# Molecular Mechanisms of Synaptogenesis

# Molecular Mechanisms of Synaptogenesis

Alexander Dityatev
Alaa El-Husseini
Editors

Springer

Alexander Dityatev
University Medical Center Hamburg-Eppendorf
52 Martinistrasse
Hamburg 20246
Germany

Alaa El-Husseini
University of British Columbia
2255 Wesbrook Mall
Vancouver, British Columbia
Canada V6T 2A1

*Cover illustrations: Front Cover* — Numerous Excitatory and Inhibitory Contacts Received by a Single Hippocampal Neuron in Culture. The structure of neurons and location of synapses were visualized by immunostaining against the dendritic microtubule-associated protein MAP2 (blue) and markers of presynaptic excitatory and inhibitory terminals, vesicular glutamate transporter VGLUT1 (green), and vesicular GABA transporter VGAT (red), respectively. Courtesy of Drs. Alaa El-Husseini and Joshua Levinson, University of British Columbia. *Back Cover Bottom Right* — Ultrastructure of Excitatory and Inhibitory Synapses. Several structures are highlighted: a dendrite and postsynaptic spines (blue), presynaptic excitatory boutons (green), and an inhibitory bouton (red). Courtesy of Drs. Daniel Nicholson and Yuri Geinisman, Northwestern University. *Back Cover Top Left* — Model Structure of the Neuroligin/β-Neurexin Trans-Synaptic Complex Together with their Putative Intracellular Binding Partners PSD-95 and CASK. The presynaptic compartment with CASK (yellow) and β-neurexin (red) is shown in green. The postsynaptic region containing PSD-95 (magenta) and neuroligin (green) is shown in blue. For more details, see Chapter 7, Figure 2. Courtesy of Drs. Markus Missler and Cartsen Reissner, University of Göttingen.

Library of Congress Control Number: 2006920793

ISBN-10: 0-387-32560-3 e-ISBN 0-387-32562-X
ISBN-13: 978-0387-32560-6

Printed on acid-free paper.



Printed in the United States of America. (SPI/EB)

9 8 7 6 5 4 3 2 1

springer.com

# Preface

The brains of humans and all other animals rely on communication between nerve cells (or neurons). Communication occurs at synapses, highly specialized junctions between neurons, where one neuron can secrete a transmitter to convey a signal to another neuron. In humans, the majority of synapses form during early prenatal and postnatal development, until about 1 year after birth. It is now widely accepted that changes in synapse function hold the key to our eventual understanding of how the brain encodes the major events responsible for development from birth to adulthood, and how neural events related to learning and memory are controlled.

Synapse specificity and plasticity provide the structural and functional basis for the formation and maintenance of the complex neural network comprising the brain. The number, location, and type of synapses formed are tightly controlled, as evidenced by the fact that synaptic circuits are formed in a highly reproducible way. This implies the existence of cellular and molecular properties that determine the thousands of connections formed by each of the 100 billion neurons in the human nervous system.

In the last decade, recent advances in molecular and cellular biology, combined with the development of sophisticated fluorescence microscopy tools to visualize synapses in live neurons, have revealed many intriguing and unexpected findings regarding the dynamics of synapse formation. Studies by a number of researchers have identified several critical protein components of synapses and have shown the time course of their arrival at the synapse. Several classes of molecules, including cell adhesion molecules, as well as scaffolding and signaling proteins, appear to serve as factors maintaining synaptic contacts between nerve cells. These protein–protein interactions act to bring about the early changes in morphology and content of sites of contact between neurons, and determine which contacts are initially stabilized. In addition, there is evidence that these molecules can act later in life to determine whether the synapse becomes potentiated or depressed and that this process contributes to diverse learning paradigms.

Several stimulating meetings were held in 2004–2005 whose goals were to discuss the recent advances in synaptogenesis. These included symposiums on *Spinogenesis and Synaptic Plasticity* (Westerburg, Germany) and *Synapse Function and Plasticity* (Vancouver, Canada) as well as a mini-symposium on *Cell Adhesion Molecules in Synapse Formation* (San Diego, USA). The idea of writing a book on molecular mechanisms of synaptogenesis was initiated by the latter conference, but scientific programs and contacts established during all three meetings were pivotal in helping us to formulate the concept of the book and recruit leading experts actively working on analysis of synaptogenesis to contribute chapters to the book. The excellence of their research and their enthusiastic support of our initiative have made it possible to share our excitement about the rapid evolution of the synaptogenesis field.

It is clear that synaptogenesis, like long-term potentiation, is a long-term problem for neuroscientists. The complexity of synaptogenesis as well as major questions needing to be addressed are highlighted in the Overview section. The first group of chapters in *Molecular Mechanisms of Synaptogenesis* introduces and examines several experimental models (from neuromuscular junctions of simple organisms such as Drosophila to hippocampal cultures of mammalian species). This section of the book also includes a discussion of the advantages of these models and summarizes the most significant results gained through their use.

Part II deals mainly with cell adhesion molecules, which have recently received a great deal of attention due to the multitude of their roles in synapse development. These molecules appear to be prominent players in all stages of synapse assembly, from contact initiation to stabilization and modification. Recent investigation into the roles of cell adhesion and extracellular matrix molecules in synapse formation has brought much insight into the basic principles governing the formation of glutamatergic and GABAergic synapses, as well as the neuromuscular junction. However, considering the existence of numerous cell adhesion molecules, it is clear that our understanding of the potential these molecules have for regulating the formation of synapses has barely scratched the surface. Moreover, the adhesion systems governing the formation of many other types of synapses, such as dopaminergic and synapses remain largely unknown. Indeed, we hope that this book will stimulate research in this fascinating area.

Parts III and IV deal with transport of synaptic components and the roles of cytoskeletal proteins and signaling molecules in assembly of synapses. These chapters demonstrate that through continued work to discover the many interactions and signaling cascades involved in these processes, a deeper understanding of the nuances governing synapse formation is gained. After synapses are formed and stabilized, their function is further shaped by life experience.

Part V describes learning-induced changes in the structure and efficacy of synapses in different brain regions and focuses on trafficking of glutamate receptors as one of the central mechanisms underlying changes in synaptic strength. In this section the importance of synaptic modifications for learning and memory associated paradigms is discussed.

Although the pursuit of a more profound understanding of the world around us has been enough to nourish the curious nature of scientists for centuries, the goal to alleviate the detrimental effects associated with mental diseases affecting people becomes possible only recently due to the drastic development of neuroscience. In this spirit, it has recently been determined that some of the critical molecules involved in building neuronal contacts are affected in psychiatric disorders such as autism and some forms of mental retardation. Also, a reduction in synapse number in specific brain regions has been found in patients suffering from various brain disorders. Thus, imbalance in synaptic contact formation may lead to abnormal neuronal circuitry underlying the aberrant behaviors manifested in these diseases. These recent findings therefore imply that major loss of neurons or neuronal populations may only be a secondary event reflecting problems in improper neuronal communication. Part VI provides a new compilation of information that links changes in basic synapse structure to brain diseases.

The question remains, however: how can one exploit this knowledge to treat such diseases? Since a specific family of genes affected in psychiatric disorders has been now identified, one possible approach might be gene therapy. This would involve the introduction of an undamaged version of affected genes into neurons to either repair damage or prevent it before it occurs. Recent studies showed that specific secreted proteins and short peptides mimicking the function of neural cell adhesion molecules can significantly enhance the formation of synapses in the brain. Peptides that interfere with the function of receptors implicated in memory formation have been used recently to disrupt memories associated with addiction. These recent advances in basic research may lay the necessary scientific groundwork to develop pharmacological treatments targeting faulty synaptogenesis and allow neurobiologists to improve the lives of people affected by brain disorders.

Thus, *Molecular Mechanisms of Synaptogenesis* will not only be useful for researchers, as well as graduate and undergraduate students in neuroscience, biology, and biochemistry, but it will also benefit students in medicine and nursing programs, expanding their knowledge of fundamental cellular mechanisms involved in the formation of synaptic contacts and communication between neuronal cells and how they are affected in brain diseases associated with abnormal neural wiring.

Before you begin the journey of synaptogenesis laid out in this book, we leave you with three images shown below and in Colorplate 1: A contact is required for exchange of information between brain cells (image on left by Catherine Gauthier-Campbell), as it is for exchange of emotion between people (as presented by *Promenade* of Marc Chagall [1917]; image in middle). This concept can also be expressed in an abstract form (as in the painting *Synapses and Genes, the Building Blocks of Life* by Alaa El-Husseini [2004]; image on the right). This book is for you to explore the hidden dimensions of synapses. It may also help you one day see the Chinese concept of yin and yang intricately woven into an image of the perforated synapse.

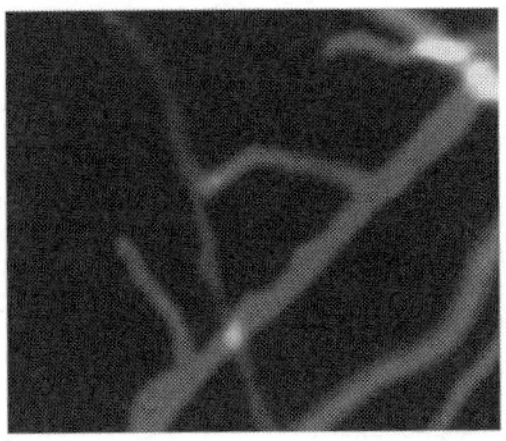

**Dedication**

We extend our greatest gratitude to the chapter authors for their outstanding contribution and timely efforts. We greatly appreciate the efforts of all colleagues who provided us with valuable information and beautiful images for the book cover. We are in debt to all members of our groups and, in particular, Galina Dityateva, Joshua Levinson, Kimberly Gerrow, and Marie-France Lisé for stimulating discussions on synaptogenesis. Many thanks to the kind and astute influences that Drs. Nazeeh Khalidi, Musa Khalidi, Hans Jacobs, Robert Shiu, Jean Paterson, Steven Vincent, David Bredt, and Roger Nicoll had on Alaa El-Husseini's research interests and career development, and as many thanks to Drs. Alexander Bart, Valery Kohzanov, Vladimir Levchenko, Peter Clamann, and Melitta Schachner, for their generous support of Alexander Dityatev. Thanks to organizations, such as the Brain Research Centre at the University of British Columbia, which have helped us to realize the significance of this project, to our families and close friends for their generous support, and to Claire Wynperle and Veronika Gorbacheva who served as Editorial Assistants during the long and arduous process of assembling this book.

Alaa El-Husseini
Alexander Dityatev

Vancouver, British Columbia
Hamburg, Germany

# Contents

# Contributors

Tonya R. Anderson, Mount Sinai School of Medicine, New York, NY, 10029, USA

Alasdair M. Barr, University of British Columbia, Vancouver, BC, V5Z 1L8, Canada

Deanna L. Benson, Mount Sinai School of Medicine, New York, NY,10029, USA

Thomas Biederer, Yale University, New Haven CT, 06520-8220, USA

Karen Brebner, University of British Columbia, Vancouver, BC, V6T 2A1, Canada

Robert W. Burgess, Jackson Laboratory, Bar Harbor, ME, 04609, USA

Daniel Choquet, Université Bordeaux, Bordeaux, 33077, France

Brian R. Christie, University of British Columbia, Vancouver, BC, V6T 1Z4, Canada

Laurent Cognet, Université Bordeaux, Bordeaux, 33077, France

Ann Marie Craig, University of British Columbia, Vancouver, BC, V6T 2B5, Canada

Alexander Dityatev, University Medical Center Hamburg-Eppendorf, Hamburg, 20246, Germany

Thomas Dresbach, University of Heidelberg, Heidelberg, 69120, Germany

Alaa El-Husseini, University of British Columbia, Vancouver, BC, V6T 1Z4, Canada

José A. Esteban, University of Michigan, Ann Arbor, MI, 48109, USA

Richard Fairless, Georg-August University, Göttingen, 37073, Germany

Anna Fejtová, Leibniz Institute for Neurobiology, Magdeburg, 39118, Germany

Yuri Geinisman, Northwestern University Feinberg School of Medicine, Chicago, IL, 60611-3008, USA

Laurent Groc, Université Bordeaux, Bordeaux, 33077, France

Eckart D. Gundelfinger, Leibniz Institute for Neurobiology, Magdeburg, 39118, Germany

Kurt Haas, University of British Columbia, Vancouver, BC, V6T 2B5, Canada

Martin Heine, Université Bordeaux, Bordeaux, 33077, France

Fritz A. Henn, Central Institute for Mental Health, Mannheim, 68159, Germany

Rochelle M. Hines, University of British Columbia, Vancouver, BC, V6T 1Z4, Canada

William G. Honer, University of British Columbia , Vancouver, BC, V5Z 1L8, Canada

Fumitoshi Irie, Burnham Institute, La Jolla, CA, 92037, USA

Hyun-soo Je, National Institute of Child Health and Human Development, Bethesda, MD, 20892-3714, USA

Yishi Jin, University of California, Santa Cruz, CA, 95064, USA

Ridha Joober, Douglas Hospital Research Centre, Montreal, QC, H4H 1R3, Canada

Yunhee Kang, University of British Columbia, Vancouver, BC, V6T 2B5; Canada and Korea University College of Medicine, Seoul, 110-744, South Korea

Matthias Kneussel, University Medical Center Hamburg-Eppendorf, Hamburg, 20246, Germany

Imbritt König, Leibniz Institute for Neurobiology, Magdeburg, 39118, Germany

Hans-Jürgen Kreienkamp, University Medical Center Hamburg-Eppendorf, Hamburg 20246, Germany

Michael R. Kreutz, Leibniz Institute for Neurobiology, Magdeburg, 39118, Germany

Brahim Lounis, Université Bordeaux, Bordeaux, 33077, France

Bai Lu, National Institute of Child Health and Human Development, Bethesda, MD, 20892-3714, USA

Jochen C. Meier, Charité–University Medicine Berlin, Berlin, 10117, Germany

Austen J. Milnerwood, Raymond University of British Columbia, Vancouver, BC, V6T 1X7, Canada

Markus Missler, Georg-August University Göttingen, 37073, Germany

Daniel A. Nicholson, Northwestern University Feinberg School of Medicine, Chicago, IL, 60611-3008, USA

Anthony G. Phillips, University of British Columbia, Vancouver, BC, V6T 2A1, Canada

Giovanni Piccoli, University of Milan, Milan, 20129, Italy

Lynn A. Raymond, University of British Columbia, Vancoumer, BC, V6T 1X7, Canada

Carsten Reissner, George-August University Göttingen, 37073, Germany

Stefano Romorini, University of Milan, Milan, 20129, Italy

Carlo Sala, University of Milan, Milan, 20129, Italy

Ken Sawada, Kochi Medical School, Kochi, 780-8520, Japan

Stephen W. Scheff, University of Kentucky, Lexington, KY, 40536-0230, USA

Thomas C. Südhof, University of Texas Southwestern Medical Center at Dallas, TX, 75390-9111, USA

Naweed I. Syed, University of Calgary, Calgary, AB, T2N 1N4, Canada

Erik M. Ullian, University of California, San Francisco, CA, 94143-07730, USA

Heather Van Epps, University of California, Santa Cruz, CA, 95064, USA

Barbara Vollmayr, Central Institute for Mental Health, Mannheim, 68159, Germany

Yu Tian Wang, University of British Columbia, Vancouver, BC, V6T 2A1, Canada

Philip E. Washbourne, University of Oregon, Eugene, OR, 97403, USA

Alina Webber,University of British Columbia, Vancouver, BC, V6T 1Z4, Canada

Joshua A. Weiner, University of Iowa, Iowa City, IA, 52242, USA

Ryanne Wiersma-Meems, University of Calgary, Calgary, AB, T2N 1N4, Canada

Tak Pan Wong,University of British Columbia, Vancouver, BC, V6T 2AI, Canada

Newton H. Woo, National Institute of Child Health and Human Development, Bethesda, MD, 20892-3714, USA

Yu Yamaguchi, Burnham Institute, La Jolla, CA, 92037, USA

Clint E. Young, University of British Columbia, Vancouver, BC, V5Z 1L8, Canada

Min Zhuo, University of Toronto, Toronto, ON, M5S 1A8, Canada

Mathias Zink, Central Institute for Mental Health, Mannheim, 68159, Germany

# Overview

## SYNAPTOGENESIS: WHEN LONG-DISTANCE RELATIONS BECOME INTIMATE

Thomas C. Südhof*

### 1. SUMMARY

Neurons in brain talk to each other at synapses which connect neurons into vast communicating synaptic circuits. Synapses are specialized intercellular junctions that are diverse and dynamic. The number, locations, and distinct functional properties of synapses confer onto synaptic circuits an enormous complexity that is essential for information processing by these circuits. Insight into how synaptic connections in such circuits are specified represents a multifaceted problem that includes four interrelated questions: 1. How does a neuron identify the correct target neurons for synapse formation? 2. How does a neuron form a synapse on specific parts of that target neuron, e.g., distal dendrites or axon hillocks? 3. How is the decision reached, whether to keep or to discard a given synapse after it has been formed? 4. How is the functional diversity of synapses generated and controlled? Clearly synapse formation means more than just establishing contacts, and includes specification of the dynamics and types of these synaptic contacts. Although much remains to be clarified, the available data suggest that axonal pathfinding is a major component in establishing synaptic specificity, that initial formation of synapses is fueled by mechanisms that involve multiple cell adhesion molecules, and that the development of synaptic properties and use of a synapse are crucial in the decision about whether or not a synapse survives.

*Department of Molecular Genetics, Center for Basic Neuroscience, and Howard Hughes Medical Institute, University of Texas Southwestern Medical Center, 6000 Harry Hines Boulevard, Dallas, TX 75390-9111, USA; thomas.sudhof@utsouthwestern.edu

## 2. THE SCOPE OF THE SYNAPSE FORMATION PROBLEM

The brain contains more than $10^{11}$ neurons, each of which forms and receives many synapses – from a few hundred to more than 100,000. Neurons fall into different types and subtypes that are defined by a characteristic pattern of synaptic connectivity. Here, the 'pattern of synaptic connectivity' refers to which downstream target neurons a neuron forms synaptic contacts with, and which upstream neurons in turn form synapses on a particular neuron. Moreover, the pattern of synaptic connectivity that is characteristic for a given type of neuron includes the properties of these synaptic contacts (location on the target neurons, properties of synaptic transmission). For a given type of neuron, the pattern of synaptic connectivity is very reproducible from one neuron to the next; moreover, for a given type of neuron the average number of inputs and outputs is always the same, i.e., there are no "better" or "worse" neurons in terms of how connected a neuron is.

A good example of these basic synaptic circuit principles is the CA1 region of the hippocampus, one of the best studied brain areas (reviewed in ref. 1; see Figure 0.1; Colorplate 1). The CA1 region contains an apparently homogeneous set of excitatory pyramidal neurons that are arranged in a single layer (stratum pyramidale), and that elaborate extensive apical and basal dendrites. CA1-region pyramidal neurons receive five different types of excitatory inputs: In addition to axon collaterals from the CA1-region pyramidal neurons themselves, these inputs are derived from CA3-region pyramidal neurons, and from entorhinal, amygdala, and thalamic inputs. Furthermore, the pyramidal neurons receive inhibitory inputs from at least 12 types of local interneurons. Finally, four additional types of interneurons in the CA1 region only synapse onto other interneurons. The complex excitatory and inhibitory inputs onto CA1-region pyramidal neurons are precisely targeted to restricted parts of the pyramidal neurons – nothing appears to be left to chance. The inputs from CA3-region pyramidal neurons that are closest to the CA1 region innervate only dendrites in the stratum oriens, whereas inputs from CA3-region pyramidal neurons that are further away from the CA1 region innervate only dendrites in the stratum radiatum. Thalamic inputs, in contrast, are restricted to stratum lacunosum-moleculare dendrites, while axon collateral inputs from the CA1-region pyramidal neurons and inputs from the amygdala are only made on dendrites of the distal stratum oriens (Figure 0.1). The inputs from the different types of interneurons exhibit a similar laminar-specific innervation pattern on pyramidal neurons, and these inhibitory neurons themselves in turn receive spatially well-organized inputs from excitatory and inhibitory neurons that are also differentially formed on the soma or dendrites of the interneurons.

This pattern of synaptic connectivity in the hippocampal CA1 region, exemplary for many other brain regions, results in a vast neural network that would already be enormously complex if only two types of synapses existed (excitatory and inhibitory). However, an additional element of complexity is generated by characteristic differences in synaptic transmission at synapses between defined types of neurons. Thus not only the number and location of synapses are characteristic of synaptic connections formed between two types of neurons, but also the properties of these synapses (e.g., whether they are facilitating or depressing, exhibit NMDA-receptor dependent plasticity or other forms of synaptic plasticity, and so on).

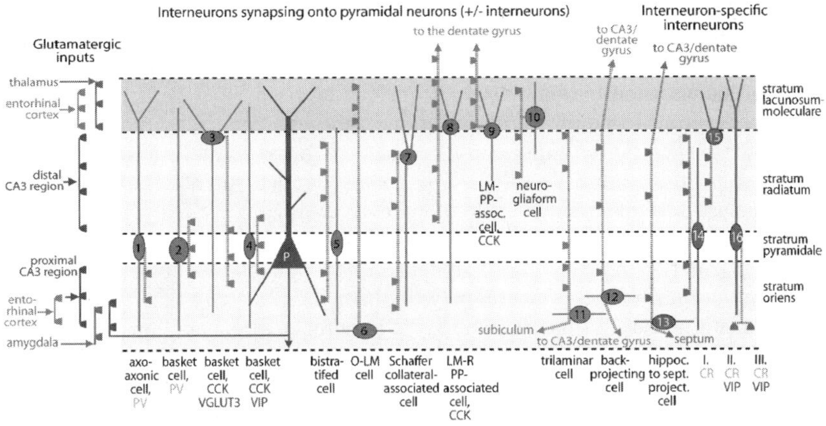


**Figure 0.1.** Schematic Diagram of the Innervation Patterns of a Pyramidal Cell in the CA1 Region of the Hippocampus by 12 Types of GABAergic Interneurons. The main laminar-specific glutamatergic inputs are indicated on the left. The somata and dendrites of interneurons innervating pyramidal cells (green) are shown in orange, those innervating mainly or exclusively other interneurons are shown in lilac (see Colorplate 1). The main termination zones of GABAergic synapses are shown by trapeziform symbols. The proposed names of neurons, some of them abbreviated, are under each schematic cell and a minimal list of molecular cell markers is given, which in combination with the axonal patterns help the recognition and characterization of each class. Note that one molecular cell marker may be expressed by several distinct cell types. The number of interneurons shown is not exhaustive or complete. Note the association of the output synapses of different sets of cell types with the perisomatic region, and either the Schaffer collateral, commissural, or the entorhinal pathway termination zones, respectively. CB, calbindin; CR, calretinin; LM-PP, lacunosum-moleculare–perforant path; LM-R-PP; lacunosum-moleculare–radiatum–perforant path; m2, muscarinic-receptor type 2; NPY, neuropeptide tyrosine; PV, parvalbumin; SM, somatostatin; VGLUT3, vesicular glutamate transporter 3. Modified with permission from ref. 1.

Synapses formed by a single neuron onto different target neurons can have dramatic differences in properties, depending on the target neuron with which these synapses are formed. A well-studied example for such differences is the synapses formed by pyramidal neurons in layer 2/3 of the cortex. Among others, these pyramidal neurons form synapses onto two types of interneurons, bitufted and multipolar neurons, with very different properties[2] (Figure 0.2). The synaptic properties suggest a large difference in presynaptic release probability; as a result, the pyramidal synapses on bitufted interneurons are very ineffective but facilitating, whereas the synapses on multipolar interneurons are reliable but depressing. These differences corresponded to differences in $Ca^{2+}$-influx[3]. The differences between presynaptic release from terminals formed by the same presynaptic neuron onto different postsynaptic neurons suggest that the postsynaptic target neuron instructs the presynaptic neuron what kind of synapse to form, indicative of a retrograde *trans*-synaptic signaling process.

These examples illustrate that the specificity of connectivity in synaptic networks not only consists of the specificity of neuronal synaptic partners (i.e., the question of which neurons communicate with each other pre- and postsynaptically), but also of the specificity of where on these neurons synapses are formed, and what functional properties these synapses have. Understanding these three facets of synaptic connectivity is of central importance for understanding the wiring diagram of the brain. Thus, the problem of synapse formation not only consists of the question of how each individual neuron receives specific inputs from a series of neurons, and forms specific outputs onto another series of neurons. In fact, phrasing this question in this manner is misleading because there is little evidence that there is point-to-point specificity; rather, the evidence suggests that there is class-to-class specificity, with neurons of class A on average forming X numbers of synapses with neurons of class B. This means that as long as classes of neurons have mechanisms of finding each other and achieving an average number of connections, there is no need for a precise blueprint of connectivity.

Based on this consideration, understanding synaptic connectivity requires insight into a fourth facet of neural network formation, namely the mechanism that ensures that a neuron forms the same average number of connections with a given class of target neuron. The assumption that connectivity is based on random connections between classes of neurons simplifies any mechanism, and suggests a plausible explanation for the massive amount of synapse elimination that occurs during development[4]. According to this hypothesis, synapses form between classes of neurons, and are then 'averaged' out, possibly in an activity-dependent manner, so that the correct average connectivity is achieved.

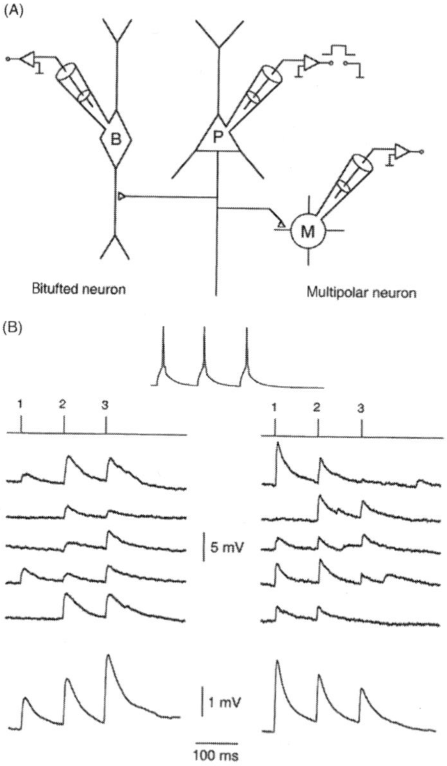


**Figure 0.2.** Synapses Formed by a Pyramidal Neuron (P) of Layer 2/3 in the Cortex with Two Different Types of Interneurons, Bitufted (B) and Multipolar neurons (M), Exhibit Distinct Release Properties. (A) Schematic drawing of the recording configuration used for the traces shown in (B). The presynaptic pyramidal neuron was stimulated by brief intracellular current injections, and unitary EPSPs were recorded simultaneously from a bitufted and a multipolar postsynaptic interneuron. (B) Representative traces of five consecutive EPSPs recorded in bitufted and multipolar interneurons. The numbers above the records indicate the time of presynaptic action potentials. The lowermost traces represent averages of 100 sweeps. The efficacy of unitary EPSPs, measured as their amplitude, was 0.92 ± 0.49 and 3.3 ± 1.9 mV for bitufted and multipolar interneurons, respectively; the reliability, measured as the failure rate, was 42 ± 18% and 1.6 ± 3.5% for bitufted and multipolar interneurons; the mean paired pulse ratios were 1.95 ± 0.59 and 0.53 ±0.12 for bitufted and multipolar interneurons, respectively; and the mean EPSP latency was 2.97 ± 0.42 and 1.17 ± 0.19 for bitufted and multipolar interneurons, respectively. Reproduced with permission from ref. 2.

## 3. THE NATURE OF THE SYNAPTIC CONNECTION

Synapses are specialized intercellular junctions that connect pre- and postsynaptic neurons (and presynaptic neurons to postsynaptic effector cells such as muscle cells). As intercellular junctions, synapses are typical in that they display membrane specializations on both sides of the junction, and in that a uniformly spaced cleft separates the two cells at the junction. Synapses are different from other intercellular junctions in that synaptic junctions are highly asymmetric, with clusters of vesicles on the presynaptic side[5]. Intercellular junctions (and junctions between cells and the extracellular matrix) usually have three functions, to couple two cells mechanically to each other (for example, in creating the architecture of a tissue), to signal between cells, and to organize the spatial organization of intracellular membrane traffic in participating cells. Synapses are no different. Their signaling and intracellular membrane trafficking functions have grown enormously compared to other types of intercellular junction, but they still also act to mechanically connect two cells, for example, in fixing axons and dendrites in space.

At a synapse, a fast signal is transferred from the presynaptic to the postsynaptic neuron in the form of a chemical neurotransmitter that is released from the presynaptic nerve terminal, and recognized by postsynaptic receptors. In addition, several slow anterograde and retrograde signals regulate synapse properties and size. The probability of synaptic transmission for each individual synaptic contact is always less than 1, sometimes considerably less. As a result, whenever a synaptic connection has to always elicit a postsynaptic response for every presynaptic action potential, neurons elaborate a multitude of individual contacts on the target cell (instead of a single larger contact). For example, in the Calyx of Held synapse a presynaptic terminal forms ~600 individual contacts with the same postsynaptic neuron; approximately a third of these contacts transmit a synaptic signal for each action potential, thereby ensuring that ~200 synaptic inputs on the postsynaptic neuron are activated per action potential[6].

Apart from the typical differences between excitatory and inhibitory synapses, no major structural differences exist between typical central synapses that would allow prediction of their often quite dramatically different functional properties. This was for example shown for cerebellar parallel fiber and climbing fiber synapses that in spite of very different release probabilities exhibit quantitatively similar ultrastructural features[7]. Although the size of synapses varies, it only varies over a relatively small range. It should be noted that these statements only to standard central synapses; the structure of the neuromuscular junction[8] or of ribbon synapses[9] is dramatically different, as are their properties.

## 4. AXONAL PATHFINDING VERSUS SYNAPTIC CELL ADHESION

The specificity of synaptic connections is established by two consecutive processes: axonal pathfinding and synaptic cell adhesion[10]. A simple consideration shows that of these two processes, axonal pathfinding is more important than synaptic cell adhesion, although obviously both are essential. This consideration is that synaptic cell adhesion can only operate at a distance of $\ll 100$ nm, which even on the scale of the densely packed neuropil is a very short distance. Whereas over short distances axons and dendrites have equal roles in establishing synaptic specificity, over long distances axons alone mediate the specificity of connections. Thus most of the specificity of synaptic connectivity must depend on guiding an axon to its correct target, with the actual formation of synaptic connections being secondary.

Axonal pathfinding is well studied, and much has been learned about its complex molecular determinants (e.g., see refs. 11–13). Axonal guidance may even contribute to the determination of the dendritic domain to which an axonal input is directed. Many key mechanisms of axonal guidance have been established, providing the molecular basis for Sperry's pioneering chemoaffinity hypothesis[10].

After axonal guidance is completed, a growth cone enters the target neuropil and becomes competent to form a synapse. In the densely populated neuropil, most axons and dendrites are not destined to form synaptic connections with each other, but chance encounters between these axons and dendrites must be extremely frequent. How does the emerging nerve terminal select the right target neuron and the right dendritic domain in the densely packed neuropil? For example, how does a thalamic input in the CA1 region of the hippocampus select between pyramidal cell dendrites in the stratum lacunosum or the dendrites of at least ten different types of interneurons (Figure 0.1; Colorplate 1)? Two not mutually exclusive hypotheses can be proposed to explain synapse selection by axons that are primed for synapse formation. First, specific recognition molecules may trigger synapse formation. Second, neurons may form promiscuous synaptic connections and then eliminate the wrong connections, i.e., act by a sampling mechanism that establishes and dissolves synapses constantly as the axon moves along. It seems likely that a combination of both hypotheses is correct.

An important clue to the nature of synapse formation comes from artificial *in vitro* experiments. When a single hippocampal or cortical neuron is plated on microislands of glia cells, it forms hundreds of synapses, the so-called autapses, onto themselves. Although autapses do occur physiologically, they are rare *in vivo* (e.g., in cortical pyramidal neurons, autapses account for <1% of the total synapses; ref. 14). Thus in the absence of appropriate targets, a neuron forms synapses onto itself by default, whereas in the presence of appropriate targets, it does not. This finding implies that synapse formation is a competitive process: when a neuron cannot find a better partner, it prefers connecting to itself instead of remaining unconnected. Thus synapse formation is not necessarily governed by a strict cell adhesion code, but at least partly operates by a mechanism that selects stable synapses based on a competition between targets (i.e., a sampling mechanism).

It is clear, however, that 'sampling and selection' of synapses is not sufficient to explain the generation of patterns of synaptic connectivity. If all neurons randomly formed synapses that were then largely eliminated by a selection mechanism, this would simply shift the specificity issue from the stage of initial synapse formation to the stage of synapse elimination. Whereas activity-dependent synapse elimination is a plausible mechanism to achieve a homogeneous distribution of synaptic inputs and outputs between two classes of neurons, it is not plausible as a mechanism that eliminates a huge number of random synapses. Apart from theoretical considerations, an experimental result that demonstrates this point is the observation that in Munc18-1 deficient mice, all synaptic activity is absent but the pattern of synaptic connectivity is nevertheless not randomized[15]. It seems likely that the major role of the sampling mechanism is to equalize synaptic inputs and outputs for a given class of neuron.

## 5. ROLES OF SYNAPTIC CELL ADHESION MOLECULES

Many cell adhesion molecules were shown to functionally affect synapses. These include homophilic cell adhesion molecules such as cadherins or NCAM, and heterophilic cell adhesion molecules such as neurexins/neuroligins and Ephrins/Eph receptors. For none of these cell adhesion molecules an unequivocal role has been demonstrated. This deficiency is probably partly due to the fact that demonstrating such a role is difficult, but may also be caused by an intrinsic functional ambiguity of such cell adhesion molecules. Specifically, many synaptic cell adhesion molecules do not only have a function at the synapse, but also in other developmental processes. Moreover, many synaptic cell adhesion molecules may be part of an ensemble of cell adhesion molecules that together determines the overall process of synapse formation and maintenance.

In principle, synaptic cell adhesion molecules could potentially be important for the initial formation of synapses, for their mechanical stabilization, and for *trans*-synaptic signaling that mediates, among others, specification of synaptic properties. An intriguing assay to probe for a role in initial synapse formation was described by Serafini and colleagues[16]. In this assay, non-neuronal cells expressing a candidate synaptic cell adhesion molecule were co-cultured with neurons, and the formation of synapses by the neuron on the non-neuronal cell was examined. These experiments revealed that neuroligins were potent inducers of artificial synapse formation[16,17], presumably by binding to neurexins. In the same assay, another unrelated synaptic cell adhesion molecule, SynCAM, was active, whereas many other proteins were not[18]. However, several observations suggest that this assay is unlikely to directly report on synapse formation, and more likely monitors the general process of *trans*-synaptic signaling. First, it is unlikely that initial synapse formation is mediated by multiple very different molecules. Second, in transfected neurons these same molecules had very different effects (e.g., see ref. 19). Third, these molecules are expressed by all neurons; it is hard to imagine how specificity of synapse formation would be generated if all neurons were triggered by these molecules to form synapses. A more likely interpretation is that the assay simply reflects activation of a *trans*-synaptic signaling cascade, and that activation of such a cascade is sufficient to stabilize a transient synaptic contact.

Synaptic cell adhesion molecules perform a minimum of four roles: 1. a function in the initial establishment of a synaptic contact; 2. a function in mechanical cohesion of synaptic contacts, thereby aligning pre- and postsynaptic specializations with each other; 3. a function in specifying the synaptic properties of a synaptic contact (e.g., the target-induced differences in presynaptic release properties of L2/3 cortical neurons, which could be referred to as stable *trans*-synaptic signaling; see Figure 0.2), and 4. a function in transient *trans*-synaptic signaling events. Assaying for these functions is difficult because both *in vitro* and *in vivo* assays have limitations, and because many synaptic cell adhesion molecules likely perform multiple functions. Nevertheless, given the apparent promiscuity of initial synapse formation, it seems likely that the majority of synaptic cell adhesion molecules will be more involved in role #2–4, and that these roles are more important for synaptic circuit formation than the initial formation of synapses.

## 6. PUTTING EVERYTHING TOGETHER

Whereas enormous progress has been made in the functional identification of axonal guidance molecules, the study of synaptogenesis and synaptic cell adhesion remains far behind. The reason for the relatively slow progress in studying synaptic cell adhesion appears to be two-fold. First, the molecules involved in synaptic cell adhesion likely have less clear-cut functions, but act in ensembles which among others determine the strength of a synaptic contact and specify its properties. Second, assaying for the functions in synaptogenesis and synaptic cell adhesion is much less straightforward than assaying for molecules with a function in axonal guidance. Nevertheless, the principles outlined above provide general ideas for a molecular framework of synaptogenesis that is illustrated as a model with candidate molecules, as far as identified, in Figure 0.3[20].

Briefly, it seems almost certain that all molecules involved in synaptogenesis will be multifunctional, but no molecule will be a master regulator of all other molecules. Moreover, it seems likely that synaptogenesis is a complex process that involves signaling on both sides of the synapse-to-be. Thus we postulate that initial synapse formation is due to a transient and rather nonspecific contact that may be mediated by the combined action of multiple cell adhesion molecules, possibly including protocadherins and other cadherins[21]. Afterward, synapses are validated and stabilized by a large number of activity-dependent processes, including the action of neurexins/neuroligins. It is likely that this 'stabilization' involves more than one particular process, and several different types of subreactions. For this step, neurexins and neuroligins are the best candidates because deletions of a subset of neurexins do not abolish synapse formation, but dramatically alter synapse specification[22], and because increases in synapse density that are induced in transfected neurons by expression of neuroligin 1 depend on activation of NMDA receptors (ref. 20; see Figure 0.3). The variability of neurexins and neuroligins produced by multiple genes and alternative spicing might contribute to

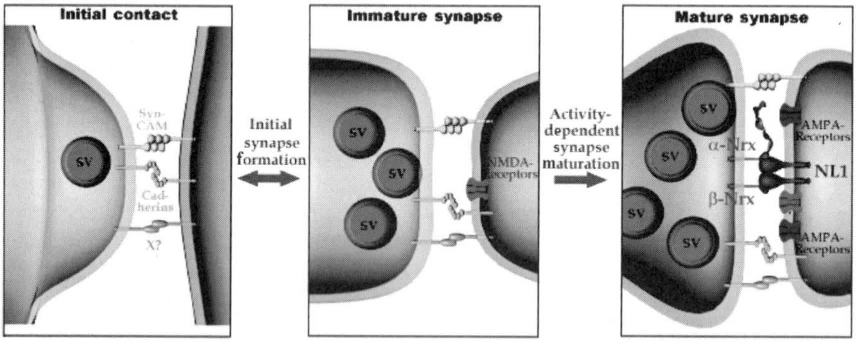


**Figure 0.3.** Model of the Role of Cell Adhesion Molecules in Synapse Formation and Maturation. The initial synaptic contact between neurons is proposed to involve multiple cell adhesion molecules, including SynCAM and cadherins, which might impart specificity on synaptic contacts. The resulting immature synapses are functional, but are stabilized by activity-dependent processes. The model suggests that neuroligin 1 mediates the activity-dependent stabilization of transient synaptic contacts, and depends on the simultaneous activation of NMDA receptors. In promoting activity-dependent synapse stabilization, postsynaptic neuroligin 1 likely transduces a *trans*-synaptic signal triggered by binding of its extracellular esterase-like domain to presynaptic neurexins. For more details, see ref. 20.

synapse specification[23]. However, it is also probable that neurexins and neuroligins are not the only molecules that act here; for example, ephrins and Eph receptors might have additional major roles (e.g., see ref. 24). Uncovering the roles of various candidate synaptic cell adhesion molecules will be a major challenge for years to come.

## 7. REFERENCES


1. Somogyi, P., and Klausberger, T. (2005) *J Physiol* **562**, 9–26.
2. Rozov, A., Burnashev, N., Sakmann, B., and Neher, E. (2001) *J Physiol* **531**, 807–826.
3. Koester, H.J., and Johnston, D. (2005) *Science* **308**, 863–866.
4. Hua, J.Y., and Smith, S.J. (2004) *Nat Neurosci* **7**, 327–332.
5. Peters, A., Palay, S.L., and Webster, H.deF. (1991) The Fine Structure of the Nervous System. Neurons and Their Supporting Cells. 3rd ed., Oxford University Press, New York.
6. Satzler, K., Sohl, L.F., Bollmann, J.H., Borst, J.G., Frotscher, M., Sakmann, B., and Lubke, J.H. (2002) *J Neurosci* **22**, 10567–10579.
7. Xu-Friedman, M.A., Harris, K.M., and Regehr, W.G. (2001) *J Neurosci* **21**, 6666–6672.
8. Harlow, M.L., Ress, D., Stoschek, A., Marshall, R.M., and McMahan, U.J. (2001) *Nature* **409**, 479–484.
9. Sterling, P., and Matthews, G. (2005) *Trends Neurosci* **28**, 20–29.
10. Sperry, R.W. (1963) *Proc Natl Acad Sci U S A* **50**, 703–710.
11. Skutella, T., and Nitsch, R. (2001) *Trends Neurosci* **24**, 107–113.
12. Tessier-Lavigne, M. (2002–2003) *Harvey Lect* **98**, 103–143.
13. Charron, F., and Tessier-Lavigne, M. (2005) *Development* **132**, 2251–2262.
14. Lubke, J., Markram, H., Frotscher, M., and Sakmann, B. (1996) *J Neurosci* **16**, 3209–3218.
15. Verhage, M., Maia, A.S., Plomp, J.J., Brussaard, A.B., Heeroma, J.H., Vermeer, H., Toonen, R.F., Hammer, R.E., van den Berg, T.K., Missler, M., Geuze, H., and Südhof, T.C. (2000) *Science* **287**, 864–869.
16. Scheiffele, P., Fan, J., Choih, J., Fetter, R., and Serafini, T. (2000) *Cell* **101**, 657–669.
17. Chubykin, A.A., Liu, X., Comoletti, D., Tsigelny, I., Taylor, P., and Südhof, T.C. (2005) *J Biol Chem* **280**, 22365–22374.
18. Biederer, T., Sara, Y., Mozhayeva, M., Atasoy, D., Liu, X., Kavalali, E.T., and Südhof, T.C. (2002) *Science* **297**, 1525–1531.
19. Sara, Y., Biederer, T., Atasoy, D., Mozhayeva, M.G., Chubykin, A., Südhof, T.C., and Kavalali, E.T. (2005) *J Neurosci* **25**, 260–270.
20. Chubykin et al. submitted.
21. Weiner, J.A., Wang, X., Tapia, J.C., and Sanes, J.R. (2005) *Proc Natl Acad Sci U S A* **102**, 8–14.
22. Missler, M., Zhang, W., Rohlmann, A., Kattenstroth, G., Hammer, R.E., Gottmann, K., and Sudhof, T.C. (2003) *Nature* **423**, 939–948.
23. Boucard, A., Chubykin, A.A., Comoletti, D., Taylor, P., and Südhof, T.C. (2005) *Neuron* **20**, 229–236.
24. Penzes, P., Beeser, A., Chernoff, J., Schiller, M.R., Eipper, B.A., Mains, R.E., and Huganir, R.L. (2003) *Neuron* **37**, 263–274.

# Part I

## EXPERIMENTAL MODELS OF SYNAPTOGENESIS

# 1

# THE FORMATION OF THE VERTEBRATE NEUROMUSCULAR JUNCTION: ROLES FOR THE EXTRACELLULAR MATRIX IN SYNAPTOGENESIS

Robert W. Burgess*

## 1. SUMMARY

The vertebrate neuromuscular junction (NMJ) is one of the best studied models of synapse formation. This chapter summarizes the cellular and molecular lessons learned from these studies. It also specifies where these conclusions can be directly generalized to other models of synaptogenesis and where other systems may be using analogous but distinct mechanisms. In particular, it focuses on the extracellular matrix (ECM) of the NMJ, which has essential roles in its formation. Nerve-derived agrin is a component of the synaptic ECM and is an essential signal for postsynaptic differentiation in the muscle. Laminins are also ECM proteins that play a role in organizing the subsynaptic structure of the junction. Together these proteins, as well as other less well-defined mechanisms, direct the formation of the specialized point or cell–cell contact that matures into a highly organized functional NMJ.

## 2. INTRODUCTION

There are a number of reasons, both historical and practical, for the popularity of the vertebrate neuromuscular junction (NMJ) in studies of synaptogenesis. Many of the seminal studies on the quantal nature of synaptic transmission were performed at the NMJ, establishing a battery of techniques for analysis and allowing a correlation of synaptic structure and function. The NMJ is also very accessible, with motor neuron input contacting each muscle approximately in the middle, and each muscle fiber receiving input from a single motor axon under normal circumstances in the adult. Thus, reliably finding a well-defined NMJ for

---

*The Jackson Laboratory, 600 Main Street, Bar Harbor, ME 04609, USA; robert.burgess@jax.org

experiments is trivial compared to repeatedly finding a well-defined population of central nervous system (CNS) synapses. The NMJ is also very large, and each nerve terminal is composed of many active zones (Figure 1.1), which is a benefit to physiological, ultrastructural, and imaging studies. Finally, and importantly, related synapses were amenable to biochemical analyses. Specifically, the electric organs of rays such as *Torpedo californica* are essentially huge stacks of cholinergic synapses that functionally and molecularly most closely resemble NMJs. Several key proteins found at cholinergic NMJs, including agrin, rapsyn, and even acetylcholine receptors (AChRs) themselves, were first characterized in such preparations.

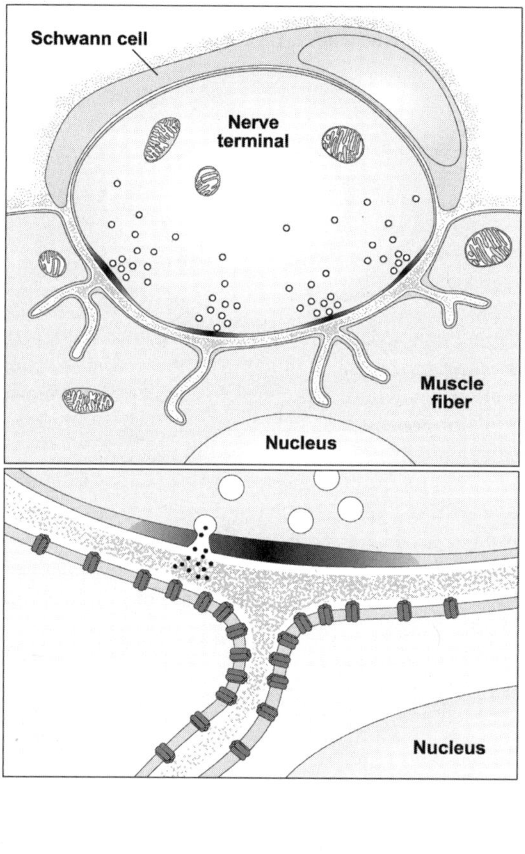


**Figure 1.1.** The Neuromuscular Junction. The NMJ is the synapse between motor neurons and their target muscle fibers. In vertebrates, the motor neuron has a terminal ending with multiple presynaptic release sites. Synaptic vesicles containing the neurotransmitter acetylcholine, as well as ATP, are polarized in the terminal and cluster near thickenings in the presynaptic membrane, the active zones. The nerve terminal is recessed into the muscle cell membrane, and the entire structure is capped by a nonmyelinating glial cell, the terminal Schwann cell. The muscle fiber has invaginations, junctional folds, which are in direct apposition to the presynaptic active zones. An extracellular matrix runs through the entire synaptic cleft. At individual release sites, vesicles fuse with the presynaptic membrane and release neurotransmitter into the cleft. The acetylcholine receptors in the postsynaptic membrane are at their highest concentration near the crests of the junctional folds. The extracellular matrix contains many signaling and structural molecules, as well as acetylcholinesterase, the enzyme that degrades ACh and thus curtails synaptic transmission.

As a synapse, the NMJ is quite specialized. Most CNS neurons function as integrators, balancing excitatory and inhibitory synaptic inputs, and rarely firing an action potential in response to a single stimulus. The NMJ, however, functions as a failsafe synapse, where an action potential in the motor neuron is guaranteed to produce sufficient depolarization in the muscle fiber to reach the threshold for a muscle action potential, leading to contraction. To ensure this is the case, there is an extreme accumulation of AChRs, which function as ligand-gated cationic channels, and voltage-gated sodium channels in the postsynaptic membrane (Figure 1.1). In addition, the multiple release sites of the presynaptic terminal have a high cumulative probability of vesicular release in response to an action potential, and the amount of acetylcholine (ACh) released locally saturates postsynaptic receptors. Transmission is limited spatially and temporally by acetylcholinesterase (AChE) in the synaptic cleft, which degrades the transmitter, terminating the signal.

There are structural correlates to the functional requirements described above (Figure 1.1). Each of the many presynaptic active zones, the sites of vesicular release, is directly aligned with a postsynaptic specialization called a junctional fold. The folds are invaginations into the muscle plasma membrane. The highest concentration of AChRs is found near the tops of these folds, while the voltage-gated sodium channels are found deeper in the folds. This intimate arrangement of channels and increased membrane surface area provided by the folds helps ensure large synaptic currents and threshold depolarization. Another important structural feature of the NMJ is the thick basal lamina of extracellular matrix (ECM) that runs through the cleft. As mentioned above, this matrix contains AChE, which is important for synaptic function. In addition, it houses many of the signaling and structural molecules that are now known to be essential players in the development and stability of the NMJ.

From the studies of NMJ formation, a number of principles have become clear. First, there is a great deal of specialization and organization of both the pre- and postsynaptic components. This level of organization is not an absolute prerequisite for synaptic transmission, but is required for optimized synaptic function. Second, there is reciprocal signaling between the pre- and postsynaptic cells to achieve this functional optimization. This reciprocal signaling occurs during the formation and maturation of the NMJ, and also in adulthood, in cases of plasticity due to changes in synaptic efficacy or motor neuron loss. Third, components of the ECM serve critical signaling and structural roles during the formation and maturation of the synapse. Of these principles, the first two clearly apply to CNS synapses also. The role of the ECM may be somewhat specific for the NMJ, but analogous secreted signaling factors and transmembrane adhesion molecules serve these purposes in the brain (see Chapter 11).

## 3. SIGNALING BY THE ECM IN SYNAPTOGENESIS: AGRIN

Studies on synaptogenesis in regenerating nerve and muscle preparations have illustrated that the cue or cues directing synapse formation are to be found in the ECM, or "basal lamina ghosts"[1,2]. Many putative signaling molecules have been tested, but agrin quickly emerged as the strongest candidate. Agrin was identified and named based upon its ability to aggregate AChRs into clusters on cultured myotubes, forming sites of postsynaptic differentiation that closely resembled the sites seen on developing muscle fibers *in vivo*[3]. Although this AChR clustering activity was detectable in protein homogenates from many sources, it was ultimately purified by a combination of biochemical and antibody approaches using ECM prepared from the *T. californica* electric organ. The cDNA sequence was determined shortly thereafter[4,5]. In 1990, Dr. U. J. McMahan put forward the "Agrin hypothesis," stating that agrin is the nerve-derived organizer of postsynaptic differentiation at the NMJ[6].

The agrin hypothesis has been beset by various challenges in the last 15 years, but has largely proven correct. Initial alarm was caused by the finding that agrin was made by muscle as well as motor neurons, thus making it unclear how it could be the *nerve-derived* organizer of the NMJ. This quandary was quickly resolved with the discovery of alternatively spliced forms of agrin (Figure 1.2).

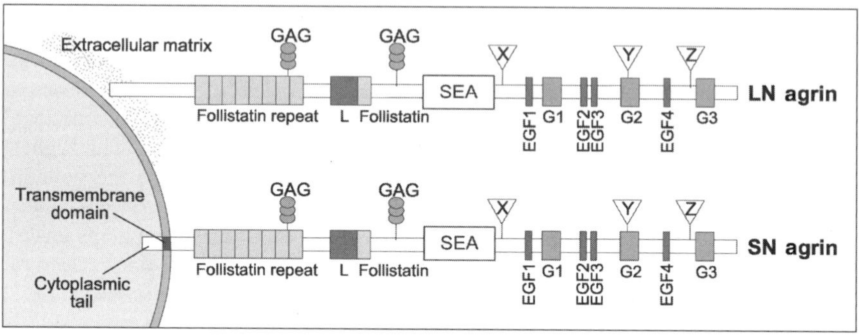


**Figure 1.2.** The Agrin Protein. Agrin is a heparan sulfate proteoglycan of almost 2,000 amino acids. At its N-terminus, agrin has nine follistatin-like repeats that also have homology with Kazal-type protease inhibitor domains. Also in the N-terminal half are the sites of glycosaminoglycan (GAG) addition. This post-translational modification is usually heparan sulfate, but may sometimes include chondroitin sulfate. The SEA domain is proposed to mediate interactions with other extracellular O-linked glycoproteins. In the C-terminal half of the protein are four EGF-like repeats and three laminin-type globular G-domains (G). Agrin is alternatively spliced in its C-terminal half. The X-splice site is an alternative splice acceptor site in exon 20 of the agrin gene of unknown significance. The Y-splice site includes an exon of just 12 base pairs, encoding four amino acids. Inclusion of these amino acids confers heparin-binding activity to agrin. The Z-splice site involves two exons of 24 and 33 bp, resulting in the inclusion of eight, 11, or the combined 19 amino acids. The inclusion of these amino acids is specific for neuronally expressed isoforms of this protein, and these amino acids are necessary for MuSK activation and the AChR clustering activity of agrin. The N-terminus of agrin has two variants resulting from alternative transcriptional start sites and thus different translational start sites. The LN form of agrin is encoded by two distinct exons before joining the common sequence, resulting in an isoform of the protein that has a signal peptide for secretion from the cell and a domain that interacts with the γ1 chain of laminins for assembly into the ECM. This is the predominant non-neuronal form of agrin, and it is also expressed by motor neurons, where it becomes anchored into the ECM of the NMJ. The other isoform, SN agrin, is encoded by a single unique exon before its transcript rejoins the common sequence at the same point as LN agrin. SN agrin is the primary form of agrin found in the brain. The protein is a type-2 transmembrane protein that remains associated with the cell surface. Its role in the brain and on the surface of neurons is under investigation.

Isoforms containing an insertion at a C-terminal alternative splice site (called the Z+ splice forms) were found to be active at inducing AChR clusters in cultured muscle fibers, while isoforms that did not include these amino acids (Z− isoforms) were inactive[7]. Furthermore, the active isoforms were made only in the nervous system, while muscle and other non-neuronal tissues made the inactive isoforms[8,9]. Thus, while agrin is made by both nerves and muscles, only nerves made isoforms of the protein that were active in AChR clustering, refining but not disproving the agrin hypothesis.

### 3.1. Genetic Tests of the Agrin Hypothesis

The agrin hypothesis has also endured *in vivo* testing using genetic approaches in mice. Mice that lack agrin die at birth with nonfunctional NMJs[10]. During embryonic mouse development, motor neurons contact muscle fibers at approximately embryonic day 12–13 (E12–13). Examination of NMJs as early as E15 reveals that postsynaptic differentiation is nearly absent in mice lacking agrin.

This phenotype was initially described in mice in which the nervous-system-specific exons encoding the Z+ isoforms (the exons that make agrin active in AChR clustering assays) were deleted by homologous recombination. These mice also had severely reduced expression of all other agrin isoforms, including those forms expressed by muscle. The same phenotype was seen in other engineered alleles of *agrin*. Mice lacking all agrin after the fifth N-terminal follistatin repeat (Figure 1.2)[11] and mice lacking only the Z-exons with normal expression of all other agrin isoforms[12] have the same phenotype, a marked reduction in AChR clusters and postsynaptic differentiation, and an overgrowth of motor neurons well beyond their normal domain in the central end-plate band of the muscle.

A fourth allele of *agrin* confirmed the importance of agrin as a component of the ECM for NMJ formation[13]. Agrin has two alternative N-termini that result from transcripts with distinct transcriptional and translational start sites. One of these forms (LN agrin) is secreted and binds the γ1 chains of laminin trimers. This is the form made by most non-neuronal cell types, and also by motor neurons. The second agrin isoform is a type-2 transmembrane protein that is the predominant form in most of the CNS[14,15]. The LN isoform of agrin was eliminated by a gene-trap insertion between the LN encoding exons and the SN exon. Therefore LN transcripts were selectively intercepted without disrupting the expression of the SN isoform. Mice homozygous for this gene-trap insertion lack agrin in basal laminae throughout the body and show an NMJ phenotype identical to that of other *agrin* mutants. However, SN agrin is still expressed in the CNS of these animals, and homogenates of CNS tissue are still active in AChR receptor clustering assays, while homogenates from mice lacking the bulk of the *agrin* coding sequence are not. These results indicate that transmembrane SN agrin is capable of inducing AChR clusters *in vitro*, but it is insufficient *in vivo*.

The function of SN agrin in the CNS is under investigation. Targeting the SN encoding exon results in mice that are viable despite an 80% reduction of agrin in the brain (our unpublished data). However, these mice have normal NMJ junction morphology and show no behavioral signs of neuromuscular dysfunction, indicating that transmembrane SN agrin is not necessary for NMJ formation, while the ECM-bound LN isoform is essential.

### 3.2. Agrin's Postsynaptic Mechanism

The postsynaptic signal transduction mechanisms for agrin have also been examined (Figure 1.3). Muscle specific kinase (MuSK) is a receptor tyrosine kinase that is essential for agrin signal transduction[16]. Mice that lack MuSK show a phenotype similar to *agrin*-knockout mice, with a complete lack of AChR clustering (even more severe than the *agrin*-knockout phenotype) and a similar presynaptic phenotype[17]. In addition, muscle fibers cultured from *MuSK*-knockout mice do not respond to agrin, and agrin is capable of activating MuSK in a variety of heterologous *in vitro* systems. However, direct binding of agrin and MuSK has never been achieved, leading to the hypothesized existence of an accessory factor termed MASC[16].

In addition to MuSK, agrin acts through an intracellular effector protein, rapsyn. Rapsyn serves to scaffold together AChRs in the plasma membrane and anchor them to the actin cytoskeleton[18,19]. In this scaffolding role, rapsyn functions similar to gephryn at glycinergic synapses or PDZ-domain proteins at other CNS synapses. Targeted elimination of the *rapsyn* gene in mice also results in a complete failure of AChR clustering, although the receptors still appear to be more concentrated, albeit diffusely, in the central portion of the muscle and the extent of the presynaptic overgrowth is less severe in *rapsyn* knockouts than in *agrin* and *MuSK* knockouts[20].

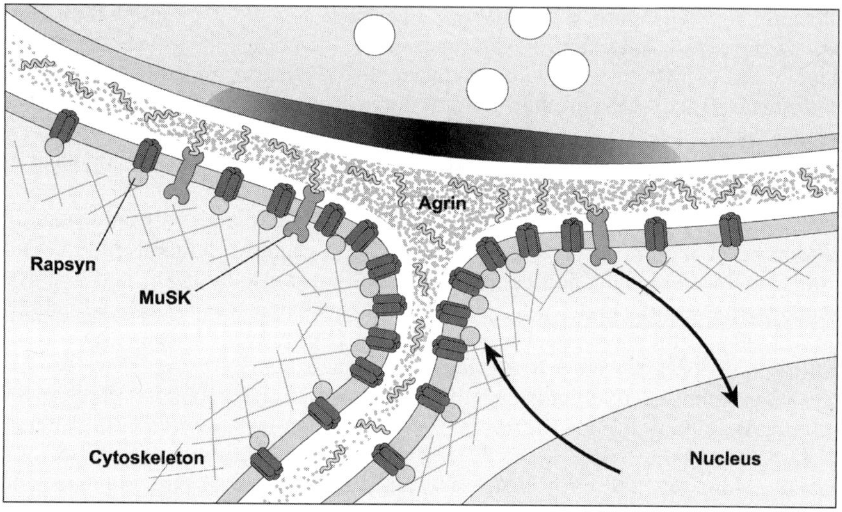


**Figure 1.3.** Agrin at the Neuromuscular Junction. Agrin is an extracellular matrix molecule that is essential for neuromuscular junction development. Agrin is made by both the motor neuron and the muscle fiber, but due to cell-type-specific alternative splicing, different protein isoforms are produced. Only those isoforms made by neurons are active in NMJ development. Agrin's signal is transduced by a receptor tyrosine kinase, Muscle Specific Kinase (MuSK). MuSK is also essential for postsynaptic differentiation. MuSK signaling influences postsynaptic gene expression and the protein has direct physical roles in the aggregation of AChRs. Rapsyn is an intracellular effector of agrin/MuSK signaling. It scaffolds AChRs and other postsynaptic proteins into clusters at the synapse and links these to the actin cytoskeleton. The loss of any of these proteins results in a failure of NMJ formation.

The *rapsyn*-knockout mice also illustrate that there is more to postsynaptic differentiation than AChR clustering. In particular, the myonuclei that fall directly below the nerve terminal (synaptic nuclei) have a different pattern of gene expression than nuclei in the rest of the syncytial muscle fiber (extrasynaptic nuclei). The synaptic nuclei are located near the plasma membrane and are closely associated with other components of the postsynaptic apparatus. In *rapsyn*-knockout mice, the pattern of synaptic gene expression for transcripts such as AChR subunits is not as severely disrupted as it is in *agrin* or *MuSK* knockouts[21]. These results suggest that agrin signaling through MuSK has an effect on postsynaptic gene expression as well as receptor clustering, while rapsyn is a more downstream effector, more important for the physical aggregation of the receptors than for signal transduction.

The complete path from agrin and MuSK to the nucleus has not been determined. Intracellular kinases including src, fyn, yes, and abl have been shown to localize to the NMJ and to have a role in signaling there[22,23]. The src, fyn, and yes kinases are more important for the maintenance of the NMJ than for its formation, whereas the abl family kinases may be directly downstream of MuSK and necessary for its signal amplification. Furthermore, transcription factors such as the Ets-domain protein, GABP, are important for postsynaptic gene expression and the promoter sequences of several synapse-specific genes such as *utrophin*, AChE, and subunits of the AChR, contain a six-base-pair upstream regulatory element, the N-box sequence[24,25]. This transcriptional mechanism is stimulated by both agrin and neuregulin signaling and relies on MAPK- and JNK-signaling pathways. Synapse-specific gene expression in synaptic nuclei of muscle fibers may be analogous to synapse-specific translation of dendritically localized mRNAs in the CNS, providing a mechanism for local synthesis of proteins important for the formation, maintenance and plasticity of the synapse.

Therefore, agrin may still represent the best example of a synaptogenic molecule to date. *In vitro*, purified or recombinant agrin is capable of inducing sites of postsynaptic differentiation on cultured muscle fibers. *In vivo*, expression of Z+ forms of agrin in muscle causes AChR aggregation independent of nerves [26,27]. Furthermore, the disruption of the *agrin* gene results in a loss of postsynaptic differentiation in muscle[10,12,13]. However, recent results suggest that it is too simplistic to conclude that agrin induces postsynaptic differentiation.

### 3.3. Induction versus Stabilization of Postsynaptic Differentiation by Agrin

In mice with mutations that result in a complete absence of motor innervation to a muscle, such as *HB9* or *topoisomerase2β* mutations, a pattern of postsynaptic differentiation and AChR clustering appears in the embryonic muscle at about the same time a motor neuron would normally be making contact (E12–13 in the mouse)[28,29]. This pattern persists until birth, although it is not as well defined spatially or as robust as in wild-type mice. However, it is dramatically better than in agrin-knockout mice. Therefore, the complete *absence* of a nerve leads to more persistent postsynaptic differentiation in muscle than one sees in the *presence* of a nerve lacking agrin. Interestingly, in double-knockout experiments, this postsynaptic differentiation was shown to still be dependent on MuSK, but to be completely independent of agrin[11,30]. There are a number of conclusions to be drawn from this work. First, muscle appears to have an intrinsic program of differentiation that is not dependent on cues from the nerve. Second, this program of differentiation does depend on MuSK, either through activation by a non-agrin, non-neuronal ligand, or through autoactivation of the receptor, as suggested by the strong inhibitory effects of genetic dosage seen in *MuSK* heterozygous animals that also lacked innervation. Third, the presence of a nerve that lacks agrin is in some way able to disperse the sites of postsynaptic differentiation assembled by the muscle.

Acetylcholine itself has emerged as at least one possible signal for the dispersal of muscle-derived sites of innervation. In mice lacking choline acetyltransferase (ChAT), the enzyme that synthesizes ACh, innervated NMJs form even in the absence of cholinergic transmission[31,32]. However, abnormalities are present, including a large number of noninnervated AChR clusters. In *agrin/ChAT* double-knockout mice, postsynaptic differentiation is significantly restored compared to *agrin* knockouts [33,34]. However, the pattern is still not as extensive as in muscle that completely lacks innervation, suggesting additional signals may be involved. The intracellular kinase CDK5 appears to be a part of the mechanism by which muscle fibers disperse aneural receptor clusters, and the conclusions *in vivo* are supported by *in vitro* analyses, in which agrin-induced AChR clusters were more stable in the face of ACh application than spontaneously arising clusters.

Thus, agrin's precise role in NMJ formation may be to stabilize muscle-derived sites of postsynaptic differentiation against dispersal activities of cholinergic activity, and possibly other signals, rather than inducing them *de novo* upon nerve contact. However, the fact that recombinant Z+ isoforms agrin are capable of inducing postsynaptic differentiation, both *in vitro* and when expressed in muscle *in vivo*, cannot be ignored. The final answer may therefore be that agrin from the motor neuron has both activities: stabilizing a muscle-derived site of AChR accumulation if it is encountered, and inducing a new site if a pre-existing site is not available.

Recent studies in zebrafish using time-lapse visualization of both postsynaptic receptor clusters and presynaptic motor neurons support this idea[35]. In early-forming synapses, motor neurons found and stabilized sites of AChR clustering already initiated by the muscle. However, in later-forming synapses, AChR clusters appeared at points of nerve contact, presumably induced by neuronally derived agrin.

### 3.4. Extrapolation to Agrin in the CNS

How completely the signaling function of agrin at the NMJ will apply to CNS synaptogenesis remains to be determined. While CNS synapses have comparable pre- and postsynaptic specializations, CNS synapses are typically much smaller and may not require the degree of signal amplification that the MuSK cascade provides at the NMJ. As subsequent chapters of this book will demonstrate, many CNS synaptogenesis mechanisms are based on adhesion molecules and direct contact of transmembrane proteins spanning the narrower synaptic cleft (Chapters 4–10), although the role of ECM molecules in formation and function of certain subtypes of central synapses is emerging (Chapter 11). Agrin clearly has effects on neurons[36–38] (Chapter 11); however, whether agrin is functioning in CNS synaptogenesis is ambiguous. Several *in vitro* studies in which agrin is acutely perturbed suggest a role in synaptogenesis[39,40], while other studies, including those in knockout mice and in primary cultures from those mice, suggest little effect[41,42]. The notion of agrin stabilizing synaptic connections against the disruptive effects of synaptic activity may also require a reassessment of the anticipated synaptic phenotype for agrin in the CNS. Recent work suggests that agrin may mediate a switch from gap junction connectivity to chemical synapses[43], and cholinergic synapses in sympathetic ganglia are impaired but not abolished in *agrin*-knockout mice[44]. The presence of the CNS-specific transmembrane form of agrin is also provocative. However, additional experiments will be required to determine the extent to which this NMJ signaling molecule may also function in the brain.

## 4. SUBSYNAPTIC ARCHITECTURE OF THE NMJ: LAMININS

The ECM is also clearly essential for establishing and maintaining the subsynaptic architecture of the NMJ. Details such as the alignment of presynaptic active zones with postsynaptic junctional folds and AChRs require *trans*-synaptic coordination of the molecular components of these complexes. Often this coordination will rely on cell-surface receptors, but it also requires an appropriate and precise subsynaptic organization of the ECM. There are many specialized components of the synaptic ECM, but the laminins are among the most interesting and have been well characterized for their role in synaptogenesis.

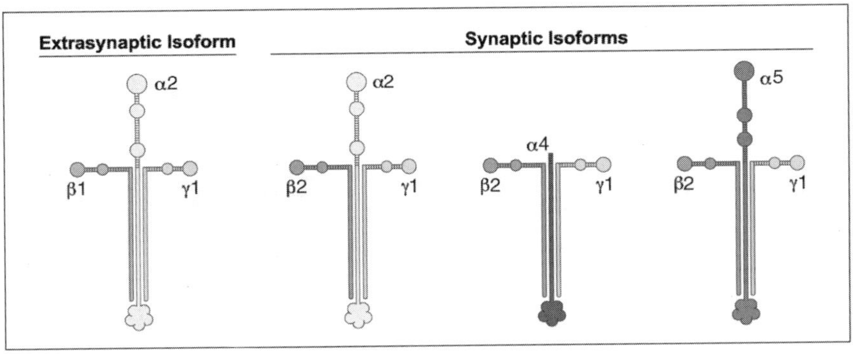


**Figure 1.4.** Laminin Protein Structure. Laminins are trimeric extracellular matrix molecules comprising an alpha chain, a beta chain, and a gamma chain. The total molecular weight of the trimers varies from 500 to 1000 kD. Alpha chains have N-terminal globular domains (domains IV and VI) separated by EGF-like repeats. This structure varies amongst the alphas, with α4 being truncated at its N-terminus and α5 being the longest, with additional EGF repeats present. The beta and gamma chains share this globular domain/EGF repeat structure at their N-termini. The central domain of the alpha chains and the C-terminal portion of the beta and gamma chains are coiled coil domains that are responsible for the assembly of the trimers. The alpha chains have an additional five globular domains (G domains) at their C-terminus. The laminins interact with many other ECM proteins. The N-terminal and C-terminal globular domains of the alpha chain both mediate interactions with integrins and heparin. Dystroglycan also binds the C-terminal G domains of the alpha chains. The γ1 chain mediates interactions with nidogen/entactin and agrin.

### 4.1. Laminin Proteins

Laminins are trimeric molecules consisting of an alpha chain, a beta chain, and a gamma chain, arranged in a cross-like structure with their C-termini coiled (Figure 1.4).[45] Mammals express at least five alpha chains, three beta chains, and three gamma chains; however, not all of these are present at the NMJ. Distinct genes encode each subunit, and the expression patterns are often dynamic during development. The laminin trimers have several domains in common, including N-terminal globular domains, EGF repeats, coiled-coil domains, and C-terminal globular domains (Figure 1.4). The laminins are known to act as ligands for integrins and other cell-surface receptors such as dystroglycan.

### 4.2. Laminin Localization

Laminins are dynamically and precisely localized on muscle fibers, and even within the synaptic cleft (Figure 1.5). In embryonic muscle, one finds trimers of α5, β1, and γ1. However, during late embryogenesis, α5 is gradually replaced by α2 in most of the muscle, but α5 expression persists at the NMJ and the protein is localized throughout the synapse[46,47]. Similarly, laminin trimers containing β2 (or S-laminins for synaptic laminins) are found only at the NMJ in muscle[48]. Trimers containing laminin α4 are also restricted to the NMJ, and show even greater specificity in their localization, flanking the active zone and junctional folds, effectively surrounding the actual sites of synaptic transmission[49]. Such specific localizations suggest synaptic functions for these laminins. Additional biochemical and genetic data have proved this to be true.

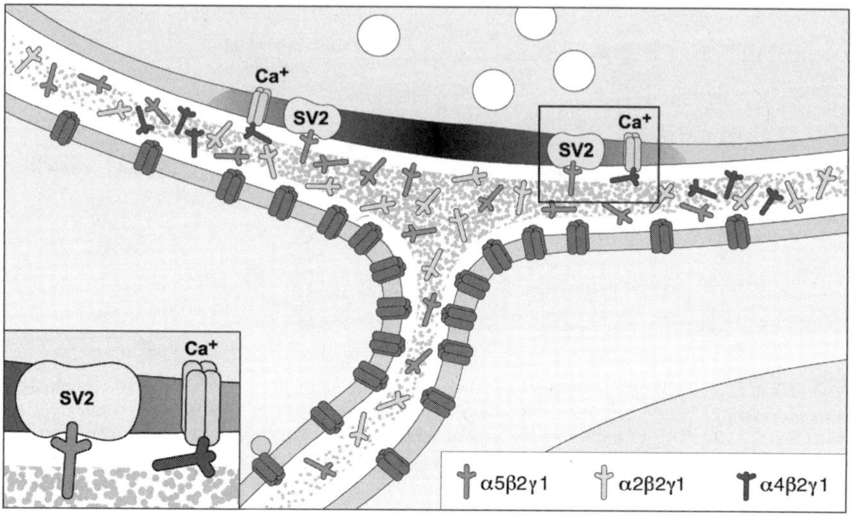


**Figure 1.5.** Laminin within the Neuromuscular Junction. Laminins have differential localizations within the muscle ECM and the neuromuscular junction. In extrasynaptic portions of the muscle ECM, the predominant laminin trimer is composed of α2, β1, and γ1 chains. Within the synapse, the β1 chain is largely replaced by β2. α2-containing trimers are also found throughout the ECM of the synapse, as are α5-containing trimers. α4-containing trimers are found only in the region of the synapse flanking the active zone. The laminins interact with many other proteins, but of particular interest at the NMJ are interactions of α5 with the presynaptic protein SV2, and β2 with presynaptic voltage-gated calcium channels. Through such interactions, the laminins organize the subsynaptic architecture of the NMJ.

## 4.3. Roles for Laminins in Synaptic Organization

Mutations in *laminin α2* (*Lama2*) cause muscular dystrophies and myelination defects in humans and mice. The α2 chain is the primary alpha chain in most of the extra-synaptic muscle ECM, but it is also found throughout the synapse, superimposed on the laminin α5 localization. In addition to myelination defects and muscular dystrophy, *laminin α2* mutant mice (the spontaneous *dy* mutation, and also engineered alleles) show mild defects in synaptic morphology and function, including a loss of junctional folds[50]. The other nerve and muscle phenotypes of these mice make the impact of the mutation on NMJ function difficult to assess, but transmission failures do occur in these mutant junctions[51].

Genetic analysis of laminin α4 function in mice has also confirmed its role as a synaptic organizer. Mice with a targeted mutation in the *laminin α4* gene (*Lama4*) no longer have active zones that are in direct apposition to the junctional folds of the muscle fiber[49]. Also, these active zones are sometimes "split," as though components of the presynaptic density are no longer anchored in the presynaptic membrane. Essentially, the pre- and postsynaptic specializations are no longer in precise register. This phenotype is subtle, requiring ultrastuctural analysis to detect the differences, but it provides an excellent example of how ECM components such as laminin α4 are responsible for establishing the subsynaptic organization of the NMJ.

Laminin β2 has been even more extensively studied, and its role in synaptogenesis at the NMJ is broader than that of laminin α4. The gene was targeted in mice, and the resulting homozygotes show NMJ abnormalities[52]. Unlike in agrin mutants, the NMJs form, but they fail to mature into an adult morphology, the terminal never becomes polarized, with vesicles remaining diffusely localized throughout the terminal, and very few active zones form. The terminals eventually become functionally compromised. A contributing factor for this may be the invasion of the synaptic cleft by the terminal Schwann cell[53], decreasing the efficacy of the synaptic contact. Consistent with this, β2-containing laminin-11 (α5, β2, γ1) presents a nonpermissive substrate for Schwann cell processes *in vitro*.

Like β2, the α5 chain is distributed throughout the synaptic cleft, and some of the effects of laminin β2 are likely to be mediated, at least in part, by trimers that also contain α5 (laminin-11). Purified preparations of laminin-11 cause neurites to stop and differentiate *in vitro*[54,55]. This effect is partially recapitulated by purified β2 alone, but α5 knockout mice (*Lama5*) have embryonic nerve-terminal defects that are more severe than those seen in the β2 knockout, suggesting that the α5 chain is also playing an active role in this process.

The importance of laminins has also been illustrated by biochemical analyses of their protein–protein interactions. Laminins containing the α5 chain can bind to SV2[56], and β2-containing laminins bind to presynaptic voltage-gated $Ca^{2+}$ channels [57]. These β2-containing trimers also contain the α4 chain. The disruption of the interaction between β2 and the $Ca^{2+}$ channels results in a disassembly of the presynaptic active zone, similar to that seen in the autoimmune Lambert–Eaton myasthenia[58]. Through such interactions, the laminins may functionally organize the presynaptic membrane components within the active zone and insure their alignment with postsynaptic specializations in the muscle.

## 5. PRESYNAPTIC DIFFERENTIATION FACTORS

The laminins bind to presynaptic proteins and mutations in laminin genes cause defects in presynaptic differentiation. However, these mutations do not cause a total failure of the motor axon to stop and differentiate. Interestingly, the only mutations where a nearly complete failure in *presynaptic* differentiation is seen are mice lacking agrin or MuSK, or even rapsyn, all proteins primarily implicated in *postsynaptic* differentiation. Agrin can serve directly as a signal for stopping neurite outgrowth and presynaptic differentiation *in vitro*[59–61]. However, in mutant mice that lack only the alternatively spliced "active" Z exons and express normal levels of muscle agrin, the presynaptic overgrowth phenotype is still present and is as severe as the null allele of *agrin*[12]. This suggests that agrin in the muscle ECM is not directly responsible for presynaptic differentiation of motor-nerve terminals *in vivo*. Indeed, the possibility of rapsyn, an intracellular scaffolding protein, interacting directly with nerve terminals is unlikely. Therefore, the conclusion to be reached is that presynaptic differentiation fails in these mutant mice secondary to the failure of postsynaptic differentiation. Presumably, a retrograde signal (or signals) that is generated in the muscle is no longer produced when the agrin-signaling pathway is perturbed.

While the laminins may be contributing components to this presynaptic differentiation signal, they are clearly not the only components. Furthermore, it is not clear whether this signal is a diffusible factor acting non-cell-autonomously, or an adhesion molecule acting directly on individual nerve terminals and muscle fibers, or a combination of such factors.

### 5.1. Ubiquitination Pathways

Candidate factors do exist for such a signal. In inter-neuronal synapses, signals including FGFs, WNTs, neurotrophins, and neurexin/neuroligin interactions have been implicated; however, it is unclear to what extent these factors play a role in synaptogenesis at the NMJ. Another system implicated in the determination of presynaptic morphology is ubiquitination[62]. This was first reported in *Drosophila* and *C. elegans*, where mutations in *highwire* and *rpm1* (regulator of presynaptic morphology 1), respectively, cause defects in presynaptic differentiation, leading to ineffective synaptic transmission[63–65]. The *highwire* and *rpm1* genes encode E3 ring finger ubiquitin ligases. The mammalian ortholog of this gene is protein associated with Myc (*PAM*)[66]. An analogous role in mammalian presynaptic differentiation was indicated by an analysis of mice lacking *Phr1* (for PAM-highwire-rpm1), generated by combining narrowly overlapping deficiencies[67]. These mice die at birth with apparently ineffective NMJ synaptic transmission, and defects in both sensory and motor-nerve terminal morphology. How ubiquitination influences presynaptic morphology and synaptic function is intriguing, but largely unanswered.

## 6. OTHER MECHANISMS AT THE NMJ

A wide variety of other mechanisms have been suggested to influence the formation of the NMJ. To some extent, data support all of them, indicating the complexity of forming such a specialized structure; however, none are as molecularly well-defined as agrin signaling for the formation of the NMJ or the laminins for NMJ maturation, maintenance, and microarchitecture.

### 6.1. ARIA/Neuregulin

Prominent among these other mechanisms is neuregulin, or acetylcholine receptor inducing activity (ARIA). Neuregulin induces the expression of AChR genes and plays an important role in the patterning of postsynaptic gene expression[68]. The neuregulin signal is transduced by ErbB receptors, of which ErbB2, ErbB3, and ErbB4 are present at the NMJ[69–71]. The function of neuregulin, however, is not in the early stages of synapse formation, but rather in maintaining the pattern of postsynaptic gene expression in the muscle following NMJ formation. This is supported by the phenotype of mice lacking isoforms of neuregulin containing the cysteine-rich domains, and by the simultaneous elimination of ErbB2 and ErbB4 in muscle[72,73]. In both cases, the early stages of NMJ formation are normal but motor-nerve terminals eventually degenerate. In both cases, the degeneration is consistent with a loss of Schwann cells, and not a primary defect in neuromuscular connectivity. These results demonstrate that synaptogenesis proceeds even with severely perturbed neuregulin signaling; they also highlight the importance of Schwann cells in NMJ maturation and maintenance.

### 6.2. Schwann Cells

The role of Schwann cells in NMJ stability should not be overlooked. The Schwann cells migrate along the motor axons and arrive at the synapse immediately after the axons themselves contact the muscle. If Schwann cells fail to migrate, the early connections formed by the motor axons quickly degenerate, and the axons retract[74,75]. This presumably reflects a critical trophic role for Schwann cells in maintaining the viability of the axon. In adulthood, Schwann cells are also essential for NMJ plasticity. Sprouts from the terminal Schwann cell of a denervated NMJ seek neighboring active synapses and stimulate the neighboring motor neurons to sprout and reinnervate the lost junction[76,77]. Thus, Schwann cells are essential for the process as a whole, but the intimate interactions between motor axons and the glia make it difficult to determine whether the effects are directly attributable to the Schwann cell, or whether the Schwann cell is indirectly enabling the motor axon to form and maintain its connections. A role for glia in synapse formation, stabilization, and plasticity is undoubtedly a shared feature of NMJs and CNS synapses[78].

Numerous other systems are just beginning to be explored. Integrins are present at the NMJ and may well play a role in its development[79], but their analysis is complicated by their function in many developmental pathways, which often obscure their effect on synaptogenesis. Similarly, synapse-specific glycosylation is also intriguing as a possible mechanism through which synapses may be organized or stabilized[80]. Studies of this are underway, but are technically challenging. The importance of glycosyltransferases for the neuromuscular system is established by diseases such as Fukuyama's muscular dystrophy, muscle–eye–brain disease, and the *Large* mutations in mice[81–83]. While some aspects of these diseases may be attributed to improper glycosylation of dystroglycan, a comprehensive understanding of the substrates and mechanisms of these enzymes remains to be achieved.

## 7. CONCLUSIONS

The vertebrate NMJ provides a powerful model system for studies in synaptogenesis. The large number of factors that influence NMJ formation demonstrates the complexity of establishing such as specialized site of cell–cell contact. However, roles for the ECM, and particularly agrin signaling and laminins, have been well characterized. The extent to which these functions will be preserved in the CNS remains to be seen, but analogous processes are at work, and the principles described in work on the NMJ provide a valuable context in which to evaluate other synaptogenic mechanisms.

## 8. REFERENCES


1. Sanes, J.R., Marshall, L.M., and McMahan, U.J. (1978) *J Cell Biol* **78**, 176–198.
2. Burden, S.J., Sargent, P.B., and McMahan, U.J. (1979) *J Cell Biol* **82**, 412–425.
3. Godfrey, E.W., Nitkin, R.M., Wallace, B.G., Rubin, L.L., and McMahan, U.J. (1984) *J Cell Biol* **99**, 615–627.
4. Tsim, K.W., Ruegg, M.A., Escher, G., Kroger, S., and McMahan, U.J. (1992) *Neuron* **8**, 677–689.
5. Rupp, F., Payan, D.G., Magill-Solc, C., Cowan, D.M., and Scheller, R.H. (1991) *Neuron* **6**, 811–823.
6. McMahan, U.J. (1990) *Cold Spring Harb Symp Quant Biol* **55**, 407–418.
7. Ferns, M., Hoch, W., Campanelli, J.T., Rupp, F., Hall, Z.W., and Scheller, R.H. (1992) *Neuron* **8**, 1079–1086.
8. Ferns, M., Campanelli, J.T., Hoch, W., Scheller, R.H., and Hall, Z.M. (1993) *Neuron* **11**, 491–502.
9. Hoch, W., Ferns, M., Campanelli, J.T., Hall, Z.W., and Scheller, R.H. (1993) *Neuron* **11**, 479–490.
10. Gautam, M., Noakes, P.G., Moscoso, L., Rupp, F., Scheller, R.H., Merlie, J.P., and Sanes, J.R. (1996) *Cell* **85**, 525–535.
11. Lin, W., Burgess, R.W., Dominguez, B., Pfaff, S.L., Sanes, J.R., and Lee, K.F. (2001) *Nature* **410**, 1057–1064.
12. Burgess, R.W., Nguyen, Q.T., Son, Y.J., Lichtman, J.W., and Sanes, J.R. (1999) *Neuron* **23**, 33–44.
13. Burgess, R.W., Skarnes, W.C., and Sanes, J.R. (2000) *J Cell Biol* **151**, 41–52.
14. Burgess, R.W., Dickman, D.K., Nunez, L., Glass, D.J., and Sanes, J.R. (2002) *J Neurochem* **83**, 271–284.
15. Neumann, F.R., Bittcher, G., Annies, M., Schumacher, B., Kroger, S., and Ruegg, M.A. (2001) *Mol Cell Neurosci* **17**, 208–225.
16. Glass, D.J. et al. (1996) *Cell* **85**, 513–523.
17. DeChiara, T.M. et al. (1996) *Cell* **85**, 501–512.
18. Froehner, S.C., Luetje, C.W., Scotland, P.B., and Patrick, J. (1990) *Neuron* **5**, 403–410.
19. Phillips, W.D., Kopta, C., Blount, P., Gardner, P.D., Steinbach, J.H., and Merlie, J.P. (1991) *Science* **251**, 568–570.
20. Gautam, M., Noakes, P.G., Mudd, J., Nichol, M., Chu, G., Sanes, J.R., and Merlie, J. (1995) *Nature* **377**, 232–236.
21. Gautam, M., DeChiara, T.M., Glass, D.J., Yancopoulos, G.D., and Sanes, J.R. (1999) *Brain Res Dev Brain Res* **114**, 171–178.
22. Smith, C.L., Mittaud, P., Prescott, E.D., Fuhrer, C., and Burden, S.J. (2001) *J Neurosci* **21**, 3151–3160.
23. Finn, A.J., Feng, G., and Pendergast, A.M. (2003) *Nat Neurosci* **6**, 717–723.
24. Briguet, A. and Ruegg, M.A. (2000) *J Neurosci* **20**, 5989–5996.
25. Schaeffer, L., de Kerchove d'Exaerde, A., and Changeux, J.P. (2001) *Neuron* **31**, 15–22.
26. Cohen, I., Rimer, M., Lomo, T., and McMahan, U.J. (1997) *Mol Cell Neurosci* **9**, 237–253.
27. Meier, T., Hauser, D.M., Chiquet, M., Landmann, L., Ruegg, M.A., and Brenner, H.R. (1997) *J Neurosci* **17**, 6534–6544.
28. Yang, X., Li, W., Prescott, E.D., Burden, S.J., and Wang, J.C. (2000) *Science* **287**, 131–134.
29. Thaler, J., Harrison, K., Sharma, K., Lettieri, K., Kehrl, J., and Pfaff, S.L. (1999) *Neuron* **23**, 675–687.
30. Yang, X., Arber, S., William, C., Li, L., Tanabe, Y., Jessell, T.M., Birchmeier, C., and Burden, S.J. (2001) *Neuron* **30**, 399–410.
31. Misgeld, T., Burgess, R.W., Lewis, R.M., Cunningham, J.M., Lichtman, J.W., and Sanes, J.R. (2002) *Neuron* **36**, 635–648.
32. Brandon, E.P., Lin, W., D'Amour, K.A., Pizzo, D.P., Dominguez, B., Sugiura, Y., Thode, S., Ko, C.P., Thal, L.J., Gage, F.H., and Lee, K.F. (2003) *J Neurosci* **23**, 539–549.
33. Lin, W., Dominguez, B., Yang, J., Aryal, P., Brandon, E.P., Gage, F.H., and Lee, K.F. (2005) *Neuron* **46**, 569–579.
34. Misgeld, T., Kummer, T.T., Lichtman, J.W., and Sanes, J.R. (2005) *Proc Natl Acad Sci USA* **102**, 11088–11093.
35. Flanagan-Steet, H., Fox, M.A., Meyer, D., and Sanes, J.R. (2005) *Development* **132**, 4471–4481.
36. Mantych, K.B. and Ferreira, A. (2001) *J Neurosci* **21**, 6802–6809.
37. Ji, R.R., Bose, C.M., Lesuisse, C., Qiu, D., Huang, J.C., Zhang, Q., and Rupp, F. (1998) *J Neurosci* **18**, 9695–9702.
38. Hilgenberg, L.G., Hoover, C.L., and Smith, M.A. (1999) *J Neurosci* **19**, 7384–7393.
39. Bose, C.M., Qiu, D., Bergamaschi, A., Gravante, B., Bossi, M., Villa, A., Rupp, F., and Malgaroli, A. (2000) *J Neurosci* **20**, 9086–9095.

40. Ferreira, A. (1999) *J Cell Sci* **112**, 4729–4738.
41. Li, Z., Hilgenberg, L.G., O'Dowd, D.K., and Smith, M.A. (1999) *J Neurobiol* **39**, 547–557.
42. Serpinskaya, A.S., Feng, G., Sanes, J.R., and Craig, A.M. (1999) *Dev Biol* **205**, 65–78.
43. Martin, A.O., Alonso, G., and Guerineau, N.C. (2005) *J Cell Biol* **169**, 503–514.
44. Gingras, J., Rassadi, S., Cooper, E., and Ferns, M. (2002) *J Cell Biol* **158**, 1109–1118.
45. Colognato, H. and Yurchenco, P.D. (2000) *Dev Dyn* **218**, 213–234.
46. Patton, B.L., Miner, J.H., Chiu, A.Y., and Sanes, J.R. (1997) *J Cell Biol* **139**, 1507–1521.
47. Miner, J.H., Patton, B.L., Lentz, S.I., Gilbert, D.J., Snider, W.D., Jenkins, N.A., Copeland, N.G., and Sanes, J.R. (1997) *J Cell Biol* **137**, 685–701.
48. Hunter, D.D., Shah, V., Merlie, J.P., and Sanes, J.R. (1989) *Nature* **338**, 229–234.
49. Patton, B.L., Cunningham, J.M., Thyboll, J., Kortesmaa, J., Westerblad, H., Edstrom, L., Tryggvason, K., and Sanes, J.R. (2001) *Nat Neurosci* **4**, 597–604.
50. Tremblay, J.P., Gregoire, L., Sasseville, R., Guay, G., and Belhumeur, C. (1988) *Synapse* **2**, 148–156.
51. Edwards, J.P., Hatton, P.A., and Wareham, A.C. (1998) *Brain Res* **788**, 262–268.
52. Noakes, P.G., Gautam, M., Mudd, J., Sanes, J.R., and Merlie, J.P. (1995) *Nature* **374**, 258–262.
53. Patton, B.L., Chiu, A.Y., and Sanes, J.R. (1998) *Nature* **393**, 698–701.
54. Patton, B.L., Miner, J.H., Chiu, A.Y., and Sanes, J.R. (1997) *J Cell Biol* **139**, 1507–1521.
55. Son, Y.J., Patton, B.L., and Sanes, J.R. (1999) *Eur J Neurosci* **11**, 3457–3467.
56. Son, Y.J., Scranton, T.W., Sunderland, W.J., Baek, S.J., Miner, J.H., Sanes, J.R., and Carlson, S.S. (2000) *J Biol Chem* **275**, 451–460.
57. Sunderland, W.J., Son, Y.J., Miner, J.H., Sanes, J.R., and Carlson, S.S. (2000) *J Neurosci* **20**, 1009–1019.
58. Nishimune, H., Sanes, J.R., and Carlson, S.S. (2004) *Nature* **432**, 580–587.
59. Campagna , J.A., Ruegg, M.A., and Bixby, J.L. (1995) *Neuron* **15**, 1365–1374.
60. Campagna, J.A., Ruegg, M.A., and Bixby, J.L. (1997) *Eur J Neurosci* **9**, 2269–2283.
61. Chang, D., Woo, J.S., Campanelli, J., Scheller, R.H., and Ignatius, M.J. (1997) *Dev Biol* **181**, 21–35.
62. DiAntonio, A. and Hicke, L. (2004) *Annu Rev Neurosci* **27**, 223–246.
63. Wan, H.I., DiAntonio, A., Fetter, R.D., Bergstrom, K., Strauss, R., and Goodman, C.S. (2000) *Neuron* **26**, 313–329.
64. Schaefer, A.M., Hadwiger, G.D., and Nonet, M.L. (2000) *Neuron* **26**, 345–356.
65. Zhen, M., Huang, X., Bamber, B., and Jin, Y. (2000) *Neuron* **26**, 331–343.
66. Guo, Q., Xie, J., Dang, C.V., Liu, E.T., and Bishop, J.M. (1998) *Proc Natl Acad Sci USA* **95**, 9172–9177.
67. Burgess, R.W., Peterson, K.A., Johnson, M.J., Roix, J.J., Welsh, I.C., and O'Brien, T.P. (2004) *Mol Cell Biol* **24**, 1096–1105.
68. Falls, D.L., Rosen, K.M., Corfas, G., Lane, W.S., and Fischbach, G.D. (1993) *Cell* **72**, 801–815.
69. Zhu, X., Lai, C., Thomas, S., and Burden, S.J. (1995) *EMBO J* **14**, 5842–5848.
70. Moscoso, L.M., Chu, G.C., Gautam, M., Noakes, P.G., Merlie, J.P., and Sanes, J.R. (1995) *Dev Biol* **172**, 158–169.
71. Altiok, N., Bessereau, J.L., and Changeux, J.P. (1995) *EMBO J* **14**, 4258–4266.
72. Escher, P., Lacazette, E., Courtet, M., Blindenbacher, A., Landmann, L., Bezakova, G., Lloyd, K.C., Mueller, U., and Brenner, H.R. (2005) *Science* **308**, 1920–1923.
73. Wolpowitz, D., Mason, T.B., Dietrich, P., Mendelsohn, M., Talmage, D.A., and Role, L.W. (2000) *Neuron* **25**, 79–91.
74. Morris, J.K., Lin, W., Hauser, C., Marchuk, Y., Getman, D., and Lee, K.F. (1999) *Neuron* **23**, 273–283.
75. Lin, W., Sanchez, H.B., Deerinck, T., Morris, J.K., Ellisman, M., and Lee, K.F. (2000) *Proc Natl Acad Sci USA* **97**, 1299–1304.
76. Son, Y.J. and Thompson, W.J. (1995) *Neuron* **14**, 133–141.
77. Son, Y.J. and Thompson, W.J. (1995) *Neuron* **14**, 125–132.
78. Ullian, E.M., Sapperstein, S.K., Christopherson, K.S., and Barres, B.A. (2001) *Science* **291**, 657–661.
79. Martin, P.T., Kaufman, S.J., Kramer, R.H., and Sanes, J.R. (1996) *Dev Biol* **174**, 125–139.
80. Martin, P.T. (2003) *J Neurocytol* **32**, 915–929.
81. Martin, P.T. and Freeze, H.H. (2003) *Glycobiology* **13**, 67R–75R.
82. Michele, D.E., Barresi, R., Kanagawa, M., Saito, F., Cohn, R.D., Satz, J.S., Dollar, J., Nishino, I., Kelley, R.I., Somer, H., Straub, V., Mathews, K.D., Moore, S.A., and Campbell, K.P. (2002) *Nature* **418**, 417–422.
83. Levedakou, E.N., Chen, X.J., Soliven, B., and Popko, B. (2005) *Mol Cell Neurosci* **28**, 757–769.

# 2

# SYNAPSE FORMATION BETWEEN IDENTIFIED MOLLUSCAN NEURONS: A MODEL SYSTEM APPROACH

Ryanne Wiersma-Meems and Naweed I. Syed*

## 1. SUMMARY

From simple reflexes to complex motor patterns and learning and memory, all nervous system functions hinge upon the precise synaptic connectivity that is orchestrated during early development. The synaptogenic program does not stop with the cessation of development, rather it continues well into adulthood, forming the basis for synaptic plasticity that underlies learning and memory. Despite extensive advances in the field of neurodevelopment, the precise cellular and molecular mechanisms of synapse formation between central neurons remain largely unknown, due primarily to the complexity of developing mammalian brain and the rate at which the synaptogenic program proceeds (20,000–30,000synapses/ 10 min). Here we report on the utility of various molluscan models whereby various steps underlying synapse formation can be investigated at the level of individual neurons and synapses. In these models, synaptogenesis can be examined both *in vivo* during regeneration and following single-cell transplantation, as well as *in vitro* through a variety of cell culture approaches.

## 2. INTRODUCTION

Imagine $10^{15}$ neurons – all wired and interconnected through synapses into one motherboard – the brain, which in turn determines all that we do throughout our lives. Imagine also all those synaptic connections in different areas of this organ, being concurrently active, either transmitting or receiving information in a highly ordered manner. Also imagine the immaculate orchestration that would be required to connect this organ appropriately and the dire consequences if this wiring were to

*Departments of Anatomy and Physiology, Health Sciences Center, 3330 Hospital Dr. N.W., Calgary, AB, Canada T2N 4N1; nisyed@acs.ucalgary.ca

go haywire. Considering all this, it is then not so difficult to envisage challenges confronting our resolve to break the connectivity code of mammalian nervous system – let alone understand its functionality. The sheer numbers of mammalian neurons, their smaller sizes, and the rate at which the synaptic connectivity proceeds, all render it rather impossible to study synapse formation between defined sets of pre- and postsynaptic neurons. Invertebrates, on the other hand, and particularly mollusks, are endowed with relatively simple nervous systems, consisting of some 20,000–30,000 neurons. The molluscan neurons are often behaviorally well defined and have large somata that are readily identifiable on the basis of their size (50 μm–1 mm), position, and coloration. Moreover, injured adult molluscan neurons have the innate propensity to recapitulate their developmental patterns of connectivity with remarkable accuracy[1]. Although the snail models leg behind their fly and worm counterparts regarding the genetic know-how of synaptic connectivity, they are nevertheless amenable to direct electrophysiological analysis at the level of single pre- and postsynaptic neurons.

## 3. SYNAPSE FORMATION: LESSONS LEARNED FROM VARIOUS MOLLUSCAN MODELS

Among the molluscan species used extensively to define cellular and synaptic mechanisms of neurite outgrowth, target cell selection, and specific synapse formation, are the pond snails *Lymnaea stagnalis*[2,3] and *Helisoma trivolvis*[4], the land snail *Helix pomatia* and the sea hare *Aplysia californica*[5,6]. This chapter specifically focuses on various *in vivo* and *in vitro* techniques that are being used to reveal cellular and molecular mechanisms underlying synaptic connectivity.

### 3.1. *In Vivo* Regeneration and Synaptic Connectivity

In contrast with vertebrate and other invertebrate species (*Drosophila* and *C. elegans*) where the mechanisms underlying specific synapse formation are generally investigated in developing animals, the adult mollusks with their inherent regenerative capacity are most favored for similar studies. Behaviorally defined, injured neurons from a variety of molluscan species such as *Melampus*[7], *Lymnaea*[8] (Figure 2.1A; Colorplate 2) *Helisoma*[9,10], and *Aplysia*[11] have been shown to regenerate their axonal projections, not only *in vitro*, but also in the intact animals. In most instances, regeneration from injured central neurons is complete and results in functional recovery[7]. However, the mechanisms underlying specific synapse formation could still not be elucidated in these intact preparations, due mainly to the intricacies of the intact brain and the lack of knowledge regarding various intrinsic and extrinsic factors that facilitate synapse formation *in vivo*. A variety of *in vitro* cell culture techniques were thus developed that enabled the extraction of uniquely identifiable neurons from the CNS/central ring ganglia. These approaches have since revealed that individually isolated neurons not only regenerate their neuritic processes in cell culture, but also develop specific synapses which are similar to those seen *in vivo*. This innate propensity of molluscan neurons to reconnect *in vitro* has enabled a number of laboratories to define various steps and mechanisms underlying synapse formation.

### 3.2. The Molluscan Cell Culture Techniques

Large molluscan neurons, such as those of *Lymnaea* (Figure 2.1A; Colorplate 2), are often clearly discernable in the central ring ganglia, even under a dissection microscope. Neuronal identification is facilitated by their size, coloration (white to red), and unique position within the ganglia. Following enzymatic treatment, which softens the connective tissue surrounding the ganglia, and subsequent mechanical removal of the inner sheath that encapsulates the neurons, individual somata can be extracted by applying gentle pressure through a suction pipette attached to a micromanipulator (Figure 2.1B). The isolated neurons (somata with their intact axon stumps) are then plated either on plastic or poly-L-lysine-coated glass coverslips attached to tissue culture dishes (Figure 2.1C). Whereas *Lymnaea* and *Helisoma* brain conditioned medium (CM – contains trophic factors), prepared by incubating central ring ganglia (2 brains/ml) in defined medium (DM) over a period of several days, is used to induce neurite outgrowth from their respective neurons, most *Aplysia* neurons are first cultured in defined medium (DM – does not contain trophic factors), with hemolymph (*Aplysia* blood) added later to promote sprouting. Within hours of neuronal plating in growth permissive medium, growth cones (Figure 2.1D) emerge either from the axon stump or from the soma itself (Figure 2.1E). Although these molluscan growth cones are much larger in size (50–100 μM), they are structurally and functionally similar to their vertebrate counterparts. When cultured in CM, molluscan neurons exhibit extensive outgrowth (Figure 2.1F) and develop synapses that are similar to those seen *in vivo*.

### 3.3. *In Vitro* Reconstruction of Neuronal Networks

Several studies have shown that specific synaptic connections between pairs of pre- and postsynaptic neurons can reform in cell culture. True synapse specificity is, however, best tested at the network level, where neurons are concurrently challenged with multiple partners. It is also imperative to demonstrate that functionally defined networks of neurons are able to generate patterned activity in a manner similar to that of *in vivo*. The molluscan models were the first in which networks of functionally defined circuits were reconstructed in culture. Specifically, the neural network mediating gill withdrawal reflex in *Aplysia* was reconstituted in cell culture where the neurons not only re-established specific connections with their select partners but also exhibited synaptic plasticity that underlies learning and memory in the intact animals[12]. Similarly, the *Lymnaea* preparation was the first in which a three-cell network, comprising the respiratory central pattern generator (CPG) was reconstructed in culture. The co-cultured respiratory neurons not only re-established their specific synapses *in vitro*, but they also generated rhythmical patterned activity that was similar to that seen *in vivo*[13]. This *in vitro* reconstructed CPG has since been used to define fundamental mechanisms of synapse formation as well as respiratory rhythm generation. These studies underscore the importance of molluscan models for elucidating mechanisms underlying not only network connectivity but also how such interconnected neurons generate rhythmical patterns in a manner similar to those seen in the intact brain.

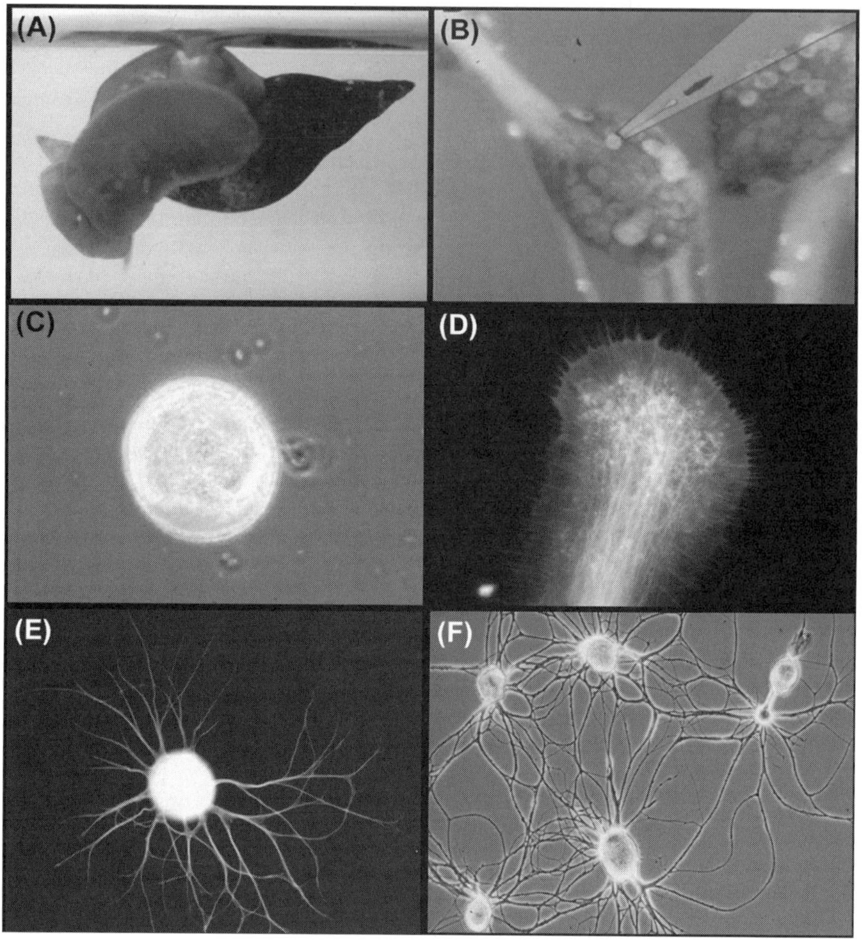


**Figure 2.1.** Cell Culture Techniques. (A) The pond snail *Lymnaea stagnalis*. (B) A cell extraction pipette is positioned to isolate individual neuronal somata. The arrow depicts neuronal extraction in progress. (C) An acutely isolated *Lymnaea* neuron in culture. (D) A *Lymnaea* growth cone fluorescently labeled with actin (red) and tubulin (green) antibodies. (E) An isolated neuron exhibiting neurite outgrowth overnight. (F) Interspecies synapse formation between *Lymnaea* and *Helisoma* neurons. A neuronal "hybrid" network reconstructed from two different snail species. Original copyright notice: Figure 2.1A reproduces Figure 3A on page 543 in *J Comp Physiology A*, volume 169, from the article by Syed, N.I., Harrison, D., and Winlow, W. (1991) Respiratory behavior in the pond snail *Lymnaea stagnalis*. Figures 2.1B, C and E reproduce Figures 2B, C (page 363), and 4C (page 372), respectively, of the book on *Modern Techniques in Neuroscience Research* (1999) Eds: Windhorst, U. and Johansson, H. Authors: Syed, N.I., Zaidi, H., and Lovell, P. Chapter 12: *In vitro* reconstruction of neuronal networks: a simple model system approach. All figures are reproduced with kind permission of Springer Science and Business Media. See Colorplate 2.

### 3.4. Transmitter–Receptor Interactions: A Mechanism for Synapse Specificity

When co-cultured with their synaptic partners, most molluscan neurons recapitulate their patterns of synaptic connections[1]. Intracellular recordings, concomitant with time lapse imaging of growth cones, revealed that synapses form within an hour of contact between pre- and postsynaptic neurons[14]. The process of specific target cell selection and subsequent synapse formation was shown to be facilitated by transmitter–receptor interactions between the approaching growth cones. Perturbation of either presynaptic transmitter release (dopamine) or postsynaptic neurotransmitter receptors, affected target cell selection but did not completely block synapse formation. It is important to note that transmitter release from the growth cone of an identified presynaptic *Lymnaea* neuron (Right Pedal Doral 1 = RPeD1) not only attracted growth cones from synaptic partners, but it also repelled nonpartner growth cones, thus avoiding synapse formation with inappropriate (synapses that do not exist *in vivo*) targets[14]. More recently, blocking cholinergic receptors during soma–axon pairing in *Lymnaea* was also shown to perturb synapse formation in cell culture[15]. It therefore seems safe to suggest that transmitter–receptor interactions between regenerating molluscan neurons play important roles in target cell selection and synapse formation, which in turn may define the early patterns of synaptic connectivity.

### 3.5. Synaptic Hierarchy: A Putative Mechanism for Determining Synapse Specificity

During early synapse formation, target cell contact is known to bring about specific changes in presynaptic transmitter release, and these changes can range from coupling of the secretory machinery to action potentials in snails[16], to switching of transmitter phenotype in rats[17]. The *Lymnaea* model has uncovered yet another novel interaction that occurs between reciprocally connected neurons during early synapse formation. Specifically, the two respiratory neurons visceral dorsal 4 (VD4) and RPeD1 paired in culture establish mutual inhibitory synaptic connections within 24 h. However, when examined during early stages of synapse formation (12–18 h), the cell VD4 wins over RPeD1 by being the first to establish inhibitory synapses with RPeD1. This is achieved through VD4-induced suppression of transmitter release from RPeD1. This suppression is transient and it involves peptide release from the VD4[18]. Once VD4 has fully established synapses with RPeD1, its suppressive control over RPeD1's secretory machinery is lifted, thus enabling RPeD1 to release transmitter and hence establish its inhibitory synapses with VD4. This study provides a unique example of "synaptic hierarchy" whereby neurons such as VD4 which control higher order behaviors (such as cardio-respiratory) outcompete other neurons for target occupancy. Together, these studies from *Lymnaea* show that transmitter–receptor interactions can serve many important developmental roles in mollusks, ranging from target cell selection, synapse formation, to establishing the synaptic hierarchy.

### 3.6. Interspecies Synaptogenesis between Molluscan Neurons

The molluscan preparations were the first in which it was demonstrated that the mechanisms underlying specific synapse formation are conserved across different snail species. Specifically, the giant dopaminergic neurons located in the right and left pedal ganglia of the snails *Lymnaea* and *Helisoma*, respectively, are considered homologs and innervate a variety of target cells in their respective species. To test whether the giant dopamine cells were indeed homologs, the *Lymnaea* dopamine cell (RPeD1) was paired with the postsynaptic targets of the *Helisoma* dopamine cell. The cells were allowed to exhibit outgrowth, which resulted in the formation of synapses between *Lymnaea* and *Helisoma* neurons[19,20] (Figure 2.1F). This study involving "neuronal hybrids" not only provided unequivocal evidence regarding the homologous nature of these cells but also demonstrated that the mechanisms of target cell selection and specific synapse formation are most likely conserved across the two related molluscan species. Therefore, it could be suggested that specific presynaptic neurons serving similar functions in a variety of molluscan species, may follow a common synaptogenic program to an extent that neuronal mixing and matching does not interfere with the mechanisms regulating synapse specificity in two different snails.

### 3.7. The Soma–Soma Synapse Model

Even in simpler model systems such as mollusks, synapses that develop between neurites are located at some distance from the somata and are thus inaccessible for direct morphological and electrophysiological analysis. Moreover, the precise timing and the numbers of synapses in these preparations are also difficult to define fully. Therefore, molluscan preparations were explored further to develop synapses between the somata, in the absence of outgrowth – an approach that was pioneered in leech[21,22] and subsequently refined in the snail *Helisoma*[23]. Specifically, Haydon[23] used identified neurons from *Helisoma* and juxtaposed their somata in culture. Indeed, chemical synapses developed between the somata in the absence of neurite outgrowth. Although the synapses developed between soma–soma paired *Helisoma* neurons did not exist in the intact brain, this model did nevertheless provide important insights into mechanisms of synapse formation, at a resolution that had not been attained before. Feng and colleagues[24] paired functionally well-defined, pre-and postsynaptic neurons from *Lymnaea* in a soma–soma configuration. The soma–soma paired *Lymnaea* neurons (Figure 2.2A; Colorplate 3) developed appropriate inhibitory, cholinergic connections overnight, and these synapses were target cell specific and both structurally (Figure 2.2B) and electrophysiologically similar to those seen *in vivo*[24]. In addition, voltage-induced, $Ca^{2+}$ hotspots were shown to develop between soma–soma paired cells during synapse formation and these $Ca^{2+}$ specializations were both target cell and contact site specific[25]. Using FM1-43 dye, which is taken up at the active synaptic site through endocytosis, the presynaptic secretory machinery was demonstrated to be specialized at the contact site between the paired cells[26,27]. Together, these data show that soma–soma synapses are both structurally and functionally similar to their neurite–neurite and *in vivo* counterparts, and are thus suitable for further studies on synapse formation.

### 3.8. Trophic Factors, Synapse Formation, and Synaptic Plasticity

*In vivo*, the processes of neurite outgrowth and synapse formation rely upon the availability of various trophic factors. Because neurite outgrowth precedes synapse formation, the involvement of growth factors in synaptogenesis, independent of neurite outgrowth, cannot be studied directly.

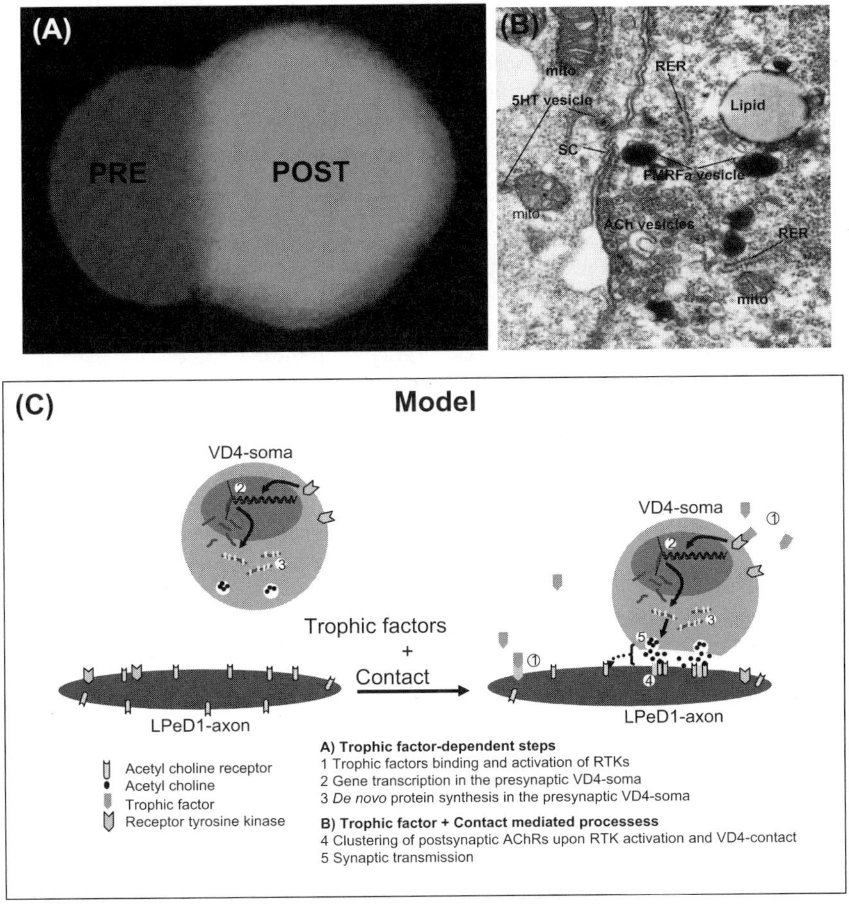


**Figure 2.2.** Mechanisms Regulating Synapse Formation in Various Model Preparations. (A) *Lymnaea* presynaptic neuron (visceral dorsal 4 – VD4 – fluorescently labeled with red dye, sulforhodamine) and postsynaptic neuron (left pedal dorsa 1 – LPeD1 – injected with Lucifer yellow) were soma–soma paired. In this configuration, most molluscan neurons develop appropriate excitatory and inhibitory synapses similar to those seen *in vivo*. (B) An electron micrograph showing the nature of synaptic contacts between soma–soma paired neurons. Vesicles dock at presynaptic site juxtaposed against the postsynaptic cell (Figure courtesy of Dr. Matthias Amrein, University of Calgary). (C) Model depicting steps and mechanisms underlying trophic factor and target cell contact-induced synapse formation between *Lymnaea* neurons in a soma–axon configuration. The model predicts that both target cell contact and extrinsic trophic support are required for appropriate, excitatory synapse formation. See Colorplate 3.

However, the soma–soma model permitted the study of trophic factor's effects on synapse formation in the absence of neurite outgrowth. When paired in DM, inhibitory synapses between the identified neurons VD4 and RPeD1 developed[24]. However, attempts to reconstruct excitatory synapses between VD4 and its other partner left pedal dorsal 1 (LPeD1) failed in DM. Under these experimental conditions, the neurons established inhibitory synapses, which were inappropriate and do not exist *in vivo*. In contrast, pairing in CM enabled appropriate excitatory synapses to develop between VD4 and LPeD1. This trophic factor-induced formation of excitatory synapses was mediated through receptor tyrosine kinases[28,29]. Similarly, the addition of trophic factors to pairs that developed inappropriate inhibitory synapses in DM resulted in a switch to appropriate excitatory synapses. This synapse switching also required receptor tyrosine kinase activity[29,30]. These findings have since been confirmed in the land snail *Helix* where appropriate, excitatory synapses were shown to rely upon the availability of specific trophic factors as well[31]. Utilizing novel synapses between soma–axon pairs (presynaptic soma paired with a somaless postsynaptic axon) from *Lymnaea*, it was subsequently shown that the trophic factor-induced excitatory synapse formation involves mobilization of excitatory, postsynaptic acetylcholine receptor from extrasynaptic to synaptic sites[32]. Together, these studies underscore the importance of trophic factor-mediated signaling in synapse formation and synaptic plasticity.

The precise identity of synapse-specific trophic molecules in *Lymnaea* is yet to be determined. However, the growth-promoting effects of human EGF (hEGF) on *Lymnaea* neurons led to the search for an EGF homolog in *Lymnaea*[33]. The *Lymnaea* albumen gland was found to be a rich source of *Lymnaea* EGF (L-EGF) and, when added to neurons *in vitro*, L-EGF exerted distinct growth-promoting effects on select neurons[33]. Addition of L-EGF to DM containing soma–soma[28], and soma–axon pairs[32] resulted in the formation of excitatory synapses, which were similar to those seen in CM. These results strongly indicate that L-EGF may be a major component of the CM-derived trophic factors that mediate excitatory synapse formation between paired neurons. However, L-EGF induces synapse formation between only 40% of the paired neurons[28], thus the search continues for the remaining complement of synapse-specific factors present in CM.

### 3.9. Synaptogenic Program Suppresses Neurite Outgrowth

It is generally believed that contact between growth cones from pre- and post-synaptic neurons results in the cessation of neurite outgrowth and the formation of specific synaptic contacts. It is therefore generally believed that neurite outgrowth and synapse formation may be two reciprocally inhibitory programs. Nowhere else is it better illustrated than in the soma–soma model. Single identified neurons plated in CM exhibit extensive sprouting, whereas identical neurons maintained in CM and paired with a specific synaptic partner failed to extend processes[34]. This suppression of neurite outgrowth from soma–soma paired cells is however transient and as the synapse matures fully, the neurons begin extending process in a manner similar to their single counterparts[34]. The snail model thus provides a clear evidence for direct, inhibitory interactions between neurite outgrowth and synapse formation.

### 3.10. Synapse-Specific Protein Synthesis, Gene Induction, and Synaptic Plasticity

#### *3.10.1. Gene Transcription*

Because soma–soma paired neurons failed to extend neurites as compared to their single unpaired counterparts, a search began to identify genes in both single and paired neurons that might be differentially regulated under these two different experimental conditions. Quantitative polymerase chain reaction (QPCR) techniques were utilized and this approach resulted in the identification of a molluscan homolog of the multiple endocrine neoplasia type 1 (*MEN1*) tumor suppressor gene, which encodes the transcription factor *menin*. This specific gene was found to be upregulated in soma–soma paired neurons during synapse formation[35]. To demonstrate the significance of *menin* in synapse formation, *MEN1* mRNA was selectively knocked down with antisense in paired neurons *in vitro*. MEN1 perturbations completely blocked both excitatory and inhibitory synapse formation. Interestingly, cell-specific knock-down of *MEN1* mRNA revealed that *menin* expression was required only in the postsynaptic neuron[35]. The *MEN1* gene is the first such gene that has been shown to be essential for synapse formation between *Lymnaea* neurons. Consistent with the notion that neurite outgrowth and synapse formation are mutually exclusive, the BERP gene that is thought to be involved in neurite outgrowth in vertebrates[36] was found to be upregulated in single, regenerating *Lymnaea* neurons and downregulated in soma–soma paired cells. Therefore, the BERP gene may regulate neurite outgrowth from *Lymnaea* neurons[37], though its inhibitory effects on the synaptogenic program remain to be determined.

### *3.10.2. Protein Synthesis*

In addition to transcription of specific genes, synapse formation between soma–soma pairs also requires *de novo* protein synthesis[24,28,29]. However, the precise site where this protein synthesis is required (pre- versus postsynaptic cell) remained unknown. A step toward resolving this issue was taken by Meems and colleagues[32] who used soma–axon pairs, consisting of a presynaptic cell body paired with a postsynaptic axon severed from its respective cell body. The removal of postsynaptic cell body alone did not affect synapse formation between the pairs. However, when presynaptic somata were removed, the synapses failed to develop in cell culture[32]. These data suggest that during early synapse formation between soma–axon pairs, the gene transcription and protein synthesis is required in presynaptic soma but not the postsynaptic axon. All that is needed of the postsynaptic axon is a redistribution of neurotransmitter receptors[32]. Specifically, presynaptic target cell contact facilitates the mobilization of postsynaptic cholinergic receptors from extrasynaptic to synaptic sites, independent of gene transcription or protein synthesis, although it does require trophic factors (Figure 2.2C; Colorplate 3). Taken together these studies underscore the importance of trophic factors and site-specific gene transcription and protein synthesis in synapse formation.

### *3.10.3. Synaptic Plasticity and Synapse Formation*

In *Lymnaea*, the synaptic plasticity-induced formation of long-term memory (LTM) in the intact animals is both transcription and translation dependent[38]. Interestingly, the locus for this transcription and translation-dependent process is confined to one single neuron termed RPeD1. Ablation of its cell body in the intact animals completely blocks memory formation[39]. Similarly, long-term facilitation (LTF), which is thought to be the underlying mechanism for LTM in *Aplysia*, has also been extensively studied and a variety of molecules and underlying mechanisms identified. One of the key issues is to define the precise locus for new protein synthesis during plasticity. For instance, how is a single synaptic connection subjected to modification in a neuron that has multiple synaptic connections? To address this question, a single bifurcated sensory neuron with two branches was allowed to make synapses *in vitro* with two spatially separated motor neurons. In this configuration individual synapses could be selectively exposed to plasticity-inducing stimuli. LTF was induced by repeated stimulation with the neuro-transmitter serotonin (5-HT) at one of the synaptic sites, whereas the other synapses were left untreated. This paradigm revealed modification only of the treated synapse and indicated that a single neuron can undergo branch-specific LTF. Furthermore, this branch-specific, synaptic modification is dependent upon gene transcription and local (at the synapse) protein synthesis. Interestingly, protein synthesis was only necessary in the presynaptic but not the postsynaptic terminals[40]. While altered gene transcription likely serves all synaptic connections formed by any given neuron, the local protein synthesis occurs only at synapses that are subjected to plasticity-specific stimuli. The products of altered gene expression and local protein synthesis thus collectively account for the changes that underlie LTF. Consistent with this idea are studies in which cell bodies from *Aplysia* neurons were removed after synapse formation to determine the contributions of local protein synthesis in synaptic plasticity. The synapses were exposed to LTF-inducing stimuli and changes in synaptic efficacy at synapses either with or without presynaptic somata were compared, showing that somaless axons exhibit LTF similar to their intact counterparts[41]. However, the LTF induced in the somaless configuration appeared to be transient. It was thus suggested that local protein synthesis in axons accounts only for initial changes in synaptic efficacy, whereas gene transcription was required for the maintenance of LTF[41]. These results underscore the importance of gene transcription and protein synthesis in synapse formation as well as in long-term changes in synaptic efficacy.

The search for the molecular machinery that contributes to gene transcription and protein synthesis-dependent LTF has led to the identification of a variety of well-known kinases. In *Helix* for instance, synapsin (a synaptic vesicle-associated phosphoprotein) was found important for increased efficiency of neurotransmitter release in neurons that were otherwise cultured under low-release conditions[42]. Furthermore, a *Helix* synapsin ortholog cloned from *Aplysia* (ApSyn) was mutated on its phosphorylation sites in search of ApSyn substrates that are involved in synaptic plasticity. ApSyn appeared to be an excellent substrate for cAMP-dependent protein kinase. Injection of wild-type ApSyn in identified *Helix* neurons cultured under low-release conditions resulted in increased neurotransmitter release, whereas injection of mutant ApSyn failed to do so[43]. These data indicated that cAMP-dependent protein kinase may be an essential player involved in the induction of synaptic plasticity. Earlier, *Aplysia* protein kinases A and C were shown to be essential for the formation of 5-HT-induced LTF[44–47], whereas more recently mitogen-activated protein kinase (MAPK) has also been implicated in the induction of increased synaptic efficacy[48,49]. Moreover, MAP kinase was shown to translocate to the nucleus[50], suggesting that the site for MAPK action could reside within the nucleus, where it may be involved in transcribing various mRNAs and the translation of their encoded proteins. Taken together, a variety of kinases appear to be involved in the cellular and molecular changes underlying synaptic efficacy, although the exact order of these events remains to be elucidated.

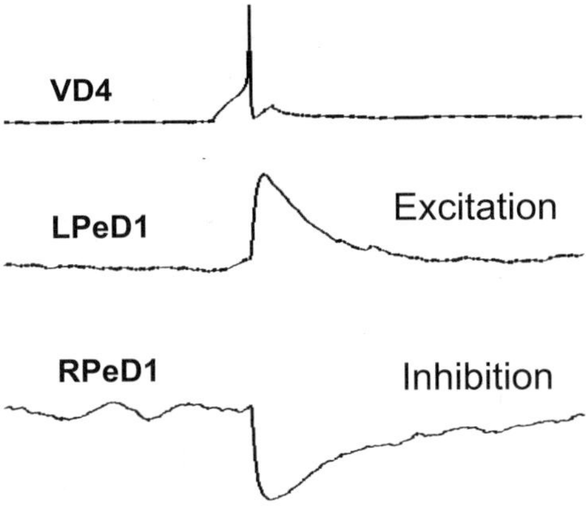


**Figure 2.3.** Synapse Specificity Between Multiple *Lymnaea* Neurons. (A) Presynaptic neuron VD4 paired simultaneously with its inhibitory (RPeD1) and excitatory (LPeD1) partners recapitulates appropriate synapses in cell culture. An action potential in VD4 simultaneously induces 1:1 excitatory (LPeD1) and inhibitory (RPeD1) postsynaptic potentials (PSPs) in synaptic targets. Both synaptic responses involve cholinergic transmission between VD4 and its target neurons.

## 3.11. Regulation of Synapse Number and Synaptic Scaling

An important unanswered question in the field of neurodevelopment and regeneration is the following: How is synapse number and efficacy regulated? That is, how does a neuron "know" that it has acquired its full complement of synapses that would be required for functionality? It is hypothesized that the synaptic efficacy is regulated globally and enhances synaptic output nonselectively[51,52]. Alternatively, synaptic efficacy could also be modulated locally and selectively at specific synaptic sites[40,53]. Similarly, the number of synapses that any given neuron establishes could also be regulated either globally or locally. To this end, single isolated LPeD1-axons were paired with two identical presynaptic VD4 neurons. Electrophysiological recordings demonstrated that in this configuration only one VD4 formed synaptic connections with the isolated axon[15]. Since the isolated axon is devoid of its cell body, the synapse numbers are likely regulated locally by the axon itself. To address this issue further, we isolated a single, presynaptic neuron from the *Lymnaea* central ring ganglia and challenged it with two identical postsynaptic targets. Specifically, a single VD4 neuron was simultaneously soma–soma paired with two postsynaptic LPeD1 neurons. In the intact brain VD4 connects with only one LPeD1. Interestingly, under these experimental conditions only one postsynaptic cell received innervation from VD4 during the first 12–18 h of cell pairing. In contrast, when VD4 was paired with two different postsynaptic neurons, LPeD1 and RPeD1, it formed excitatory and inhibitory synapses, respectively, with both neurons (Figure 2.3). Similarly, when the LPeD1–VD4–LPeD1 triplet was examined during early stages of synapse formation (4 h), both LPeD1 cells were innervated by VD4. However, the efficacy of each individual synapse under these experimental conditions was a fraction of the monosynaptic strength exhibited among pairs and involved the cAMP–PKA-dependent pathway. Indeed, experimental activation of the cAMP–PKA pathway resulted in reduced synaptic efficacy, whereas inhibition of this cascade generated hyperinnervation and an enhancement of synaptic strength[54]. These data show that cAMP–PKA-dependent signaling plays a novel role in controlling synaptic efficacy and thus regulating single or multiple innervations. In mollusks, this may thus serve as one of the mechanisms that ensure a balance between neuronal input and output capacities.

### 3.12. In Vivo Synapse Formation and Behavioral Recovery Following Single-Cell Transplantation

It can always be argued that the *in vitro* approach only offers an artificial environment, which bears little resemblance to the *in vivo* milieu that is pivotal for normal development. To determine whether the data obtained *in vitro* can be extrapolated to derive conclusions about fundamental principles governing synapse formation *in vivo*, single-cell transplantation techniques were developed in *Lymnaea*. Specifically, ablation of the single respiratory neuron VD4 in the intact animal rendered the snail unable to exhibit normal respiratory behavior[55]. This behavioral deficit was however restored by transplanting a VD4 neuron from a donor animal. The transplanted neuron subsequently recapitulated its pattern of synaptic connectivity and restored functional contacts with its target cells as well. Interestingly, when a VD4 was transplanted into the host in the presence of its native VD4, the newly transplanted cell failed to innervate its appropriate target and began making contacts with inappropriate target cells. These findings have been confirmed further by transplanting RPeD1 into the right parietal ganglia[56], a location that is different from its native habitat (right pedal ganglia). Consistent with our previous study, the transplanted neuron would only connect with its targets if they were deprived of synaptic input form the host cell. Together, these data suggest that the neuronal ability to regenerate and recognize appropriate target neurons in cell culture involves fundamental mechanisms that are likely operative in the intact brain.

## 4. THE FUTURE OF MOLLUSCAN MODELS: FROM PROTEINS AND GENES TO SILICON CHIPS

The ability to manipulate single molluscan neurons holds tremendous potential toward the identification and characterization of various genes and their encoded proteins regulating synapse formation. One possibility for future exploration will be the identification and characterization of a full complement of both pre- and postsynaptic proteins involved in synapse formation. This functional proteomics approach has already enabled the identification and characterization of a whole array of proteins within specific compartments of select mammalian neurons[57,58]. Although these mammalian neurons hold tremendous potential for identifying synapse-specific proteins, the molluscan neurons due to their larger somata, are perhaps better suited for such functional proteomics approaches. For instance, Jimenez and colleagues[59] have recently successfully identified and characterized a full complement of neuronal neuropeptides at the level of a single cell. This approach can now be employed to identify various proteins during synapse formation – at the level of single pre- and postsynaptic neurons. These proteins can then be manipulated experimentally to determine their exact involvement in synapse formation and synaptic plasticity.

Finally, a new approach to investigate connectivity of multiple neurons concurrently and noninvasively is the utilization of silicon chip technologies. The neuron and silicon chip interfacing allows noninvasive examination of large neuronal ensembles during synapse formation (Figure 2.4). Specifically, the

*Lymnaea* model was recently used to successfully interface individual neurons with silicon chips and a bidirectional communication was established between brain cells and this electronic device[60]. This approach now provides an unprecedented opportunity to examine the role of activity-dependent mechanisms in synapse formation and synaptic plasticity at a resolution that has never been attained before.

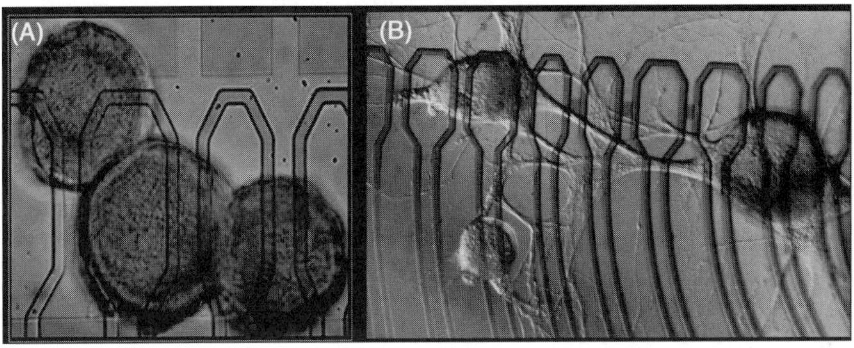


**Figure 2.4.** Synapse Formation as Revealed Through Brain–Chip Neuron Interfacing. (A) Identified *Lymnaea* neurons can be successfully interfaced with silicon chips (Kaul, Syed, Fromherz, unpublished data). The cultured cells can either be soma–soma paired (A) or allowed to extend neurites to develop networks (B). This approach has been used to reconstruct specific synapse on the chip which was subsequently used to stimulate and record synaptic activity and plasticity.

## 5. REFERENCES

1. Bulloch, A.G., and Syed, N.I. (1992) *Trends Neurosci* **15**, 422–427.
2. Munno, D.W., and Syed, N.I. (2003) *J Physiol* **552**, 1–11.
3. Wiersma-Meems, R., Van Minnen, J., and Syed, N.I. (2005) *Neuroscientist* **11**, 228–237.
4. Haydon, P.G., and Drapeau, P. (1995) *Trends Neurosci* **18**, 196–201.
5. Ghirardi, M., Casadio, A., Santarelli, L., and Montarolo, P.G. (1996) *Invert Neurosci* **2**, 41–49.
6. Cibelli, G., Ghirardi, M., Onofri, F., Casadio, A., Benfenati, F., Montarolo, P.G., and Vitiello, F. (1996) *Eur J Neurosci* **8**, 2530–2543.
7. Moffett, S., and Snyder, K. (1985) *J Neurobiol* **16**, 193–209.
8. Benjamin, P.R. and. Allison., P. (1987) *Growth and Plasticity of Neural Connections* (Winlow, W., McCrohan, C.R., Eds.), Manchester University Press, Manchester.
9. Kruk, P.J., and Bulloch, A.G. (1992) *J Neurosci Res* **31**, 401–412.
10. Murphy, A.D., and Kater, S.B. (1980) *Brain Res* **186**, 251–272.
11. Schacher, S., and Proshansky, E. (1983) *J Neurosci* **3**, 2403–2413.
12. Rayport, S.G., and Schacher, S. (1986) *J Neurosci* **6**, 759–763.
13. Syed, N.I., Bulloch, A.G., and Lukowiak, K. (1990) *Science* **250**, 282–285.
14. Spencer, G.E., Lukowiak, K., and Syed, N.I. (2000) *J Neurosci* **20**, 8077–8086.
15. Wiersma-Meems, R. (2005) Thesis: at Department of Molecular and Cellular Neurobiology, Free University, Amsterdam.
16. Zoran, M.J., Doyle, R.T., and Haydon, P.G. (1991) *Neuron* **6**, 145–151.
17. Landis, S.C. (1990) *Trends Neurosci* **13**, 344–350.
18. Lovell, P., McMahon, B., and Syed, N.I. (2002) *J Neurophysiol* **88**, 1328–1338.
19. Syed, N.I., Lukowiak, K., and Bulloch, A.G. (1992) *Neuroreport* **3**, 793–796.
20. Lukowiak, K., Syed, N.I., Ridgway, R.L., and Bulloch, A.G. (1993) *Acta Biol Hung* **44**, 115–119.
21. Fuchs, P.A., Nicholls, J.G., and Ready, D.F. (1981) *J Physiol* **316**, 203–223.
22. Fuchs, P.A., Henderson, L.P., and Nicholls, J.G. (1982) *J Physiol* **323**, 195–210.
23. Haydon, P.G. (1988) *J Neurosci* **8**, 1032–1038.
24. Feng, Z.P., Klumperman, J., Lukowiak, K., and Syed, N.I. (1997) *J Neurosci* **17**, 7839–7849.

25. Feng, Z.P., Grigoriev, N., Munno, D., Lukowiak, K., MacVicar, B.A., Goldberg, J.I., and Syed, N.I. (2002) *J Physiol* **539**, 53–65.
26. Woodall, A.J., Naruo, H., Prince, D.J., Feng, Z.P., Winlow, W., Takasaki, M., and Syed, N.I. (2003) *J Neurophysiol* **90**, 2232–2239.
27. Naruo, H., Onizuka, S., Prince, D., Takasaki, M., and Syed, N.I. (2005) *Anesthesiology* **102**, 920–928.
28. Hamakawa, T., Woodin, M.A., Bjorgum, M.C., Painter, S.D., Takasaki, M., Lukowiak, K., Nagle, G.T., and Syed, N.I. (1999) *J Neurosci* **19**, 9306–9312.
29. Woodin, M.A., Hamakawa, T., Takasaki, M., Lukowiak, K., and Syed, N.I. (1999) *Learn Mem* **6**, 307–316.
30. Woodin, M.A., Munno, D.W., and Syed, N.I. (2002) *J Neurosci* **22**, 505–514.
31. Fiumara, F., Leitinger, G., Milanese, C., Montarolo, P.G., and Ghirardi, M. (2005) *Neuroscience* **134**, 1133–1151.
32. Meems, R., Munno, D., van Minnen, J., and Syed, N.I. (2003) *J Neurophysiol* **89**, 2611–2619.
33. Hermann, P.M., van Kesteren, R.E., Wildering, W.C., Painter, S.D., Reno, J.M., Smith, J.S., Kumar, S.B., Geraerts, W.P., Ericsson, L.H., Smit, A.B., Bulloch, A.G., and Nagle, G.T. (2000) *J Neurosci* **20**, 6355–6364.
34. Feng, Z.P., Hasan, S.U., Lukowiak, K., and Syed, N.I. (2000) *J Neurobiol* **42**, 357–369.
35. van Kesteren, R.E., Syed, N.I., Munno, D.W., Bouwman, J., Feng, Z.P., Geraerts, W.P., and Smit, A.B. (2001) *J Neurosci* **21**, RC161.
36. El-Husseini, A.E., and Vincent, S.R. (1999) *J Biol Chem* **274**, 19771–19777.
37. van Kesteren, R.E. (personal communication).
38. Sangha, S., Scheibenstock, A., McComb, C., and Lukowiak, K. (2003) *J Exp Biol* **206**, 1605–1613.
39. Scheibenstock, A., Krygier, D., Haque, Z., Syed, N.I., and Lukowiak, K. (2002) *J Neurophysiol* **88**, 1584–1591.
40. Martin, K.C., Casadio, A., Zhu, H., Yaping, E., Rose, J.C., Chen, M., Bailey, C.H., and Kandel, E.R. (1997) *Cell* **91**, 927–938.
41. Liu, K., Hu, J.Y., Wang, D., and Schacher, S. (2003) *J Neurobiol* **56**, 275–286.
42. Fiumara, F., Onofri, F., Benfenati, F., Montarolo, P.G., and Ghirardi, M. (2001) *Neuroscience* **104**, 271–280.
43. Fiumara, F., Giovedi, S., Menegon, A., Milanese, C., Merlo, D., Montarolo, P.G., Valtorta, F., Benfenati, F., and Ghirardi, M. (2004) *J Cell Sci* **117**, 5145–5154.
44. Sossin, W.S., Sacktor, T.C., and Schwartz, J.H. (1994) *Learn Mem* **1**, 189–202.
45. Wu, F., Friedman, L., and Schacher, S. (1995) *J Neurosci* **15**, 7517–7527.
46. Manseau, F., Sossin, W.S., and Castellucci, V.F. (1998) *J Neurophysiol* **79**, 1210–1218.
47. Sacktor, T.C., Kruger, K.E., and Schwartz, J.H. (1988) *J Physiol (Paris)* **83**, 224–231.
48. Purcell, A.L., Sharma, S.K., Bagnall, M.W., Sutton, M.A., and Carew, T.J. (2003) *Neuron* **37**, 473–484.
49. Sharma, S.K., and Carew, T.J. (2004) *Learn Mem* **11**, 373–378.
50. Martin, K.C., Michael, D., Rose, J.C., Barad, M., Casadio, A., Zhu, H., and Kandel, E.R. (1997) *Neuron* **18**, 899–912.
51. Turrigiano, G.G., Leslie, K.R., Desai, N.S., Rutherford, L.C., and Nelson, S.B. (1998) *Nature* **391**, 892–896.
52. Leslie, K.R., Nelson, S.B., and Turrigiano, G.G. (2001) *J Neurosci* **21**, RC170.
53. Casadio, A., Martin, K.C., Giustetto, M., Zhu, H., Chen, M., Bartsch, D., Bailey, C.H., and Kandel, E.R. (1999) *Cell* **99**, 221–237.
54. Munno, D.W., Prince, D.J., and Syed, N.I. (2003) *J Neurosci* **23**, 4146–4155.
55. Syed, N.I., Ridgway, R.L., Lukowiak, K., and Bulloch, A.G. (1992) *Neuron* **8**, 767–774.
56. Lee, T.K., and Syed, N.I. (2004) *Synapse* **51**, 186–193.
57. Li, K.W. et al. (2004) *J Biol Chem* **279**, 987–1002.
58. Jimenez, C.R., Stam, F.J., Li, K.W., Gouwenberg, Y., Hornshaw, M.P., De Winter, F., Verhaagen, J., and Smit, A.B. (2005) *Mol Cell Proteomics* **4**, 120–132.
59. Jimenez, C., Spijker, S., de Schipper, S., Lodder, J.C., Janse, C., Geraerts, W.P.M., van Minnen, J., Syed, N.I., Burlingame, A.L., Smit, A.B., and Li, K.W. (2006) *J Neurosci* **26**, 518–529.
60. Merz, M. (2004) Thesis at Max-Planck Institute for Biochemistry, Technical University Munchen, Munchen.

# 3

# DEVELOPMENT OF THE *DROSOPHILA* AND *C. ELEGANS* NEUROMUSCULAR JUNCTIONS

Heather Van Epps† and Yishi Jin*

## 1. SUMMARY

Proper development of the neuromuscular junction (NMJ) is essential for synaptic transmission and hence coordinated muscle function. The complex, yet precise, organization of molecular machinery is necessary for proper presynaptic neurotransmitter release and for postsynaptic detection. The NMJ is an excellent model to dissect mechanisms of synaptogenesis, because it has the experimental advantages of being simple and accessible. Many principles that control NMJ formation have proven to be relevant in other types of synapses. Studies in the genetic model organisms, *Drosophila melanogaster* and *Caenorhabditis elegans*, have led to fundamental advances in our molecular understanding of *in vivo* NMJ development. In this chapter, we first describe the architecture of the *Drosophila* and *C. elegans* NMJs. We then briefly summarize the methodology used to analyze the invertebrate NMJs. Finally, we review specific findings in synaptic target recognition, synaptic assembly, and growth.

## 2. INTRODUCTION

Nervous system function depends on the intricate molecular architecture of the synapse. The NMJ is composed of a presynaptic terminal and a postsynaptic muscle target (Figure 3.1). Like all chemical synapses, the NMJ presynaptic terminal is characterized by a presynaptic density surrounded by a cluster of synaptic vesicles (SVs). This is frequently termed the "active zone", referring to the "active" site of SV fusion and neurotransmitter release. Surrounding the active zone is the region that accommodates reserve SVs and endocytosis and may act to

†H. Van Epps, Department of Molecular Cell and Developmental Biology, University of California, Santa Cruz, California, USA; vanepps@biology.ucsc.edu
*Y. Jin, Department of Molecular Cell and Developmental Biology, and Howard Hughes Medical Institute, University of California, Santa Cruz, California, USA; jin@biology.ucsc.edu

separate synaptic domains from nonsynaptic cytoplasm. This region is frequently termed the "periactive zone." The postsynaptic site is defined by its juxtaposition against the electron-dense presynaptic site. Clustered neurotransmitter receptors at the postsynaptic site detect and then transduce the neuronal signal.

Synapse formation therefore involves the recruitment of a specific ensemble of molecules at the active zone, periactive zone, vesicle pool, and postsynaptic site. Once mature, the synapse can be maintained for the life of the organism, and in some cases is capable of remodeling in response to input. A number of synaptic genes involved in synapse formation have been identified and characterized at the invertebrate NMJ. The NMJs in *C. elegans* and *Drosophila* provide a well-defined genetic framework in which synaptogenesis can be studied *in vivo* at single-cell and single-synapse resolution.

## 3. THE *DROSOPHILA* AND *C. ELEGANS* NMJs

### 3.1. *C. elegans*

*C. elegans* NMJs form *en passant* at discrete regions along the axon between selective nerve processes and muscles (Figure 3.1C)[2]. The muscles extend specialized arms to meet the presynaptic axon. NMJs use several types of neurotransmitters, including γ-amino butyric acid (GABA), acetylcholine (ACh), and serotonin. GABA and ACh receptors are clustered in postsynaptic sites of body wall muscles. The localization of serotonin receptors is not yet known. Typical clear SV (35–45 nm) and large-dense core vesicles (LDCV) (40–53 nm) are found at the presynaptic terminal. The size of axonal swellings, the number of SVs, and the size of the presynaptic density define the size of the presynaptic region. The presynaptic region can vary considerably, even among the same types of synapses or among synapses from the same neuron. The presynaptic density in worms is relatively small and does not have elaborate ultrastructural characteristics. The postsynaptic site has no visible density. The cholinergic NMJ synapses on to both muscle and neuron, whereas the GABAergic NMJ synapses only on to muscle (Figure 3.1C).

### 3.2. *Drosophila*

*Drosophila* NMJs form at axon terminals, and multiple motor neurons innervate muscle fibers (Figure 3.1D)[3]. Glutamate is the primary excitatory neurotransmitter in the fly NMJ. The individual synaptic units, termed "boutons," resemble beads on a string and decorate the terminal nerve branches. The presynaptic active zone appears as an electron-dense T-shaped bar surrounded by SVs (Figure 3.1B).

NMJ development spans 8 h from the time of initial neuromuscular contact in the embryo to the formation of a mature synapse in the hatching larva. Following motor axon guidance and neuronal exploration of the muscle target, the nerve terminals become restricted to the synaptic site. Soon afterward, the glutamate receptors cluster postsynaptically in response to glutamate released from the immature presynaptic terminal. Two to three hours later mature boutons, containing T-bars and SVs, are apparent. Throughout larval development there is a

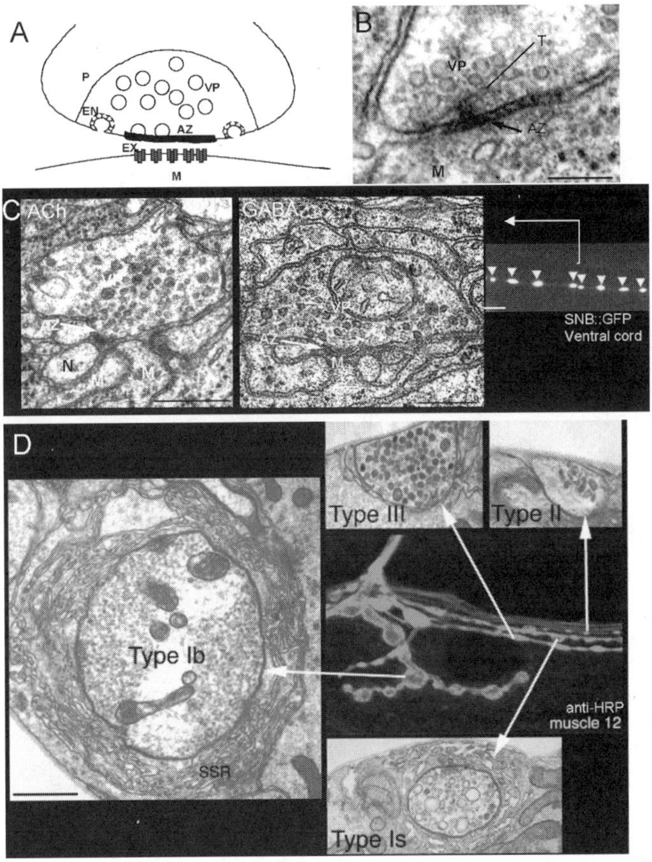


**Figure 3.1.** The *C. elegans* and *Drosophila* NMJs. (A) Generic schematic of an NMJ. The active zone (AZ) is denoted by the presynaptic density (dark bar) and surrounding vesicles (circles). The regions of endocytosis (EN) and the vesicle pools (VP) lie near to the active zone. The periactive zone (P) surrounds the active zone. The active zone is juxtaposed against the postsynaptic receptors in the muscle (M) membrane. Extracellular matrix and signaling molecules (Ex) occupy the space between pre- and postsynaptic sites. (B) Electron micrograph of a *Drosophila* synapse (Type II) (courtesy of ref. 1). The presynaptic *Drosophila* T-bar specialization (T) can be seen at the active zone protruding into the presynaptic cytoplasm. Scale bar = 200 nm. (C) Electron micrographs of the *C. elegans* NMJ reveals ultrastructural characteristics of cholinergic and GABAergic NMJs. Cholinergic NMJs are smaller and synapse on both muscle (M) and neuron (N). GABAergic neurons are larger and synapse solely on muscle. Vesicle pools (VP) surround the electron-dense presynaptic density, which marks the active zone (AZ). Scale bar = 200 nm. Presynaptic vesicle pools (arrow heads) at NMJs along the nerve cord are visualized with SNB::GFP in *C. elegans*. These pools of vesicles correspond to regions of synapses along the nerve cord. Scale bar = 5 μm. (D) Electron micrographs of the *Drosophila* NMJ reveal ultrastructural characteristics of three types of neuromuscular junctions. Type I axon terminals are short and contain 1–5 μm diameter boutons. The subtype Ib is larger than Is. Type II axon terminals are long and contain less than 2 μm boutons. Type III axon terminals comprise a small class of intermediate terminals. Scale bar = 1 μm. Individual boutons (arrows) are visualized with anti-HRP and appear like beads on a string (3.1.D courtesy of V. Budnik).

ten-fold increase in bouton number and active zone/bouton ratio. Boutons grow in part by budding or splitting from existing boutons[4].

Based on the size of the synaptic boutons and the anatomy of the arbors, the NMJs fall into three classes (Figure 3.1D)[3]. Type I NMJs are characterized by short-terminal branches and large presynaptic release sites. Type II NMJs are characterized by long thin branches, numerous small neurotransmitter release sites, and contain a variety of vesicles including LDCV. Type III axon terminals comprise a small class of intermediate terminals. In larval stages, type I NMJs contain a convoluted specialization of the postsynaptic membrane opposed to the active zone called the subsynaptic reticulum (SSR). Glutamate receptors cluster in the muscle membrane at the postsynaptic site. Although glutamate is the major neurotransmitter and is present in all types of *Drosophila* NMJs, octopamine, proctolin, leukokinin I, and insulin have been implicated as additional neurotransmitters and can be found at a subset of *Drosophila* NMJs.

## 4. METHODS FOR STUDYING *DROSOPHILA* AND *C. ELEGANS* NMJ

There are many advantages to studying the invertebrate NMJ as a model synapse. The cell lineages of motor neurons and muscles in *C. elegans* and *Drosophila* are well defined and form reproducible neuromuscular patterns. Most of the circuitry is anatomically and physiologically well characterized. Although simple, both invertebrates have an array of specific movements that can be used to assess neuromuscular function. Finally, most synapses can be readily imaged in living animals. Such features, in combination with genetic screens, cell biology, electrophysiology, and biochemistry, have aided the study of synapse development at a unique level of resolution.

### 4.1. Visualizing Synapses

The use of fluorescent proteins (such as Green Fluorescent Protein (GFP) and its derivatives) has revolutionized the study of the cellular biology of synaptogenesis. Fusion proteins between synaptic components and GFP allow *in vivo* analysis of the synapse in translucent fly larvae and worm. These fusion proteins have also served as powerful screening tools for mutants that affect the distribution of synaptic components.

#### *4.1.1. C. elegans*

A key reagent used to examine *C. elegans* NMJs is the Synaptobrevin::GFP fusion reporter[5] (Figure 3.1C). Synaptobrevin (SNB-1) is an integral membrane protein of SVs. SV pools are visualized by SNB::GFP, whose pattern correlates with that of most synapses. A SYD-2::GFP (SYnapse Defective::GFP) fusion reporter is also used to visualize active zones in multiple types of NMJs[7].

SYD-2 is a Liprin-$\alpha$ protein important for active zone assembly (see Section 5.2.3)[6]. Several fluorescently tagged presynaptic proteins have also recently been used to visualize synaptic contacts[8].

### *4.1.2. Drosophila*

In contrast to live imaging approaches used to visualize the developing synapse in *C. elegans*, visualization of *Drosophila* NMJs has traditionally relied on a bank of monoclonal antibodies[9]. For example, anti-HRP is commonly used to label neurons (Figure 3.1D). However, several GFP reagents have also been developed for live observations. CD8::GFP::Sh is used to visualize the postsynaptic muscle of boutons. This protein contains the extracellular and transmembrane domains of the human T lymphocyte protein (CD8), GFP, and the cytoplasmic C-terminal sequence of the Shaker potassium channel (Sh). The CD8::Sh protein is expressed specifically at the postsynaptic muscle[10]. In addition, proteins such as synaptophysin::GFP and synaptotagmin::GFP (SV-associated proteins) as well as SNB::GFP are commonly used to visualize presynaptic terminals. These fluorescent molecules have enabled characterization of target recognition and synapse formation.

## 4.2. Genetic Approaches for Synaptogenesis

### *4.2.1. C. elegans*

Worms move in a stereotypical sinusoidal wave pattern. The excitatory cholinergic and inhibitory GABAergic NMJs coordinate the wave propagation. Disruption of NMJ development and function leads to a wide range of uncoordinated behaviors. Early screens for *C. elegans* uncoordinated behavior yielded a large number of genes important for nervous system development and function, known as the "*unc*" genes[11]. Subsequent decades of molecular analyses have revealed that many of these genes are important players at the NMJ, a few of which contribute to synapse development.

Direct visual inspection of fluorescently tagged synapse molecules has allowed for targeted identification of synapse genes. Genetic screens using SNB::GFP have played a pivotal role in the identification of new genes specific to synaptogenesis. Most screens have used neuron-type specific promoters to drive SNB::GFP expression in type D motor neurons[6,12], mechanosensory neurons[13], chemosensory neurons[14], and hermaphrodite-specific neurons (HSNs)[15]. Mutations isolated using this approach generally have wide effects on many or all synapses, but often cause few behavioral abnormalities. Moreover, the particular synaptic phenotype differs in a synapse-type specific manner. Although these mutations were isolated based on abnormal patterns of the SV component SNB::GFP, the characterizations thus far have shown that most of these genes do not encode integral SV components, but rather encode a wide range of proteins involved in synapse development and function, validating the SNB::GFP screening approach. Recently, SYD-2::GFP has been used to screen for mutants defective in active zone assembly[7]. This screen yielded alleles of genes previously identified in SNB::GFP screens, and also identified new genes involved in active zone formation.

In addition to traditional forward genetic screens, RNA interference (RNAi) is a powerful reverse genetic approach useful for characterizing gene function. The reverse genetic approach of RNAi allows for the targeted analysis of specific genes or unbiased whole genome screens. However, poorly understood mechanisms reduce RNAi effectiveness in neurons. This problem was overcome by the identification of a number of *C. elegans* mutants that are more susceptible to RNAi. These include mutants in the 5′exonuclease, *eri-1,* the RNA-dependent RNA polymerase, *rrf-3,* and a subset of retinoblastoma mutants[16,17]. These mutants have enabled potent RNAi effects on neuronal targets. A large-scale screen using an RNAi-sensitized strain has led to the identification of 132 genes that were not previously implicated in synaptic development and transmission[8].

### *4.2.2. Drosophila*

Initial *Drosophila* screens for synapse target recognition defects used antibodies specific to a subset of NMJs. Many individual terminal branches and synapses are clearly visible in preparations stained by this method. These antibody screens identified a number of mutants with normal gross CNS morphology, but defective outgrowth or target recognition[9].

The generation of a modular misexpression system in *Drosophila* allowed screening based on gain-of-function effects at specific times and in specific cells. The modular misexpression system is based on Gal4 transactivation of a mobile enhancer and promoter that "targets" random endogenous genes for expression[18]. This approach was used to assess high-level expression of genes in moto neurons[19]. Defects in axon guidance or synaptogenesis in *Drosophila* larvae were analyzed using a pan neuronal GFP reporter. This screen identified new genes in a wide range of classes, including kinases and phosphatases, GTPases and their regulatory proteins, RNA-binding proteins, and transcriptional regulators.

The reverse genetic RNAi approach is used in *Drosophila* to analyze gene function. RNAi was used to identify genes involved in stabilization of the *Drosophila* NMJ[20]. Synaptic sites develop the SSR postsynaptic specialization only in the presence of a presynaptic terminal. If the terminal retracts, the SSR is temporarily maintained leaving a "footprint" that contains SSR markers, but no presynaptic markers. By screening for synaptic footprints in RNAi-treated flies, this screen successfully identified necessary cytoskeletal components for synapse stabilization. A GAL4/UAS RNAi expression system has been developed to allow tissue specific knockdown of target genes[21]. This approach allows dissection of presynaptic versus postsynaptic effects at specific synapses.

## 4.3. Assessment of Physiology

### *4.3.1. Electrophysiology*

Electrophysiology is another invaluable tool in the analysis of neuronal function. *Drosophila* was one of the first few genetically malleable organisms amenable to electrophysiology, and was used to study the NMJ as early as the mid-1970s[22]. The small size of *C. elegans* neurons initially appeared prohibitive to electrophysiology, but such limitations were overcome in the 1990s[23]. Spontaneous, pharmacologically stimulated, or nerve-evoked muscle responses can now be recorded in both *Drosophila* and *C. elegans*.

### *4.3.2. Pharmacology*

Drugs that impair synaptic transmission are powerful tools for studying *C. elegans* NMJ function. The ACh esterase inhibitor, aldicarb, can be used to measure the steady-state ACh release in the live worm. Aldicarb causes accumulation of ACh at the NMJ leading to paralysis and death. Mutants that have reduced release of ACh are aldicarb resistant and mutants that have increased release are aldicarb hypersensitive. Aldicarb has been used successfully to analyze and isolate many synaptic transmission mutants[24]. A second drug, levamisole, is a nematode-specific nicotinic agonist that activates an ACh receptor in muscle[25]. Acute exposure to high concentrations of levamisole leads to paralysis and death. Worms are resistant to levamisole if levamisole-sensitive ACh receptors or their downstream effectors are defective. Aldicarb and levamisole can be used to determine if specific structural defects lead to detrimental effects on the physiology of neurotransmission.

### *4.3.3. Calcium Imaging*

Neuronal excitation causes transient changes in calcium levels at the synapse, which control neurotransmitter release. These calcium transients can be detected using genetically encoded fluorescent intracellular calcium sensors. These calcium indicators can be targeted to specific cells and structures. A number of sensors have been developed[26]. The selection of an appropriate indicator and the correct interpretation of the optical signals depend on the sensor's unique properties. It is possible to analyze and compare electrically evoked calcium transients in individual connections made by a single neuron[27].

## 5. MOLECULAR MECHANISMS OF *DROSOPHILA* AND *C. ELEGANS* NMJ DEVELOPMENT

Synapse formation involves a complex host of adhesion, signaling, and structural molecules. A developing neurite must grow toward its target (axon guidance), recognize its target, and stabilize the connection (target recognition and stabilization). Synapse-specific molecules must then be recruited and organized to create functional pre- and postsynaptic sites (assembly). Below, we discuss the *Drosophila* and *C. elegans* contributions to the molecular understanding of target recognition, target stabilization, and synapse assembly. Table 1 lists the major molecules involved in these processes.

**Table 1.** *Drosophila* and *C. elegans* Synapse Development Genes.

| Synapse Development | Cellular Process | Fly | Worm | Mammalian |
|---|---|---|---|---|
| **Target Recognition** | Wnt signaling | *wingless*[81] | *cwn-1,cwn-2,egl-20, lin-44, mom-2* | *wnt ligands* |
| | | *dishevelled* | *dsh-1,dsh-2,mig-5* | *dishevelled* |
| | | *β-catenin* | *hmp-2, bar-1, wrm-1* | *β-catenin* |
| | | *frizzled* | *cfz-2, lin-17, mom-5, mig-1* | *frizzled* |
| | | *shaggy*[82] | *gsk-3* | *gsk-3* |
| | Receptor protein tyrosine phosphatase | *dlar*[31-33] | *ptp-3*[34] | *lar* |
| | Cell adhesion | *irrecC and kirre/DUF* | *syg-1*[15,36] | *neph1* |
| | | *sticks and stones and hibris* | *syg-2*[36] | *nephrin* |
| | | *flamingo*[30] *(retina)* | *fmi-1* | *flamingo* |
| | | *N-cad*[35] *(retina)* | *hmr-1* | *N-cadherin* |
| | | *semaphorin II*[38] | *smp-1* | *semaphorin* |
| | | *fasciclin II*[3] | | *Ncam* |
| | | *sidestep*[28] | | |
| | | *fasciclin III*[3] | | |
| | TGF signaling | *wishful thinking*[75,76] *thick vein*[78,79] *saxophone*[78,79] | *daf-4, daf-1* | *bmpr2, bmpr1, acvr1* |
| | | *glass bottom boat*[77] | *daf-7* | *bmp7* |
| | Netrin signaling | *netrin*[38] | *unc-6* | *netrin 1* |
| | | *frazzled*[38] | *unc-40* | *netrin receptor* |
| | RhoGAP | *rho GAP100F* | *syd-1*[41] | |
| **SV Localization and Clustering** | Ser/Thr kinase | *CG6114-PA* | *sad-1*[14,40] | *sad-A, sad-B*[102] |
| | Kinesins | *khc*[43] | *unc-116*[44] | *kinesin heavy chain* |
| | | *unc104* | *unc-104*[42] | *kif1a* |
| | JNK scaffolding | *sunday driver*[45] | *unc-16*[44] | *jsap1/jip3* |
| | Liprin | *D-liprin-a*[47,56] | *syd-2*[6,7] | *liprin-a* |
| **Presynaptic Assembly and Stabilization** | Liprin | *D-liprin-a*[47,56] | *syd-2*[6,7] | *liprin-α* |
| | Ubiquitin regulators | *highwire*[68,79] | *rpm-1*[12,69] | *pam* |
| | | *CG4643-PB* | *fsn-1*[70] | *fbxo45* |
| | | *fat facets*[71] | T24B8.7 | *USP9x* |
| | TGF beta regulator | *spinster*[80] | Y111B2A.19, C13C4.5, C39E9.10, F09A5.1 | *spinl* |
| | Receptor tyrosine kinase-like orphan receptor | *ror-PA* | *cam-1*[65] | *ror1 and musk* |
| | Cell adhesion | *fasiclin II*[3] | *ncam-1* | *ncam* |
| | | *dlg-1*[10] | *dlg-1* | *dlg1* |
| | | *toll*[37] | *tol-1* | *toll like receptor* |
| | | *connectin*[3] | *rbc-1, rbc-2* | |
| | Basal lamina | *nidogen* | *nid-1*[59] | *nidogen* |
| | | *CG33171-PE* | *cle-1*[59] | *collagen a XVIII* |
| | | *dap160*[60,61] | *itsn-1* | *intersectin* |
| | Cytoskeleton | *b-spectrin*[21] | *unc-70* | *sptbn1* |
| | | *a-spectrin*[21] | *spc-1* | *sptan1* |
| | | *nervous wreck*[85] | *tag-13* | *FCHSD2* |
| | | *Dfxr*[89] | | *fmrp* |
| | | *futsch*[87] | | *map1b* |
| | | *still life*[90] | | |
| | Dynactin, dynein regulatory complex | *centractin*[20] | | *centractin* |
| | | *glued*[20] | *dnc-1* | *p150glued* |

| | | | | |
|---|---|---|---|---|
| **Receptor Clustering** | Glutamate or GABA biosynthetic pathway | *gad*[61] | **unc-25*[63] | *glutamic acid decarboxylase 1* |
| | | *got2*[61] | *C14F11.1* | *glutamate oxaloacetate transaminase 2* |
| | | *gs2*[61] | *4Q934* | *glutamine synthetase 1* |
| | Anaphase promoting complex E3 ubiquitin ligase | *Dcdc27,morula, and fizzy-related*[72] | *mat-1, N/A, and fzr-1*[73] | *cdc27, apc2, and cdh1 respectively* |
| | Receptor tyrosine kinase-like orphan receptor | *ror-PA* | *cam-1*[65] | *ror1, or musk* |
| | Rho type GEF | *pix*[67] | | |
| | Transmembrane | *CG31149* | *lev-10*[66] | |
| | Translation | *pum*[92] | many | *pum* |

*in contrast to its fly homologue, *unc-25* is NOT involved in synapse development

## 5.1. Synaptic Target Recognition

Selecting synaptic partners is a crucial step in neural circuit formation. Synapses are formed *in vivo* at specific locations, in part, under the control of target recognition molecules. Evidence shows that target selection is mediated not by a point-to-point system, but rather a combinatorial system. Localized attractants, repellants, and stabilization molecules act in combination to create precise neuromuscular contacts essential for muscle function.

### *5.1.1. Localized Attraction and NMJ Stabilization*

Cell adhesion molecules (CAMs) are crucial for stabilizing synaptic connections. Changes in adhesive interactions can promote or restrict changes in synapse strength. *Drosophila* Fasciclin II (Fas II) is an immunoglobulin family CAM with sequence similarity to vertebrate Neural Cell Adhesion Molecule (NCAM; Chapter 6). Recent findings indicate that Fas II stabilizes the motor axon innervation pattern[3]. *Fas II (lf)* mutants show reduced bouton number and size, while overexpression of Fas II causes synaptic overgrowth[3]. Fas II and the voltage-gated potassium channels bind to, and are clustered by, Discs Large (Dlg). Dlg is a membrane-associated guanylate kinase, a member of the Postsynaptic Density-95 (PSD-95) family[10]. The synaptic localization of Fas II by Dlg probably contributes to the stabilization of neuronal contacts and also reduces the probability of ectopic synapse formation.

Fasciclin III (Fas III), another homophilic CAM, is expressed on both the growth cone and the synaptic site on the muscle during NMJ development. Fas III acts as a sufficient, but not essential, synaptic target recognition molecule[3]. Ectopic expression of Fas III in muscle is sufficient to transform the muscle into an acceptable synaptic target. Target recognition is controlled by Fas III dosage and spatio-temporal expression.

Other CAMs shown to regulate target recognition include Sidestep, Connectin, and cadherins. Sidestep is an immunoglobulin CAM necessary for synaptic target recognition. Embryonic muscles express Sidestep when motor axons need to extend onto them. In *sidestep* mutants, axons fail to extend onto muscles, but instead they extend along the motor neurons[28]. Ectopic expression of Sidestep results in extensive and prolonged motor axon contact with inappropriate muscle targets. Connectin, a transmembrane glycoprotein that has homophilic cell adhesion properties, is expressed in a subset of muscles and the motor neurons that innervate them. Motor neurons inappropriately innervate neighboring nontarget muscle that ectopically express connectin[3]. Furthermore, the ectopic synapse formation is dependent on the endogenous connectin expression on the motor neurons. These results show that connectin can function as an attractive and homophilic target recognition molecule *in vivo*. Cadherins are implicated in target recognition or synaptic stabilization in various synapses. A role for cadherins at the NMJ is well characterized in vertebrates, but has not been shown in flies and worms. Most invertebrate studies have been conducted in the *Drosophila* retina. In *N-cadherin* mutant fly eyes, the characteristic spatial arrangement of axon terminals is disrupted as synaptogenesis proceeds. Although synapses form, underlying cytoplasmic structures are not fully specialized at both pre- and postsynaptic terminals and SVs abnormally accumulate[29]. Loss of the protocadherin, Flamingo, also disrupts the local pattern of synaptic terminals in the retina[30]. These studies suggest that the combinatorial effects of several cell adhesion molecules may be critical for target recognition.

Receptor protein tyrosine phosphatases (RPTP) regulate cell–cell adhesion at the synapse directly via homophilic binding or indirectly by association with other known CAMs. In flies, multiple RPTP genes, including *dlar* and *dptp69d*, participate in target selection both in the NMJ and in the retina[31–33]. *C. elegans* has a single Leukocyte common Antigen Related (LAR)-like RPTP gene, *ptp-3* that produces at least two isoforms. The *ptp-3b* isoform affects axon outgrowth and targeting, while the *ptp-3a* isoform exhibits specific effects on synapse patterning[34]. In fly retina, loss of function in *N*-cadherin or in Dlar results in nearly identical target recognition defects[35], suggesting Dlar and *N*-cadherin may act in the same pathway.

Heterophilic interactions between immunoglobulin CAMs also play an important role in the target selection of *C. elegans* HSNs. HSNs form NMJs on vulval muscles and stimulate egg laying. HSN synapse formation is guided by the epithelial cells of the developing vulva. SYG-1 (SYnaptoGenesis abnormal) and SYG-2 are members of the immunoglobulin superfamily[15,36]. SYG-1 functions cell autonomously in HSNs. SYG-2 is expressed in vulval epithelial cells. In *syg-2* mutants, the SYG-1::GFP fusion protein fails to cluster at HSN synapses[36]. Thus, SYG-2 expressed in vulva epithelium acts as a ligand for neuronal SYG-1 to determine the site of the HSN synapse. The localized expression of SYG-2 may be regulated during vulva epithelium specification. These studies provide the first molecular mechanism for guidepost cells in synaptic target recognition. The homologs of SYG-1 are the vertebrate NEPH1 and *Drosophila* IrrecC and Kirre/DUF. The homologs of SYG-2 are vertebrate Nephrin and *Drosophila* Sticks and Stones (SNS) and Hibris (HIB). Both vertebrate homologs were identified for their roles in myoblast fusion. The neuronal functions of the SYG-1/SYG-2 homologs remain to be determined.

The above-described CAMs and RPTPs all act as attractants and adhesion molecules in synaptic target recognition. However, it is poorly understood how these various molecules are regulated in time and cellular space.

### *5.1.2. Localized Inhibition and Synapse Destabilization*

Localized inhibition and destabilization work in concert with localized attraction and stabilization to control synapse development. The transmembrane protein, Toll, is expressed in *Drosophila* embryo muscles and acts locally to inhibit synapse formation. In *toll* null mutants, the neuron innervates incorrect muscle cells, including those that normally express Toll[37]. Delayed expression of Toll leads to normal target recognition but delayed synaptic assembly. These studies show that both the temporal and spatial control of Toll expression is crucial for synapse formation.

Semaphorins are repulsive guidance molecules that signal through plexins to affect axon guidance and target recognition. Further evidence for repulsive signals in the body wall come from misexpression studies of *Drosophila* Semaphorin II (Sema II)[38]. *Drosophila* Sema II is transiently expressed in muscle during neurite outgrowth and synapse formation. *Sema II (lf)* mutants and ectopically or overexpressed Sema II reveal that it functions *in vivo* as a selective target-derived signal that inhibits synapse formation. A target recognition role for another axon guidance molecule, Netrin, has also been reported. Netrin B can also act as an inhibitor of target recognition, but does not require the Netrin receptor, Frazzled[38].

These studies reveal that target recognition, like axon guidance, is based on measurement and response to a specific dynamic and malleable balance of repulsive and attractive forces. The targeting system uses combinatorial interactions between broadly and specifically expressed, temporal and spatial, repulsive and attractive forces. The specific combinations of these molecules create unique recognition signals. Future studies are needed to understand the precise regulation of the combinations of adhesive and repulsive molecules.

## 5.2. NMJ Assembly

Synaptic assembly involves coordinated action of several interdependent events. These include: (1) formation of the electron-dense presynaptic site at the membrane opposite the post-synaptic cell, (2) clustering of the SVs, (3) establishment of the active zone with the proper ratio of docked SVs to presynaptic density, (4) localization of channels and regulatory proteins involved in neurotransmission, and (5) clustering of postsynaptic receptors. The localization of the major players in NMJ development is diagrammed in Figure 3.1A.

A large part of synaptic development occurs via integrated retrograde and anterograde signaling at the synapse, as well as interconnected synaptic molecule recruitment. Early developmental events take place in muscle fibers that are not innervated, including the expression of CAMs. The *Drosophila* neurotransmitter, glutamate, and its receptor are also expressed in motor neurons prior to the arrival of the neuron at the synaptic site[3]. However, the development of the mature functional NMJ requires the presence of both the nerve and muscle.

### 5.2.1. *Links Between Target Recognition and Synaptic Assembly*

The molecular link between target recognition and assembly of synaptic components is largely unknown. Conceivably, proteins important for both target recognition and synaptic component recruitment may coordinate these two processes. For example, the guanine exchange factor, Trio, interacts with Dlar and may transduce LAR signaling to the synaptic cytoskeleton via Enabled[39]. *C. elegans* SAD-1 (Synapses of Amphids Defective) defines a novel conserved family of protein Ser/Thr kinases with the kinase domain closely related to that of PAR-1 (abnormal embryonic PARtitioning of cytoplasm) and the MARK (Microtubule Affinity Regulating Kinase) family. *sad-1 (lf)* mutants display mis-localized SNB::GFP in dendritic processes. Overexpression of SAD-1 can include clustering of SV markers or vesicle precursors at nonsynaptic regions. Thus, SAD-1 functions to couple neuronal polarity determination with SV clustering[14,40]. *C. elegans* SYD-1 is a novel presynaptic active zone protein containing PDZ, C2, and rhoGAP-like domains that also functions to regulate synapse assembly. Mutations in *syd-1* cause mislocalization of synatic proteins and ectopically formed presynaptic terminals in dendritic compartments[41]. The ectopic expression of synaptic components in *syd-1* and *sad-1* mutants suggests that they participate in linking target recognition with synaptic assembly.

### *5.2.2. Vesicle Clustering*

SV precursors are generated at the cell body and transported to the terminals along microtubules. The UNC-104/KIF1A kinesin is the major motor that transports SV precursors. In *unc-104* mutant animals very few mature SVs are detected at synapses, whereas SV-like vesicles are retained in the cell bodies[42]. The conventional Kinesin-1 is a heterotetramer composed of two heavy chains and two light chains. The processive kinesin motor can walk continuously along a microtubule for several micrometers. Mutations in Kinesin-1 disrupt SV trafficking in flies and worms[43,44]. A conserved protein family, including *Drosophila* Sunday Driver, *C. elegans* UNC-16, and mammalian JIPs (c-Jun N-terminal kinase (JNK)-Interacting Protein) act as cargo adaptors for Kinesin-1[44,45]. These proteins appear to affect SV trafficking by linking JNK scaffolding to the Kinesin-1 motor. Mutations in components of the retrograde motor, dynein, cause defective trafficking of the SV proteins, SNB, and synaptotagmin, but not other synaptic proteins[46]. In *Drosophila liprin-α* mutants, motor axons accumulate SV markers (Synaptotagmin and SNB::GFP) and clear-core vesicles[47]. Direct visualization of SNB::GFP transport in *Dliprin-α* mutants shows a decrease in anterograde processivity and an increase in retrograde transport initiation. This study suggests a role for Liprin-*α* in promoting the delivery of synaptic material by an increase in Kinesin processivity and an indirect suppression of Dynein activation. In addition to transport, a major contributor to the SV pool is synaptic membrane recycling. *Drosophila* and *C. elegans* mutants defective in SV endocytosis such as *unc-26/synaptojanin*[48,49], *unc-57/endophilin*[50,51], and *unc-11/AP180*[52,53], often lead to a severe depletion of SVs at synapses. This shows endocytosis is a significant contributor to the SV pool.

### *5.2.3. Formation of the Active Zone*

The active zone is the region directly juxtaposed to the postsynaptic density, and includes the docked and primed SV, SV release machinery, and cytomatrix at the active zone (CAZ). The precise architecture of this region is important for proper neurotransmission. Liprins are an important family of proteins implicated in regulating the assembly of the active zone and are localized to the presynaptic active zone and the postsynaptic density in a variety of synapses. Liprins were originally identified through binding to the intracellular domain of LAR-RPTP. Liprins bind active zone proteins, RIM (Rab3 Interacting Molecule) and

ERC (ELKs-Rab6-interacting protein-CAST)[54,55] family protiens. *syd-2* and *Dliprin-α* encode the worm and fly liprin-α, respectively, mutants of which display abnormal active zone morphology[6,56]. Thus, liprins are important organizers of the presynaptic density.

*Dlar* mutants have abnormal active zone morphology similar to *Dliprin-α*[56]. Mutations in the *C. elegans* LAR isoform, *ptp-3A*, show synaptic pattern defects similar to, but weaker than, *syd-2(lf)*[34]. These studies are consistent with a role of LAR and Liprin-α acting together to pattern presynaptic assembly. The extracellular portion of LAR-like receptors can bind Nidogen-Laminin and Collagen XVIII[57,58]. Mutations in the *C. elegans* homologs *nid-1* (*nidogen*) and *cle-1* (*collagen XVIII*) cause distinct changes in synapse morphology[59]. Mutations in *syd-2* are epistatic to mutations in *nid-1* and *ptp-3*, but not *cle-1*[59]. These analyses hint at the possibility that the presynaptic organizing activity of SYD-2 may be modulated through LAR-Nidogen, which, together with Laminin, secure the synaptic site.

### *5.2.4. Scaffolding Synaptic Machinery*

Formation and stabilization of the synapse often relies on large multidomain scaffolding/signaling molecules. Dap160, the *Drosophila* homolog of Intersectin, is such an adaptor protein, which localizes endocytic machinery[60,61]. *dap160* mutant synapses have decreased levels of endocytic proteins, including Dynamin, Endophilin, Synaptojanin, and AP180, while other active zone and periactive zone markers remain unaltered. *dap160* mutants display abundant, but small and severely deformed synaptic boutons. Dap160 is proposed to scaffold endocytic machinery and synaptic signaling systems to the periactive zone. The ~400 kDa multidomain *C. elegans* RPM-1 (Regular of Presynaptic Morphology) and *Drosophila* Highwire proteins (discussed further in Section 5.3.1.) could possibly scaffold synaptic machinery. However, this capacity has not yet been characterized.

### *5.2.5 Postsynaptic Receptor Clustering*

Clustering of neurotransmitter receptors is crucial for effective neurotransmission and is dependent on the presynaptic neuron. *Drosophila* postsynaptic glutamate receptor fields do not form unless induced by the presynaptic neuron[62]. In muscles with delayed or ectopic innervation, both receptor clustering and receptor synthesis are delayed or ectopic. Furthermore, mutations altering glutamate levels at the synapse alter the glutamate receptor field in *Drosophila.* Mutations in enzymes that affect levels of glutamate all affect the glutamate receptor field. For example, decreased glutamate levels increase glutamate receptor expression and clustering. The postsynaptic glutamate receptor fields, once formed, may be limited in size by nonvesicular glutamate release from the presynaptic terminal[63]. In contrast to *Drosophila* glutamate receptors, postsynaptic clustering of *C. elegans* GABA receptors requires the presynaptic neuron, but does not require GABA neurotransmission[64].

In vertebrate NMJs, agrin activates a receptor complex in muscles that includes the Muscle-Specific Kinase (MuSK) and the rapsyn- and dystrophin-associated complexes, which in turn induce clustering of ACh receptors. The *C. elegans* receptor tyrosine kinase CAM-1 is similar to MuSK. *cam-1* is expressed in motor neurons and is required for nicotine-sensitive ACh neurotransmission. It regulates the localization and stabilization of a postsynaptic

ACh receptor, *acr-16*[65]. Notably, *cam-1* mutants also exhibit defects in the presynaptic site, suggesting that retrograde signaling is important for presynaptic development or maintenance. In worms, LEV-10 is a transmembrane protein required in the body wall muscle for clustering ACh receptors[66]. *lev-10* mutants display resistance to levamisole and have reduced density of levamisole-sensitive ACh receptors. The LEV-10 extracellular region is sufficient to rescue ACh receptor aggregation in *lev-10* mutants, suggesting extracellular protein–protein interactions are involved in ACh receptor clustering. *Drosophila* Dlg stabilizes glutamate receptors within the clustered synaptic receptor field. Dlg, Fas II, and glutamate receptors are regulated by *Drosophila* PIX, a Rho-type guanine exchange factor[67]. *Dpix* mutations lead to decreased synaptic levels of Dlg, Fas II, and the glutamate receptor subunit GluRIIA and a complete reduction in the SSR. PIX and the Rho-type effector kinase (Pak) together regulate postsynaptic structure. It is poorly understood how the cell adhesion complexes, signaling pathways, muscle innervation, and neurotransmitter levels coordinate receptor expression and clustering.

## 5.3. Regulatory Mechanisms of Synapse Development

### *5.3.1. Ubiquitination*

The covalent attachment of ubiquitin is a powerful mechanism for controlling protein activity and localization. Ubiquitination is a reversible process conducted by ubiquitin ligases and reversed by deubiquitinating proteases. Controlled protein ubiquitination has become a major regulatory theme in synapse development in recent years. Figure 3.2 summarizes many of the known ubiquitination players at the invertebrate NMJ.

*C. elegans* RPM-1 and *Drosophila* Highwire are large conserved proteins with vertebrate homologs known as Esrom (zebrafish), Phr1 (mouse), and Pam (human) [12,68]. The RPM-1/Highwire family has several predicted functional modules, one of which is a RING-H2 finger ubiquitin E3 ligase domain. *rpm-1* mutant NMJs display severe reduction of SV numbers per synapse and the presence of several presynaptic densities within one synaptic terminal[12,69]. NMJs of *highwire* mutants have greatly expanded bouton numbers and length of branches[12,68]. The synapse displays normal ultrastructure, but has reduced quantal content. A synaptic quanta is the smallest unit of neurotransmitter release and corresponds to the amount of neurotransmitter molecules contained within a single SV. RPM-1 binds to an F-box domain containing protein, FSN-1, which associates with homologs of Cullin and Skp. This SCF (Skp, Cullin, F-box) ubiquitin ligase complex is required in presynaptic neurons for the restriction and/or maturation of synapses[70]. Two targets for this SCF complex have recently been identified. One is a p38 Mitogen-Activated Protein Kinase (MAPK) cascade, with the upstream MAPKKK DLK-1/a dual-leucine zipper MAPK as a direct substrate[69]. Another is an ALK-like receptor tyrosine kinase[70]. The functional connection between the ALK kinase and the DLK-1 pathway remains to be investigated. Neuronal overexpression of the deubiquitinating protease Fat Facets at the *Drosophila* NMJ severely disrupts synaptic growth control and thus causes an increase in the number of synaptic boutons[71]. Mutations in *fat facets* enhance the NMJ defects in *highwire* mutants. Together, these results support an evolutionarily conserved function of the RPM-1/Highwire and Fat Facet proteins in modulating ubiquitin to control NMJ development.

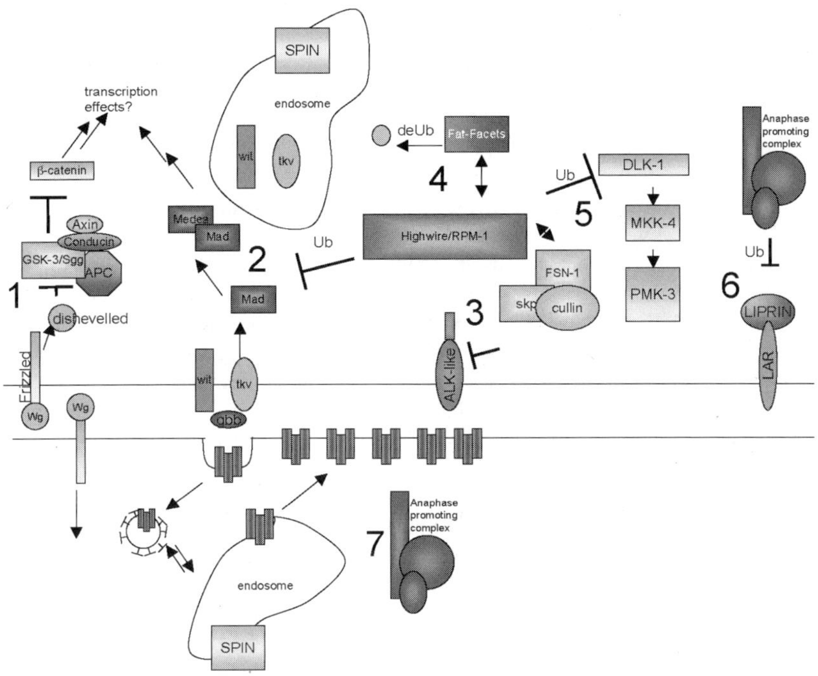


**Figure 3.2.** Ubiquitination and Cell Signaling Mechanisms in Invertebrate NMJ Development. (1) The Wnt ligand Wg signaling is required for presynaptic formation of the active zone and for postsynaptic development of the SSR. The Wnt pathway has been well described in early development. Wg may function via Sgg, but the Wg Wnt pathway at the invertebrate NMJ is not well understood. (2) The TGF-beta signaling pathway is important for proper synapse development. Highwire downregulates the TGF-beta signaling pathway and directly binds the transcription factor Medea. Spinster downregulates TGF-beta signaling, possibly by affecting endosomal recycling of BMP receptors. (3) RPM-1 binds the F-box protein, FSN-1, which forms a novel SCF ubiquitin ligase complex. This complex regulates the ALK tyrosine kinase. (4) Highwire binds the deubiquitinating enzyme, Fat Facets. (5) RPM-1 also downregulates the MAPK pathway. (6) The APC downregulates Liprin to control active zone assembly and also (7) regulates neurotransmitter receptors possibly via an endocytic mechanism.

The Anaphase Promoting Complex (APC), an E3 ubiquitin ligase that was originally identified through its function in regulating the cell cycle, has recently been shown to control synapse development and function in multiple contexts. In fly NMJs, the APC is localized at synaptic boutons and controls synaptic size by regulating levels of Liprin-$\alpha$ through ubiquitination[72]. In fly and worm, the APC also regulates the synaptic abundance of postsynaptic receptors[72,73]. In worm, the regulation of the postsynaptic receptor concentration and abundance possibly occurs via an endocytic mechanism[73]. Mutations in APC subunits increase the number of glutamate receptor subunits. Mutations that block clathrin-mediated endocytosis block the APC mutation effects, suggesting that the APC regulates glutamate receptor recycling.

### *5.3.2. Wnt and BMP Signaling*

Proper synapse development requires positional information, which coordinates and maintains the juxtaposition of pre- and postsynaptic elements.

Secreted retrograde and anterograde signals aid in this aspect of synapse development. Wnt and TGF-beta signaling pathways, which are best known for providing positional information during morphogenesis, have fundamental roles in synapse development[74]. Wnts and TGF-betas are secreted by either the pre- or postsynaptic cell and provide crucial signals that regulate the coordinated development of synaptic specializations.

The bone morphogenetic proteins (BMPs) are secreted TGF-beta family proteins. Their diverse functions include primary neural induction, dorsal–ventral patterning of the neural tube, axon guidance, as well as synapse development. TGF-beta ligand binding causes heterodimerization of type I and II receptors. The type II receptor then activates the type I receptor by phosphorylation. The type I receptor then phosphorylates R-Smads (Mad proteins in *Drosophila* and Sma and Daf proteins in *C. elegans*), which transduce the signal to the nucleus via a co-Smad. The R-Smad/co-Smad complex binds transcription factors to regulate gene expression. *Drosophila* Wishful Thinking (Wit) is a type II BMP receptor and is required for adequate synaptic growth and synaptic transmission[75,76]. In *wit* mutant NMJs, pre- and postsynaptic membranes are detached and the T-bar structure is separated from the presynaptic membrane. This effect appears to involve changes in CAMs, as Fas II is downregulated at the *wit* mutant NMJ. The Wit ligand, TGF-beta Glass Bottom Boat (Gbb), signals retrogradely from the muscle to affect synaptic growth[77]. *gbb* and *wit* mutants show similar ultrastructural and physiological defects. Similarly, mutants in the type I BMP receptors *thick veins (tkv)* and *saxophone (sax),* in the R-Smad *mothers against dpp (mad)*, or the Co-Smad *medea* all have defective synapses and neurotransmission in embryonic NMJs[78,79]. Together, these findings demonstrate that an entire TGF-beta pathway is essential for synapse development.

The regulation of the Wit TGF-beta pathway appears to involve both the E3 ubiquitin ligase, Highwire, and an endosome pathway[79,80]. *spinster (spin)* encodes a protein that may help to regulate this growth factor signaling via an endosomal/lysosomal pathway. It is localized to the late endosome/lysosome compartment in the pre- and postsynaptic terminals[80]. In the *spin* null mutant, NMJs overgrow by more than 200% compared to wild type. Genetic evidence in *spin* mutants shows that the synaptic overgrowth is due, at least in part, to enhanced TGF-beta signaling. Late endosomes have dramatically changed architecture in *spin* mutants. This disrupted architecture could potentially cause changes in endosomal function and thus misregulation of growth factor signaling. An understanding of TGF-beta pathway regulation in developmental time and space is needed.

Wnt signaling molecules are essential regulators of cell fate, cell polarity, cell proliferation, tissue patterning, and synaptogenesis. Many aspects of the Wnt signaling pathway have been defined. Wnts signal through the membrane receptor Frizzled and the co-receptor (*Drosophila* Arrow). In the absence of Wnt signaling, β-catenin (*Drosophila* Armadillo/*C. elegans* HMP-2) is phosphorylated by casein kinase 1 and glycogen synthase kinase 3β (GSK-3β) (*Drosophila* Shaggy/*C. elegans* GSK3β). It is then ubiquitinated and degraded. This process requires the formation of a complex with the scaffolding protein Axin and the tumor suppressor protein Adenomatous Polyposis Coli (APC). Activation of the Wnt pathway leads to phosphorylation of Dishevelled, which prevents β-catenin degradation, which ultimately leads to transcription changes in Wnt responsive genes.

The *Drosophila* Wnt signaling molecule, Wingless (Wg), acts as an anterograde regulator of synaptic growth[81]. Wg is expressed in motor neurons. It is required for presynaptic formation of the active zone, for postsynaptic development of the SSR, and for localization of Fas II, Dlg, and glutamate receptors. Wg binds Frizzled 2. The Wnt downstream effector, GSK-3/Shaggy (Sgg), is expressed presynaptically in type I boutons at the NMJ[82]. *sgg* loss of function causes an increase in synapse number, while sgg overexpression causes a reduction in synapse number. Wg may function via Sgg to regulate synapse development. However, the Wnt signaling pathway is poorly understood at the invertebrate NMJ. The specific molecules in the Wnt pathway, how the pathway is regulated, and the down stream effectors are not yet known. The human APC–β-catenin complex was shown to bind the human homolog of the *Drosophila* DLG, suggesting the APC– DLG complex may regulate neuronal function[83].

Genetic and cell biological studies of Wnts and TGF-betas have revealed a synapse growth and development role for these well-known early development genes. However, the mechanism, regulation, and downstream effectors of the Wnt and TGF-beta pathways, which mediate cross talk between pre- and postsynaptic sites, is not understood. Further analysis of these pathways, their effectors, and the feedback mechanisms that impinge on them will further our understanding of synapse development.

### *5.3.3. Synaptic Cytoskeletal Rearrangement*

Remodeling of the cytoskeleton allows for the control of an amazing diversity of cell shapes and dynamic intracellular behaviors. The NMJ cytoskeleton consists principally of actin and microtubule protein filaments. Polymerized actin (F-actin) is abundantly distributed throughout the synaptic terminal and is concentrated primarily around the active zone and SV clusters. Microtubules primarily run close to, but not within, the synapse. The cytoskeleton plays an important role in NMJ development and maintenance. Figure 3.3 diagrams the major known cytoskeletal regulators at the invertebrate NMJ.

A number of experimental observations support the view that dynamic actin filaments are a prerequisite for synapse formation[84]. First, pharmacological perturbation of actin inhibits synaptogenesis. Second, actin clusters at the presynaptic site before functional synapses form, possibly acting as a spatial marker for synaptogenesis. Third, disruption of signaling pathways involved in actin reorganization results in synaptogenesis defects. The first *in vivo* evidence that regulation of the actin cytoskeleton can directly affect synaptogenesis in *Drosophila* comes from the studies of the periactive zone protein, Nervous Wreck. Nervous Wreck contains an N-terminal Fes/CIP4 Homology (FCH) domain and two SH3 domains. The FCH domain is implicated in actin binding. Nervous Wreck activates Wasp, which promotes F-actin assembly in growth cones. Mutations in either *nervous wreck* or *wasp* cause an increase in the number of synaptic boutons and branch formation[85]. Mutant boutons are reduced in size and have fewer active zones, associated with a reduction in synaptic transmission.

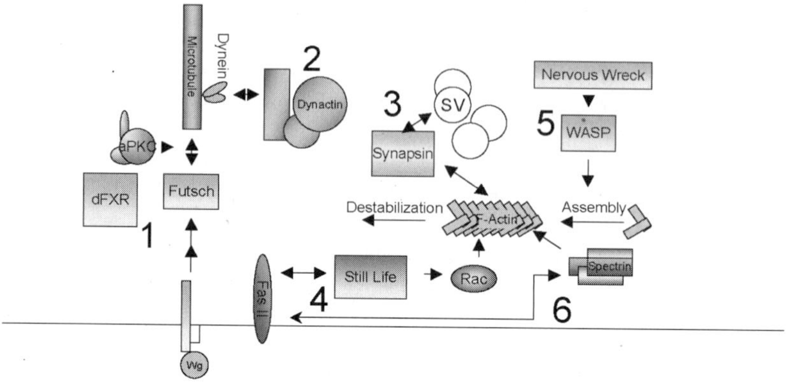


**Figure 3.3.** Cytoskeleton Regulators Involved in Invertebrate NMJ Development. (1) Futsch regulates microtubules. dFXR represses translation of Futsch. Wg localizes Futsch. aPKC promotes the association of Futsch with microtubules. (2) Dynactin regulates the Dynein microtubule motor. (3) Synapsins tether vesicles to actin. (4) Fas II localizes the RacGEF Still Life[90]. Still Life regulates actin dynamics via Rac. (5) Nervous Wreck activates Wasp, which promotes F-actin assembly. (6) Spectrin heterotetramers interact with short actin filaments and stabilize NMJ cell adhesion molecules.

Actin clusters SVs within a presynaptic vesicle pool. Inducing actin depolymerization with cytochalasin-D at the *Drosophila* NMJ causes activity-dependent depletion of SVs[86]. This study corroborates a model in which actin plays a facilitatory role in SV endocytosis and mobilization from SV pools. Synapsins are a family of actin-associated phosphoproteins that are required for maintaining the SV reserve pool[84]. They act by tethering the SVs to the actin cytoskeleton. Synapsins are present in fly (Syn) and worm (SNN-1), but their functions are not yet reported. Spectrins are actin-bundling proteins essential for embryonic development and important for synapse stabilization at the adult NMJ. β-Spectrin and α-Spectrin form heterotetramers that interact with an actin network localized to the plasma membrane. The genomes of *C. elegans* and *Drosophila* have a single gene for each subunit. In *Drosophila,* depletion of presynaptic spectrins, using transgenic RNAi, revealed a role of spectrin in synapse stability[21]. The mutant NMJs show altered axonal transport and disrupted synaptic microtubules, reminiscent of synapse retraction. Loss of presynaptic Spectrin causes a disorganization and elimination of synaptic CAMs, Fas II, and Neuroglian. Spectrin may thus have a role in linking synaptic cell adhesion with the stabilization of the cytoskeleton.

Regulation of the cytoskeleton via Microtubule-Associated Protein (MAP), affects synaptic formation and synaptic growth *in vivo* at the *Drosophila* NMJ[87]. The MAP1B-like protein, Futsch, is necessary for microtubule organization during synapse development and growth[87]. The stable microtubule loop at the axon terminal undergoes rearrangement to allow bouton division. Futsch colocalizes with the stable, looped microtubule ends, but adopts a punctate, diffuse appearance during bouton division[87]. *futsch* mutants display disrupted terminal microtubule loops, increased bouton size, and decreased bouton number. Futsch appears to be subject to several regulations. Atypical protein kinase C (aPKC) promotes the association of Futsch with microtubules[88]. Fragile X mental retardation protein

(FMRP) is an RNA-binding protein that regulates translation. The *Drosophila* FMRP homolog, Fragile X-related protein (dFXR), represses translation of Futsch[89]. Thus, dFXR controls the microtubule cytoskeleton to maintain both synaptic strength and function. The Wnt family protein, Wg, localizes Futsch. *wg* mutants have an increase in the number of boutons that show splayed microtubules[81]. The opposite occurs in sgg mutants in which looped boutons are in excess[82].

In addition to the microtubules themselves, regulation of the microtubule motor, Dynein, affects NMJ development. Dynactin is a multiprotein complex thought to mediate Dynein function in the cell. The cytoskeletal Dynactin protein complex, including Centractin and Glued, is necessary for synapse stabilization[20].

*5.3.4. Localized Synaptic Protein Translation*

Synaptic proteins were originally thought to be synthesized in the cell body and then trafficked to the synapse. However, local protein translation is an elegant way to obtain fast and efficient synapse specific localization of new proteins. There is increasing evidence that local protein translation modifies active synapses and thus affects synaptic plasticity. Localized translation is important for the efficacy and morphology of the *Drosophila* NMJ. Glutamate receptors are locally translated in the postsynaptic muscle. Mutation of translation factors, including eIF4E and poly(A)-binding protein, alters the abundance of receptors in the postsynaptic membrane[91]. Large aggregates of translational components as well as messenger RNA of the postsynaptic glutamate receptor subunit DGluR-IIA are localized to the larval NMJ. Genetic manipulations using the translation initiation factors eIF4E and poly(A)-binding protein cause an increase in synaptic translation aggregates. This aggregation is associated with a significant increase in postsynaptic DGluR-IIA and a reduction in the cell-adhesion molecule Fas II. In addition, the NMJ size and neurotransmission efficacy is significantly increased.

Pumilio (Pum) negatively regulates expression of the translation factor eIF-4E at the NMJ[92]. It is localized to the postsynaptic side of the NMJ in third instar larvae and is also expressed in larval neurons. GluRIIa is upregulated in *pum* mutants. Neuronal Pum also regulates synaptic growth. *Pum (lf)* mutant NMJ boutons are larger and fewer in number, while Pum overexpression increases bouton number and decreases bouton size.

Unlike postsynaptic sites, presynaptic terminals appear to lack ribosomes and translation machinery. It is thus surprising that several RNA-binding proteins appear to be present in axons and presynaptic terminals at the *Drosophila* NMJ. The precise function of these RNA-binding proteins awaits future studies.

## 5.4. Activity Dependence of Synapse Assembly and Growth

Synaptic activity plays a crucial role in shaping the *Drosophila* NMJ and has recently been implicated in shaping the *C. elegans* NMJ as well. In *Drosophila,* synaptic activity influences synaptic connectivity, size, and functional homeostasis[3]. Hyperexcitable and hypoexcitable mutants display abnormal axon and synapse morphology. For example, hyperexcitable Shaker (Sh) and ether a go-go (eag) mutants synthetically interact to cause an increase in the number of higher-order axonal branches and synaptic swellings on the neurites[93]. This effect is mediated by presynaptic changes in excitability. *Drosophila* glutamate receptors preferentially cluster opposite the largest and most physiologically active sites[94]. Elevated expression of muscle glutamate receptors leads to significant NMJ expansion[94,95]. These results suggest retrograde control of motor neuron growth is influenced by synaptic receptor activation. Presynaptic effects on postsynaptic glutamate receptor activity require the activity of Wit, mediated through $Ca^{2+}$/calmodulin-dependent protein kinase II (CaMKII)[96]. CaMKII is regulated by activity-dependent calcium changes, and thus may provide an additional mechanism for activity-dependent NMJ changes. Localization of DLG and Fas II to the NMJ is regulated by the activity of CaMKII[97]. Whether synaptic excitability affects the NMJ via these signaling molecules has yet to be examined.

Fas II and the cAMP Response Element-Binding Protein (CREB) modulate activity-dependent NMJ changes[98]. Increased activity causes an increase in cAMP levels that leads to a reduction of Fas II and a concurrent increase of CREB activity. Decreased Fas II levels promote expansion of the NMJ. Transcription of CREB responsive genes ultimately increases neurotransmitter release. In the fly NMJ, the Ras MAPK cascade functions through Fas II cell adhesion to control synapse growth[99].

In *C. elegans,* mutants with reduced cholinergic synaptic transmission show abnormal synapse sprouting in a set of cholinergic motor neurons[100]. These effects are mediated by an unknown retrograde signal from the postsynaptic cell that affects formation of synaptic connections. UNC-122 is a postsynaptic transmembrane protein that affects neuromuscular signaling[101]. In unc-122 mutants, impaired synaptic transmission causes axon-sprouting defects in a subset of GABAergic neurons. Although synaptic activity is implicated in invertebrate synapse assembly and growth, the mechanism is poorly understood.

## 6. CONCLUSIONS

Few other systems are as highly genetically malleable as the worm and fly. The unbiased forward genetic screening approach has led to the identification of a number of molecules involved in synapse development. Advances in cell biology, *in vivo* imaging, and electrophysiology have led to the characterization of identified synapse development molecules. With these approaches, studies in *C. elegans* and *Drosophila* have shed light on many fundamental questions. For example, the use of genetics has revealed entire *in vivo* signaling mechanisms involved in synapse development, such as the MAPK and TGF-beta pathways. *Drosophila* and *C. elegans* cell biology and *in vivo* imaging have allowed the site-specific dissection of synapse effects. The first large-scale RNAi screen in neuroscience was conducted in *C. elegans*[8]. This screen has revealed the immense promise of a functional genomics approach at the synapse. Direct observation of synapse formation in real time would assist our understanding of integrated developmental processes. The valuable *Drosophila* and *C. elegans* experimental systems can now be used to integrate the individual processes of synapse development. With great promise, future studies will network the mechanistic interactions and signaling pathways of synapse development.*

*We would like to apologize for not including all relevant studies due to space constraints. We would like to thank Vivian Budnik for the kind contribution of the *Drosophila* images in Figure 3.1. We would like to thank the following people for helpful comments on the manuscript: C. Weaver, H. Brown, K. Chan, P.C. Spiegel, B. Ackley, C. Suh, and C. Pfeiffenberger. The work in Y.J.'s laboratory is supported by grants from NIH, NSF, and HHMI. Y.J. is an investigator of HHMI.

## 7. REFERENCES


1. Broadie, K.S., and Richmond, J.E. (2002) *Curr Opin Neurobiol* **12**, 491–498.
2. White, J.S., Thomson J.E., and Brenner S. (1986) *Phil Trans Royal Soc Lond Ser B, Biol Scien* **314**, 1–340.
3. Keshishian, H., Broadie, K., Chiba, A., and Bate, M. (1996) *Annu Rev Neurosci* **19**, 545–575.
4. Zito, K., Parnas, D., Fetter, R.D., Isacoff, E.Y., and Goodman, C.S. (1999) *Neuron* **22**, 719–729.
5. Nonet, M.L. (1999) *J Neurosci Methods* **89**, 33–40.
6. Zhen, M., and Jin, Y. (1999) *Nature* **401**, 371–375.
7. Yeh, E., Kawano, T., Weimer, R.M., Bessereau, J.L., and Zhen, M. (2005) *J Neurosci* **25**, 3833–3841.
8. Sieburth, D., Ch'ng, Q., Dybbs, M., Tavazoie, M., Kennedy, S., Wang, D., Dupuy, D., Rual, J.F., Hill, D.E., Vidal, M., Ruvkun, G., and Kaplan, J.M. (2005) *Nature* **436**, 510–517.
9. Van Vactor, D., Sink, H., Fambrough, D., Tsoo, R., and Goodman, C. (1993) *Cell* **73**, 1137–1153.
10. Zito, K., Fetter, R.D., Goodman, C.S., and Isacoff, E.Y. (1997) *Neuron* **19**, 1007–1016.
11. Brenner, S. (1974) *Genetics* **77**, 71–94.
12. Zhen, M., Huang, X., Bamber, B., and Jin, Y. (2000) *Neuron* **26**, 331–343.
13. Schaefer, A.M., Hadwiger, G.D., and Nonet, M.L. (2000) *Neuron* **26**, 345–356.
14. Crump, J.G., Zhen, M., Jin, Y., and Bargmann, C.I. (2001) *Neuron* **29**, 115–129.
15. Shen, K., and Bargmann, C.I. (2003) *Cell* **112**, 619–630.
16. Wang, D., Kennedy, S., Conte, D., Jr., Kim, J.K., Gabel, H.W., Kamath, R.S., Mello, C.C., and Ruvkun, G. (2005) *Nature* **436**, 593–597.
17. Simmer, F., Tijsterman, M., Parrish, S., Koushika, S.P., Nonet, M.L., Fire, A., Ahringer, J., and Plasterk, R.H. (2002) *Curr Biol* **12**, 1317–1319.
18. Rorth, P. (1996) *Proc Natl Acad Sci U S A* **93**, 12418–12422.
19. Kraut, R., Menon, K., and Zinn, K. (2001) *Curr Biol* **11**, 417–430.
20. Eaton, B.A., Fetter, R.D., and Davis, G.W. (2002) *Neuron* **34**, 729–741.
21. Pielage, J., Fetter, R.D., and Davis, G.W. (2005) *Curr Biol* **15**, 918–928.
22. Jan, L.Y., and Jan, Y.N. (1976) *J Physiol* **262**, 215–236.
23. Francis, M.M., Mellem, J.E., and Maricq, A.V. (2003) *Trends Neurosci* **26**, 90–99.
24. Nguyen, M., Alfonso, A., Johnson, C.D., and Rand, J.B. (1995) *Genetics* **140**, 527–535.
25. Lewis, J.A., Wu, C.H., Berg, H., and Levine, J.H. (1980) *Genetics* **95**, 905–928.
26. Reiff, D.F., Ihring, A., Guerrero, G., Isacoff, E.Y., Joesch, M., Nakai, J., and Borst, A. (2005) *J Neurosci* **25**, 4766–4778.
27. Guerrero, G., Rieff, D.F., Agarwal, G., Ball, R.W., Borst, A., Goodman, C.S., and Isacoff, E.Y. (2005) *Nat Neurosci* **8**, 1188–1196.
28. Sink, H., Rehm, E.J., Richstone, L., Bulls, Y.M., and Goodman, C.S. (2001) *Cell* **105**, 57–67.
29. Iwai, Y., Hirota, Y., Ozaki, K., Okano, H., Takeichi, M., and Uemura, T. (2002) *Mol Cell Neurosci* **19**, 375–388.
30. Lee, R.C., Clandinin, T.R., Lee, C.H., Chen, P.L., Meinertzhagen, I.A., and Zipursky, S.L. (2003) *Nat Neurosci* **6**, 557–563.
31. Desai, C.J., Gindhart, J.G., Jr., Goldstein, L.S., and Zinn, K. (1996) *Cell* **84**, 599–609.
32. Sun, Q., Bahri, S., Schmid, A., Chia, W., and Zinn, K. (2000) *Development* **127**, 801–812.
33. Wills, Z., Bateman, J., Korey, C.A., Comer, A., and Van Vactor, D. (1999) *Neuron* **22**, 301–312.
34. Ackley, B.D., Harrington, R.J., Hudson, M.L., Williams, L., Kenyon, C.J., Chisholm, A.D., and Jin, Y. (2005) *J Neurosci* **25**, 7517–7528.
35. Lee, C.H., Herman, T., Clandinin, T.R., Lee, R., and Zipursky, S.L. (2001) *Neuron* **30**, 437–450.
36. Shen, K., Fetter, R.D., and Bargmann, C.I. (2004) *Cell* **116**, 869–881.
37. Rose, D., Zhu, X., Kose, H., Hoang, B., Cho, J., and Chiba, A. (1997) *Development* **124**, 1561–1571.
38. Winberg, M.L., Mitchell, K.J., and Goodman, C.S. (1998) *Cell* **93**, 581–591.
39. Bateman, J., Shu, H., and Van Vactor, D. (2000) *Neuron* **26**, 93–106.
40. Kishi, M., Pan, Y.A., Crump, J.G., and Sanes, J.R. (2005) *Science* **307**, 929–932.
41. Hallam, S.J., Goncharov, A., McEwen, J., Baran, R., and Jin, Y. (2002) *Nat Neurosci* **5**, 1137–1146.
42. Hall, D.H., and Hedgecock, E.M. (1991) *Cell* **65**, 837–847.
43. Hurd, D.D., and Saxton, W.M. (1996) *Genetics* **144**, 1075–1085.
44. Byrd, D.T., Kawasaki, M., Walcoff, M., Hisamoto, N., Matsumoto, K., and Jin, Y. (2001) *Neuron* **32**, 787–800.
45. Bowman, A.B., Kamal, A., Ritchings, B.W., Philp, A.V., McGrail, M., Gindhart, J.G., and Goldstein, L.S. (2000) *Cell* **103**, 583–594.

46. Koushika, S.P., Schaefer, A.M., Vincent, R., Willis, J.H., Bowerman, B., and Nonet, M.L. (2004) *J Neurosci* **24**, 3907–3916.
47. Miller, K.E., DeProto, J., Kaufmann, N., Patel, B.N., Duckworth, A., and Van Vactor, D. (2005) *Curr Biol* **15**, 684–689.
48. Verstreken, P., Koh, T.W., Schulze, K.L., Zhai, R.G., Hiesinger, P.R., Zhou, Y., Mehta, S.Q., Cao, Y., Roos, J., and Bellen, H.J. (2003) *Neuron* **40**, 733–748.
49. Harris, T.W., Hartwieg, E., Horvitz, H.R., and Jorgensen, E.M. (2000) *J Cell Biol* **150**, 589–600.
50. Schuske, K.R., Richmond, J.E., Matthies, D.S., Davis, W.S., Runz, S., Rube, D.A., van der Bliek, A.M., and Jorgensen, E.M. (2003) *Neuron* **40**, 749–762.
51. Verstreken, P., Kjaerulff, O., Lloyd, T.E., Atkinson, R., Zhou, Y., Meinertzhagen, I.A., and Bellen, H.J. (2002) *Cell* **109**, 101–112.
52. Bao, H., Daniels, R.W., MacLeod, G.T., Charlton, M.P., Atwood, H.L., and Zhang, B. (2005) *J Neurophysiol* **94**, 1888–1903.
53. Nonet, M.L., Holgado, A.M., Brewer, F., Serpe, C.J., Norbeck, B.A., Holleran, J., Wei, L., Hartwieg, E., Jorgensen, E.M., and Alfonso, A. (1999) *Mol Biol Cell* **10**, 2343–2360.
54. Ko, J., Na, M., Kim, S., Lee, J.R., and Kim, E. (2003) *J Biol Chem* **278**, 42377–42385.
55. Schoch, S., Castillo, P.E., Jo, T., Mukherjee, K., Geppert, M., Wang, Y., Schmitz, F., Malenka, R.C., and Sudhof, T.C. (2002) *Nature* **415**, 321–326.
56. Kaufmann, N., DeProto, J., Ranjan, R., Wan, H., and Van Vactor, D. (2002) *Neuron* **34**, 27–38.
57. O'Grady, P., Thai, T.C., and Saito, H. (1998) *J Cell Biol* **141**, 1675–1684.
58. Aricescu, A.R., McKinnell, I.W., Halfter, W., and Stoker, A.W. (2002) *Mol Cell Biol* **22**, 1881–1892.
59. Ackley, B.D., Kang, S.H., Crew, J.R., Suh, C., Jin, Y., and Kramer, J.M. (2003) *J Neurosci* **23**, 3577–3587.
60. Marie, B., Sweeney, S.T., Poskanzer, K.E., Roos, J., Kelly, R.B., and Davis, G.W. (2004) *Neuron* **43**, 207–219.
61. Koh, T.W., Verstreken, P., and Bellen, H.J. (2004) *Neuron* **43**, 193–205.
62. Broadie, K.S. (1994) *J Physiol Paris* **88**, 123–139.
63. Featherstone, D.E., Rushton, E., and Broadie, K. (2002) *Nat Neurosci* **5**, 141–146.
64. Jin, Y., Jorgensen, E., Hartwieg, E., and Horvitz, H.R. (1999) *J Neurosci* **19**, 539–548.
65. Francis, M.M., Evans, S.P., Jensen, M., Madsen, D.M., Mancuso, J., Norman, K.R., and Maricq, A.V. (2005) *Neuron* **46**, 581–594.
66. Gally, C., Eimer, S., Richmond, J.E., and Bessereau, J.L. (2004) *Nature* **431**, 578–582.
67. Parnas, D., Haghighi, A.P., Fetter, R.D., Kim, S.W., and Goodman, C.S. (2001) *Neuron* **32**, 415–424.
68. Wan, H.I., DiAntonio, A., Fetter, R.D., Bergstrom, K., Strauss, R., and Goodman, C.S. (2000) *Neuron* **26**, 313–329.
69. Nakata, K., Abrams, B., Grill, B., Goncharov, A., Huang, X., Chisholm, A.D., and Jin, Y. (2005) *Cell* **120**, 407–420.
70. Liao, E.H., Hung, W., Abrams, B., and Zhen, M. (2004) *Nature* **430**, 345–350.
71. DiAntonio, A., Haghighi, A.P., Portman, S.L., Lee, J.D., Amaranto, A.M., and Goodman, C.S. (2001) *Nature* **412**, 449–452.
72. van Roessel, P., Elliott, D.A., Robinson, I.M., Prokop, A., and Brand, A.H. (2004) *Cell* **119**, 707–718.
73. Juo, P., and Kaplan, J.M. (2004) *Curr Biol* **14**, 2057–2062.
74. Packard, M., Mathew, D., and Budnik, V. (2003) *Nat Rev Neurosci* **4**, 113–120.
75. Marques, G., Bao, H., Haerry, T.E., Shimell, M.J., Duchek, P., Zhang, B., and O'Connor, M.B. (2002) *Neuron* **33**, 529–543.
76. Aberle, H., Haghighi, A.P., Fetter, R.D., McCabe, B.D., Magalhaes, T.R., and Goodman, C.S. (2002) *Neuron* **33**, 545–558.
77. McCabe, B.D., Marques, G., Haghighi, A.P., Fetter, R.D., Crotty, M.L., Haerry, T.E., Goodman, C.S., and O'Connor, M.B. (2003) *Neuron* **39**, 241–254.
78. Rawson, J.M., Lee, M., Kennedy, E.L., and Selleck, S.B. (2003) *J Neurobiol* **55**, 134–150.
79. McCabe, B.D., Hom, S., Aberle, H., Fetter, R.D., Marques, G., Haerry, T.E., Wan, H., O'Connor, M.B., Goodman, C.S., and Haghighi, A.P. (2004) *Neuron* **41**, 891–905.
80. Sweeney, S.T., and Davis, G.W. (2002) *Neuron* **36**, 403–416.
81. Packard, M., Koo, E.S., Gorczyca, M., Sharpe, J., Cumberledge, S., and Budnik, V. (2002) *Cell* **111**, 319–330.
82. Franco, B., Bogdanik, L., Bobinnec, Y., Debec, A., Bockaert, J., Parmentier, M.L., and Grau, Y. (2004) *J Neurosci* **24**, 6573–6577.
83. Matsumine, A., Ogai, A., Senda, T., Okumura, N., Satoh, K., Baeg, G.H., Kawahara, T., Kobayashi, S., Okada, M., Toyoshima, K., and Akiyama, T. (1996) *Science* **272**, 1020–1023.

84. Dillon, C., and Goda, Y. (2005) *Annu Rev Neurosci* **28**, 25–55.
85. Coyle, I.P., Koh, Y.H., Lee, W.C., Slind, J., Fergestad, T., Littleton, J.T., and Ganetzky, B. (2004) *Neuron* **41**, 521–534.
86. Kuromi, H., and Kidokoro, Y. (1998) *Neuron* **20**, 917–925.
87. Roos, J., Hummel, T., Ng, N., Klambt, C., and Davis, G.W. (2000) *Neuron* **26**, 371–382.
88. Ruiz-Canada, C., Ashley, J., Moeckel-Cole, S., Drier, E., Yin, J., and Budnik, V. (2004) *Neuron* **42**, 567–580.
89. Zhang, Y.Q., Bailey, A.M., Matthies, H.J., Renden, R.B., Smith, M.A., Speese, S.D., Rubin, G.M., and Broadie, K. (2001) *Cell* **107**, 591–603.
90. Sone, M., Suzuki, E., Hoshino, M., Hou, D., Kuromi, H., Fukata, M., Kuroda, S., Kaibuchi, K., Nabeshima, Y., and Hama, C. (2000) *Development* **127**, 4157–4168.
91. Sigrist, S.J., Thiel, P.R., Reiff, D.F., Lachance, P.E., Lasko, P., and Schuster, C.M. (2000) *Nature* **405**, 1062–1065.
92. Menon, K.P., Sanyal, S., Habara, Y., Sanchez, R., Wharton, R.P., Ramaswami, M., and Zinn, K. (2004) *Neuron* **44**, 663–676.
93. Mosca, T.J., Carrillo, R.A., White, B.H., and Keshishian, H. (2005) *Proc Natl Acad Sci U S A* **102**, 3477–3482.
94. Sigrist, S.J., Reiff, D.F., Thiel, P.R., Steinert, J.R., and Schuster, C.M. (2003) *J Neurosci* **23**, 6546–6556.
95. Sigrist, S.J., Thiel, P.R., Reiff, D.F., and Schuster, C.M. (2002) *J Neurosci* **22**, 7362–7372.
96. Haghighi, A.P., McCabe, B.D., Fetter, R.D., Palmer, J.E., Hom, S., and Goodman, C.S. (2003) *Neuron* **39**, 255–267.
97. Koh, Y.H., Popova, E., Thomas, U., Griffith, L.C., and Budnik, V. (1999) *Cell* **98**, 353–363.
98. Davis, G.W., Schuster, C.M., and Goodman, C.S. (1996) *Neuron* **17**, 669–679.
99. Koh, Y.H., Ruiz-Canada, C., Gorczyca, M., and Budnik, V. (2002) *J Neurosci* **22**, 2496–2504.
100. Zhao, H., and Nonet, M.L. (2000) *Development* **127**, 1253–1266.
101. Loria, P.M., Hodgkin, J., and Hobert, O. (2004) *J Neurosci* **24**, 2191–2201.
102. Kishi, M., Pan, Y.A., Crump, J.G., and Sanes, J.R. (2005) *Science* **307**, 929–932.

# 4

# MECHANISMS THAT REGULATE NEURONAL PROTEIN CLUSTERING AT THE SYNAPSE

Rochelle M. Hines and Alaa El-Husseini*

## 1. SUMMARY

There is considerable evidence that synaptic contact formation and remodeling is regulated by assembly of several important classes of proteins, including neurotransmitter receptors, adhesiòn molecules, and scaffolding proteins. This process is regulated by intricate mechanisms that regulate protein trafficking and clustering at the synapse. In this chapter we discuss some of the newly discovered mechanisms that govern protein targeting to the synapse and how this process contributes to clustering of ion channels and adhesion molecules at contact sites, with emphasis on the role of scaffolding proteins in this process.

## 2. INTRODUCTION

Brain function hinges on the clustering of appropriate neurotransmitter receptors and associated signaling molecules at the postsynaptic membrane in perfect opposition to the presynaptic active zone where the release of a specific neurotransmitter occurs. Amazingly, this organization occurs along the dendrites, long distances away from the cell body where proteins are synthesized. Understanding how neurons organize this intricate network of synapses formed in early development is integral to understanding central nervous system function. In general, the formation of a synapse is thought to occur in three successive steps beginning with the synthesis of new proteins in the soma, followed by protein sorting and trafficking to axons and dendrites, and then by their assembly at nascent neuronal contacts (Figure 4.1A). At postsynaptic sites, the intense clustering of neurotransmitter

*University of British Columbia, Vancouver, BC, , Canada V6T 1Z3; alaa@interchange.ubc.ca

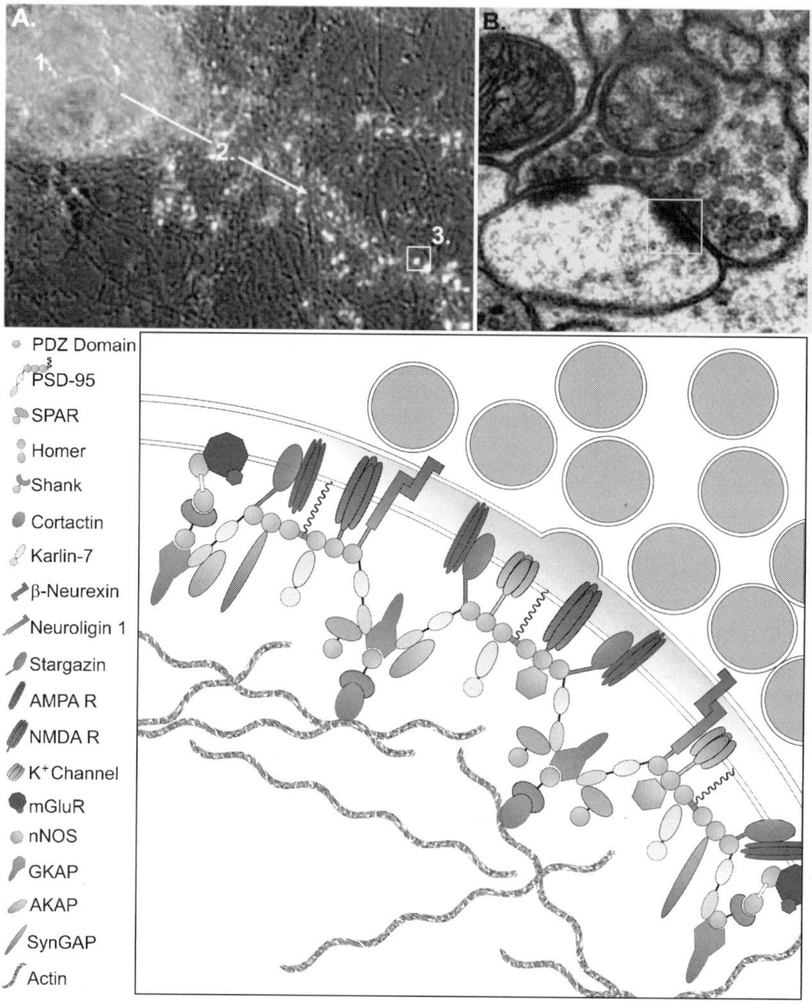


**Figure 4.1.** Protein Complexes at Excitatory Synapses. (A) An image of a neuron stained with the presynaptic marker synaptophysin (green), to identify synaptic contacts. This panel illustrates steps involved in the assembly of proteins at contact sites. Synapse formation is generally thought to involve three basic steps which include production of proteins in the cell soma (A-1), transport of these proteins to early sites of contact between axons and dendrites (A-2), and assembly of protein complexes at synapses (A-3). (B) The intense clustering of proteins seen at the PSD of excitatory synapses is highlighted in the electron micrograph shown in (B). A schematic diagram of this region is blown up in, illustrating the role of scaffolding molecules such as PSD-95 in assembly of large protein complexes. PSD-95 forms the core of the protein network, which is associated with the membrane through palmitoylation, and anchored within the postsynaptic compartment by several proteins that associate with actin. Coupling of PSD-95 to adhesion molecules such as neuroligins allows for *trans*-synaptic signaling. See Colorplate 4.

receptors, adhesion molecules, and signaling proteins at the postsynaptic density (PSD) can be clearly seen in electron micrographs of excitatory synapses (Figure 4.1B). Remarkably, the content and morphology of inhibitory postsynaptic sites is fundamentally different from that of excitatory contacts (see Chapter 19). A schematic diagram depicting some of the proteins identified at postsynaptic sites of excitatory synapses emphasizes the complexity and specificity of these structures (Figure 4.1). What drives clustering and assembly of diverse proteins at these specific locations? In this chapter we discuss some of the mechanisms implicated in protein trafficking, especially those driven by scaffolding molecules that induce protein clustering through multiple protein–protein interaction domains and protein multimerization.

Much of our understanding of protein targeting and clustering at synaptic sites has been gleaned from studies in cell culture. Cell culture systems have been used in scientific research for centuries, and have been the basis for the development of many biomedical advances such as vaccines[1]. Provided with the appropriate support, cells in culture display many of the properties of the differentiated cells from which they originated[2]. In general, cell culture systems can be divided into two categories, primary cell cultures that are prepared from fresh tissues and cell lines which can be split and repeatedly divide over long periods of time. Cell culture systems offer a well-defined, simplified homogenous system that avoids much of the inherent complexity *in vivo* and gains direct control over the environment of the cells. Another advantage of cultured cell systems is their receptiveness to gene transfer techniques. Gene transfer techniques (transfection) allow investigators to monitor and manipulate gene expression levels in cells, revealing information on the function of specific genes. The strength of this technique is amplified by the fusion of fluorescent probes or tags to the DNA before transfection, which allows for live imaging or immunohistochemical detection.

Although cultured cells provide a clean and powerful system for dissecting out mechanisms underlying protein assembly and contact formation, they may be unable to provide definitive answers about how synapses are formed *in vivo*. Because the natural surroundings of the cells have been removed, processes that depend on interaction with neighboring cells and the natural extracellular matrix may be altered *in vitro*. Unfortunately, along with the natural cellular environment *in vivo* comes inherent complexity. *In vivo* studies present more challenges for labeling, imaging, and manipulating individual cells amidst millions of others. However, as we discuss below, many of the techniques pioneered *in vitro* generated much of the current knowledge on how synapses assemble and cluster ion channels, cell adhesion molecules, and associated signaling and scaffolding proteins.

## 3. COMPLEX ASSEMBLY AND CLUSTERING OF SYNAPTIC PROTEINS

Recent studies revealed that PDZ domain-containing scaffolding proteins serve an important role in clustering of diverse molecules into large complexes[3]. The membrane-associated guanylate kinase (MAGUK) family of scaffolding proteins is of central importance in regulation of protein clustering at the synapse. MAGUKs are defined by the presence of a domain homologous to the yeast guanylate kinase (GK) domain, which is catalytically inactive. The GK domain is preceded by an Src-homology-3 (SH3) domain. In addition, MAGUKs contain PDZ domains, which are named for the original proteins in which the domains were identified (PSD-95, discs large, and zona occludens 1). PDZ domains bind the carboxyl terminus of proteins or form dimers with other PDZ domain-containing proteins[4]. PSD-95 (postsynaptic density protein of 95 kDa) is the prototypical MAGUK, and extensive research has been conducted on this protein. PSD-95 family is encoded by four genes – PSD-95/synapse-associated protein 90 (SAP90), PSD protein of 93 kDa (PSD-93)/chapsyn-110, synapse-associated protein 102 (SAP102), and synapse-associated protein 97 (SAP97), which are characterized by three PDZ domains, in addition to the SH3 and GK domain characteristic of MAGUKs[3]. The presence of several protein–protein interaction domains, combined with the observation that their accumulation at the synapse precedes several other known synaptic proteins, suggested an important role for MAGUKS in the assembly of a large complex at the synapse. In addition to multiple protein–protein interaction domains, protein multimerization appears to be critical for clustering proteins into large complexes[3,5]. For instance, PSD-95 multimerization, mediated through the palmitoylated N-terminal cysteines, is essential for ion channel clustering. Thus, protein multimerization serves as a mechanism by which scaffolding proteins can act to increase the number of molecules assembled at specific sites, such as the PSD.

*In vitro* assays utilizing expression of proteins in heterologous cells provided the first direct evidence that MAGUKs have the ability to cluster ion channels and associated proteins. Early studies showed that transfection of PSD-95 into heterologous cell lines, such as COS cells, leads to a mainly diffuse cytoplasmic distribution of PSD-95[6–8]. In contrast, when PSD-95 is co-transfected with other known binding partners, both proteins form distinct clusters, or patches on the cell membrane. This phenomenon was first demonstrated with the shaker subclass of voltage-gated potassium ($K^+$) channels[6]. Heterologous expression of shaker-type subunits with PSD-95 results in the co-clustering of the two proteins. Further, subunit binding and clustering by PSD-95 is blocked by a mutation in the carboxyl terminus of the shaker subunit Kv1.4. This heterologous system for analyzing protein interaction and trafficking is highly effective, and has since been used to further study the dynamics of PSD-95 interaction and clustering. In a second set of experiments the specific PDZ domain requirements of PSD-95 for clustering of Kv1.4 were examined[8]. Missense and deletion mutations were introduced into the PDZ1 and/or the PDZ2 domains of full length PSD-95. Kv1.4 co-expression with the PDZ2 mutant showed a significant disruption in clustering, whereas PDZ1 mutation had little effect on channel clustering. Interestingly, a mutant containing inverted PDZ1 and PDZ2 domains also reduces clustering of the Kv1.4 channel subunits when co-expressed. Thus, the PDZ2 domain must be both intact and in the correct position in order for proper protein interaction and clustering[8]. Similar to clustering of $K^+$ channel subunits, PSD-95 has also been shown to cluster several other proteins *in vitro*. Co-expression with PSD-95 leads to the clustering of semaphorin 4B, an integral membrane protein that participates in axon and dendrite guidance[9]. Studies of a similar design in COS cells show that PSD-95 induces clustering of the adhesion molecule neuroligin-1 through PDZ-mediated interactions (Figure 4.2)[7].

In addition to PSD-95, this assay has been useful in the assessment of the clustering activity of other scaffolding proteins. For example, Shank has been shown to cluster mGluR5 in heterologous cells in the presence of Homer[10]. Further, Shank was shown to mediate the co-clustering of Homer with a complex containing PSD-95 and GKAP. Thus, Shank may provide a link between mGluR5/Homer and PSD-95/NMDA receptor complexes at the PSD[10]. Further, the experiments in heterologous cell lines reveal the general role of scaffolding proteins, acting to assemble and localize protein complexes.

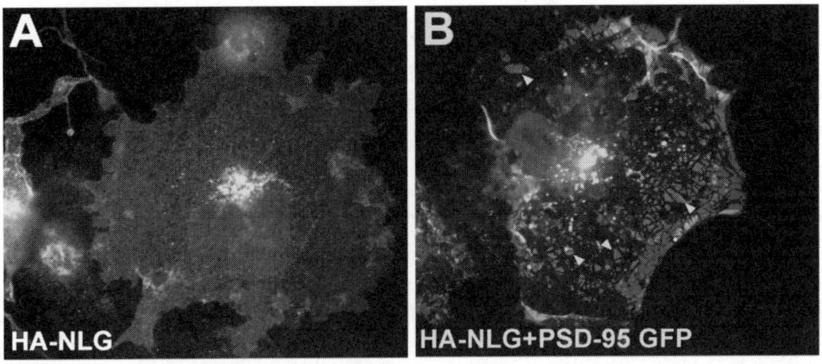


**Figure 4.2.** Clustering of the Postsynaptic Adhesion Molecule, neuroligin-1 (NLGI), by PSD-95 in COS Cells. COS cells were transfected with either (A) HA-tagged NLG1 (HA-NLG1) alone, or with (B) HA-NLG1 and PSD-95 GFP. When co-expressed, these two proteins co-localize into large clusters or patches (white arrow heads).

Overexpression studies in neurons show that PSD-95 accelerates clustering of several proteins, including neuroligin-1 and the glutamate receptor subunit GluR1 (Figure 4.3). Co-expression of neuroligin-1 and PSD-95 also results in an increase in the clustering of several presynaptic proteins, including synaptophysin. Taken together, these findings indicate that clustering of postsynaptic proteins controls the size and function of the PSD and presynaptic terminals.

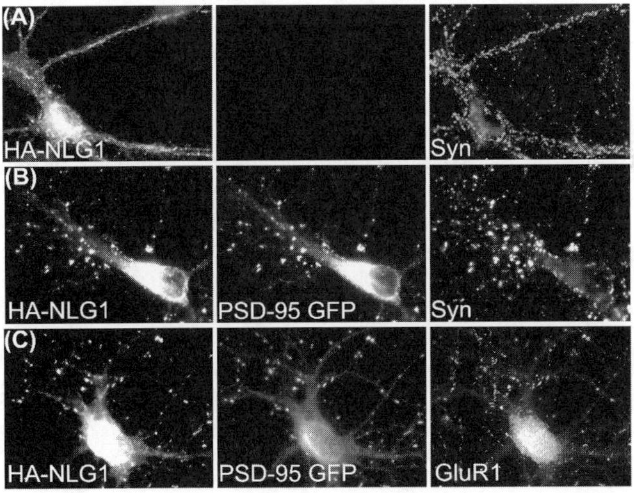


**Figure 4.3.** Enhanced Expression of Neuroligin-1 and PSD-95 in Hippocampal Neurons Results in the Clustering of both Pre- and Postsynaptic Proteins. Hippocampal neurons transfected with (A) HA-tagged neuroligin-1 (HA-NLG1) alone or with (B,C) GFP-tagged PSD-95 (PSD-95 GFP). (A) Expression of HA-NLG1 enhances the number of presynaptic sites positive for synaptophysin (Syn). (B,C) Co-expression of HA-NLG1 with PSD-95 results in an increase in the size of clusters of (B) synaptophysin and (C) the postsynaptic proteins, HA-NLG1, and the glutamate receptor subunit GluR1. Note the relative increase in the size of HA-NLG1 clusters in cells expressing HA-NLG1 and PSD-95 when compared to cells transfected with HA-NLG1 alone.

Another important contributor to protein clustering is the presence of filamentous actin (F-actin) at the synapse. F-actin acts as an anchor that stabilizes multiple protein complexes within the PSD. For instance, coupling to actin is thought to assemble a complex containing PSD-95, GKAP (guanylate kinase-associated protein), shank, and cortactin[11]. Cytoskeletal components of the PSD are also thought to trap transmembrane proteins and signaling molecules in domains specified by adhesion molecules[12,13]. This retention concentrates receptors and effectors in a region directly opposed to the presynaptic active zone, allowing for rapid signal transduction between the closely associated proteins and preventing lateral diffusion[13]. Another mechanism thought to contribute to retention of the clustered receptors at the synapse is the formation of puncta adherentia, small junctions formed by actin cytoskeleton in conjunction with adhesion molecules. These junctions are characterized by dense thickenings on both sides of the junction and an absence of synaptic vesicles. Puncta adherentia stabilize connections between adjacent cells at sites lateral to the synapse[14]. Thus, cytoskeletal elements are important for stabilizing cell junctions, assembly of protein complexes, and maintaining their clustering at the PSD and in turn optimize their function at the synapse.

## 4. REGULATION OF PROTEIN SORTING AND CLUSTERING BY LIPID MODIFICATIONS

In addition to PDZ-dependent protein clustering, modification of proteins with long-chain fatty acids has recently emerged as an important mechanism for regulating protein trafficking and clustering. Multiple lipid modifications have been characterized, including the co-translational addition of myristic acid to the amino-terminal glycine (myristoylation), and the post-translational attachment of prenyl groups to carboxy-terminal cysteine-containing motifs (prenylation)[15–17]. In palmitoylation, palmitate is linked through thioester bonds to cysteine residues[15,16]. This lipid modification increases hydrophobicity, thus facilitating protein interaction with the membrane. Palmitoylation, in contrast to stable myristoylation and prenylation, is a reversible modification allowing for dynamic regulation of protein targeting[16,17].

Topinka and colleagues demonstrated that PSD-95 is a major palmitoylated protein in neuronal cells[18]. Importantly, palmitoylated PSD-95 was found to be exclusively associated with the membrane. Moreover, dual palmitoylation allows PSD-95 to associate with a specific perinuclear vesiculotubular compartment[19]. This is presumably required for the formation of transport intermediates destined for postsynaptic sites. Through systematic mutagenesis, these experimenters were also able to determine that palmitoylation of PSD-95 occurs at conserved N-terminal cysteines 3 and 5. Palmitoylation-deficient mutants of PSD-95 were not found associated with membranes and were also unable to cluster Kv1.4 channel subunits in transfected cell lines[18]. Palmitoylation has profound effects on protein interactions and trafficking. Using the heterologous cell assay discussed above, it has been shown that PSD-95 mediates cell surface ion channel clustering[20]. Remarkably, palmitoylation of a subset of members of the PSD-95 family differentially regulates certain functions mediated by these proteins. For example, studies in heterologous cells showed that only the palmitoylated members are capable of clustering ion channels and associated proteins. Although the amino termini of PSD-93, PSD-95 and SAP-102 all contain cysteines, only PSD-93 and PSD-95 are palmitoylated and cluster ion channels. The amino terminus of SAP-102 lacks surrounding hydrophobic amino acids that are required for palmitoylation[16]. SAP-97 lacks amino-terminal cysteines, is not palmitoylated, and does not act to cluster ion channels[16]. Indeed, attaching the amino terminus of palmitoylated PSD-95 to SAP-97 instigates postsynaptic clustering of the chimera, further demonstrating the importance of this domain in the function of MAGUKs[20]. The differential palmitoylation of specific isoforms of the glutamate–receptor-interacting protein ABP/GRIP provides another mechanism for regulation of AMPA receptor clustering at specific subcellular sites. Taken together, these studies elucidate an important role for palmitoylation in regulating neuronal protein clustering and function at the synapse.

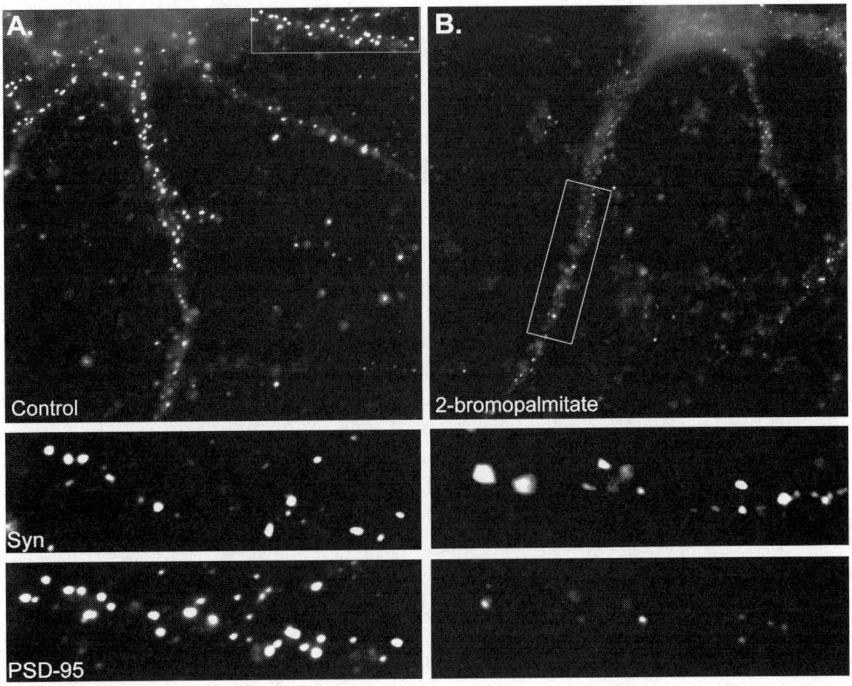


**Figure 4.4.** Palmitoylation is Required for PSD-95 Clustering at the Synapse. Hippocampal neurons were either given (A) no treatment (control) or (B) 2-bromopalmitate treatment for 8 h to block palmitoylation. Neurons were then fixed and stained for endogenous PSD-95 and synaptophysin (Syn). Lower panels showing dendrites of control neurons with many synaptophysin positive puncta that colocalize with PSD-95. In contrast, 2-bromopalmitate treated neurons show dramatic reduction of PSD-95 clustering at sites apposed to synaptophysin positive puncta.

Similar to protein phosphorylation, palmitoylation is a reversible process that is dynamically regulated by specific cellular stimuli. In cultured hippocampal neurons, blocking palmitoylation using 2-bromopalmitate disperses PSD-95 clusters (Figure 4.4) and causes a reduction in the surface expression of AMPA receptors[21]. This result suggests that PSD-95 regularly alternates between palmitoylated and nonpalmitoylated states, and that PSD-95 palmitoylation is necessary for both PSD-95 and AMPA receptor clustering. Prolonged synaptic activity accelerates depalmitoylation of PSD-95, resulting in enhanced AMPA receptor endocytosis[19,21]. In mature neurons, defects in PSD-95 palmitoylation block specific forms of synaptic plasticity associated with the regulated delivery of AMPA receptor subunits to the synapse[22,23]. Moreover, expression of a palmitoylation-deficient form of PSD-95 dampens dopamine-mediated effects on AMPA receptor phosphorylation and surface expression[24]. More recently, the AMPA-type glutamate receptors have been shown to be palmitoylated and this process regulates activity-dependant internalization of these proteins[25]. Thus, palmitoylation appears to regulate diverse aspects of protein clustering, assembly and function of postsynaptic proteins.

In addition to postsynaptic sites, palmitoylation plays a major role in the control of trafficking of numerous presynaptic proteins associated with the synaptic vesicle fusion machinery and neurotransmitter release. Prime examples include SNAP-25 and members of the synaptotagmin family. Palmitoylation of SNAP-25 is required for efficient SNARE-complex dissociation and the regulation of vesicle exocytosis, but not for membrane targeting[26]. In contrast, palmitoylation of specific members of the synaptotagmin family contributes to protein sorting and clustering at presynaptic terminals[27]. For example, the palmitoylated amino terminus of synaptotagmin I, but not synaptotagmin VII, is required for targeting to presynaptic sites. Palmitoylation also regulates synaptotagmin I sequestration from the presynaptic plasma membrane to synaptic vesicles[27]. In the SNARE protein Ykt6, palmitoylation provides a mechanism for regulating the rate of intracellular membrane flow and vesicle fusion[28]. In addition to these proteins, several other components of the synaptic vesicle machinery, including cysteine-string protein, α-SNAP (α-soluble *N*-ethylmaleimide-sensitive-factor-attachment protein), and VAMP (vesicle-associated membrane protein) are palmitoylated, however the role of this modification in these proteins remains unclear. The modification of a large number of presynaptic proteins with palmitate points to a critical role for this modification in regulating various aspects of neurotransmitter release. The sorting of enzymes involved in neurotransmitter synthesis also requires palmitoylation. For instance, presynaptic clustering of glutamic acid decarboxylase 65 kDa (GAD65) requires palmitoylation. Palmitoylation is required for trafficking of GAD65 from Golgi membranes to Rab5-regulated endosomes for delivery to presynaptic sites[29].

Enzymes that regulate addition and removal of palmitate may serve an important role in regulating protein sorting and clustering at the synapse. However, the diversity of proteins modified by palmitate combined with the lack of a common consensus sequence for palmitoylation hindered the identification of enzymes involved in this process[30]. Characterization of the first enzymes involved in palmitoylation came from work done in flies which revealed that *skinny hedgehog* (*Ski*) and *porcupine* (*porc*) are palmitoyl acyl-transferases (PATs) that mediate palmitoylation of secreted factors hedgehog and Wnt-1. This lipid modification is critical for hedgehog and Wnt-1 function in neuronal differentiation[31–33]. However, Ski and Porc are luminal proteins that palmitoylate substrates within the lumen of the secretory pathway, and thus are not likely involved in palmitoylation of cytosolic proteins. A major advance in the field came from genetic analysis of a family of yeast enzymes that contains a cysteine-rich domain (CRD). This domain also harbors a conserved DHHC (Asp-His-His-Cys) motif. These include Erf2p/Erf4p protein complex which palmitoylates Ras[34] and Akr1p which modifies the yeast enzyme casein kinase II[35].

The existence of 23 mammalian proteins that contain the DHHC domain prompted investigations to analyze the role of these proteins in neuronal protein palmitoylation. Investigation by Bredt and colleagues characterized the role of the existing 23 PATs named DHHC-1 through -23 in the palmitoylation of a subset of neuronal proteins[22]. Among them, DHHC-2, -3, -7, and -15 showed the strongest specificity for PSD-95. In a parallel work, *in vitro* palmitoylation assays revealed that the huntingtin-interacting protein 14 (HIP14)[36], also known as DHHC-17, palmitoylates numerous neuronal proteins which include huntingtin, PSD-95, SNAP-25, synaptotagmin I, and GAD65[37]. The identified enzymes utilize the intermediate palmitoyl-acyl CoA for palmitate transfer to specific substrates. These studies also revealed that activity of the purified enzymes does not rely on auxiliary subunits. Overexpression experiments as well as loss of function analyses in neuronal cells demonstrated that most of the identified DHHC proteins modulate palmitoylation-dependent protein trafficking[22]. For instance, knockdown of HIP14 reduces synaptic clustering of PSD-95 and GAD65[37].

Although the exact subcellular locations for palmitoylation remain ambiguous, the enrichment of several of the identified DHHC proteins in the Golgi suggests that this compartment is a major site for palmitoylation[37–39]. Studies on HIP14, however, revealed that this enzyme also associates with tubulovesicular organelles and recycling endosomes present in dendrites, axons, and spine necks[37]. Mobile HIP14-positive vesicular structures were rapidly transported from a perinuclear compartment to various subcellular locations. Thus, at least some of the identified enzymes are involved in the cytosolic modification of both cytoplasmic and integral membrane proteins at numerous subcellular locations. In the future it will be important to determine whether the efficacy of these enzymes is subject to changes in neuronal activity and whether other binding partners regulate their function. The availability of new advanced imaging techniques and sensitive assays for detecting palmitoylated proteins may help clarify mechanisms that regulate the rate of turnover of palmitate. These tools may elucidate some of the mechanisms that control protein clustering at the synapse.

## 5. MONITORING ASSEMBLY AND CLUSTERING OF SYNAPTIC PROTEINS IN LIVE CULTURED NEURONS

Synaptogenesis involves clustering of presynaptic and postsynaptic components at sites of initial contact[40–47]. This includes recruitment of the presynaptic release machinery, postsynaptic neurotransmitter receptors, and associated signaling molecules (see Chapters 14 and 16). Clustering of different proteins to presynaptic and postsynaptic sites suggests involvement of heterotypic *trans*-synaptic adhesion molecules, and their associated scaffolding and signaling proteins. These *trans*-synaptic signals would also require specificity, the ability to recruit the appropriate neurotransmitter on the presynaptic side and their cognate receptors on the postsynaptic side, in order to avoid mismatching.

In the late 1970s Banker and colleagues developed a culture system for the study of isolated hippocampal neurons. Examination of hippocampal neurons in culture reveals that they rapidly attach to the artificial extracellular matrix and begin to extend processes within 24–48 h[48,49]. A large proportion of the cells in culture mature to resemble typical hippocampal pyramidal cells *in vivo*, bearing polarized processes emerging from their triangular-shaped cell bodies. Cultured neurons provide many advantages over intact specimens for microscopic imaging. Dissociated cell cultures are optically accessible, with little to no overlap between individual cell components, allowing for easier interpretation of results. In addition, cultured cells are more accessible for labeling or loading with markers for imaging. Imaging in cell culture was revolutionized by the advent of fluorescent proteins, such as green fluorescent protein (GFP). Many live time-lapse imaging studies of synaptogenesis take advantage of direct fusion of a fluorescent marker to their protein of interest, revealing the time course and organization of molecules in synapse formation. Further understanding of the molecules involved in synapse formation has been gained by fixing the culture after the time-lapse imaging and conducting retrospective immunohistochemistry. Likewise, the ultrastructure of newly formed or developing synapses can be examined using retrospective electron microscopy. Additional tools useful in live imaging are styryl dyes (FM-4-64) which permit visualization of vesicle turnover, an important indicator of a functional synapse. These dyes are washed into the culture media, and are taken up when vesicles fuse with the cell membrane and are re-endocytosed. Following uptake, when the dye has been washed out, presynaptic compartments that contain recycled synaptic vesicles will remain labeled, signifying that they are active. In addition, live imaging studies allowed observation of initial contact events between outgrowing axons and dendritic filopodia. These approaches have given us insights into which compartments initiate contact, and what structures or events developmentally precede synapse maturation. Such approaches also allowed visualization of the recruitment of transport packets containing presynaptic proteins such as synaptophysin at initial sites of contact (Figure 4.5).

Intense study has focused on the order of synaptic protein recruitment and whether proteins are recruited as preassembled complexes or as individual molecules. Studies from several laboratories have suggested that vesicular delivery plays a prominent role in this process[50]. The recruitment of vesicles carrying structural components of the presynaptic active zone, such as piccolo and bassoon, have been observed shortly after initial contact between an axon and a dendrite[46]. Another population of precursor presynaptic vesicles containing proteins important for active neurotransmitter release, such as VAMP and synaptophysin, has also been observed to fuse shortly after initial contact[51]. In a live time-lapse imaging study, FM 4-64 dye loading was used to study the time course of both pre- and postsynaptic protein recruitment to new sites of axo-dendritic contact[40,42]. Once the new synapses were labeled with the FM dye, different periods of time were allowed to elapse before fixation and retrospective immunohistochemistry. Using this paradigm, it was demonstrated that new presynaptic boutons capable of vesicle recycling form within 30 min of initial contact between dendrites and axons. The presynaptic scaffold bassoon was found at all new contact sites, whereas postsynaptic proteins, such as PSD-95, gradually accumulated at the apposed sites[40,42].

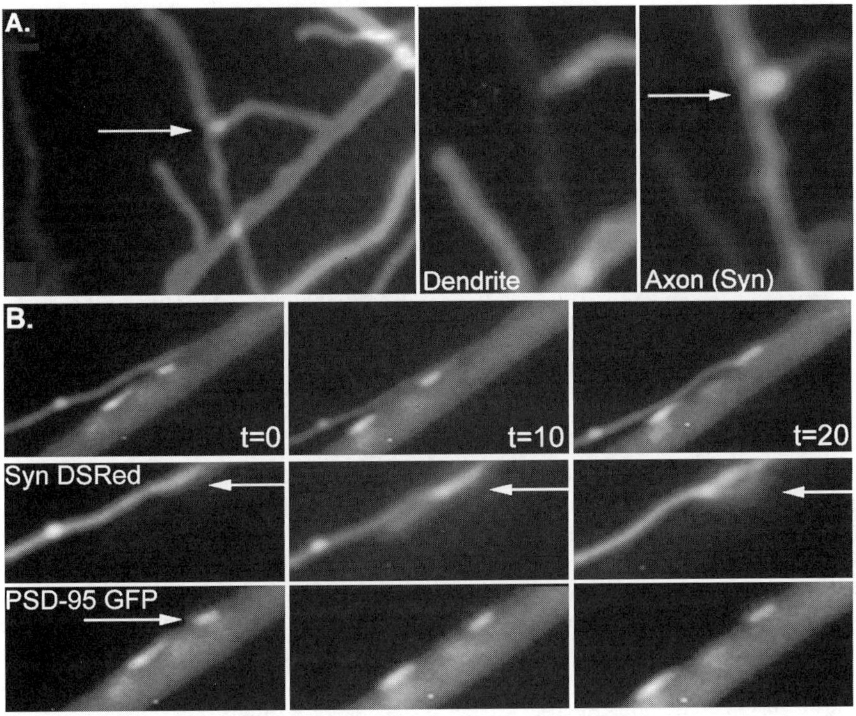


**Figure 4.5.** Recruitment of Clusters of Presynaptic and Postsynaptic Proteins at Early Sites of Contact Between Axons and Dendrites. (A) Shows an accumulation of a synaptophysin cluster at a contact site between dendritic filopodia of a cell transfected with a membrane targeted GFP and an axon from a neuron transfected with synaptophysin tagged with DsRed (SYN DsRed). (B) Time-lapse images showing accumulation of SYN DsRed at a site apposed to an existing PSD-95 GFP cluster, occurred over a time (*t*) period of 20 min. See Colorplate 5.

However, a recent study by our group in young hippocampal neurons has shown that clusters of PSD-95 can also be found associated with neuroligin-1 in the absence of an active presynaptic terminal. In these young neurons, protein complexes exist in two distinct subpopulations which differ in their content, mobility, and involvement in synapse formation. (One subpopulation of clusters is mobile and rely on actin transport to nascent and existing synapses. Further, the majority of mobile clusters, containing the scaffolding proteins PSD-95, GKAP, and Shank, lack neuroligin-1. The second subpopulation consist of stationary nonsynaptic scaffold complexes. These complexes contain neuroligin-1 and recruit synaptophysin-containing vesicles to the opposing axonal site of contact. Importantly, these sites were shown to become functional presynaptic contacts through FM 4-64 dye loading. In this case, the clustering of postsynaptic proteins such as neuroligin-1 may facilitate recruitment of presynaptic proteins important for vesicular release. Another study looking at new contacts made by axon filopodia demonstrated the rapid recruitment of NMDA-type glutamate receptor clusters independent of PSD-95, and before the establishment of an active presynaptic terminal (see Chapter 14)[47]. Although critical for synapse maturation, it remains unknown how clustering of these proteins is developmentally and spatially regulated, and what adhesion systems are involved.

Recent *in vitro* studies showed that several of the identified cell adhesion molecules modulate synaptic contact number, morphology, and function. However, strong evidence that any of these molecules is indispensable for synapse formation *in vivo* is lacking, suggesting a redundancy in their function. Thus, synapse formation and maturation may rely on assembly of several adhesion systems. Whether the numerous adhesion families act in parallel or in a hierarchical manner is unknown, and future studies will be required to tease apart the nuances of how these adhesion systems work together in the establishment and function of the synapse. To study the ability of cell adhesion molecules to cluster presynaptic proteins at contact sites, a clever assay has been developed using co-culture of transfected heterologous cells with neurons. In a series of experiments by Scheiffele's group, HEK cells were transfected with DNA encoding the adhesion molecule neuroligin-1 and co-cultured with developing neurons[52]. Remarkably, the expression of this postsynaptic adhesion molecule in heterologous cells resulted in the differentiation of presynaptic terminals at sites of axon–-HEK cell contact. Furthermore, these contacts were found to have not only morphological but functional characteristics of actual presynaptic contacts, with the accumulation of synaptic vesicles[52]. In hippocampal neurons, overexpression of neuroligins increased the number of both excitatory and inhibitory presynaptic terminals[53–57]. Similar assays have been used to show that SYNCAM can also induce clustering of presynaptic proteins at contact sites. Conversely β-neurexin, a binding partner of neuroligins, presented to dendrites via heterologous cells or beads has been demonstrated to cluster postsynaptic proteins (see Chapter 19)[58,59]. These studies suggest that initial interaction between adhesion molecules initiates[53,55] clustering of postsynaptic proteins and recruitment of neurotransmitter receptors to newly formed neuronal contacts[52,58,59]. Thus, early events that involve clustering and assembly of protein complexes appear to regulate not only initial contact formation and stabilization but also drive synapse maturation.

## 6. CONCLUSIONS

Researchers are now beginning to understand the many intricate mechanisms that control the trafficking and clustering of proteins important for synaptic function. Advances in imaging techniques revealed that in some cases proteins may be trafficked to contact sites as preformed complexes as opposed to individual protein molecules. Other studies showed that scaffolding proteins serve to cluster neurotransmitter receptors, and adhesion molecules, spatially restricting them to precise positions and optimizing their function. In mature neurons, scaffolding proteins may serve to modulate clustering of ion channels and other proteins, facilitating control of synaptic function and strength in response to activity, through mechanisms such as protein phosphorylation, palmitoylation, and/or degradation. Another critical finding revealed from recent investigations is that clustering of cell adhesion molecules and scaffolding proteins contributes to the specific sorting of neurotransmitter receptors that determine synapse function. Further investigation of the functional, temporal, and spatial attributes of scaffolding proteins and adhesion molecules will reveal how they work in a cooperative and integrated manner in synapse formation and at mature synapses.*

## 7. REFERENCES

1. Hilleman, M.R. (2000) *Vaccine* **18**, 1436–1447.
2. Geller, H.M., Quinones-Jenab, V., Poltorak, M., and Freed, W.J. (1991) *J Cell Biochem* **45**, 279–283.
3. Kim, E., and Sheng, M. (2004) *Nat Rev Neurosci* **5**, 771–781.
4. Hsueh, Y.P., Kim, E., and Sheng, M. (1997) *Neuron* **18**, 803–814.
5. Fanning, A.S., and Anderson, J.M. (1999) *Curr Opin Cell Biol* **11**, 432–439.
6. Kim, E., and Sheng, M. (1996) *Neuropharmacology* **35**, 993–1000.
7. Irie, M., Hata, Y., Takeuchi, M., Ichtchenko, K., Toyoda, A., Hirao, K., Takai, Y., Rosahl, T.W., and Sudhof, T.C. (1997) *Science* **277**, 1511–1515.
8. Imamura, F., Maeda, S., Doi, T., and Fujiyoshi, Y. (2002) *J Biol Chem* **277**, 3640–3646.
9. Burkhardt, C., Muller, M., Badde, A., Garner, C.C., Gundelfinger, E.D., and Puschel, A.W. (2005) *FEBS Lett* **579**, 3821–3828.
10. Tu, J.C., Xiao, B., Naisbitt, S., Yuan, J.P., Petralia, R.S., Brakeman, P., Doan, A., Aakalu, V.K., Lanahan, A.A., Sheng, M., and Worley, P.F. (1999) *Neuron* **23**, 583–592.
11. Naisbitt, S., Kim, E., Tu, J.C., Xiao, B., Sala, C., Valtschanoff, J., Weinberg, R.J., Worley, P.F., and Sheng, M. (1999) *Neuron* **23**, 569–582.
12. Beck, K.A., and Nelson, W.J. (1996) *Am J Physiol* **270**, C1263–1270.
13. Baines, A.J., Keating, L., Phillips, G.W., and Scott, C. (2001) *Cell Mol Biol Lett* **6**, 691–702.
14. Spacek, J., and Harris, K.M. (1998) *J Comp Neurol* **393**, 58–68.
15. Resh, M.D. (2004) *Subcell Biochem* **37**, 217–232.
16. El-Husseini, A., and Bredt, D.S. (2002) *Nat Rev Neurosci* **3**, 791–802.
17. Huang, K., and El-Husseini, A. (2005) *Curr Opin Neurobiol* **15**, 527–535.
18. Topinka, J.R., and Bredt, D.S. (1998) *Neuron* **20**, 125–134.
19. El-Husseini, A.E., Craven, S.E., Chetkovich, D.M., Firestein, B.L., Schnell, E., Aoki, C., and Bredt, D.S. (2000) *J Cell Biol* **148**, 159–172.
20. El-Husseini, A.E., Topinka, J.R., Lehrer-Graiwer, J.E., Firestein, B.L., Craven, S.E., Aoki, C., and Bredt, D.S. (2000) *J Biol Chem* **275**, 23904–23910.
21. El-Husseini A., Schnell, E., Dakoji, S., Sweeney, N., Zhou, Q., Prange, O., Gauthier-Campbell, C., Aguilera-Moreno, A., Nicoll, R.A., and Bredt, D.S. (2002) *Cell* **108**, 849–863.
22. Fukata, M., Fukata, Y., Adesnik, H., Nicoll, R.A., and Bredt, D.S. (2004) *Neuron* **44**, 987–996.
23. Ehrlich, I., and Malinow, R. (2004) *J Neurosci* **24**, 916–927.
24. Swayze, R.D., Lise, M.F., Levinson, J.N., Phillips, A., and El-Husseini, A. (2004) *Neuropharmacology* **47**, 764–778.
25. Hayashi, T., Rumbaugh, G., and Huganir, R.L. (2005) *Neuron* **47**, 709–723.
26. Washbourne, P., Cansino, V., Mathews, J.R., Graham, M., Burgoyne, R.D., and Wilson, M.C. (2001) *Biochem J* **357**, 625–634.
27. Kang, R., Swayze, R., Lise, M.F., Gerrow, K., Mullard, A., Honer, W.G., and El-Husseini, A. (2004) *J Biol Chem* **279**, 50524–50536.
28. Fukasawa, M., Varlamov, O., Eng, W.S., Sollner, T.H., and Rothman, J.E. (2004) *Proc Natl Acad Sci U S A* **101**, 4815–4820.
29. Kanaani, J., Diacovo, M.J., El-Husseini Ael, D., Bredt, D.S., and Baekkeskov, S. (2004) *J Cell Sci* **117**, 2001–2013.
30. Linder, M.E., and Deschenes, R.J. (2004) *J Cell Sci* **117**, 521–526.
31. Nusse, R. (2003) *Development* **130**, 5297–5305.
32. Chamoun, Z., Mann, R.K., Nellen, D., von Kessler, D.P., Bellotto, M., Beachy, P.A., and Basler, K. (2001) *Science* **293**, 2080–2084.
33. Zhai, L., Chaturvedi, D., and Cumberledge, S. (2004) *J Biol Chem* **279**, 33220–33227.
34. Lobo, S., Greentree, W.K., Linder, M.E., and Deschenes, R.J. (2002) *J Biol Chem* **277**, 41268–41273.
35. Roth, A.F., Feng, Y., Chen, L., and Davis, N.G. (2002) *J Cell Biol* **159**, 23–28.
36. Singaraja, R.R., Hadano, S., Metzler, M., Givan, S., Wellington, C.L., Warby, S., Yanai, A., Gutekunst, C.A., Leavitt, B.R., Yi, H., Fichter, K., Gan, L., McCutcheon, K., Chopra, V., Michel, J., Hersch, S.M., Ikeda, J.E., and Hayden, M.R. (2002) *Hum Mol Genet* **11**, 2815–2828.

*A.E.H. research is supported by funds from CIHR, EJLB, MSFHR, and Neuroscience Canada.

37. Huang, K., Yanai, A., Kang, R., Arstikaitis, P., Singaraja, R.R., Metzler, M., Mullard, A., Haigh, B., Gauthier-Campbell, C., Gutekunst, C.A., Hayden, M.R., and El-Husseini, A. (2004) *Neuron* **44**, 977–986.
38. Keller, C.A., Yuan, X., Panzanelli, P., Martin, M.L., Alldred, M., Sassoe-Pognetto, M., and Luscher, B. (2004) *J Neurosci* **24**, 5881–5891.
39. Mukai, J., Liu, H., Burt, R.A., Swor, D.E., Lai, W.S., Karayiorgou, M., and Gogos, J.A. (2004) *Nat Genet* **36**, 725–731.
40. Bresler, T., Ramati, Y., Zamorano, P.L., Zhai, R., Garner, C.C., and Ziv, N.E. (2001) *Mol Cell Neurosci* **18**, 149–167.
41. Bresler, T., Shapira, M., Boeckers, T., Dresbach, T., Futter, M., Garner, C.C., Rosenblum, K., Gundelfinger, E.D., and Ziv, N.E. (2004) *J Neurosci* **24**, 1507–1520.
42. Friedman, H.V., Bresler, T., Garner, C.C., and Ziv, N.E. (2000) *Neuron* **27**, 57–69.
43. Garner, C.C., Zhai, R.G., Gundelfinger, E.D., and Ziv, N.E. (2002) *Trends Neurosci* **25**, 243–251.
44. Shapira, M., Zhai, R.G., Dresbach, T., Bresler, T., Torres, V.I., Gundelfinger, E.D., Ziv, N.E., and Garner, C.C. (2003) *Neuron* **38**, 237–252.
45. Ziv, N.E., and Garner, C.C. (2001) *Curr Opin Neurobiol* **11**, 536–543.
46. Ziv, N.E., and Garner, C.C. (2004) *Nat Rev Neurosci* **5**, 385–399.
47. Washbourne, P., Bennett, J.E., and McAllister, A.K. (2002) *Nat Neurosci* **5**, 751–759.
48. Banker, G.A., and Cowan, W.M. (1977) *Brain Res* **126**, 397–342.
49. Banker, G.A., and Cowan, W.M. (1979) *J Comp Neurol* **187**, 469–493.
50. Waites, C.L., Craig, A.M., and Garner, C.C. (2005) *Annu Rev Neurosci* **28**, 251–274.
51. Ahmari, S.E., Buchanan, J., and Smith, S.J. (2000) *Nat Neurosci* **3**, 445–451.
52. Scheiffele, P., Fan, J., Choih, J., Fetter, R., and Serafini, T. (2000) *Cell* **101**, 657–669.
53. Prange, O., Wong, T.P., Gerrow, K., Wang, Y.T., and El-Husseini, A. (2004) *Proc Natl Acad Sci U S A* **101**, 13915–13920.
54. Chih, B., Engelman, H., and Scheiffele, P. (2005) *Science* **307**, 1324–1328.
55. Levinson, J.N., Chery, N., Huang, K., Wong, T.P., Gerrow, K., Kang, R., Prange, O., Wang, Y.T., and El-Husseini, A. (2005) *J Biol Chem* **280**, 17312–17319.
56. Song, J.Y., Ichtchenko, K., Sudhof, T.C., and Brose, N. (1999) *Proc Natl Acad Sci U S A* **96**, 1100–1105.
57. Varoqueaux, F., Jamain, S., and Brose, N. (2004) *Eur J Cell Biol* **83**, 449–456.
58. Nam, C.I., and Chen, L. (2005) *Proc Natl Acad Sci U S A* **102**, 6137–6142.
59. Graf, E.R., Zhang, X., Jin, S.X., Linhoff, M.W., and Craig, A.M. (2004) *Cell* **119**, 1013–1026.

# Part II

## ROLES OF CELL ADHESION AND SECRETED MOLECULES IN SYNAPTIC DIFFERENTIATION

# 5

# CADHERIN-MEDIATED ADHESION AND SIGNALING DURING VERTEBRATE CENTRAL SYNAPSE FORMATION

Tonya R. Anderson and Deanna L. Benson*

## 1. SUMMARY

The events that encompass synaptogenesis—growth cone guidance, axonal and dendritic arborization, dendritic spine growth, even neurotransmitter release, and receptor dynamics—all are calcium-dependent processes. As calcium-dependent molecules, the members of the cadherin superfamily are particularly well suited to mediate shifting requirements for adhesion during these dynamic events. These molecules, which are thought to bind homophilically across synapses, couple to the actin-based cytoskeleton as well as synaptic vesicles, ion channels, and neurotransmitter receptors. Given these biochemical features, it is not surprising that manipulating cadherin functions compromises normal synapse maturation and plasticity alike. Although the contributions of individual family members to synaptogenesis appear to differ to varying degrees, their fundamental roles in calcium-dependent junctional maintenance are ontogenetically and phylogenetically conserved.

## 2. INTRODUCTION

Synapses are adhesive junctions. This is a widely acknowledged and accepted fact, but most early studies on the mechanisms of CNS synapse development focused on the events and components that contribute directly to neurotransmission. For several years, the synaptic cleft was ignored. The discovery that members of the cadherin family of cell-adhesion molecules were concentrated at synapses[1,2] renewed interest in an old debate that had focused on the morphological similarities between adherens junctions, which are generated by

*Fishberg Department of Neuroscience, Mount Sinai School of Medicine, 1425 Madison Avenue Box 1065, New York, NY 10029, USA; tonya.anderson@mssm.edu and deanna.benson@mssm.edu

cadherin adhesion and synaptic junctions[3,4]. Cadherins were not the first adhesion proteins to be found at synapses, but they were the first that were known to be capable of generating cell–cell junctions. They also cluster at synapses early in synaptogenesis and different cadherins delineate functionally distinct synapse types. These findings collectively suggested that cadherins could be critical components for synapse assembly and could potentially guide synapse specificity[5]. Experimental findings now support that cadherin-based adhesion is important for synapse assembly, synapse stabilization, dendritic spine morphology, and functional plasticity.

## 3. STRUCTURE OF CADHERINS AND BINDING INTERACTIONS

Cadherins mediate calcium dependent, mostly homophilic adhesion. They were first identified in vertebrates, but the family is evolutionarily conserved and members can be found in echinoderms, arthropods, and chordates. Between phyla, there is substantial structural diversity, but cadherins have in common a cadherin domain—a ~110 amino acid sequence that is usually one of several tandem repeats separated by a calcium-binding domain. Current estimates indicate that the cadherin superfamily has as many as 300 members that can be parsed into several subfamilies[6] (Figure 5.1). By far the most is known about classic cadherins, which in vertebrates have five cadherin domains, but in nonchordates can have up to 15 (e.g., DN-cadherin in *Drosophila*). Their highly homologous intracellular domains bind strongly to the actin cytoskeleton via the catenin proteins: cadherins bind directly to armadillo repeat-containing proteins, ß or γ catenin, which bind in turn to the unrelated but similarly named α-catenin. Alpha-catenin binds to F-actin directly or indirectly via α-actinin. Classic cadherins and protocadherins (see Chapter 9) are the only cadherins that have been localized to synapses thus far.

Classic cadherins (usually referred to simply as "cadherins") mediate strong adhesion at junctions like adherens junctions or central nervous system (CNS) synapses but can also engage in weaker, nonjunctional adhesion such as that mediating neurite extension. A single cadherin extending from the plasma membrane of one cell engages its partner in *trans* in an interaction that is specified by the first cadherin domain, and requires at least the first two cadherin domains for adhesion[7–9]. The adhesive force generated by a single binding interaction in *trans* would be expected to be weak but could be important for traction during neurite extension. Within the plane of the plasma membrane, cadherins can also form *cis*-dimers with one another, and crystal structures of the *N*-cadherin N-terminal EC1 domain support that a series of *cis*-dimerized cadherins could engage similarly dimerized cadherins in *trans*, forming an "adhesion zipper"[7]. This would be expected to confer very strong adhesion, and consistent with this idea, many junctional cadherins are resistant to trypsin digestion[10]. More recent work on the crystal structure of the entire extracellular domain of *Xenopus C*-cadherin suggests that the *cis*- and *trans*-binding interfaces may be shared and therefore must alternate binding with one another[11]. Thus, while it is not clear whether *cis* dimers form *in vivo*, a great deal of experimental evidence supports that *cis*-clustering augments *trans* adhesion. *Cis* interactions are modulated by proteins binding to the juxtamembrane domain: armadillo repeat-containing proteins p120 catenin and δ-catenin/NPRAP, and presenilin-1 which also participates in the ε-cleavage of cadherins, a process that releases a C-terminal cadherin fragment into the cytosol[12,13] (Figure 5.2).

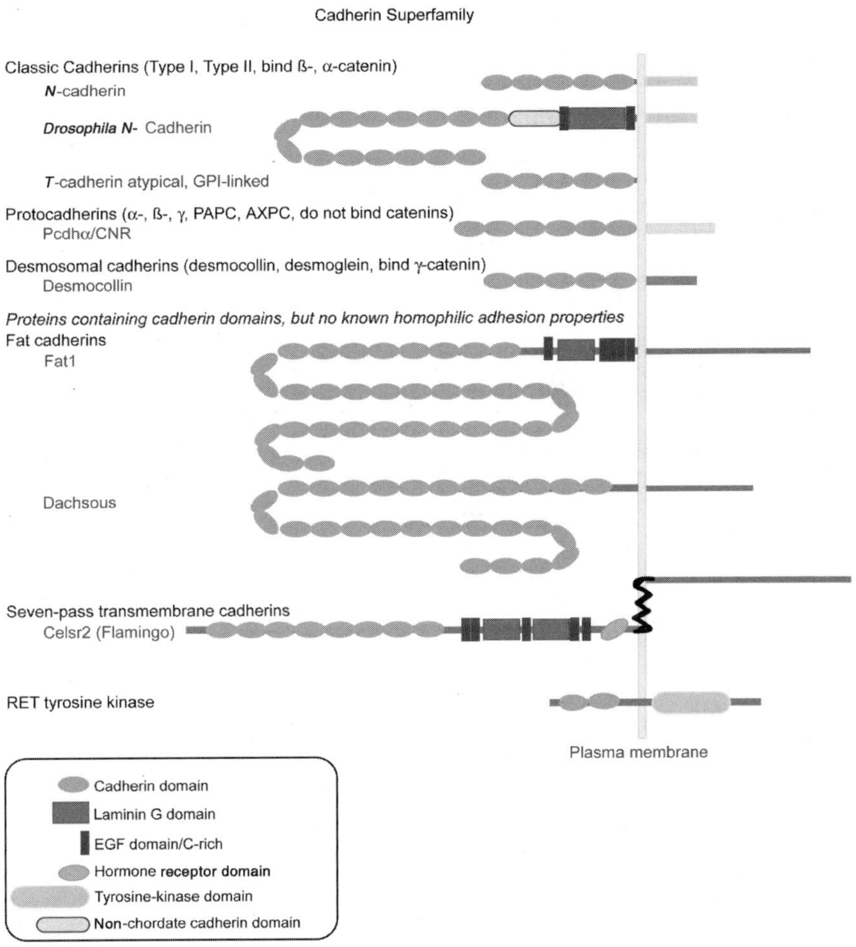


**Figure 5.1.** Cadherins in the Nervous System. All have cadherin-type extracellular repeats. Data are repeats. Data are taken from refs 6,18,68. *Abbreviations*: CNR: cadherin-related neuronal receptor, GPI: glycosylphosphatidylinositol, RET: rearranged during transfection.

Most of the data on cadherin actions and the consequences of cadherin binding and signaling arise from experiments using the expression of cadherin mutant proteins or peptides mimicking key domains. Traditional genetic ablation has been less successful because many cadherins are essential for early stages stages of embryonic development prior to brain development, and experiments

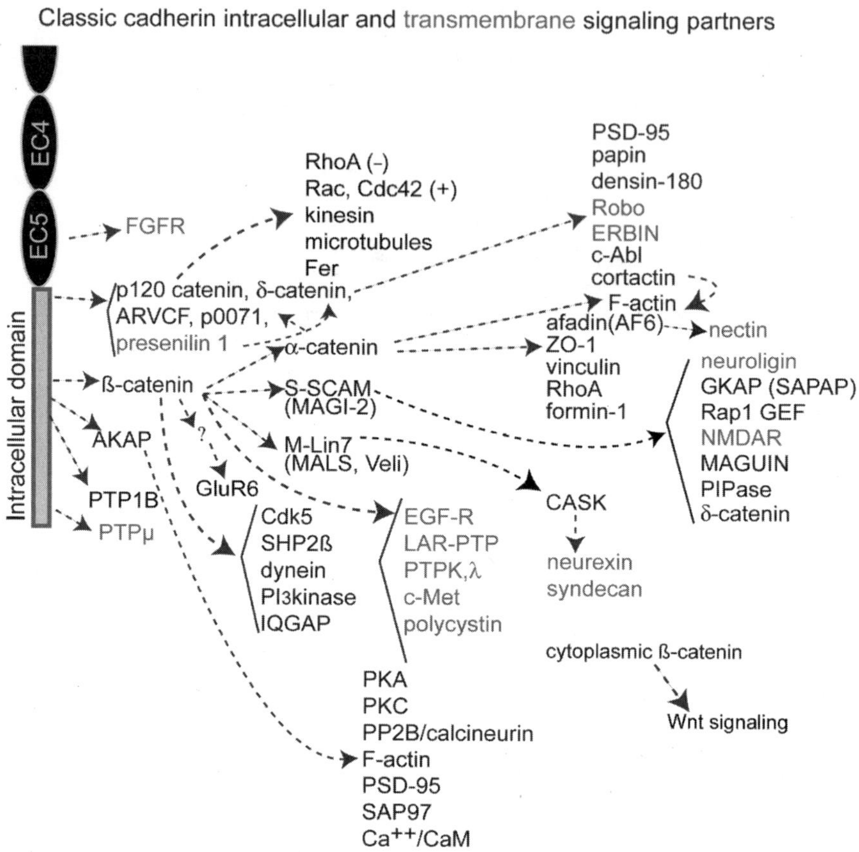


**Figure 5.2.** Classic Cadherin Extracellular and Intracellular Binding Interactions. This list, while not exhaustive, illustrates the enormous variety of signaling interactions and regulatory mechanisms associated with cadherins. Transmembrane proteins are written in light gray and cytosolic proteins, in black. Some secondary and tertiary interactions are included to indicate known and potential mechanisms for cross-talk between cadherins and other systems. Data are taken principally from refs 18,69. *Abbreviations*: AKAP: A-kinase anchoring protein, ARVCF: armadillo repeat gene deletes in velocardiofacial syndrome, $Ca^{2+}$/CaM:calcium–calmodulin, c-Abl: Abelson tyrosine kinase, CASK: calcium/calmodulin-dependent serine protein kinase, Cdk: cyclin-dependent kinase, ERBIN: ErbB2-interacting protein, Fer: Fer kinase, FGFR: fibroblast growth factor receptor, GEF: GTP exchange factor, GKAP: G kinase anchoring protein, IQGAP: GTPase activating protein with IQ motifs, MAGI: membrane-associated guanylate kinase, MAGUIN: membrane-associated guanylate kinase-interacting protein, MALS: mammalian LIN-7, PKA: protein kinase A, PKC: protein kinase C, PP2B: protein phosphatase 2B, PSD95: postsynaptic density protein 95 (also SAP90), PTP: protein tyrosine phosphatase, SAPAP: SAP90/PSD-95-associated protein, SAP: synapse-associated protein, Shc: SRC homology 2 domain containing (transforming protein-1), SHP2: SRC homology 2 domain-containing tyrosine phosphatase, S-SCAM: synaptic scaffolding molecule, Veli: vertebrate homolog of LIN-7, Wnt: wingless type, ZO-1: zona occludens 1.

which key cadherin-binding partners have been delated selectively in the nervous system or late in development show normal cadherin distribution surprisingly undisturbed at synapses[14,15]. Binding between cadherins can be blocked or attenuated by exposure to peptides containing HAV, highly conserved sequence in the EC1 domain of all type I classic cadherins. Some crystal structure data support that the HAV motif is at the *trans*-binding interface[7], while more recent work has called this into question[11]. HAV-containing peptides are more effective at blocking *N*-cadherin-mediated axon extension for a variety of neuron types and systems than the assembly of cell–cell junctions[16,17], suggesting that the cadherin interactions supporting outgrowth and junctional adhesion are likely to be different and may explain some of the apparent contradictions in the literature. Membrane-targeted cadherin mutants lacking the extracellular domain act as pan-cadherin-interfering proteins *in vivo* and in culture. The effects of such mutants appear to be potent and quite specific for classic cadherins. Mutants lacking the intracellular domain act as dominant-interfering proteins for individual cadherins, but there are some data that the extracellular domain can be adhesive on its own[18,19]. While it is clear that loss of all intracellular interactions abrogates many cadherin functions[20], the contributions of particular family members are likely to become clearer with the increased use of RNAi to selectively decrease the levels of single cadherins. Cell-permeant peptides directed against particular intracellular domains appear to act as sinks for the normal cadherin-binding partners and have been used to help dissect differences in the actions of proteins binding the juxtamembrane or ß-catenin binding domains[21]. Their actions are immediate but cell wide, affecting both cytoplasmic and membrane-bound pools of proteins.

Extracellular binding provokes cadherin binding to the actin cytoskeleton in a Rac1-dependent fashion in non-neuronal cells[22] and this inhibits RhoA[23]. The srength of adhesion can be modulated by binding interactions with the catenin proteins which in the case of ß-catenin is negatively regulated by its phosphorylation (Figure 5.2). ß-catenin interacts with an enormous variety of scaffolding, signaling, and transmembrane proteins in addition to its interactions with cadherins and actin, thereby linking adhesion and ion channel activity at the cell surface to intracellular signaling pathways. For example, the LAR receptor protein tyrosine phosphatase co-immunoprecipitates with the cadherin-ß–catenin complex as well as with the AMPA receptor-binding protein, GRIP. Mutations in LAR disrupting either its function or binding interactions reduce the synaptic localization of ß-catenin as well as the surface concentration of AMPA receptors, suggesting that cadherin–catenin interactions may regulate AMPA receptor concentration[24]. Changes in activity can also modulate cadherin interactions. When neurons are exposed to NMDA, interactions between *N*-cadherin and the PKA-binding protein, AKAP79/150, are disrupted while *N*-cadherin homophilic binding is strengthened[25]. Since AKAP79/150 also binds to PSD95/SAP90 which binds directly to NMDA receptors, NMDA exposure may sever the connection between cadherins and NMDA receptors[26] (Figure 5.2). These findings and others indicate that cadherin function is coupled actively to neurotransmission.

## 4. CELLULAR AND SUBCELLULAR LOCALIZATION AND TRAFFICKING DURING TERMINAL OUTGROWTH AND SYNAPTOGENESIS

Several individual cadherins are expressed reciprocally by afferent and target regions[5]. These findings have lent support to the concept of a cadherin code whereby cadherins, acting individually or in combination, impart specificity to synapse targeting and maturation. This notion is supported further by cases of mutual exclusion, such as the localization of *N*-cadherin and cadherin-8 to synapses in developing rat somatosensory cortex according to their respective thalamic nuclei of origin, and *N*- and *E*-cadherin immunolabeling in adult hippocampus[1,27]. The time course for such actions may be protracted, as *N*-cadherin is initially present at all synapses between hippocampal neurons and gradually becomes excluded from GABAergic sites as neurons mature[28].

As the rat hippocampus develops, immunogold-labeled cadherins are widely and regularly distributed along the synaptic cleft of young synapses, supporting the idea that cadherins are major contributors to adhesion at young synaptic junctions. The even distribution also suggests a possible role in matching pre- to postsynaptic size. Over the course of maturation, cadherins become concentrated in clusters that can be localized within or just outside of the active zone[29]. In adult mouse cerebellum, immunogold labeling for α*N*- and β-catenins, presumably bound to the cytoplasmic tail of cadherins, is concentrated in clusters similar to those seen in hippocampus, but localized exclusively at the edges of synaptic active zones suggesting that the adult distribution of cadherins within synapses may differ slightly between brain areas[2]. More importantly, the data suggest that the functions of cadherins differ between newly formed synaptic junctions and more mature sites.

Most studies of the synaptic localization and trafficking of individual cadherins have focused on *N*-cadherin. Prior to synaptogenesis, *N*-cadherin is concentrated in vesicle-like particles in axonal and dendritic cytoplasm[28]. At least some of these particles correspond to the large dense-core vesicles that also transport Bassoon, RIM, and other essential components to nascent synapses [30,31]. Live images of fluorescently tagged *N*-cadherin in developing zebrafish show the deposition of protein from an intracellular pool at nascent axon terminals of spinal Rohon–Beard neurons. Deletion of the extracellular domain, but not the cytoplasmic tail, occludes this synaptic accumulation[32], suggesting that *N*-cadherin becomes targeted to presynaptic terminals during exocytosis or by lateral membrane diffusion and trapping. As with *E*-, and *P*-cadherins, *N*-cadherin has been identified in postsynaptic fractions purified from adult rat forebrain[33–35]. The presence of *N*-cadherin at both sides of the synapse is consistent with the presumed *trans*-synaptic binding and function of cadherins.

## 5. ROLES IN AXONAL TARGETING AND TERMINATION, DENDRITIC ARBORIZATION, AND SPINE GROWTH

*N*-cadherin appears to influence axonal targeting by signaling lamina-specific termination and promoting outgrowth and fasciculation[36–39]. Specific adhesive interactions also may target axons, in a less traditional sense, by instructing the encroaching axons of projection neurons and interneurons to form boutons onto distinct subcellular regions of target neurons. For instance, basket cell axons do not restrict their termination to Purkinje cell axon initial segments in mice lacking an axosomatic gradient of the cell-adhesion molecule neurofascin[40]. Cadherins do not appear to serve in this capacity, since long-term expression of an extracellular domain deletion mutant (see Section 3) in hippocampal neurons does not alter the normal distribution of inhibitory and excitatory terminals onto target neurons[41]. However, transient transfections at either early or later stages of maturation do reduce terminal density[15], indicating that cadherins contribute to normal synapse formation, in part, by promoting or stabilizing the initial contacts between axons and dendrites.

Cadherins appear to instruct morphogenesis and innervation of differentiating neurites concomitantly, since changes in cadherin expression also affect dendritic differentiation. *N*-cadherin overexpression in hippocampal neurons promotes branching, albeit modest, of the dendritic arbor, while arborization is reduced by overexpression of the soluble intracellular domain[42]. This reduced branching may result from the retraction of destabilized dendritic arbors, as appears to occur after knocking down the expression of Celsr2, the 7-pass transmembrane cadherin, in Purkinje cells[43]. Additionally, overexpressing extracellular domain deletion mutants causes dendritic spines to appear more elongated or spiky, similar to spines lacking F-actin[15,44,45].

## 6. ROLES IN ASSEMBLY, RETENTION, AND FUNCTION OF PRE- AND POSTSYNAPTIC COMPONENTS

*N*-cadherin expressing HEK293 cells fail to induce vesicle clusters in apposing hippocampal axons in heterologous cultures[46]. This finding, along with its targeting to nascent but pre-existing Rohon–Beard synapses[32], suggests that *N*-cadherin is present from early stages but does not initiate synapse assembly by itself. However, several lines of evidence indicate that, beyond their morphogenetic actions, cadherins may dynamically influence the composition and function of maturing synapses. Overexpression of an extracellular domain deletion mutant in dissociated hippocampal neurons grown in culture for a week blocks synapse formation and retention completely. In more mature neurons transient expression of the same mutant reduces presynaptic terminal density generally, but also reduces GABAergic innervation to a lesser but significant extent[15]. Both excitatory and inhibitory terminal densities recover to normal levels with longer blockades, but active vesicle recycling, as visualized by styryl dye reuptake, is reduced[41]. Notably, mEPSCs recorded from these transfected neurons occur with lower frequency, indicating that presynaptic activity in untransfected neurons also is affected by the postsynaptic disruption and further supporting the notion that postsynaptic cadherins can act retrogradely via homophilic adhesion[47,48]. Postsynaptically, PSD95 clusters are smaller and fewer with transient pan-cadherin functional blockades in hippocampal neurons or following perturbation of cadherin-6B function in chick retinal neurons[15,49]. These latter observations imply that individual cadherins may either cluster or anchor particular synaptic components. Consistent with this potential role, beads coated with the *N*-cadherin extracellular domain can cluster GluR6-containing kainate receptors and *E*-cadherin can recruit the MAGUK SAP97 to the cortical cytoskeleton in non-neuronal cell lines[50,51], although it remains to be demonstrated whether these functions are served in neurons. In a separate vein, infusion of embryonic chick ciliary neurons with the intercellular juxtamembrane domain of *N*-cadherin (sJMD) attenuates the amplitude of high-voltage activated calcium currents[52]. Thus, in addition to simply clustering synaptic components, cadherin interactions interactions may modulate channel surface expression or activity.

## 7. CADHERIN LOCALIZATION AND FUNCTION DURING PLASTICITY

Given their apparent influence on spine maturation and the trafficking and function of synaptic components, cadherin synaptogenic functions may be intimately tied to nascent activity patterns. These functions also have obvious implications for synaptic plasticity at later stages. Indeed, CA1 LTP induced by tetanic stimulation is augmented in hippocampal slices from adult cadherin-11 deficient mice relative to wild-type controls[53]. Although cadherin-11 function was chronically absent throughout the development of these animals, acute perturbations support more immediate roles for individual cadherins in mediating plasticity. Manipulating *E*- or *N*-cadherin function, with function-perturbing antibodies or adhesion-blocking HAV motif-containing peptides, can prohibit the maintenance of long-term potentiation (LTP) or even occlude its induction in the CA1 region of hippocampal slices[47]. Cadherins can also mediate morphological plasticity at mature synapses. For instance, transient spine head expansion following KCl-mediated depolarization in three-week cultured neurons is restricted by the expression of *N*-cadherin mutants either lacking the extracellular domain or carrying a point mutation that abrogates homophilic binding[54]. As with maturation, individual cadherins may coordinate these functional and morphological changes at mature synapses.

In some sense, the calcium-dependent adhesion conferred by cadherins in these cases may serve as both a structural index of synaptic strength at a given time and an activity-dependent modulator of synapse composition. As such, cadherins themselves appear to be dynamically modulated by activity. Hippocampal neurons maintained for four to five weeks in culture show a transient lateral dispersion of both *N*-cadherin and synaptophysin following a brief depolarization of by KCl application[25], which is thought to produce concerted neurotransmitter vesicle fusion. A corresponding transient lateral diffusion of *N*-cadherin is seen postsynaptically within the expanding spine heads of three-week cultured neurons with this manipulation, consistent with a functional coupling of pre- and postsynaptic *N*-cadherin following neurotransmitter release[25,54]. Biochemical evidence demonstrates an activity-dependent increase in *N*-cadherin dimerization and an increased resistance to trypsin digestion that is NMDA receptor dependent[25]. Hippocampal slices in which LTP has been induced also show increased levels of *N*-cadherin dimers[48]. Thus, cadherins appear to be stabilized at active glutamatergic synapses. These findings also suggest that cadherins could become sensitive to the actions of endogenous proteases, particularly at sites with active neurotransmitter release and relatively little NMDA receptor activation, such as some GABAergic synapses or depressed glutamatergic synapses. This possibility may explain the loss of *N*-cadherin from GAD65-positive hippocampal synapses or thalamic terminals in somatosensory cortex during development[28,55]. It also may partly explain the persistent loss of *E*-cadherin expression in nonpeptidergic C-fibers following spinal axotomy[56]. Potentiation, on the other hand, produces a protein synthesis-dependent increase in synaptic *N*-cadherin in juvenile CA1[41], consistent with its deployment to strengthened sites.

## 8. MOLECULAR MECHANISMS OF CADHERIN ACTION DURING SYNAPTOGENESIS AND SYNAPTIC FUNCTION

How cadherins mediate synapse maturation and function is not well understood, but their associations with β- and δ-catenin and, thereby, the actin cytoskeleton, are considered prime modes of action (Figure 5.2). F-actin, like the cadherins, is required to assemble and maintain synapses in hippocampal neurons during the first week in culture[45]. By contrast, actin depolymerization with latrunculin A in mature cultures, where synaptogenesis is virtually complete, has no apparent effect on terminal density. PSD-95 clusters are not affected by depolymerization at any stage of maturation, but NMDA and AMPA receptor number and synaptic localization are susceptible to this treatment to varying degrees[44,45], supporting a general requirement for actin-linked cell-adhesion molecules in maintaining synapse composition independent of PSD-95 scaffolding. Long-term blockade of cadherin-based adhesion occludes the F-actin independence of synapse maintenance normally attained in the third week of culture[15,41]. Stabilizing actin, in contrast, prohibits rundown and long-term depression (LTD) of NMDA receptor-mediated responses at maturing hippocampal synapses[57,58]. Cadherins and other cell-adhesion molecules conceivably achieve tight control of these F-actin dependent physiological properties under normal conditions. Cadherins also have been implicated in modulating glutamate receptor-dependent, actin-mediated spine dynamics[54,59], perhaps in coordination with physiological properties.

Overexpression of either β-catenin or α*N*-catenin, as with *N*-cadherin, enhances dendritic arborization and spine maturation[42,60]. Conversely, the dendritic arbors of 3-week old hippocampal neurons from α*N*-catenin deficient mice are studded with elongated or spiky protrusions rather than spines[15,41,42]. While these and other observations implicate these catenins as direct mediators of cadherin function, care must be taken when dissecting the individual contributions of the several molecules that bind the cadherin intracellular domain (Figure 5.2) or ascribing their actions to cadherins. For instance, although the effects of β-catenin deletion on vesicle clustering are quite similar to those seen with cadherin blockade, its effect on synaptophysin trafficking does not appear to involve cadherins, as cadherins remain clustered at affected sites following β-catenin deletion[14]. Additionally, the principal effects of ß-catenin on dendritic arborization appear to be mediated via the Wnt pathway and not through classic cadherins[42]. Modulation of high-voltage activated (HVA) calcium channels by the sJMD is counteracted by acute inactivation of either RhoA GTPase- or Rho-associated kinase, which similarly implicates these latter proteins as effectors of sJMD function, particularly through their functional interactions with p120 catenin[23,52]. Still, other binding partners of the sJMD, either cadherin-associated or in cytosolic pools, could contribute to its modulation of HVA channels. To dissect the functional contributions of individual binding partners unequivocally, conditional ablation of cadherins will be need to be combined with the conditional expression of mutated cadherins.

Other studies suggest that cadherins can influence synapse maturation and function independent of their interactions with the catenins and actin. *N*-, *E*-, and *P*-cadherins along with actin, several catenins, and the A-kinase anchoring protein AKAP79/150 previously have been found in large multiprotein complexes with the NMDA receptor[34]. Under more stringent conditions, the membrane-targeted

AKAP79/150, which also forms a complex with PSD-95, the GluR1 subunit of the AMPA receptor and the protein phosphatase calcineurin (PP2B), has been shown to bind noncompetitively with β-catenin to a site within the cadherin intracellular domain[26,61,62] (Figure 5.2). Given the requirement of AKAP79 for PKA phosphorylation of the AMPA receptor, these biochemical findings collectively imply that mobilizing cadherins can have immediate effects on glutamate receptor function during developmental and adult plasticity. A presenilin-1-generated cleavage fragment of intracellular *N*-cadherin, which can bind CREB-binding protein and inhibit CREB-mediated transcription in non-neuronal cells[12], may effect an additional, delayed modulation of synaptic strength in neurons.

## 9. CADHERIN FUNCTIONS IN OTHER SYSTEMS

As with vertebrates, *N*-cadherin synaptogenic functions are the most widely characterized among cadherin family members in *Drosophila*, mainly based on analyses of mosaic fly embryos. Initially, all cells in these flies carry one copy of a gene of interest that has been disrupted by insertion of a reporter gene that also has recombination elements. Cells within a distinct subset also carry a second mutation that allows heat-inducible mitotic recombination such that, upon transferring these embryos to the permissive temperature, homozygous mutations are generated in a particular class of cells. This manipulation allows distinct neurons to be mutated with some temporal control and these mutant neurons to be visualized in a comparatively normal (i.e. "wild-type") background.

*N*-cadherin function has been examined by this type of mosaic analysis foremost in the fly olfactory system. Olfactory receptor neurons (ORNs) in the sensory neuropil send primary axons to the dendrites of projection neurons (PNs) at glomeruli in the antenna lobe. The dendritic arbors of individual PNs—which, at first, only loosely demarcate a protoglomerulus—become confined to form a single, more distinct glomerulus upon innervation by several ORNs, which themselves express the same, unique odorant receptor (OR) but are broadly distributed[63]. Normally, both ORN axons and PN dendrites express *N*-cadherin, but individual, mosaic PN dendrites homozygous for a null *N*-cadherin mutation fail to refine their arbors in response to wild-type ORN innervation or neighboring dendritic arbors[63]. By comparison, OR-specific ORN axons homozygous for either a null or inactivating *N*-cadherin mutation properly co-fasciculate and target the antenna lobe but fail to restrict their terminals to dendrites within a single protoglomerulus[63,64]. These findings suggest that, as in rat hippocampal neurons, *N*-cadherin does not encode target selectivity *per se*, but may mediate the selective strengthening of certain synapses within a network in response to activity.

This role also may hold true for the fly visual system. Normally, photoreceptors 1 through 6 (R1–6) from each ommatidium project according to subtype to stereotyped columns of laminar neurons—first the "column of origin," then the target column—within the optic neuropil. Instead, photoreceptors largely fail to project beyond their specific column of origin to lamina neurons in the target column when either lacks *N*-cadherin[65]. Although mosaic flies mutant for *E*-cadherin and other nonclassical cadherins had no obvious phenotypes in these contexts[63,65], these latter family members are likely to serve other aspects of synapse formation or maturation in this system.

## 10. CONCLUSIONS

The level of reduction in synapse number seen with transient cadherin blockade in the first week of culture is on par with that reported following shRNA knockdown of neuroligins-1, -2, and -3 at similar maturational stages[15,66]. Although the morphological criteria used to identifying synapses in these studies differ, these findings collectively suggest that cadherins have some synaptogenic capacities. To date, only SynCAM and members of the neurexin/neuroligin families are thought to be capable of initiating synapse formation, based on their abilities to cluster axonal vesicles and induce glutamatergic transmission between neurons and non-neuronal cells in heterologous cultures[46,67] (see Chapters 7 and 8). Notably, the non-neuronal HEK293 cells used in these studies express endogenous cadherins. Thus, it is possible that cadherins act coordinately with these and other cell-adhesion molecules[*] to form normal synapses. Beyond general synapse formation, the synaptogenic properties of cadherins appear to have been appropriated in adulthood to impart networks with dynamic control of synapse maintenance, strengthening and signaling. Recent work suggests this function is likely to be critical for maintaining normal cognitive capacity. Presenilin-1 can regulate the strength of cadherin adhesion and participates in the $\gamma$-secretase mediated $\varepsilon$-cleavage of *N*-cadherin. The liberated intracellular fragment can enter the nucleus and negatively regulate the activity of the transcription factor CREB (cyclic AMP response element binding protein) which controls the expression of certain genes critical for nervous system function and plasticity. Mutations in presenilin-1 that are known to be involved in early onset familial Alzheimer's disease prevent *N*-cadherin cleavage suggesting a link between the regulation of cadherin adhesion, transcriptional regulation and cognition[12].[†]

## 11. REFERENCES


1. Fannon, A.M., and Colman, D.R. (1996) *Neuron* **17**, 423–434.
2. Uchida, N., Honjo, Y., Johnson, K.R., Wheelock, M.J., and Takeichi, M. (1996) *J Cell Biol* **135**, 767–779.
3. Landis, D.M., and Reese, T.S. (1974) *J Comp Neurol* **155**, 93–125.
4. Peters, A., Palay, S.L., and Webster, H.D. (1991) The Fine Structure of the Nervous System, Oxford University Press, New York.
5. Benson, D.L., Colman, D.R., and Huntley, G.W. (2001) *Nat Rev Neurosci* **2**, 899–909.
6. Tepass, U., Truong, K., Godt, D., Ikura, M., and Peifer, M. (2000) *Nat Rev Mol Cell Biol* **1**, 91–100.
7. Shapiro, L., Fannon, A.M., Kwong, P.D., Thompson, A., Lehmann, M.S., Grubel, G., Legrand, J.F., Als-Nielsen, J., Colman, D.R., and Hendrickson, W.A. (1995) *Nature* **374**, 327–337.
8. Overduin, M., Harvey, T.S., Bagby, S., Tong, K.I., Yau, P., Takeichi, M., and Ikura, M. (1995) *Science* **267**, 386–389.
9. Chappuis-Flament, S., Wong, E., Hicks, L.D., Kay, C.M., and Gumbiner, B.M. (2001) *J Cell Biol* **154**, 231–243.
10. Takeichi, M. (1977) *J Cell Biol* **75**, 464–474.


---

[*]It is well established that $\alpha E\beta 7$ integrin and *E*-cadherin bind directly in leukocytes (70). Interactions between $\alpha 2\beta 1$ and *E*-cadherin in fibrosarcoma cells suggest such heterophilic binding may be more widespread (71), but it has not yet been examined in the nervous system.

[†]We thank Vanja Nagy for helpful discussion and Dr. Ioana Carcea for her critical reading of this chapter.

11. Boggon, T.J., Murray, J., Chappuis-Flament, S., Wong, E., Gumbiner, B.M., and Shapiro, L. (2002) *Science* **296**, 1308–1313.
12. Marambaud, P., Shioi, J., Serban, G., Georgakopoulos, A., Sarner, S., Nagy, V., Baki, L., Wen, P., Efthimiopoulos, S., Shao, Z., Wisniewski, T., and Robakis, N.K. (2002) *EMBO J* **21**, 1948–1956.
13. Yap, A.S., Niessen, C.M., and Gumbiner, B.M. (1998) *J Cell Biol* **141**, 779–789.
14. Bamji, S.X., Shimazu, K., Kimes, N., Huelsken, J., Birchmeier, W., Lu, B., and Reichardt, L.F. (2003) *Neuron* **40**, 719–731.
15. Togashi, H., Abe, K., Mizoguchi, A., Takaoka, K., Chisaka, O., and Takeichi, M. (2002) *Neuron* **35**, 77–89.
16. Blaschuk, O.W., Sullivan, R., David, S., and Pouliot, Y. (1990) *Dev Biol* **139**, 227–229.
17. Renaud-Young, M., and Gallin, W.J. (2002) *J Biol Chem* **277**, 39609–39616.
18. Gumbiner, B.M. (2005) *Nat Rev Mol Cell Biol* **6**, 622–634.
19. Vestal, D.J., and Ranscht, B. (1992) *J Cell Biol* **119**, 451–461.
20. Nagafuchi, A., and Takeichi, M. (1988) *EMBO J* **7**, 3679–3684.
21. Arregui, C., Pathre, P., Lilien, J., and Balsamo, J. (2000) *J Cell Biol* **149**, 1263–1274.
22. Lambert, M., Choquet, D., and Mege, R.M. (2002) *J Cell Biol* **157**, 469–479.
23. Noren, N.K., Arthur, W.T., and Burridge, K. (2003) *J Biol Chem* **278**, 13615–13618.
24. Dunah, A.W., Hueske, E., Wyszynski, M., Hoogenraad, C.C., Jaworski, J., Pak, D.T., Simonetta, A., Liu, G., and Sheng, M. (2005) *Nat Neurosci* **8**, 458–467.
25. Tanaka, H., Shan, W., Phillips, G.R., Arndt, K., Bozdagi, O., Shapiro, L., Huntley, G.W., Benson, D.L., and Colman, D.R. (2000) *Neuron* **25**, 93–107.
26. Gorski, J.A., Gomez, L.L., Scott, J.D., and Dell'Acqua, M.L. (2005) *Mol Biol Cell* **16**, 3574–3590.
27. Gil, O.D., Needleman, L., and Huntley, G.W. (2002) *J Comp Neurol* **453**, 372–388.
28. Benson, D.L., and Tanaka, H. (1998) *J Neurosci* **18**, 6892–6904.
29. Elste, A.M., and Benson, D.L. (2006) *J Comp Neurol* **495**, 324–335.
30. Shapira, M., Zhai, R.G., Dresbach, T., Bresler, T., Torres, V.I., Gundelfinger, E.D., Ziv, N.E., and Garner, C.C. (2003) *Neuron* **38**, 237–252.
31. Zhai, R.G., Vardinon-Friedman, H., Cases-Langhoff, C., Becker, B., Gundelfinger, E.D., Ziv, N.E., and Garner, C.C. (2001) *Neuron* **29**, 131–143.
32. Jontes, J.D., Emond, M.R., and Smith, S.J. (2004) *J Neurosci* **24**, 9027–9034.
33. Beesley, P.W., Mummery, R., and Tibaldi, J. (1995) *J Neurochem* **64**, 2288–2294.
34. Husi, H., Ward, M.A., Choudhary, J.S., Blackstock, W.P., and Grant, S.G. (2000) *Nat Neurosci* **3**, 661–669.
35. Phillips, G.R., Florens, L., Tanaka, H., Khaing, Z.Z., Fidler, L., Yates, J.R., 3rd, and Colman, D.R. (2005) *J Neurosci Res* **81**, 762–775.
36. Doherty, P., Skaper, S.D., Moore, S.E., Leon, A., and Walsh, F.S. (1992) *Development* **115**, 885–892.
37. Inoue, A., and Sanes, J.R. (1997) *Science* **276**, 1428–1431.
38. Treubert-Zimmermann, U., Heyers, D., and Redies, C. (2002) *J Neurosci* **22**, 7617–7626.
39. Poskanzer, K., Needleman, L.A., Bozdagi, O., and Huntley, G.W. (2003) *J Neurosci* **23**, 2294–2305.
40. Ango, F., di Cristo, G., Higashiyama, H., Bennett, V., Wu, P., and Huang, Z.J. (2004) *Cell* **119**, 257–272.
41. Bozdagi, O., Valcin, M., Poskanzer, K., Tanaka, H., and Benson, D.L. (2004) *Mol Cell Neurosci* **27**, 509–521.
42. Yu, X., and Malenka, R.C. (2003) *Nat Neurosci* **6**, 1169–1177.
43. Shima, Y., Copeland, N.G., Gilbert, D.J., Jenkins, N.A., Chisaka, O., Takeichi, M., and Uemura, T. (2002) *Dev Dyn* **223**, 321–332.
44. Allison, D.W., Gelfand, V.I., Spector, I., and Craig, A.M. (1998) *J Neurosci* **18**, 2423–2436.
45. Zhang, W., and Benson, D.L. (2001) *J Neurosci* **21**, 5169–5181.
46. Scheiffele, P., Fan, J., Choih, J., Fetter, R., and Serafini, T. (2000) *Cell* **101**, 657–669.
47. Tang, L., Hung, C.P., and Schuman, E.M. (1998) *Neuron* **20**, 1165–1175.
48. Bozdagi, O., Shan, W., Tanaka, H., Benson, D.L., and Huntley, G.W. (2000) *Neuron* **28**, 245–259.
49. Honjo, Y., Nakagawa, S., and Takeichi, M. (2000) *Genes Cells* **5**, 309–318.
50. Coussen, F., Normand, E., Marchal, C., Costet, P., Choquet, D., Lambert, M., Mege, R.M., and Mulle, C. (2002) *J Neurosci* **22**, 6426–6436.
51. Reuver, S.M., and Garner, C.C. (1998) *J Cell Sci* **111**, 1071–1080.
52. Piccoli, G., Rutishauser, U., and Bruses, J.L. (2004) *J Neurosci* **24**, 10918–10923.

53. Manabe, T., Togashi, H., Uchida, N., Suzuki, S.C., Hayakawa, Y., Yamamoto, M., Yoda, H., Miyakawa, T., Takeichi, M., and Chisaka, O. (2000) *Mol Cell Neurosci* **15**, 534–546.
54. Okamura, K., Tanaka, H., Yagita, Y., Saeki, Y., Taguchi, A., Hiraoka, Y., Zeng, L.H., Colman, D.R., and Miki, N. (2004) *J Cell Biol* **167**, 961–972.
55. Huntley, G.W., and Benson, D.L. (1999) *J Comp Neurol* **407**, 453–471.
56. Brock, J.H., Elste, A., and Huntley, G.W. (2004) *J Neurosci* **24**, 8806–8817.
57. Li, B., Chen, N., Luo, T., Otsu, Y., Murphy, T.H., and Raymond, L.A. (2002) *Nat Neurosci* **5**, 833–834.
58. Morishita, W., Marie, H., and Malenka, R.C. (2005) *Nat Neurosci* **8**, 1043–1050.
59. Fischer, M., Kaech, S., Wagner, U., Brinkhaus, H., and Matus, A. (2000) *Nat Neurosci* **3**, 887–894.
60. Abe, K., Chisaka, O., Van Roy, F., and Takeichi, M. (2004) *Nat Neurosci* **7**, 357–363.
61. Colledge, M., Dean, R.A., Scott, G.K., Langeberg, L.K., Huganir, R.L., and Scott, J.D. (2000) *Neuron* **27**, 107–119.
62. Dell'Acqua, M.L., Dodge, K.L., Tavalin, S.J., and Scott, J.D. (2002) *J Biol Chem* **277**, 48796–48802.
63. Zhu, H., and Luo, L. (2004) *Neuron* **42**, 63–75.
64. Hummel, T., and Zipursky, S.L. (2004) *Neuron* **42**, 77–88.
65. Prakash, S., Caldwell, J.C., Eberl, D.F., and Clandinin, T.R. (2005) *Nat Neurosci* **8**, 443–450.
66. Chih, B., Engelman, H., and Scheiffele, P. (2005) *Science* **307**, 1324–1328.
67. Biederer, T., Sara, Y., Mozhayeva, M., Atasoy, D., Liu, X., Kavalali, E.T., and Sudhof, T.C. (2002) *Science* **297**, 1525–1531.
68. Nollet, F., Kools, P., and van Roy, F. (2000) *J Mol Biol* **299**, 551–572.
69. Hirano, S., Suzuki, S.T., and Redies, C.M. (2003) *Front Biosci* **8**, D306–355.
70. Kutlesa, S., Wessels, J.T., Speiser, A., Steiert, I., Muller, C.A., and Klein, G. (2002) *J Cell Sci* **115**, 4505–4515.
71. Whittard, J.D., Craig, S.E., Mould, A.P., Koch, A., Pertz, O., Engel, J., and Humphries, M.J. (2002) *Matrix Biol* **21**, 525–532.

# 6

# SYNAPTIC FUNCTIONS OF THE NEURAL CELL ADHESION MOLECULE, NCAM

Alexander Dityatev*

## 1. SUMMARY

The neural cell adhesion molecule, NCAM, belongs to the immunoglobulin superfamily of cell adhesion molecules, and is well recognized as an important regulator of different morphogenetic events, including neural cell proliferation and migration, as well as axonal fasciculation and outgrowth. This chapter highlights synaptic functions of NCAM. At early stages of synaptogenesis in primary hippocampal cultures, clusters of NCAM at the cell surface – linked via spectrin to trans-Golgi network (TGN)-derived organelles – translocate along growing neurites to sites of neurite-to-neurite contacts within several minutes of initial contact formation. There, NCAM mediates an anchoring ("synaptic trap") of the intracellular organelles. At later stages of synaptogenesis, the relative levels of postsynaptic NCAM expression control both the number and strength of synapses in an activity-dependent manner. This process requires polysialylation of NCAM and activity of FGF and NMDA receptors. In mature brains, NCAM is important for induction of NMDA receptor-dependent long-term potentiation (LTP) and depression (LTD) in the CA1 area of the hippocampus, LTP-associated increase in the number of perforated synapses, NMDA receptor-independent mossy fiber LTP, several forms of learning and memory and morphological plasticity in the hypothalamo-neurohypophysial system. Additionally, NCAM was found to be essential for synaptic vesicle trafficking at the neuromuscular junction (NMJ) and for catecholamine release from neuroendocrine chromaffin cells. Thus, NCAM is important for formation and plasticity of synapses and plays a basic role in transmitter release. The significance of NCAM is further underscored by the efficacy of NCAM-derived compounds to affect these processes, and emerging links between NCAM and schizophrenia.

---

*Department of Neurophysiology and Center for Molecular Neurobiology, University Medical Center Hamburg-Eppendorf, Martinistr. 52, 20246 Hamburg, Germany; dityatev@zmnh.uni-hamburg.de

## 2. INTRODUCTION

The neural cell adhesion molecule (NCAM), originally described by Jorgensen and Bock in 1974 as a synaptic membrane glycoprotein termed D2[1], was the first vertebrate molecule to be identified and characterized by the Edelman group as a cell adhesion molecule[2]. This glycoprotein is widely expressed in the central and peripheral nervous systems, and decades of intensive research have disclosed a great deal of information on structure, interaction partners, signaling pathways, and functions of NCAM. Interestingly, in the mammalian brain, NCAM is a unique carrier of the unusual polyanionic carbohydrate, polysialic acid (PSA). Functions of NCAM appear to depend on PSA, providing probably one of the most impressive examples in neurobiology of how glycosylation may modify the activity of a protein. In this chapter, I discuss emerging multiple mechanisms by which NCAM and associated PSA affect synaptogenesis, transmitter release, and synaptic plasticity.

## 3. BASIC CHARACTERISTICS OF NCAM

### 3.1. Structure of NCAM

NCAM is a molecule belonging to the immunoglobulin (Ig) superfamily (for a review, see ref. 3). The extracellular part of NCAM contains five Ig-like domains and two fibronectin type III-like repeats (Figure 6.1). NCAM sequences from vertebrates, ranging from human to frog, share residue identities of 70–98%. More distantly related are fasciclin II of *Drosophila melonagaster* and the *Aplysia californica* cell adhesion molecule, apCAM, which have 25% identity to vertebrate NCAM proteins. NCAM exists in several isoforms, due to alternative splicing of a single gene consisting of at least 26 exons. In the mouse, NCAM exists in three major membrane-bound isoforms derived from one gene: NCAM120, NCAM140, and NCAM180. The last two forms are integral membrane proteins differing in the size of their intracellular domains. NCAM120, on the other hand, is linked to the membrane via a GPI anchor. GPI-linked proteins appear to be relatively free to move within the plane of the lipid bilayer, in contrast to membrane-spanning proteins that may interact with cytoskeletal components and are thus more rigidly inserted in the membrane. Additionally, several soluble forms of NCAM are present in the brain[4]. Some isoforms contain (on their fifth Ig-like domain) PSA in an unusual $\alpha 2,8$ linkage in chains which can be up to 200 residues long. Attachment of the highly negatively charged PSA to NCAM results in a large hydration sphere and modulates the function of the molecule. For example, homophilic binding of nonpolysialylated NCAM is much stronger than of PSA-NCAM.

### 3.2. Extracellular Binding Partners of NCAM

The domains involved in homophilic NCAM–NCAM recognition remain a matter of debate. Early data suggested antiparallel association of the five Ig-like domains of one molecule with five Ig-like domains of another NCAM molecule[5], with the highest affinity found in that between the third Ig-like domains. More recently, an interaction involving the first two Ig-like domains has been put forward[6]. Additionally, more complex models of *trans*-interaction between dimers of *cis*-interacting NCAM molecules have been proposed on the basis of crystal structure of the first three NCAM Ig domains (reviewed in ref. 7). However, two recent studies using surface force apparatus and atomic force microscopy[8,9] have suggested that NCAM preferentially adheres in two fashions: either in a fully, antiparallel alignment requiring Ig3, or in a separate, weaker antiparallel interaction between the Ig1–Ig2 domains.

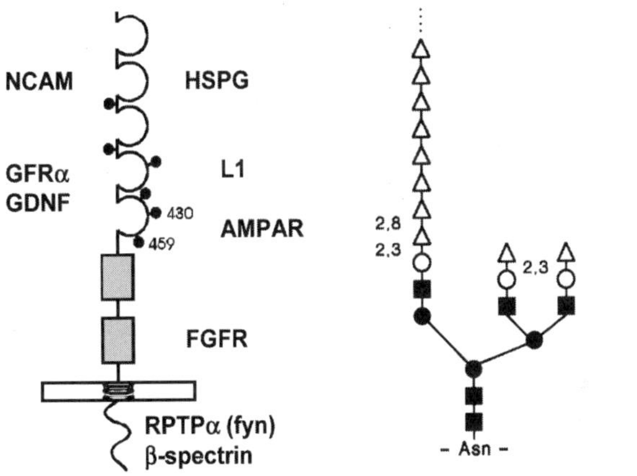


**Figure 6.1.** Structure of NCAM and PSA. The extracellular domain of NCAM contains five Ig-like subdomains (semicircles) and two FNIII repeats (rectangles). Glycosylation sites are indicated by black spots; two numbered sites indicate amino acid positions to which PSA may be attached. Several NCAM-binding proteins are listed around NCAM. PSA (on the right) is found on di- and triantennary *N*-glycan chains attached to at least two N-glycosylation sites within the fifth Ig-like domain of all three NCAM isoforms. A large number of α2,8-linked sialic acid residues (empty triangles) are found attached to one antenna through α2,3-linked sialic acid (empty circles). *N*-acetylglucosamine and mannose residues are shown as filled squares and circles, respectively. For more details, see ref. 10.

NCAM is also engaged in a number of heterophilic interactions. The fourth Ig-like domain of NCAM contains a carbohydrate recognition site for oligomannosidic glycans that is important for its *cis*-association with another adhesion molecule, termed L1. This *cis*-interaction has been suggested to enhance the homophilic *trans*-binding of L1[11]. Using short synthetic peptides, a heterophilic-binding region for heparin, consisting of two clusters of basic amino acid residues, has been allocated to the second Ig-like domain of NCAM[12]. Structural analysis suggests that the heparin and chondroitin sulfate-binding sites in Ig2 coincide, and that this site overlaps with the homophilic-binding site[13]. Other extracellular ligands of NCAM are neurocan and phosphacan[14], soluble nervous tissue-specific chondroitin sulfate proteoglycans (CSPGs). The FGF receptor is another major binding partner of NCAM[15]. Interaction between molecules may involve a direct interaction between the first two fibronectin type III domains of NCAM and the Ig2 and Ig3 immunoglobulin domains of the FGF receptor[7]. NCAM also associates with GFR1, a GPI-anchored receptor for GDNF. This interaction downregulates NCAM-mediated cell adhesion and promotes high-affinity binding of GDNF to NCAM140, resulting in rapid activation of the cytoplasmic protein tyrosine kinases, fyn and FAK[16]. Among other binding partners of NCAM are prion protein, TAG-1/axonin-1, and collagens.

As mentioned above, presence of PSA on NCAM inhibits NCAM–NCAM mediated interactions. PSA also inhibits adhesive interactions mediated by L1, cadherins, and integrins[17], and thus appears as a rather universal inhibitor of adhesion. Recent molecular force measurements directly show that NCAM polysialylation increases the range and magnitude of intermembrane repulsion[18]. On the other hand, PSA promotes binding between HSPGs and the NCAM heparin-binding domain[19], as well as signaling mediated by brain-derived neurotrophic factor (BDNF), platelet-derived growth factor (PDGF), and glutamate receptors[20–22]. Thus, polysialylation of NCAM may function as a 'switch,' determining whether NCAM is engaged in cell adhesion or in signaling.

### 3.3. Intracellular Signaling Mediated by NCAM

NCAM may influence different second messenger systems. Originally, it was found that polyclonal antibodies against L1 and NCAM reduced intracellular levels of the inositol phosphates, $IP_2$ and $IP_3$, while the intracellular level of cAMP was unaffected. The antibodies also reduced intracellular pH and increased intracellular $Ca^{2+}$ by opening $Ca^{2+}$-channels in a manner which was sensitive to inhibition by pertussis toxin[23]. Further insight came from a study showing that NCAM-dependent neurite outgrowth was selectively inhibited in cultures of neurons from fyn knockout mice[24]. In wild-type cells, fyn was constitutively associated with NCAM140, the focal adhesion kinase, FAK, and became recruited to the NCAM140-fyn complex in response to stimulation with antibodies against the extracellular region of NCAM[25]. The NCAM140 isoform appears to directly interact with the intracellular domain of the receptor-type protein tyrosine phosphatase RPTPα, and this interaction is important for NCAM triggered activation of fyn[26].

The surface mobility of NCAM140 is higher than that of NCAM180[27], suggesting an association of the latter with the cytoskeleton or other stabilizing factors. NCAM180 is accumulated at sites of cell-to-cell contacts, where the cytoskeleton–membrane linker protein, spectrin, and actin also accumulate. Heteromeric spectrin (αIβI) binds to the intracellular domain of NCAM180, whereas isolated spectrin subunits bind to both NCAM180 and NCAM140. NCAM140/ NCAM180-βI spectrin complexes are located both in lipid rafts and raft-free membrane domains. Activation of NCAM enhances formation of complexes between PKCβ2 and NCAM140/NCAM180-spectrin, and results in their redistribution to lipid rafts. This process requires activity of FGF receptors and is necessary for NCAM-mediated neurite outgrowth[28]. These data support a model in which NCAM-mediated outgrowth requires co-signaling via raft-associated kinases and FGF receptors[29] (for a recent review see ref. 30).

## 4. ROLES OF NCAM IN SYNAPTOGENESIS

### 4.1 Roles of NCAM Homologs in *Aplysia* and *Drosophila*

After synapse formation in *Drosophila,* the NCAM-like molecule, fasciclin II (Fas II), is localized in both pre- and postsynaptic structures, where it controls synapse stabilization. In Fas II null mutants, synapse formation is normal, but boutons then retract during larval development. Synapse elimination and resulting lethality are rescued by transgenes that drive Fas II expression both pre- and postsynaptically. Driving Fas II expression on either side alone is insufficient[31]. The increase in number of synapses evoked by enhanced synaptic activity in *eag Shaker* and *dunce* mutants is accompanied by a 50% decrease of presynaptic

Fas II expression. This decrease is necessary and sufficient for presynaptic sprouting: Fas II mutants that have decreased Fas II levels by approximately 50% have sprouting similar to *eag Shaker* and *dunce*, while transgenes that maintain wild-type synaptic Fas II levels suppress sprouting in *eag Shaker* and *dunce*[31]. However, Fas II mutants that cause a 50% increase in bouton number do not alter synaptic strength; rather evoked release from single boutons has a reduced quantal content, suggesting that the wild-type amount of release machinery is distributed throughout more boutons[32].

In *Aplysia*, the serotonin-induced long-term facilitation of synaptic efficacy is accompanied by the growth of new synaptic connections and downregulation of apCAM. Within 1 h, serotonin led to a more than 50% reduction of apCAM due to its internalization[33]. Despite the overall decline in apCAM, there is an increase of apCAM expression at postsynaptic sites where new varicosities form following treatment with serotonin. Antibodies to apCAM block these structural and functional changes. Internalization of apCAM at the surface membrane of the sensory neuron depends on MAPK phosphorylation, which, thus, represents an early regulatory step in the growth of new synaptic connections that accompanies long-term facilitation[34].

## 4.2. Role of NCAM in Mammalian Synaptogenesis

The role of NCAM in formation of hippocampal synapses was established in 2000[35]. Since NCAM is expressed both pre- and postsynaptically, we used heterogenotypic co-cultures of wild-type (NCAM+/+) and NCAM-deficient (NCAM–/–) neurons to dissect roles that pre- and postsynaptic NCAM may play in synaptic functions. Using this system, double-cell patch-clamp recordings of unitary excitatory postsynaptic currents (uEPSCs) in NCAM+/+ and NCAM–/– neurons (evoked by intracellular stimulation of NCAM+/+ or NCAM–/– presynaptic neurons) were performed. Comparison of the mean amplitudes of uEPSPs in synaptic connections with different patterns of NCAM expression revealed that the presence of NCAM presynaptically did not influence synaptic strength, whereas postsynaptic expression of NCAM increased synaptic strength by a factor of 2. Paired-pulse facilitation and amplitude of miniature EPSPs were normal in NCAM–/– neurons, suggesting that the probability of release and postsynaptic response elicited by release of a single vesicle are not affected by the absence of NCAM. Also, no changes in size of NCAM–/– neurons were found. However, analysis of synaptophysin immunoreactivity associated with NCAM–/– and NCAM+/+ neurons revealed a 2-fold higher synaptic coverage of NCAM+/+ cells, measured as the number of synaptophysin-rich puncta or as the mean intensity of synaptophysin immunofluorescence. This was observed only in heterogenotypic cultures, i.e., under conditions when growing axons have a choice which postsynaptic target to select: NCAM+/+ or NCAM–/–. There was no difference between NCAM–/– and NCAM+/+ neurons in synaptic coverage in homogenotypic cultures. Thus, expression of NCAM dictates where to form synapses, but is not required for synapse formation. Evidently, the absence of NCAM in NCAM–/– cultures can be compensated by other molecules. Since expression of NCAM and PSA in the CNS is regulated in an activity-dependent manner[36–39] an increase in NCAM/PSA-NCAM expression may promote experience-dependent synaptogenesis in stimulated neurons and/or dendritic subdomains (Figure 6.2; Colorplate 5).

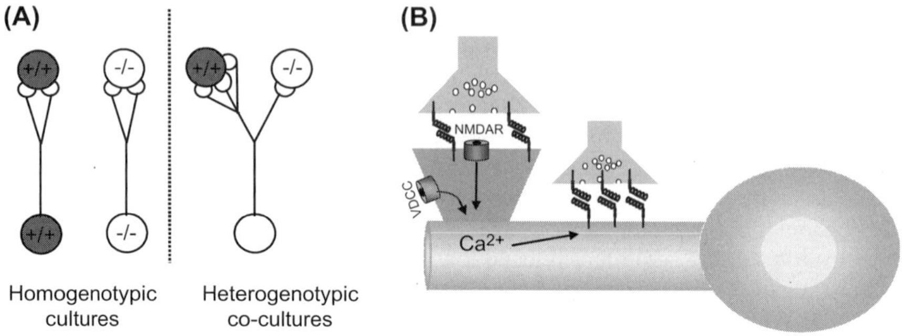


**Figure 6.2.** NCAM Promotes Synaptogenesis in a Choice Situation. (A) A schematic diagram depicting normal synaptic coverage of NCAM-deficient (–/–) neurons, as compared to wild-type neurons (+/+) in homogenotypic cultures, and of reduced synaptic coverage in heterogenotypic co-cultures. (B) A hypothetical model according to which synaptic activity and activation of NMDA receptors (NMDAR) and voltage-dependent $Ca^{2+}$ channels (VDCC) during induction of LTP may lead to increases in NCAM expression in stimulated neurons or in neighborhood of stimulated synapses, which may promote synaptogenesis in these cells/subcellular domains. See Colorplate 5.

Does NCAM act as a ligand or a receptor? Transfection of NCAM-deficient neurons with either of three major NCAM isoforms, GPI-linked NCAM120 or transmembrane domain-containing NCAM140 or NCAM180, stimulated preferential synapse formation on all NCAM isoform-expressing neurons[40]. These experiments suggest that the extracellular domain of NCAM has synaptogenic activity. To investigate the involvement of PSA, cultures were treated with endoneuraminidase N (endo-N) which removes PSA linked to protein. Intriguingly, this treatment completely abolished preferential formation of synapses in NCAM-expressing cells. Enzymatic removal of heparan sulfates from cultured neurons, a mutation in the heparin-binding domain (HBD) of NCAM, and application of recombinant soluble extracellular domains of NCAM and PSA-NCAM similarly diminished synaptogenic activity of neuronally expressed PSA-NCAM, suggesting that interaction of NCAM with heparan sulfate proteoglycans mediates this activity. PSA-NCAM-driven synaptogenesis was also blocked by antagonists to fibroblast growth factor (FGF) receptor and the NMDA subtype of glutamate receptors, but not by blockers of non-NMDA glutamate receptors and voltage-dependent $Na^+$ channels. Enzymatic removal of PSA and heparan sulfates also suppressed the increase in the number of perforated spine synapses associated with NMDA receptor-dependent LTP in the CA1 region of organotypic hippocampal slice cultures[40]. Thus, neuronal PSA-NCAM in complex with heparan sulfate proteoglycans promotes synaptogenesis and activity-dependent remodeling of synapses (Figure 6.3).

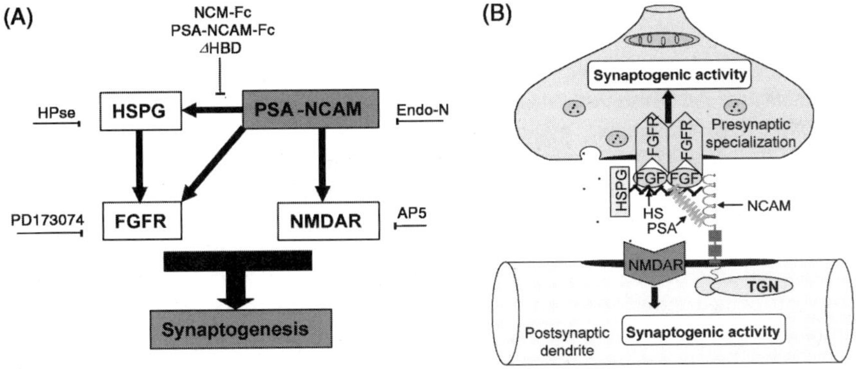


**Figure 6.3.** Signaling of PSA-NCAM via FGF and NMDA Receptors Promotes Synaptogenesis. (A) Summary of genetic and pharmacological experiments showing that blockade of FGF or NMDA receptors, removal of PSA by endo-N or heparan sulfates (HS) by heparinase (HPse), addition of soluble NCAM-Fc and PSA-NCAM-Fc, or deletion of heparin-binding domain (ΔHBD) inhibits the synaptogenic activity of NCAM. (B) A hypothetical scheme showing a complex of PSA-NCAM and HSPG that may function to enhance signaling via presynaptic FGF receptors. For more explanation, see text and ref. 40.

Which HSPGs cooperate with NCAM in hippocampal neurons is currently unknown. It is striking, however, that the presynaptically secreted heparan sulfate, agrin, induces postsynaptic differentiation at the NMJ (Chapters 1 and 11) and is involved in NCAM-mediated heterophilic adhesion[19,41]. Heparan sulfate proteoglycans of the syndecan and glypican families may also be candidates for mediating PSA-NCAM synaptogenic activity. Agrin binds to FGF-2, and syndecans and glypicans are known to promote FGF-2 triggered signaling. Several studies have shown that heparan sulfate chains serve to facilitate dimerization of FGF and FGF receptors (e.g., ref. 42), which activates the tyrosine kinase domain of these receptors through autophosphorylation. FGF receptors may also be activated by NCAM via direct interaction between these molecules[7]. In agreement with our present study, FGF-2 and a synthetic peptide, FGL, which corresponds to the site of binding between NCAM and FGF receptor 1, stimulated formation of synapses in hippocampal cultures[43]. Particularly pertinent to the many signaling mechanisms described for FGF receptor function is activation of neurite outgrowth through generation of polyunsaturated fatty acids (e.g., arachidonic acid) which activate voltage-gated $Ca^{2+}$ channels[44]. Restricted $Ca^{2+}$ influx at sites of FGF action may lead to phosphorylation-sensitive clustering of synaptic vesicles at these sites[45]. Arachidonic acid may also affect synaptogenesis via activation of protein kinase C, which has also been shown to stimulate synaptogenesis[46]. A recent study identifies FGF-22, FGF-7, and FGF-10, the closest relatives of FGF-22, to be potent inducers of presynaptic differentiation in mossy fibers[47]. In summary, these data suggest that heparin sulfate proteoglycans in association with PSA-NCAM may assist in activating axonally localized FGF receptors, thus stimulating differentiation of presynaptic specializations (Figure 6.3B). Such mechanisms may explain the higher number of presynaptic terminals contacting onto postsynaptic neurons expressing PSA-NCAM. It is, however, also conceivable that NCAM and FGF receptors may interact postsynaptically to stabilize presynaptic contacts via retrograde messengers. More experiments are necessary to verify these hypotheses and the role of NMDA receptors in NCAM-driven synaptogenesis.

### 4.3. Role of NCAM During Initial Stages of Mammalian Synaptogenesis

To evaluate the roles of NCAM in formation and stabilization of initial contacts, we performed time-lapse recordings in dissociated hippocampal cultures[48]. NCAM was visualized by indirect immunofluorescence, and TGN-derived organelles were loaded with a fluorescent dye, FM4-64. Within minutes of establishment of physical contacts between the growth cone and target neurite, NCAM immunoreactive clusters and associated organelles that had moved along the target neurite began to accumulate at sites of contact. The clusters and associated intracellular aggregates often passed several times through the actual site of contact until one or several clusters and associated organelles were "trapped" at the contact site and remained there until the end of the recordings (10–100 min) (Figure 6.4). Only occasionally did the growth cone contact the target neurite at the site of an NCAM immunoreactive cluster. In this case, the NCAM immunoreactive clusters and associated intracellular aggregates remained at the site of contact from the moment of its formation. These data suggest that the location of initial contact is not predetermined by NCAM expression, but NCAM may be quickly recruited – together with intracellular organelles – to initial contacts in a quantal manner.

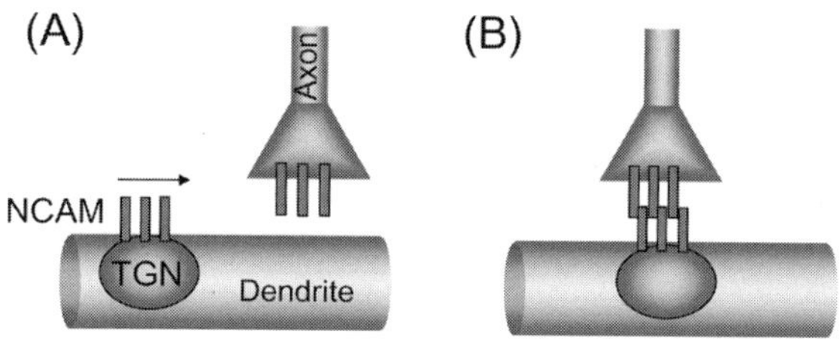


**Figure 6.4.** Synaptic Trapping of TGN-derived Intracellular Organelles by NCAM. Location of pre- and postsynaptic structures, NCAM clusters, and associated organelles are shown before (A) and a few minutes after (B) initial contact formation.

To investigate whether pre- or postsynaptic NCAM plays a role in the stabilization of organelles at sites of contact, we used the heterogenotypic co-culture model described above[35]. In co-cultures of wild-type and NCAM-deficient neurons maintained for 4 days *in vitro*, γ-adaptin immunoreactive TGN organelles were also significantly more often associated with contact sites between heterogenotypic axons and dendrites when compared to NCAM negative contacts, showing that heterophilic interactions of NCAM become apparent in a choice situation, i.e., in the presence of NCAM-expressing and NCAM-deficient neurons. TGN organelles were even more often associated with sites of contacts formed by NCAM positive axons and dendrites, implicating a possible contribution of homophilic *trans*-interactions. Time-lapse recordings showed that contacts between NCAM-deficient neurons were more often disrupted due to retraction of neurites when compared to wild-type neurons. Moreover, organelles moved away from contact points approximately four times more often in NCAM-deficient neurons compared to wild-type neurons. Thus, these data demonstrate that NCAM is important for stabilization of initial contacts and recruitment of intracellular organelles to these contact sites.

## 5. NCAM AND SYNAPTIC PLASTICITY

### 5.1. Hippocampal Synaptic Plasticity

The first direct evidence that NCAM may play a role in synaptic plasticity was provided in 1994 by a seminal study that showed that perturbation of NCAM function significantly reduced LTP in the CA1 area of the hippocampus[49]; (see Chapter 25 for introduction in synaptic plasticity). Polyclonal antibodies against NCAM, soluble oligomannosides that block interaction of NCAM with oligomannosidic carbohydrates carried by L1, and synthetic peptides from the fourth Ig-like domain of NCAM, which mediates interaction with L1, were used in these experiments. Further studies using constitutive NCAM-knockout and conditionally NCAM-deficient mice, in which the NCAM gene is ablated in neurons after cessation of major developmental events, showed impairment of CA1 LTP in both mutants, thus supporting the view that NCAM plays acute functional role in synaptic plasticity in the CA1 region[20,36,50]. Additionally, LTD in the CA1 was impaired in conditional NCAM knockout mice[50]. In the CA3 region, constitutive NCAM but not conditional NCAM-deficient mice were found to have abnormalities in lamination of mossy fiber projections and to be impaired in mossy fiber LTP, suggesting that NCAM is required for proper development and function of mossy fiber–CA3 synapses[50–52].

Since NCAM is the only carrier of PSA in mammalian brains, significant efforts were made to dissect the functions of PSA and the NCAM glycoprotein backbone. Enzymatic removal of PSA by endo-N inhibited LTP and LTP-associated formation of perforated synapses, and LTD in CA1[36,40,53]. Experiments using mice deficient in one of two polysialyltransferases ST8SiaII (STX) or ST8SiaIV (PST), enzymes required for polysialylation of NCAM in immature and mature cells, respectively, provided the first genetic evidence for the importance of this carbohydrate in synaptic plasticity in the CA1[54,55]. No involvement of these enzymes in mossy fiber LTP in the CA3 region was revealed, despite abnormal lamination of mossy fiber projections in STX-deficient mutants[55]. The first in vivo study[56] demonstrated that mice deficient in either PST or STX also exhibited normal LTP in the dentate gyrus. However, this form of LTP was impaired in NCAM-/- mice[56]. Thus, NCAM is important for synaptic plasticity in the CA1, CA3 and dentate gyrus (Table 1).

Despite this intensive research, the mechanisms mediating NCAM action in synaptic plasticity are largely unknown. Since peptides blocking interaction of proteins with the fourth immunoglobulin-like domain of NCAM reduced LTP when applied before induction of LTP but not afterward[57], a role of NCAM in LTP induction was suggested. Since impairment of LTP in NCAM-deficient mice could be rescued by elevation of extracellular $Ca^{2+}$ concentration, one may speculate that NCAM influences $Ca^{2+}$ entry via NMDA receptors[50]. Whether this is due to a reduced number of synaptic NMDA receptors or impaired function of these receptors remains to be determined. By analogy with the adhesion molecule neuroligin, which is connected indirectly via PSD-95 to NMDARs (Chapters 4 and 7), an indirect connection between the NCAM and the NMDARs seems possible. This is supported by the similar central location of NCAM180 and the NR2A within the postsynaptic density in untreated animals, and a similar redistribution of these molecules to the edges of postsynaptic density in animals after induction of LTP[58]. One of the scaffolding molecules cross-linking NCAM180 and NMDAR

Table 1. Effects of NCAM, PSA and Polysialyltransferases on Different Forms of Hippocampal Synaptic Plasticity.

| Condition | CA1 LTP | CA1 LTD | CA3 LTP | DG LTP | References |
|---|---|---|---|---|---|
| NCAM-/- | ↓ | n.d. | ↓ | ↓ | (36,52,56) |
| NCAMff+ | ↓ | ↓ | = | n.d. | (50) |
| Endo-N | ↓ | ↓ | = | n.d. | (36,20) |
| PST-/- | ↓ | ↓ | = | = | (54,56) |
| STX-/- | = | n.d. | = | = | (55,56) |

*Abbreviations:* DG, dentate gyrus; NCAMff+, conditional NCAM-deficient mice; ↓, impaired; =, normal; n.d., not determined

could be spectrin, which has been reported to bind to these two molecules and is highly expressed in postsynaptic densities[27,48,59]. Because PSA may directly potentiate opening of AMPA-type glutamate receptors[22], another interesting possibility would be that PSA influences activity of NMDA receptors via a direct interaction with the extracellular domain of receptors.

The results showing the role of NCAM in hippocampal plasticity are nicely complemented by studies of hippocampus-dependent forms of learning and memory. Perturbation of NCAM function with NCAM antibodies caused amnesia in a passive avoidance task[60] and defects in spatial learning in rats[61]. Constitutively and conditionally NCAM-deficient mice show a deficit in spatial learning in the Morris water maze paradigm[50,51]. Mutant mice either lacking NCAM, or transgenically designed to overproduce soluble extracellular fragments of NCAM, or deficient in the polysialyltransferase ST8SiaII, all show impaired contextual and cued fear memories[55,62,63]. Consistent with this, an enzymatic removal of PSA and injection of a synthetic dendropeptide C3d that binds to the first Ig domain of NCAM show impaired contextual memory in a fear-conditioning paradigm[53,64]. In contrast, the NCAM-derived peptide, FGL, corresponding to the binding site of NCAM to FGF receptors, improves long-lasting retrieval of fear memory[43].

## 5.2. Structural Plasticity in the Hypothalamo-Neurohypophysial System

The role of PSA has also been intensively investigated in the hypothalamo-neurohypophysial system. Oxytocin-secreting neurons of this system undergo reversible morphological changes during physiological stimulation. In the hypothalamus, such structural plasticity is represented by modifications in the size and shape of their somata and dendrites, as well as the extent to which their surfaces are covered by glia. Due to retraction of astrocytic processes, adjacent somata and dendrites of oxytocinergic neurons become extensively juxtaposed and receive an increased number of synapses. In the neurohypophysis, there is a parallel reduction in glial coverage of their axons together with retraction of glial processes from the perivascular basal lamina and an increase in the number and size of axon terminals. These changes occur rapidly, within a few hours, and are reversible upon termination of physiological stimulation[65].

Evidently this plasticity requires dynamic cell interactions that must bring into play cell surface and extracellular matrix molecules such as those intervening in developing neuronal systems. Indeed, neurons and glial cells of the adult hypothalamo-neurohypophysial system continue to express such molecules, including PSA-NCAM and the extracellular matrix glycoprotein tenascin-C. Furthermore, removal of PSA with endo-N prevents withdrawal of astrocytic processes and increases the number of synaptic contacts normally induced by lactation and dehydration[66]. Endo-N also prevents a significant reduction in the number of GABAergic axo-somatic synapses normally induced by estradiol in the adult female arcuate nucleus[67]. *In vitro* experiments involving stimulation of the neurohypophysis by exposure to hyperosmotic conditions or a beta-adrenergic agonist also revealed a suppression of axonal and glial changes by endo-N[68]. Thus, PSA promotes structural plasticity in the hypothalamo-neurohypophysial system. Not only PSA is important for this process but the rest of NCAM glycoprotein seems to be involved as well, since the supraoptic nucleus displays diminished astrocytic coverage and increased synaptic input in NCAM-deficient mice. This correlated with enhanced plasma and intranuclear concentrations of oxytocin and reduced anxiety-related behavior, showing hyperactivity of the oxytocin system in NCAM-deficient mice[69].

## 6. NCAM AND TRANSMITTER RELEASE

In the hippocampus, NCAM-deficient mice show normal levels of synaptic transmission and minor abnormalities in the paired-pulse facilitation, detectible only at 1.5 mM extracellular $Ca^{2+}$ but not at higher $Ca^{2+}$ concentrations[50]. However, NCAM–/– NMJs appears to be very interesting in this respect, since they lack paired-pulse facilitation and fail to maintain transmitter output with repetitive stimuli[70]. Experiments using electrical stimulation-induced loading of synaptic terminal terminals with the styryl fluorescent dye FM1-43 revealed that terminals can be loaded and destained more quickly in NCAM–/– mice than in wild-type controls[71]. Furthermore, clusters of dye-loaded vesicles were observed not only at the end plate but also at the preterminal part of axon, as was previously found in immature axons. As in the case of quantal release from immature axons in wild-type animals, which has been shown to be blocked by brefeldin A, preterminal loading of nerves in mature NCAM–/– junctions could also be blocked by this compound. Furthermore, application of brefeldin A strongly reduced cyclical periods of total transmission failures in NCAM–/– mice. This phenotype could be mimicked in NCAM+/+ mice by inhibitors of myosin light chain kinase (MLCK).

Examination of mice deficient in the NCAM180 revealed that the immature vesicle cycling–transmitter release system was downregulated at NMJs in the absence of NCAM 180 as in wild-type mice, but vesicle cycling was not confined to presynaptic active zones apposed to muscle[72]. NCAM 180 isoform-deficient synapses were also unable to sustain transmitter output with repetitive stimuli like NCAM–/– mice, despite normal $Ca^{2+}$ influx into presynaptic terminals. Similar to mice lacking all NCAM isoforms, the expression of several presynaptic molecules was reduced. Among them were the calcium channel subunit $Ca_v2.1$, syntaxin, and Rab 3-interacting molecule (RIM), a molecule involved in the modulation of synaptic transmission.

Recent evidence supports the pharmacological data on involvement of MLCK in NCAM-mediated mechanisms. A highly conserved C-terminal (KENESKA) domain on NCAM was found to be required to maintain effective transmission via a pathway involving MLCK and probably myosin light chain (MLC) and myosin II[73]. By introducing peptides into adult NMJs, the hypothesized role of proteins in this pathway was tested by competitive disruption of protein–protein interactions. The effects of the KENESKA and other peptides on MLCK and MLC activation and on transmission failures in both wild-type and NCAM180-deficient junctions supported the role for NCAM in this pathway. Furthermore, serine phosphorylation of the KENESKA peptide appeared to be critical for this NCAM function. Thus, this pathway is required to replenish synaptic vesicles during high levels of exocytosis by facilitating myosin-driven delivery of synaptic vesicles to active zones for subsequent exocytosis. Recent data also demonstrate that NCAM–/– mice exhibit deficits in catecholamine granule trafficking between the readily releasable pool and the highly release-competent immediately releasable pool[74]. In summary, NCAM appears to play a fundamental role in the transmitter release mechanism found in neuroendocrine cells, at the NMJ and, possibly in central synapse active zones.

## 7. CONCLUSIONS AND FUTURE DIRECTIONS

The data discussed in this chapter show that NCAM has "many faces" and is involved in many synaptic functions. Numerous *in vitro*, *in situ*, and *in vivo* studies have converged to demonstrate that 1) spatial and temporal patterns of NCAM and PSA expression is regulated by neuronal activity and animal experience; 2) learning and memory are affected by manipulation of NCAM and PSA; and 3) NCAM and PSA promote synaptogenesis and synaptic plasticity. The underlying mechanisms appear to be at least partially relevant to mechanisms underlying activity of NCAM during neurite outgrowth, which involve interaction with spectrin and signaling via the FGF receptor and fyn kinase. Currently, we know that spectrin is involved in stabilization of intracellular organelles during early synaptogenesis and that FGF receptors are required for synaptogenic activity of NCAM at later stages. However, the roles of FGF receptors, fyn kinase, and spectrin in NCAM-mediated synaptic plasticity have not yet been verified. Apart from these mechanisms, reported interactions between NCAM and glutamate receptors[22] and a rescue of LTP deficits in NCAM-deficient mice by exogenous BDNF application[20], suggest that new players may be specifically involved in the regulation of synaptic functions by NCAM, as compared to neuritogenesis.

Since PSA is involved in neuroplasticity, it is particularly interesting that there is a downregulation of polysialylated NCAM in the hippocampi of patients with schizophrenia[75] and that a polymorphism in the promoter region of polysialyltransferase is associated with schizophrenia[76]. Additionally, soluble forms of NCAM are elevated in cerebrospinal fluid and in the brain of schizophrenic patients, whereas expression of the membrane-associated NCAM isoforms is unaffected[77]. These findings have stimulated an interest in soluble NCAM and a transgenic mouse (NCAM-EC) was recently generated, which overexpresses soluble NCAM in the brain[63]. NCAM-EC transgenic mice exhibit a striking reduction in synaptic puncta of GABAergic interneurons in several brain regions, as shown by decreased immunolabeling of GABAergic terminals. In addition, there is a reduction in excitatory synapses, as revealed by synaptophysin staining and apical dendritic spine density of cortical pyramidal cells, and expression of numerous behavioral abnormalities relevant to schizophrenia. Thus, understanding of mechanisms by which NCAM regulates synaptogenesis may guide us in finding effective treatments to compensate for abnormalities in synaptic connectivity in diseased brains.*

## 8. REFERENCES

1. Jorgensen, O.S., and Bock, E. (1974) *J Neurochem* **23,** 879–880.
2. Thiery, J.P., Brackenbury, R., Rutishauser, U., and Edelman, G.M. (1977) *J Biol Chem* **252,** 6841–6845.
3. Brummendorf, T., and Rathjen, F.G. (1994) *Protein Profile* **1,** 951–1058.
4. Krog, L., Olsen, M., Dalseg, A.M., Roth, J., and Bock, E. (1992) *J Neurochem* **59,** 838–847.
5. Chothia, C., and Jones, E.Y. (1997) *Annu Rev Biochem* **66,** 823–862.
6. Jensen, P.H., Soroka, V., Thomsen, N.K., Ralets, I., Berezin, V., Bock, E., and Poulsen, F.M. (1999) *Nat Struct Biol* **6,** 486–493.
7. Kiselyov, V.V., Soroka, V., Berezin, V., and Bock, E. (2005) *J Neurochem* **94,** 1169–1179.
8. Johnson, C.P., Fujimoto, I., Rutishauser, U., and Leckband, D.E. (2005) *J Biol Chem* **280,** 137–145.
9. Wieland, J.A., Gewirth, A.A., and Leckband, D.E. (2005) *J Biol Chem* **280,** 41037–41046.
10. Kleene, R., and Schachner, M. (2004) *Nat Rev Neurosci* **5,** 195–208.
11. Kadmon, G., Kowitz, A., Altevogt, P., and Schachner, M. (1990) *J Cell Biol* **110,** 193–208.
12. Cole, G.J., and Akeson, R. (1989) *Neuron* **2,** 1157–1165.
13. Kulahin, N., Rudenko, O., Kiselyov, V., Poulsen, F.M., Berezin, V., and Bock, E. (2005) *J Neurochem* **95,** 46–55.
14. Margolis, R.K., Rauch, U., Maurel, P., and Margolis, R.U. (1996) *Perspect Dev Neurobiol* **3,** 273–290.
15. Saffell, J.L., Williams, E.J., Mason, I.J., Walsh, F.S., and Doherty, P. (1997) *Neuron* **18,** 231–242.
16. Paratcha, G., Ledda, F., and Ibanez, C.F. (2003) *Cell* **113,** 867–879.
17. Fujimoto, I., Bruses, J.L., and Rutishauser, U. (2001) *J Biol Chem* **276,** 31745–31751.
18. Johnson, C.P., Fujimoto, I., Rutishauser, U., and Leckband, D.E. (2005) *J Biol Chem* **280,** 137–145.
19. Storms, S.D., and Rutishauser, U. (1998) *J Biol Chem* **273,** 27124–27129.
20. Muller, D., Djebbara-Hannas, Z., Jourdain, P., Vutskits, L., Durbec, P., Rougon, G., and Kiss, J.Z. (2000) *Proc Natl Acad Sci U S A* **97,** 4315–4320.
21. Zhang, H., Vutskits, L., Calaora, V., Durbec, P., and Kiss, J.Z. (2004) *J Cell Sci* **117,** 93–103.
22. Vaithianathan, T., Matthias, K., Bahr, B., Schachner, M., Suppiramaniam, V., Dityatev, A., and Steinhauser, C. (2004) *J Biol Chem* **279,** 47975–47984.
23. Schuch, U., Lohse, M.J., and Schachner, M. (1989) *Neuron* **3,** 13–20.
24. Beggs, H.E., Soriano, P., and Maness, P.F. (1994) *J Cell Biol* **127,** 825–833.
25. Beggs, H.E., Baragona, S.C., Hemperly, J.J., and Maness, P.F. (1997) *J Biol Chem* **272,** 8310–8319.
26. Bodrikov, V., Leshchyns'ka, I., Sytnyk, V., Overvoorde, J., den Hertog, J., and Schachner, M. (2005)
*J Cell Biol* **168,** 127–139.
27. Pollerberg, G.E., Schachner, M., and Davoust, J. (1986) *Nature* **324,** 462–465.
28. Leshchyns'ka, I., Sytnyk, V., Morrow, J.S., and Schachner, M. (2003) *J Cell Biol* **161,** 625–639.
29. Niethammer, P., Delling, M., Sytnyk, V., Dityatev, A., Fukami, K., and Schachner, M. (2002) *J Cell Biol* **157,** 521–532.
30. Hinsby, A.M., Berezin, V., and Bock, E. (2004) *Front Biosci* **9,** 2227–2244.
31. Schuster, C.M., Davis, G.W., Fetter, R.D., and Goodman, C.S. (1996) *Neuron* **17,** 641–654.
32. Davis, G.W., Schuster, C.M., and Goodman, C.S. (1996) *Neuron* **17,** 669–679.
33. Bailey, C.H., Chen, M., Keller, F., and Kandel, E.R. (1992) *Science* **256,** 645–649.
34. Bailey, C.H., Kaang, B.K., Chen, M., Martin, K.C., Lim, C. S., Casadio, A., and Kandel, E.R. (1997) *Neuron* **18,** 913–924.
35. Dityatev, A., Dityateva, G., and Schachner, M. (2000) *Neuron* **26,** 207–217.
36. Muller, D., Wang, C., Skibo, G., Toni, N., Cremer, H., Calaora, V., Rougon, G., and Kiss, J.Z. (1996) *Neuron* **17,** 413–422.

* A.D. was supported by research grants from the DFG.

37. Holst, B.D., Vanderklish, P.W., Krushel, L.A., Zhou, W., Langdon, R.B., McWhirter, J.R., Edelman, G.M., and Crossin, K.L. (1998) *Proc Natl Acad Sci U S A* **95,** 2597–2602.
38. Sandi, C., Merino, J.J., Cordero, M.I., Kruyt, N.D., Murphy, K.J., and Regan, C.M. (2003) *Biol Psychiatry* **54,** 599–607.
39. Bouzioukh, F., Tell, F., Jean, A., and Rougon, G. (2001) *J Neurosci* **21,** 4721–4730.
40. Dityatev, A., Dityateva, G., Sytnyk, V., Delling, M., Toni, N., Nikonenko, I., Muller, D., and Schachner, M. (2004) *J Neurosci* **24,** 9372–9382.
41. Storms, S.D., Kim, A.C., Tran, B.H., Cole, G.J., and Murray, B.A. (1996) *Cell Adhes Commun* **3,** 497–509.
42. Loo, B.M. and Salmivirta, M. (2002) *J Biol Chem* **277,** 32616–32623.
43. Cambon, K., Hansen, S.M., Venero, C., Herrero, A.I., Skibo, G., Berezin, V., Bock, E., and Sandi, C. (2004) *J Neurosci* **24,** 4197–4204.
44. Walsh, F.S., and Doherty, P. (1997) *Annu Rev Cell Dev Biol* **13,** 425–456.
45. Dai, Z., and Peng, H.B. (1995) *J Neurosci* **15,** 5466–5475.
46. Hama, H., Hara, C., Yamaguchi, K., and Miyawaki, A. (2004) *Neuron* **41,** 405–415.
47. Umemori, H., Linhoff, M.W., Ornitz, D.M., and Sanes, J.R. (2004) *Cell* **118,** 257–270.
48. Sytnyk, V., Leshchyns'ka, I., Delling, M., Dityateva, G., Dityatev, A., and Schachner, M. (2002) *J Cell Biol* **159,** 649–661.
49. Lüthi, A., Laurent, J.P., Figurov, A., Muller, D., and Schachner, M. (1994) *Nature* **372,** 777–779.
50. Bukalo, O., Fentrop, N., Lee, A.Y., Salmen, B., Law, J.W., Wotjak, C.T., Schweizer, M., Dityatev, A., and Schachner, M. (2004) *J Neurosci* **24,** 1565–1577.
51. Cremer, H., Lange, R., Christoph, A., Plomann, M., Vopper, G., Roes, J., Brown, R., Baldwin, S., Kraemer, P., Scheff, S. et al. (1994) *Nature* **367,** 455–459.
52. Cremer, H., Chazal, G., Carleton, A., Goridis, C., Vincent, J.D., and Lledo, P.M. (1998) *Proc Natl Acad Sci U S A* **95,** 13242–13247.
53. Becker, C.G., Artola, A., Gerardy-Schahn, R., Becker, T., Welzl, H., and Schachner, M. (1996) *J Neurosci Res* **45,** 143–152.
54. Eckhardt, M., Bukalo, O., Chazal, G., Wang, L., Goridis, C., Schachner, M., Gerardy-Schahn, R., Cremer, H., and Dityatev, A. (2000) *J Neurosci* **20,** 5234–5244.
55. Angata, K., Long, J.M., Bukalo, O., Lee, W., Dityatev, A., Wynshaw-Boris, A., Schachner, M., Fukuda, M., and Marth, J.D. (2004) *J Biol Chem* **279,** 32603–32613.
56. Stoenica, L., Senkov, O., Weinhold, B., Gerady-Schahn, R., Schachner, M., and Dityatev A. (2006) *Eur J Neurosci* **23,** 2255–2264.
57. Staubli, U., Chun, D., and Lynch, G. (1998) *J Neurosci* **18,** 3460–3469.
58. Fux, C.M., Krug, M., Dityatev, A., Schuster, T., and Schachner, M. (2003) *Mol Cell Neurosci* **24,** 939–950.
59. Wechsler, A., and Teichberg, V.I. (1998) *EMBO J* **17,** 3931–3939.
60. Doyle, E., Nolan, P.M., Bell, R., and Regan, C.M. (1992) *J Neurochem* **59,** 1570–1573.
61. Arami, S., Jucker, M., Schachner, M., and Welzl, H. (1996) *Behav Brain Res* **81,** 81–87.
62. Stork, O., Welzl, H., Wolfer, D., Schuster, T., Mantei, N., Stork, S., Hoyer, D., Lipp, H., Obata, K., and Schachner, M. (2000) *Eur J Neurosci* **12,** 3291–3306.
63. Pillai-Nair, N., Panicker, A.K., Rodriguiz, R.M., Gilmore, K.L., Demyanenko, G.P., Huang, J.Z., Wetsel, W.C., and Maness, P.F. (2005) *J Neurosci* **25,** 4659–4671.
64. Cambon, K., Venero, C., Berezin, V., Bock, E., and Sandi, C. (2003) *Neuroscience* **122,** 183–191.
65. Theodosis, D.T., Piet, R., Poulain, D.A., and Oliet, S.H. (2004) *Neurochem Int* **45,** 491–501.
66. Theodosis, D.T., Bonhomme, R., Vitiello, S., Rougon, G., and Poulain, D.A. (1999) *J Neurosci* **19,** 10228–10236.
67. Hoyk, Z., Parducz, A., and Theodosis, D.T. (2001) *Eur J Neurosci* **13,** 649–656.
68. Monlezun, S., Ouali, S., Poulain, D.A., and Theodosis, D.T. (2005) *Mol Cell Neurosci* **29,** 516–524.
69. Theodosis, D.T., Schachner, M., and Neumann, I.D. (2004) *Eur J Neurosci* **20,** 3270–3280.
70. Rafuse, V.F., Polo-Parada, L., and Landmesser, L.T. (2000) *J Neurosci* **20,** 6529–6539.
71. Polo-Parada, L., Bose, C.M., and Landmesser, L.T. (2001) *Neuron* **32,** 815–828.
72. Polo-Parada, L., Bose, C.M., Plattner, F., and Landmesser, L.T. (2004) *J Neurosci* **24,** 1852–1864.
73. Polo-Parada, L., Plattner, F., Bose, C., and Landmesser, L.T. (2005) *Neuron* **46,** 917–931.
74. Chan, S.A., Polo-Parada, L., Landmesser, L.T., and Smith, C. (2005) *J Neurophysiol* **94,** 1037–1047.
75. Barbeau, D., Liang, J.J., Robitalille, Y., Quirion, R., and Srivastava, L.K. (1995) *Proc Natl Acad Sci U S A* **92,** 2785–2789.
76. Arai, M., Yamada, K., Toyota, T., Obata, N., Haga, S., Yoshida, Y., Nakamura, K., Minabe, Y., Ujike, H., Sora, I., Ikeda, K., Mori, N., Yoshikawa, T., and Itokawa, M. (2005) *Biol Psychiatry.*
77. Vawter, M.P., Cannon-Spoor, H.E., Hemperly, J.J., Hyde, T.M., VanderPutten, D.M., Kleinman, J.E., and Freed, W.J. (1998) *Exp Neurol* **149,** 424–432.

# 7

# ROLE OF NEUROLIGIN BINDING TO NEUREXINS IN SYNAPTIC ORGANIZATION

Richard Fairless, Carsten Reissner, and Markus Missler*

## 1. SUMMARY

The complex formed by the synaptic cell adhesion molecules neuroligin and β-neurexin has been implicated in synaptogenesis due to the molecular asymmetry of their heterophilic binding, reflecting the asymmetric nature of the synapse. During the past 5 years, their role in synapse formation has been explored *in vitro*, yielding exciting, though sometimes conflicting results. In this chapter, we focus on the biochemical and functional aspects of the neuroligin/β-neurexin complex, with an emphasis on the recent contributions coming from various cell culture approaches. In addition, we point out important unresolved issues with respect to their binding properties, their proposed function at excitatory versus inhibitory synapses, and their putative involvement in psychiatric diseases such as autism spectrum disorders.

## 2. INTRODUCTION

The first demonstration that neuroligins could bind to neurexins came in 1995 with their discovery by affinity chromatography on immobilized β-neurexin[1]. It was shown that all three rat neuroligins interacted with the extracellular domain of neurexin 1β in a calcium-dependent manner, and that this interaction was also dependent upon the splice variation of β-neurexin. Neurexins 1β, 2β, and 3β were all able to interact with neuroligins in biochemical and cell culture-binding assays, but only when β-neurexins were lacking an insert in splice site #4 (1,2; but see ref. 3 for binding of a particular splice variant of neuroligin to all neurexins). Since its discovery, the neuroligin/β-neurexin complex has been suggested to play a role in

*Center for Physiology, Georg-August University Göttingen, Humboldtallee 23, 37073 Göttingen, Germany; mmissle1@gwdg.de

synapse formation and/or maturation. This was subsequently supported by the complex' ability to mediate cell–cell adhesion[4], the postsynaptic location[5] and interactions of neuroligin[6], and by high expression levels of neuroligin and β-neurexin mRNAs during rapid synaptogenesis in the brain[6].

## 3. BIOCHEMICAL ASPECTS OF THE INTERACTION BETWEEN NEUREXIN AND NEUROLIGIN

The binding between the extracellular domains of neuroligin and β-neurexin is characterized by its dependence upon alternative splicing of both molecules, the glycosylation and dimerization of neuroligin, and the presence of $Ca^{2+}$ ions. Intracellularly, both transmembrane proteins interact with different binding partners in the pre- and postsynaptic compartments, respectively.

### 3.1. Isoform-Dependent Binding

Neurexins form a large family of cell-surface proteins[7,8]. Each of the three vertebrate neurexin genes has two alternative promoters giving rise to a long mRNA transcript encoding α-neurexins, and a small mRNA transcript encoding β-neurexins. Both α- and β-neurexins consist of a conserved, intracellular C-terminus with a type II PDZ-recognition motif, a transmembrane region, a short serine-/threonine-rich sequence, and varying numbers of laminin-like (LNS) domains with interspersed EGF-like sequences. α-Neurexins contain six LNS repeats, whereas β-neurexins only have one[9,10]. There are at least six principal neurexin isoforms but due to extensive alternative splicing at five (α-neurexins) or two (β-neurexins) conserved splice sites (referred to as #1–5), over 1,000 variants in total may result[7,8]. α- and β-neurexins are neuron specific, but different neurons may express different combinations of neurexins[8]. Although they share most of their sequences, α-neurexins play an essential role in neurotransmission that cannot be replaced by β-neurexins[11,12], and partly different extracellular binding partners exist for α-neurexins[13–15] and β-neurexins[13].

Neuroligins, in turn, consist of a short intracellular sequence ending in a type I PDZ-recognition motif, a transmembrane region, and a long extracellular sequence, including an α/β hydrolase domain. This latter domain is homologous to members of the esterase family (e.g., acetylcholine esterase, AChE), although it is catalytically inactive due to the lack of an active site serine[1]. Like the neurexins, neuroligin mRNA is susceptible to splicing (at two positions referred to as A and B), although it did not appear to affect binding to neurexins until a recent study identified a splice variant (lacking the insert in B) that binds indiscriminately to all β-neurexins and presumably all α-neurexins[3]. However, the abundance and distribution in brain of the newly investigated neuroligin isoform have yet to be explored before this surprise finding can be fully appreciated.

### 3.2. Glycosylation and Dimerization

Sequence analysis of both β-neurexin and neuroligin cDNA has highlighted glycosylation sites, including an O-linked carbohydrate-rich domain just N-terminal to the transmembrane regions of both, as well as multiple scattered consensus sequences for N-linked glycosylation[1,9]. Experimentally, neuroligin obtained from the cell lysate (preglycosylated form) had a size of 98 kDa, whereas protein obtained from the cell surface, had a larger apparent molecular weight of 126 kDa due to glycosylation[16]. Likewise, glycohydrolase treatment of β-neurexins revealed that they are extensively O-glycosylated, with only some N-glycosylation[9]. The function of their glycosylation is not fully understood yet, but deglycosylation of neuroligin increased its affinity for β-neurexin as measured by surface plasmon resonance, whereas deglycosylation of β-neurexin had no effect[16].

Sedimentative equilibrium studies and analysis of the hydrodynamic properties of neuroligin by size exclusion chromatography have suggested that neuroligin forms dimers in solution[16]. This was supported by chemical cross-linking of purified recombinant neuroligin, resulting in covalently linked dimers and tetramers[17]. Further evidence for neuroligin dimerization included the presence of higher molecular weight species of neuroligin 1 following native gel electrophoresis. In addition, dimerization appears to be functionally necessary[17].

### 3.3. $Ca^{2+}$-Binding Sites

From the initial purification of neuroligins by affinity chromatography using the extracellular domain of β-neurexin, it has been noted that $Ca^{2+}$ is a necessary component for adhesion between β-neurexin and neuroligin[1]. $Ca^{45}$ overlay blotting revealed that neuroligin is capable of binding $Ca^{2+}$, but β-neurexin was not[16]. In addition, circular dichroism techniques demonstrated that incubation with $Ca^{2+}$ did not cause any observable structural changes in β-neurexin, nor did it affect β-neurexin stability as assessed by temperature-dependent denaturation[16]. It appeared from these studies that neuroligin is the $Ca^{2+}$ binding partner, supported by the identification of a degenerate EF-hand motif in neuroligins[18].

We have summarized the available biochemical evidence in a structural model of the neuroligin/β-neurexin complex (Figure 7.1; Colorplate 6). Based on a crystal structure for β-neurexin[19], and on homology modeling of neuroligin, this model sufficiently explains the steric hindrance between inserts in splice site #4 of neurexins and splice site B of neuroligins. In addition, the position of dimerization and glycosylation domains of neuroligin is consistent with experimental data. However, the exact whereabouts of the putative $Ca^{2+}$-binding sites remain obscure because the positions proposed thus far appear very distant from the contact interface of both proteins[18], and AChE folds may be too rigid for a calcium-induced conformational change.

### 3.4. Intracellular Binding Partners

In addition to the extracellular interaction between β-neurexin and neuroligin, their respective intracellular binding partners have also been investigated (see Figure 7.2 and Colorplate 7 for an overview model). All neuroligin isoforms are able to bind to a common postsynaptic density protein, PSD-95[6]. PSD-95 is a post-synaptic scaffolding protein which employs PDZ domains to scaffold proteins together synaptic sites[20,21]. Using PSD-95 and neuroligin mutants it was established that the C-terminus of neuroligin binds to the third PDZ domain of PSD-95[6]. The first two PDZ domains mediate PSD-95 interactions with the NR2 subunit of the *N*-methyl D-aspartate (NMDA) glutamate receptor (NMDAR)[20,21], thus demonstrating the potential for PSD-95 to couple neuroligin to other postsynaptic components. In addition, PSD-95 has been shown to be responsible for the recruitment of α-amino-3-hydroxy-5-methylisoxazole-4-propionate (AMPA) receptor (AMPAR) subunits[22], despite interacting only indirectly via the protein Stargazin[23]. β-Neurexin, in turn, has been shown to bind to CASK which is enriched in brain at synaptic plasma membranes[24]. CASK, like PSD-95, is a member of the membrane-associated guanylate kinase (MAGUK) superfamily. The guanylate kinase-like motif lacks catalytic activity, but along with the PDZ and SH3 domains, may mediate protein–protein interactions. The type II PDZ domain of CASK has been shown biochemically to mediate binding to the C-terminus of neurexins[24], and pulls down reliably all α- and β-neurexins. Intracellularly, CASK forms a complex with both Mint-1 and Veli[26–28] which has been proposed to link neurexin-mediated cell adhesion events to components of the exocytosis machinery and/or to synaptic vesicle trafficking. Along this line, other CASK-binding partners include protein 4.1[29], calmodulin[24], rabphilin 3a,[30], and calcium channels[31]. In addition, CASK has even been proposed as a transcriptional regulator[32]. Given the promiscuous nature of the PDZ domain interactions and conflicting published results with respect to binding partners and localization[27,33], more work will be needed to establish the physiological significance of CASK for the neuroligin/β-neurexin complex.

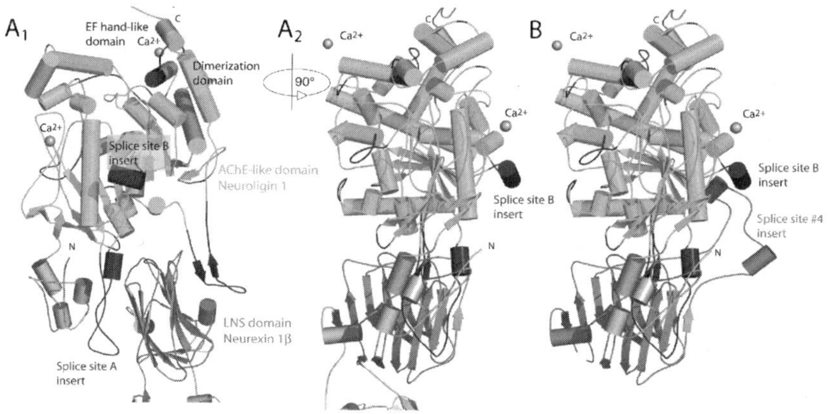


**Figure 7.1.** Model Structure of the Extracellular Domains of Neuroligin 1 and Neurexin 1β. An interaction in a head-to-head *trans*-complex is proposed because it best fits the available biochemical data. The AChE-like domain of neuroligin (green) interacts with the laminin (LNS)-like domain of neurexin (yellow) when the latter contains no insert in splice site #4 (panels $A_1$ and $A_2$, the latter showing the structure at a different angle as indicated). Several loops (red) in neurologin 1 are much longer than in other proteins with an AChE-like fold, including insertions in the two splice sites A and B. This interaction is severely hindered when β-neurexin contains an insert at splice site #4 (panel B), presumably by sterical interfering with the position of the insert B in neuroligin. This model is supported by a recent finding that lack of insert B allows binding to all neurexins irrespective of their splice combination[3]. While this interaction should be less efficient when neuroligin is glycosylated at the splice site insert B, their $Ca^{2+}$-dependence is more difficult to explain based on this model structure: the proposed major calcium binding site at a degenerate EF-hand-like motif is located at the C-terminus, and a second potential site is close to splice site B (best seen in panel $A_2$). The structures have been built using coordinates from protein data bank (PDB, www.rcsb.org) entries 1c4r for neurexin, and 1mah & 1fss for neuroligin. The structures of the N-terminal sequences as well as those regions linking the domains have been predicted using software programs SPDBViewer and Threader, and the web service Hmmstr/Rosetta. See Colorplate 6.

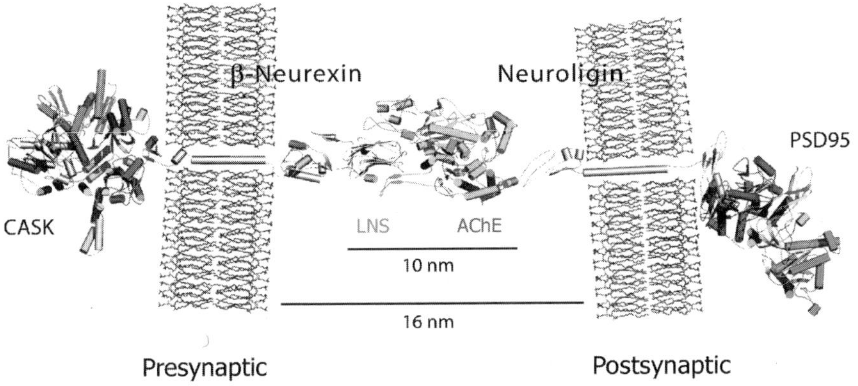


**Figure 7.2.** Model Structure of the Neuroligin/β-Neurexin *Trans*-Synaptic Complex Together with Their Putative Intracellular Binding Partners PSD-95 and CASK. The size of the complex formed by the laminin (LNS)-like domain (yellow) of β-neurexin with the AChE-like domain (green) of neuroligin is about 10 nm (compare also Figure 7.1) and for itself not sufficient to span the synaptic cleft. Each of these domains is linked to the adjacent membrane by a sequence that may build structurally flexible domains (gray) and would have to expand to the required lengths, building a bridge of at least 16 nm over the synaptic cleft. The model is based on the structures of the LNS domain from β-neurexin, the PDZ and the guanylate kinase domain from CASK, and the three PDZ domains from PSD-95 that have been solved by X-ray crystallography. The structures of the other domains have been obtained by homology modeling using coordinates from the protein data bank (PDB, www.rcsb.org), i.e., entries 1kgd, 1kwa, 1jxm, 1y74, and 1rso for CASK; 1c4r for neurexin; and 1mah & 1fss for neuroligin; 1be9, 1iu0, 1jxm, 1qlc, and 1v1t for PSD-95. Predictions were made using programs SPDBViewer and Threader, and the web service Hmmstr/Rosetta, and represent most compact forms[25]. The bio-informatical work for this review was supported by grant SFB406-C9 (to MM). See Colorplate 7.

## 4. ROLE OF NEUROLIGIN AND β-NEUREXIN IN SYNAPSE FORMATION AND FUNCTION *IN VITRO*

The importance of a heterotypic adhesion complex such as the one formed by neuroligin and β-neurexin arises from the asymmetric nature of the synapse with presynaptic and postsynaptic specializations. Recently, experimental support for a role of neuroligin and β-neurexin in synapse formation and/or function has been provided by *in vitro* assays.

### 4.1. Effects on Synapse Formation

In 2000, a chimeric culture model between neuronal and non-neuronal cells was first used to address the role of neuroligin in synapse formation by Scheiffele and colleagues[34], where pontine explants were co-cultured with non-neuronal HEK293 cells. This model has the advantage that the non-neuronal cell can be engineered to express proteins of interest by conventional transfection, and the resultant effect can be monitored to elucidate the function of these proteins. The chimeric synapses can also be easily distinguished from endogenous synapses between neurons due to the absence of other synaptic markers within the non-neuronal cell. Presenting neuroligin 1 on a non-neuronal cell was sufficient to instigate synapse formation, resulting in the recruitment and clustering of synapsin at the neuronal presynaptic contact surface with the transfected HEK293 cell. This was a specific effect, not elicited by other candidate molecules, such as *N*-cadherin, ephrinB1, TAG-1, agrin, or L1[34]. More recently, analyses of the homophilic cell adhesion molecule SynCAM have revealed comparable synapse-inducing properties for an adhesion molecule with a structure completely unrelated to neuroligin domains[35,36]. Following the initial demonstration of neuroligin's potential to induce presynaptic specializations, the so-called reverse chimeric system has been used to demonstrate that β-neurexin can induce postsynaptic differentiation. β-Neurexin-expressing COS cells induced the clustering of several postsynaptic markers, including PSD-95 and gephyrin, as well as γ aminobutyric acid $(GABA)_A$ and NMDA receptor subunits[37]. In addition, HEK293 cells co-transfected with both neuroligin and PSD-95 revealed that neuroligin was sufficient to anchor the postsynaptic PSD-95 with neuronal presynaptic synapsin, indicating that the neuroligin/β-neurexin complex can bridge the synaptic cleft[38]. Neuroligin and β-neurexin may nucleate the formation of both pre- and postsynaptic elements and are therefore a promising candidate pair for organizing the asymmetric nature of the synapse.

Consistent with Scheiffele's initial observations[34], overexpression of neuroligin 1 caused an increase in the number and size of presynaptic terminals[41]. In other studies, neuroligin 1 overexpression resulted in increased postsynaptic differentiation and additional dendritic spines[40], and a splice variant of neuroligin 1 (without an insert in B, see also Figure 7.1; Colorplate 6) predominantly caused changes in the size of spines and presynaptic terminals[3]. It appears that the actual effect observed not only depends on the molecules and their splice variants over-expressed but also on the availability and amount of their binding partner PSD-95, and possibly yet unknown other interacting molecules. It was hypothesized that limiting endogenous pools of PSD-95 may be responsible for the lack of postsynaptic changes in some cases, and co-transfection of PSD-95 with neuroligin did result in an enlargement and increased number of postsynaptic structures[41]. Although data from several recent investigations agree that neuroligins robustly modulate presynaptic maturation, these approaches and their most important results, which are compiled in Table 1, demonstrate that similar experiments sometimes yield quite different results on the regulation of postsynaptic protein clustering.

Differences between recent studies also exist with respect to the postsynaptic recruitment of neurotransmitter receptors (NTRs), which is an essential aspect of the differentiation of functional synapses (Table 1). Upon presentation of β-neurexin on COS cells[37] or PC12 cells[39], the NR1 subunit of NMDAR was detected within the resulting postsynaptic clusters, but no AMPAR subunit recruitment took place. However, upon overexpression of neuroligin in neurons, one study saw clustering of both NR1 and some AMPAR subunits, GluR2/3[40], whereas another did not observe any recruitment of the NTR subunits analyzed[41]. Interestingly, as with the postsynaptic effects of neuroligin addressed above, this latter study found that transfection of neurons with PSD-95 resulted in the recruitment of the AMPAR subunit, GluR1. Nam and Chen eventually demonstrated that in neurons, overexpression of β-neurexin only elicited the formation of AMPAR-negative postsynaptic clusters[39] which subsequently could be induced to recruit the AMPAR subunit GluR1 by treatment with glutamate. The suggestion was that since these synapses already contained NMDAR glutamate receptors, glutamate treatment may represent an activity-dependent mechanism for initiating further maturation of the synapse. This mechanism has been proposed to occur

**Table 1.** Comparison of Overexpression Studies *In Vitro* to Analyze the Role of the Neuroligin/β-Neurexin Complex in Synapse Formation.

| Method | Result | Reference |
|---|---|---|
| **Overexpression in non-neuronal cell of** | | |
| β-neurexin | Induces postsynaptic clustering in neurons | |
| | Neuroligins 1 and 2, PSD-95, gephyrin, NMDA (NR1) and $GABA_A$ receptor subunits | (37) |
| | PSD-95, NMDAR (NR2) | (39) |
| Neuroligin 1 or 2 | Induces presynaptic clustering in neurons | |
| | Synapsin | (34) |
| Neuroligin 1 (mutated to abolish binding to β-neurexin | Fails to induce presynaptic clustering in neurons | |
| | Synapsin | (58) |
| **Overexpression in neurons of** | | |
| Neuroligin 1 | Induces presynaptic clustering | |
| | VGLUT1 | (37,40,41) |
| | GAD65/VGAT | (37,41) |
| | β-neurexin, synaptobrevin | (17) |
| | Induces postsynaptic clustering | |
| | PSD-95 | (17,40) |
| | NONE – requires PSD-95 co-expression | (42,41) |
| Neuroligin 2 | Induces presynaptic clustering | |
| | VGLUT1 | (37,42) |
| | GAD65/VGAT | (37,42) |
| C-terminally truncated neuroligin 1 | Induces presynaptic clusters | (34) |
| | Decreases postsynaptic clustering of PSD-95 | (39–41) |
| | No effect on postsynaptic recruitment of NMDAR | (40) |
| Splice variant of neuroligin 1 (lacking insert in B) | Increases mostly size of dendritic spines and Presynaptic terminals | (3) |

*Abbreviations:* GABA, γ-aminobutyric acid; GAD65, glutamic acid decarboxylase-65; NMDAR, *N*-methyl-D-aspartate receptors; NR1/2, NMDA receptor subunit 1/2; PSD-95, postsynaptic density-95; VGLUT1, vesicular glutamate transporter 1; VGAT, vesicular GABA transporter.

when so-called silent synapses become conductive during synapse development, through activation by a long-term potentiation-like process involving calmodulin--dependent protein kinase II (CaMKII)[43]. Consistent with this proposal, transfection of these neurons with constitutively active CaMKII also resulted in the recruitment of AMPAR. It can therefore be concluded from these studies that neuroligin/β-neurexin-induced synapses are able to recruit NMDAR subunits, and that AMPAR subunits can be recruited subsequently depending upon the activation status of the cell which resembles developmental processes[43,44].

## 4.2. The Neuroligin/β-Neurexin Complex and Synapse Function

Following the demonstration that neuroligin and β-neurexin are capable of triggering formation of both pre- and postsynaptic specializations, it was necessary to show that these resulting synapses are not just synaptic protein aggregates but rather represent functional synapses. This was first achieved by demonstrating that neuroligin-induced presynaptic elements in chimeric cultures were capable of undergoing synaptic vesicle turnover following depolarization in a similar manner to endogenous synapses[34]. Subsequently, electrophysiological measurements of these artificial synapses confirmed that the synapses were actually capable of transmission, using HEK293 cells co-transfected with neuroligin and subunits of NMDAR and AMPAR that allowed direct measurement of currents in the heterologous cells[38]. β-Neurexin was first shown to be essential for synapse function through the application of soluble β-neurexin in neuronal cultures, which disrupted the formation of synapsin-positive clusters at chimeric and endogenous synapses[34]. This study further demonstrated that the synapse-inducing effects of neuroligin were mediated through its β-neurexin partner. In another approach to disrupt the complex, neuroligin 2 was overexpressed in neurons at very high levels such that it mislocalized to the entire dendritic surface[37], resulting in disrupted postsynaptic receptor clustering of PSD-95, gephyrin, and NR1, and a reduction of synaptic transmission. Finally, the role of the complex was also explored using small inhibitory RNA (siRNA) against different neuroligin isoforms, whereby a reduction in neuroligin expression resulted in disrupted synaptic transmission[40].

An alternative strategy to study their role at synapses has been to employ mutated versions of β-neurexin and neuroligin. Extracellularly, the AChE domain of neuroligin 1 was shown to be necessary for both β-neurexin binding and the induction of presynaptic terminal formation[17,58]. In agreement with this, upon truncating the extracellular domain of neuroligin, El-Husseini and colleagues demonstrated a loss in its ability to induce presynaptic clustering in neurons[41]. Likewise for β-neurexin, removal of the LNS domain resulted in an impaired ability to induce postsynaptic differentiation[37]. The glycosylation-rich domain was also found to be important for β-neurexin function since its deletion disrupted the ability to instigate synapse formation, however, it could not induce synaptogenesis by itself, suggesting that it may be required for structural positioning of the LNS domain. The synaptogenic activity of the neuroligin/β-neurexin complex appears to result from *trans*-synaptic aggregation, since aggregation of neuroligin by application of β-neurexin attached to beads, or by antibodies, results in the same clustering as co-culturing[37]. β-Neurexin may therefore exert its effects through induced clustering of neuroligin. As described above, oligomerization of neuroligin is essential in this process[16] since mutants which are still capable of binding β-neurexin, but can no longer oligomerize with each other, fail to induce presynaptic terminal formation[17]. In addition, loop-exchange mutants of neuroligin 1 with acetylcholinesterase demonstrated that the capacity to bind β-neurexin is necessary for its ability to induce synapse formation[58]. Intracellularly, the PDZ-recognition sequence at the C-terminus of neuroligin is responsible for the interaction with PSD-95. Thus, truncation of the C-terminal region resulted in a failure to recruit PSD-95 to postsynaptic clusters[39–41], reflecting the loss of interaction between neuroligin and PSD-95 but surprisingly did not affect the targeting of neuroligin itself to synapses[45]. This mutant also resulted in a loss of recruitment of AMPAR subunits and in reduced excitatory transmission, whilst inhibitory transmission was unaltered[39]. In addition, a mutant in which the intracellular and transmembrane domains were replaced by a GPI anchor (allowing membrane sequestering without any direct intracellular signalling) was still able to induce presynaptic specializations[34], demonstrating that the extracellular domain was sufficient. However, its activity was less than for the full-length neuroligin, suggesting that there may therefore still be a need for a feedback system, or cross-communication.

Taken together, the discrepancies observed on postsynaptic differentiation are most likely due to differences in the levels of overexpression of these proteins. Despite these inconsistencies, however, these results hint at a mechanism by which induction of both neurexin and neuroligin clustering results in recruitment of additional pre- and post-synaptic proteins. Future studies on the time course of recruitment and clustering of endogenous neurexins and neuroligins at the synapse may help clarify some of these issues.

## 5. THE NEUROLIGIN/β-NEUREXIN COMPLEX AT EXCITATORY VERSUS INHIBITORY SYNAPSES

Immunohistochemical studies of neuroligin 1 in brain tissue have revealed a localization at excitatory synapses, based on its co-localization with GluR2/3 receptors, rather than $GABA_A$ receptors[5]. It has been proposed that neuroligin 2, in turn, is localized exclusively to inhibitory GABAergic synapses[46], leading to the hypothesis that different neuroligin isoforms may determine which type of synapse is formed. Analyses of endogenous neuroligin in neuronal cultures have revealed a similar picture, i.e., neuroligin 1 is almost always associated with excitatory synapse markers[37,42], whereas neuroligin 2 is predominantly associated with inhibitory synaptic markers such as gephyrin[37] or vesicular GABA transporter[40,42]. This distribution may change, however, when cultured neurons overexpress neuroligins[37,40–42], indicating that presumably all neuroligin isoforms have the potential to associate at both excitatory and inhibitory synapses under certain experimental conditions. Although this behavior makes interpretation difficult, a common picture emerges that the neuroligin/β-neurexin complex may be responsible for the ratio between excitatory and inhibitory synapses[47].

A possible mechanism for determining to which type of synapses neuroligins are localized came from an investigation of PSD-95. PSD-95 co-expression with neuroligin 1 was found to enhance the size and number of postsynaptic clusters, and resulted in a change of neuroligin localization from both excitatory and inhibitory synapses to solely excitatory[41]. In support of this, siRNA was used to reduce PSD-95 expression within the neuron, resulting in an increase in VGAT-positive presynaptic contacts, presumably through a shift of neuroligin 1 to inhibitory synapses, indicating that the level of PSD-95 determines neuroligin 1 localization. Surprisingly, the same mechanism seems to regulate neuroligin 2 location: co-expression of PSD-95 with neuroligin 2 also promoted neuroligin 2 to associate predominantly with excitatory synapses, instead of with both excitatory and inhibitory synapses in the case of overexpressed neuroligin 2 alone[42], or instead of solely inhibitory synapses in the case of endogenous neuroligin 2[37]. In line with these data, a C-terminally truncated neuroligin 1 resulted in impaired excitatory synaptic transmission, whereas inhibitory transmission was unaffected[39]. Although all neuroligin isoforms contain a PDZ-domain recognition motif necessary for PSD-95 binding, it may be that neuroligin 1 has a greater affinity for PSD-95 than neuroligin 2, or alternatively, there may be other intracellular binding partners that compete with PSD-95 for the C-terminus of neuroligin. The different approaches and resultant effects of the neuroligin/β-neurexin complex on excitatory versus inhibitory synapses are summarized in Table 2.

β-Neurexin expression has also been detected in both excitatory and inhibitory neurons[8]. More recently, Graf et al. found that β-neurexin-expressing non-neuronal cells induce both excitatory and inhibitory postsynaptic specializations at chimeric synapses[37]. Moreover, this process involved the recruitment of different neuroligins, i.e., neuroligins 1 and 2, presumably to mediate the specificity of induced PSD-95 or gephyrin clusters. Antibody-induced aggregation of neuroligin 1 was sufficient to co-aggregate PSD-95 (excitatory postsynaptic clusters) but not gephyrin (inhibitory postsynaptic clusters). Conversely, direct aggregation of neuroligin 2 clustered predominantly gephyrin, although PSD-95 clustering could also be found[37]. Application of soluble β-neurexin, in turn, decreased the number of inhibitory synapses induced by overexpression of neuroligins 1 and 2, while increasing the number of PSD-95 co-clustering with neuroligin[42], indicating that β-neurexin binding induces the aggregation of PSD-95 with both neuroligins 1 and 2.

Interfering with the neuroligin/β-neurexin complex *in vitro* not only alters the clustering and recruitment of binding partners but also appears to change dramatically the balance between excitatory and inhibitory synaptic transmission. Application of soluble β-neurexin resulted in an increased overall ratio of excitatory to inhibitory miniature postsynaptic currents[42]. Since this affected mini frequencies, and not amplitudes, the effect was most likely caused by disruption of the presynaptic machinery, rather than postsynaptic differentiation. Suppression of any one of the three neuroligins by siRNA treatment, in turn, reduced both excitatory and inhibitory synapse numbers[40]. In contrast, suppression of all three neuroligin isoforms together resulted in the disruption of mainly inhibitory synaptic transmission, with only a moderate effect on excitatory transmission. Since suppression of a single isoform had such a dramatic effect, it supports the idea that deletion of any of the neuroligins severely disrupts the balance between excitation and inhibition. Synaptic activity was subsequently shown to be completely restored by the addition of only one human neuroligin isoform[40], suggesting that its overexpression is sufficient to restore a full and functional neuroligin population. It has to be emphasized, however, that all these studies relied on measuring asynchronous spontaneous events rather that action potential-evoked synaptic transmission. Evoked postsynaptic responses are more difficult to record in primary neuronal cultures but such recordings are mandatory before definitive conclusions can be made on the functional aspects of the synaptic contacts induced by the neuroligin/neurexin complex *in vitro*.

## 6. CLINICAL ASPECTS OF THE NEUROLIGIN/β-NEUREXIN COMPLEX

Disturbances in the ratio of excitatory and inhibitory synapses may have implications for neurodevelopmental disorders such as autism and mental retardation[48]. Indeed, shortly after the *human* neuroligin family was cloned[49], revealing the existence of a fourth human neuroligin, the genes encoding neuroligins 3 and 4 were found to be present within X-linked loci (Xp22.3 and Xq13) associated with autism[50]. Upon screening for neuroligin mutations within families with autistic members, a frame shift mutation was identified in neuroligin 4 at position 396 (D396X), leading to a truncation, and a substitution mutation within neuroligin 3 (R451C). Both mutations were absent from unaffected family

**Table 2.** Differential Effects of the Neuroligin/β-Neurexin Complex on Excitatory versus Inhibitory Synapses.

| Method | Result | Reference |
|---|---|---|
| ***In situ/in vitro* localization of** | | |
| Neuroligin 1 | Excitatory synapses | |
| | *in situ* | (5) |
| | *in vitro* | (37,42) |
| Neuroligin 2 | Inhibitory synapses | |
| | *in situ* | (46) |
| | *in vitro* | (37,40,42) |
| **Localization after overexpression of** | | |
| Neuroligin 1 | Induces both excitatory and inhibitory presynaptic clusters | (40–42) |
| | Induces mainly excitatory clusters | (37) |
| Neuroligin 2 | Induces both excitatory and inhibitory presynaptic clusters | (41,42) |
| | Induces mainly inhibitory, with some excitatory clusters | (37,40) |
| **Shift in localization following PSD-95 overexpression** | | |
| Neuroligin 1 (overexpressed) | From both excitatory and inhibitory to excitatory contacts | (41,42) |
| Neuroligin 2 (overexpressed and endogenous) | From both excitatory and inhibitory to excitatory contacts | (37,42) |
| **Knockdown/ Disruption of** | | |
| Neuroligin 1 | C-terminal mutant (no PSD-95 binding) | |
| | Still recruits NMDAR – but not PSD-95 | (40) |
| | RNAi – also against neuroligins 2 and 3 | |
| | Reduces inhibitory synapses | (40) |
| | Dominant-negative neuroligin 1 | |
| | Impairs excitatory but not inhibitory synapses | (39) |
| Neuroligin 2 | Overexpression of mislocalized neuroligin 2 | |
| | Disrupts both excitatory and inhibitory synapses | (37) |
| **Role of β-neurexin** | Interfering with neuroligin (1 and 2) function by addition of soluble β-neurexin – causes clustering of PSD-95 and reduces excitatory and inhibitory synapses | (42) |
| | Causes clustering of PSD-95 | (37) |

*Abbreviations:* NMDAR, *N*-methyl-D-aspartate receptor; PSD-95, postsynaptic density protein 95; RNAi, RNA interference.

members, and all unrelated controls analyzed. Support for the involvement of neuroligin 4 with autism followed from a large study of autistic individuals, where 4 missense mutations in neuroligin 4 were identified[51]. In agreement, a two base pair deletion in neuroligin 4, leading to a truncation in the expressed protein (position 429), was identified in a family containing both autistic and nonautistic but mentally retarded family members[52]. This suggested that neuroligin-associated disorders include both autism and mental retardation, and thus that these two syndromes may involve a common synaptic mechanism. However, controversy has arisen over the association of autism with neuroligin gene mutations. In two other studies involving larger populations, neuroligin mutations were not identified in association with autism[53,54]. Although these studies do not rule out a role for neuroligin, they suggest that it may only be responsible for a small percentage of the heterogeneous population suffering from autism[55] and mental retardation[56].

To further the investigation into the role that neuroligin mutations may play in nervous system disorders, constructs of neuroligins 3 and 4 encoding the previously identified disease-associated mutations R451C and D396X[50], were generated and transfected into heterologous cells[57,59]. Only low levels of neuroligin 3 R451C (and the corresponding mutation R471C in rat neuroligin) and no neuroligin 4 D396X were targeted to the cell surface, and co-labeling revealed that both mutant proteins were retained in the endoplasmic reticulum. Upon transfection into hippocampal neurons, the mutants failed to promote presynaptic differentiation[57]. In a related study using an autism mutant of neuroligin 3 and a comparable mutation site in neuroligin 1, transfections resulted in less efficient trafficking to the cell surface, despite retaining some synaptogenic activity[58]. Although much work remains to be done to establish its role in disease, the disruption of the neuroligin/neurexin complex may significantly alter the ratio between excitatory and inhibitory neurotransmission, manifesting itself in an impaired cognitive development, and possibly psychiatric disorders such as autism.

## 7. CONCLUSIONS

Neuroligin and neurexin represent an ideal candidate pair for a role in the formation and/or maturation of synapses due to their heterotypic binding, regulated by alternative splicing and $Ca^{2+}$ dependence. At least *in vitro*, neuroligin and β-neurexin induce pre- and postsynaptic specializations, respectively, including recruitment of relevant synaptic molecules to newly formed synaptic contacts. Through its interaction with PSD-95, the complex seems to be of particular importance for the balance between excitatory and inhibitory synapses. The newly induced synaptic contacts appear functional due to the detection of neurotransmitter release, although this has only been tested by miniature postsynaptic current recordings. Disruption of the β-neurexin/neuroligin complex *in vitro*, in turn, results in the abrogation of synaptic structure and function. Based on identified mutations, neuroligins and the neuroligin/β-neurexin complex are putative candidates for the molecular basis of cognitive diseases such as autism and mental retardation.[*]

## 8. REFERENCES

1. Ichtchenko, K., Hata, Y., Nguyen, T., Ullrich, B., Missler, M., Moomaw, C., and Südhof, T.C. (1995) *Cell* **81**, 435–443.
2. Ichtchenko, K., Nguyen, T., and Südhof, T.C. (1996) *J Biol Chem* **271**, 2676–2682.
3. Boucard, A.A., Chubykin, A., Comoletti, D., Taylor, P., and Südhof, T.C. (2005) *Neuron* **48**, 229–236.
4. Nguyen, T., and Südhof, T.C. (1997) *J Biol Chem* **272**, 26032–26039.
5. Song, J.Y., Ichtchenko, K., Südhof, T.C., and Brose, N. (1999) *Proc Natl Acad Sci U S A* **96**, 1100–1105.
6. Irie, M., Hata, Y., Takeuchi, M., Ichtchenko, K., Toyoda, A., Hirao, K., Takai, Y., Rosahl, T.W., and Südhof, T.C. (1997) *Science* **277**, 1511–1515.

[*]The bioinformatic work for this review was supported by grant SFB406-C9 (to M.M.).

7. Missler, M., and Südhof, T.C. (1998) *Trends Genet* **14**, 20–26.
8. Ullrich, B., Ushkaryov, Y.A., and Südhof, T.C. (1995) *Neuron* **14**, 497–507.
9. Ushkaryov, Y.A., Petrenko, A.G., Geppert, M., and Südhof, T.C. (1992) *Science* **257**, 50–56.
10. Ushkaryov, Y.A., Hata, Y., Ichtchenko, K., Moomaw, C., Afendis, S., Slaughter, C.A., and Südhof, T.C. (1994) *J Biol Chem* **269**, 11987–11992.
11. Missler, M., Zhang, W., Rohlmann, A., Kattenstroth, G., Hammer, R.E., Gottmann, K., and Sudhof, T.C. (2003) *Nature* **424**, 939–948.
12. Zhang, W., Rohlmann, A., Sargsyan, V., Aramuni, G., Hammer, R.E., Sudhof, T.C., and Missler, M. (2005) *J Neurosci* **25**, 4330–4342.
13. Sugita, S., Saito, F., Tang, J., Satz, J., Campbell, K., and Sudhof, T.C. (2001) *J Cell Biol* **154**, 435–445.
14. Beglopoulos, V., Montag-Sallaz, M., Rohlmann, A., Piechotta, K., Ahmad, M., Montag, D., and Missler, M. (2005) *Mol Cell Biol* **25**, 7278–7288.
15. Missler, M., and Südhof, T.C. (1998) *J Neurosci* **18**, 3630–3638.
16. Comoletti, D., Flynn, R., Jennings, L.L., Chubykin, A., Matsumura, T., Hasegawa, H., Sudhof, T.C., and Taylor, P. (2003) *J Biol Chem* **278**, 50497–50505.
17. Dean, C., Scholl, F.G., Choih, J., DeMaria, S., Berger, J., Isacoff, E., and Scheiffele, P. (2003) *Nat Neurosci* **6**, 708–716.
18. Tsigelny, I., Shindyalov, I.N., Bourne, P.E., Südhof, T.C., and Taylor, P. (2000) *Protein Sci* **9**, 180–185.
19. Rudenko, G., Nguyen, T., Chelliah, Y., Südhof, T.C., and Deisenhofer, J. (1999) *Cell* **99**, 93–101.
20. Kim, E., Niethammer, M., Rothschild, A., Jan, Y.N., and Sheng, M. (1995) *Nature* **378**, 85–88.
21. Kornau, H.C., Schenker, L.T., Kennedy, M.B., and Seeburg, P.H. (1995) *Science* **269**, 1737–1740.
22. El Husseini, A.E., Schnell, E., Chetkovich, D.M., Nicoll, R.A., and Bredt, D.S. (2000) *Science* **290**, 1364–1368.
23. Leonard, A.S., Davare, M.A., Horne, M.C., Garner, C.C., and Hell, J.W. (1998) *J Biol Chem* **273**, 19518–19524.
24. Hata, Y., Butz, S., and Südhof, T.C. (1996) *J Neurosci* **16**, 2488–2494.
25. Nakagawa, T., Futai, K., Lashuel, H.A., Lo, I., Okamoto, K., Walz, T., Hayashi, Y., and Sheng, M. (2004) *Neuron* **44**, 453–467.
26. Borg, J., Straight, S.W., Kaech, S.M., Taddeo-Borg, M., Kroon, D.E., Karnak, D., Turner, R.S., Kim, S.K., and Margolis, B. (1998) *J Biol Chem* **273**, 31633–31636.
27. Butz, S., Okamoto, M., and Südhof, T.C. (1998) *Cell* **94**, 773–782.
28. Biederer, T., and Südhof, T.C. (2000) *J Biol Chem* **275**, 39803–39806.
29. Biederer, T., and Sudhof, T.C. (2001) *J Biol Chem* **276**, 47869–47876.
30. Zhang, Y., Luan, Z., Liu, A., and Hu, G. (2001) *FEBS Lett* **487**, 99–102.
31. Maximov, A., Sudhof, T.C., and Bezprozvanny, I. (1999) *J Biol Chem* **274**, 24453–24456.
32. Hsueh, Y.P., Wang, T.F., Yang, F.C., and Sheng, M. (2000) *Nature* **404**, 298–302.
33. Hsueh, Y.P., Yang, F.C., Kharazia, V., Naisbitt, S., Cohen, A.R., Weinberg, R.J., and Sheng, M. (1998) *J Cell Biol* **142**, 139–151.
34. Scheiffele, P., Fan, J., Choih, J., Fetter, R., and Serafini, T. (2000) *Cell* **101**, 657–669.
35. Biederer, T., Sara, Y., Mozhayeva, M., Atasoy, D., Liu, X., Kavalali, E.T., and Sudhof, T.C. (2002) *Science* **297**, 1525–1531.
36. Sara, Y., Biederer, T., Atasoy, D., Chubykin, A., Mozhayeva, M.G., Sudhof, T.C., and Kavalali, E.T. (2005) *J Neurosci* **25**, 260–270.
37. Graf, E.R., Zhang, X., Jin, S.X., Linhoff, M.W., and Craig, A.M. (2004) *Cell* **119**, 1013–1026.
38. Fu, Z., Washbourne, P., Ortinski, P., and Vicini, S. (2003) *J Neurophysiol* **90**, 3950–3957.
39. Nam, C.I., and Chen, L. (2005) *Proc Natl Acad Sci U S A* **102**, 6137–6142.
40. Chih, B., Engelman, H., and Scheiffele, P. (2005) *Science* **307**, 1324–1328.
41. Prange, O., Wong, T.P., Gerrow, K., Wang, Y.T., and El Husseini, A.E. (2004) *Proc Natl Acad Sci U S A* **101**, 13915–13920.
42. Levinson, J.N., Chery, N., Huang, K., Wong, T.P., Gerrow, K., Kang, R., Prange, O., Wang, Y.T., and El-Husseini, A. (2005) *J Biol Chem* **280**, 17312–17319.
43. Durand, G.M., Kovalchuk, Y., and Konnerth, A. (1996) *Nature* **381**, 71–75.
44. Rao, A., Kim, E., Sheng, M., and Craig, A.M. (1998) *J Neurosci* **18**, 1217–1229.
45. Dresbach, T., Neeb, A., Meyer, G., Gundelfinger, E.D., and Brose, N. (2004) *Mol Cell Neurosci* **27**, 227–235.
46. Varoqueaux, F., Jamain, S., and Brose, N. (2004) *Eur J Cell Biol* **83**, 449–456.
47. Levinson, J.N., and El Husseini, A.E. (2005) *Neuron* **48**, 171–174.

48. Rubenstein, J.L., and Merzenich, M.M. (2003) *Genes Brain Behav* **2**, 255–267.
49. Bolliger, M.F., Frei, K., Winterhalter, K.H., and Gloor, S.M. (2001) *Biochem J* **356**, 581–588.
50. Jamain, S., Quach, H., Betancur, C., Rastam, M., Colineaux, C., Gillberg, I.C., Soderstrom, H., Giros, B., Leboyer, M., Gillberg, C., and Bourgeron, T. (2003) *Nat Genet* **34**, 27–29.
51. Yan, J., Oliveira, G., Coutinho, A., Yang, C., Feng, J., Katz, C., Sram, J., Bockholt, A., Jones, I.R., Craddock, N., Cook Jr, E.H., Vicente, A., and Sommer, S.S. (2005) *Mol Psychiatry* **10**, 329–335.
52. Laumonnier, F., Bonnet-Brilhault, F., Gomot, M., Blanc, R., David, A., Moizard, M.P., Raznaud, M., Ronce, N., Lemonnier, E., Calvas, P., Laudier, B., Chelly, J., Fryns, J.P., Ropers, H.H., Hamel, B.C., Andres, C., Barthelemy, C., Moraine, C., and Briault, S. (2004) *Am J Hum Genet* **74**, 552–557.
53. Vincent, J.B., Kolozsvari, D., Roberts, W.S., Bolton, P.F., Gurling, H.M., and Scherer, S.W. (2004) *Am J Med Genet B Neuropsychiatr Genet* **129**, 82–84.
54. Gauthier, J., Bonnel, A., St-Onge, J., Karemera, L., Laurent, S., Mottron, L., Fombonne, E., Joober, R., and Roleau, G.A. (2005) *Am J Med Genet B Neuropsychiatr Genet* **132**, 74–75.
55. Risch, N. et al. (1999) *Am J Hum Genet* **65**(2), 493–507.
56. Chiurazzi, P., Hamel, B.C., and Neri, G. (2001) *Eur J Hum Genet* **9**, 71–81.
57. Chih, B., Afridi, S.K., Clark, L., and Scheiffele, P. (2004) *Hum Mol Genet* **13**, 1471–1477.
58. Chubykin, A.A., Liu, X., Comoletti, D., Tsigelny, I., Taylor, P., and Sudhof, T.C. (2005) *J Biol Chem* **280**, 22365–22374.
59. Comoletti, D., De Jaco, A., Jennings, L.L., Flynbn, R.E., Tsigelny, I., Ellisman, M.H., and Taylor, P. (2004) *J Neurosci* **24**, 4889–4893.

# 8

# SynCAM IN FORMATION AND FUNCTION OF SYNAPTIC SPECIALIZATIONS

Thomas Biederer*

## 1. SUMMARY

SynCAM 1 (Synaptic Cell Adhesion Molecule 1), a member of the immunoglobulin (Ig) superfamily of proteins, is an intercellular adhesion molecule at synapses in the central nervous system (CNS). It mediates interactions that bridge the synaptic cleft between pre- and postsynaptic membranes. SynCAM 1 has an active role in synaptic differentiation and induces formation of new presynaptic terminals. In keeping with this activity, SynCAM 1 is expressed throughout the developing brain, indicating a general role in synaptogenesis. SynCAM 1 is a member of a family of four genes found solely in vertebrates. The domain structure of all SynCAM family members is well defined. Three Ig-like domains constitute their extracellular sequence, followed by a single transmembrane region and a short cytosolic tail. Their intracellular sequences display an interaction motif for PDZ domain-containing adaptor molecules and for components of the actin cytoskeleton. These cytosolic protein interaction motifs are highly conserved among all four SynCAM proteins, underscoring their important role in mediating membrane differentiation.

## 2. Ig-SUPERFAMILY MEMBERS IN SYNAPTIC DIFFERENTIATION

The first step of synaptic differentiation is initiated by axo-dendritic contact. This physical contact between incoming axons and postsynaptic targets initiates clustering of synaptic vesicles indicative of the formation of presynaptic terminals[1]. Two key roles of adhesion molecules in this process can be distinguished. First, they confer the specificity of synaptic contacts. Second, they can affect formation of synaptic specializations.

*Department of Molecular Biophysics and Biochemistry, Yale University Medical School, 333 Cedar Street, New Haven, CT 06520, USA; thomas.biederer@yale.edu

Studies of these processes have highlighted various functions of the Ig superfamily in synaptic differentiation. Ig superfamily members have diverse roles in extracellular recognition and perform important functions in the vertebrate and invertebrate CNS[2,3]. They have critical roles in axon guidance[4–6] and affect synaptic plasticity (see Chapters 3, 5, and 6). The roles of Ig superfamily members in synaptic differentiation were illustrated by studies of axon fasciculation in insects, which identified the membrane protein fasciclin II (Fas II)[7,8]. Fas II contains five extracellular Ig-like domains followed by two fibronectin repeats. It engages in a homophilic interaction at *Drosophila* neuromuscular junctions, and determines growth, maturation, and morphology of neuromuscular synapses[9,10]. Importantly, muscle-cell expressed Fas II controls the patterning of synaptic input[11], and affects both specificity and maturation of neuromuscular synapses in *Drosophila*. Fas II has similar functions at central synapses in *Drosophila*[12].

Important roles of Ig superfamily members in the spatial definition of specific synaptic sites were recently identified in the nematode *C. elegans* and in the vertebrate retina. In *C. elegans*, the motor neurons that express the Ig superfamily member SYG-1 make synapses only with certain neurons and muscle cells. This specificity is directed by epithelial guidepost cells expressing the SYG-2 protein, which interacts in a heterophilic interaction with the neuronal SYG-1[13,14]. In the vertebrate retina, an analogous function in target recognition is exerted by the Sidekicks proteins, synaptic membrane proteins with six extracellular Ig-like domains, and 13 fibronectin repeats. Sidekicks are differentially expressed within retinal ganglion cells, and their homophilic interactions specify synaptic connectivity in distinct retinal laminae[15]. Together, these studies demonstrate that target-derived Ig superfamily members can control synaptic specificity and differentiation.

## 3. SPECIFIC ADHESION SYSTEMS INCLUDING THE Ig-SUPERFAMILY MEMBER SynCAM 1 INDUCE SYNAPSE FORMATION

Studies in vertebrates identified selected adhesion molecules that not only regulate synaptic differentiation, but also demarcate future synaptic sites and directly drive formation of new synaptic membrane specializations. Examples of cell adhesion molecules with both adhesive and inducing functions at synapses are the neurexin/neuroligin proteins (see Chapters 4 and 7, and Figure 8.1). The postsynaptic neuroligins induce presynaptic specializations via binding to neurexins, their presynaptic partners[16–18]. Different neuroligin family members vary in their ability to induce excitatory and inhibitory presynaptic terminals[19–21]. Neurexins themselves also can signal the formation of excitatory and inhibitory postsynaptic specializations[19].

However, biochemical evidence indicated that other adhesion systems exist at central synapses. Biochemical analysis showed that the interaction of neurexin and neuroligin is $Ca^{2+}$-dependent[16], as is the homophilic interaction of cadherins, synaptic adhesion molecules that promote synapse maturation (see Chapter 5). Yet, preparations of nerve terminals remain attached to each other in the absence of $Ca^{2+}$, indicating that additional adhesion systems exist at the cleft of central synapses.

Recent studies of synaptogenesis identified the vertebrate Ig superfamily member SynCAM 1 (Synaptic Cell Adhesion Molecule 1; human gene symbol *IGSF4*). SynCAM proteins constitute a distinct, $Ca^{2+}$-independent adhesion system that connects pre- and postsynaptic membranes (Figure 8.1). Importantly, SynCAM combines this adhesive function with the activity to induce presynaptic specializations in the vertebrate CNS.

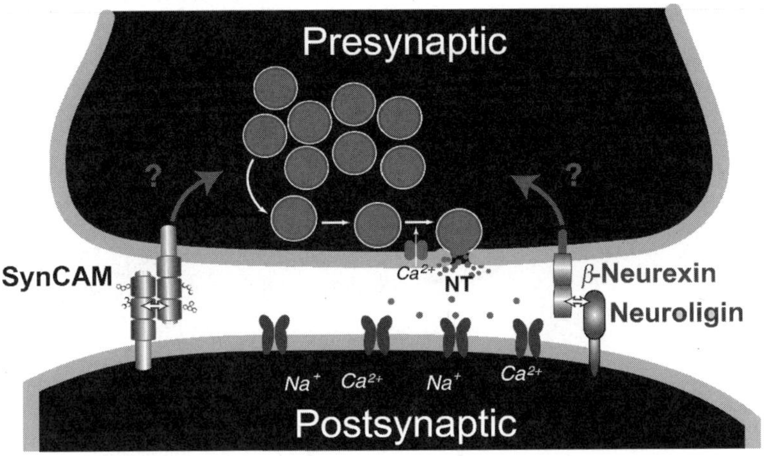


**Figure 8.1.** Induction Pathways of Synaptic Terminals at Central Synapses. The postsynaptic adhesion molecules SynCAM 1 and neuroligin interact with their presynaptic partner molecules SynCAM and β-neurexin, respectively, in interactions that bridge the synaptic cleft. Presynaptic proteins that interact with SynCAM and β-neurexin then trigger the formation of presynaptic specializations through unidentified signaling pathways (arrows). Both postsynaptic SynCAM 1 and neuroligin are sufficient to induce functional presynaptic specializations containing actively recycling synaptic vesicles that release neurotransmitter (NT) as indicated.

## 4. IDENTIFICATION OF SynCAM 1

SynCAM 1 was identified in a bioinformatic search for brain-expressed adhesion molecules belonging to the Ig superfamily that contain intracellular PDZ-domain interaction motifs[22]. PDZ domains characterize a set of adaptor molecules and typically allow them to bind to the carboxyl termini of membrane proteins, serving important functions in membrane protein scaffolding and targeting[23]. These PDZ domain-containing adaptor molecules are particularly common at synaptic sites, where they form submembraneous molecular scaffolds[24,25]. These scaffolds are probably critical for the initial assembly and organization of synaptic membrane specializations. SynCAM 1 is one of a comparatively small number of Ig superfamily members containing a PDZ binding consensus motif. This group of molecules also includes all the invertebrate Ig superfamily members mentioned above that are involved in synaptic differentiation. This analogy of extracellular domains and intracellular sequence motifs drew particular attention to SynCAM 1, and aided its identification as a synaptic adhesion molecule in vertebrates.

In addition, SynCAM 1 had been independently identified by other approaches. It was mapped and cloned as a potential tumor suppressor gene[26,27]. This role was confirmed using SynCAM 1-expressing tumor cells in immunocompromised mice[28]. These studies indicate that SynCAM 1 could regulate migration, proliferation, and differentiation of non-neuronal cells. SynCAM 1 was also cloned as an Ig domain containing protein expressed in testis during spermatogenesis[29] and as retinoic acid-inducible gene from PC12 cells, which implicates it in neuronal differentiation[30].

## 5. DOMAIN ORGANIZATION AND MOTIFS OF SynCAM 1

SynCAM 1 is a single-spanning membrane protein (Figure 8.2). SynCAM 1 transcripts are alternatively spliced in the extracellular region (see paragraph 9), and diverse transcripts encode SynCAM 1 isoforms whose lengths vary from 416 to 455 amino acids. The predicted signal peptide of SynCAM 1 is 41 amino acids long. Sequence analysis predicts a molecular weight of 42.0–47.9 kDa for SynCAM 1 protein products of the shortest and longest alternatively spliced transcripts after signal peptidase cleavage, respectively.

The extracellular domain of SynCAM 1 is composed of three Ig-like domains, which belong to the V-set, C1-set, and I-set subclasses of Ig-like domains, respectively[31] (Figure 8.2).

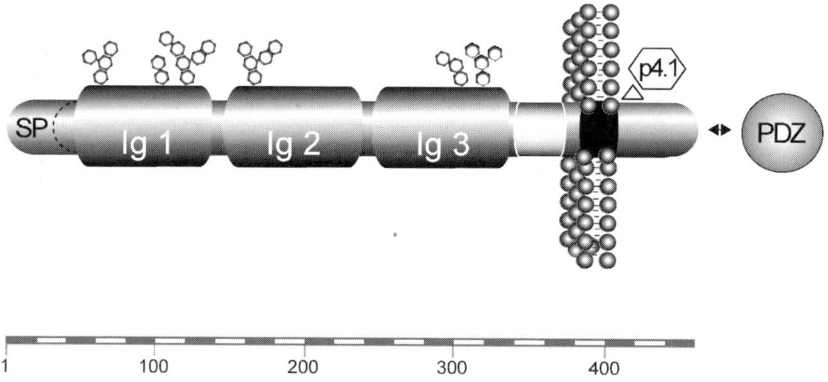


**Figure 8.2.** Domain Organization of Human SynCAM 1. The figure depicts the product of the longest human SynCAM 1 splice variant. The broken line at left indicates the signal peptidase cleavage site, and hexagons show the predicted N-glycosylation sites in the extracellular domain. The light gray segment marks the extracellular sequence stretch that consists mainly of putative O-glycosylation sites, which varies in length depending on alternative splicing of this sequence. The following black segment surrounded by an indicated lipid bilayer depicts the single transmembrane region. The cytosolic tail contains sequences interacting with the cytoskeletal protein 4.1 and PDZ-domain containing adaptor molecules as indicated. The model is drawn to scale, and the bottom scale bar indicates amino acid numbers. SP, signal peptide; Ig, Ig-like domain; PDZ, PDZ-domain containing protein.

Six predicted N-glycosylation sites lie within these three extracellular Ig-like domains. The Ig-like domains of SynCAM 1 are followed by an alternatively spliced O-glycosylation stalk, which can contain different numbers of predicted O-glycosylation sites depending on alternative splicing of the exons encoding this region (see paragraph 9). Splicing is conserved between human and mouse orthologs[31], but the functions of this conserved splicing in the O-glycosylation stalk are not yet known. In agreement with these predicted glycosylation sites, SynCAM 1 protein is post-translationally modified with carbohydrates in brain, and N-glycosylation at multiple sites in the Ig-like domains increases its apparent molecular weight. Additionally, SynCAM 1 is O-glycosylated in agreement with predicted sites present in different splice variants[32].

The single transmembrane region is followed by a short, 47 amino acid long cytosolic sequence. This cytosolic sequence of SynCAM 1 contains two identified protein interaction motifs (Figure 8.2). First, a juxtamembraneous intracellular sequence is predicted to interact with members of the protein 4.1 family, which stabilize the assembly of the actin/spectrin cytoskeleton[33]. This sequence was first analyzed in the erythrocyte membrane protein glycophorin C[34–36]. Notably, this motif is also present in the neurexin family of synaptic adhesion molecules[37]. Second, SynCAM 1 displays a sequence at the extreme carboxyl terminus predicted to interact with type II class PDZ domains[38]. This prediction has been confirmed, and SynCAM 1 is known to interact with the intracellular PDZ-domain containing adaptor molecules CASK, Mint1, and syntenin[22].

## 6. SynCAM 1 EXPRESSION IN THE VERTEBRATE BRAIN

SynCAM 1 homologs are present throughout vertebrate genomes from puffer fish to humans, whereas no evolutionarily related genes could be identified in invertebrates[31]. It is encoded by the *IGSF4* (*Ig superfamily member 4*) gene, which is highly conserved both on the exon/intron level and in its coding sequence[31]. SynCAM 1 orthologs display high sequence similarity throughout vertebrate genomes. The high degree of conservation across species is exemplified by the fact that in humans and puffer fish, 71% of the amino acids are identical or highly similar for their two SynCAM 1 orthologs. Human and mouse SynCAM 1 proteins are even more closely related and share 98% sequence identity.

SynCAM 1 transcripts are abundant in brain, present in testis, and weakly detected in other tissues[29,32,39]. During early postnatal development, SynCAM 1 is expressed in all brain regions analyzed so far. Its expression levels in mice increase during the first 2 weeks after birth, the peak period of synaptogenesis in rodents[40,41]. An antibody raised against the carboxyl terminus of SynCAM 1 shows synaptic staining in immunohistochemistry and immunoelectron microscopy[22]. This expression of SynCAM 1 throughout the brain indicates a general role for this adhesion molecule at synapses in the developing CNS.

## 7. SynCAM 1 DRIVES SYNAPSE FORMATION

SynCAM 1 is sufficient to induce neurons to form functional presynaptic specializations and its expression in neurons promotes synaptic transmission[22,42]. The function of SynCAM 1 in synaptic differentiation was demonstrated in a co-culture assay of non-neuronal cells with dissociated hippocampal neurons (Figure 8.3). In this experimental approach, HEK293 cells are seeded atop the dissociated neurons at postnatal day 6–7, a time when intense synaptogenesis begins to occur in culture[43]. Using HEK293 cells transfected with SynCAM 1, the protein is presented at the surface of these non-neuronal cells to cultured hippocampal neurons. Analysis of synapse induction atop the surface of HEK293 cells is conducted 1–2 days after seeding. Three different observations demonstrate that SynCAM 1 induces presynaptic specializations.

First, immunostaining of these co-cultures for presynaptic marker proteins demonstrates that the neurons develop specializations containing synaptic vesicle proteins at contact sites with SynCAM 1-expressing HEK293 cells (Figure 8.3). Quantitative analysis confirms that this effect is specific[22].

Second, optical imaging studies of live co-cultures of SynCAM 1-expressing HEK293 cells with cultured neurons show that the specializations induced by SynCAM 1 contain actively recycling vesicles[22,42]. These experiments utilize FM styryl dyes to monitor synaptic vesicle endo- and exocytosis[44]. The quantum yield of FM dye fluorescence is highly increased when the dye partitions into a hydrophobic environment such as the membrane of a synaptic vesicle. In these live cell imaging experiments, FM dyes are added to dissociated neuronal cultures. Simultaneously, synaptic vesicle release is stimulated by depolarization. FM dyes are endocytosed within the lumen of synaptic vesicles, where their fluorescence is detected. Consequently, presynaptic terminals containing actively recycling synaptic vesicles are labeled. This approach allows optical imaging of synaptic vesicles and estimation of the size of synaptic vesicle pools in presynaptic specializations. Subsequent to dye loading, the kinetics of synaptic vesicle recycling can be determined in dye release experiments[45]. Optical FM dye imaging studies reveal that the specializations formed by neurons atop SynCAM 1-expressing HEK293 cells contain recycling vesicles that endocytose FM dyes during stimulation of synaptic vesicle release[22,42]. Significant amounts of FM dye are loaded into these specializations, indicating that their vesicle pool size is considerable. Furthermore, the neuronal specializations induced on top of SynCAM 1-expressing HEK293 cells have synaptic vesicle recycling properties remarkably similar to presynaptic terminals formed between neurons. SynCAM 1 is therefore able to induce presynaptic specializations containing actively recycling synaptic vesicles with properties highly similar to those found at normal synapses formed between neurons.

Third, the function of SynCAM 1 in synaptic differentiation is shown by the reconstitution of synaptic transmission *in vitro*. The optical imaging studies described above strongly indicate that the presynaptic specializations induced by SynCAM 1 release neurotransmitter onto the surface of the HEK293 cells, as they contain synaptic vesicles undergoing exocytosis. This was tested and confirmed using a modified co-culture assay. This assay employs HEK293 cells expressing both SynCAM 1, to induce presynaptic specializations, and the glutamate receptor subunit GluR2, to detect neurotransmitter released from excitatory presynaptic terminals. Patch-clamp recordings from these HEK293 cells in co-culture with neurons show that evoked responses can be detected under conditions of SynCAM 1 and GluR2 co-expression[22]. These experiments were the first to reconstitute synaptic transmission. This reconstituted system furthermore allows detection of excitatory miniature potentials[42].

Together, these co-culture experiments show that target-derived SynCAM 1 induces hippocampal neurons to develop functional presynaptic specializations upon axonal contact. This activity is limited to a few neuronal cell surface proteins. Among multiple cell adhesion and signaling molecules tested, only SynCAM 1 and neuroligins induce presynaptic specializations[17,22,42].

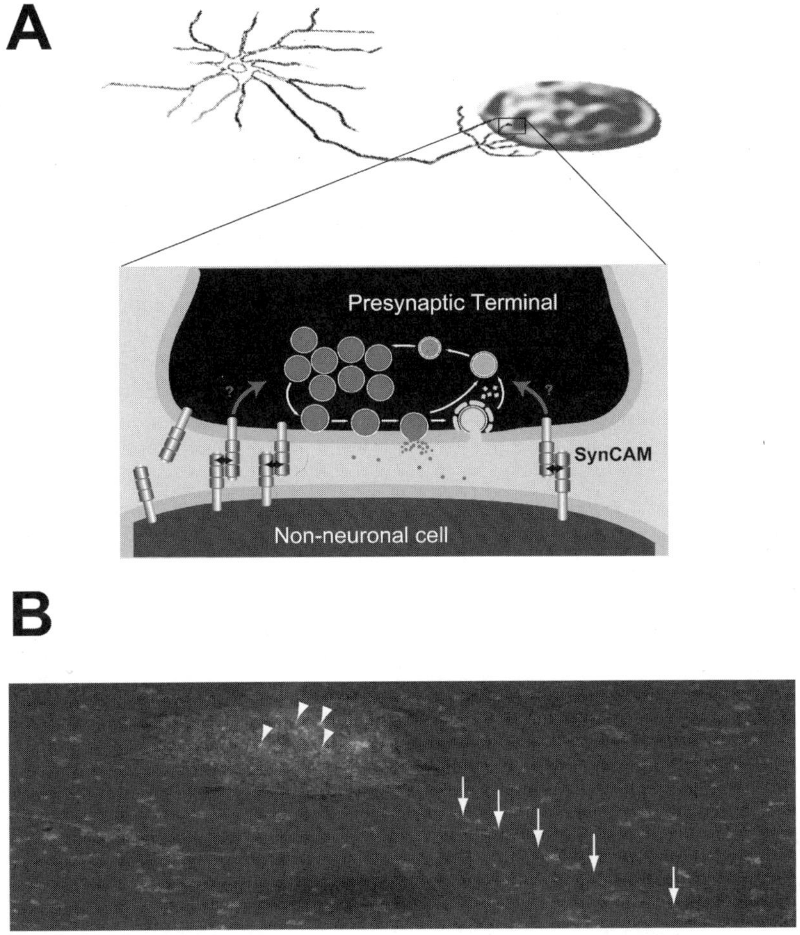


**Figure 8.3.** Induction of Presynaptic Specializations by SynCAM 1. (A) Model of the co-culture assay of hippocampal neurons with HEK293 cells expressing SynCAM 1. (B) Three-dimensional reconstruction of a confocal image of a HEK293 cell in co-culture with hippocampal neurons. The HEK293 cell is identified by co-expression of ECFP (gray), and co-cultures were stained for synaptophysin (white) as presynaptic vesicle marker. The image shows an axon displaying diffuse synaptophysin staining (arrows) that traverses a HEK293 cell. Atop this cell, the axon forms multiple presynaptic specializations (arrowheads).

In agreement with these results, SynCAM 1 affects excitatory neurotransmission. When overexpressed in hippocampal neurons, SynCAM 1 increases the frequency of miniature excitatory currents recorded from these neurons 2–3 fold[22,42]. This experimental design is analogous to the co-culture experiments described above, inasmuch as SynCAM 1 levels are increased on the postsynaptic side (contributed either by a HEK293 cell or a transfected neuron). The fact that the amplitude distribution of these responses is unaltered[22] indicates an effect by postsynaptic SynCAM 1 on the presynaptic inputs received by the SynCAM 1-overexpressing neuron. This effect of SynCAM 1 on mini frequencies is dependent on neuronal development. It is detectable at the peak of synaptogenesis, but not at later stages when synapses have matured further[42]. Spontaneous neurotransmission in these cultures increases after the peak of synapse formation (reflected in an increase of mini frequency over time[42]). These results may indicate a developmental time window during which SynCAM 1 exerts its activity. Alternatively, neurons at earlier stages may be more responsive so that SynCAM 1 can make a difference, and less so as they mature. Notably, SynCAM 1 enhances excitatory neurotransmission of hippocampal neurons but does not affect GABAergic transmission, indicating specific roles in the formation and/or function of glutamatergic synapses[42].

These results provide insights into the functions of postsynaptically expressed SynCAM 1. In addition, a dominant negative approach permitted identification of presynaptic functions of SynCAM 1 in cultured neurons. Overexpression of the cytosolic tail of SynCAM 1 in neurons, which can be expected to titrate out its endogenous binding partners, results in structural and functional synaptic changes[22]. Expression of the full-length cytosolic sequence of SynCAM 1 reduces the density of presynaptic puncta in overexpressing cultured hippocampal neurons, and interferes with synaptic vesicle recycling. These experiments demonstrate that the intracellular binding partners of SynCAM 1 have important roles in synapse differentiation and function.

## 8. STRUCTURE/FUNCTION ANALYSIS OF SynCAM 1 IN SYNAPSE INDUCTION

The structure/function analysis of SynCAM 1 provides important insights into the domains and protein interaction partners mediating SynCAM activities. All physiological functions of SynCAM 1 that have been identified in neurons depend on its extracellular Ig-like domains, as a mutant lacking the Ig domains is inactive[22,42]. This mutant of SynCAM 1 lacking the three Ig-like domains does not promote synaptic transmission and does not induce neurons to form presynaptic specializations[22,42]. These Ig-like domains of SynCAM 1 mediate its homophilic interactions as shown in cell aggregation assays and confirmed by biochemical approaches[22]. This interact of SynCAM 1 extracellular domains with each other is independent of $Ca^{2+}$. These results clearly demonstrate that the extracellular adhesive interactions of SynCAM 1 are required to mediate its synaptic activities.

The importance of the intracellular tail of SynCAM 1 was revealed by mutational analysis. Overexpression of the full-length tail in neurons interferes with presynaptic differentiation and function as described above. Mutant constructs lacking the last three carboxyl terminal amino acids of the SynCAM 1 cytosolic tail, however, are inactive[22]. This mutation abolishes binding of PDZ domain-containing adaptor molecules to SynCAM 1 as expected[22,32]. Several PDZ domain-containing adaptors are known to bind the cytosolic sequence of SynCAM 1, including CASK, Mint1, and syntenin[22]. CASK and Mint1 also bind to neurexins and are part of a tripartite complex that could link synaptic adhesion and the synaptic vesicle recycling machinery[46,47]. Syntenin binds to several other membrane and signaling proteins in addition to SynCAM 1. It is not known which of these molecules are the main interaction partners of SynCAM 1.

Several parallels between SynCAM and neurexins exist at the molecular level. Both proteins presumably drive presynaptic differentiation through interactions of their cytosolic tails (see Figure 8.1). Interestingly, their carboxyl terminal sequences are highly similar, and neurexins bind and recruit similar proteins to the plasma membrane, including Mint1, and CASK, itself binding several other adaptor molecules[46,48]. One of the molecules recruited by CASK is protein 4.1[37], a component of the cytoskeleton that promotes formation of actin/spectrin filaments[33]. A physical connection to cytoskeletal components is a feature common to adhesion molecules, and the protein 4.1 link can confer this property to SynCAM and neurexins alike. The parallels in the intracellular protein binding properties of neurexins and SynCAM make it likely that both engage overlapping presynaptic signaling pathways. Their extracellular domains, however, are not related, providing an example of how different proteins can perform analogous functions by means of shared protein interactions.

## 9. THE SynCAM FAMILY COMPRISES FOUR MEMBERS

Bioinformatic analysis revealed the SynCAM family as a new group of vertebrate-specific immunoglobulin superfamily members that comprises four genes[31]. The four SynCAM genes encode highly similar single-spanning membrane proteins with the same overall domain organization (Table 1). All SynCAM family members contain three extracellular immunoglobulin-like domains of the V-set, C1-set, and I-set subclasses. Their extracellular domains are related to each other but somewhat diverse in sequence. The intracellular sequences of all four SynCAM proteins, however, are very highly conserved. This indicates that the intracellular protein–protein interaction motifs play an important role for their functions. As SynCAM genes have been identified in different physiological contexts, several protein names can be found in the literature for the members of this family. Genes and splice products encoding the SynCAM 1-4 proteins have now been clarified in a unified nomenclature[31].

**Table 1.** Sequence Similarities of Human SynCAM Proteins. Numbers indicate percentage of amino acids identical or highly similar between family members. Similarities were calculated as described[31].

| | SynCAM 2 (%) | SynCAM 3 (%) | SynCAM 4 (%) |
|---|---|---|---|
| SynCAM 1 | 62 | 56 | 59 |
| SynCAM 2 | – | 70 | 59 |
| SynCAM 3 | – | – | 58 |

Both SynCAM 1 and 2 are alternatively spliced in the exons encoding the extracellular O-glycosylation stalk, the juxtamembraneous region rich in predicted O-glycosylation sites[31]. Splicing generates SynCAM forms that either lack or contain a stalk with up to 21 predicted O-glycosylation sites. This analysis of variable splice patterns draws attention to the O-glycosylation of SynCAM 1 and 2, but functional roles are not understood. Overall, the high similarity among members of the SynCAM family to SynCAM 1 and the fact that they are also expressed in brain[32] indicates that they could share functions with SynCAM 1 in the developing nervous system. This question is currently under investigation.

## 10. CONCLUSIONS AND FUTURE DIRECTION

The Ig superfamily member SynCAM 1 functions in synaptic differentiation during early postnatal developmental stages and is sufficient to drive the formation of presynaptic specializations. Concurrently, SynCAM 1 promotes neurotransmission between neurons and exerts this activity specifically at glutamatergic synapses in the CNS. Only the neurexin/neuroligin adhesion system shares similar functions. Thus, these synaptic adhesion molecules not only bridge the synaptic cleft, but can also initiate subsequent steps of synapse formation. SynCAM 1 has been identified in additional physiological contexts, which could provide new directions in understanding this gene family. For example, SynCAM 1 may have neuroimmunological roles, as indicated by the fact that it engages in adhesion of neurites to mast cells[49]. SynCAM 1 also appears to have tumor suppressor activity[28], which could be due to roles of this adhesion molecule in cell migration and differentiation. However, the best understood functions of SynCAM 1 to date are in synaptic differentiation.

Three key questions are the focus of current studies of SynCAM 1 in synaptogenesis. First, what are the intracellular interaction partners of SynCAM 1 that signal and organize the formation of synaptic membrane specializations? Several intracellular candidate proteins are known to interact with SynCAM 1, but the extent to which they function downstream remains to be investigated. Furthermore, it is likely that other adaptor and signaling molecules bind SynCAM 1, and their identification and characterization will prove critical for understanding SynCAM 1 function. Second, the extracellular interactions of SynCAM 1 need to be characterized in molecular detail. These studies will identify the extracellular partners of SynCAM 1 beyond its own homophilic binding. Candidate extracellular interacting proteins include the other SynCAM family members. Future questions will also include whether these Ig domain-mediated interactions are regulated by N- and O-glycosylation of the SynCAM 1 extracellular domain. Third, the physiological functions of SynCAM 1 in the brain will have to be further characterized using a combination of *in vitro* and *in vivo* approaches, notably mouse genetics.

These studies can lead to insights not only into the roles of SynCAM 1 in synapse induction, but also in other aspects of synaptic differentiation. The possibility that SynCAM 1 exerts functions subsequent to synapse induction is indicated by the fact that it remains expressed in adult animals after the peak of synaptogenesis has occurred. It is conceivable that SynCAM 1 promotes maturation and stability of synapses after they are formed. SynCAM 1 therefore could affect synaptic plasticity subsequent to triggering synapse formation. Since SynCAM 1 is the representative member of a small family of four highly conserved vertebrate genes expressed in the developing brain, their characterization may provide further insights into mechanisms of target recognition, synaptic specificity, and synapse differentiation mediated by these Ig superfamily members. Future studies are expected to throw light on the roles of SynCAM proteins at developing synapses and on the extent to which they affect synaptic function.

## 11. REFERENCES

1. Fletcher, T.L., Cameron, P., De Camilli, P., and Banker, G. (1991) *J Neurosci* **11**, 1617–1626.
2. Williams, A.F., and Barclay, A.N. (1988) *Ann Rev Immunol* **6**, 381–405.
3. Rougon, G., and Hobert, O. (2003) *Ann Rev Neurosci* **26**, 207–238.

4. Tessier-Lavigne, M., and Goodman, C.S. (1996) *Science* **274**, 1123–1133.
5. Walsh, F.S., and Doherty, P. (1997) *Ann Rev Cell Developl Biol* **13**, 425–456.
6. Yu, T.W., and Bargmann, C.I. (2001) *Nat Neurosci* **4**, 1169–1176.
7. Bastiani, M.J., Harrelson, A.L., Snow, P.M., and Goodman, C.S. (1987) *Cell* **48**, 745–755.
8. Harrelson, A.L., and Goodman, C.S. (1988) *Science* **242**, 700–708.
9. Schuster, C.M., Davis, G.W., Fetter, R.D., and Goodman, C.S. (1996) *Neuron* **17**, 641–654.
10. Stewart, B.A., Schuster, C.M., Goodman, C.S., and Atwood, H.L. (1996) *J Neurosci* **16**, 3877–3886.
11. Davis, G.W., Schuster, C.M., and Goodman, C.S. (1997) *Neuron* **19**, 561–573.
12. Baines, R.A., Seugnet, L., Thompson, A., Salvaterra, P.M., and Bate, M. (2002) *J Neurosci* **22**, 6587–6595.
13. Shen, K., and Bargmann, C.I. (2003) *Cell* **112**, 619–630.
14. Shen, K., Fetter, R.D., and Bargmann, C.I. (2004) *Cell* **116**, 869–881.
15. Yamagata, M., Weiner, J.A., and Sanes, J.R. (2002) *Cell* **110**, 649–660.
16. Nguyen, T., and Südhof, T.C. (1997) *J Biol Chem* **272**, 26032–26039.
17. Scheiffele, P., Fan, J., Choih, J., Fetter, R., and Serafini, T. (2000) *Cell* **101**, 657–669.
18. Dean, C., Scholl, F.G., Choih, J., DeMaria, S., Berger, J., Isacoff, E., and Scheiffele, P. (2003) *Nature Neurosci* **6**, 708–716.
19. Graf, E.R., Zhang, X., Jin, S.-X., Linhoff, M.W., and Craig, A.M. (2004) *Cell* **119**, 1013–1026.
20. Chih, B., Engelman, H., and Scheiffele, P. (2005) *Science*, 1324–1328.
21. Levinson, J.N., Chery, N., Huang, K., Wong, T.P., Gerrow, K., Kang, R., Prange, O., Wang, Y.T., and El-Husseini, A. (2005) *J Biol Chem* **280**, 17312–17319.
22. Biederer, T., Sara, Y., Mozhayeva, M., Atasoy, D., Liu, X., Kavalali, E.T., and Südhof, T.C. (2002) *Science* **297**, 1525–1531.
23. Hung, A.Y., and Sheng, M. (2002) *J Biol Chem* **277**, 5699–5702.
24. Montgomery, J.M., Zamorano, P.L., and Garner, C.C. (2004) *Cell Molec Life Sci* **61**, 911–929.
25. Kim, E., and Sheng, M. (2004) *Nat Rev Neurosci* **5**, 771–781.
26. Gomyo, H., Arai, Y., Tanigami, A., Murakami, Y., Hattori, M., Hosoda, F., Arai, K., Aikawa, Y., Tsuda, H., Hirohashi, S., Asakawa, S., Shimizu, N., Soeda, E., Sakaki, Y., and Ohki, M. (1999) *Genomics* **62**, 139–146.
27. Pletcher, M.T., Nobukuni, T., Fukuhara, H., Kuramochi, M., Maruyama, T., Sekiya, T., Sussan, T., Isomura, M., Murakami, Y., and Reeves, R.H. (2001) *Gene* **273**, 181–189.
28. Kuramochi, M., Fukuhara, H., Nobukuni, T., Kanbe, T., Maruyama, T., Ghosh, H.P., Pletcher, M., Isomura, M., Onizuka, M., Kitamura, T., Sekiya, T., Reeves, R.H., and Murakami, Y. (2001) *Nat Genet* **27**, 427–430.
29. Wakayama, T., Ohashi, K., Mizuno, K., and Iseki, S. (2001) *Molec Reprod Develop* **60**, 158–164.
30. Urase, K., Soyama, A., Fujita, E., and Momoi, T. (2001) *Neuroreport* **12**, 3217–3221.
31. Biederer, T. (2006) *Genomics* **87**, 139–150.
32. Biederer, T., and colleagues, unpublished observation.
33. Hoover, K.B., and Bryant, P.J. (2000) *Curr Opin Cell Biol* **12**, 229–234.
34. Marfatia, S.M., Leu, R.A., Branton, D., and Chishti, A.H. (1995) *J Biol Chem* **270**, 715–719.
35. Hemming, N.J., Anstee, D.J., Staricoff, M.A., Tanner, M.J., and Mohandas, N. (1995) *J Biol Chem* **270**, 5360–5366.
36. Workman, R.F., and Low, P.S. (1998) *J Biol Chem* **273**, 6171–6176.
37. Biederer, T., and Südhof, T.C. (2001) *J Biol Chem* **276**, 47869–47876.
38. Songyang, Z., Fanning, A.S., Fu, C., Xu, J., Marfatia, S.M., Chishti, A.H., Crompton, A., Chan, A.C., Anderson, J.M., and Cantley, L.C. (1997) *Science* **275**, 73–77.
39. Fukami, T., Satoh, H., Fujita, E., Maruyama, T., Fukuhara, H., Kuramochi, M., Takamoto, S., Momoi, T., and Murakami, Y. (2002) *Gene* **295**, 7–12.
40. Fiala, J.C., Feinberg, M., Popov, V., and Harris, K.M. (1998) *J Neurosci* **18**, 8900–8911.
41. Harris, K.M., Jensen, F.E., and Tsao, B. (1992) *J Neurosci* **12**, 2685–2705.
42. Sara, Y., Biederer, T., Atasoy, D., Chubykin, A., Mozhayeva, M.G., Südhof, T.C., and Kavalali, E.T. (2005) *J Neurosci* **25**, 260–270.
43. Fletcher, T.L., De Camilli, P., and Banker, G. (1994) *J Neurosci* **14**, 6695–6706.
44. Rizzoli, S.O., Richards, D.A., and Betz, W.J. (2003) *J Neurocyt* **32**, 539–549.
45. Mozhayeva, M.G., Sara, Y., Liu, X.R., and Kavalali, E.T. (2002) *J Neurosci* **22**, 654–665.
46. Biederer, T., and Südhof, T.C. (2000) *J Biol Chem* **275**, 39803–39806.
47. Butz, S., Okamoto, M., and Südhof, T.C. (1998) *Cell* **94**, 773–782.
48. Tabuchi, K., Biederer, T., Butz, S., and Südhof, T.C. (2002) *J Neurosci* **22**, 4264–4273.
49. Furuno, T., Ito, A., Koma, Y.-I., Watabe, K., Yokozaki, H., Bienenstock, J., Nakanishi, M., and Kitamura, Y. (2005) *J Immunol* **174**, 6934–6942.

# 9

# PROTOCADHERINS AND SYNAPSE DEVELOPMENT

Joshua A. Weiner*

## 1. SUMMARY

Three families of protocadherin (Pcdh) genes, *Pcdh-α, -β, and -γ,* are clustered in a tandem array on a single chromosome in mammals. Members of these *Pcdh* families are expressed in an overlapping, combinatorial manner in the nervous system and the proteins they encode are present at some synaptic sites. This has led to the hypothesis that protocadherins may provide an adhesive code involved in the specification of synaptic patterning. Mice in which the 22-gene *Pcdh-γ* cluster has been deleted or disrupted die shortly after birth with synaptic defects that may be consistent with such a function. This chapter reviews these and other recent data on the organization, expression, and function of the clustered *Pcdh* genes and enumerates several unanswered questions that will be important to address in future work.

## 2. INTRODUCTION

Synaptic transmission between neurons in a circuit is the fundamental organizing principle of brain function. It is generally assumed, therefore, that disruptions in the formation, patterning, or maintenance of synapses underlie a wide range of debilitating neurological and psychiatric disorders. As tautologous as this assumption may be, its usefulness for understanding and, eventually, treating such disorders will require identification of the molecules that control synapse development in the central nervous system (CNS). Much progress has been made in the past decade, with many of the candidates thus far identified being synaptic adhesion molecules, including the "classical" cadherins (see Chapter 5); immunoglobulin superfamily members such as the sidekicks[1], NCAM (see Chapter 6), SynCAM (see Chapter 8), and the SYGs[2,3]; and the neuroligins/ neurexins (see Chapter 7; for recent reviews, see refs. 4–6). The critical roles identified for such a diverse group of synaptic adhesion molecules provide experimental support for the

*Department of Biological Sciences and Neuroscience Graduate Program, University of Iowa, 143 Biology Building, Iowa City, IA 52242, USA; joshua-weiner@uiowa.edu

classic "chemoaffinity" hypothesis of Sperry[7], which suggested that the specificity of synaptic patterning could be mediated by cell-surface tags interacting in a "lock-and-key" manner to join pre- and postsynaptic partners.

The ~60 protocadherin adhesion molecules encoded by the *Pcdh-α, -β,* and *-γ* gene families have, since their discovery[8], been widely discussed as leading candidates for mediating the molecular recognition of synaptic partners due to their diversity, combinatorial neuronal expression, and partial synaptic localization. In this chapter, I review what is known about the organization, expression, and function of the clustered protocadherin genes. Particular attention is paid to the γ-protocadherins, for which our recent work has identified critical functions in synapse development and neuronal survival[9,10].

## 3. CLONING OF PROTOCADHERIN GENES AND ORGANIZATION OF THE *Pcdh-α, -β,* AND *-γ* GENE CLUSTERS

The term "protocadherin" was coined in 1993 by Shintaro Suzuki and co-workers[11] to refer to two novel cDNAs isolated in a degenerate oligonucleotide PCR screen for sequences homologous to the "classical" cadherins. Since then, nearly 70 protocadherin genes have been identified, all of which exhibit varying numbers of ectodomains similar to those of the classical cadherins that mediate $Ca^{2+}$-dependent cell–cell adhesion. This makes the protocadherins the largest and most diverse group in the cadherin superfamily, which also includes the type I and II classical cadherins (e.g., *N*-cadherin, *E*-cadherin, cadherin-7) that form adherens junctions, the desmocollins and desmogleins that form desmosomes, the 27- or 34-ectodomain proteins Dachsous and Fat, and the 7 transmembrane-domain Flamingo cadherins (reviewed in refs. 12,13; see Chapter 5).

The genes isolated by Sano et al.[11], initially termed *protocadherin-42/Pcdh1* and *protocadherin-43/Pcdh2*, encoded transmembrane proteins with six or seven cadherin ectodomains. The cytoplasmic domains of these protocadherins, however, differed significantly from those of the classical cadherins, apparently lacking their characteristic catenin-binding sites. These and several other protocadherin genes subsequently isolated by Suzuki's group[14–18] and by that led by Takeshi Yagi[19] were found to be widely expressed in the brain and, in some cases, localized to synapses, providing the first hint of the importance of protocadherins for nervous system function.

A major advance in the field came in 1999, when Wu and Maniatis discovered over 50 genes encoding three families of protocadherins (termed *Pcdh-α, -β,* and *-γ*) clustered in a tandem array spanning ~750 kilobase on human chromosome 5q31[8]. An orthologous array with a similar arrangement was subsequently found on mouse chromosome 18[20] and, with some variation, in chimpanzee[21], rat[21,22], chicken[23], zebrafish[21,24,25], and even coelacanth[26]. These observations extended, and gave coherence to, a number of previous reports of novel protocadherin genes that are now known to be members of the *Pcdh-α, -β,* or *-γ* loci: for example, *protocadherin-43/Pcdh2*[11,15] and *protocadherin2C*[18] reside in the *Pcdh-γ* locus, *Pcdh3*[14] is a member of the *Pcdh-β* family, and the eight genes reported as cadherin-related neuronal receptors (CNRs ; ref. 19) make up more than half of the *Pcdh-α* family.

The organization of the *Pcdh-α, -β,* and *-γ* gene clusters is reminiscent of the immunoglobulin and T-cell receptor genomic loci, in that multiple "variable" (V) exons encoding variant isoforms are arrayed in tandem and, for the *Pcdh-α and -γ* clusters, are spliced to three short, invariant "constant" (C) exons (Figure 9.1; the *Pcdh-β* locus is similar but lacks any constant exons; refs. 8,20,21). Each large V exon encodes the six cadherin ectodomains, the transmembrane domain, and the proximal cytoplasmic domain of a single Pcdh isoform, with the C-terminal having 125–150 amino acids of the α- and γ-Pcdhs encoded by their respective C exons (Figure 9.1). The resulting α- and γ-Pcdh proteins are thus exactly the kind of molecules that one might predict would underlie a Sperry-like "chemoaffinity" mechanism: their divergent extracellular domains could engage in a variety of specific homo- or heterophilic interactions at the cell surface, while their shared cytoplasmic domain could be involved in conserved signaling mechanisms leading to the assembly or stabilization of the synaptic apparatus. In mice, there are 14 *Pcdh-α*, 22 *Pcdh-β*, and 22 *Pcdh-γ*V exons (20; see Figure 9.1). The *Pcdh-γ* genes can be further grouped into three subfamilies, termed *a* (12 genes), *b* (7 genes), and *c* (3 genes), based on sequence similarity (Figure 9.1). Two of the *Pcdh-α* genes are referred to as *c1* and *c2* because their similarity to the *Pcdh-γ*-c subfamily (*c3, c4,* and *c5*) is greater than their similarity to the other 12 *Pcdh-α* genes[8,21].

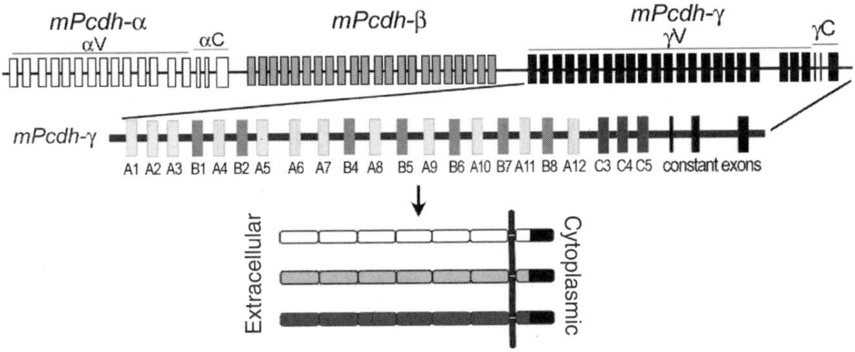


**Figure 9.1.** The Murine *Pcdh-α, -β,* and *-γ* Gene Clusters. Top: a schematic representation of the three clustered *Pcdh* loci on ~900 kilobases of mouse chromosome 18. V, variable exon; C, constant exon. Middle: an expanded view of the *Pcdh-γ* locus, showing the three subfamilies (A, B, and C) of V exons. Bottom: Each V exon, which encodes most of a single γ-Pcdh isoform, is expressed from its own promoter and subsequently spliced to the C exons, which encode a shared C-terminal cytoplasmic domain.

This intriguing genomic organization raised the possibility that the *Pcdh* genes could be substrates for a long-hypothesized somatic DNA recombination mechanism in the CNS, similar to that employed by the immune system to generate the vast repertoire of antibodies and T-cell receptors[27–30]. Subsequent detailed analyses of *Pcdh-α* and *-γ* gene expression, however, have failed to provide support for such a mechanism, demonstrating instead that expression is controlled by promoter choice and alternative V exon splicing[31,32]. The sequence upstream of each *Pcdh* V exon contains its own promoter region, including an approximately 20-base pair conserved sequence element that is required for expression. Through mechanisms that remain obscure, a given V exon promoter is "chosen" and transcription through the remaining portion of the *Pcdh-α* or *-γ* cluster proceeds; intervening V exons are then removed when the 5′ V exon is *cis*-spliced to the downstream C exons[31,32]. Although the *Pcdh-β* cluster does not contain its own C exons, each *Pcdh-β* V exon harbors a consensus 5′ splice site near its end[8], suggesting the possibility of splicing to the C exons of other clusters. Such intercluster spliced transcripts, while rare, are apparently present in neurons, as are low levels of αV/γC and γV/αC hybrid transcripts produced by *trans*-splicing between separate pre-mRNA intermediates[31–33]. Even given that such intercluster splicing is rare, it has the potential to greatly increase the already-significant diversity of the proteins encoded by the *Pcdh-α*, *-β*, and *-γ* clusters.

Although the adhesive capability and specificities of most of the clustered protocadherins remain unknown, the extant studies on individual β- or γ-Pcdh isoforms[11,14,15,34] suggest that they mediate calcium-dependent, homophilic adhesion that is weaker than that of the classical cadherins. For one γ-Pcdh isoform (γC3/Pcdh2), homophilic adhesion could be strengthened by replacing its cytoplasmic domain with that of *E*-cadherin[15]. Interestingly, existing data suggest that the α-Pcdhs may not be homophilic adhesion molecules; instead, an α-Pcdh/CNR isoform (α4/CNR1) was found to mediate heterophilic adhesion to β1 integrins via an RGD site in its first ectodomain[35]. It has also been reported that several α-Pcdh/CNR proteins can bind reelin, an extracellular matrix-associated protein critical for lamination of the cerebral cortex, also in an RGD-dependent manner[36]. A function as reelin receptors would, in fact, be consistent with expression studies showing juxtaposition of *CNR*-positive and *reelin*-positive cells in the cortex and spinal cord[36,37]. Other researchers have, however, been unable to replicate the binding of α-Pcdhs to reelin[38,39]; this, combined with extensive validation of the lipoprotein receptors ApoER2 and VLDLR as reelin receptors[40], makes it unclear at present whether α-Pcdhs play any role in reelin signaling.

## 4. EXPRESSION AND LOCALIZATION OF THE α-, β-, AND γ-PROTOCADHERINS IN THE NERVOUS SYSTEM

The initial studies of individual protocadherin genes now known to reside within the *Pcdh-α*, *-β*, or *-γ* clusters found that their expression was essentially limited to the nervous system[11,14,16,18,19]. This has been confirmed by more recent studies examining the expression of entire clusters by using riboprobes against the *Pcdh-α* or *-γ* constant exons; essentially all neurons express *Pcdh* genes, and outside of the nervous system, only weak lung and muscle expression has been found[10,11,34]. Expression begins during the embryonic period, and in most cases has been found to peak in the early postnatal period, a time during which many synapses are being formed[19,34]. *In situ* hybridization using probes specific for individual V exons has found that they exhibit distinct but apparently overlapping expression patterns within the CNS. Most experiments have relied on side-by-side comparisons of individual expression patterns to reach this conclusion[16,18,34], but limited studies employing double-label *in situ* hybridization with pairs of *Pcdh-α*[19] or *Pcdh-γ*[10]; J.A.W., unpublished data) probes have confirmed that neurons can express more than one gene from each cluster. An elegant and technically impressive recent study[41] used single-cell RT-PCR of neurons and polymorphisms that can distinguish between two mouse strains to show that expression of *Pcdh-α* V exons is monoallelic but combinatorial. Single cerebellar Purkinje cells were found to express 2-4 *Pcdh-α* V exons; while the gene *clusters* of both alleles were active in a given neuron, transcripts of each *individual* V exon derived solely from one allele[41]. This differs from previously reported forms of monoallelic gene expression and, given the presence of nonsynonymous single-nucleotide polymorphisms in human and mouse *Pcdh-α* V exons[33,42,43], suggests yet another mechanism that may increase the diversity of protocadherin interactions.

The hypothesis that such interactions might play a role in the establishment of synaptic connections is supported by studies using antibodies raised against the constant domains of α- or γ-Pcdhs. Subcellular fractionation of brain lysates followed by Western blotting has consistently found that α-Pcdh[19,44] and γ-Pcdh[10,44,45] proteins are concentrated in synaptic membrane and post-synaptic density (PSD) fractions. In addition, proteomic methods have identified γ-Pcdhs as components of the "presynaptic particle web" that provides a scaffold for organizing neurotransmitter release[46]. Consistent with this biochemical analysis, immunostaining of brain sections with antibodies against the constant domains of α- or γ-Pcdhs yields a punctate, neuropil-like pattern[19,44]. Our studies using a mouse line in which the gene for green fluorescent protein (GFP) is fused to the third *Pcdh-γ* constant exon[10], found a similar pattern of fusion protein localization and further indicated that γ-Pcdhs are highly concentrated in synapse-rich zones such as the retinal inner and outer plexiform layers. Within these zones, some γ-Pcdh puncta overlap with puncta labeled by antibodies against presynaptic (SV2, synaptophysin) and postsynaptic (PSD-95, gephyrin) proteins, but there is clearly substantial extrasynaptic γ-Pcdh as well[10]. This suggests that the γ-Pcdhs may localize to particular types of synapses, which is consistent with immunostaining studies of cultured hippocampal neurons. In hippocampal cultures, the γ-Pcdhs are highly expressed by inhibitory interneurons, but in an apparently nonsynaptic pattern, while the half of the spiny excitatory neurons that express γ-Pcdhs localize them to only a portion of their synapses[45].

These data have been confirmed and extended by Phillips et al.[45], who performed immunogold electron microscopy with anti-γ-Pcdh constant domain antibodies. This work clearly showed that while some hippocampal synapses contain γ-Pcdh protein at both the pre- and postsynaptic membranes, much of the labeling was in "tubulovesicular" structures within axonal terminals and dendritic profiles[45]. Although the precise function of these structures is not yet clear, they may represent endosomal compartments that could mediate dynamic insertion of γ-Pcdhs at synapses, perhaps in response to activity or during synaptic maturation. An α-Pcdh isoform (α4/CNR1) was also detected at the synapse by immunogold electron microscopy[19]; although α-Pcdhs are clearly located extrasynaptically as well, it is not yet known whether they are found in "tubulovesicular" structures as are the γ-Pcdhs. In the chick ciliary ganglion, α-Pcdhs have been reported to be present in axons and perisynaptic sites, rather than directly at the synapse[47]. There does appear to be substantial co-localization of α- and γ-Pcdhs in hippocampal neurons both *in vitro* and *in vivo*, and results of a recent study[44] suggest that this is more than coincidental. In many cell lines transfected with a construct expressing a full-length α-Pcdh isoform, the exogenous protein failed to localize efficiently to the cell surface[35]. Murata et al.[44] present evidence that efficient surface expression of α-Pcdhs requires a *cis*-interaction with γ-Pcdhs. Antibodies against the α-Pcdhs can co-immunoprecipitate γ-Pcdhs in brain lysates, and affinity-tag pull-down assays with truncated Pcdh proteins suggest that both the extracellular and cytoplasmic domains of α- and γ-Pcdhs can interact[44]. Such interactions likely underlie the increased cell-surface delivery of an exogenous α-Pcdh isoform when HEK293T cells are co-transfected with constructs encoding γ-Pcdh isoforms[44]. Whether interactions between α- and γ-Pcdhs are important for their surface expression in neurons remains to be seen, but such a mechanism could contribute to the dynamic modulation of protocadherin-mediated adhesion at the synapse.

## 5. GENETIC ANALYSIS OF γ-PROTOCADHERIN FUNCTION

Despite the great interest in the clustered protocadherins, we still know relatively little about their function. The multiplicity of *Pcdh-α*, *-β*, and *-γ* genes makes genetic redundancy likely, such that ablation of any single isoform by traditional gene knockout methods might not produce any detectable phenotype. In addition, because there are few antibodies against individual protocadherin isoforms available, we know almost nothing about the neuronal subsets that express each particular isoform and thus about the types of connections they make. This would make it difficult to know where to look for defects in any single *Pcdh* gene knockout animals, should such defects be subtle. Conversely, since multiple *Pcdh* genes are expressed by (probably) every neuron in the nervous system, forced over- or mis-expression of any single isoform might not have any observable effect on neuronal function. Our own work has avoided these problems by taking advantage of mouse mutants, produced in the course of gene expression studies[31], in which the entire *Pcdh-γ* gene cluster has been deleted or otherwise disrupted[9,10].[*] Analysis of these *Pcdh-γ* mutant mice has provided the first genetic evidence that at least some of the clustered protocadherins play critical roles in nervous system development and function.

The *Pcdh-γ* deletion mutant was produced in mouse embryonic stem (ES) cells by sequential targeting to insert loxP DNA sequences both upstream of V exon γA1 and within C exon 3[10,31]. Subsequent expression of Cre recombinase in the double-targeted ES cells led to excision of the entire ~300 kb *Pcdh-γ* locus. The resulting ES cells were used to generate chimeric mice, which were then bred to obtain germline transmission. Homozygous *Pcdh-γ* deletion mice were born alive in the expected Mendelian ratio, but died within hours after birth. Mutant pups exhibited multiple signs of severe neurological defects: a hunched posture, repetitive limb tremor, lack of voluntary movements, and nearly nonexistent reflexes and withdrawal behaviors. No obvious defects were found in non-neural organs, and analysis of intramuscular nerves and neuromuscular junctions suggested that they had formed during embryonic development and were grossly normal. Within the CNS, however, severe neuronal defects were readily apparent[10].

[*]The work described in this section and published in refs. (9) and (10) was performed while the author was a postdoctoral fellow in the laboratory of Dr. Joshua R. Sanes (Washington University School of Medicine; now at Harvard University), in collaboration with Dr. Xiaozhong Wang (Baylor College of Medicine; now at Northwestern University).

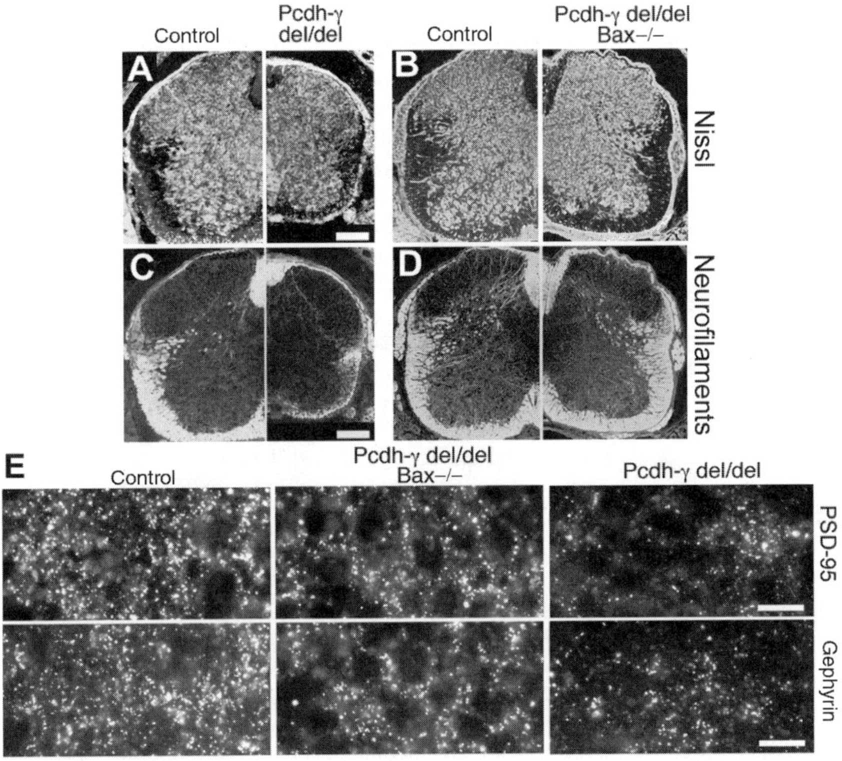


**Figure 9.2.** Genetically Dissociable Neurodegeneration and Synaptic Defects in Mice Lacking the *Pcdh-γ* Locus. (A–D) Sections through the spinal cords of P0 control, *Pcdh-γ* deletion homozygotes (del/del), and *Pcdh-γ* deletion/Bax knockout double mutants (del/del; Bax –/–), stained for Nissl substance (A,B) to reveal cell bodies or with antibodies to neurofilaments (C,D) to label axonal fiber tracts. The reduction in spinal cord size due to neuron and fiber tract loss observed in *Pcdh-γ* deletion mice (A,C) is rescued by additional deletion of Bax (B,D). Double mutants still, however, display severe reductions in both excitatory (PSD-95+) and inhibitory (gephyrin +) synaptic puncta in spinal cord sections (E). These reductions are not quite as severe as those in the *Pcdh-γ* deletion mutants, in which the death of neurons exacerbates synapse loss, but are still very significant, approaching 50% (see ref. 9). Bar = 100 μm in (A–D), 12 μm in (E).

The spinal cords of *Pcdh-γ* deletion homozygotes were greatly reduced in size and neuronal density compared to control littermates, despite the fact that the pups themselves were of normal stature (Figure 9.2A). Analyses of *Pcdh-γ* deletion mutants at multiple embryonic stages demonstrated that spinal cord neurons proliferated, migrated out of the ventricular zone, extended axons and dendrites and expressed cell-type specific markers normally in the mutant embryos. At late embryonic stages, however, mutant interneurons, particularly those in the intermediate gray and ventral horn, underwent apoptotic cell death, as detected by TUNEL staining, immunoreactivity for cleaved caspase-3, and the prevalence of pyknotic nuclei at both the light and electron microscopic levels[10]. Consistent with this, mutant spinal cords exhibited multiple signs of neurodegeneration: loss of neurofilament-stained fiber tracts in the lateral and ventral funiculi (Figure 9.2C), reactive gliosis, and macrophage invasion. Similar evidence of normal early development followed by neurodegeneration was found in the brain, with the affected regions being those that are most mature at birth, when the mutants died. Together, these data suggested a previously unhypothesized role for the γ-Pcdhs in promoting the survival of particular neuronal populations.

So what about synapses? Electron microscopic analysis uncovered synaptic profiles in thin sections of neonatal mutant spinal cord that were grossly normal in appearance, with a well-defined synaptic cleft, docked vesicles, and a PSD[10]. This demonstrates that γ-Pcdhs are not strictly required for all synapse formation *per se*. However, such synaptic profiles were much rarer in mutant spinal cord as compared to controls, and immunostaining for presynaptic (synaptophysin) and postsynaptic (PSD-95, gephyrin) markers confirmed a 60–70% reduction in synaptic density (Figure 9.2E). Because so many neurons had been lost by the time of birth, it remained possible that the paucity of synapses was entirely a secondary effect. That this was not the case was, however, suggested by the fact that when synapses were quantified as the number of PSD-95- or gephyrin-positive puncta *per neuron*, a reduction of similar magnitude was still observed[10]. Nevertheless, our initial analysis of *Pcdh-γ* deletion mutants, while demonstrating the critical importance of these genes for neural development, could not definitively identify a synaptic function.

This uncertainty has been resolved recently through the use of two novel genetic approaches[9]. First, mice harboring the *Pcdh-γ* deletion mutation were crossed with those lacking the gene for Bax, a pro-apoptotic member of the Bcl-2 family in the absence of which much neuronal apoptosis is blocked[48]. As predicted, this results in a complete rescue of the neurodegenerative phenotype: Double-mutant spinal cords were normal in size, neuron number, and fiber tract density, and lacked any morphological or biochemical evidence of apoptosis (Figure 9.2B,D). However, *Pcdh-γ* deletion/Bax knockout mice still died within hours of birth and exhibited neurological defects similar to, if less severe than, those of mice harboring the deletion allele alone. This is apparently due to the fact that, even in the absence of neurodegeneration, loss of γ-Pcdh function leads to severe reductions in synaptic density: PSD-95-positive puncta were reduced by over 35%, and gephyrin or glycine receptor-positive puncta were reduced by 50% in spinal cord sections (Figure 9.2E)[9].

A possible caveat in these results is that neurons in the double mutants could be "living dead"; that is, unhealthy in some way, but artificially able to complete the apoptotic program due to the absence of Bax. A second approach using mice harboring a hypomorphic *Pcdh-γ* allele has, however, obviated this concern. The *Pcdh-γ* truncation allele encodes γ-Pcdhs lacking the C-terminal 57 amino acids due to insertion of an internal ribosome entry site (IRES)-GFP-LacZ reporter cassette into the third C exon[9]. This reporter cassette proved to be nonfunctional in the resulting mice, but serendipitously results in a hypomorph that produces only very low levels of γ-Pcdh proteins. These *Pcdh-γ* truncation mice display a dissociation of neurodegenerative and synaptic phenotypes similar to that observed in the *Pcdh-γ* deletion/Bax knockout double mutants. Their spinal cords appear normal in structure and neuronal density, and levels of apoptosis are low. Nevertheless, *Pcdh-γ* truncation mice die within hours of birth, exhibiting neurological defects due to a reduction of 40–50% in the density of both excitatory and inhibitory spinal cord synapses[9]. Unfortunately, the fact that the truncation mutants are likely hypomorphic for all γ-Pcdh proteins makes it impossible to conclude whether the observed synaptic defects are due to the C-terminal truncation that these mice harbor. These results do, however, suggest that the

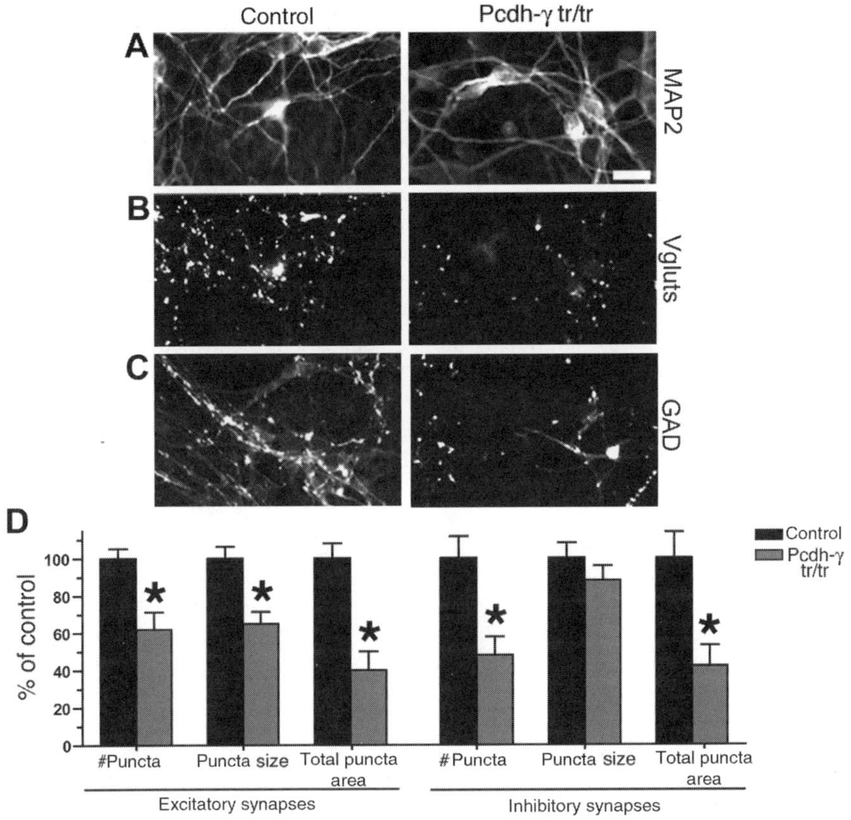


**Figure 9.3.** Spinal Neurons Deficient in γ-Pcdhs Make Fewer Synapses *in Vitro*. (A–C) Spinal neuron cultures from control and homozygous *Pcdh-γ* truncation embryos (tr/tr) cultured for 9 days and immunostained for the somato-dendritic protein MAP2 (A) and markers of excitatory (VGluts, vesicular glutamate transporters; B) or inhibitory (GAD, glutamic acid decarboxylase; C) presynaptic terminals. Despite normal neuronal differentiation in mutant cultures (see ref. 9 for further data), the numbers of both excitatory and inhibitory synapses are significantly reduced. Quantification confirms this reduction and further demonstrates a significant reduction in the size of excitatory terminals (D; $^{*}p < 0.001$, mean ± SEM of data from four cultures per genotype). Bar = 30 μm. The micrographs in A–C show the same fields within each genotype.

missing cytoplasmic region is not strictly required for potential γ-Pcdh functions in the promotion of spinal interneuron survival.

To further substantiate the synaptic defects observed in *Pcdh-γ* mutant mice, as well as to establish a model system in which the mechanism of these defects could be elucidated, spinal cord neurons were cultured from *Pcdh-γ* truncation mutant embryos[9]. Mutant cultures exhibited normal neuronal differentiation with no evidence of excessive apoptosis, as expected from our analysis of the intact spinal cord. Immunostaining of mature cultures (>8 days in vitro) for synaptic markers, however, revealed that *Pcdh-γ* mutant neurons made 40–50% fewer synapses than did control neurons. The number of both excitatory (vesicular glutamate transporter-positive) and inhibitory (glutamic acid decarboxylase-positive) terminals were significantly reduced, as was the size of excitatory terminals (Figure 9.3). Patch-clamp recordings of individual neurons in these cultures revealed that *Pcdh-γ* mutant neurons exhibited normal resting membrane potentials, and were able to conduct action potentials at normal depolarization thresholds. However, the amplitude of both excitatory and inhibitory spontaneous synaptic currents was reduced by ~50% in mutant neurons[9]. The reduced current amplitudes in mutant neurons, along with the decrease in the size of some synaptic terminals, suggest that γ-Pcdhs may be important for the maturation of synaptic contacts into fully functioning synapses. The significant reduction in the number of immunostained synaptic puncta could indicate a role for γ-Pcdh function in either initial synaptogenesis or in synaptic maintenance.

Together, our data provide the first genetic analysis of any of the clustered protocadherins, demonstrating that the γ-Pcdhs are critically required for both synapse development and neuronal survival in the spinal cord. While confirming that these molecules are indispensable for neural development is an important first step, much remains unknown about the mechanisms by which protocadherins exert their functions. The following section highlights some of the critical issues to be addressed and discusses progress made by several recent studies.

## 6. UNANSWERED QUESTIONS AND FUTURE DIRECTIONS

Within the clustered protocadherin families, functional data have as yet only been reported for the γ-Pcdhs. Regarding the *Pcdh-β* locus, the lack of a shared constant domain has likely been a hindrance to both expression studies and the design of gene knockout strategies. Of all the protocadherins, the α-Pcdhs have been the most intensively studied, and we have a good deal of information about their molecular evolution, gene expression, protein localization, and adhesive interactions, due in large part to the efforts of Takeshi Yagi and colleagues[19,22–24,35,41–44,49–51]. Genetic analyses of the *Pcdh-α* cluster remain unreported as of this writing, though it seems likely that results of such experiments are imminent. In this regard, it is interesting that a 16.7-kilobase deletion in the human *Pcdh-α* locus that truncates one isoform and removes two others causes no obvious abnormalities in individuals from several populations[52]. This suggests that strategies similar to those pursued for the *Pcdh-γ* locus, in which multiple isoforms are mutated, may be required to obtain detectable phenotypes.

As for the γ-Pcdhs, several aspects of the mutant phenotypes remain to be examined in detail, and there are many unanswered questions, three of which I discuss here. First, why are only specific neuronal types affected? In the *Pcdh-γ* deletion mice, only interneurons in the spinal cord, and a subset of neurons in the brain, undergo excessive apoptosis. Major projection neuron populations, such as sensory neurons of the dorsal root ganglia and motor neurons of the ventral spinal cord, appear to be spared, although there could be subtle alterations in their synaptic connectivity[9,10]. The ubiquitous neuronal expression of the *Pcdh-γ* genes makes this specificity difficult to understand. A clue may be found in the fact that spinal interneurons only become affected late in their development, once they have migrated and begun to make synaptic connections. If γ-Pcdh function only becomes critical as neurons mature, the lack of obvious defects in neurons, such as those in the forebrain, that are relatively immature at the time the mutant neonates die would not be surprising. A timing difference, however, cannot explain why sensory and motor neurons appear spared in the spinal cord, as these neurons develop very early. Thus, some of the specificity of phenotype between neuronal populations likely reflects deeper differences in the repertoire of *Pcdh-γ* genes they express, or in the presence or absence of (as yet unidentified) associated signaling partners. Our current work is aimed at clearing up this issue by using conditional mutants to disrupt γ-Pcdh function in isolated neuronal subsets and/or at different stages of development.

Second, for which stages of synapse development are the γ-Pcdhs required? Most of our results support a role for γ-Pcdh function in synaptic maturation and/or maintenance rather than initial synaptogenesis. However, a time-course analysis of synapse development in mutant neurons compared to that of control neurons will be necessary before this can be determined. Also important to clarify in this regard are the dynamics of γ-Pcdh protein localization at the synapse. The substantial extrasynaptic localization of γ-Pcdhs and their presence in axonal and dendritic tubulovesicular compartments[10,45] suggest that they are shuttled to and from synapses, and the regulation of such trafficking might be coordinated with synaptic maturation. It is also entirely possible that γ-Pcdhs exert their functions through intracellular signaling partners, and thus influence synapse development "at a distance" rather than being exclusively involved in adhesion across the synaptic cleft.

Third, are γ-Pcdhs involved in specifying the pattern of synaptic connections? The phenotype of *Pcdh-γ* mutant mice is certainly not inconsistent with such a notion, as synapses that form between "incorrect" partners are likely to be smaller, weaker, and less stable than those found in the "correct" circuit (see references in ref. 9). However, the existing data cannot exclude the possibility that γ-Pcdhs are important for more "generic" synaptic functions and that individual *Pcdh-γ* genes are, to some extent, interchangeable. The differential and combinatorial expression of *Pcdh-γ* isoforms is, however, highly suggestive of a role in synaptic specificity, and the idea is so attractive that it is likely to remain the default assumption. Conclusive data in support of this will probably require the development of new knockout and transgenic mouse lines in which the repertoire of *Pcdh-γ* genes is altered. Correlating the expression of particular isoforms to neuronal subsets for which the pattern of connectivity is known may also help us to answer this question, particularly if misexpression of another isoform can be shown to shift such connectivity in an observable way (for an example of such an approach, see ref. 1). A corollary to this question is whether apoptosis in *Pcdh-γ* mutants is a consequence of inappropriate synapse formation or is due to a separate role of γ-Pcdhs.

In parallel to neurobiological studies of protocadherin function at the synapse, it will be important to address mechanisms of intracellular signaling. The *CNR* genes (*Pcdh-α* members) were initially isolated in a yeast two-hybrid screen for binding partners of Fyn[19], a multifunctional tyrosine kinase that has been implicated in neuronal and glial biology[53,54]. Mice lacking *Fyn* display several neurological abnormalities that could reflect disrupted signaling downstream of α-Pcdhs[55,56]. Interestingly, although the organization of the *Pcdh-γ* C exons is very similar to those of the *Pcdh-α* cluster, there is little amino acid homology between the domains they encode, and the γ-Pcdh C-terminus lacks the multiple SH3-binding motifs through which the α-Pcdhs likely interact with Fyn[8]. The signaling partners that may be activated by γ-Pcdhs are, therefore, unclear, as are those that might interact with the constant domain-lacking β-Pcdhs.

Three recent reports have begun to shed some light on the issue of γ-Pcdh signaling. Gayet et al.[57] performed a yeast two-hybrid screen for proteins that interacted with the V exon-encoded portion of the γ-Pcdh B1 cytoplasmic domain. This screen identified the stathmin homolog SCG10, which has been implicated as a microtubule destabilizing protein[58]. GST pull-down and co-immunoprecipitation experiments confirmed this interaction, and further suggested that SCG10 can interact with two other "B" subfamily γ-Pcdh isoforms, but not with the divergent C4 isoform[57]. While such an interaction is intriguing, it is not yet clear whether it is of functional significance *in vivo*. Also unknown are the proteins that may interact with the constant cytoplasmic domain that is shared by all 22 γ-Pcdh isoforms.

Two similar reports have, however, suggested that this domain can undergo presenilin-dependent cleavage followed by translocation to the nucleus[59,60]. In HEK293 or COS cells transiently or stably overexpressing individual *Pcdh-γ* cDNAs, it was found that the γ-Pcdh ectodomain was cleaved at the cell surface, and that this cleavage could be blocked by inhibitors of matrix metalloproteinases. A second cleavage event near the cytoplasmic face of the membrane released an approximately 20 kDa C-terminal fragment that was found to localize to the nucleus. This C-terminal cleavage event was carried out by presenilin, as it was blocked by specific inhibitors and was absent in fibroblasts cultured from mice lacking both presenilin genes[59,60]. Using a luciferase reporter assay in transfected HEK293 cells, Hambsch et al.[59] present evidence that nuclear translocation of the γ-Pcdh intracellular domain can lead to transactivation of *Pcdh-γ*V exon promoter regions. Surprisingly, such transactivation did not depend on the "conserved sequence element" that was shown previously to be required for *Pcdh-γ* expression[31,32]. Because these experiments were conducted using non-neuronal cell lines overexpressing exogenous constructs, the relevance of these mechanisms to neuronal development *in vivo* is not entirely clear. In support of such relevance, Hambsch et al. show that mice in which the first *Pcdh-γ*C exon has been deleted express very low levels of γ-Pcdhs[59], a situation similar to that in the *Pcdh-γ* truncation line that we have described[9]. However they might work in developing neurons, the mechanisms discovered by Hambsch et al. and Haas et al.[59,60] are remarkably similar to those that result in the processing of *N*-cadherin (reviewed in ref. 61). These recent data suggest that γ-Pcdhs could influence synapse development "indirectly" via intracellular signaling pathways in addition to their presumed "direct" role in adhesion across the synaptic cleft. Further work toward identifying protocadherin signaling mechanisms should greatly enhance our understanding of the roles played by these molecules in the development of the nervous system.

## 7. REFERENCES


1. Yamagata, M., Weiner, J.A., and Sanes, J. R. (2002) *Cell* **110**, 649–660.
2. Shen, K., and Bargmann, C.I. (2003) *Cell* **112**, 619–630.
3. Shen, K., Fetter, R.D., and Bargmann, C.I. (2004) *Cell* **116**, 869–881.
4. Yamagata, M., Sanes, J.R., and Weiner, J.A. (2003) *Curr Opin Cell Biol* **15**, 621–632.
5. Scheiffele, P. (2003) *Annu Rev Neurosci* **26**, 485–508.
6. Washbourne, P., Dityatev, A., Scheiffele, P., Biederer, T., Weiner, J.A., Christopherson, K.S., and El-Husseini, A. (2004) *J Neurosci* **24**, 9244–9249.
7. Sperry, R.W. (1963) *Proc Natl Acad Sci U S A* **50**, 703–710.
8. Wu, Q., and Maniatis, T. (1999) *Cell* **97**, 779–790.
9. Weiner, J.A., Wang, X., Tapia, J.C., and Sanes, J.R. (2005) *Proc Natl Acad Sci U S A* **102**, 8–14.

10. Wang, X., Weiner, J.A., Levi, S., Craig, A.M., Bradley, A., and Sanes, J.R. (2002) *Neuron* **36**, 843–854.
11. Sano, K., Tanihara, H., Heimark, R.L., Obata, S., Davidson, M., St John, T., Taketani, S., and Suzuki, S. (1993) *Embo J* **12**, 2249–2256.
12. Yagi, T., and Takeichi, M. (2000) *Genes Dev* **14**, 1169–1180.
13. Nollet, F., Kools, P., and van Roy, F. (2000) *J Mol Biol* **299**, 551–572.
14. Sago, H., Kitagawa, M., Obata, S., Mori, N., Taketani, S., Rochelle, J.M., Seldin, M.F., Davidson, M., St John, T., and Suzuki, S.T. (1995) *Genomics* **29**, 631–640.
15. Obata, S., Sago, H., Mori, N., Rochelle, J.M., Seldin, M.F., Davidson, M., St John, T., Taketani, S., and Suzuki, S.T. (1995) *J Cell Sci* **108**, 3765–3773.
16. Obata, S., Sago, H., Mori, N., Davidson, M., St John, T., and Suzuki, S.T. (1998) *Cell Adhes Commun* **6**, 323–333.
17. Hirano, S., Yan, Q., and Suzuki, S.T. (1999) *J Neurosci* **19**, 995–1005.
18. Hirano, S., Ono, T., Yan, Q., Wang, X., Sonta, S., and Suzuki, S.T. (1999) *Biochem Biophys Res Commun* **260**, 641–645.
19. Kohmura, N., Senzaki, K., Hamada, S., Kai, N., Yasuda, R., Watanabe, M., Ishii, H., Yasuda, M., Mishina, M., and Yagi, T. (1998) *Neuron* **20**, 1137–1151.
20. Wu, Q., Zhang, T., Cheng, J.F., Kim, Y., Grimwood, J., Schmutz, J., Dickson, M., Noonan, J.P., Zhang, M.Q., Myers, R.M., and Maniatis, T. (2001) *Genome Res* **11**, 389–404.
21. Wu, Q. (2005) *Genetics*, in press.
22. Yanase, H., Sugino, H., and Yagi, T. (2004) *Genomics* **83**, 717–726.
23. Sugino, H., Yanase, H., Hamada, S., Kurokawa, K., Asakawa, S., Shimizu, N., and Yagi, T. (2004) *Biochem Biophys Res Commun* **316**, 437–445.
24. Tada, M.N., Senzaki, K., Tai, Y., Morishita, H., Tanaka, Y.Z., Murata, Y., Ishii, Y., Asakawa, S., Shimizu, N., Sugino, H., and Yagi, T. (2004) *Gene* **340**, 197–211.
25. Noonan, J.P., Grimwood, J., Schmutz, J., Dickson, M., and Myers, R.M. (2004) *Genome Res* **14**, 354–366.
26. Noonan, J.P., Grimwood, J., Danke, J., Schmutz, J., Dickson, M., Amemiya, C.T., and Myers, R.M. (2004) *Genome Res* **14**, 2397–2405.
27. Yagi, T. (2003) *Genes Cells* **8**, 1–8.
28. Chun, J.J., Schatz, D.G., Oettinger, M.A., Jaenisch, R., and Baltimore, D. (1991) *Cell* **64**, 189–200.
29. Chun, J. (1999) *Curr Biol* **9**, R651–654.
30. Dreyer, W.J., and Roman-Dreyer, J. (1999) *Genetica* **107**, 249–259.
31. Wang, X., Su, H., and Bradley, A. (2002) *Genes Dev* **16**, 1890–1905.
32. Tasic, B., Nabholz, C.E., Baldwin, K.K., Kim, Y., Rueckert, E.H., Ribich, S.A., Cramer, P., Wu, Q., Axel, R., and Maniatis, T. (2002) *Mol Cell* **10**, 21–33.
33. Hirayama, T., Sugino, H., and Yagi, T. (2001) *Genes Cells* **6**, 151–164.
34. Frank, M., Ebert, M., Shan, W., Phillips, G.R., Arndt, K., Colman, D.R., and Kemler, R. (2005) *Mol Cell Neurosci* **29**, 603–616.
35. Mutoh, T., Hamada, S., Senzaki, K., Murata, Y., and Yagi, T. (2004) *Exp Cell Res* **294**, 494–508.
36. Senzaki, K., Ogawa, M., and Yagi, T. (1999) *Cell* **99**, 635–647.
37. Carroll, P., Gayet, O., Feuillet, C., Kallenbach, S., de Bovis, B., Dudley, K., and Alonso, S. (2001) *Mol Cell Neurosci* **17**, 611–623.
38. Jossin, Y., Ignatova, N., Hiesberger, T., Herz, J., Lambert de Rouvroit, C., and Goffinet, A.M. (2004) *J Neurosci* **24**, 514–521.
39. Homayouni, R., Rice, D.S., and Curran, T. (2001) *Biochem Biophys Res Commun* **289**, 539–547.
40. Tissir, F., and Goffinet, A.M. (2003) *Nat Rev Neurosci* **4**, 496–505.
41. Esumi, S., Kakazu, N., Taguchi, Y., Hirayama, T., Sasaki, A., Hirabayashi, T., Koide, T., Kitsukawa, T., Hamada, S., and Yagi, T. (2005) *Nat Genet* **37**, 171–176.
42. Miki, R., Hattori, K., Taguchi, Y., Tada, M. N., Isosaka, T., Hidaka, Y., Hirabayashi, T., Hashimoto, R., Fukuzako, H., and Yagi, T. (2005) *Gene,* **349**, 1–14.
43. Taguchi, Y., Koide, T., Shiroishi, T., and Yagi, T. (2005) *Mol Biol Evol,* **22**, 1433–1443.
44. Murata, Y., Hamada, S., Morishita, H., Mutoh, T., and Yagi, T. (2004) *J Biol Chem* **279**, 49508–49516.
45. Phillips, G.R., Tanaka, H., Frank, M., Elste, A., Fidler, L., Benson, D.L., and Colman, D.R. (2003) *J Neurosci* **23**, 5096–5104.
46. Phillips, G.R., Huang, J.K., Wang, Y., Tanaka, H., Shapiro, L., Zhang, W., Shan, W.S., Arndt, K., Frank, M., Gordon, R.E., Gawinowicz, M.A., Zhao, Y., and Colman, D.R. (2001) *Neuron* **32**, 63–77.
47. Blank, M., Triana-Baltzer, G.B., Richards, C.S., and Berg, D.K. (2004) *Mol Cell Neurosci* **26**, 530–543.

48. Deckwerth, T.L., Elliott, J.L., Knudson, C.M., Johnson, E.M., Jr., Snider, W.D., and Korsmeyer, S.J. (1996) *Neuron* **17**, 401–411.
49. Sugino, H., Hamada, S., Yasuda, R., Tuji, A., Matsuda, Y., Fujita, M., and Yagi, T. (2000) *Genomics* **63**, 75–87.
50. Morishita, H., Kawaguchi, M., Murata, Y., Seiwa, C., Hamada, S., Asou, H., and Yagi, T. (2004) *Eur J Neurosci* **20**, 2843–2847.
51. Morishita, H., Murata, Y., Esumi, S., Hamada, S., and Yagi, T. (2004) *Neuroreport* **15**, 2595–2599.
52. Noonan, J.P., Li, J., Nguyen, L., Caoile, C., Dickson, M., Grimwood, J., Schmutz, J., Feldman, M.W., and Myers, R.M. (2003) *Am J Hum Genet* **72**, 621–635.
53. Yagi, T. (1999) *Biochem Pharmacol* **57**, 845–850.
54. Umemori, H., Sato, S., Yagi, T., Aizawa, S., and Yamamoto, T. (1994) *Nature* **367**, 572–576.
55. Yuasa, S., Hattori, K., and Yagi, T. (2004) *Neuroreport* **15**, 819–822.
56. Miyakawa, T., Yagi, T., Tateishi, K., and Niki, H. (1996) *Neuroreport* **7**, 2723–2726.
57. Gayet, O., Labella, V., Henderson, C.E., and Kallenbach, S. (2004) *FEBS Lett* **578**, 175–179.
58. Grenningloh, G., Soehrman, S., Bondallaz, P., Ruchti, E., and Cadas, H. (2004) *J Neurobiol* **58**, 60–69.
59. Hambsch, B., Grinevich, V., Seeburg, P.H., and Schwarz, M.K. (2005) *J Biol Chem,* **280**, 15888–15897.
60. Haas, I.G., Frank, M., Veron, N., and Kemler, R. (2005) *J Biol Chem* **280**, 9313–9319.
61. Junghans, D., Haas, I.G., and Kemler, R. (2005) *Curr Opin Cell Biol*, **17**, 446–452.

# 10

# EPHRINS AND EPH RECEPTORS IN SPINOGENESIS AND SYNAPTIC PLASTICITY

Yu Yamaguchi and Fumitoshi Irie*

## 1. SUMMARY

Originally identified as repulsive axon guidance cues, Eph receptors and their cell surface-bound ligands ephrins have been increasingly implicated in the development and functional modulation of synapses. Two synaptic functions of ephrins and Eph receptors have now been well established by *in vitro* and *in vivo* studies, namely the induction and maintenance of dendritic spines and the regulation of synaptic plasticity. This chapter summarizes recent progress in this rapidly expanding area of ephrin–Eph research. An emerging view is that Eph receptors regulate spine morphology by activating multiple intracellular signaling cascades that involve Rho family GTPases. In contrast, genetic studies suggest that regulation of synaptic plasticity does not require extensive intracellular signaling.

## 2. INTRODUCTION

Eph receptors comprise the largest family of receptor tyrosine kinases, with 16 genes encoding Eph receptors having been identified in mammalian species. Eph receptors are classified into two groups, namely ten EphA receptors and six EphB receptors. The preferential ligands for EphA receptors are GPI-anchored A-ephrins, whereas EphB receptors are preferentially bound by transmembrane B-ephrins. There are six A-ephrin and three B-ephrin genes in human genome (reviewed in refs. 1–3). Binding of ephrin ligands causes multimerization of Eph receptors, leading to activation of their cytoplasmic tyrosine kinase domain. Activated Eph receptors phosphorylate cytoplasmic substrates, including themselves (autophosphorylation)[4,5]. Autophosphorylated tyrosine residues within Eph receptors also serve as the binding sites for cytoplasmic SH2 domain-

*Developmental Neurobiology Program, Burnham Institute for Medical Research, La Jolla, CA, USA; yyamaguchi@burnham.org and firie@burnham-inst.org

containing proteins (Figure 10.1). In addition to these classical signaling mechanisms, the C-terminal tail of Eph receptors can bind PDZ domain-containing proteins[6,7]. Guanine nucleotide exchange factors (GEFs) are another class of molecules that bind the cytoplasmic domain of Eph receptors independent of SH2 domains[8–10].

Although traditionally regarded as ligands, ephrins themselves can act as receptors that transduce signals into cells on which they are expressed. In this case, binding by Eph receptors triggers this "reverse signaling"[11]. Since ephrins do not possess kinase domains, reverse signaling does not involve stimulation of intrinsic kinase activity. Yet B-ephrins are phosphorylated by cytosolic Src family kinases and phosphorylated B-ephrins can be recognized by the SH2 domain-containing adaptor protein Grb4[12,13]. Like Eph receptors, the cytoplasmic tail of B-ephrins binds PDZ domain-containing proteins[14].

The feature that makes the ephrin–Eph system so unique is the fact that the system acts as a signaling system based on both cell contact and tyrosine phosphorylation. Furthermore, its mode of action is generally asymmetrical (i.e., ephrins on one side and Eph receptors on the other sides) and bidirectional (i.e., forward and reverse signaling). In other words, the system seems to be created to mediate exchange of signals between two cells interacting asymmetrically, such as the growth cone of axons and target neurons, and sprouting endothelial cells and surrounding mesenchymal cells (reviewed in ref. 3). Such situations are abundant during development, as crosstalk of two cell types and germ layers is often the basis of morphogenesis. Then how about synapses? Although both cells forming a synapse are neurons (except for the case of neuromuscular synapses where postsynaptic cells are muscle cells), many synapses are formed between axons and dendrites and pre- and postsynaptic membranes are differentially specialized. Thus, it seems reasonable that studies during the last 5 years increasingly reveal functional roles of the ephrin–Eph system in synapse formation and function[15]. In this chapter, we review the recent progress in our understanding of the role of these remarkable molecules in spine formation and synaptic plasticity (see also Chapters 17 and 21).

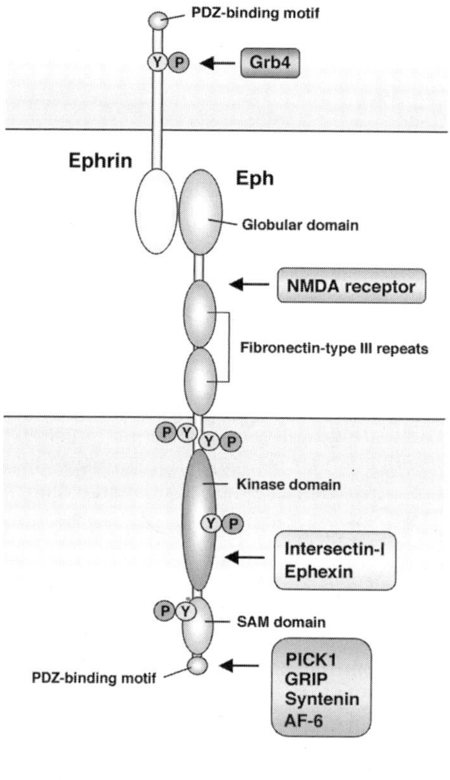


**Figure 10.1.** Domain Structures of Ephrins and Eph Receptors. Transmembrane ephrinB is shown as a representative ligand. Molecules known to interact with ephrins and Eph receptors are shown in rectangles on the right. Y, tyrosine residues; P, phosphate.

## 3. EXPRESSION AND LOCALIZATION OF EPHRINS AND Eph RECEPTORS IN SYNAPSES

Although ephrins and Eph receptors have been studied in the context of the nervous system from the beginning, earlier studies were almost solely focused on axon guidance and segmental patterning during embryonic development (reviewed in ref. 1). The discovery of ephrins and Eph receptors in synapses was rather incidental. Torres et al.[6] screened for binding partners of the C-terminal tail of EphA7 and EphB2 by the yeast two-hybrid system and identified two synaptic PDZ domain-containing proteins, namely PICK1 and GRIP. This immediately prompted immunocytochemical localization of EphB2 and ephrinB in synapses in cultured hippocampal neurons. Now it is known that many ephrins and Eph receptors are expressed in the adult brain, especially in synapses. Because both PICK1 and GRIP were believed to be predominantly concentrated in postsynaptic sites, it was natural to speculate that Eph receptors were present in postsynaptic sites, whereas ephrin ligands were present in presynaptic sites. Indeed, subcellular localization studies have identified EphA4, EphA7, EphB2, and EphB3 in postsynaptic sites[7,16], and B-ephrins are in axons and presynaptic sites[17,18]. However, the *in vivo* picture is not so simple. Along with a panel of Eph receptors, ephrinB2 is also present in the postsynaptic site[16]. B-ephrin localization in synapses differs depending on the site in the brain and the type of synapses. Further complicating the issue, there is evidence that the A-ephrin/EphA combination is expressed at the site of neuron–astrocyte contacts. Here, ephrinA3 is localized in astrocytic surfaces surrounding synapses, whereas EphA4 is localized on spines[19]. It should be noted that these observations regarding synaptic localization of ephrins and Eph receptors have been made mainly in glutamatergic synapses. EphB receptors were not found in GABAergic synapses formed directly on dendritic shafts[8,20].

## 4. Eph RECEPTORS IN THE REGULATION OF SPINE FORMATION AND MORPHOLOGICAL PLASTICITY

Although there is still some controversy about the cellular process how dendritic spines are formed, whether or not they are derived directly from dendritic filopodia that have made contacts with axons, it is clear that some cell surface recognition events between presumptive pre- and postsynaptic neurons play a key role in the morphogenesis of spines. Moreover, the formation of such specialized membrane appendages almost certainly involves localized signaling that leads to local reorganization of cellular (membrane, cytoskeleton) architectures. Being a cell contact-triggered signal transduction system, the ephrin–Eph system appears to possess necessary properties for carrying out such a role.

The function of EphB receptors in spine formation was first demonstrated by Ethell et al.[20]. Transfection of kinase-inactive EphB2, which acts in a dominant negative manner to signaling by all EphB receptors, inhibited spine formation in cultured hippocampal neurons. Dendritic protrusions in neurons transfected with kinase-inactive EphB2 remained as immature, filopodium-like protrusions even at 21 days *in vitro*. The number of AMPA receptor clusters was markedly reduced as a result, whereas the number of synapses formed directly on dendritic shafts (mainly GABA receptors) was unchanged. Thus these results indicate that kinase-dependent signaling by EphB receptors is required for developmental spine formation.

Subsequent studies demonstrated that activation of EphB receptor kinase signaling by exogenous B-ephrins (ephrinB1 and ephrinB2) is sufficient to induce spine formation in hippocampal and cortical neurons[18,21]. Furthermore, the observation made in EphB receptor knockout mice confirmed the physiological relevance of EphB receptor signaling in spine formation *in vivo*. In $EphB1^{-/-}$,$EphB2^{-/-}$,$EphB3^{-/-}$ triple knockout mice, a significant reduction in spine density and the prevalence of spines with no heads and of abnormally small size were demonstrated[18]. Interestingly, such an abnormal spine development was recapitulated in transgenic mice expressing a kinase-inactive, truncated EphB2 mutant on $EphB2^{-/-}$ background, indicating that cell-autonomous forward signaling is responsible for spine formation[18]. Together, these *in vitro* and *in vivo* data essentially confirmed the requirement of EphB kinase signaling in dendritic spine formation. The requirement of kinase activity is the crucial difference between the effect of EphB receptors on spine morphology and that on NMDA-dependent synaptic plasticity (see below).

Meanwhile, much less is known about which ephrin(s) is involved in spine formation *in vivo*. Considering complicated patterns of ephrinB/EphB receptor expression, this question will need detailed analyses. At least in the case of spines in the CA3 pyramidal neurons, ephrinB3 and ephrinA4, both of which can activate EphB receptors, are expressed in the dentate gyrus neurons, which project to CA3 pyramidal neurons. This suggests that these ephrins are good candidates as physiological Eph ligands in these synapses.

The observations described above have been made in the context of axon–dendrite contacts, which is presumably the major factor that drives spine development. Yet there is an indication that spine morphology is also regulated by astrocytes. Astrocytes are an integral component of mature synapses *in vivo*, with their processes closely enwrapping synapses[22]. At least one ephrin, ephrinA3, is expressed by astrocytes in the adult mouse hippocampus while the EphA4 receptor is highly expressed on the dendritic spines of hippocampal pyramidal neurons[19]. Furthermore, stimulation of hippocampal slices with ephrinA3 causes shortening of spines, and spines in $EphA4^{-/-}$ neurons are irregularly shaped and longer than normal[19]. Thus ephrinA–EphA signaling may be functionally involved in regulation of the morphological plasticity of spines through the crosstalk between astrocytic processes and extrasynaptic regions of spines.

## 5. EPHRINS AND Eph RECEPTORS IN THE REGULATION OF SYNAPTIC PLASTICITY

There is increasing evidence that ephrin–Eph signaling has a significant influence on synaptic plasticity[15]. The first observation suggesting the role for Eph receptors in synaptic plasticity was already made in the late 1990s. Gao et al.[23] showed that extracellular application of soluble EphA5-Fc fusion protein impairs activity-induced changes in long-term potentiation (LTP) in mouse hippocampal slices, whereas ephrinA5-Fc induces LTP-like enhancement of basal synaptic transmission. Furthermore, infusion of EphA5-Fc into hippocampus resulted in impaired performance in behavioral paradigms, whereas infusion of ephrinA5-Fc enhanced performance[24] and improved anesthesia-induced memory loss[25]. Then in 2001, two articles simultaneously reported changes in synaptic plasticity in EphB2 knockout mice[17,26]. These mice show deficits in NMDA-dependent, hippocampal synaptic plasticity and minor defects in spatial memory. Curiously, these phenotypes in synaptic plasticity and behavior were not found in *EphB2$^{lacZ}$* knock-in mice, in which the wild-type *EphB2* allele was replaced with truncated EphB2 lacking the tyrosine kinase domain[26]. Thus, the kinase activity of EphB2 is not required for its function in NMDA-dependent synaptic plasticity, which is in stark contrast to the function of EphB2 in spine morphogenesis. This, in turn, suggests that the underlying molecular mechanism for the regulation of NMDA-dependent plasticity may be the interaction of EphB2 with the NR1 subunit of NMDA receptors through their extracellular domains[27].

NMDA-independent synaptic plasticity is also regulated by EphB receptors. Contractor et al.[28] showed that extracellular application of ephrinB1-Fc fusion protein impairs mossy fiber-CA3 LTP, which is believed to be dependent on presynaptic changes in glutamate release probability, whereas EphB2-Fc fusion protein enhances basal excitatory transmission and occludes subsequent LTP. Blocking of the interaction between EphB2 and the AMPA receptor-binding protein GRIP impairs mossy fiber LTP. Thus, the ephrinB–EphB receptor system is thought to exert a retrograde, *trans*-synaptic effect, which leads to presynaptic expression of long-term changes in glutamate release probability. Since GRIP binds both AMPA receptor (the C-terminus of the GluR2 subunit) and EphB2, a natural interpretation of these results is that the GRIP-mediated association between EphB2 and AMPA receptors might be the key to this effect on LTP. In view of recent findings on the molecular interaction of GRIP, however, the role of GRIP in this phenomenon may need further clarification. Fukata et al.[29] showed by a proteomic approach that only a negligible fraction of GRIP in the brain associates with AMPA receptors. Meanwhile, the GRIP–EphB2 interaction has been reported to be critical for delivery of EphB to dendrites and morphogenesis of dendritic arbors[30].

The above observations generally point to postsynaptic Eph receptors in the regulation of synaptic plasticity. However, in CA3–CA1 synapses, ephrinB2 and ephrinB3 are postsynaptically localized[16]. What is the function of these B-ephrins? Recent data revealed that B-ephrins also play a role in the regulation of synaptic plasticity. In brain-specific ephrinB2 conditional knockout mice, CA3–CA1 hippocampal LTP and long-term depression (LTD) are impaired, whereas basal parameters of synaptic transmission are unaffected[16]. Similar changes were found in ephrinB3 knockout mice. EphA4 is considered a candidate receptor for postsynaptic ephrinB2 and ephrinB3, as EphA4 knockout mice display similar impairment in CA3–CA1 LTP[16]. The LTP defects in EphA4 knockout mice are independent of the EphA4 cytoplasmic domain.

While the observations described above were all made in the context of synaptic plasticity in the cerebrum, there is evidence that ephrin–Eph receptor signaling is involved in other types of synaptic plasticity. Central sensitization is a form of synaptic plasticity that occurs in somatosensory neurons in the spinal cord following intense peripheral noxious stimuli, tissue injury, or nerve damage[31]. NMDA receptors are key mediators for central sensitization in the dorsal spinal cord. Battaglia et al.[32] showed that EphB1 and B-ephrins (ephrinB1, ephrinB2, and ephrinB3) are expressed in the sensory system, and that injection of ephrinB2-Fc induces central sensitization. In the spinal cord, EphB activation by exogenous ephrinB2-Fc induces Src activation, which is postulated to be a mechanism that eventually leads to NMDA receptor activation[32].

## 6. Eph RECEPTOR DOWNSTREAM SIGNALING MECHANISMS IN SYNAPSES AND SPINES

### 6.1. Effects on Actin Cytoskeleton

The cytoskeleton of dendritic spines is composed predominantly of actin, and the motility of spines is governed by dynamic reorganization of actin cytoskeleton[33–35]. A wealth of evidence now indicates a crucial role for Rho family GTPases in spine formation (see ref. 36 for review). For instance, transgenic mice expressing constitutively active Rac1 have smaller but more numerous spines in Purkinje cell dendrites[37]. Hippocampal slices transfected with dominant negative Rac1 exhibit progressive elimination of spines[38]. In Drosophila, loss-of-function mutation of Cdc42 causes a reduction in spine density in the vertical system neurons[39]. Rnd1, a new member of Rho family GTPases, also affects spine morphology[40]. Thus it was tempting to speculate that EphB receptor signaling somehow connects to actin polymerization through Rho family GTPases.

Irie and Yamaguchi[9] demonstrated a functional linkage between EphB2 and Cdc42 via the GEF intersectin-l. Treatment of cultured hippocampal neurons with ephrinB2-Fc causes activation of Cdc42, which is temporally in parallel with EPhB2 activation. EphB2 associates with intersectin-l and its association enhances the intersectin's GEF activity in concert with N-WASP, which also binds intersectin-1 (Figure 10.2; Colorplate 7). Dominant negative forms of Cdc42, intersectin-l, and N-WASP all inhibit spine formation in cultured hippocampal neurons[9]. These results indicate that Arp2/3-mediated actin polymerization plays a role in spine formation. The role of Arp2/3 in spine formation has also been suggested by a cortactin knockdown study[41]. Penzes et al.[21] demonstrated a Kalirin-Rac1-PAK pathway as another signaling cascade downstream of EphB2 (see Chapter 17). Here ephrinB2-mediated EphB receptor activation induces synaptic translocation of Kalirin, which is, like intersectin, a GEF. Activated Kalirin then activates the Rac1-PAK pathway. Dominant negative forms of Rac1 and Kalirin-7, as well as a fusion protein of the PAK1 inhibitory domain, block spine formation in cultured hippocampal neurons[21]. Antisense-mediated reduction of Kalirin expression in hippocampal slices results in a reduction in spine density, with dispersion of postsynaptic density markers and elimination of presynaptic endings[42].

It is noteworthy that these effects of ephrinB–EphB signaling on spine formation require the cytoplasmic domain of EphB receptors. In fact, several proteins present in synapses are phosphorylated by EphB receptors. Such proteins include syndecan-2[20], Kalirin-5 and Kalirin-7[21], Src[43], and synaptojanin 1 (44; also see below). Phosphorylation of syndecan-2 is implicated in its clustering in postsynaptic sites, which is a critical step for this cell surface proteoglycan to induce spine formation in immature neurons[20,45]. Phosphorylation of Src is implicated in NMDA receptor-dependent calcium influx through Src-dependent phosphorylation of NMDA receptor[43]. Other downstream target molecules of Eph receptors, such as Abl and Arg, may have implications in spine formation, as they can affect actin polymerization via Ena/VASP proteins[46]. In Drosophila, Ena promotes the formation of both dendritic branches and actin-rich spine-like protrusions in dendritic arborization sensory neurons[47].

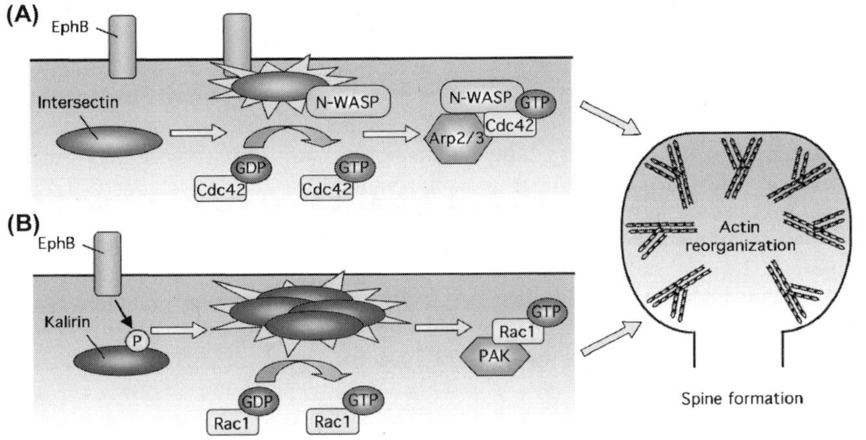


**Figure 10.2.** EphB Downstream Signaling Pathways Leading to Actin Polymerization and Spine Morphogenesis. (A) The Cdc42 GEF intersectin-l is recruited to the site of ephrinB–EphB signaling through its interaction with EphB receptors. Association of EphB2 and N-WASP has a synergistic effect to promote the GEF activity of intersectin-l, which in turn causes local activation of Cdc42. Together with N-WASP, activated Cdc42 triggers Arp2/3-mediated actin polymerization into a branched network[9]. (B) EphB receptors phosphorylate the Racl GEF Kalirin and induce its translocation to synapses and dustering. Activated Kalirin then stimulates the Racl-PAK cascade, resulting in actin rearrangement[21].

It is possible that other parallel pathways linking Eph signaling and actin polymerization exist in spines. For instance, the GEF Tiaml plays a role in ephrin–Eph-mediated neurite outgrowth[10]. Another GEF, Vav2, has been reported to have a critical function during ephrinA1-induced axon repulsion. Binding of ephrinA1 to EphA receptors on the growth cone triggers Vav2-dependent endocytosis of the ligand–receptor complex[48]. Although the role of these novel pathways has not been investigated in the context of spine formation, it is likely that the list of GTPases and GEFs involved in spine formation will grow.

## 6.2. Endocytosis and Intracellular Trafficking

Transport of neurotransmitter receptors to and from postsynaptic membranes is of critical importance in the regulation of synaptic function and efficacy. In excitatory synapses, AMPA-class glutamate receptors undergo dynamic trafficking between the postsynaptic membrane and intracellular site[49,50]. In this vein, it is interesting that there is increasing evidence indicating the role for the ephrin–Eph system in endocytosis and intracellular vesicle trafficking.

One of the remarkable findings in this regard is bidirectional internalization of the ephrinB–EphB complex upon contact between cells expressing EphB receptors and those expressing B-ephrin ligands[51,52]. This process is similar to phagocytosis, clearly distinct from clathrin-mediated endocytosis, and requires the cytoplasmic domain of both receptor (EphB2) and ligand (ephrinB1). Treatment with cytochalasin D or perturbation of Racl activity inhibits this bidirectional internalization[51]. These findings provide an explanation how an initial, contact-based ephrin–Eph interaction results in a physical separation of two cells during growth cone repulsion. It is interesting to see whether a similar mechanism operates in synapses and spines, and if so, how such a mechanism plays a role in the regulation of morphological plasticity of spines.

Another recent finding highlights a direct role for EphB receptors in the regulation of AMPA receptor endocytosis. In the search for novel EphB kinase substrates, Irie and colleagues[44] found that synaptojanin 1, a key accessory protein for clathrin-mediated endocytosis with phosphatidylinositol-phosphatase activity[53], is phosphorylated in its proline-rich domain by EphB2. This domain binds another endocytic protein, endophilin (Figure 10.3). Both synaptojanin 1 and endophilin are essential for clathrin uncoating, a late step in clathrin-mediated endocytosis[54]. EphB2-dependent tyrosine phosphorylation of synaptojanin 1 inhibits its interaction with endophilin. In EphB2-expressing cells, treatment with ephrinB2 elevates the level of phosphatidylinositol 4,5-bisphosphate (PIP2) via inhibition of phosphatidylinositol-phosphatase activity of synaptojanin 1 and promotes clathrin-mediated endocytosis, whereas both a kinase-inactive EphB2 mutant and a phosphorylation site mutant of synaptojanin 1 neutralize the effect of ephrinB2 on endocytosis[44]. Thus these results indicate that ephrinB–EphB signaling regulates clathrin-mediated endocytosis by influencing protein interactions and phosphoinositide turnover through tyrosine phosphorylation of synaptojanin 1. In the postsynaptic neuron, AMPA receptors are endocytosed by clathrin-mediated mechanisms [55,56]. Both the kinase-inactive EphB2 and phosphorylation site mutant of synaptojanin 1 block ligand-induced AMPA receptor endocytosis in hippocampal neurons[44]. Thus, ephrinB–EphB receptor signaling may regulate synaptic plasticity by altering endocytic activity in the postsynaptic site. This does not necessarily mean that EphB–synaptojanin 1 signaling should occur in synchrony with the endocytic cycle. Enhanced levels of synaptojanin 1 phosphorylation may provide a long-lasting condition in which AMPA receptor endocytosis occurs more efficiently. An important unanswered question is whether and how dephosphorylation of synaptojanin 1 is regulated. If such a mechanism (which likely involves tyrosine phosphatases) exists, it could also affect synaptic plasticity by modulating AMPA receptor endocytosis in a direction opposite to EphB receptor activation.

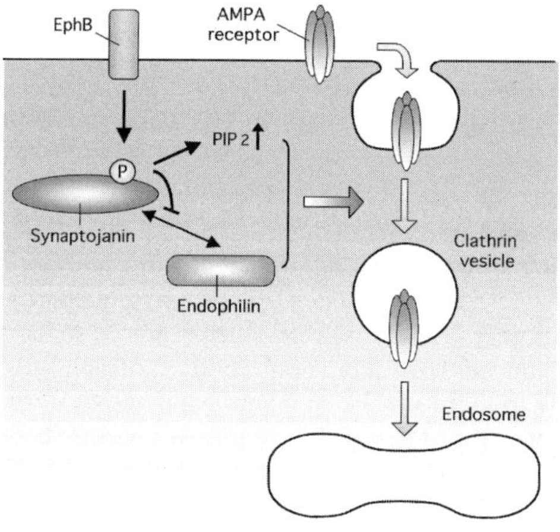


**Figure 10.3.** Model for the Role of EphB Receptor Signaling in AMPA Receptor Endocytosis. Activated EphB2 phosphorylates synaptojanin 1 in its proline-rich domain. This inhibits the phosphatidylinositol-phosphatase activity of synaptojanin 1 and binding to endophilin, leading to the elevation of cellular PIP2 levels and the stimulation of the initial phase of clathrin-mediated endocytosis of cell surface molecules, including AMPA receptors[44].

Finally, a recent article by Cottrell et al.[57] provides a possible link between the EphB receptor effect on spine formation and that on endocytosis. CPG2 is a synapse-specific protein playing a role in endocytosis of AMPA receptors. Interestingly, RNAi-mediated CPG2 knockdown causes not only a reduction in AMPA receptor endocytosis but also a reduction in spine size. This result indicates that supply of membranes from internal membrane organelles may be one of the critical elements that determine the size of spines. Dendritic spines contain endosomal compartments, which serve as a store of recycling membrane[58]. The plasma membrane as well as receptor proteins are delivered to endosomes by endocytic pathway[59]. Thus, ephrinB–EphB signaling might be involved in spine formation and plasticity by not only the effect on actin cytoskeleton but also the effect on membrane trafficking.

## 7. CONCLUSIONS AND FUTURE PROSPECTS

Our understanding on the role of ephrins and Eph receptors in synapses has grown considerably during the last 5 years. Not only have these studies affirmed the early speculation that the ephrin–Eph signaling system operates in synapses, they have also exposed its naïveness; these molecules turned out to function in a way much more complicated and diverse than that had initially been speculated. Accordingly, it would not be surprising to see yet other novel physiological and neurobiological effects of these molecules and additional downstream signaling pathways. With regard to downstream signaling pathways, it will be necessary to determine which pathway is more physiologically relevant to a given phenomenon than the others.

More long-term, and medically important question is whether dysfunction of ephrins and Eph receptors may underlie some neuropsychiatric and mental disorders. Abnormal spines have been reported in patients of mental retardation, autism, and other cognitive disorders[60–63]. Dysfunction of glutamatergic synapses has been implicated in autism and schizophrenia[64,65]. Since many of ephrins and Eph receptors play important developmental roles, null mutations are likely to result in severe developmental phenotypes or even fetal death, as seen in several knockout mouse models. Therefore, epigenetic dysfunction of ephrins and Eph receptors may be more clinically relevant. In this vein, it is interesting to note that administration of drugs of abuse induces a significant elevation in the expression of ephrinB2 in the nucleus accumbens[66] and EphB1 in the striatum[67]. These drugs also cause long-lasting changes in the structures of dendrites and dendritic spines[68,69]. The possible role of ephrins and Eph receptor in the development of drug addiction and other nongenetic disorders would thus be one of the important research topics in the near future.[*]

## 8. REFERENCES

1. Flanagan, J.G., and Vanderhaeghen, P. (1998) *Annu Rev Neurosci* **21**, 309–345.
2. Kullander, K., and Klein, R. (2002) *Nat Rev Mol Cell Biol* **3**, 475–486.

* Work in the author's laboratory is supported by NIH grants HD25938, NS32717, NS41332, and a research grant from the Mizutani Foundation for Glycoscience.

3. Pasquale, E.B. (2005) *Nat Rev Mol Cell Biol* **6**, 462–475.
4. Kalo, M.S., and Pasquale, E.B. (1999) *Biochemistry* **38**, 14396–14408.
5. Kalo, M.S., Yu, H.H., and Pasquale, E.B. (2001) *J Biol Chem* **276**, 38940–38948.
6. Torres, R., Firestein, B.L., Dong, H., Staudinger, J., Olson, E.N., Huganir, R.L., Bredt, D.S., Gale, N.W., and Yancopoulos, G.D. (1998) *Neuron* **21**, 1453–1463.
7. Buchert, M., Schneider, S., Meskenaite, V., Adams, M.T., Canaani, E., Baechi, T., Moelling, K., and Hovens, C.M. (1999) *J Cell Biol* **144**, 361–371.
8. Shamah, S.M., Lin, M.Z., Goldberg, J.L., Estrach, S., Sahin, M., Hu, L., Bazalakova, M., Neve, R.L., Corfas, G., Debant, A., and Greenberg, M.E. (2001) *Cell* **105**, 233–244.
9. Irie, F., and Yamaguchi, Y. (2002) *Nat Neurosci* **5**, 1117–1118.
10. Tanaka, M., Ohashi, R., Nakamura, R., Shinmura, K., Kamo, T., Sakai, R., and Sugimura, H. (2004) *EMBO J* **23**, 1075–1088.
11. Cowan, C.A., and Henkemeyer, M. (2002) *Trends Cell Biol* **12**, 339–346.
12. Cowan, C.A., and Henkemeyer, M. (2001) *Nature* **413**, 174–179.
13. Palmer, A., Zimmer, M., Erdmann, K.S., Eulenburg, V., Porthin, A., Heumann, R., Deutsch, U., and Klein, R. (2002) *Mol Cell* **9**, 725–737.
14. Lu, Q., Sun, E.E., Klein, R.S., and Flanagan, J.G. (2001) *Cell* **105**, 69–79.
15. Yamaguchi, Y., and Pasquale, E.B. (2004) *Curr Opin Neurobiol* **14**, 288–296.
16. Grunwald, I.C., Korte, M., Adelmann, G., Plueck, A., Kullander, K., Adams, R.H., Frotscher, M., Bonhoeffer, T., and Klein, R. (2004) *Nat Neurosci* **7**, 33–40.
17. Henderson, J.T., Georgiou, J., Jia, Z.P., Robertson, J., Elowe, S., Roder, J.C., and Pawson, T. (2001) *Neuron* **32**, 1041–1056.
18. Henkemeyer, M., Itkis, O.S., Ngo, M., Hickmott, P.W., and Ethell, I.M. (2003) *J Cell Biol* **163**, 1313–1326.
19. Murai, K.K., Nguyen, L.N., Irie, F., Yamaguchi, Y., and Pasquale, E.B. (2003) *Nat Neurosci* **6**, 153–160.
20. Ethell, I.M., Irie, F., Kalo, M.S., Couchman, J.R., Pasquale, E.B., and Yamaguchi, Y. (2001) *Neuron* **31**, 1001–1013.
21. Penzes, P., Beeser, A., Chernoff, J., Schiller, M.R., Eipper, B.A., Mains, R.E., and Huganir, R.L. (2003) *Neuron* **37**, 263–274.
22. Vesce, S., Bezzi, P., and Volterra, A. (1999) *Cell Mol Life Sci* **56**, 991–1000.
23. Gao, W.Q., Shinsky, N., Armanini, M.P., Moran, P., Zheng, J.L., Mendoza-Ramirez, J.L., Phillips, H.S., Winslow, J.W., and Caras, I.W. (1998) *Mol Cell Neurosci* **11**, 247–259.
24. Gerlai, R., Shinsky, N., Shih, A., Williams, P., Winer, J., Armanini, M., Cairns, B., Winslow, J., Gao, W., and Phillips, H.S. (1999) *J Neurosci* **19**, 9538–9549.
25. Gerlai, R., and McNamara, A. (2000) *Behav Brain Res* **108**, 133–143.
26. Grunwald, I.C., Korte, M., Wolfer, D., Wilkinson, G.A., Unsicker, K., Lipp, H.P., Bonhoeffer, T., and Klein, D. (2001) *Neuron* **32**, 1027–1040.
27. Dalva, M.B., Takasu, M.A., Lin, M.Z., Shamah, S.M., Hu, L., Gale, N.W., and Greenberg, M.E. (2000) *Cell* **103**, 945–956.
28. Contractor, A., Rogers, C., Maron, C., Henkemeyer, M., Swanson, G.T., and Heinemann, S.F. (2002) *Science* **296**, 1864–1869.
29. Fukata, Y., Tzingounis, A.V., Trinidad, J.C., Fukata, M., Burlingame, A.L., Nicoll, R.A., and Bredt, D.S. (2005) *J Cell Biol* **169**, 399–404.
30. Hoogenraad, C.C., Milstein, A.D., Ethell, I.M., Henkemeyer, M., and Sheng, M. (2005) *Nat Neurosci* **8**, 906–915.
31. Ji, R.R., Kohno, T., Moore, K.A., and Woolf, C.J. (2003) *Trends Neurosci* **26**, 696–705.
32. Battaglia, A.A., Sehayek, K., Grist, J., McMahon, S.B., and Gavazzi, I. (2003) *Nat Neurosci* **6**, 339–340.
33. Landis, D.M., and Reese, T.S. (1983) *J Cell Biol* **97**, 1169–1178.
34. Matus, A. (2000) *Science* **290**, 754–758.
35. Bonhoeffer, T., and Yuste, R. (2002) *Neuron* **35**, 1019-1027
36. Govek, E.E., Newey, S.E., and Van Aelst, L. (2005) *Genes Dev* **19**, 1–49.
37. Luo, L., Hensch, T.K., Ackerman, L., Barbel, S., Jan, L.Y., and Jan, Y.N. (1996) *Nature* **379**, 837–840.
38. Nakayama, A.Y, Harms, M.B., and Luo, L. (2000) *J Neurosci* **20**, 5329–5338.
39. Scott, E.K., Reuter, J.E., and Luo, L. (2003) *J Neurosci* **23**, 3118–3123.
40. Ishikawa, Y., Katoh, H., and Negishi, M. (2003) *J Neurosci* **23**, 11065–11072.
41. Hering, H., and Sheng, M. (2003) *J Neurosci* **23**, 11759–11769.
42. Ma, X.M., Huang, J., Wang, Y., Eipper, B.A., and Mains, R.E. (2003) *J Neurosci* **23**, 10593–10603.
43. Takasu, M.A., Dalva, M.B., Zigmond, R.E., and Greenberg, M.E. (2002) *Science* **295**, 491–495.
44. Irie, F., Okuno, M., Pasquale, E.B., and Yamaguchi, Y. (2005) *Nat Cell Biol* **7**, 501–509.
45. Ethell, I.M., and Yamaguchi, Y. (1999) *J Cell Biol* **144**, 575–586.

46. Krause, M., Dent, E.W., Bear, J.E., Loureiro, J.J., and Gertler, F.B. (2003) *Annu Rev Cell Dev Biol* **19**, 541–564.
47. Li, W., Li, Y., and Gao, F.B. (2005) *Dev Dyn* **234**, 512–522.
48. Cowan, C.W., Shao, Y.R., Sahin, M., Shamah, S.M., Lin, M.Z., Greer, P.L., Gao, S., Griffith, E.C., Brugge, J.S., and Greenberg, M.E. (2005) *Neuron* **46**, 205–217.
49. Malinow, R., and Malenka, R.C. (2002) *Annu Res Neurosci* **25,** 103–126.
50. Bredt, D.S., and Nicoll, R.A. (2003) *Neuron* **40**, 361–379.
51. Marston, D.J., Dickinson, S., and Nobes, C.D. (2003) *Nat Cell Biol* **5**, 879–888.
52. Zimmer, M., Palmer, A., Kohler, J., and Klein, R. (2003) *Nat Cell Biol* **5**, 869–878.
53. McPherson, P.S., Garcia, E.P., Slepnev, V.I., David, C., Zhang, X., Grabs, D., Sossin, W.S., Bauerfeind, R., Nemoto, Y., and De Camilli, P. (1996) *Nature* **379**, 353–357.
54. Slepnev, V.I., and De Camilli, P. (2000) *Nat Rev Neurosci* **1**, 161–172.
55. Man, H.Y., Lin, J.W., Ju, W.H., Ahmadian, G., Liu, L., Becker, L.E., Sheng, M., Wang, Y.T. (2000) *Neuron* **25**, 649–662.
56. Lee, S.H., Liu, L., Wang, Y.T., and Sheng, M. (2002) *Neuron* **36**, 661–674.
57. Cottrell, J.R., Borok, E., Horvath, T.L. and Nedivi, E. (2004) *Neuron* **44**, 677–690.
58. Cooney, J.R., Hurlburt, J.L., Selig, D.K., Harris, K.M. and Fiala, J.C. (2002) *J Neurosci* **22**, 2215–2224.
59. Maxfield, F.R., and McGrow, T.E. (2004) *Nat Rev Mol Cell Biol* **5**, 121–132.
60. Rudelli, R.D, Brown, W.T., Wisniewski, K., Jenkins, E.C., Laure-Kaminowska, M., Connell F., and Wisniewski, H.M. (1985) *Acta Neuropathol (Berl)* **67**, 289–295.
61. Hinton, V.J, Brown, W.T., Wisniewski, K., and Rudelli, R.D. (1991) *Am J Med Genet* **41**, 289–294.
62. Glantz, L.A., and Lewis, D.A. (2000) *Arch Gen Psychiatry* **57**, 65–73.
63. Kaufmann, W.E., and Moser, H.W. (2000) *Cerebral Cortex* **10**, 981–991.
64. Goff, D.C., and Coyle, J.T. (2001) *Am J Psychiatry* **158**, 1367–1377.
65. Purcell, A.E., Jeon, O.H., Zimmerman, A.W., Blue, M.E., and Pevsner, J. (2001) *Neurology* **57**, 1618–1628.
66. Yue, Y., Widmer, D.A.J., Halladay, A.K., Ceretti, D.P., Wagner, G.C., Dreyer, J.-L., and Zhou, R. (1999) *J Neurosci* **19**, 2090–2101.
67. Halladay, A.K., Yue, Y., Michna, L., Widmer, D.A.J., Wagner, G.C., and Zhou, R. (2000) *Mol Brain Res* **85**, 171–178.
68. Nestler, E.J. (2001) *Nat Rev Neurosci* **2**, 119–128.
69. Robinson, T.E., and Kolb, B. (2004) *Neuropharmacology* **47**, Suppl 1, 33–46.

# 11

# EXTRACELLULAR MATRIX MOLECULES AND FORMATION OF CNS SYNAPSES

Erik M. Ullian* and Alexander Dityatev†

## 1. SUMMARY

Intracellular scaffolding proteins play important roles in synaptogenesis by linking major components of pre- and postsynaptic machineries with pre- and postsynaptic cell adhesion molecules. Molecules of the extracellular matrix (ECM), secreted from neurons and glial cells, also form scaffolds, although in the extracellular space. These scaffolds are involved in recruitment and stabilization of synaptic components. For instance, ECM molecules may cluster postsynaptic glutamate receptors via direct interactions with the extracellular domains of these receptors in a synapse-specific manner, or accumulate presynaptic $Ca^{2+}$ channels and growth factors. Furthermore, binding of ECM molecules to cell surface receptors, such as integrins and apolipoprotein E receptor 2, may trigger intracellular signaling cascades resulting in maturation and plasticity of synapses. Additionally, thrombospondins (TSPs) and tenascin-R have prominent effects on formation of synapses via not-yet-identified mechanisms. In this chapter we review available data on synaptogenic activities of ECM molecules, suggesting that these molecules are important for many aspects of synaptic function in the central nervous system (CNS).

## 2. INTRODUCTION

Tremendous advances have been made in our understanding of the proteins and signals required for synaptogenesis in the CNS. An array of both intrinsic neuronal proteins and extrinsic signals from other cells – such as glia – influence

* Department of Ophthalmology and Physiology, University of California, San Francisco, CA 94143-07730, USA; ulliane@vision.ucsf.edu

† Department of Neurophysiology and Center for Molecular Neurobiology, University Medical Center Hamburg-Eppendorf, Martinistr. 52, 20246 Hamburg, Germany; dityatev@zmnh.uni-hamburg.de

aspects of CNS synapse formation and function. An emerging picture is now forming of which of these proteins and signals are necessary and sufficient for synaptogenesis. Thus, it appears that transmembrane proteins such as the neurexin–neuroligin proteins and SynCAM are sufficient to induce formation of CNS synapses[1–3] (Chapters 4,7,8). However, these signals do not act alone. Aspects of synapse formation, stabilization, elimination, and plasticity are influenced by other signals that are either intrinsic or extrinsic to neurons. For example, cadherins, ephrins, neuroligan, Wnts, TGFβ family members, neuronal pentraxins, FGF-22, and the ECM molecules have been identified as helping to promote pre- or postsynaptic differentiation in a variety of invertebrate and vertebrate preparations[2,4–7]. Here, we focus on the role of the ECM in aspects of CNS synapse formation and stabilization. Several recent reviews have provided insight into the ECM's role in synaptic plasticity[8,9]. While ECM molecules have an established and important role in PNS synaptogenesis, the function of ECM in CNS synaptogenesis is relatively poorly understood. Nevertheless, several interesting studies have recently implicated ECM in different aspects of CNS synapse formation and function.

## 3. THROMBOSPONDINS

### 3.1. The Thrombospondin Family

TSPs form a highly conserved family of molecules[10] and constitute a gene family of five members in vertebrates. All TSPs are extracellular multimeric, multidomain calcium-binding glycoproteins that function at cell surfaces and in the ECM[10–13]. Their best known functions are to promote cell attachment and to regulate the cytoskeleton, migration, and angiogenesis. Subgroup A TSPs (TSP1 and TSP2) are homotrimeric, whereas subgroup B TSPs (TSPs 3, 4, and 5) are homopentameric molecules. Members of each subgroup are very closely related. TSP5 is only found in cartilage, but there is evidence that the other four forms are expressed at varying times during development in the nervous system. TSP1 is the best studied of the five TSPs. Structurally, TSP1 consists of a heparin-binding N-terminal domain (HBD), a linker with homology to procollagen, three TSP-type-1 (properdin) repeats, three TSP-type-2 (EGF) repeats, seven TSP-type-3 (calcium binding) repeats, and a cell-binding carboxyl-terminal domain (CBD) (Figure 11.1). Its C-terminal domain is associated with adhesive functions, whereas its amino-terminal domain is implicated in de-adhesive functions. TSP2 has a very similar structure. TSP3, 4, and 5 lack the N-terminal domains, procollagen homology linker, and the TSP-type-1 repeats, and display much more restricted expression patterns. Importantly, each of the TSP domains binds to different receptors.

In the nervous system, TSP1 and TSP2 have been found to be expressed by astrocytes only during postnatal development[14,15]. TSP3 has been observed predominantly in embryonic brain. TSP4 is expressed by CNS neurons in adulthood rather than during postnatal development[16] and is highly concentrated in synaptic layers in the adult retina and brain, as well as at the mature NMJ. The functions of these TSP isoforms in the nervous system are not known.

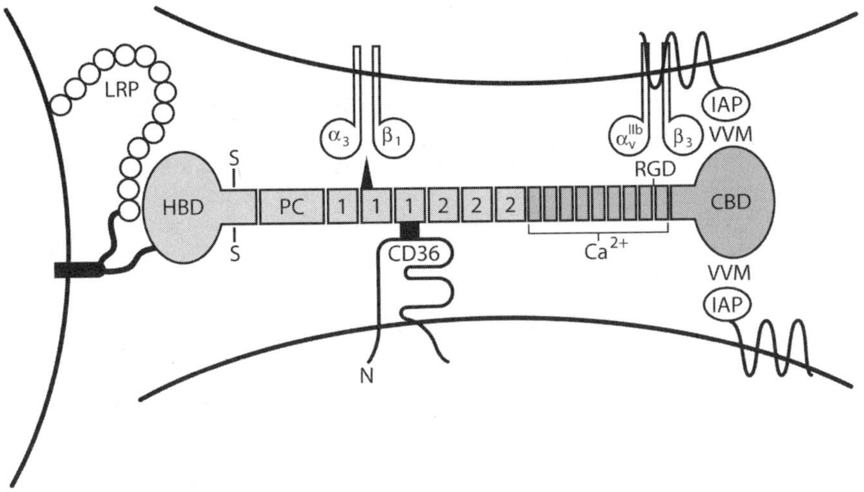


**Figure 11.1.** Structure of TSP1 and 2. The molecules comprise distinct domains that in non-neuronal cells have been found to bind to distinct receptors and mediate different functions of the proteins. TSP1 and 2 consist of a heparin-binding N-terminal domain (HBD), a linker with homology to procollagen, three TSP-type-1 (properdin) repeats (labeled as 1), three TSP- type-2 (EGF) repeats (labeled as 2), seven TSP type-3 ($Ca^{2+}$ binding) repeats, and a cell-binding carboxyl-terminal domain (CBD). VVM-containing motifs in the CBD of TSP bind to integrin-associated protein (IAP). The HBD in involved in interaction with low-density lipoprotein receptor-related protein (LRP) and other proteins (see ref. 10).

## 3.2. Thrombospondin Receptors Are Highly Localized to CNS Synapses

Among the TSP receptors that are highly localized to synapses in the CNS are multiple integrins and integrin-associated protein (IAP/CD47). Integrins comprise a large family of cell adhesion molecules that mediate interactions between the extracellular environment and the cytoplasm[17–19]. Integrins are expressed as cell surface heterodimers consisting of α and β subunits. There are 16 different α and eight different β mammalian integrin subunits, which associate to form 22 recognized α β heterodimers. Each integrin recognizes specific ligands, often ECM molecules such as laminin and fibronectin, or other cell surface receptors such as cell adhesion molecules. Recently, it has been found that integrins are localized to synapses, where they participate in synaptic development, function, and plasticity[20–24]. CD47/IAP is also highly localized to synaptic regions throughout the brain and retina[25,26], where its function is unknown.

## 3.3. Induction of CNS Synapses by Thrombospondins

Thrombospondin has recently been identified as a component of the CNS ECM that is both necessary and sufficient to form synapses *in vitro* and plays a role in synapse formation *in vivo*. Using purified cultures of neurons and glia, numerous studies have found that neurons in the absence of other cell types can survive and extend robust processes, but they do not form many synapses. Interestingly, another cell type associated with synaptic structures *in vivo*, the astrocytes, produces soluble signals that can increase the number of synapses on neurons nearly 10-fold[27–29]. This increase in synapse number is seen when media conditioned over a feeding layer of astrocytes (astrocyte conditioned media) is added to neurons, indicating that the signal is secreted from astrocytes, much as ECM molecules are secreted. What is the identity of this astrocyte-derived synapse promoting factor(s)? One astrocyte-derived synapse promoting signal turns out to be TSP.

Evidence that TSP is the astrocyte-derived signal that increases synapses on purified neurons comes from the recent work of Christopherson and co-workers[30]. These authors found that astrocytes *in vivo* and *in vitro* express high levels of TSP1 and TSP2 mRNA. Interestingly, *in vitro*, TSP2 seems to be the main protein expressed. Is TSP2 necessary and sufficient to increase the synapse number *in vitro*? Addition of purified recombinant TSP2 to cultures of neurons resulted in a dramatic increase in synapse number, similar to the effects of astrocyte-conditioned media[30], indicating that TSP2 is indeed sufficient to increase synapse number (Figure 11.2A; Colorplate 8). To determine if TSP2 is a necessary signal from astrocytes for their synapse-promoting ability, TSP2 was removed from astrocyte-conditioned media with a TSP2-specific antibody. TSP2 depletion almost entirely eliminated the synapse-inducing activity of astrocyte-conditioned media. These findings taken together indicate that TSP2 is sufficient and necessary for astrocyte-conditioned media to induce synaptogenesis between retinal ganglion cells *in vitro*. TSP1 had a similar synapse-promoting effect as TSP2 and appeared to be expressed *in vivo*[30]. Therefore, it is likely that both TSPs play a role in synaptogenesis in vivo. To determine what role TSP1 and TSP2 play in synaptogenesis *in vivo*, the recently generated TSP1 and TSP2 double knockout (KO) mice were analyzed.

### 3.4. TSP1 and TSP2 Double Knockout Mice Have A Reduced Number of Synapses

Analysis of double KO brains found a 30% decrease in synapse number at early postnatal ages (P5–6) and this reduction in synapses is maintained into adulthood[30]. Interestingly, no significant effect on synapse number was seen in the TSP1 or TSP2 single KOs. This indicates that both TSP1 and TSP2 are required *in vivo* for normal synapse formation or stabilization. Because of the high homology of the TSP1 and TSP2 orthologs, it is likely that they can serve redundant functions in the nervous system, each being able to regulate synaptogenesis.

How do TSP1 and TSP2 regulate synaptogenesis *in vitro* and *in vivo*? One interesting property of TSP is its ability to regulate function of other adhesion molecules. This raises the possibility that TSP functions by regulating the adhesiveness or signaling of other proteins already implicated in synapse formation, such as neurexin–neuroligin or SynCAM. Whether TSP1 and TSP2 function by regulating such molecules and how these TSPs send signals to neurons has yet to be determined.

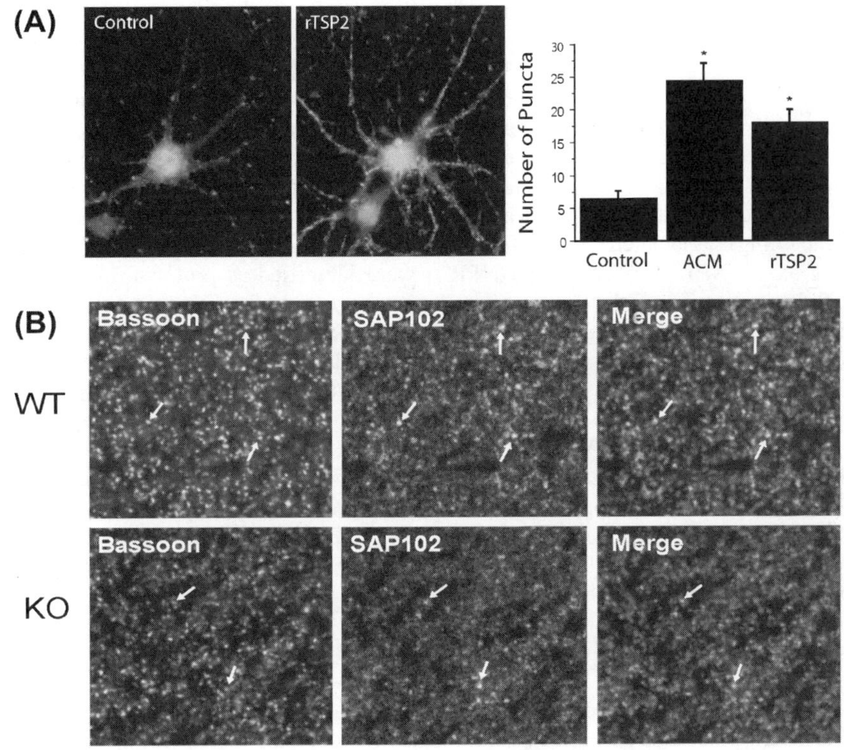


**Figure 11.2.** TSP is Necessary and Sufficient for Synapse Formation. (A) Purified CNS neurons do not form synapses under control condition (Control) as seen by the absence of synaptic puncta after staining for presynaptic synaptotagmin and postsynaptic PSD-95. Addition of recombinant TSP2 (rTSP2) induces a dramatic increase in the number of colocalized presynaptic puncta and postsynaptic puncta (yellow puncta in Colorplate 8). Graph shows a nearly 4-fold increase in synapse number after addition of rTSP2 that is comparable to the increase elicited by the astrocyte-conditioned medium (ACM). (B) Post-natal day 21 wild-type (WT) and TSP1/2 double knockout (KO) mice brains were sectioned and stained for presynaptic marker Bassoon and postsynaptic marker SAP102. KO animals show a reduction in the number of colocalized synaptic puncta (arrows). Reprinted from ref. 30, Copyright (2005) with permission of Elsevier.

## 4. NEURONAL PENTRAXINS

### 4.1. The Pentraxin Family

The pentraxin family is a large family of proteins defined by a conserved motif, the pentraxin domain. Some pentraxins are structurally characterized by the arrangement of subunits in a pentamer. The pentraxin family is further broken up into two main types of pentraxins, the short pentraxins and the long pentraxins. The short pentraxins consist of C-reactive protein (CRP) and serum amyloid P. The long pentraxins consist of pentraxin 3 and 4 (PTX3 and PTX4), neuronal pentraxin 1 and 2 (NP1 and NP2, also known as Narp), and neuronal pentraxin receptor (NPR)[31].

The first identified role for pentraxins was a role in innate immunity. Innate immunity is characterized by the recognition of pathogens and damaged tissue by pattern recognition receptors. The first pentraxin to be identified, CRP, is an example of such a receptor. This pentraxin binds to the *C*-polysaccharide of *Streptococcus pneumoniae* and mediates the immune attack and kills sequence for complement-mediated lysis of these pathogenic bacteria. Similarly, serum amyloid protein also binds a variety of pathogens and mediates the innate immune response to pathogens. In addition, pentraxins appear to play an important role in tissue remodeling associated with fertility and inflammation[32–34]. The neuronal pentraxins, NP1[35], NP2[36–39], and NPR[40], were identified as proteins that bind to affinity columns of the presynaptic snake venom toxin, taipoxin, and/or the lumenal calcium-binding protein TCBP49[41]. NP1, NP2, and NPR protein sequences are 50% identical to each other and their carboxy terminal half is 20–30% identical to classical pentraxins such as the acute phase proteins, CRP[42], serum amyloid P protein[43], and the long pentraxin, PTX3[44]. Neuronal pentraxins form heteromultimers[45]. While NP1 and NP2 are secreted when expressed alone, they can also be tethered to the cell membrane when expressed as heteromultimers with NPR. NP2 has also been identified as a synaptic protein that is rapidly and dramatically upregulated by neuronal activity[39]. This places NP2 in the family of immediate early genes. The function of NP1 and NP2 *in vivo* is unknown, although both NPs have been implicated in regulating glutamate receptor clustering *in vitro*.

## 4.2. Pentraxin's Role in Glutamate Receptor Clustering

One of the best studied functions of NPs is the clustering of AMPA receptors (AMPARs). Both NP1 and NP2 are capable of clustering AMPARs *in vitro*. Initially, NP2 was found to cluster AMPARs in culture[46]. This study reported that NP2 was enriched at excitatory synapses both on hippocampal and spinal neuron cultures, and the expressed Myc-tagged NP2 was delivered to synapses. Strikingly, NP2 molecules co-clustered and co-immunoprecipitated with AMPA receptor subunits GluR1–3, but not with the AMPA receptor subunit GluR4, NMDA receptor subunits NR1 and NR1/2A, or kainate receptor subunit GluR6. Expression of a dominant-negative NP2 mutant protein, which prevented secretion of endogenous NP2, inhibited NP2 accumulation at synapses and affected clustering of AMPARs. Because it is not entirely clear that the dominant-negative NP2 exerts its effects only through NP2 secretion, these results must be interpreted somewhat cautiously. It is possible that endogenous levels of NP2 are not directly involved in AMPAR clustering, but this work certainly implies that NP2 plays some role in this process.

Recent work has suggested that NP2 may also indirectly cluster NMDA-type receptors in spinal neurons. Mi and co-workers[47] found that NP2 may act through the AMPAR subunit GluR2 to cluster NMDARs containing the NR2A or NR2B subunits. This clustering depended both on the GluR2 C-terminal tail and on the C-terminal tail of stargazin, a protein found to cluster AMPARs at synapses[48]. NP2 does not directly bind to either NMDARs or stargazin, implying that there may be an interaction between clustered AMPARs and NMDARs through stargazin in nonspiny neurons. Interestingly, these authors also found that stargazin DC (Stargazin without the C-terminal domain) does not affect surface GluR1 clusters in transfected nonspiny motor neurons and hippocampal interneurons, despite having a potent effect on reducing GluR1 clusters in spiny hippocampal pyramidal neurons[48]. Because NP2 can cluster AMPARs on these types of neurons, one interpretation of this result is that NP2 takes the place of stargazin for AMPAR clustering. However, stargazin is one member of a family of proteins termed Transmembrane AMPA Receptor Proteins (TARPs) with AMPAR binding and surface expression and synaptic targeting functions[49]. Thus, it is quite possible that different members of the TARP family are required for AMPAR clustering in different types of synaptic connections. Nevertheless, based on the previous work, it appears that NPs play a role in the clustering and synaptic distribution of AMPARs, especially at nonspiny synapses. Because NP1 shares about 50% sequence identity with NP2 and has also been shown to cluster AMPARs, it is thought that these two proteins may function in concert to cluster AMAPRs at nonspiny synapses[50–52].

### 4.3. Pentraxin's Role in Synaptogenesis

Overexpression of NP2 not only induced clustering of AMPARs but also increased the number of excitatory synapses in hippocampal cultures[46]. Furthermore, NP2 expression in transfected HEK293 cells was sufficient to induce AMPAR clustering on neuronal dendrites that contacted the transfected cells. Interestingly, these overexpression studies revealed that NP2 can be secreted to synapses both from the presynaptic terminal as well as from the postsynaptic dendrite, although it is not clear whether under normal conditions there is significant dendritic expression of NP2.

In subsequent studies it was found that dominant-negative NP2 not only reduced the number of GluR1 subunit AMPAR clusters at contact sites between a dominant-negative NP2 expressing neuron and a postsynaptic neuron, but also induced a slight, although significant, decrease in the number of excitatory synaptic contacts in this connection[50]. No effect was found on the number of inhibitory synapses as assessed by gephyrin or GAD clusters. These studies led to a hypothesis that NP2 in the CNS may act analogously to agrin clustering of nicotinic acetylcholine receptor clustering in the PNS (see Chapter 1), being necessary and sufficient to cluster AMPARs at excitatory synapses and thereby leading to increases in the number of synapses found in culture. Both NP1 and NP2 are able to cluster AMPARs, but pentamers comprising both NP1 and NP2 have greater clustering activity than pentamers of either alone[52]. Interestingly, while secretion of NP2 is activity dependent, secretion of NP1 is not[39,52]. This finding leads to a model in which neurons have a low level of receptor clustering activity through NP1 that is dramatically upregulated by activity and subsequent release of NP2. Thus, this mechanism appears to contribute to both activity-independent and activity-dependent excitatory synaptogenesis[52]. What roles NP1 and NP2 play *in vivo* are completely unknown. NP1, NP2, and NP1/2 double KO mice must be carefully analyzed to understand the role of this neuronal ECM protein in synapse formation and function.

## 5. TENASCIN-R

### 5.1 The Tenascin Family

Tenascins are multimeric ECM glycoproteins. In vertebrates, five tenascins are found in various tissues, tenascins-C, -N, -R, -X, and -W. Of these, tenascins-C and -R are not only found in the CNS but also implicated in modulating different aspects of synaptic plasticity[8,53]. Structurally, tenascins-C and -R contain EGF-like repeats, fibronectin type III repeats, and a fibrinogen domain. Tenascin-R is one of major components of the perineuronal nets that surround a subset of neurons in the brain and spinal cord[54]. These nets – known already to Golgi and Cajal – have recently received renewed interest because of several studies suggesting that they may have inhibitory modulatory effects on regeneration and structural plasticity[55,56].

### 5.2. Role of Tenascin-R in Formation of GABAergic Synapses

Electron microscopic analysis of tenascin-R KO mice revealed a reduction in the number of inhibitory somatic synapses on CA1 pyramidal neurons. In addition, existing synapses displayed both shorter active zones and reduced number of predocked vesicles[57]. Overall, there was a 30–40% decrease in the number of inhibitory active zones per unit length of pyramidal neuron. Such a dramatic decrease would be expected to have severe functional consequences and indeed, the tenascin-R KO mice showed a two-fold decrease in $GABA_A$ receptor-mediated unitary evoked inhibitory postsynaptic current (IPSC) amplitudes and increase in the number of release failures. These effects of tenascin-R deficiency on evoked release occurred with no changes in the size of spontaneous miniature IPSCs, but with changes in their frequency, indicating that the effect is likely to be presynaptic[58].

### 5.3. Tenascin-R and $GABA_B$ Receptors

The mechanism underlying synaptogenic activity of tenascin-R possibly involves a carbohydrate HNK-1 (originally found on human natural killer cells, hence the name) carried by tenascin-R. Antibodies to this carbohydrate inhibited perisomatic IPSCs, although not dendritic IPSCs or EPSCs[57]. The effects of HNK-1 antibodies were blocked by antagonists to $GABA_B$ receptors and postsynaptic Kir channels coupled to these receptors. Moreover, HNK-1 was found to bind and inhibit recombinant $GABA_B$ receptors[58]. These data led to a model in which tenascin-R via HNK-1 inhibits postsynaptic $GABA_B$ receptor activation under normal conditions. In the tenascin-R KO animals, there is excessive activation of $GABA_B$ receptors, resulting in a postsynaptic activation of Kir channels and the accumulation of $K^+$ in the extracellular space. This in turn may lead to a chronic depolarization of presynaptic terminals, an increase in spontaneous transmitter release, and a reduction in release probability and IPSC amplitude[58]. An increase in postsynaptic $GABA_B$ receptor-mediated currents in the tenascin-R KO animals supports this model[56]. Further pharmacological analysis or tenascin-R KO may help to verify whether tenascin-R regulates the number of inhibitory synapses via activation of postsynaptic $GABA_B$ receptors.

## 6. AGRIN

### 6.1. Structure and Binding Partners of Agrin

The best-studied synaptic organizing ECM molecule is agrin, which is required for formation of the vertebrate skeletal neuromuscular junction (see Chapter 1). Agrin is a large (200 kDa), complex protein that may bear two heparan sulfate chains and therefore is considered to be a heparan sulfate proteoglycan. Agrin is expressed as a secreted protein with the longer N-terminal (LN) or as a transmembrane protein with the shorter N-terminal (SN) domain. Following the SN or LN domains is a set of eight follistatin repeats related to those that bind growth factors and inhibit proteases. The central region of the protein contains two repeats homologous to domain III of laminin chains, a ninth follistatin repeat, two serine/threonine-rich segments. The C-terminal part of agrin, which is responsible for the molecule's known signaling functions, contains four EGF repeats and three so-called G domains homologous to those found in laminin α chains, neurexins, and slits. This C-terminal portion also contains three sites of alternative splicing, called in mammals as x, y, and z. Agrin activates or binds to several membrane- or matrix-associated proteins, including muscle-specific receptor tyrosine kinase, MuSK; laminin γ1 chain; β1 integrins; dystrogycan; NCAM; heparan sulfates; and thrombospondins.

### 6.2. Agrin's Role in Synaptogenesis

The nerve-derived alternatively spliced z+ isoform of agrin proved to be necessary for postsynaptic differentiation of neuromuscular junction[59]. Synaptogenic function of agrin is mediated by its activation of MuSK. A crucial effector downstream of tyrosine phosphorylation activity mediated by MuSK is the cytoplasmic scaffolding protein rapsyn that induces clustering of acetylcholine receptors, thus providing one prototypic mechanism of ECM action on postsynaptic differentiation (see Chapter 1).

Recent findings indicate that agrin is important for formation of synapses between cholinergic preganglionic axons and sympathetic neurons in the superior cervical ganglion[60] and for formation of splanchnic nerve–chromaffin cell cholinergic synapses in rat adrenal acute slices[61]. In the latter, agrin decreases gap junction-mediated electrical coupling that precedes an increase in nicotinic synaptic transmission. This developmental switch from predominantly electrical to chemical communication is fully operational within 1 hour and depends on the activation of Src family-related tyrosine kinases.

Also in hippocampal cultures, the density of presynaptic boutons and vesicular turnover was reduced when agrin expression or function was suppressed by antisense oligonucleotides and specific antibodies[62,63]. However, synaptogenesis occurred normally in primary hippocampal and cortical neurons derived from agrin-deficient mice[64,65], suggesting the possibility of functional redundancy of agrin and activation of compensatory mechanisms during development in agrin-deficient cultures. Application of recombinant agrin to cultured cortical neurons induces multiple events that involve tyrosine kinase activity and result in modulation of intracellular $Ca^{2+}$ levels and activity of MAPK and CaMKII, phosphorylation of CREB and induction of expression of the immediate early gene c-fos[66–68]. Conversely, agrin mutants show reduced c-fos expression and resistance to excitotoxicity and seizures. In summary, in several types of synaptic connections, particularly in cholinergic synapses, agrin stimulates activities of tyrosine kinases that affect important aspects of synaptic organization and signaling.

## 7. LAMININS

### 7.1. The Laminin Family

Laminins are extracellular glycoproteins that form heterotrimers of various combinations of α, β, and γ subunits that in vertebrates combine to form at least 15 distinct laminins (see ref. 69 and Chapter 1). These heterotrimers assemble into cross-shaped structures consisting of a long α and short β and γ chains, making up each arm of the cross[70–72]. Generally the laminins have conserved domains with either matrix assembly or cell-binding activities. The short arms are often involved in cell matrix assembly and binding while the long α chain can bind to cell surface receptors by either the N- or C-terminal end of the cross. The long α chain may also bind and assemble matrix through an N-terminal globular domain that is present in many, but not all, members of the α chain family. Laminins play multiple roles in vertebrate development, being expressed at very early time points in the embryo. They mediate potent effects on neurite outgrowth, are required for the overall organization of the cortical layers, and play a role in many stages of the brain formation and development[69,73,74]. Because of these early and strong effects of laminins, it had been difficult to tease out their role in CNS synaptogenesis.

### 7.2. Synaptogenic Activity of Laminins

Like agrin, laminins are known to play an essential role in proper synaptogenesis at the neuromuscular junction[75–78]. There, β2 chain-containing laminins bind directly to the presynaptic P/Q- and N-type calcium channels and induce their clustering, which in turn recruit other presynaptic components. Surprisingly, recruitment occurs independent of $Ca^{2+}$ influx via the channels[75]. Perturbation of laminin-channel interaction *in vivo* results in disassembly of neurotransmitter release sites, active zones, resembling defects previously observed in an autoimmune neuromuscular disorder, Lambert–Eaton myasthenic syndrome. These abnormalities correlate with physiological defects in quantal release and use-dependent modulation of synaptic transmission at the NMJ[79]. Earlier *in vitro* study of ciliary ganglion neurons also showed induction of presynaptic structures by recombinant β2 fragments in the absence of a postsynaptic cell[80]. Thus, it appears that the β2 laminin is capable of initiating a complex set of changes in axons that leads to presynaptic differentiation. Since the N- and P/Q-type calcium channels, to which β2 laminins bind, are widely expressed in CNS, laminins are good candidates to initiate assembly of presynaptic zones in central synapses.

In mice lacking another laminin, α4, active zones and junctional folds of the NMJ form in normal numbers, but not precisely apposed to each other[77], suggesting that laminins may be important players in alignment of pre- and postsynaptic machineries. Abnormalities in rod photoreceptor synapses in laminin β2 chain-deficient mice[81] also suggest that laminins are important for alignment of pre- and postsynaptic sites. In the normal retina, rod photoreceptors make synapses, called triads, in which three postsynaptic elements invaginate into the base of the photoreceptor such that two horizontal cell dendrites lie laterally, and one bipolar cell dendrite lies centrally. However, in the β2 chain-deficient animals, another type of synaptic configurations, dyads, with only one or two horizontal cell processes apposed to the presynaptic specialization (ribbon), are most common. Also unusually common are floating synapses, wherein a fully formed ribbon, is seen without any postsynaptic element apposed.

## 8. INTEGRIN'S ROLE IN FORMATION AND MATURATION OF SYNAPSES

Further evidence for the role of laminins in the CNS comes from studies of one of the main laminin receptor classes, the integrins. While there are various laminin receptors including dystroglycan, lectins, and proteoglycans, integrins are the most prevalent and well-studied receptors. As mentioned in Section 3.2, integrins are a family of α and β subunit heterodimeric cell surface receptors expressed throughout the nervous system[74] that mediate a variety of neuronal developmental and functional processes.

Integrins α8, αvβ8, and α3 have all been directly localized to synaptic sites in CNS[82,83]. Since both RGD peptides that block interaction between integrins and their ligands and antibodies to β1 integrins interfered with formation of synapses in organotypic slice cultures, these integrins have been suggested to play a role in synaptogenesis[23]. More evidence on synaptogenic activity of integrins came from a recent work of Hama and colleagues [84] that shows changes in the synaptogenic state of a neuron in response to local contact with astrocytes. Astrocyte contact triggers integrin-mediated activation of protein kinase C throughout the neuron (see Chapter 21). This mechanism may provide a global enhancement of synaptogenesis. This study suggests that integrins may not directly mediate synaptogenesis by acting as cell adhesion molecules at the synapse itself, but rather may act as signaling molecules that change a neuron from either a quiescent or outgrowth state to a synaptogenic state. Such a role would be consistent with some of the known functions of ECM molecules as modifiers of cell function and adhesion through alteration of other adhesion proteins and cell signaling pathways. This study and others[85–88] begin to address a critical, yet poorly understood aspect of CNS synaptogenesis. Namely whether the timing of synapse formation is intrinsically determined by properties of interacting neurons or it may also depend on extrinsic factors such as the ECM. These studies strongly suggest that environmental or ECM signals can have important effects on the timing of CNS synaptogenesis.

Interestingly, during later stages of synaptic development, integrins proved to be necessary for maturation of excitatory hippocampal synapses, converting immature hippocampal synaptic contacts, which express the NR2B subunit of NMDA receptors and have a high glutamate release probability, into mature synapses, which lack NR2B and have lower release probability[21]. Furthermore, numerous studies demonstrated the important role of integrins in synaptic plasticity (for review, see ref. 8).

## 9. REELIN

### 9.1. Reelin and Its Receptors

Reelin is a secreted ECM glycoprotein that binds transmembrane receptors on neurons and activates tyrosine kinase signaling cascades[89]. Animals with mutations in Reelin (Reeler mice) strikingly display abnormal neuronal migration such that cortical neurons reverse their normal layering pattern[90]. Reelin signals through two main receptors, the very-low-density lipoprotein receptor (VLDLR) and apolipoprotein E receptor 2 (ApoER2). Both of these receptors appear to signal through the disabled (Dab-1) signaling pathway to mediate Reelin's effects on neuronal migration[91]. In addition, there is evidence that Reelin can signal through $\alpha3\beta1$ integrin-dependent pathways as well[92].

### 9.2. Reelin's Role in Synaptogenesis

In addition to an effect on migration, studies have also linked Reelin to synapse formation in the hippocampus. Although hippocampal layering is largely unaffected in the Reeler mouse, defects are seen both in axonal pathfinding and synaptogenesis in the hippocampus[93,94]. Hippocampal entorhinal projections are topographically correct in Reeler mice, but more fibers make aberrant projections, there are fewer branches of entorhinal axons, and there is a nearly 50% reduction in the number of synapses at P2 and a 30% reduction at P12[94]. Also heterozygous Reeler mice exhibit a decrease of dendritic spine density on cortical and CA1 pyramidal neurons of the hippocampus[95]. It is difficult to verify if Reelin is playing a direct role in hippocampal synaptogenesis or if the reduction is due to poorly developed dendritic trees or the reduced branching and misrouting of the entorhinal projections. Multiple mechanisms for Reelin in synaptogenesis are likely because there are fewer synaptic varicosities per length of axon in Reeler mice[94], Reelin is expressed in CNS synapses and affects synaptic properties[96] as well as dendritic development[97].

Another intriguing example of a possible role for Reelin in synaptogenesis comes from studies of the retina. Reelin is expressed by retinal ganglion cells, amacrine cells, and rod bipolar cells in the developing and adult retina[98]. Reelin is also strongly expressed in the synaptic inner plexiform layer of the retina and as this layer matures into its distinct ON/OFF sublaminae (see ref. 99 for review), Reelin becomes localized predominantly to the ON layer[92]. In the Reeler mouse, a subset of Reelin expressing rod bipolar cells terminate in the incorrect layer of the retina and do not synapse in the ON sublamina. Thus, Reelin may play a role in the establishment of appropriate synaptic circuitry in the retina perhaps by modulating the relative adhesiveness of proteins involved in synaptic specificity.

### 9.3. Reelin and Synaptic Maturation

Several studies suggest that Reelin affects synaptic functions via VLDL and apoER2 receptors which may induce mDab1 tyrosine phosphorylation and activation of nonreceptor tyrosine kinases of the Src family. This idea is supported by recent data on maturation of synapses in hippocampal cultures[96]. In this system, chronic blockade of the function of Reelin with antisense oligonucleotides or the function-blocking antibody prevented the decrease of NR1/2B-mediated whole-cell currents, which is a hallmark of synaptic maturation. Conversely, exogenously added recombinant Reelin accelerated the maturational changes in NMDA-evoked currents. Importantly, the change in NMDAR subunits was also blocked by chronic treatment with an inhibitor of the Src kinase signaling pathway or an antagonist of the LDL receptors, receptor associated protein[96]. Even stronger evidence supports a link between Reelin, apoER2, tyrosine phosphorylation of NMDA receptors, and long-term potentiation[8]. Thus, signaling via apoER2/VLDL receptors and tyrosine phosphorylation appeared to be critical in mediating effects of Reelin on synaptic functions. Whether the same mechanisms are operant during stabilization of nascent synapses remains to be investigated. The importance of the above-described findings is underscored by reduction of Reelin expression in psychotic patients and similarities in some cellular and synaptic abnormalities observed in these patients and Reeler mice.

## 10. CONCLUSIONS

The findings discussed in this chapter suggest that ECM molecule may shape synaptogenesis via several mechanisms. On the one hand, they may aggregate and serve as scaffolds for accumulation of presynaptic $Ca^{2+}$ channels, postsynaptic glutamate receptors and growth factors (Figure 11.3).

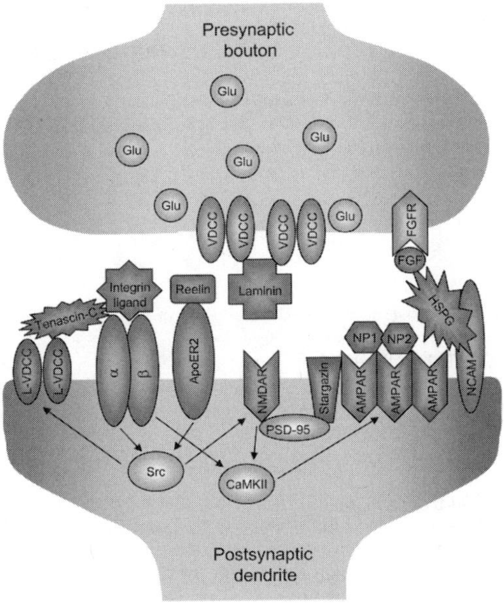


**Figure 11.3.** Hypothetical Mechanisms by which ECM Molecules Shape Synaptogenesis. Neuronal pentraxins, NP1 and NP2, may form scaffolds in the extracellular space, thus promoting clustering of postsynaptic AMPA subtype of glutamate (Glu) receptors (AMPARs) and associated proteins, such as Stargazin, PSD-95, and NMDA receptors (NMDARs). Scaffold-containing laminins cluster presynaptic voltage-dependent $Ca^{2+}$ channels (VDCC). Similarly, heparan sulfate proteoglycans (HSPGs) in association with NCAM serve to accumulate FGF and amplify signaling via FGF receptors. Reelin and laminin may bind to their cell surface receptors (ApoER2 and integrins) and trigger intracellular cascades resulting in activation of tyrosine kinases of the Src family and tyrosine phosphorylation of effectors, including NMDA receptors (NMDAR).

Data on laminin and neuronal pentraxins spectacularly illustrate this mechanism. Additionally, a recent study on synaptogenic activity of the neural cell adhesion molecule NCAM shows that it is related to formation of a complex between polysialylated NCAM and heparan sulfate proteoglycans and signaling via fibroblast growth factor receptor (see Chapter 6). On the other hand, ECM molecules may serve as ligands capable of activating intracellular signaling cascades via binding to cell surface receptors. A common feature of these signaling events is tyrosine phosphorylation of effectors involved in synaptic transmission and plasticity.

The ECM molecules and receptors mentioned above are by no means the only ones involved in synapse formation and function. An increasing number of ECM signals are now being implicated in these processes such as aggrecan, neurocan, brevican, versican and HB-GAM (see ref. 100 for review). In addition, ECM components such as F-spondin may modify protease function and lead to altering processing and breakdown of neuronal proteins implicated in synapse formation and function, such as amyloid precursor protein[100,101]. Thus, it is becoming clear that ECM may play a role in synaptogensis by modifying the adhesion of neuronal synaptogenic proteins, diffusional parameters of the extracellular space[102], the response of neurons to growth factors [73], and the synaptogenic state of neurons. How ECM does this and what receptors and signaling pathways are involved are questions still mostly waiting for answers.[*]

## 11. REFERENCES


1. Biederer, T., Sara, Y., Mozhayeva, M., Atasoy, D., Liu, X., Kavalali, E. T., and Sudhof, T.C. (2002) *Science* **297**, 1525–1531.
2. Scheiffele, P. (2003) *Annu Rev Neurosci* **26**, 485–508.
3. Waites, C.L., Marie Craig, A., and Garner, C.C. (2005) *Annu Rev Neurosci* **28**, 251–274.
4. Hall, A.C., Lucas, F.R., and Salinas, P.C. (2000) *Cell* **100**, 525–535.
5. Dalva, M.B., Takasu, M.A., Lin, M.Z., Shamah, S.M., Hu, L., Gale, N.W., and Greenberg, M.E. (2000) *Cell* **103**, 945–956.
6. Packard, M., Mathew, D., and Budnik, V. (2003) *Nat Rev Neurosci* **4**, 113–120.
7. Umemori, H., Linhoff, M.W., Ornitz, D.M., and Sanes, J.R. (2004) *Cell* **118**, 257–270.
8. Dityatev, A., and Schachner, M. (2003) *Nat Rev Neurosci* **4**, 456–468.
9. Pavlov, I., Lauri, S., Taira, T., and Rauvala, H. (2004) *Birth Defects Res C Embryo Today* **72**, 12–24.
10. Adams, J.C. (2001) *Annu Rev Cell Dev Biol* **17**, 25–51.
11. Adams, J.C., and Tucker, R.P. (2000) *Dev Dyn* **218**, 280–299.
12. Lawler, J. (2000) *Curr Opin Cell Biol* **12**, 634–640.
13. Bornstein, P. (2001) *J Clin Invest* **107**, 929–934.
14. Liska, D.J., Hawkins, R., Wikstrom, K., and Bornstein, P. (1994) *J Cell Physiol* **158**, 495–505.
15. Scott-Drew, S., and ffrench-Constant, C. (1997) *J Neurosci Res* **50**, 202–214.
16. Arber, S., and Caroni, P. (1995) *J Cell Biol* **131**, 1083–1094.
17. Hynes, R.O. (1992) *Cell* **69**, 11–25.
18. Sheppard, D. (2000) *Matrix Biol* **19**, 203–209.
19. Milner, R., and Campbell, I.L. (2002) *J Neurosci Res* **69**, 286–291.
20. Rohrbough, J., Grotewiel, M.S., Davis, R.L., and Broadie, K. (2000) *J Neurosci* **20**, 6868–6878.
21. Chavis, P., and Westbrook, G. (2001) *Nature* **411**, 317–321.
22. Kramar, E.A., Bernard, J.A., Gall, C.M., and Lynch, G. (2002) *Neuroscience* **110**, 29–39.
23. Nikonenko, I., Toni, N., Moosmayer, M., Shigeri, Y., Muller, D., and Sargent Jones, L. (2003) *Brain Res Dev Brain Res* **140**, 185–194.
24. Lin, B., Arai, A.C., Lynch, G., and Gall, C.M. (2003) *J Neurophysiol* **89**, 2874–2878.
25. Huang, A.M., Wang, H.L., Tang, Y.P., and Lee, E.H. (1998) *J Neurosci* **18**, 4305–4313.
26. Mi, Z.P., Jiang, P., Weng, W.L., Lindberg, F.P., Narayanan, V., and Lagenaur, C.F. (2000) *J Comp Neurol* **416**, 335–344.
27. Ullian, E.M., Christopherson, K.S., and Barres, B.A. (2004) *Glia* **47**, 209–216.
28. Slezak, M., and Pfrieger, F.W. (2003) *Trends Neurosci* **26**, 531–535.
29. Pfrieger, F.W. (2003) *Biochim Biophys Acta* **1610**, 271–280.
30. Christopherson, K.S., Ullian, E.M., Stokes, C.C., Mullowney, C.E., Hell, J.W., Agah, A., Lawler, J., Mosher, D.F., Bornstein, P., and Barres, B.A. (2005) *Cell* **120**, 421–433.


[*]A.D. was supported by the DFG grants Di 702/4-1,2,3.

31. Garlanda, C., Bottazzi, B., Bastone, A., and Mantovani, A. (2005) *Annu Rev Immunol* **23**, 337–366.
32. Varani, S., Elvin, J.A., Yan, C., DeMayo, J., DeMayo, F.J., Horton, H.F., Byrne, M.C., and Matzuk, M.M. (2002) *Mol Endocrinol* **16**, 1154–1167.
33. Salustri, A., Garlanda, C., Hirsch, E., De Acetis, M., Maccagno, A., Bottazzi, B., Doni, A., Bastone, A., Mantovani, G., Beck Peccoz, P., Salvatori, G., Mahoney, D.J., Day, A.J., Siracusa, G., Romani, L., and Mantovani, A. (2004) *Development* **131**, 1577–1586.
34. Wisniewski, H.G., and Vilcek, J. (2004) *Cytokine Growth Factor Rev* **15**, 129–146.
35. Schlimgen, A.K., Helms, J.A., Vogel, H., and Perin, M.S. (1995) *Neuron* **14**, 519–526.
36. Hsu, Y.C., and Perin, M.S. (1995) *Genomics* **28**, 220–227.
37. Noland, T.D., Friday, B.B., Maulit, M.T., and Gerton, G.L. (1994) *J Biol Chem* **269**, 32607–32614.
38. Reid, M.S., and Blobel, C.P. (1994) *J Biol Chem* **269**, 32615–32620.
39. Tsui, C.C., Copeland, N.G., Gilbert, D.J., Jenkins, N.A., Barnes, C., and Worley, P.F. (1996) *J Neurosci* **16**, 2463–2478.
40. Dodds, D.C., Omeis, I.A., Cushman, S.J., Helms, J.A., and Perin, M.S. (1997) *J Biol Chem* **272**, 21488–21494.
41. Dodds, D., Schlimgen, A.K., Lu, S.Y., and Perin, M.S. (1995) *J Neurochem* **64**, 2339–2344.
42. Whitehead, A.S., Zahedi, K., Rits, M., Mortensen, R.F., and Lelias, J.M. (1990) *Biochem J* **266**, 283–290.
43. Dowton, S.B., and McGrew, S.D. (1990) *Biochem J* **270**, 553–556.
44. Breviario, F., d'Aniello, E.M., Golay, J., Peri, G., Bottazzi, B., Bairoch, A., Saccone, S., Marzella, R., Predazzi, V., Rocchi, M. et al. (1992) *J Biol Chem* **267**, 22190–22197.
45. Kirkpatrick, L.L., Matzuk, M.M., Dodds, D.C., and Perin, M.S. (2000) *J Biol Chem* **275**, 17786–17792.
46. O'Brien, R.J., Xu, D., Petralia, R.S., Steward, O., Huganir, R.L., and Worley, P. (1999) *Neuron* **23**, 309–323.
47. Mi, R., Sia, G.M., Rosen, K., Tang, X., Moghekar, A., Black, J.L., McEnery, M., Huganir, R.L., and O'Brien, R.J. (2004) *Neuron* **44**, 335–349.
48. Chen, L., Chetkovich, D.M., Petralia, R.S., Sweeney, N.T., Kawasaki, Y., Wenthold, R.J., Bredt, D.S., and Nicoll, R.A. (2000) *Nature* **408**, 936–943.
49. Tomita, S., Chen, L., Kawasaki, Y., Petralia, R.S., Wenthold, R.J., Nicoll, R.A., and Bredt, D.S. (2003) *J Cell Biol* **161**, 805–816.
50. O'Brien, R., Xu, D., Mi, R., Tang, X., Hopf, C., and Worley, P. (2002) *J Neurosci* **22**, 4487–4498.
51. Mi, R., Tang, X., Sutter, R., Xu, D., Worley, P., and O'Brien, R.J. (2002) *J Neurosci* **22**, 7606–7616.
52. Xu, D., Hopf, C., Reddy, R., Cho, R.W., Guo, L., Lanahan, A., Petralia, R.S., Wenthold, R.J., O'Brien, R.J., and Worley, P. (2003) *Neuron* **39**, 513–528.
53. Chiquet-Ehrismann, R., and Chiquet, M. (2003) *J Pathol* **200**, 488–499.
54. Bruckner, G., Grosche, J., Schmidt, S., Hartig, W., Margolis, R.U., Delpech, B., Seidenbecher, C.I., Czaniera, R., and Schachner, M. (2000) *J Comp Neurol* **428**, 616–629.
55. Bradbury, E.J., Moon, L.D., Popat, R.J., King, V.R., Bennett, G.S., Patel, P.N., Fawcett, J.W., and McMahon, S.B. (2002) *Nature* **416**, 636–640.
56. Pizzorusso, T., Medini, P., Berardi, N., Chierzi, S., Fawcett, J.W., and Maffei, L. (2002) *Science* **298**, 1248–1251.
57. Nikonenko, A., Schmidt, S., Skibo, G., Bruckner, G., and Schachner, M. (2003) *J Comp Neurol* **456**, 338–349.
58. Saghatelyan, A.K., Dityatev, A., Schmidt, S., Schuster, T., Bartsch, U., and Schachner, M. (2001) *Mol Cell Neurosci* **17**, 226–240.
59. Sanes, J.R., and Lichtman, J.W. (2001) *Nat Rev Neurosci* **2**, 791–805.
60. Gingras, J., Rassadi, S., Cooper, E., and Ferns, M. (2002) *J Cell Biol* **158**, 1109–1118.
61. Martin, A.O., Alonso, G., and Guerineau, N.C. (2005) *J Cell Biol* **169**, 503–514.
62. Ferreira, A. (1999) *J Cell Sci* **112**, 4729–4738.
63. Bose, C.M., Qiu, D., Bergamaschi, A., Gravante, B., Bossi, M., Villa, A., Rupp, F., and Malgaroli, A. (2000) *J Neurosci* **20**, 9086–9095.
64. Li, Z., Hilgenberg, L.G., O'Dowd, D.K., and Smith, M.A. (1999) *J Neurobiol* **39**, 547–557.
65. Serpinskaya, A.S., Feng, G., Sanes, J.R., and Craig, A.M. (1999) *Dev Biol* **205**, 65–78.
66. Ji, R.R., Bose, C.M., Lesuisse, C., Qiu, D., Huang, J.C., Zhang, Q., and Rupp, F. (1998) *J Neurosci* **18**, 9695–9702.
67. Hilgenberg, L.G., Hoover, C.L., and Smith, M.A. (1999) *J Neurosci* **19**, 7384–7393.
68. Hilgenberg, L.G., and Smith, M.A. (2004) *J Neurobiol* **61**, 289–300.
69. Miner, J.H., and Yurchenco, P.D. (2004) *Annu Rev Cell Dev Biol* **20**, 255–284.

70. Beck, K., Dixon, T.W., Engel, J., and Parry, D.A. (1993) *J Mol Biol* **231**, 311–323.
71. Sasaki, T., Fassler, R., and Hohenester, E. (2004) *J Cell Biol* **164**, 959–963.
72. Hallmann, R., Horn, N., Selg, M., Wendler, O., Pausch, F., and Sorokin, L.M. (2005) *Physiol Rev* **85**, 979–1000.
73. Venstrom, K.A., and Reichardt, L.F. (1993) *FASEB J* **7**, 996–1003.
74. Clegg, D.O., Wingerd, K.L., Hikita, S.T., and Tolhurst, E.C. (2003) *Front Biosci* **8**, d723–750.
75. Nishimune, H., Sanes, J.R., and Carlson, S.S. (2004) *Nature* **432**, 580–587.
76. Marangi, P.A., Wieland, S.T., and Fuhrer, C. (2002) *J Cell Biol* **157**, 883–895.
77. Patton, B.L., Cunningham, J.M., Thyboll, J., Kortesmaa, J., Westerblad, H., Edstrom, L., Tryggvason, K., and Sanes, J.R. (2001) *Nat Neurosci* **4**, 597–604.
78. Huh, K.H., and Fuhrer, C. (2002) *Mol Neurobiol* **25**, 79–112.
79. Knight, D., Tolley, L.K., Kim, D.K., Lavidis, N.A., and Noakes, P.G. (2003) *J Physiol* **546**, 789–800.
80. Son, Y.J., Patton, B.L., and Sanes, J.R. (1999) *Eur J Neurosci* **11**, 3457–3467.
81. Libby, R.T., Lavallee, C.R., Balkema, G.W., Brunken, W.J., and Hunter, D.D. (1999) *J Neurosci* **19**, 9399–9411.
82. Einheber, S., Schnapp, L.M., Salzer, J.L., Cappiello, Z.B., and Milner, T.A. (1996) *J Comp Neurol* **370**, 105–134.
83. Nishimura, S.L., Boylen, K.P., Einheber, S., Milner, T.A., Ramos, D.M., and Pytela, R. (1998) *Brain Res* **791**, 271–282.
84. Hama, H., Hara, C., Yamaguchi, K., and Miyawaki, A. (2004) *Neuron* **41**, 405–415.
85. Pfrieger, F.W., and Barres, B.A. (1997) *Science* **277**, 1684–1687.
86. Ullian, E.M., Sapperstein, S.K., Christopherson, K.S., and Barres, B.A. (2001) *Science* **291**, 657–661.
87. Mauch, D.H., Nagler, K., Schumacher, S., Goritz, C., Muller, E.C., Otto, A., and Pfrieger, F.W. (2001) *Science* **294**, 1354–1357.
88. Nagler, K., Mauch, D.H., and Pfrieger, F.W. (2001) *J Physiol* **533**, 665–679.
89. Rice, D.S., and Curran, T. (2001) *Annu Rev Neurosci* **24**, 1005–1039.
90. D'Arcangelo, G., Miao, G.G., Chen, S.C., Soares, H.D., Morgan, J.I., and Curran, T. (1995) *Nature* **374**, 719–723.
91. Benhayon, D., Magdaleno, S., and Curran, T. (2003) *Brain Res Mol Brain Res* **112**, 33–45.
92. Rice, D.S., Nusinowitz, S., Azimi, A.M., Martinez, A., Soriano, E., and Curran, T. (2001) *Neuron* **31**, 929–941.
93. Del Rio, J.A., Heimrich, B., Borrell, V., Forster, E., Drakew, A., Alcantara, S., Nakajima, K., Miyata, T., Ogawa, M., Mikoshiba, K., Derer, P., Frotscher, M., and Soriano, E. (1997) *Nature* **385**, 70–74.
94. Borrell, V., Del Rio, J.A., Alcantara, S., Derer, M., Martinez, A., D'Arcangelo, G., Nakajima, K., Mikoshiba, K., Derer, P., Curran, T., and Soriano, E. (1999) *J Neurosci* **19**, 1345–1358.
95. Liu, W.S., Pesold, C., Rodriguez, M.A., Carboni, G., Auta, J., Lacor, P., Larson, J., Condie, B.G., Guidotti, A., and Costa, E. (2001) *Proc Natl Acad Sci U S A* **98**, 3477–3482.
96. Sinagra, M., Verrier, D., Frankova, D., Korwek, K.M., Blahos, J., Weeber, E.J., Manzoni, O.J., and Chavis, P. (2005) *J Neurosci* **25**, 6127–6136.
97. Niu, S., Renfro, A., Quattrocchi, C.C., Sheldon, M., and D'Arcangelo, G. (2004) *Neuron* **41**(1), 71–84.
98. Rice, D.S., and Curran, T. (2000) *J Comp Neurol* **424**, 327–338.
99. Tian, N. (2004) *Vision Res* **44**, 3307–3316.
100. Rauch, U. (2004) *Cell Mol Life Sci* **61**, 2031–2045.
101. Ho, A., and Sudhof, T.C. (2004) *Proc Natl Acad Sci U S A* **101**, 2548–2553.
102. Sykova, E., Vorisek, I., Mazel, T., Antonova, T., and Schachner, M. (2005) *Eur J Neurosci* **22**, 1873–1880.

# 12

# ROLE OF NEUROTROPHINS IN THE FORMATION AND MAINTENANCE OF SYNAPSES

Newton H. Woo, Hyun-soo Je, and Bai Lu*

## 1. SUMMARY

During development, the establishment of precise synaptic connections between neurons and specific target cells is critical for normal function of the nervous system. Several attributes of neurotrophins, particularly their activity-dependent expression and secretion, render these secreted proteins excellent extracellular signals that regulate the formation of synapses. Here we review experimental evidence for the potential role of neurotrophins in synapse formation. Members of the neurotrophin family have been shown to regulate the maturation of both excitatory and inhibitory synapses and also have been demonstrated to impact refinement of synaptic connections. Here, we discuss the multitude of signaling cascades that allow neurotrophins to achieve such diverse effects on CNS development and elaborate on the neurotrophin functions in the regulation of synaptogenesis in several important regions of the mammalian brain.

## 2. INTRODUCTION

A key function of neurotrophins is to regulate the formation and refinement of synaptic connections in the central nervous system (CNS). Since the initial discovery of nerve growth factor (NGF), a member of the neurotrophin family, extensive efforts have been made to identify and elucidate the cellular actions of neurotrophins. Like other secreted proteins, neurotrophins arise from precursors, pro-neurotrophins (30–35 kDa), which are proteolytically cleaved to produce mature proteins (12–13 kDa). In the mammalian brain, four members of the neurotrophin family have been identified: NGF, brain-derived neurotrophic factor (BDNF), neurotrophin 3 (NT-3), and neurotrophin 4 (NT-4). These closely related molecules act by binding to two distinct classes of transmembrane receptors: the

---

*Section on Neural Development & Plasticity, NICHD; Genes, Cognition and Psychosis Program (GCAP), NIMH, National Institutes of Health, Bethesda, MD 20892-3714; bailu@mail.nih.gov

p75 pan-neurotrophin receptor ($p75^{NTR}$) and the Trk family of receptor tyrosine kinases. All neurotrophins bind $p75^{NTR}$ with similar affinity. In contrast, Trk receptor kinases interact selectively with a different neurotrophin: BDNF and NT-4 bind to TrkB, NGF to TrkA and NT-3 to TrkC. This diversity in receptor and ligand interactions, as well as the distinct patterns of expression in several critical regions of the brain, underscores the importance of neurotrophins in regulating many facets of neural development.

Normal function of the brain requires the formation of complex neuronal networks built on numerous synaptic connections that are appropriately formed and maintained. This is the result of a multistep process occurring throughout development that involves neurogenesis, survival, differentiation, and dendritic and axonal growth. The final step culminates in the formation of synapses, also known as synaptogenesis. Not limited to development, synaptogenesis also occurs during normal cognitive functions in the adult brain such as learning and memory, as well as during regeneration after brain injury. It is now recognized that synaptogenesis itself is a multistep process that includes initial synaptic contact, maturation, and stabilization. In addition, excessive and inappropriate synapses also need to be eliminated, and this process requires co-incident activity between pre- and postsynaptic partners.

Recent studies have shown that neurotrophins influence various aspects of synaptogenesis. In this chapter we highlight the diverse signal transduction cascades utilized by neurotrophins to regulate synapse formation, and discuss the functional roles of neurotrophins at different stages of synaptogenesis by focusing on several regions of the brain. We start out with an overview of the signaling cascades that are activated by neurotrophins. We next discuss the role of neurotrophins in synaptogenesis in a simple system: the neuromuscular junction (NMJ). Subsequently, we examine the formation of synapses in the developing hippocampus and cerebellum. We then discuss how neurotrophins regulate the formation of ocular dominance during visual cortex (VC) development, an example of how neurotrophins refine synaptic connections. Finally we provide evidence for neurotrophic regulation of synaptic connections in the adult or mature organism as observed in the case of the barrel cortex.

## 3. NEUROTROPHIN SIGNALING

Neuronal activity controls the synthesis and secretion of neurotrophins during critical periods (CPs) of synaptic development and remodeling[1–3]. Cumulative evidence suggests that neurotrophins regulate structure and function of synapses, both pre- and postsynaptically[4–6]. These results make neurotrophins particularly attractive candidates as regulators of synaptogenesis. Nevertheless, studies of neurotrophic regulation of synaptogenesis are still at a descriptive stage, and relatively little is known about the underlying mechanisms or functional significance in living animals. This is due largely to the ever-changing landscape of research in neurotrophin signaling. Novel interacting proteins and new signaling cascades are continuously being discovered. Thus, it is important to update the current understanding of neurotrophin signaling (Figure 12.1).

A general scheme (Figure 12.1A), largely based on studies of NGF and TrkA in PC12 cells, in that binding of neurotrophins to their cognate Trk receptors causes receptor dimerization and phosphorylation of tyrosine residues in their intracellular domains (ICDs)[7,8]. Phosphorylation of these tyrosines creates docking sites for several adaptor proteins, which in turn activate three key signaling pathways (Figure 12.1A): the PLC-γ, Ras-Erk/MAPK, and PI3K/Akt cascades[9,10]. Active PLC-γ hydrolyzes phosphatidylinositol 4,5-biphosphate [PI(4,5)P2] to generate inositol trisphosphate (IP3) and diacylglycerol (DAG), which in turn induces $Ca^{2+}$ release from internal stores and activates protein kinase C, respectively. Activation of Shc, through interaction with Grb2 and SOS, triggers a series of phosphorylation reactions that include Raf, MEK, and mitogen-associated protein kinase (MAP kinase). Through Gab-1, Shc also activates phosphoinositide 3-kinase (PI3 kinase).

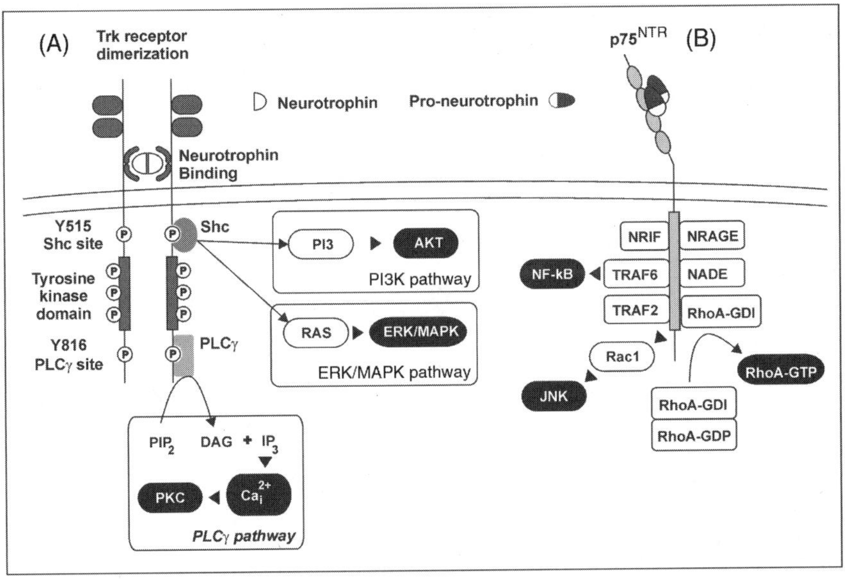


**Figure 12.1.** Neurotrophin Signaling Through Trk and $p75^{NTR}$ Receptors. (A) Homodimeric neurotrophins (NGF, BDNF, NT-3, NT4/5) induces the dimerization of their cognate Trk tyrosine kinase receptors. Autophosphorylation of tyrosine residues in the kinase domain primes subsequent phosphorylation events at other tyrosine residues, Y515 and Y816, and creates binding sites for Shc and PLCγ, respectively. As a result, three major signaling cascades can be activated: PI3K pathway, ERK/MAPK pathway, and PLCγ pathway. (B) Pro-neurotrophins and to a lesser extent mature neurotrophins bind $p75^{NTR}$ receptors. Upon activation, a signaling complex that includes NRAGE, NRIF, NADE, and TRAF can be assembled and induce Rac1 and consequent JNK activation. In a different context, $p75^{NTR}$ can also mediate RhoA activation. Rho-GDI is associated with inactive Rho-GDP, however, ligand binding to a receptor complex, consisting of $p75^{NTR}$ and NogoR and LINGO-1, induces a conformational shift that allows Rho-GDI to bind to the intracellular domain of $p75^{NTR}$ and releases RhoA. Once released, RhoA can exchange GDP to GTP and achieve its active conformation.

Although this general scheme is well accepted, each Trk receptor may trigger a unique set of signaling events. Thus, signaling pathways involved in specific synaptic functions in specific brain regions remain to be worked out. For example, acute potentiation of synaptic transmission at the developing NMJ is mediated by concomitant activation of PLC-γ-IP3 and PI3 kinase pathways, but not MAP kinase pathway[11]. In sympathetic neurons, activation of TrkA by NGF stimulates cell survival predominantly through the PI3 kinase pathway, whereas BDNF suppresses apoptosis through TrkB activation of both MAP kinase and PI3 kinase pathways[12]. Moreover, there are many ways that neurotrophin signal transduction is regulated. TrkA activated at the axonal terminals, through endocytosis of NGF–TrkA complex and retrograde transport of signaling endosomes, can induce MAP kinase Erk5 activation[13]. In contrast, direct stimulation of TrkA on the cell soma activates Erk1/2, but not Erk5.

Signal transduction mechanisms induced by acute or long-term exposure of neurotrophins are dramatically different[5]. Acute effects of neurotrophins depend on a transient activation of the Ras-MAP kinase and PI3 kinase but not its downstream effector Akt. Long-term effects of neurotrophins appear to be mediated by endocytosis of Trk receptors and sustained activation of Rap1-MAP kinase as well as PI3 kinase-Akt. For synaptogenesis, the most important and relevant attribute of neurotrophin signaling is the activity-dependent regulation of Trk receptors[14]. Recent studies demonstrated that neuronal activity facilitates the insertion of TrkB onto the cell surface[15,16], and also promotes the endocytosis of BDNF–TrkB complexes[17]. BDNF is secreted from active synapses and neurons recruit TrkB from extrasynaptic sites into lipid rafts, microdomains of membrane that are enriched at synapses[18]. Postsynaptic rises in cAMP concentrations facilitate translocation of TrkB into the postsynaptic density[19]. Futures studies are required to dissect the molecular cascades involved in each individual step of synapse formation and maintenance and also identify which critical steps are regulated by neurotrophins.

All neurotrophins are capable of binding to the pan-neurotrophin receptor ($p75^{NTR}$). It has now been widely accepted that prime ligands for $p75^{NTR}$ are not the mature form but rather the unprocessed, pro-neurotrophins[20]. Recent studies support a "yin–yang" hypothesis that interaction of pro-neurotrophins with $p75^{NTR}$ often elicits biological effects opposite to those elicited by mature neurotrophin–Trk interactions[21]. While $p75^{NTR}$ facilitates Trk-mediated cell survival in certain cases, a major function of $p75^{NTR}$, especially when it is expressed alone, is to promote apoptosis[22,23]. The pro-apoptotic effect of $p75^{NTR}$ is believed to involve the activation of Rac, a small GTP-binding protein, and the activation of Jun N-terminal kinase (JNK). Proximal events that couple $p75^{NTR}$ and Rac are less clear, but the interaction between $p75^{NTR}$ cytoplasmic domain and several adaptor proteins, including TRAF-6, NRAGE, NADE and NRIF, appears to be important (Figure 12.1B). Another important function of $p75^{NTR}$ is to regulate axonal growth and regeneration, which is primarily mediated by the interaction of $p75^{NTR}$ with RhoA, another small GTP-binding protein (Figure 12.1B). The third and paradoxical role of $p75^{NTR}$ is to facilitate cell survival through TRAF-6 and NF-κB (Figure 12.1B). In addition to the complexities of signaling events associated with $p75^{NTR}$, two recent advances have dramatically changed the landscape of $p75^{NTR}$ research. First, Nykjaer et al. (2004) discovered that the pro-domain of unprocessed neurotrophins binds to the extracellular domain of sortilin, and such binding is required for pro-neurotrophin-induced cell death[24]. Based on this study, a model was proposed that pro-neurotrophins (e.g., proNGF) simultaneously bind sortilin (via the pro-domain) and $p75^{NTR}$ (via the mature domain). Interaction of the pro-domain and sortilin has also been shown to be important for intracellular trafficking and regulated secretion of proBDNF[25] The second major advance is that $p75^{NTR}$ is similar to amyloid precursor protein, Notch and ErB4, in the sense that its extracellular domain can also undergo proteolytic cleavage by α-secretase. This is followed by a second cleavage by γ-secretase within the transmembrane domain to generate an ICD, which is translocated to the nucleus. It remains to be established whether the production and translocation of ICD controls gene transcription.

Advances in $p75^{NTR}$ research have generally lagged behind that of Trk receptors. This may explain why virtually all functional studies on neurotrophins thus far have concentrated on the effects mediated by Trk receptors. Given the breakthroughs made in the $p75^{NTR}$ signaling in the last few years, we expect more studies will reveal synaptic functions of neurotrophins mediated by $p75^{NTR}$. Indeed, a recent study has shown that secreted proBDNF, through activation of $p75^{NTR}$, plays a key role in NMDA receptor-dependent long-term depression (LTD) at hippocampal synapses[26].

## 4. NEUROMUSCULAR JUNCTION (NMJ)

Due to simplicity and easy accessibility, much of our understanding of synapse formation and stabilization derives from studies utilizing the NMJ as a model system[27]. The formation of neuromuscular synapses can be roughly divided into three stages (Figure 12.2). The first stage is the initial synaptic contact. Axonal growth cones come in contact with the postsynaptic target cells, the myocytes (Figure 12.2A). The second stage is synapse maturation. Through cell–cell interaction, the growth cones are differentiated into presynaptic terminals. The quantum secretion machinery gradually matures, and synaptic vesicles become clustered near the presynaptic membranes to form active zones. This process occurs in parallel with the differentiation of postsynaptic myocytes (Figure 12.2B). A hallmark of this stage is the aggregation of acetylcholine receptors (AChRs). The final stage is synapse maintenance, where appropriate synaptic connections are stabilized and incorrect ones are eliminated.

Neurotrophins are synthesized and secreted from the muscle cells in an activity-dependent manner[28,29]. Their cognate receptors, such as TrkB and TrkC, are expressed in the presynaptic spinal neurons and postsynaptic muscle cells[28,30–32]. Experiments over the last decades have clearly established the role of neurotrophins in the first and second stages. There is also evidence that NT-4 is involved in the maintenance of effective NMJ in the adult[32], but whether neurotrophins also play a role in activity-dependent synaptic elimination remains to be investigated.

The role of neurotrophins in the first stage of synaptogenesis was examined using the *Xenopus* nerve–muscle co-cultures. During the earliest stages, exposure to BDNF induces a rapid collapse of growth cones[33]. The morphology of growth cones changed within minutes following treatment with BDNF. This may be necessary to convert a dynamic growth cone into a differentiated terminal. At the stage where initial synaptic contact occurs, acute application of BDNF or NT-3 elicits a marked potentiation of transmitter release from presynaptic terminals[34]. It appears that BDNF and NT-3 elicit their acute effects through different intracellular mechanisms. BDNF acts almost exclusively at the presynaptic terminals[35], and this effect requires $Ca^{2+}$ influx[36]. In contrast, NT-3 is thought to act on the cell body[37], and its effect is mediated by $Ca^{2+}$ from intracellular $Ca^{2+}$ stores[38].

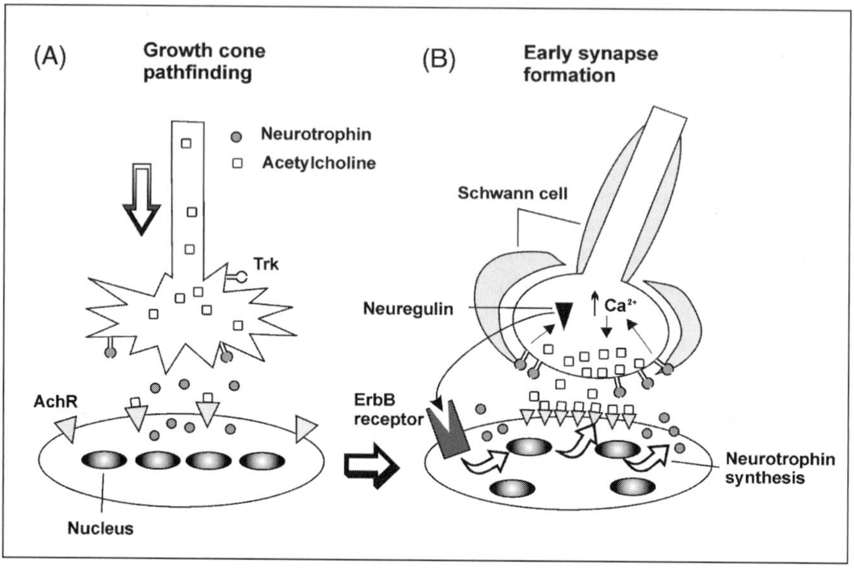


**Figure 12.2.** Outline of Synaptogenesis at the Neuromuscular Junction. (A) The growth cone of a motor neuron approaches a newly formed myotube by recognizing neurotrophin. Near the area of contact, the growth cone of a motor neuron differentiates into a nerve terminal that is specialized for neurotransmitter release. Note that synaptic vesicles in the nerve terminal are not strictly localized at the synaptic sites. Acetylcholine receptors (AChRs) are initially present throughout the myotube surface. (B) As synapses are forming, AChRs are clustered on the postsynaptic membrane and extrasynaptic AChRs are diminishing due to redistribution of AChR proteins and local synthesis of AChRs. The local synthesis results from enhanced transcription of AChR genes by subsynaptic nuclei. Neurotrophins secreted from myotubes enhance transmitter release by increasing $[Ca^{2+}]_{in}$. Also, Trk signaling upon neurotrophin binding triggers neuregulin expression in the presynaptic neuron and the secreted neuregulin from presynaptic neuron binds to ErbB receptor in the myotubes and increases AChR transcription. Furthermore, increased presynaptic activity induces gene transcription and synthesis of neurotrophins. This positive feedback loop between presynaptic and postsynaptic cells may stabilize the newly synthesized synapse.

Much more extensive work has been carried out on the role of neurotrophins in synapse maturation. Two effects have been observed when *Xenopus* nerve–muscle co-cultures were treated with BDNF or NT3 for a prolonged period of time (e.g., 48 h) after initial synaptic contacts are made. First, electrophysiology experiments demonstrated that chronic exposure to BDNF or NT-3 resulted in more mature and uniformed quanta[39–41]. Failure rate and variability of impulse-evoked synaptic transmission were decreased. Treatment with NT-3 also increased the excitability of the presynaptic motor neurons[42]. Second, BDNF or NT-3 also elicited dramatic morphological changes at the presynaptic terminals. Immunohistochemistry and FM dye experiments showed that the number of synaptic varicosities, which reflect clusters of synaptic vesicles, were increased in response to chronic treatment of BDNF and NT-3[39]. Biochemically, BDNF or NT-3 enhanced the expression of synaptic vesicle proteins such as synapsin-I and synaptophysin at the motor nerve terminals[39]. Synapsins are capable of clustering synaptic vesicles in the presynaptic terminals[43]. Synapsin I has been shown to promote synapse maturation at the developing NMJ[44,45]. It is therefore conceivable that neurotrophins promote the structural and functional maturation of the presynaptic motor terminals by the up-regulation of synapsin I.

Endogenous neurotrophins involved in specific aspects of synapse development remain to be established. Different neurotrophins are expressed in muscle cells at different stages of development. The levels of BDNF and NT-3 mRNAs are high in embryonic and neonatal muscle cells but low in those of adult rat[32]. Moreover, the expression of NT-3 in developing muscle cells is activity dependent[28]. Membrane depolarization or repetitive electrical stimulation specifically increased the levels of NT-3 mRNA rapidly. NT-3 gene expression was also enhanced by ACh, the neurotransmitter that causes muscle membrane depolarization. Thus, innervation and synaptic transmission at the NMJ enhance postsynaptic NT-3 production. Remarkably, conditioned medium from depolarized myocytes enhanced synaptic transmission at the developing NMJ, and such effect was blocked by the NT-3 scavenger TrkC-IgG[28]. These results suggest that NT-3 may serve as a retrograde messenger, derived from postsynaptic myocytes to stimulate presynaptic maturation at the NMJ. In the adult, NT-4 is the major neurotrophin expressed in muscle cells. Muscle expression of NT-4 in adult rodent can be regulated by synaptic transmission and muscle activity[32]. These results raise the possibility that NT-4 is responsible for the effects of exercise and electrical stimulation on neuromuscular performance.

In addition to facilitating presynaptic maturation, neurotrophins appear to be involved in the regulation of postsynaptic targets during synaptogenesis by regulating AChR density and/or clustering at the postsynaptic membrane[29,46,47]. As the muscle becomes innervated, ACh-evoked responses become much larger at the site of innervation and the responses at extrasynaptic regions decline, suggesting the increased numbers of receptors at synaptic sites. Both *in vitro* and *in vivo* imaging with α-bungarotoxin (α-Btx) labeling techniques confirmed that AChR clusters become highly localized to the synapses as synaptogenesis proceeds. Nancy and colleagues demonstrated that muscle-derived NT-3 enhances AChR clustering on the myotubes by using co-cultures of NG108 neuroblastoma cells, which express TrkC and C2C12 myotubes transfected with various neurotrophins[46]. In addition to this *in vitro* finding, NT-4 knockout mice display enlarged and fragmented NMJs with disassembled postsynaptic AChR clusters *in vivo*, suggesting muscle derived NT-4 is necessary for the maintenance of postsynaptic AChR clustering[48]. Interestingly, when TrkB signaling in postnatal muscle cells was disrupted by expressing dominant negative TrkB, AChR clusters on the postsynaptic muscle cells were disassembled *in vivo*[47]. These results indicate that NT-4/TrkB signaling is required for the maintenance of AChR clustering in postsynaptic cells. It is still not clear how NT-4/TrkB signaling affects AChR clustering, but growing evidence suggests that NT-4, through TrkB, could possibly activate Fyn kinase, which is implicated in the formation and stability of AChR clusters[49–51].

Neurotrophins can also modulate postsynaptic development indirectly through the regulation of neuregulin[52]. A major function of neuregulin is to stimulate the transcription of the AChR gene in the "synaptic nucleus" of polynucleus myotubes[53]. Recent data suggest that neurotrophins upregulate neuregulin mRNA expression in the spinal neurons and this in turns increase the number of AChRs at synaptic sites[52]. In support for this, blocking synaptic activity with curare reduces synaptic neuregulin expression in chick spinal neurons, but neuregulin expression and AChR clustering are restored when exogenous BDNF or NT-3 is added to this nerve–muscle culture[52]. Taken together, it is clear that neurotrophins modulate synapse formation at NMJ by affecting both presynaptic and postsynaptic sites either directly or indirectly.

## 5. HIPPOCAMPUS AND CEREBELLUM

Compared to the NMJ, relatively less work has been done on synaptogenesis in the brain. But recent technological advances such as real-time imaging and transgenic animals have provided some insights into the molecular and cellular processes that underlie the formation of central synapses. An emerging concept is that the multiple steps involved in synapse formation observed in the NMJ also occur in the CNS (Figure 12.3).

Early studies using dissociated cultures of hippocampal neurons provided key evidence that neurotrophins play an important role for the formation of both inhibitory and excitatory synapses[54]. Hippocampal neurons isolated from embryonic day 16 (E16) rats do not generally develop spontaneous synaptic activity, however, treatment with neurotrophins for 3 days elicited a marked increase in the number of functional synapses[54]. More specifically, BDNF induced formation of both excitatory and inhibitory synapses, whereas NT-3 induced formation of only excitatory synapses. When NT-4 gene was knocked into the BDNF locus, functional synapses as measured by FM dye labeling in cultured hippocampus formed much earlier, compared with those in wild-type mice[55]. Since NT-4 binds TrkB with similar affinity and that application of BDNF and NT-4 to wild-type cultures elicited similar effects on synapse formation[55], it is likely that differential expression (soma versus dendrites or axons) or secretion (constitutive versus regulated) of BDNF and NT-4 may contribute to the differential effects.

Do neurotrophins regulate hippocampal synaptogenesis *in vivo*? In neonatal hippocampus (postnatal day 12–13 in rats) in which endogenous BDNF expression is low, CA1 synapses are unable to express long-term potentiation (LTP), primarily due to the inability of these synapses to follow high-frequency, tetanic stimulation (HFS) commonly used to induce LTP. Application of exogenous BDNF to developing hippocampal slices facilitates LTP by enhancing the CA1 synapses to respond to HFS[56,57]. These results suggest that BDNF promotes the maturation of hippocampal CA1 synapses. Indeed, in neonatal hippocampi derived from TrkB–/– and TrkC–/– mice, there is a significant decrease in many synaptic proteins and in the total number of synaptic vesicles per nerve terminal, as well as a reduction in the number of synaptic contacts[58]. Conversely, transgenic mice overexpressing BDNF not only increased the number of both excitatory and inhibitory synapses but also elicited earlier maturation of these synapses[59].

Using three different lines of conditional knockout mice that deleted TrkB gene in pre- and postsynaptic cells at different developmental stages, Parada and colleagues elegantly demonstrated that TrkB signaling has a cell-autonomous role required for normal development of both presynaptic and postsynaptic components of the Schaffer collateral synapses[60]. Presynaptically, BDNF–TrkB signaling appears to promote synapse maturation by facilitating synaptic vesicle docking at the active zone. Electron microscopic studies showed that deletion of BDNF gene reduced the number of vesicles docked on the presynaptic membranes of CA1 synapses[61]. Similar vesicle docking deficit was observed at the parallel fibers to Purkinje cell synapses in the cerebellum of BDNF–/– mice[62]. Conversely, treatment of neonatal hippocampal slices with BDNF promoted synaptic vesicle docking[63,64]. Changes in the levels and/or distribution of synaptic vesicle proteins may contribute to BDNF regulation of vesicle docking (Figure 12.3). In synaptosomes prepared from BDNF knockout mice, there was a selective reduction of synaptophysin as well as synaptobrevin, proteins involved in docking and fusion of the vesicles[61]. Treatment of slice cultures derived from neonatal hippocampus with BDNF increased the expression of several synaptic vesicle proteins including synaptobrevin[63].

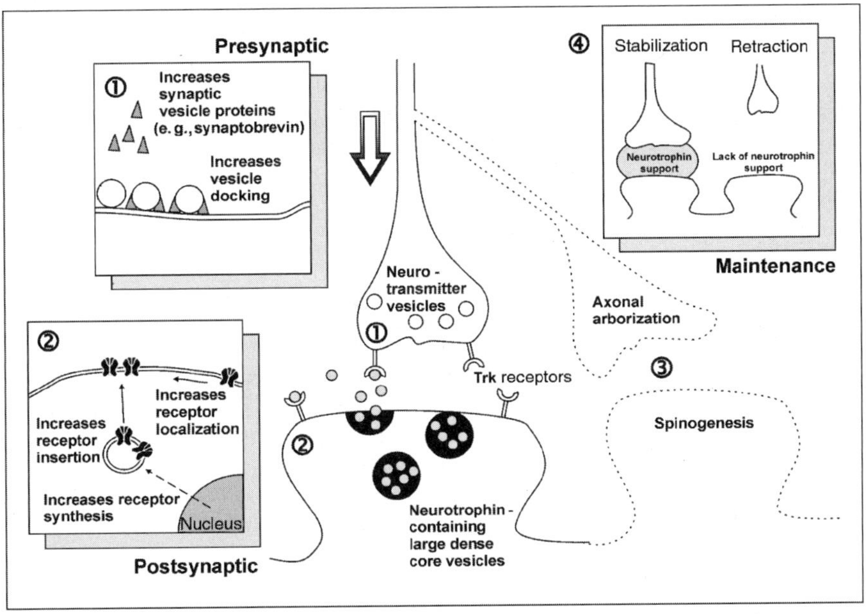


**Figure 12.3.** Diverse Actions of Neurotrophins during Synaptogenesis. Activity-dependent release of neurotrophins promotes synaptogenesis by stimulating the maturation of pre- and postsynaptic cells and also by inducing profound morphological changes. ① Presynaptically, neurotrophins enhance vesicle docking to the active site and also increase the expression of several key proteins involved in neurotransmitter release, including synaptobrevin, synaptophysin and synaptotagmin. ② In addition, neurotrophins elicit several postsynaptic effects that play an important role in synaptogenesis. These include the enhancement of receptor synthesis, as well as the insertion and localization of various neurotransmitter receptors at synaptic sites. ③ In terms of morphological alterations, neurotrophins enhance axonal arborization and spinogenesis of both excitatory and inhibitory neurons. ④ Once new synapses are formed, neurotrophins stabilize these connections and prevent axonal retraction.

It appears that the regulatory effects of neurotrophins are not limited to the hippocampus. In cerebellar slice cultures, BDNF or NT-4 promoted the development of inhibitory axosomatic synapses in Purkinje cells when neuronal activity was blocked, suggesting that neurotrophins underlie this activity-dependent form of regulation [65]. Conversely, blocking neurotrophins with antibodies greatly reduced synaptogenesis in inhibitory neurons[66]. Further evidence comes from a study using conditional TrkB knockout mice. This study demonstrated that TrkB is required for the establishment of cerebellar inhibitory synapses [67].

Postsynaptically, neurotrophins control synaptogenesis through the regulation of neurotransmitter receptor at synapses (Figure 12.3). In cultured neurons, application of BDNF facilitates the insertion of AMPA receptor into the postsynaptic membranes through an NSF-dependent mechanism[68,69]. Interestingly,

BDNF can trigger PDZ protein expression at synaptic sites in developing cortical neurons[70]. Therefore, BDNF may promote the synaptic expression of AMPA receptor during development by facilitating the interaction between AMPA receptor and proteins involved in receptor trafficking such as NSF and PDZ proteins (SAP97, Grip, PICK1). A recent study demonstrated that TrkB-mediated signaling affects the number and synaptic localization of postsynaptic NMDA and $GABA_A$ receptor clusters[71]. In this study, BDNF treatment increased the number of receptor clusters and also the proportion of clusters apposed to presynaptic terminals. It should be pointed out that these results were obtained using cultured neurons and further work is necessary to confirm whether this holds true *in vivo*.

In addition to receptor clustering, neurotrophins promote the formation of postsynaptic dendritic spines (Figure 12.3). Most excitatory synapses are localized to dendritic spines. In cerebellar cultures, BDNF, together with co-cultured granule cells, enhanced spinogenesis in Purkinje cells without affecting dendritic complexity[75]. Similarly, long-term treatment of hippocampal slice cultures with BDNF increases the spine density in apical dendrites of CA1 pyramidal neurons, without affecting dendritic length and branching[64]. A recent study suggests that BDNF regulation of spine formation in hippocampal neurons is controlled by cAMP, through two different mechanisms[19]. One is "cAMP gating," where cAMP promotes BDNF signaling by potentiating TrkB phosphorylation. In addition, cAMP facilitates translocation of TrkB into the postsynaptic density. Both mechanisms work together to mediate BDNF regulation of spine formation.

Although these findings provide preliminary insights, the exact step(s) of synaptogenesis that are regulated by neurotrophins, and the mechanisms downstream of neurotrophin receptor activation, are not well understood. Further work is required to determine the necessity and sufficiency of neurotrophins at each particular step in synapse formation.

## 6. OPTIC TECTUM

Studies using the optic tectum have significantly contributed to our understanding of the many roles of neurotrophins in synaptogenesis[76,77]. The retinal ganglion cells (RGC) extend a single axon to the optic tectum in lower vertebrates or the lateral geniculate nucleus (LGN) in higher vertebrates. A remarkable early finding was that injection of BDNF into the optic tectum of live *Xenopus* tadpoles increased the arborization of RGC axon terminals, which express TrkB[78]. BDNF also regulates RGC dendritic growth, but this regulation is more complex. Target (tectum)-derived BDNF increased dendritic arbor complexity while local (retina)-derived BDNF inhibited dendritic arborization[79,80].

However, an enhancement in axon terminal arborization does not necessarily lead to an increase in synapse number. By labeling presynaptic sites with GFP-synaptobrevin and axonal terminals with DsRed, Alsina et al. showed that injection of BDNF enhanced RGC terminal arbor complexity as well as the density of GFP-synaptobrevin identified at presynaptic sites *in vivo*[81]. Conversely, inhibition of endogenous tectal BDNF by function-blocking antibodies significantly enhanced GFP-synaptobrevin cluster elimination, and this was in parallel with an increase in branch elimination[82].

Neuronal/synaptic activity also regulates axon terminal arborization, as well as synapse number. Global activity such as action potentials may preferentially affect terminal arborization. Time-lapse imaging showed that retinal activity blockade by tetrodotoxin (TTX) led to an increase in both addition and elimination of terminal branches, reducing the stabilization of axon terminals. In contrast, BDNF promoted axonal arborization by increasing branch addition and lengthening, without affecting branch elimination, and this effect is activity independent[83]. Local, synaptic activity may preferentially affect synapse number. Tectal injection of the NMDA receptor antagonists APV or MK801 reduced synapse number, as reflected by GFP-synaptobrevin cluster, but did not influence axon branch addition or elimination. Co-injection of NMDA receptor antagonists with BDNF prevented synaptic de-stabilization[82]. Therefore BDNF influences RGC synaptogenesis by not only promoting morphological maturation of RGC axons but also by stabilizing synapses.

Taken together, these studies provide direct correlations between structural changes and synapse formation, both of which are regulated by BDNF. However, it is unclear whether newly formed or stabilized synapses are physiologically active. Moreover, the signaling cascades mediating morphological changes induced by neurotrophins are poorly understood. Recent studies that examine actin cytoskeletal dynamics suggest that F-actin stabilization and/or polymerization may be a candidate since pre- and postsynaptic changes of actin directly influences the stability of developing synapses[84,85].

## 7. VISUAL CORTEX

Visual information, initially encoded by the RGCs, is propagated to morphologically distinct layers in the LGN of the thalamus. By way of the geniculocortical pathway, LGN neurons send terminal arbors that form dense clusters in layer 4 of the VC. The development of thalamic–cortical synapses exhibit several characteristic features: (1) ocular dominance (OD): pyramidal neurons, particularly those in layer 4, respond preferentially to inputs from one eye. In cat and primates, cells that respond to one eye are organized into alternating, eye-specific columns; (2) activity-dependent synaptic competition: in cells that receive inputs from both eyes, the strength of the synaptic connection depends on the activity of the thalamic inputs from a specific eye; and (3) critical period (CP): this activity-dependent modulation of cortical synapse formation occurs in a defined period of time during development. These features have allowed scientists to use two powerful experimental approaches. One is monocular deprivation (MD), namely manipulations that block the neuronal activity of one eye (e.g., eyelid suture) during CP. A striking effect of MD is to cause a shift of OD toward the nondeprived eye. The second commonly used approach is dark rearing (DR), namely to block the activity of both eyes. This manipulation delays the formation of OD and therefore postpones CP.

These interesting features of VC synapses strongly suggest that RGC inputs from the two eyes compete for factors produced in the postsynaptic cells in an activity-dependent manner. To determine whether a factor is critically involved in the development of OD, several criteria should be fulfilled. This factor should be produced in a limited amount in the cortex, and its production (or responsiveness) should be controlled by the electrical activity of input eyes. Thalamic afferents should be responsive (express receptors) to the factor. Experimentally, application (or overexpression) of the factor should rescue the OD shift induced by MD, and enhance synapse formation in the VC and shorten CP.

Studies by Maffei and his colleagues suggest that NGF is the critical factor in the rat VC. Intraventricular infusion of NGF during the CP abolished the physiological shift in OD toward the nondeprived eye following MD[86] as well as prevented the shrinkage of LGN cell bodies[87,88]. Conversely, administration of antibodies against NGF extended the CP[89] and also induced shrinkage of LGN cells[90]. NGF also prevented some of the deficits induced by DR, allowing normal development of VC synapses[91,92]. However, NGF is poorly expressed in the VC. Furthermore the NGF receptor TrkA is only expressed in cholinergic afferents that stem from the basal forebrain, but not in the axons of geniculate neurons[93]. Collectively, these observations weaken the argument that NGF is a factor that directly controls OD formation.

More compelling evidence supports TrkB ligands as the critical factor in the VC. Several studies highlight a role for BDNF and perhaps NT-4, through activation of TrkB, in controlling the competitive interactions between geniculate inputs to the VC. TrkB receptor is highly expressed by LGN neurons during early postnatal life[94–97]. Both BDNF[98] and NT-4[99] mRNAs are detected in layer 4 pyramidal neurons. Moreover, BDNF expression increases rapidly after eye opening[98,100]. Lid suture or intravitreal injection of TTX induced a downregulation of BDNF expression[94,98]. Thus, BDNF mRNA levels correlate with the activity level of postsynaptic visual cortical neurons. More importantly, when TrkB ligands were infused into kitten VC during the CP, LGN axons failed to segregate into OD columns near the site of infusion[101]. However when BDNF infusion was performed on adult cats, OD formation was unaffected[102]. Therefore the sensitivity of visual cortical cells to BDNF appears to be restricted to the CP. Administration of NT-4 into VC prevented the OD shift[103] as well as the shrinkage of LGN neurons normally associated with the deprived eye after MD[104]. Conversely, administration of TrkB-IgG, a scavenger for BDNF/NT-4, inhibited segregation of LGN axons near the site of infusion[105]. These experiments support a model in which activity-dependent expression and/or uptake of TrkB ligands, driven by eye-specific LGN axons, underlies the maturation and stabilization of eye-specific synapses, and therefore the segregation of OD.

In addition to its direct role in the formation of excitatory synapses between LGN axons and pyramidal neurons, BDNF also regulates GABAergic synapses. Mice deficient for GAD-65, an enzyme required for GABA synthesis, which does not have fast synaptic inhibition, were not sensitive to MD[106]. In these animals, the OD shift in favor of the open eye normally observed after monocular deprivation did not occur but was reversed by the administration of diazepam, a GABAergic use-dependent agonist[106]. These results suggest an important role for GABAergic interneurons in the development of OD. In the cortex, GABAergic neurons express TrkB but not BDNF. Application of BDNF enhances GABA release and promotes the differentiation of GABAergic neurons. Mice that lack BDNF have significantly reduced GABAergic markers, such as parvalbumin, in the cortex[107]. On the other hand, maturation of GABAergic inhibition is accelerated in the VC of BDNF-overexpressing mice[108]. As a result, these BDNF over-expressing mice have an earlier and shorter CP that is sensitive to MD.

## 8. BARREL CORTEX

Analogous to the visual system, neurotrophins play an important role in the development of the barrel cortex. This distinct structure in the somatosensory cortex receives sensory information derived from the whiskers of rodents. Incoming sensory information passed to the thalamus originate from axons of the ventral posterior medial (VPM) nucleus, which then segregate into distinct rings termed barrels. Each barrel consists of a set of neurons that receives sensory information from a single whisker. In TrkB mutant mice that lack the kinase domain receptor but express the truncated receptor, a developmental delay in barrel formation was observed[109]. It was later identified that BDNF and not NT-4 was the endogenous ligand for TrkB involved in thalamocortical axon segregation and barrel formation[110].

BDNF also plays an important role in synaptogenesis occurring during adulthood. In mature animals, altered sensory experience results in robust physiological changes in the barrel cortex[111], which occur in parallel with structural changes[112,113]. By using two-photon microscopy, Trachtenberg and colleagues demonstrated that formation of synapses is influenced by dendritic spine growth and retraction as a result of altered sensory input. Remarkably, stimulation of a single whisker results in an increase in BDNF gene expression in the barrel corresponding to that whisker[114]. The same whisker stimulation induces the formation of inhibitory synapses[112], raising the possibility that the synaptogenesis observed in the barrel cortex is mediated by activity-dependent expression of BDNF. Indeed, structural plasticity that normally occurs in wild-type animals does not occur in mice heterozygous for BDNF[115].

## 9. CONCLUSIONS AND FUTURE PERSPECTIVES

Synaptogenesis is a multistep process that underlies the formation of precise synaptic connections. Because neurotrophins regulate structure and function of both pre- and postsynaptic elements, this class of secreted proteins are attractive candidates as regulators of synaptogenesis. Firstly, neurotrophins regulate axonal and dendritic morphological changes that may aid in establishing initial contact between synaptic partners. Secondly, neurotrophins induce changes in pre- and postsynaptic elements associated with the formation of new synapses, such as maturation of presynaptic release properties and postsynaptic receptor clustering. Finally, neurotrophins regulate the stabilization and/or elimination of newly formed synapses to form appropriate networks. Taken together, neurotrophins critically regulate several key steps in synaptogenesis (Figure 12.3).

However, there are many outstanding issues and problems. The underlying signaling mechanisms important for the regulation of synapse formation by neurotrophins remain to be elucidated. In addition, $p75^{NTR}$ signaling, highlighted in many recent studies of neurite growth, has not been rigorously examined in synaptogenesis. Multidisciplinary approaches such as *in vivo* imaging and electrophysiological recording, together with the use of genetically engineered animals, may help us answer these unresolved questions and issues.

Recent breakthroughs in stem cell technology have provided an optimistic view that this technology may be applied in the near future. However, this form of therapy would not be practical if newly generated neurons cannot make appropriate functional synapses. Neurotrophins may conceivably be used to promote synapse formation in both pre-existing and newly generated neurons.

Therefore, it is imperative to understand the diverse mechanisms utilized by neurotrophins to regulate synaptogenesis. This is important not only for our understanding of proper nervous system development but may also aid in the development of neurotrophin-based treatments that will facilitate the regeneration of neurons in the face of neurodegenerative diseases and injury.

## 10. REFERENCES


1. Katz, L.C., and Shatz, C.J. (1996) *Science* **274**, 1133–1138.
2. Lu, B. (2003) *Learn Memory* **10**, 86–98.
3. Thoenen, H. (1995) *Science* **270**, 593–596.
4. McAllister, A.K., Katz, L.C., and Lo, D.C. (1999) *Annu Rev Neurosci* **22**, 295–318.
5. Lu, B. (2004) *Progr Brain Res* **146**, 137–150.
6. Poo, M.M. (2001) *Nat Rev Neurosci* **2**, 24–32.
7. Kaplan, D.R., and Stephens, R.M. (1994) *J Neurobiol* **25**, 1404–1417.
8. Huang, E.J., and Reichardt, L.F. (2003) *Annu Rev Biochem* **72**, 609–642.
9. Segal, R.A., and Greenberg, M.E. (1996) *Annu Rev Neurosci* **19**, 463–489.
10. Kaplan, D.R., and Miller, F.D. (2000) *Curr Opin Neurobiol* **10**, 381–391.
11. Yang, F., He, X., Feng, L., Mizuno, K., Liu, X., Russell, J., Xiong, W., and Lu, B. (2001) *Nature Neurosci* **4**, 19–28.
12. Atwal, J.K., Massie, B., Miller, F.D., and Kaplan, D.R. (2000) *Neuron* **27**, 265–277.
13. Watson, F.L., Heerssen, H.M., Bhattacharyya, A., Klesse, L., Lin, M.Z., Segal, R.A., Moheban, D.B., Sauvageot, C.M., and Pomeroy, S.L. (2001) *Nat Neurosci* **4**, 981–988.
14. Nagappan, G., and Lu, B. (2005) *Trends Neurosci* **28**, 464–471.
15. Meyer-Franke, A., Wilkinson, G.A., Kruttgen, A., Hu, M., Munro, E., Hanson, M.G., Jr., Reichardt, L.F., and Barres, B.A. (1998) *Neuron* **21**, 681–693.
16. Du, J., Feng, L., Yang, F., and Lu, B. (2000) *J Cell Biol* **150**, 1423–1434.
17. Du, J., Feng, L., Zaitsev, E., Je, H.S., Liu, X.W., and Lu, B. (2003) *J Cell Biol* **163**, 385–395.
18. Suzuki, S., Numakawa, T., Shimazu, K., Koshimizu, H., Hara, T., Hatanaka, H., Mei, L., Lu, B., and Kojima, M. (2004) *J Cell Biol* **167**, 1205–1215.
19. Ji, Y., Pang, P.T., Feng, L., and Lu, B. (2005) *Nat Neurosci* **8**, 164–172.
20. Lee, R., Kermani, P., Teng, K.K., and Hempstead, B.L. (2001) *Science* **294**, 1945–1948.
21. Lu, B., Pang, P.T., and Woo, N.H. (2005) *Nat Rev Neurosci* **6**, 603–614.
22. Chao, M.V. (2003) *Nat Rev Neurosci* **4**, 299–309.
23. Dechant, G., and Barde, Y.A. (2002) *Nat Neurosci* **5**, 1131–1136.
24. Nykjaer, A., Lee, R., Teng, K.K., Jansen, P., Madsen, P., Nielsen, M.S., Jacobsen, C., Kliemannel, M., Schwarz, E., Willnow, T.E., Hempstead, B.L., and Petersen, C.M. (2004) *Nature* **427**, 843–848.
25. Chen, Z.Y., Ieraci, A., Teng, H.K., Dall, H., Meng, C.X., Herrera, D.G., Nykjaer, A., Hempstead, B.L., and Lee, F.S. (2005) *J Neurosci*, **25**, 6156–6166.
26. Woo, N.H., Teng, H.K., Siao, C.J., Chiaruttini, C., Pang, P.T., Milner, T.A., Hempstead, B.L., and Lu, B. (2005) *Nat Neurosci* **8**, 1069–1077.
27. Sanes, J.R., and Lichtman, J.W. (1999) *Annu Rev Neurosci* **22**, 389–442.
28. Xie, K., Wang, T., Olafsson, P., Mizuno, K., and Lu, B. (1997) *J Neurosci* **17**, 2947–2958.
29. Wang, X.H., and Poo, M.M. (1997) *Neuron* **19**, 825–835.
30. Henderson, C.E., Camu, W., Mettling, C., Gouin, A., Poulsen, K., Karihaloo, M., Rullamas, J., Evans, T., McMahon, S.B., Armanini, M.P., Berkemeier, L., Phillips, H.S., and Rosenthal, A. (1993) *Nature* **363**, 268–270.
31. Koliatsos, V.E., Clatterbuck, R.E., Winslow, J.W., Cayouette, M.H., and Price, D.L. (1993) *Neuron* **10**, 359–367.
32. Funakoshi, H., Belluardo, N., Arenas, E., Yamamoto, Y., Casabona, A., Persson, H., and Ibanez, C.F. (1995) *Science* **268**, 1495–1499.
33. Wang, Q., and Zheng, J.Q. (1998) *J Neurosci* **18**, 4973–4984.
34. Lohof, A.M., Ip, N.Y., and Poo, M.M. (1993) *Nature* **363**, 350–353.
35. Stoop, R., and Poo, M.M. (1995) *Science* **267**, 695–699.
36. Stoop, R., and Poo, M.M. (1996) *J Neurosci* **16**, 3256–3264.
37. Chang, S., and Popov, S.V. (1999) *Proc Natl Acad Sci U S A* **96**, 4095–4100.
38. He, X., Yang, F., Xie, Z., and Lu, B. (2000) *J Cell Biol* **149**, 783–792.
39. Wang, T., Xie, K.W., and Lu, B. (1995) *J Neurosci* **15**, 4796–4805.
40. Liou, J.C., Yang, R.S., and Fu, W.M. (1997) *J Physiol* **503**, 129–139.

41. Liou, J.C., and Fu, W.M. (1997) *J Neurosci* **17**, 2459–2468.
42. Nick, T.A., and Ribera, A.B. (2000) *Nat Neurosci* **3**, 142–149.
43. Benfenati, F., Valtorta, F., Rossi, M.C., Onofri, F., Sihra, T., and Greengard, P. (1993) *J Cell Biol* **123**, 1845–1855.
44. Lu, B., Czernik, A., Papov, S., Wang, T., Poo, M.-M., and Greengard, P. (1996) *Neurosci* **74**, 1087–1097.
45. Lu, B., Greengard, P., and Poo, M.-M. (1992) *Neuron* **8**, 521–529.
46. Fu, A.K., Ip, F.C., Lai, K.O., Tsim, K.W., and Ip, N.Y. (1997) *Neuroreport* **8**, 3895–3900.
47. Gonzalez, M., Ruggiero, F.P., Chang, Q., Shi, Y.J., Rich, M.M., Kraner, S., and Balice-Gordon, R.J. (1999) *Neuron* **24**, 567–583.
48. Belluardo, N., Westerblad, H., Mudo, G., Casabona, A., Bruton, J., Caniglia, G., Pastoris, O., Grassi, F., and Ibanez, C.F. (2001) *Mol Cell Neurosci* **18**, 56–67.
49. Mohamed, A.S., Rivas-Plata, K.A., Kraas, J.R., Saleh, S.M., and Swope, S.L. (2001) *J Neurosci* **21**, 3806–3818.
50. Marangi, P.A., Wieland, S.T., and Fuhrer, C. (2002) *J Cell Biol* **157**, 883–895.
51. Smith, C.L., Mittaud, P., Prescott, E.D., Fuhrer, C., and Burden, S.J. (2001) *J Neurosci* **21**, 3151–3160.
52. Loeb, J.A., Hmadcha, A., Fischbach, G.D., Land, S.J., and Zakarian, V.L. (2002) *J Neurosci* **22**, 2206–2214.
53. Fischbach, G.D., and Rosen, K.M. (1997) *Annu Rev Neurosci* **20**, 429–458.
54. Vicario-Abejon, C., Collin, C., McKay, R.D., and Segal, M. (1998) *J Neurosci* **18**, 7256–7271.
55. Fan, G., Egles, C., Sun, Y., Minichiello, L., Renger, J.J., Klein, R., Liu, G., and Jaenisch, R. (2000) *Nat Neurosci* **3**, 350–357.
56. Figurov, A., Pozzo-Miller, L., Olafsson, P., Wang, T., and Lu, B. (1996) *Nature* **381**, 706–709.
57. Gottschalk, W., Pozzo-Miller, L.D., Figurov, A., and Lu, B. (1998) *J Neurosci* **18**, 6830–6839.
58. Martinez, A., Alcantara, S., Borrell, V., Del Rio, J.A., Blasi, J., Otal, R., Campos, N., Boronat, A., Barbacid, M., Silos-Santiago, I., and Soriano, E. (1998) *J Neurosci* **18**, 7336–7350.
59. Aguado, F., Carmona, M.A., Pozas, E., Aguilo, A., Martinez-Guijarro, F.J., Alcantara, S., Borrell, V., Yuste, R., Ibanez, C.F., and Soriano, E. (2003) *Development* **130**, 1267–1280.
60. Luikart, B.W., Nef, S., Virmani, T., Lush, M.E., Liu, Y., Kavalali, E.T., and Parada, L.F. (2005) *J Neurosci* **25**, 3774–3786.
61. Pozzo-Miller, L., Gottschalk, W.A., Zhang, L., McDermott, K., Du, J., Gopalakrishnan, R., Oho, C., Shen, Z., and Lu, B. (1999) *J Neurosci* **19**, 4972–4983.
62. Carter, A.R., Chen, C., Schwartz, P.M., and Segal, R.A. (2002) *J Neurosci* **22**, 1316–1327.
63. Tartaglia, N., Du, J., Neale, E., Tyler, W.J., Pozzo-Miller, L., and Lu, B. (2001) *J Biol Chem* **276**, 37585–37593.
64. Tyler, W.J., and Pozzo-Miller, L.D. (2001) *J Neurosci* **21**, 4249–4258.
65. Seil, F.J. (1999) *Brain Res* **818**, 561–564.
66. Seil, F.J., and Drake-Baumann, R. (2000) *J Neurosci* **20**, 5367–5373.
67. Rico, B., Xu, B., and Reichardt, L.F. (2002) *Nat Neurosci* **5**, 225–233.
68. Narisawa-Saito, M., Carnahan, J., Araki, K., Yamaguchi, T., and Nawa, H. (1999) *Neuroscience* **88**, 1009–1014.
69. Narisawa-Saito, M., Iwakura, Y., Kawamura, M., Araki, K., Kozaki, S., Takei, N., and Nawa, H. (2002) *J Biol Chem* **277**, 40901–40910.
70. Jourdi, H., Iwakura, Y., Narisawa-Saito, M., Ibaraki, K., Xiong, H., Watanabe, M., Hayashi, Y., Takei, N., and Nawa, H. (2003) *Dev Biol* **263**, 216–230.
71. Elmariah, S.B., Crumling, M.A., Parsons, T.D., and Balice-Gordon, R.J. (2004) *J Neurosci* **24**, 2380–2393.
72. Fiala, J.C., Feinberg, M., Popov, V., and Harris, K.M. (1998) *J Neurosci* **18**, 8900–8911.
73. Dunaevsky, A., Tashiro, A., Majewska, A., Mason, C., and Yuste, R. (1999) *Proc Natl Acad Sci U S A* **96**, 13438–13443.
74. Ziv, N.E., and Smith, S.J. (1996) *Neuron* **17**, 91–102.
75. Shimada, A., Mason, C.A., and Morrison, M.E. (1998) *J Neurosci* **18**, 8559–8570.
76. Cohen-Cory, S., and Lom, B. (2004) *Int J Dev Biol* **48**, 947–956.
77. Cohen-Cory, S. (2002) *Science* **298**, 770–776.
78. Cohen-Cory, S., and Fraser, S.E. (1994) *Neuron* **12**, 747–761.
79. Lom, B., and Cohen-Cory, S. (1999) *J Neurosci* **19**, 9928–9938.
80. Lom, B., Cogen, J., Sanchez, A.L., Vu, T., and Cohen-Cory, S. (2002) *J Neurosci* **22**, 7639–7649.
81. Alsina, B., Vu, T., and Cohen-Cory, S. (2001) *Nat Neurosci* **4**, 1093–1101.
82. Hu, B., Nikolakopoulou, A.M., and Cohen-Cory, S. (2005) *Development* **132**, 4285–4298.
83. Cohen-Cory, S., and Lom, B. (1999) *J Neurosci* **19**, 9996–10003.
84. Dillon, C., and Goda, Y. (2005) *Annu Rev Neurosci* **28**, 25–55.
85. Zhang, W., and Benson, D.L. (2001) *J Neurosci* **21**, 5169–5181.

86. Maffei, L., Berardi, N., Domenici, L., Parisi, V., and Pizzorusso, T. (1992) *J Neurosci* **12**, 4651–4662.
87. Domenici, L., Parisi, V., and Maffei, L. (1992) *Neuroscience* **51**, 19–24.
88. Domenici, L., Cellerino, A., and Maffei, L. (1993) *Proc R Soc Lond B Biol Sci* **251**, 25–31.
89. Domenici, L., Cellerino, A., Berardi, N., Cattaneo, A., and Maffei, L. (1994) *Neuroreport* **5**, 2041–2044.
90. Berardi, N., Cellerino, A., Domenici, L., Fagiolini, M., Pizzorusso, T., Cattaneo, A., and Maffei, L. (1994) *Proc Natl Acad Sci U S A* **91**, 684–688.
91. Fagiolini, M., Pizzorusso, T., Porciatti, V., Cenni, M., and Maffei, L. (1997) *Eur J Neurosci* **9**, 102–112.
92. Pizzorusso, T., Porciatti, V., Tseng, J.L., Aebischer, P., and Maffei, L. (1997) *Neuroscience* **80**, 307–311.
93. Gibbs, R.B., and Pfaff, D.W. (1994) *J Comp Neurol* **341**, 324–339.
94. Bozzi, Y., Pizzorusso, T., Cremisi, F., Rossi, F.M., Barsacchi, G., and Maffei, L. (1995) *Neuroscience* **69**, 1133–1144.
95. Cabelli, R.J., Allendoerfer, K.L., Radeke, M.J., Welcher, A.A., Feinstein, S.C., and Shatz, C.J. (1996) *J Neurosci* **16**, 7965–7980.
96. Schoups, A.A., Elliott, R.C., Friedman, W.J., and Black, I.B. (1995) *Brain Res Dev Brain Res* **86**, 326–334.
97. Allendoerfer, K.L., Cabelli, R.J., Escandon, E., Kaplan, D.R., Nikolics, K., and Shatz, C.J. (1994) *J Neurosci* **14**, 1795–1811.
98. Lein, E.S., and Shatz, C.J. (2000) *J Neurosci* **20**, 1470–1483.
99. Bozzi, Y., and Borrelli, E. (1999) *Eur J Neurosci* **11**, 1275–1284.
100. Castren, E., Zafra, F., Thoenen, H., and Lindholm, D. (1992) *Proc Natl Acad Sci USA* **89**, 9444–9448.
101. Cabelli, R.J., Hohn, A., and Shatz, C.J. (1995) *Science* **267**, 1662–1666.
102. Galuske, R.A., Kim, D.S., Castren, E., and Singer, W. (2000) *Eur J Neurosci* **12**, 3315–3330.
103. Gillespie, D.C., Crair, M.C., and Stryker, M.P. (2000) *J Neurosci* **20**, 9174–9186.
104. Riddle, D.R., Lo, D.C., and Katz, L.C. (1995) *Nature* **378**, 189–191.
105. Cabelli, R.J., Shelton, D.L., Segal, R.A., and Shatz, C.J. (1997) *Neuron* **19**, 63–76.
106. Hensch, T.K., Fagiolini, M., Mataga, N., Stryker, M.P., Baekkeskov, S., and Kash, S.F. (1998) *Science* **282**, 1504–1508.
107. Jones, K.R., Farinas, I., Backus, C., and Reichardt, L.F. (1994) *Cell* **76**, 989–999.
108. Huang, Z.J., Kirkwood, A., Pizzorusso, T., Porciatti, V., Morales, B., Bear, M.F., Maffei, L., and Tonegawa, S. (1999) *Cell* **98**, 739–755.
109. Vitalis, T., Cases, O., Gillies, K., Hanoun, N., Hamon, M., Seif, I., Gaspar, P., Kind, P., and Price, D.J. (2002) *J Neurosci* **22**, 4987–5000.
110. Lush, M.E., Ma, L., and Parada, L.F. (2005) *Int J Dev Neurosci* **23**, 523–536.
111. Welker, E., Rao, S.B., Dorfl, J., Melzer, P., and van der Loos, H. (1992) *J Neurosci* **12**, 153–170.
112. Knott, G.W., Quairiaux, C., Genoud, C., and Welker, E. (2002) *Neuron* **34**, 265–273.
113. Trachtenberg, J.T., Chen, B.E., Knott, G.W., Feng, G., Sanes, J.R., Welker, E., and Svoboda, K. (2002) *Nature* **420**, 788–794.
114. Rocamora, N., Welker, E., Pascual, M., and Soriano, E. (1996) *J Neurosci* **16**, 4411–4419.
115. Genoud, C., Knott, G.W., Sakata, K., Lu, B., and Welker, E. (2004) *J Neurosci* **24**, 2394–2400.

# Part III

## TRANSPORT OF SYNAPTIC PROTEINS

# 13

# MOTOR–CARGO INTERACTIONS INVOLVED IN TRANSPORT OF SYNAPTIC PROTEINS

Matthias Kneussel*

## 1. SUMMARY

Intracellular transport is fundamental for synaptogenesis, neuronal transmission, and synaptic plasticity. Proteins destined for the presynaptic active zone or the postsynaptic specialization are selectively transported to either axons or dendrites. In addition, certain mRNAs are transported for local protein synthesis. Molecular motors of the kinesin, dynein, and myosin superfamily mediate cargo recruitment along cytoskeletal elements: microtubules and actin filaments. Each motor participates in a transport complex typically consisting of a heavy chain combined with accessory light chains. Directionality of transport requires mechanisms that mediate both the recognition of individual tracks and cargoes as well as the sorting of transport complexes to specific subcellular compartments. Depending on the cargoes involved, both selective transport and selective retention have been described. A number of motor–cargo adaptors, which mediate transport specificity at intracellular sites, also locate at postsynaptic densities of mature synapses, where they participate in neurotransmitter receptor-scaffold interactions. Upon delivery to the neuronal surface, receptors enter the plasma membrane at extrasynaptic sites, followed by lateral diffusion and/or active transport between extrasynaptic and synaptic sites. The delivery and the removal of receptors that are available for synaptic transmission at a certain time are thought to participate in the regulation of synaptic strength.

*Zentrum für Molekulare Neurobiologie, ZMNH, Universität Hamburg, Falkenried 94, 20251 Hamburg, Germany; matthias.kneussel@zmnh.uni-hamburg.de

## 2. INTRODUCTION

Neurons are highly polarized cells, typically consisting of a cell body, several dendrites, and a long, thin axon. The soma or perikaryon contains the nucleus, endoplasmic reticulum, mitochondria, the Golgi apparatus, and a large number of cytoskeletal fibers. The axon and the dendrites are processes, which arise from the cell body and differ in terms of shape, cytoskeletal organization, and membrane protein composition. The region of the cell body from which axons originate is known as the axon hillock, a narrow structure that tapers off to the axon initial segment. In contrast to axons, dendrites have a larger diameter and taper off gradually. Proximal dendrites are abundant in endoplasmic reticulum and ribosomes and the cytoskeletal architecture of proximal dendrites and the cell body is similar. Distal dendrites also contain endoplasmic reticulum and ribosomes and are thought to synthesize individual proteins. For local protein synthesis of individual proteins, specific mRNAs are also transported to distal dendrites. A key problem in intracellular transport is the fact that neurites extend distances that are hundreds or thousands of times the diameter of the cell body (Figure 13.1). Therefore, efficient transport mechanisms are required to recruit mRNAs and a large number of newly synthesized proteins to neurites.

Long-distance transport is mainly mediated by microtubules, cytoskeletal fibers that serve as tracks for the transport of molecules. Microtubules are long hollow cylinders and their polarity differs between axons and dendritic compartments. Significantly, neurons have multiple compartments that are devoid of microtubules where transport is still seen to occur. These areas are rich in other cytoskeletal polymers such as actin filaments, which are typically used as tracks for short-distance transport. Actin filaments also display a polarity in the polymer with a pointed- and a barbed-end and are highly enriched at the cellular cortex[1].

## 3. ORGANIZATION AND POLARITY OF MICROTUBULES, THE TRACKS FOR LONG-DISTANCE TRANSPORT

Microtubules are polymers of $\alpha$ and $\beta$ tubulins with a fast-growing (plus) end and an opposite slow-growing (minus) end[2]. Microtubules associate with microtubule-associated proteins (MAPs) and are highly crosslinked by meshworks of fine filamentous structures. The spacing of microtubules in axons is about 20 nm, whereas in dendrites it is typically 65 nm. Some MAPs selectively localize either to the axon or the dendrite. Whereas the MAP tau decorates microtubules in the axon, MAP2 binds to microtubules in dendrites.

Microtubules generally have a radial organization in many cell types, with plus ends typically oriented to the cell periphery and minus ends anchored in a microtubule-organizing center (MTOC). With respect to neurons, this uniformity of microtubule polarity is found in axons but not in dendrites. Axonal microtubules are directed with their plus ends away from the cell body toward the growth cone. In contrast, dendritic microtubules show a mixed orientation. In proximal dendritic regions, about 75 μm from the cell body, roughly equal proportions of microtubules are oriented with plus ends directed either toward the growth cone or toward the cell body. However, in distal dendritic regions, within about 15-μm distances from the growth cone, microtubule polarity orientation is similar to that in axons, thus plus ends are uniformly directed toward the growth cone. Molecular motors of the kinesin and dynein superfamilies move along microtubules. Most kinesin superfamily proteins (KIFs) represent plus-end directed motors that move toward the plus end of microtubules[3]. This direction from the cell body to axons and dendrites is considered as anterograde transport. Dyneins are minus-end directed microtubule motors that mediate retrograde transport from the axonal and dendritic terminals to the cell body[4]. In contrast, myosin motors use actin filaments as tracks for transport and most myosins move toward the barbed or plus end of the filament.

**Figure 13.1.** The Transport Problem. Schematic representation of the relationship between axon length and size of the perikaryon in a human motoneuron. Proteins that are synthesized in the neuronal perikaryon (black circle) need to travel distances that are a few orders of magnitude the diameter of the cell body. The preferential delivery to a specific subcellular compartment such as an axon, the transport over long distances, and the constant delivery of material to neurite processes or synapses is a great challenge for the cellular transport machinery.

## 4. MOLECULAR MOTORS IN NEURONS

Each motor protein complex – kinesin, dynein, and myosin – consists of multiple domains or accessory subunits. All motors are enzymes that convert chemical energy stored in adenosine triphosphate (ATP) into molecular motion, thereby producing force upon the associated cytoskeletal polymer. Typically, the

ATPase function is mediated by the heavy chain of the respective motor protein complex, whereas accessory intermediate and light chains are specialized for self-assembly or interaction with molecular cargo. However, individual heavy chains also have been shown to directly interact with cargo molecules.

Kinesins form a large protein family, known as kinesin superfamily (KIF) with 45 members in mice and humans[5]. Many KIFs are primarily expressed in the nervous system, but KIFs are generally expressed in other tissues and mediate a large variety of intracellular transport. The first kinesin to be discovered is known as conventional kinesin and corresponds to KIF5. On the basis of the positions of their motor domains, three types of kinesins have been classified: the amino (N)-terminal motor types (*N*-kinesins), the middle motor types (*M*-kinesins) and the carboxy (C)-terminal motor types (*C*-kinesins)[5]. *N*-kinesins and *M*-kinesins generally move toward the plus end of microtubules, and the latter also depolymerize microtubules. In contrast, *C*-kinesins move toward the microtubule minus end. KIFs often exist as tetramers of two heavy chains, consisting of motor, neck, stalk, and tail domains and two light chains, all of which can be involved in cargo interactions. The globular motor domain shared by all KIFs displays high degrees of homology and contains a microtubule-binding sequence and an ATPase-binding site. On the other hand, stalk and tail domains of KIFs are characterized by unique sequences. The diversity of these cargo-binding domains allows transport of numerous different cargoes. KIFs either function as monomeric proteins or form both homo- and heterodimers with other KIF superfamily members. The transport velocity of KIFs is consistent with the speed of fast axonal transport *in vivo*, which varies from 0.2 to 1.5 $\mu m\ s^{-1}$ [3]. Depending on the cargo to be transported, some KIF-containing transport complexes are recruited in the range of $\mu m\ min^{-1}$. KIF5 family members transport α-amino-3-hydroxy-5-methyl-4-isoxazole propionic acid (AMPA) receptor-containing vesicles to dendrites[6]. In this transport complex, the glutamate receptor interacting protein GRIP1 couples AMPA receptors with the carboxy (C)-terminal tail of KIF5 (Figure 13.2).

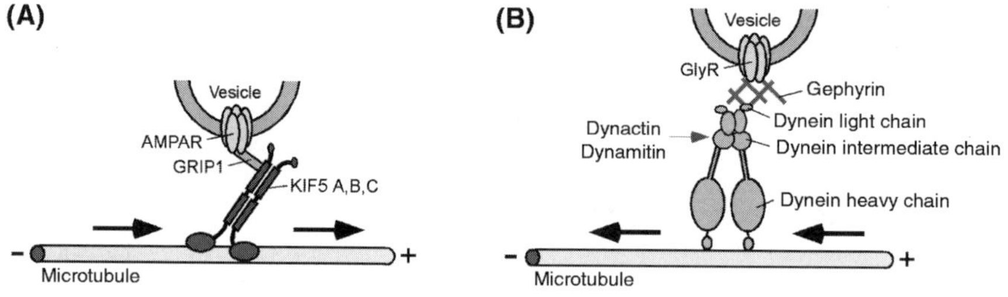


**Figure 13.2.** Schematic Representation of Microtubule-Dependent Transport Complexes. (A) Kinesin-mediated anterograde transport of excitatory AMPA-type glutamate receptors. (B) Dynein-mediated retrograde transport of inhibitory glycine receptor (GlyR). Adaptor proteins (GRIP1 and gephyrin, respectively) connect the motor complex with neurotransmitter receptors that represent integral membrane proteins of transport vesicles.

Dyneins are microtubule-based motor proteins consisting of a multisubunit complex, which contains two gigantic heavy chains of about 380 kDa and a variable number of associated polypeptides[7]. The heavy chain mediates ATP binding and microtubule attachment via a carboxy-terminal coiled coil stalk. Each heavy chain interacts with certain intermediate and light chains, allowing construction of a wide variety of different dyneins with many common features, but specialized for individual functions. Cytoplasmic dynein associates with the protein complex dynactin and associated proteins such as p150$^{Glued}$ and dynamitin. Although dynamitin often mediates the interaction between cytoplasmic dynein and its cargoes, some cargoes bind the complex by direct interaction with dynein light chains. Dynein retrogradely transports glycine receptors (GlyRs) in neurons. The transport adaptor that couples GlyR-containing vesicles with the dynein motor complex is gephyrin, a GlyR-binding protein, originally described at postsynaptic sites[8,9] (Figure 13.2).

Myosins move along actin filaments and represent a diverse protein family of more than 15 members with one or two motor domains and one or more light chains[10]. The myosin heavy chain consists of the catalytic domain, including actin and ATP-binding sites, and forms a long α-helix and a globular tail, the former of which interacts with a number of light chains that wrap around and stabilize the helical structure. Some light chains bind divalent cations, such as calmodulin, which serves as myosin light chain. Myosin V dimerizes and forms a two-headed protein that moves toward the barbed or plus end of actin filaments. Furthermore, myosin Va is one of the fastest myosins tested, which moves at a rate of about 0.4 $\mu m\ s^{-1}$[11].

## 5. CARGO RECOGNITION

Transmembrane cargo proteins do not directly bind to molecular motors but instead use adaptor/scaffolding polypeptides or protein complexes for cargo recognition and binding[6,9,12–14]. For instance the interaction between KIF17 and cargo vesicles that contain NMDA-type glutamate receptors is mediated by a triple complex of mLIN-family proteins (for: mammaliam homolog of *C. elegans* lineage abnormal) that contains mLIN10 (MINT1), mLIN2 (CASK), and mLIN7 (VELIS/MALS). The microtubule-associated motor binds to the cytoplasmic LIN10 through a PDZ domain-mediated interaction. The vesicular NR2B subunit of NMDA receptors interacts through its carboxy (C)-terminal tail with cytoplasmic mLIN7. To connect the cargo vesicle with its transport track, mLIN2 then functions as a linker between the mLIN10-motor complex and the mLIN7-receptor-vesicle complex, thereby generating a large complex that transports NR2B containing receptors toward postsynaptic sites[12]. Evidence that these interactions are important for *in vivo* function comes from a study in which transgenic overexpression of the motor protein KIF17 enhances NR2B-mediated spatial and working memory in mice[15]. All three mLIN family proteins contain PDZ domains, which are not involved to bind each other, but to recruit other proteins to the complex.

It has been discussed that kinesin motor proteins consist of a large heavy chain represented by head, neck, stalk and tail domains and also binds accessory light chains. As kinesin light chain KLC binds to the C-terminal tail of KIF5, the question arises whether cargoes bind to heavy chain sequences of KIF5 or to KLC. Different studies in neurons, including loss-of-function experiments, currently suggest that cargoes bind both to the C-terminal tail of KIF5 as well as to KLC; however binding to KLC tends to be used for transporting cargoes to axons, whereas binding to the C-terminal tail of KIF5 seems to be used for directing cargoes to dendrites[6]. This view seems to be true not only for protein transport, since also granules that contain calmodulin-dependent protein kinase II (CaMKII) or ARC mRNAs for transport to dendrites, directly bind to the C-terminal tail of KIF5[16,17].

Also if cargo binds kinesin through KLCs, the interaction of cargo and KLC is often not direct. Scaffolding proteins of the c-Jun N-terminal kinase (JNK) signaling pathway, JIP1, JIP2 or JIP3, are abundant at neuronal processes and synaptic junctions and mediate cargo binding to KLCs. For instance, JIP1 is thought to bridge the interaction of KLC with the amyloid precursor protein APP. The phosphotyrosine-binding domain of JIP1 binds APP and JIP1 is required for the interaction of APP with KLC[18,19].

In contrast to kinesins, the recognition of cargoes of cytoplasmic dynein is currently barely understood. The dynein-associated protein complex dynactin contains a short filament of actin-related protein ARP1. In addition, dynactin binds to $p150^{Glued}$ and dynamitin[7]. Although dynamitin seems to be critical for cargo binding to the dynein complex, cargo elements bind to different components of the complex. During ER to Golgi trafficking, the Golgi-associated spectrin isoform βIII spectrin directly binds to ARP1[20]. However, the transmembrane protein rhodopsin and the scaffold protein gephyrin couple to cytoplasmic dynein through binding to the dynein light chains TCTEX1 and DLC1/DLC2, respectively[8,21].

In general, a large number of cellular molecules are subject of transport by a limited number of molecular motors (Table 1). Therefore, accessory intermediate and light chains, as well as adaptor/scaffolding proteins allow a large combination of individual and unique transport complexes and thereby contribute to the specificity of cargo recognition and transport to specific subcellular domains.

**Table 1.** Presently known and putative transport complexes of neurotransmitter receptors and associated proteins

| Motor | Adaptor | | Cargo | Reference |
|---|---|---|---|---|
| KIF17 | mLin-2 (CASK)<br>mLin-7 (MALS/Velis)<br>mLin-10 (Mint/X11) | | NMDA-R 2B | (12) |
| KIF5A,B,C | GRIP1 | | AMPA-R GluR2 | (12) |
| KIF1Bα | ? ← | PSD-95<br>PSD-93<br>chapsyn-110<br>SAP-97<br>S-SCAM | → ? | (13) |
| Dynein | gephyrin | | GlyR | (8, 9) |
| Myosin V | ? ← | GKAP<br>PSD-95 | → ? | (41) |
| Myosin VI | SAP-97 | | AMPA-R GluR1 | (14) |

## 6. DIRECTIONAL SORTING AND TRANSPORT

Which mechanisms encode that a transport complex moves in a certain direction of a polar neuron? As discussed above, cargo recognition is important to form the transport complex; however the following example shows that individual molecules, which link motor and cargo, might be directly involved in the determination of transport direction. Transport vesicles containing AMPA receptors destined for the postsynaptic specialization are transported to dendrites through interaction with the AMPA receptor-interacting protein GRIP1, which binds to the C-terminal tail of KIF5 (Figure 13.2). The KIF5-binding domain of GRIP1 consists of amino acid residues 753–987 in the GRIP1 polypeptide, a region located between the sixth and seventh PDZ domain of GRIP1. Upon overexpression of this motif, endogenous KIF5 accumulates predominantly in the somatodendritic area of the neuron. In contrast, when the kinesin light chain interacting protein JIP3 is overexpressed, KIF 5 accumulates in the somatoaxonal area[6]. This indicates that cargo binding or the nature of the adaptor/cargo complex steers cargoes to specific subcellular domains, such as axons and dendrites.

It is not yet clear whether this phenomenon represents a general principle or applies to individual transport complexes; however other cellular mechanisms contribute to selectively sort and transport proteins to axons and dendrites. It has been a matter of debate whether motors are smart and would be able to distinguish between axons and dendrites, since individual motors, such as KIF21B, specifically move to dendrites[22]. However, the fact that many motors transport cargo in both compartments rather suggests that other mechanisms mainly account for the sorting of motors. For instance the polarity of microtubules differs between axons and dendrites. This difference could contribute in achieving polarized transport to different neuronal compartments.

Given the nature of the axon initial segment, it seems likely that a structural component critically contributes to sorting of material in particular to the axon. If one considers the three-dimensional structure of a neuronal soma with an axon diameter of approximately one-tenth of that of the cell body, axonally transported material needs to be sorted from the cell body in only about 0.25% of all possible directions to enter the axon[23]. Therefore, to circumvent this problem, unique components, such as for instance the nature of the tracks that extend into the axon initial segment, could direct transport complexes destined for the axon. In consistence with this view, the fusion of the KIF5 motor domain to GFP, generates a fusion protein that, when expressed in hippocampal neurons, specifically accumulates at the tips of axons. This finding indicates that the motor domain alone, lacking peripheral sequences for cargo binding, has a preference for axonal delivery[23]. Moreover, a KIF5 mutant that binds microtubules, but is neither able to translocate nor to dissociate from the tracks, accumulates in the axon initial segment, further indicating that this motor domain has a preference for microtubules particular at this subcellular region. Consequently, the nature of the tracks will have to be considered as critical for sorting processes in a polar cell, such as for instance a pyramidal neuron with many dendrites and a single axon.

Microtubules are decorated by MAPs. Some of these proteins, such as Tau, which selectively binds to axonal microtubules, are highly phosphorylated at various residues[24,25]. Given the high number of MAPs expressed in neurons in combination with the many post-translational modifications used in cells, one could consider that a large variety of individual tracks exist that contribute to the sorting of individual transport complexes.

Despite the described mechanisms to regulate sorting at the level of the track, the motor or the cargo adaptor, many proteins to be transported harbor sorting signals that can be divided into axonal and dendritic targeting sequences. Signals that have been identified for neuronal transport include dileucine-based motifs, tyrosine-based motifs or palmitoylation of cysteine residues[26–28]. The transport to dendrites is often considered to be analogous to basolateral transport in epithelial cells, in which proteins undergo polarized sorting to apical and basolateral compartments. Consistently, a GFP-tagged transferrin receptor (TfR), which is a marker for basolateral sorting in epithelial cells, localizes selectively to neuronal dendrites and rarely enters the axon. The TfR dendritic targeting signal is located in the N-terminal tail of the polypeptide and represents a tyrosine-based sequence[26]. For the postsynaptic density protein 95 (PSD-95), an N-terminal palmitoylation signal is necessary for dendritic targeting to dendrites[28]. However, in this case the motif itself is not sufficient for the targeting process, suggesting that other sequences are involved in the targeting process of PSD-95. In contrast, for the lipid-anchored peripheral protein GAP43, a palmitoylation motif within its polypeptide sequence is indeed sufficient for axonal targeting. Interestingly, when this GAP43-derived sequence is fused to PSD-95, the resulting chimera is redirected to axons. For metabotopic glutamate receptors (mGluRs), mGluR1a is targeted to dendrites, however its alternative splice transcript mGluR1b is targeted to axons[29,30]. In this case, alternative splicing generates a tripeptide motif that directs the protein to the axonal compartment.

## 7. SELECTIVE TRANSPORT AND SELECTIVE RETENTION OF CARGO

Not all proteins that accumulate in a specific subcellular compartment undergo processes of selective sorting and transport. Some proteins seem to be localized by a mechanism known as selective retention, which describes that cargoes are transported nonselectively to both axons and dendrites, but are eliminated at one side by selective endocytosis and retained at the other, where endocytosis is prevented. Prominent examples for this process are the proteins VAMP2 and NgCAM. NgCAM is sorted into carriers that preferentially deliver their cargo proteins to the axonal membrane. In contrast, VAMP2 is delivered to the surface of both axons and dendrites; however it is preferentially endocytosed from the dendritic membrane, a process, which also results in an axonal enrichment[31]. Indeed, VAMP2 harbors an endocytosis signal in its cytoplasmic domain, and mutation of this sequence consistently results in an evenly distribution of VAMP2 to cell body, dendrites, and axon. Although such process initially sounds like a waste of cellular energy, some proteins are localized by selective retention, rather than by selective sorting and transport.

## 8. TRANSPORT ADAPTORS AT POSTSYNAPTIC SCAFFOLD FORMATIONS

With focus on selective transport, components that connect motors and cargoes are of particular interest. Transmembrane cargoes, which are transported as integral membrane proteins of transport vesicles, connect to molecular motors through adaptor proteins. Different examples show that the same set of proteins, which mediate motor–cargo interactions of neurotransmitter receptors, have been previously described as components of postsynaptic densities (PSDs), where they mediate scaffolding reactions, such as for instance the clustering of receptors at axo-dendritic contacts. As discussed above, the proteins of the mLIN-family that connect the NMDA-type glutamate receptor subunit NR2B with KIF17[12] as well as

GRIP1, which couples the AMPA-type glutamate receptor subunit GluR2 to KIF5 A, B and C[6], locate and concentrate at postsynaptic sites and mediate specific protein–protein interactions at this membrane compartment. It has also been pointed out that for instance mLIN2, mLIN7, and mLIN10 harbor PDZ domains, known to interact with a variety of proteins, which enables such candidates to participate in the formation of larger protein complexes. Furthermore, the postsynaptic proteins gephyrin and PSD-95, known to participate in both transport and postsynaptic reactions, multimerize by self-interaction domains and thereby form protein scaffolds with multiple sites for protein–protein interactions[32]. In general, the use of a common set of proteins as both intracellular transport adaptors and synaptic scaffold proteins might contribute to the transport specificity and postsynaptic integration of receptors that underlie synapse formation and plasticity (Figure 13.3).

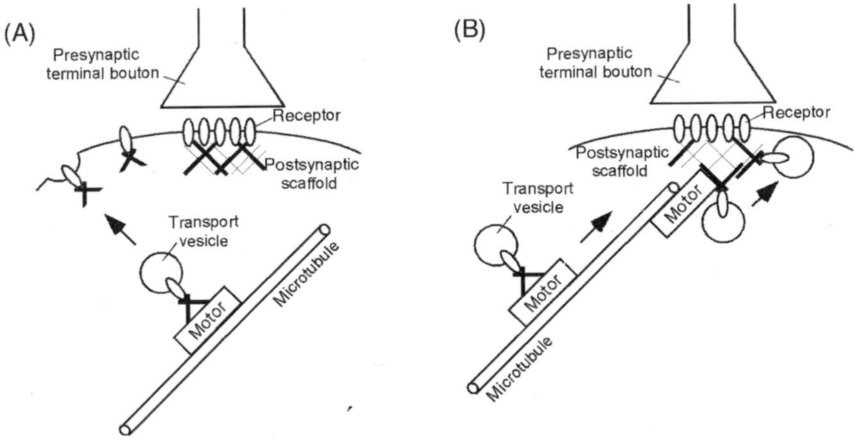


**Figure 13.3.** Proteins with Dual Functions in Motor-Protein-Dependent Transport and at Postsynaptic Membrane Specializations Contribute to the Post-Golgi Surface Membrane Delivery of Neurotransmitter Receptors. Two possibilities for the delivery of receptors are shown in (A) and (B). The currently available data favor the mechanism shown in (A).

Neurotransmitter receptors could be directly or indirectly delivered to the synapse; however the maintenance of the receptor–adaptor interaction after transport could have several advantages for the entry and stabilization of receptors at existing postsynaptic specializations. With respect to the formation of presynaptic active zones, preformed large protein complexes have been described that are subject of common transport of precursor membrane specializations[33]. Similarly, a set of proteins, which functions in both transport and membrane scaffold formation, would be suitable to mediate directed delivery of preformed protein complexes that need to be maintained. It is currently unknown whether the choice of adaptor/scaffolding proteins mediates a certain contribution of transport specificity in individual transport complexes and whether this represents a specific phenomenon or rather a general principle.

## 9. ASSOCIATION OF MICROTUBULE- AND ACTIN FILAMENT-BASED TRANSPORT SYSTEMS

Microtubules display an orientation parallel to the long axis of the axon making them ideal candidates for transport over long distances. In contrast, actin filaments in axons are primarily concentrated adjacent to the plasma membrane and have unknown orientations with respect to polarity. Remarkably, some actin filaments do associate closely with microtubules. Electron microscopy has shown that actin filaments can be interwoven with the longitudinal bundles of microtubules[34], however, it remains unclear whether this association plays a role in transport or is rather essential for maintaining the structural integrity of the microtubule bundles. It has been suggested in this respect that myosin I is likely to contribute to local transport in axons and might help that transported vesicles dissociate from the long-distance transport traffic and traverse the actin-rich cortex adjacent to the plasma membrane[10]. Myosin Va associates with axonal organelles that contain synaptic vesicle proteins[35–37] and that are transported along microtubules in axons[38]. In the absence of myosin Va activity their transport rate increases, indicating that these organelles associate with both types, microtubule-based and actin filament-based transport motors. Indeed, myosin Va has been shown to interact directly with the kinesin motor KhcU[39] and the movement of myosin Va-labeled vesicles in axons is partially microtubule dependent. These observations raise the possibility that a dual motor complex regulates certain vesicle movements along both microtubules and actin filaments[38,39].

## 10. MYOSIN FUNCTION IN DENDRITIC SPINES

Dendritic spines are unique compartments because they contain little, if any microtubules, but are rich in actin filaments. Organelles, such as smooth endoplasmic reticulum and ribososmes, as well as a large number of proteins that form the postsynaptic density need to be transported into dendritic spines. There is considerable evidence that these compartments undergo structural changes that are dependent upon actin filaments and neuronal activity[40]. Whether transport plays a role for such structural changes of spines is currently unknown. A number of studies report interactions of the myosins Va, Vb, and VI with synaptic proteins in dendritic spines. For instance, myosin Va binds indirectly to GKAP through an 8kDa protein originally identified as a dynein light chain[41]. Myosin Va further binds the synaptic protein CaMKII and contributes to its kinase activity[42]. The tail of myosin Vb was shown to indirectly interact with the M4 muscarinic acetylcholine receptor through Rab11a interactions[43] and myosin VI binds to SAP97, a binding partner for the GluR1 subunit of AMPA-type glutamate receptors[14]. Functional evidence that myosin VI is important for dendritic spine structure comes from Snell's waltzer mice, deficient in the myosin VI protein. Hippocampi from these mice exhibit a decrease in synapse number and display abnormally short dendritic spines[44]. In addition, cultured hippocampal neurons from this genetic background display decreased numbers of both synapses and spines. Since myosin VI is an actin-based motor implicated in clathrin-mediated endocytosis in non-neuronal cells, it has been suggested that endocytosis of synaptic AMPA receptors might depend on myosin VI and the loss of endocytosis induces activity-dependent changes in synaptic structure[44].

## 11. CONCLUSIONS

To generate and maintain cellular polarity with an axon and many dendrites, neurons use different cellular mechanisms. Both mRNAs and proteins are sorted and transported in protein–RNA complexes, organelles, or protein complexes, respectively. A number of motor proteins, specialized for transport of cargo material, recognize these complexes and recruit them over short or long distances within the cell. As the number of cargoes is generally much higher than the number of motors available for transport, specificity is required to recruit certain material to individual destinations. To address specificity, transport complexes not only consist of motor and cargo but also contain a variable number of adaptor or scaffolding proteins involved in motor–cargo attachment. Adaptor proteins also bind targeting signals within the polypeptide chain of cargo molecules that are thought to encode directionality of transport. Not all cargo molecules are selectively transported to their site of action. Polar cells also perform nonselective transport of material and eliminate cargo from inappropriate sites, for instance through endocytosis of transmembrane proteins. In this case, targeting signals might serve as recognition sites for the endocytic machinery to induce internalization of the respective proteins.

Molecular motors seem to be critical for all types of selective transport. Individual motors have a preference for either axons or dendrites, a phenomenon that is likely to be encoded by the molecular nature of the tracks along which motors move. Microtubule tracks are decorated by a variety of MAPs, most of which are target of post-translational modifications. Therefore, the interaction of individual motor–adaptor–cargo complexes, together with the individual nature of the tracks, generates a large combination of signals to encode directed delivery. Upon delivery of cargo at its final destination, interactions of cargo and adaptor/scaffolding proteins might remain to associate proteins for instance at subsynaptic membrane specializations such as the postsynaptic density.

Because of the large number of molecules to be transported, some transport mechanisms might be redundant. A more precise understanding of protein–protein interactions between tracks, motors, adaptors, and cargoes is required to distinguish between common principles and rather exceptional mechanisms of transport selectivity.

## 12. REFERENCES

1. Cramer, L.P. (2000) *J Cell Biol* **150**, F121–126.
2. Desai, A., and Mitchison, T.J. (1997) *Annu Rev Cell Dev Biol* **13**, 83–117.
3. Hirokawa, N. (1998) *Science* **279**, 519–526.
4. Vallee, R.B., Wall, J.S., Paschal, B.M., and Shpetner, H.S. (1988) *Nature* **332**, 561–563.
5. Miki, H., Setou, M., Kaneshiro, K., and Hirokawa, N. (2001) *Proc Natl Acad Sci U S A* **98**, 7004–7011.
6. Setou, M., Seog, D.H., Tanaka, Y., Kanai, Y., Takei, Y., Kawagishi, M., and Hirokawa, N. (2002) *Nature* **417**, 83–87.
7. Vallee, R.B., Williams, J.C., Varma, D., and Barnhart, L.E. (2004) *J Neurobiol* **58**, 189–200.
8. Fuhrmann, J.C., Kins, S., Rostaing, P., El Far, O., Kirsch, J., Sheng, M., Triller, A., Betz, H., and Kneussel, M. (2002) *J Neurosci* **22**, 5393–5402.
9. Maas, C., Tagnaouti, N., Behrend, B., Loebrich, S., Lappe-Siefke, C., and Kneussel, M. (2006) *J Cell Biol.* **172**, 441–451.
10. Bridgman, P.C. (2004) *J Neurobiol* **58**, 164–174.
11. Cheney, R.E., O'Shea, M.K., Heuser, J.E., Coelho, M.V., Wolenski, J.S., Espreafico, E.M., Forscher, P., Larson, R.E., and Mooseker, M.S. (1993) *Cell* **75**, 13–23.

12. Setou, M., Nakagawa, T., Seog, D.H., and Hirokawa, N. (2000) *Science* **288**, 1796–1802.
13. Mok, H., Shin, H., Kim, S., Lee, J.R., Yoon, J., and Kim, E. (2002) *J Neurosci* **22**, 5253–5258.
14. Wu, H., Nash, J.E., Zamorano, P., and Garner, C.C. (2002) *J Biol Chem* **277**, 30928–30934.
15. Wong, R.W., Setou, M., Teng, J., Takei, Y., and Hirokawa, N. (2002) *Proc Natl Acad Sci U S A* **99**, 14500–14505.
16. Krichevsky, A.M., and Kosik, K.S. (2001) *Neuron* **32**, 683–696.
17. Kanai, Y., Dohmae, N., and Hirokawa, N. (2004) *Neuron* **43**, 513–525.
18. Inomata, H., Nakamura, Y., Hayakawa, A., Takata, H., Suzuki, T., Miyazawa, K., and Kitamura, N. (2003) *J Biol Chem* **278**, 22946–22955.
19. Matsuda, S., Matsuda, Y., and D'Adamio, L. (2003) *J Biol Chem* **278**, 38601–38606.
20. Holleran, E.A., Ligon, L.A., Tokito, M., Stankewich, M.C., Morrow, J.S., and Holzbaur, E.L. (2001) *J Biol Chem* **276**, 36598–36605.
21. Tai, A.W., Chuang, J.Z., Bode, C., Wolfrum, U., and Sung, C.H. (1999) *Cell* **97**, 877–887.
22. Marszalek, J.R., Weiner, J.A., Farlow, S.J., Chun, J., and Goldstein, L.S. (1999) *J Cell Biol* **145**, 469–479.
23. Nakata, T., and Hirokawa, N. (2003) *J Cell Biol* **162**, 1045–1055.
24. Matenia, D., Griesshaber, B., Li, X.Y., Thiessen, A., Johne, C., Jiao, J., Mandelkow, E., and Mandelkow, E.M. (2005) *Mol Biol Cell.*
25. Johnson, G.V., and Stoothoff, W.H. (2004) *J Cell Sci* **117**, 5721–5729.
26. West, A.E., Neve, R.L., and Buckley, K.M. (1997) *J Neurosci* **17**, 6038–6047.
27. Rivera, J.F., Ahmad, S., Quick, M.W., Liman, E.R., and Arnold, D.B. (2003) *Nat Neurosci* **6**, 243–250.
28. El-Husseini Ael, D., Craven, S.E., Brock, S.C., and Bredt, D.S. (2001) *J Biol Chem* **276**, 44984–44992.
29. Francesconi, A., and Duvoisin, R.M. (2002) *J Neurosci* **22**, 2196–2205.
30. Stowell, J.N., and Craig, A.M. (1999) *Neuron* **22**, 525–536.
31. Sampo, B., Kaech, S., Kunz, S., and Banker, G. (2003) *Neuron* **37**, 611–624.
32. Kneussel, M. (2005) *EMBO Rep* **6**, 22–27.
33. Zhai, R.G., Vardinon-Friedman, H., Cases-Langhoff, C., Becker, B., Gundelfinger, E.D., Ziv, N.E., and Garner, C.C. (2001) *Neuron* **29**, 131–143.
34. Bearer, E.L., and Reese, T.S. (1999) *J Neurocytol* **28**, 85–98.
35. Prekeris, R., and Terrian, D.M. (1997) *J Cell Biol* **137**, 1589–1601.
36. Ohyama, A., Komiya, Y., and Igarashi, M. (2001) *Biochem Biophys Res Commun* **280**, 988–991.
37. Evans, L.L., Lee, A.J., Bridgman, P.C., and Mooseker, M.S. (1998) *J Cell Sci* **111**, 2055–2066.
38. Bridgman, P.C. (1999) *J Cell Biol* **146**, 1045–1060.
39. Huang, J.D., Brady, S.T., Richards, B.W., Stenolen, D., Resau, J.H., Copeland, N.G., and Jenkins, N.A. (1999) *Nature* **397**, 267–270.
40. Matus, A., Brinkhaus, H., and Wagner, U. (2000) *Hippocampus* **10**, 555–560.
41. Naisbitt, S., Valtschanoff, J., Allison, D.W., Sala, C., Kim, E., Craig, A.M., Weinberg, R.J., and Sheng, M. (2000) *J Neurosci* **20**, 4524–4534.
42. Costa, M.C., Mani, F., Santoro, W., Jr., Espreafico, E.M., and Larson, R.E. (1999) *J Biol Chem* **274**, 15811–15819.
43. Volpicelli, L.A., Lah, J.J., Fang, G., Goldenring, J.R., and Levey, A.I. (2002) *J Neurosci* **22**, 9776–9784.
44. Osterweil, E., Wells, D.G., and Mooseker, M.S. (2005) *J Cell Biol* **168**, 329–338.

# 14

# POSTSYNAPTIC TRANSPORT PACKETS

Philip E. Washbourne*

## 1. SUMMARY

To understand the molecular methods by which synapse formation occurs, it is critical to know the mechanisms by which all the synaptic components can possibly arrive at a newly forming synapse. Methods of trafficking can range from diffusion through the cytoplasm or in the plasma membrane to exo/endocytic cycling between bouts of transport along microtubules. In the case of the post-synaptic density (PSD) of glutamatergic synapses, it appears that many of the necessary proteins arrive separately and by different means. Here we examine the modes of transport and recruitment of a number of postsynaptic components to newly forming, glutamatergic synapses.

## 2. INTRODUCTION

The postsynaptic compartment is the region of the synapse specialized in receiving neurotransmitter secreted by the presynaptic terminal. This neuronal subdomain contains not only receptors which bind and sense the neurotransmitter, but also molecules required for signaling, scaffolding, and adhesion with the presynaptic neuron[1]. The best-studied postsynaptic compartment is at the glutamatergic synapse. This is probably due to the fact, that the postsynaptic specialization of glutamatergic synapses is easily recognizable in electron micrographs as a dense material under the membrane, hence the name PSD. Another reason lies in the variety of glutamate receptor types and their respective potential roles in learning and memory.

In order to understand how the glutamatergic PSD forms, it is crucial to know its composition[2,3] and where its components are synthesized and located prior to synapse formation and how they get to new sites of synapse formation. Knowledge of the trafficking of PSD proteins will ultimately lead to the comprehension of mechanisms of molecular recruitment to nascent postsynaptic sites.

*Institute of Neuroscience, University of Oregon, Eugene, OR 97403, USA; pwash@uoneuro.uoregon.edu

The largest advance in understanding molecular trafficking of synaptic proteins happened during the late 1990s with the advent of high-resolution time-lapse imaging and the tagging of synaptic components with fluorescent proteins such as green fluorescent protein[4]. Prior to these techniques, our understanding of postsynaptic molecule trafficking relied on data obtained using fixed specimens such as electron micrographs[5] and immunofluorescence images[6] or on data from electrophysiological recordings[7,8]. While the former two only provide static snapshots of protein distribution during development, the latter does not resolve the localization of receptors prior to insertion in the plasma membrane. Furthermore, electrophysiological recordings tell us nothing of the localization of intracellular postsynaptic proteins that do not directly influence channel properties. Despite the constraints of these early studies, they have shown that (1) postsynaptic molecules redistribute from a nonsynaptic dendritic localization to a synaptic localization[5–8] and (2) some postsynaptic molecules such as NMDA and AMPA-type glutamate receptors and the postsynaptic scaffolding protein PSD-95 are localized in nonsynaptic clusters prior to synapse formation[6,9].

The observation of nonsynaptic glutamate receptor and PSD-95 clusters led immediately to the question as to what these protein clusters may represent. Were they reservoirs of proteins that could then diffuse to new sites of synapse formation or did they perhaps represent a transport unit? Furthermore, the lack of colocalization of the nonsynaptic clusters of NMDA receptors, PSD-95, and AMPA receptors[6] suggested that assembly of the PSD may contrast starkly with formation of the presynaptic terminal[10,11]. It had been discovered that, at the developing presynaptic terminal, complexes containing a number of active zone and synaptic vesicle components are recruited on a rapid time scale to generate a functional presynaptic compartment[11–13] (discussed in more detail in Chapter 15). On the postsynaptic side, however, the dynamics of synaptic proteins has been controversial.

## 3. PSD-95 ACCUMULATION

PSD-95 was identified in the early 1990s by cDNA cloning of PSD proteins[14]. Since then, it has been regarded as the prototypical PSD scaffolding protein. A protein interaction domain, of which PSD-95 contains three, is seen in hundreds of such scaffolding proteins and was named after PSD-95 and its *Drosophila* homolog: the PDZ domain (PSD-95, Discs large, and ZO-1). In addition to the PDZ domains, PSD-95 presents an inactive guanylate kinase domain, making it a member of the large family of membrane-associated guanylate kinases (MAGUKs)[15]. PSD-95 has been regarded as the fail-safe marker of the mature glutamatergic synapse and by virtue of its direct and indirect interactions with NMDA and AMPA receptors, PSD-95 has been considered a key player in the development of the PSD.

Before extensive time-lapse studies were carried out, PSD-95 had been recognized as playing an important role in synaptic strength at mature synapses[16]. In fact, it was shown that for PSD-95 to function in synaptic plasticity, this essentially cytosolic molecule needs to become membrane bound by palmitoylation[17]. This interesting mechanism may also be required in its recruitment to new synapses (discussed below). The huge importance of PSD-95 allowing mature synapses to grow in response to activity[18], suggested a function in early synapse formation and maturation. With this in mind, the dynamics of PSD-95 before, after, and during synapse formation have been studied extensively.

As mentioned above, PSD-95 has been clearly seen to localize in clusters within the dendrites of hippocampal and cortical neurons prior to synapse formation[6]. This distribution was confirmed by imaging of GFP-tagged forms of PSD-95[9,18–20]. These clusters have been shown to be largely immobile[9,18–21], suggesting that they do not represent transport modules of PSD-95. EM studies have shown nonsynaptic membrane specializations or "free PSDs"[22,23], however, the nature of these structures remains obscure. Since PSD-95 has been shown to accumulate gradually at nascent synapses from a cytoplasmic pool[18–20], one may guess that these clusters act as protein reservoirs within the dendrites of neurons.

In contrast, a number of studies have shown some movement of PSD-95 clusters, albeit very slow (up to 1 μm/min)[24]. The dynamics for PSD-95 clusters include splitting of clusters[24,25], lateral movement in shafts[24], movement into and out of established spines, and movement within dendritic filopodia[24,25]. Spines represent the final maturation step of glutamatergic synapses. These stable, mushroom-shaped protrusions from the dendritic shaft are thought to act as compartments for the restriction of local second messenger systems to individual postsynaptic densities[26]. In contrast, dendritic filopodia are highly dynamic protrusions, which are thought to be involved in initiating synaptogenic contact and comprise morphological precursors of synaptic spines[27].

Since these studies did not address the colocalization of presynaptic terminals with the individual moving clusters of PSD-95[24,25], it is unclear whether they can be regarded as transport modules. It is possible that these clusters constitute assembled postsynaptic densities opposing presynaptic terminals, which, with a certain low probability, can move out of and into spines. Another possibility is that an assembled, free PSD can move during assembly and disassembly of the synapse. It remains to be determined what these PSD-95 clusters exactly represent and which other proteins are associated with them.

## 4. NMDA RECEPTOR TRAFFICKING

In contrast to the cytosolic protein PSD-95, the localization and trafficking of transmembrane proteins, such as the NMDA receptor, is intrinsically more complicated. Levels of complication are not only provided by synthesis and processing in the endoplasmic reticulum (ER) and Golgi apparatus, but also by the possibility of both vesicular trafficking and plasma membrane diffusion.

At the level of the ER, the trafficking of the NMDA receptor is highly regulated. As with other receptor complexes, exit of the NMDA receptor from the ER is determined by successful assembly of a complete receptor[28]. This is usually constituted by the interaction of the obligatory subunit, NR1, with a type 2 or 3 subunit[29]. The majority of studies have concentrated on the 2A and 2B subunits of the NMDA receptor. The NR1 subunit contains an ER retention signal, which is masked by association with NR2A or B, allowing exit from the ER[30,31]. In addition, the retention signal can be modified by alternative splicing of the NR1 messenger RNA and by phosphorylation of residues in the C-terminal tail[32]. A switch between NR2A and NR2B subunits has been shown to occur during development with NR2A being the more abundant isoform at mature synapses[33–35].

It turns out that this switch is regulated at the level of mRNA splicing and appears to be activity- and integrin-dependent[36,37].

Once NMDA receptors leave the Golgi in transport vesicles there are two possible trafficking routes. Either the transport vesicles are hauled into and through dendrites by motor proteins or, alternatively, the vesicles fuse with the plasma membrane, allowing the receptors to move laterally in the plasma membrane. There is evidence for both these possibilities.

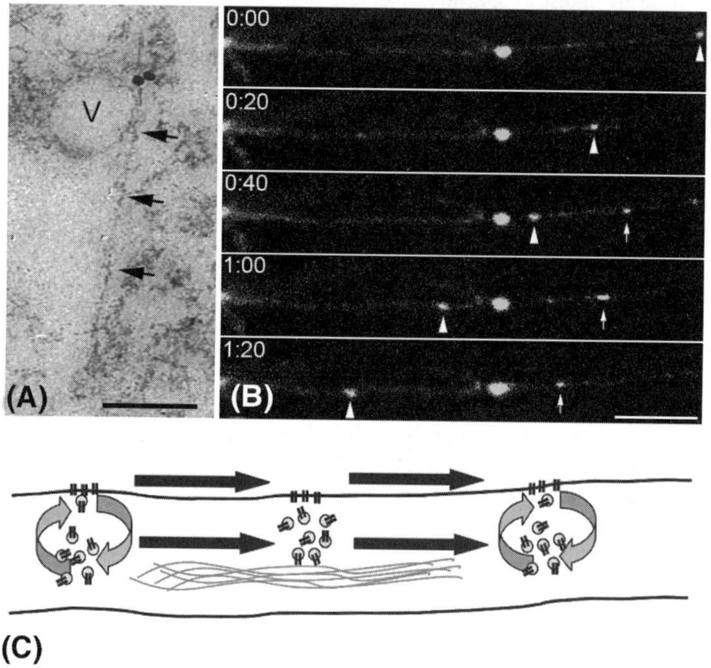


**Figure 14.1.** NMDA Receptor Transport Packets. (A) Electron micrograph of a cortical neuronal dendrite labeled with immunogold for NMDA receptor subunit 1. The gold particles clearly label a vesicle that is tightly associated with microtubules (arrows). Scale bar = 200 nm. (B) Time-lapse images of NMDA receptor clusters *in vivo*. Clusters of GFP-tagged NMDA receptor subunit 1 are mobile (arrow and arrowhead) within the dendrites of an interneuron in the spinal cord of a zebrafish embryo at 51 h post fertilization. Time in minutes and seconds; scale bar = 10 μm.(C) Model for exo/endocytic cycling of NMDA receptors with the plasma membrane during pauses of transport. NMDA receptors can travel both on vesicles within dendrites or in the plasma membrane.

Electrophysiological and single-particle tracking studies have demonstrated that NMDA receptors are present in the plasma membrane[38] and are relatively mobile prior to synaptogenesis[39,40]. In contrast, EM and live-imaging studies suggest that NMDA receptors are present in vesicles that are transported along microtubules (see Figure 14.1)[41,42]. It is possible that both intracellular vesicular transport and plasma membrane diffusion is occurring. Evidence for a fusion these two models was recently presented. During the period of synaptogenesis synaptogenesis NMDA receptors in transport vesicles fuse with the plasma membrane and rapidly re-endocytose[42]. Live imaging of GFP-tagged NMDA receptor subunit 1 together with a tagged form of clathrin light chain demonstrated that endocytosis occurs during pauses in movement[42]. As development proceeds, the amount of NMDA receptors that are endocytosed is gradually reduced and is likely due to an interaction with PSD-95[43]. At mature synapses NMDA receptors are highly stable, which contrasts the very dynamic nature of AMPA receptors[44–46].

It remains unclear why NMDA receptors undergoing transport should exhibit this dynamic flux between the plasma membrane and intracellular transport vesicles. One possibility is that Golgi-derived transport vesicles contain a number of different cargo proteins, which need to be separated into separate vesicles or retained at the plasma membrane prior to targeting to the final destination. Such a sorting mechanism was suggested by data demonstrating that the synaptic vesicle protein VAMP/synaptobrevin appears at the somatodendritic plasma membrane prior to transport along the axon[47]. On the other hand, it is possible that NMDA receptors are required at the plasma membrane for sensing glutamate but are endocytosed periodically for efficient and directed transport. A requirement for the sensing of glutamate during development is suggested by studies of neuronal migration[48] and synapse formation[49,50]. This is contrasted by pharmacological studies of neurons in culture[51–53] and knock-out mice which are unable to secrete neurotransmitter[54,55]. In the latter studies, genetic ablation of genes involved in neurotransmitter exocytosis did not reduce the ability of neurons to develop synapses in culture and *in vivo*, suggesting that the secretion of neurotransmitter is not required to initiate synapse formation. However, it is also possible that the sensitivity to glutamate release may serve to increase the likelihood or location of synapse formation within the dendritic arborization[56–59].

Live imaging of GFP-tagged NMDA receptors during development has led to differing conclusions about how they are transported. A number of studies have shown distinct clusters of NMDA receptors which are transported along microtubules[9,42]. This was corroborated by a study which demonstrated the movement of the motor protein KIF17 as discrete clusters within neuronal dendrites[41]. Retrospective immunostaining of these clusters showed clear colocalization of KIF17 with NR2B[60]. In fact, NR2B has been shown to interact with KIF17 indirectly via mLin-10, CASK, and Velis[41] (see Chapter 13). In addition, the clusters seem to move at around 8 μm/min in both anterograde and retrograde directions[9]. NMDA receptor transport clusters switch direction and stop often[9]. It appears that the clusters preferentially stop at clathrin hotspots[42]. The fact that these regions of the dendritic plasma membrane are constant sites of endocytosis[61] suggests that endo- and perhaps exocytosis of the NMDA receptor-containing vesicles occurs at these sites[42,43]. Furthermore, the clusters present NMDA receptors both at the surface and intracellularly, suggesting a tight link between intracellular and extracellular pools of the receptor[42].

In contrast to the data describing discrete mobile clusters of NMDA receptors in neuronal dendrites, another group has documented a more diffuse localization of NMDA receptors prior to synapse formation. GFP-tagged NR1 was seen to gradually accumulate at synaptic sites over time, in a similar manner to PSD-95[11]. Fluorescence recovery after photobleaching (FRAP) confirmed a gradual replacement of fluorescence at previously identified synaptic sites[11]. While both transport mechanisms may involve vesicles that move along microtubules, it is unclear why such a difference is seen between laboratories, since the experiments have been carried out in similar neuronal cultures from the same age. One potential explanation for the controversy could lie in the expression level of the GFP transgene. Regardless of whether NMDA receptors move as discrete clusters or as separate vesicles within dendrites, researchers agree that recruitment of NMDA receptors to synaptic sites is separate from other postsynaptic proteins such as PSD-95 and AMPA receptors[9,11,19,20].

Interestingly, the MAGUK protein SAP-102 is associated with NMDA receptor clusters during transport[42,62]. It is probable that SAP-102 plays a role in directing the transport of NMDA receptor-containing vesicles toward the exocyst complex[62]. This complex is involved in delivering vesicles to the plasma membrane, further lending weight to the hypothesis that exocytosis plays an important role in the transport of NMDA receptor-containing vesicles. A switch in the abundancy of SAP102 and PSD-95 at postsynaptic densities during early postnatal development of rat hippocampus, suggests that these two MAGUK proteins are exchanged as binding partners for the NMDA receptor[3,62]. This could explain the reduction of endocytosis of NMDA receptors that occurs during this same period[63] and thus begin to address a molecular mechanism for how receptors are targeted to and then maintained at synaptic sites.

## 5. AMPA RECEPTOR TRANSPORT

Due to the highly dynamic nature of AMPA receptors as a mechanism of synaptic plasticity[64–66], there exists a wealth of studies which focus on their exo- and endocytosis at mature synapses (see Chapters 22–24). A number of studies suggest that similar mechanisms governing the activity-dependent insertion of AMPA receptors may also be occurring during development[8,67,68].

AMPA receptors have long been shown to interact directly with NSF and α and β-SNAPs[69]. These proteins form part of the exocytic fusion complex and are integral to vesicle fusion with the plasma membrane. The interaction between AMPA receptors and NSF has been shown to be required for the increase in the number of AMPA receptors at synapses in response to activity[70,71]. In addition, AMPA receptors are endocytosed by a dynamin-dependent mechanism[44]. These studies all suggest that exo/endocytosis is integral to the local trafficking of AMPA receptors at mature synapses. However, can one assume that AMPA receptor trafficking prior to synapse formation also involves cycling with the plasma membrane, as seen for NMDA receptors?

AMPA receptors, like NMDA receptors, are also seen as nonsynaptic clusters in dendrites of young cortical and hippocampal neurons[9,53]. These clusters have also been described to be mobile[9] (see Figure 14.2; Colorplate 9). However, it appears that AMPA receptor trafficking may be governed by somewhat different mechanisms. The number of clusters that are seen to move at any one time is greatly reduced as compared to NMDA receptors and the velocity of their movement is also less[9]. This may be attributable to the fact that AMPA receptors are transported by the the motor protein KIF5A[72] as compared with KIF17 for NMDA receptors[41]. Interestingly, Hirokawa and colleagues demonstrated that interaction of GluR2-associated protein (GRIP1) with KIF5A preferentially directs the AMPA receptor complex into dendrites rather than axons[72].

Occasionally, clusters containing both AMPA and NMDA receptors are transported together and these appear to move with the kinetics of NMDA receptor transport packets[9]. These may represent vesicles recently derived from the Golgi apparatus prior to sorting at the plasma membrane. During development AMPA receptors are also seen at the surface of neurons at nonsynaptic sites[38,39,73,74]. It appears that AMPA receptors are found in intracellular and extracellular pools in neuronal dendrites[74]. This suggests that while AMPA receptors may be transported separately from NMDA receptors, perhaps employing a different motor protein, both receptor types may share the mechanism of cycling with the plasma membrane.

As with NMDA receptors, AMPA receptors are able to diffuse in the plasma membrane. Electrophysiological recordings[38], immunocytological methods[74], and single-particle tracking experiments[73] point to a diffuse presence of extrasynaptic glutamate receptors in the plasma membrane. Electrophysiological recordings combined with iontophoretic application of glutamate did not find large clusters of AMPA in extrasynaptic membrane[38]. Single-particle tracking experiments clearly demonstrate active transport of surface receptors and trapping of receptors at synapses, suggesting that interactions with a submembrane scaffold may be a mechanism for recruiting receptors that are already in the plasma membrane[39,40,73]. It remains to be determined which exact route AMPA receptors take to arrive at the synaptic plasma membrane.

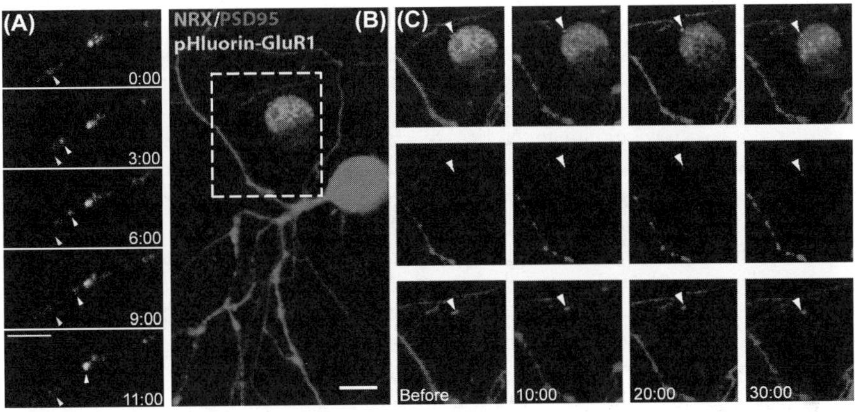


**Figure 14.2.** AMPA Receptor Transport and Recruitment. (A) Time-lapse images of AMPA receptor clusters *in vitro*. Clusters of GFP-tagged AMPA receptor subunit GluR1 are mobile (white arrow) in the dendrites of cortical neurons cultured for 4 days. Yellow arrow (see Colorplate 9) shows original location; time in minutes and seconds; scale bar = 6 μm. (B) Insertion of pH-dependent GFP-tagged GluR1 (pHluorin-GluR1) at new sites of synapse formation. Contact of a β-neurexin expressing PC12 cell (red) induces the accumulation of CFP-PSD-95 (blue). Scale bar = 10 μm. (C) The appearance of pHluorin-GluR1 at the PSD-95 cluster (arrowheads) is induced by the bath application of glutamate and glycine in brief pulses. Time in minutes and seconds. Reproduced with permission from ref. 88. Copyright (2005) National Academy of Sciences, U.S.A.

There are four different subunits of AMPA receptors, GluR1–4. A great number of studies have focused on their differential recruitment to synapses and this topic is discussed in more detail in Chapter 24. However, pertinent at this point is the observation that GluR4-only AMPA receptors are the first AMPA receptors to be incorporated at new synapses. During the development of neuronal circuits in a number of regions of the nervous system, synapses first only show NMDA receptors and are thus termed "silent"[8,67]. In response to spontaneous synaptic activity, GluR4-containing AMPA receptors are inserted at "silent" synapses[75]. Subsequently, these receptors are exchanged for GluR2-containing AMPA receptors, a process that requires little synaptic activity[75]. The consequences and reasons for such switching of AMPAR receptor subtype delivery remain largely obscure; however it has been hypothesized that long-term maintenance of a potentiated synapse requires the exchange of AMPA receptors for GluR2-containing AMPA receptors[70].

Prior to incorporation at new synaptic sites, AMPA receptors associate with a MAGUK protein early during the biosynthetic pathway. SAP-97 binds to AMPA receptors while the receptors are still in the ER or Golgi[76]. Since SAP-97 does not associate with synaptic AMPA receptors, we can assume that there is a switch in MAGUK protein association that serves to stabilize AMPA receptors at synapses. Recently, it has been hypothesized that a transmembrane protein that also associates with AMPA receptors plays a far greater role in AMPA receptor trafficking and targeting than previously surmised. Stargazin and the related transmembrane AMPA regulatory proteins (TARPs) were identified due to a naturally occurring mutation in the Stargazin gene in Stargazer mice. These mice show dystonia and often look upward, hence their name. This phenotype probably arises due to the fact that the granule cells in the cerebellum have no AMPA receptors at the plasma membrane[77]. TARPs, as this family of proteins is now known[78], have been shown to direct trafficking of AMPA receptors to the cell surface and to synapses via distinct mechanisms[77]. TARPs interact with AMPA receptors through PSD-95 and this interaction controls the number of AMPA receptors at synapses[79]. However, it appears that a direct interaction between AMPA receptors and these four transmembrane-domain proteins in their extra-cellular domains can modulate AMPA receptor gating[80]. Furthermore, quantitative immunoprecipitations from mouse brain suggest that the majority of AMPA receptors are present in an exclusive complex with TARPs, with less than a total of 1% of AMPA receptors bound to MAGUK proteins such as SAP102, SAP97, PSD-95, PICK1, and GRIP[81]. This recent study brings a large amount of controversy into the field and much work is needed to determine the overriding molecular mechanism by which AMPA receptors are transported through the biosynthetic pathway to the plasma membrane and then to synapses.

## 6. IDENTITY OF VESICULAR TRANSPORT PACKETS

The recruitment of postsynaptic proteins, such as PSD-95, NMDA, and AMPA receptors, to nascent synapses occurs separately and by distinct mechanisms. While PSD-95 accumulates from a cytosolic pool of molecules[18–20]. NMDA and AMPA receptors are present in clusters that travel along microtubules between bouts of exo- and endocytosis[9,41,42]. Despite this knowledge, it is unclear what constitutes these mobile clusters of glutamate receptors and which other proteins are associated with these clusters.

From the few studies addressing mobile NMDA receptors we know of a limited number of associated proteins. The NMDA receptors are associated with SAP102 during transport[42,62]. The NR2B subunit of the NMDA receptor is bound to the kinesin-like motor protein KIF17 via a large protein complex consisting of mLin-10 (Mint1), mLin-2 (CASK), and mLin-7 (MALS/Velis)[41]. However, it is unclear whether all NMDA receptor clusters are transported via KIF17, since transport velocities vary greatly and KIF17-purified vesicles are much smaller than the average size of NR1-labeled membrane structures in developing cortex[42]. One may then further enquire as to the nature of the NMDA receptor-containing membranous structures that are found within dendrites of neurons. These organelles are heterogeneous in size and shape; varying from 50 to 600nm in diameter, averaging around 200 nm[42]. They can be tubular in shape and are presumably of an endocytic nature, considering the high rates of exo- and endocytosis that is ongoing during development[42,43]. Indeed, the early endosomal marker EEA1 can be seen to be associated with these transport packets[42]. Similar tubulo-vesicular structures have been seen in the case of AMPA receptors and SAP-97[76], suggesting that AMPA receptors are found in a similar early endosomal compartment.

Recent data show that large tubulovesicular structures of Golgi-derived origin are transported through neuronal processes to sites of contact[82]. These structures are so large that they can be seen by using phase contrast video microscopy. These structures are somehow tethered to neural cell adhesion molecule (NCAM) which is present at the plasma membrane[82,83]. This association is thought to occur via an interaction with spectrin. The fact that these mobile organelles are Golgi derived does not necessarily preclude the possibility that they are also undergoing exo/endocytosis. It remains to be determined whether these NCAM-associated, mobile organelles are indeed the same as the NMDA and AMPA receptor transport packets.

## 7. MECHANISMS OF RECRUITMENT

While we are gradually beginning to understand mechanisms by which postsynaptic components are transported, the more important issue of how postsynaptic transport packets are trapped at nascent synaptic sites remains largely unknown. A large amount of research is currently being directed toward cell adhesion molecules as initiators of synapse formation (see Chapters 4–10). In fact, two classes of cell adhesion molecules have been found to be highly synaptogenic: Neuroligins and SynCAMs (see Chapters 7 and 8). However, the exact mechanism by which these single-transmembrane molecules transduce contact and adhesion to bring about the recruitment of pre- and postsynaptic molecules and organelles is still a mystery. Possibilities include interactions with MAGUK proteins via PDZ binding domains and the binding of actin filaments.

Even understanding the recruitment of diffusible PSD-95 to Neuroligin, which can interact with each other directly[84], presents conceptual problems or at least numerous possibilities to be tested. Presumably, PSD-95 can only bind to Neuroligin when Neuroligin is bound to its presynaptic partner, Neurexin. For this to be the case, one can imagine a change in conformation in the cytoplasmic tail of Neuroligin which could open up the C-terminal PDZ binding domain, allowing PSD-95 to bind. On the other hand, one could envisage a mechanism by which binding of Neurexin to Neuroligin transmits a signal to palmitoyl transferases, which can then palmitoylate PSD-95[85,86]. This modification anchors PSD-95 at the plasma membrane and is necessary for the localization of PSD-95 to synapses[17]. This modification could enhance binding of PSD-95 to the PDZ binding domain of Neuroligin directly by a morphological change or by reducing the diffusional space to a two-dimensional area[87].

In the case of membrane receptors being transported along microtubules in vesicles the problem of recruitment becomes even more complex. First of all we must establish whether recruitment of such transport packets occurs by directed movement or by generation of a stop signal. In the case of directed movement, we have to imagine a diffusible second messenger such as calcium or cAMP which would originate from the site of presynaptic contact and cause molecular motors to move up the gradient. The stop signal could also involve a diffusible molecule, but would need to cause motor proteins to stop once above a threshold concentration. This signal could be constituted by a kinase or a phosphatase. Another mechanism is illustrated by the study showing that intracellular domains of NCAM may provide a trapping signal via a direct interaction with proteins coating intracellular organelles, such as β1-spectrin[82]. In addition it seems that the switches in protein–protein associations between glutamate receptors and scaffolding proteins, mentioned above, could also help "stop" or "direct" transport packets to new sites of synapse formation. However, the mechanisms involved remain completely unkown.

Interestingly, it appears that a signal for the recruitment of AMPA receptors has been found. AMPA receptors appear at synapses with a delay with respect to NMDA receptors[9]. These data together with the finding that glutamatergic synapses are largely "silent," NMDA-only synapses early in development[8,67,68], spurred the hypothesis that strong activation of these synapses would recruit AMPA receptors to the postsynaptic membrane, as seen in paradigms of long-term potentiation (LTP) at mature synapses[66]. Recent evidence points to the activation of NMDA receptors by glutamate at nascent synapses as being a trigger for AMPA receptor recruitment[88]. Similarly to the induction of LTP[89], the recruitment of AMPA receptors to new synapses is mediated by calcium/calmodulin-dependent protein kinase II (CaMKII), since transfection of a constitutively active form of CaMKII increased colocalization of AMPA receptors with Neurexin-induced PSD-95 clusters[88] (see Figure 14.2; Colorplate 9). Further mechanisms of how CaMKII directs AMPA receptors to the nascent synaptic site remain to be explored.

## 8. CONCLUSIONS

We can see that while the past 5 years have increased our understanding of the transport of pre- and postsynaptic molecules, large gaps in our knowledge of this process still exist. Questions that remain to be addressed largely center around the molecular mechanisms which govern the transport of postsynaptic molecules and that are directly responsible for their recruitment to newly forming synapses. In light of the fact that the formation of synapses is of the utmost importance to brain function, where even subtle errors lead to devastating disorders of the mind, such as autism, mental retardation, and perhaps even schizophrenia, it is critical that we unravel the mysteries of protein trafficking in developing neurons.

## 9. REFERENCES


1. Kennedy, M.B. (2000) *Science* **290**, 750–754.
2. Husi, H., Ward, M.A., Choudhary, J.S., Blackstock, W.P., and Grant, S.G. (2000) *Nat Neurosci* **3**, 661–669.
3. Petralia, R.S., Sans, N., Wang, Y.X., and Wenthold, R.J. (2005) *Mol Cell Neurosci* **29**, 436–452.
4. Nakata, T., Terada, S., and Hirokawa, N. (1998) *J Cell Biol* **140**, 659–674.
5. Liu, X.B., Munoz, A., and Jones, E.G. (1998) *J Comp Neurol* **395**, 450–465.
6. Rao, A., Kim, E., Sheng, M., and Craig, A.M. (1998) *J Neurosci* **18**, 1217–1229.
7. Wu, G., Malinow, R., and Cline, H.T. (1996) *Science* **274**, 972–976.
8. Isaac, J.T., Crair, M.C., Nicoll, R.A., and Malenka, R.C. (1997) *Neuron* **18**, 269–280.
9. Washbourne, P., Bennett, J.E., and McAllister, A.K. (2002) *Nat Neurosci* **5**, 751–759.
10. Ziv, N.E., and Garner, C.C. (2001) *Curr Opin Neurobiol* **11**, 536–543.
11. Bresler, T., Shapira, M., Boeckers, T., Dresbach, T., Futter, M., Garner, C.C., Rosenblum, K., Gundelfinger, E.D., and Ziv, N.E. (2004) *J Neurosci* **24**, 1507–1520.

12. Ahmari, S.E., Buchanan, J., and Smith, S.J. (2000) *Nat Neurosci* **3**, 445–451.
13. Shapira, M., Zhai, R.G., Dresbach, T., Bresler, T., Torres, V.I., Gundelfinger, E.D., Ziv, N.E., and Garner, C.C. (2003) *Neuron* **38**, 237–252.
14. Kennedy, M.B. (1997) *Trends Neurosci* **20**, 264–268.
15. Garner, C.C., Nash, J., and Huganir, R.L. (2000) *Trends Cell Biol* **10**, 274–280.
16. El-Husseini, A.E., Schnell, E., Chetkovich, D.M., Nicoll, R.A., and Bredt, D.S. (2000) *Science* **290**, 1364–1368.
17. El-Husseini, A. et al. (2002) *Cell* **108**, 849–863.
18. Okabe, S., Kim, H.D., Miwa, A., Kuriu, T., and Okado, H. (1999) *Nat Neurosci* **2**, 804–811.
19. Bresler, T., Ramati, Y., Zamorano, P.L., Zhai, R., Garner, C.C., and Ziv, N.E. (2001) *Mol Cell Neurosci* **18**, 149–167.
20. Friedman, H.V., Bresler, T., Garner, C.C., and Ziv, N.E. (2000) *Neuron* **27**, 57–69.
21. Niell, C.M., Meyer, M.P., and Smith, S.J. (2004) *Nat Neurosci* **7**, 254–260.
22. Fiala, J.C., Feinberg, M., Popov, V., and Harris, K.M. (1998) *J Neurosci* **18**, 8900–8911.
23. Blue, M.E., and Parnavelas, J.G. (1983) *J Neurocytol* **12**, 599–616.
24. Prange, O., and Murphy, T.H. (2001) *J Neurosci* **21**, 9325–9333.
25. Marrs, G.S., Green, S.H., and Dailey, M.E. (2001) *Nat Neurosci* **4**, 1006–1013.
26. Nimchinsky, E.A., Sabatini, B.L., and Svoboda, K. (2002) *Annu Rev Physiol* **64**, 313–353.
27. Ziv, N.E., and Smith, S.J. (1996) *Neuron* **17**, 91–102.
28. Fukaya, M., Kato, A., Lovett, C., Tonegawa, S., and Watanabe, M. (2003) *Proc Natl Acad Sci U S A* **100**, 4855–4860.
29. McIlhinney, R.A., Le Bourdelles, B., Molnar, E., Tricaud, N., Streit, P., and Whiting, P.J. (1998) *Neuropharm* **37**, 1355–1367.
30. Standley, S., Roche, K.W., McCallum, J., Sans, N., and Wenthold, R.J. (2000) *Neuron* **28**, 887–898.
31. Xia, H., Hornby, Z.D., and Malenka, R.C. (2001) *Neuropharm* **41**, 714–723.
32. Scott, D.B., Blanpied, T.A., Swanson, G.T., Zhang, C., and Ehlers, M.D. (2001) *J Neurosci* **21**, 3063–3072.
33. Tovar, K.R., and Westbrook, G.L. (1999) *J Neurosci* **19**, 4180–4188.
34. Sans, N., Petralia, R.S., Wang, Y.X., Blahos, J., 2nd, Hell, J.W., and Wenthold, R.J. (2000) *J Neurosci* **20**, 1260–1271.
35. Barria, A., and Malinow, R. (2002) *Neuron* **35**, 345–353.
36. Mu, Y., Otsuka, T., Horton, A.C., Scott, D.B., and Ehlers, M.D. (2003) *Neuron* **40**, 581–594.
37. Chavis, P., and Westbrook, G. (2001) *Nature* **411**, 317–321.
38. Cottrell, J.R., Dube, G.R., Egles, C., and Liu, G. (2000) *J Neurophysiol* **84**, 1573–1587.
39. Groc, L., Heine, M., Cognet, L., Brickley, K., Stephenson, F.A., Lounis, B., and Choquet, D. (2004) *Nat Neurosci* **7**, 695–696.
40. Choquet, D., and Triller, A. (2003) *Nat Rev Neurosci* **4**, 251–265.
41. Setou, M., Nakagawa, T., Seog, D.H., and Hirokawa, N. (2000) *Science* **288**, 1796–1802.
42. Washbourne, P., Liu, X.B., Jones, E.G., and McAllister, A.K. (2004) *J Neurosci* **24**, 8253–8264.
43. Roche, K.W., Standley, S., McCallum, J., Dune Ly, C., Ehlers, M.D., and Wenthold, R.J. (2001) *Nat Neurosci* **4**, 794–802.
44. Carroll, R.C., Beattie, E.C., Xia, H., Luscher, C., Altschuler, Y., Nicoll, R.A., Malenka, R.C., and von Zastrow, M. (1999) *Proc Natl Acad Sci U S A* **96**, 14112–14117.
45. Ehlers, M.D. (2000) *Neuron* **28**, 511–525.
46. Lin, J.W., Ju, W., Foster, K., Lee, S.H., Ahmadian, G., Wyszynski, M., Wang, Y.T., and Sheng, M. (2000) *Nat Neurosci* **3**, 1282–1290.
47. Sampo, B., Kaech, S., Kunz, S., and Banker, G. (2003) *Neuron* **37**, 611–624.
48. Komuro, H., and Rakic, P. (1993) *Science* **260**, 95–97.
49. Young, S.H., and Poo, M.M. (1983) *Nature* **305**, 634–637.
50. Demarque, M., Represa, A., Becq, H., Khalilov, I., Ben-Ari, Y., and Aniksztejn, L. (2002) *Neuron* **36**, 1051–1061.
51. Craig, A.M., Blackstone, C.D., Huganir, R.L., and Banker, G. (1994) *Proc Natl Acad Sci U S A* **91**, 12373–12377.
52. Benson, D.L., and Cohen, P.A. (1996) *J Neurosci* **16**, 6424–6432.
53. Rao, A., and Craig, A.M. (1997) *Neuron* **19**, 801–812.
54. Verhage, M. et al. (2000) *Science* **287**, 864–869.
55. Washbourne, P. et al. (2002) *Nat Neurosci* **5**, 19–26.
56. Lin, S.Y., and Constantine-Paton, M. (1998) *J Neurosci* **18**, 3725–3737.
57. Lohmann, C., Myhr, K.L., and Wong, R.O. (2002) *Nature* **418**, 177–181.
58. Maletic-Savatic, M., Malinow, R., and Svoboda, K. (1999) *Science* **283**, 1923–1927.
59. Wong, W.T., and Wong, R.O. (2000) *Cur Opin Neurobiol* **10**, 118–124.
60. Guillaud, L., Setou, M., and Hirokawa, N. (2003) *J Neurosci* **23**, 131–140.
61. Blanpied, T.A., Scott, D.B., and Ehlers, M.D. (2002) *Neuron* **36**, 435–449.

62. Sans, N., Prybylowski, K., Petralia, R.S., Chang, K., Wang, Y.X., Racca, C., Vicini, S., and Wenthold, R.J. (2003) *Nat Cell Biol* **5**, 520–530.
63. Roche, K.W., Tu, J.C., Petralia, R.S., Xiao, B., Wenthold, R.J., and Worley, P.F. (1999) *J Biol Chem* **274**, 25953–25957.
64. Shi, S.H., Hayashi, Y., Petralia, R.S., Zaman, S. H., Wenthold, R.J., Svoboda, K., and Malinow, R. (1999) *Science* **284**, 1811–1816.
65. Liao, D., Scannevin, R.H., and Huganir, R. (2001) *J Neurosci* **21**, 6008–6017.
66. Sheng, M., and Lee, S.H. (2001) *Cell* **105**, 825–828.
67. Liao, D., Zhang, X., O'Brien, R., Ehlers, M.D., and Huganir, R.L. (1999) *Nat Neurosci* **2**, 37–43.
68. Petralia, R.S., Esteban, J.A., Wang, Y.X., Partridge, J.G., Zhao, H.M., Wenthold, R.J., and Malinow, R. (1999) *Nat Neurosci* **2**, 31–36.
69. Osten, P., Srivastava, S., Inman, G.J., Vilim, F.S., Khatri, L., Lee, L.M., States, B.A., Einheber, S., Milner, T.A., Hanson, P.I., and Ziff, E.B. (1998) *Neuron* **21**, 99–110.
70. Shi, S., Hayashi, Y., Esteban, J.A., and Malinow, R. (2001) *Cell* **105**, 331–343.
71. Beretta, F., Sala, C., Saglietti, L., Hirling, H., Sheng, M., and Passafaro, M. (2005) *Mol Cell Neurosci* **28**, 650–660.
72. Setou, M., Seog, D.H., Tanaka, Y., Kanai, Y., Takei, Y., Kawagishi, M., and Hirokawa, N. (2002) *Nature* **417**, 83–87.
73. Borgdorff, A.J., and Choquet, D. (2002) *Nature* **417**, 649–653.
74. Mammen, A.L., Huganir, R.L., and O'Brien, R.J. (1997) *J Neurosci* **17**, 7351–7358.
75. Zhu, J.J., Esteban, J.A., Hayashi, Y., and Malinow, R. (2000) *Nat Neurosci* **3**, 1098–1106.
76. Sans, N., Racca, C., Petralia, R.S., Wang, Y.X., McCallum, J., and Wenthold, R.J. (2001) *J Neurosci* **21**, 7506–7516.
77. Chen, L., Chetkovich, D.M., Petralia, R.S., Sweeney, N.T., Kawasaki, Y., Wenthold, R.J., Bredt, D.S., and Nicoll, R.A. (2000) *Nature* **408**, 936–943.
78. Tomita, S., Chen, L., Kawasaki, Y., Petralia, R.S., Wenthold, R.J., Nicoll, R.A., and Bredt, D.S. (2003) *J Cell Biol* **161**, 805–816.
79. Schnell, E., Sizemore, M., Karimzadegan, S., Chen, L., Bredt, D.S., and Nicoll, R.A. (2002) *Proc Natl Acad Sci U S A* **99**, 13902–13907.
80. Tomita, S., Adesnik, H., Sekiguchi, M., Zhang, W., Wada, K., Howe, J.R., Nicoll, R.A., and Bredt, D.S. (2005) *Nature* **435**, 1052–1058.
81. Fukata, Y., Tzingounis, A.V., Trinidad, J.C., Fukata, M., Burlingame, A.L., Nicoll, R.A., and Bredt, D.S. (2005) *J Cell Biol* **169**, 399–404.
82. Sytnyk, V., Leshchyns'ka, I., Delling, M., Dityateva, G., Dityatev, A., and Schachner, M. (2002) *J Cell Biol* **159**, 649–661.
83. Sytnyk, V., Leshchyns'ka, I., Dityatev, A., and Schachner, M. (2004) *J Cell Sci* **117**, 381–388.
84. Irie, M., Hata, Y., Takeuchi, M., Ichtchenko, K., Toyoda, A., Hirao, K., Takai, Y., Rosahl, T.W., and Sudhof, T.C. (1997) *Science* **277**, 1511–1515.
85. Fukata, M., Fukata, Y., Adesnik, H., Nicoll, R.A., and Bredt, D.S. (2004) *Neuron* **44**, 987–996.
86. Huang, K., Yanai, A., Kang, R., Arstikaitis, P., Singaraja, R.R., Metzler, M., Mullard, A., Haigh, B., Gauthier-Campbell, C., Gutekunst, C.A., Hayden, M.R., and El-Husseini, A. (2004) *Neuron* **44**, 977–986.
87. Washbourne, P. (2004) *Neuron* **44**, 901–902.
88. Nam, C.I., and Chen, L. (2005) *Proc Natl Acad Sci U S A* **102**, 6137–6142.
89. Poncer, J.C., Esteban, J.A., and Malinow, R. (2002) *J Neurosci* **22**, 4406–4411.

# 15

# LATERAL DIFFUSION OF EXCITATORY NEUROTRANSMITTER RECEPTORS DURING SYNAPTOGENESIS

Laurent Groc[*], Martin Heine[*], Laurent Cognet[†], Brahim Lounis[†] and Daniel Choquet[*]

## 1. SUMMARY

The understanding of how receptors accumulate within the developing synapse has captured a lot of attention. Lateral diffusion of surface neurotransmitter receptors has emerged as a key pathway to regulate receptor trafficking and surface distribution, in addition to the receptor cycling between intracellular and plasma membrane pools. The neurotransmitter receptor lateral diffusion depends on several factors, such as interactions with other proteins highly enriched within the synaptic structure. During synaptogenesis, receptors aggregate and cluster within developing synapses. As described in this chapter, lateral diffusion of receptors is likely to play an important role in such process since receptor lateral diffusion is high during synaptogenesis, providing favourable conditions for the "capture" of receptors within synaptic contacts. Indeed, the "diffusion-trap" model for receptor accumulation in developing synapses has now gained experimental support from excitatory synapses, although direct evidence to test this model is still lacking due to the absence of adequate tools to precisely control extrasynaptic receptor lateral diffusion. Within the developing synapse, it also emerges that neurotransmitter release generates surface instability of the receptors and their stabilization requires additional factors.

---

[*]CNRS UMR 5091, Université Bordeaux 2, rue Camille Saint-Saens, Bordeaux cedex 33077, France; Daniel.choquet@pcs.u-bordeaux2.fr

[†]CNRS UMR 5798, Université Bordeaux 1, Talence, France

## 2. INTRODUCTION

One key step in synapse formation is the recruitment and the stabilization of neurotransmitter receptors in the postsynaptic membrane. The neurotransmitter receptors traffic in neurons from their exit of the endoplasmic reticulum to the postsynaptic membrane. However, the routes used by the neurotransmitter receptors to reach the synapse remain unclear. Over the last decade, two paths have emerged. On one hand, neurotransmitter receptors cycle to and from the plasma membrane through exocytosis and endocytosis processes, respectively, and can be inserted to the postsynaptic membrane by vesicle exocytosis. On the other hand, neurotransmitter receptors diffuse at the surface of neurons (lateral diffusion) and may thus diffuse to the synapse from the extrasynaptic membrane.

Since it appears that most cycling events take place in the extrasynaptic membranes, both cycling and surface trafficking processes are likely involved in the regulation of the number of synaptic neurotransmitter receptors. The first report of neurotransmitter receptor lateral diffusion came from studies in the early 1970s in which acetylcholine (ACh) receptor trafficking was investigated at the surface of cultured muscle fibres[1,2]. But it was only 30 years later that the lateral trafficking of neurotransmitter receptors at the surface of neurons was reported[3–6]. Since then continuous improvement of the tracking technique using live imaging has been achieved, giving to neurobiologists the tools to localize fluorescent molecules with sub-wavelength precision and to identify receptor sub-populations. The use of nanometre-sized ligand–fluorophore complexes has even made possible to track targets within confined cellular compartments, such as the synaptic cleft. Since excitatory as well as inhibitory receptors exchange between extrasynaptic and synaptic membranes, the regulation of receptor numbers at postsynaptic sites is therefore likely to also depend on lateral diffusion. Moreover, the role of neurotransmitter receptor lateral diffusion during synaptogenesis has been proposed[1,2]. In this chapter, we discuss (i) the principles and the experimental approaches to analyze receptor lateral diffusion, (ii) the plasma membrane as a space for diffusion, (iii) the evidence that neurotransmitter receptors diffuse at the surface of neurons, and (iv) the lateral diffusion of neurotransmitter receptors during synaptogenesis.

## 3. LATERAL DIFFUSION: PRINCIPLES

From the model proposed by Singer and Nicolson in the early 1970s the plasma membrane has been envisioned as a ‘two-dimensional oriented solution of integral proteins embedded in a viscous phospholipid bilayer’[7]. However it turned out that the measured diffusion coefficients of both lipids and proteins in biological membranes were more than one order of magnitude lower than those predicted from theory or from measurements in reconstituted lipid bilayers[8]. Another expectation from the early model was that receptor size should not affect the protein diffusion within the plasma membrane, but diffusion coefficients of oligomers versus monomers dropped down significantly, arguing against the model[9]. Proteins move passively within the plasma membrane, pushed in different directions by the thermal motion of surrounding molecules, which randomly and continuously change both direction and speed. At this molecular scale, viscous forces dominate movements, so mass and inertia can be neglected. Since Brownian movements are random, one can only predict the probability of a receptor to be at a given position at a certain time point. The average distribution of large numbers of diffusing molecules therefore conforms to defined spatial and temporal laws (Fick's laws). In particular, with time each individual molecule is likely to travel farther from its starting point. This can be expressed by the function of the mean square displacement (MSD) over time t, which provides a linear function for Brownian movements. In a two-dimensional space such as the membrane:

$$\mathrm{MSD} = 4\mathrm{D}t,$$

where $D = K_bTb$ (Einstein's equation) and $K_b$ is the Boltzmann constant, $T$ the absolute temperature, and $b$ is the mobility defined as the velocity produced by a unit force. For a sphere of radius $R$, in a medium of viscosity $\mu$, $b = 1/6\pi\mu R$ (Stoke's law). Membrane proteins diffuse in a two-dimensional space and this equation can therefore not be applied. For a model in which the protein is considered as a cylinder moving in a thin sheet of fluid, its mobility varies as the logarithm of its radius[10]. For a cylinder of radius $R$ and height $h$, embedded in a viscous (lipid) layer of thickness $h$ and viscosity $\mu$, both surfaces are in contact with a less viscous (aqueous) phase with a viscosity $\mu'$, the translational diffusion coefficient of the cylindrical particle can be described as:

$$D = (KT/4\pi\mu h)(\log(\mu h/\,\mu' R) - y)$$

where $y$ is Euler's constant. The weak dependence of $D$ on $R$ has been experimentally shown[11] and this explains why clusters of receptors can diffuse at the same rate as individual receptors. However, as mentioned earlier for proteins in biological membranes, the measured values of $D$ are lower than is expected from the viscosity determined from measurements of lipid diffusion. Experimental evidence has now shown that interactions with obstacles in or beneath the membrane are mostly responsible for the unexpected low value of $D$ (see Section 5). As shown in the equation above, temperature is a critical factor that influences diffusion within membrane. Experimental evidence has indeed shown that temperature change of 10–15°C induce approximately 2-fold change in protein diffusion coefficient within plasma membrane[12,13]. Such factor is thus critical and it has thus to be taken into account for biological experiments.

## 4. LATERAL DIFFUSION OF RECEPTORS: EXPERIMENTAL APPROACHES

The first experimental approach that measured protein surface trafficking was the fluorescence recovery after photobleaching (FRAP) technique[2]. In this approach, molecules of interest are tagged with a fluorophore (e.g. fluorescent irreversible antagonist or antibody). A small area of the cell is quickly photobleached by intense excitation light and the rate of fluorescence recovery in the excited spot is monitored over time. The recovery then depends on the protein mobility from the area surrounding the bleached spot. Although this technique unravels the surface diffusion of several proteins, it only gives bulk estimates of protein mobility, as the movements of thousands of molecules are averaged and it has limited spatial resolution (~200 nm). The emergence of single particle tracking (SPT) experiments allowed real-time monitoring of the movement of individual or a small group of proteins[14]. In the SPT approach, sub-micrometre-sized particles are bound to the protein of interest, such as neurotransmitter receptors, through ligands that recognize the extracellular domain[14–16]. These ligands, such as antibodies, are either passively adsorbed or covalently linked onto the particle, which can be made of latex (0.1–1 µm diameter), gold (40 nm diameter) or silica. Particle diffusion, and therefore diffusion of the underlying receptors, is imaged by video-enhanced differential interference contrast microscopy or fluorescence imaging. The relative position of the particle is then measured with an accuracy of 1–10 nm. The size of the tag does not largely influence the movements, which are dominated by Brownian and viscous forces because membrane viscosity is approximately 500-fold greater than that of extracellular fluids. The membrane-anchored receptors by themselves slow down the particles and not the opposite. One clear disadvantage of such approach in neurons is that it allows only the tracking of extrasynaptic receptors, as the size of particles excludes the tracking of receptors within the synaptic cleft, which is approximately 20 nm wide for excitatory synapses. An alternative approach is thus to use particles of small diameter that can track for long periods of time. This can now be achieved using quantum dots (QDs) that are nanometre-sized semiconductor fluorescent particles providing long-lasting fluorescence emission (virtually no photo-bleaching). However, despite these attempts to minimize the particle size, even the smallest ones (e.g. QDs) used to track receptors within confined spaces have shown limitations, likely because of the complex size[17]. Actual improvements in the QD structure, such as reduction of the core (few nanometres) or the organic shell sizes, will likely overcome this issue. A new photothermal detection method has been developed and allows optical imaging of small gold particles (down to few nm diameter)[18] that could in addition be used to track receptors inside the synaptic cleft.

To investigate tracking of individual receptors in confined space the single molecule detection (SMD) of labelled antibodies directed against extracellular epitopes of the protein of interest has been used. SMD has emerged with the improvement of charge-coupled device (CCD) camera sensitivity and it allows imaging of single synthetic fluorophore (e.g. cyanine 3 or 5) or fluorescent proteins (e.g. GFP and DsRed) under epifluorescence[19–21]. To gather sufficient photons to image one dye molecule, the preparation must be excited at saturation through a defocused laser. However, this accelerates substantially the photobleaching, and thus single dye molecules can usually be imaged for only few seconds (at 30 Hz rate acquisition). Thus, the SMD approach provides the advantage of small probe size that enters confined space but provides the disadvantage of a limited tracking time. Finally, an alternative approach, the fluorescence correlation spectroscopy (FCS) has been developed. Using FCS, the mobility of individual fluorescent molecules is measured as their residence time in a small confocal illumination volume. The faster molecules diffuse, the shorter their residence time is in the excitation volume. As with FRAP, this approach gives measurements of the population mobility. Furthermore, it is limited to the measurement of rapidly moving molecules, as the fluorophore must remain active during the whole period of residence in the measurement volume and not photobleach.

## 5. LATERAL DIFFUSION WITHIN THE PLASMA MEMBRANE: MODELS

Using all of the approaches mentioned above, movement of membrane proteins, such as neurotransmitter receptors, can be classified into five main categories: stationary, free diffusion, directed diffusion, confined and corralled diffusion[22]. Hypotheses that the plasma membrane is compartmentalized and form microdomains have thus emerged. Such microdomains can consist of sub-membranous fences, mobile obstacles and variations in membrane fluidity in sub-compartments[23–25]. Many membrane proteins diffuse rapidly within a constrained space for short periods of time and then escape from these areas[24,26]. FRAP experiments[27] and the measurement of the barrier-free path of transmembrane protein movements by SPT[28] indicates that barriers do indeed restrict the membrane protein lateral mobility. Diffusion is more restricted for proteins with an intracellular tail that protrudes 2–3 nm below the membrane than for truncation mutants with shorter intracellular tails or glycosylphosphatidylinositol (GPI)-anchored molecules[8,29]. All together, these results have favoured the cytoskeleton-fence model[30] in which the diffusion of membrane proteins is dependent on the imperfect and fluctuating network of the membrane-associated portion of cytoskeletal proteins. Escape of proteins from these domains can be accomplished by hopping the corral fence[31]. However, the cytoskeleton-fence model cannot account for all of the experimental data since even the diffusion of GPI-anchored proteins and phospholipids (present in the outer leaflet of the plasma membrane) are compartmentalized[32,33]. It was further demonstrated that the compartmentalization of these probes depends on the actin cytoskeleton, but not on the extracellular matrix, extracellular domains of proteins or lipid composition[32]. Such confinement was interpreted as resulting from long-range interactions between membrane proteins, explaining the contradiction from the original membrane model in which the size of the probe was theoretically not influencing the lateral diffusion. The picket-barrier model[32] emerged as a variation on the general scheme of diffusion in the presence of obstacles in which transmembrane proteins anchored to the membrane skeleton are called pickets and act as posts along the membrane skeleton fence (Figure 15.1).

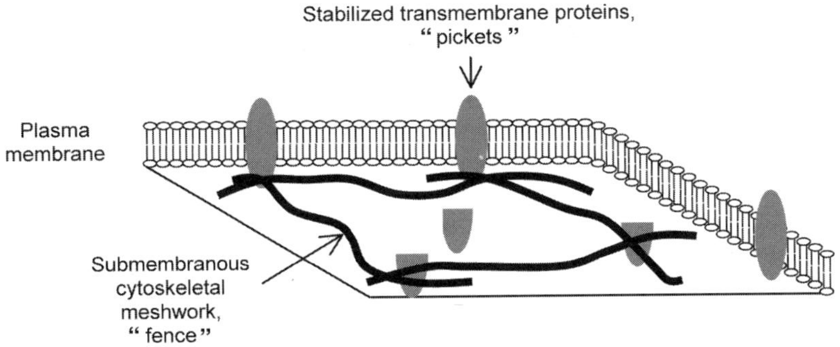


**Figure 15.1.** Schematic View of the Plasma Membrane Compartmentalization. The small compartments are made of sub-membranous cytoskeleton meshwork (e.g. actin), the so-called "fence", together with rows of stabilized transmembrane proteins, the so-called "pickets".

These pickets can then be viewed as effective diffusion barriers due to steric hindrance, and might likely affect phospholipid movements. In conclusion, the plasma membrane can be viewed as a mosaic of sub-micrometre domains, within which diffusion is as fast as expected from theory. Fences that restrict transitions from one compartment to another separate these transient domains, thereby decreasing overall diffusion. In addition, the above mechanisms cannot account for the full topological and dynamic complexity of membrane protein distribution in neurons. Interacting proteins from intra- and extracellular compartments have been proposed to anchor receptors and thus strongly affect their lateral diffusion.

## 6. LATERAL DIFFUSION OF RECEPTORS: EVIDENCES IN NEURONAL MEMBRANE

The distribution of proteins, such as the neurotransmitter receptors and voltage-gated channels, at the neuronal surface is highly compartmentalized. The most striking example of such compartmentalization in neurons is between the somatodentritic and axonal compartments. Use of FRAP and SMD has revealed that at the level of axonal initial segment, cytoskeleton-anchored proteins form a membrane "barrier" that prevent protein diffusion[34]. Thus, at mature stages, membrane proteins located in the somatodendritic compartments cannot diffuse to the axonal membrane, implying that the segregation of axonal and somatodentritic membrane proteins occurs either at early stages when the barrier is not fully formed or by intracellular trafficking. At a smaller scale, the synapse is a sub-compartment where neurotransmitter receptors are concentrated and confined within the postsynaptic membrane[13]. Studies that investigated the lateral diffusion of neurotransmitter receptors have shown that receptors diffuse laterally with different characteristics. To date, the lateral diffusion at the neuronal surface has been demonstrated for ionotropic alpha-amino-3-hydroxy-5-methyl-4-isoxazole propionic acid (AMPA)[13,17,35] (Figure 15.2; Colorplate 9), *N*-methyl-D-aspartate (NMDA)[17,36], metabotropic GluR5[37], glycine[38,39], $GABA_A$[40,41], and β2-adrenergic receptors[42]. Until recently the synaptic fluctuation in receptor number was only envisioned as cycling between surface and intracellular compartments[43,44]. However, a number of observations indicate that most membrane traffic occurs outside synapses. First, sites of endocytosis, revealed through coated pit formation, are observed mainly outside the synaptic areas using both light and electron microscopy[45,46]. Second, there are examples of synaptic glutamate receptors that are not inserted directly at synapses during constitutive or activity-dependent exocytosis[47]. Thus receptors have to laterally diffuse to the "cycling" pits (Figure 15.2A Colorplate 9). The glutamate receptor trafficking has received a lot of attention probably because changes in glutamate receptor number at the synapse form the basis for synaptic plasticity processes involved in many key cognitive tasks (e.g. learning and memory). The excitatory synaptic transmission is mostly mediated by glutamate which activates postsynaptic ionotropic (AMPARs, NMDARs, kainate receptors[48] or G protein-coupled[49] receptors. Proteomic approaches have revealed that the glutamate receptors interact both directly and indirectly with a number of intracellular proteins[50–52] concentrated within the PSD, i.e. the scaffolding proteins. Of particular interest, stargazin has been proposed to be an auxiliary sub-unit of AMPARs[53], to link PSD-95 and AMPARs[54], and to control the number of synaptic AMPARs. Glutamate itself regulates synaptic

AMPAR number by increasing their internalization[43,55]. Consistently, glutamate application increases the lateral diffusion of AMPARs which leave the synapse to the extrasynaptic membrane followed by an entry in the constitutive endocytotic pathway[13,56]. Furthermore, glutamate induces a dissociation of AMPARs from transmembrane AMPA receptor regulatory proteins (TARP). Altogether, these data point to the need for a pool of diffusive extrasynaptic receptors to regulate synaptic receptor number, and synapses may then envisioned as donors or acceptors of exchanged receptors (Figure 15.2A).

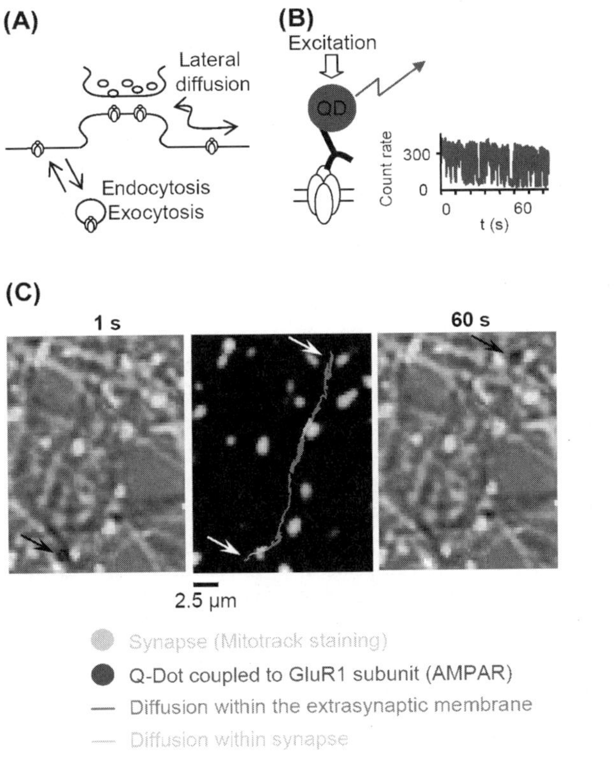


**Figure 15.2.** Receptor Lateral Diffusion at the Surface of Live Neurons Revealed by Single Particle Tracking. (A) Schematic representation of the two main pathways for neurotransmitter receptor trafficking. First, neurotransmitter receptors traffic to and from the plasma membrane by exocytosis and endocytosis, respectively. Second, neurotransmitter receptors diffuse laterally within the plasma membrane and they can even exchange between the extrasynaptic and synaptic membranes. (B) Quantum dot-based single particle tracking is based on the coupling between a single quantum dot and the receptor of interest via an antibody directed against the extracellular domain of the receptor. Note that the fluorescence emission of the quantum dot is stable over time, although blinking occurs (right panel). (C) Trajectories of a GluR1-containing AMPAR (60 s duration). Neurites are shown on right and left panels (differential interference contrast images). Note that the AMPAR diffused along a neurite and crossed two synaptic areas over this time period. See Colorplate 9.

## 7. RECEPTOR LATERAL DIFFUSION DURING SYNAPTOGENESIS

The formation of a synapse is thought to be a multi-step process initiated shortly after initial axo-dendritic contact formation and requiring the clustering of numerous partners. Over the last decades, the glutamate synapse formation has retained a lot of attention and a model has then emerged on how pre- and postsynaptic elements are sequentially pull together[57]. First, cell adhesion molecules (e.g. cadherins) form complexes, tightening together the pre- and postsynaptic membranes (see Chapters 4–10). Since surface adhesion molecules are highly diffusive[29,35], the clustering of these proteins within developing synapses is possibly due to lateral diffusion process. Second, dense-core vesicles (~80 nm) that could be precursors of the active zone are recruited to the adhesion site. Third, on the postsynaptic side, the scaffolding molecules are recruited, elaborating the PSD; the glutamate receptors being recruited later. In the following paragraphs we discuss the potential role of lateral diffusion of excitatory neurotransmitter receptor in this process.

### 7.1. Nicotinic Acetylcholine Receptor Lateral Diffusion During the Neuromuscular Junction Formation

During the neuromuscular junction formation, AChRs that are initially dispersed all over the plasma membrane become highly concentrated in the synaptic membrane. The clustering of AChRs within the synaptic cleft is mainly dependent on the extracellular glycoprotein agrin, which is released by motoneurons at the neuromuscular junction[58]. The actual model for the neuromuscular junction formation is that ACh, released by the motoneuron, destabilizes synaptic AChRs, and an important role of agrin is to counteract such "anti-synaptogenic" effect of ACh[58]. As previously mentioned, the initial reports of neurotransmitter receptor lateral diffusion described the diffusion of AChRs within the plasma membrane of developing myotubes[1,59–61]. It was shown that approximately 50% of alpha-bungarotoxin sensitive AChRs are mobile, and the mobile fraction diffuse on average at 0.1–0.01 $\mu m^2/s$ (comparable to average diffusion coefficient for other receptor types). The surface diffusion was modulated by temperature changes (e.g. switch from 35° to 22°C) consistent with the fact that receptor diffusion is due to the thermal motion of surrounding molecules[12] (see Section 3). Interestingly, AChR lateral diffusion is higher in immature myotubes when compared to mature and well-interconnected ones. This led to a developmental model in which AChRs freely diffuse in the plasma membrane and are continuously "trapped" and stabilized by agrin at the site of developing synaptic contact[12,60]. It was even shown *in vivo* that AChRs diffuse laterally away and to the neuromuscular junction on an activity-dependent manner, suggesting that the perijunctional receptor pool is both a source and a sink for the junctional receptors[62]. Recently, using the genetically screening approach in *C. elegans*, it has been reported that the clustering of ionotropic AChRs at the synapse requires a protein, lev-10[63], and the stabilization of synaptic AChRs relies on extracellular protein–protein interactions. In conclusion, it emerges that the neuromuscular junction formation relies on the synaptic clustering of laterally diffusive AChRs located in extra/perijunctional membrane. Moreover, within the synapse, AChRs are stabilized by both agrin and interactions with other proteins.

### 7.2. Focus on the Glutamatergic Synapse Maturation

How glutamatergic receptors are incorporated and stabilized into the nascent synapse has been a subject of large interest. Several cellular pathways for glutamate receptor trafficking during development have been proposed. In the developing brain, AMPARs are expressed and are functional at the surface of neuronal progenitors[64–66] before synaptogenesis takes place. Migrating and unconnected neurons express both surface AMPARs and NMDARs[67] and it has been proposed that these receptors are themselves involved in the early step of synaptogenesis by regulating dendritic filopodia movements[68]. How and when are these receptors recruited in the developing synapse? Schematically, the scaffold molecules seem to be recruited first in the postsynaptic side approximately 30 min after the formation of the presynaptic bouton, and then both AMPARs and NMDARs are recruited, suggesting that the presence of scaffold protein is necessary to cluster and retain the receptors[57]. The recruitment of NMDARs was found to be progressive, not stepwise as shown for presynaptic markers, indicating a constant accumulation of individual receptors into the nascent synapse[69]. These results have suggested that the incorporation of glutamate receptors within the postsynaptic membrane relies more on random lateral diffusion of AMPARs from extrasynaptic membrane to synapse rather than the insertion of receptor-containing vesicles. Consistently, during *in vitro* development, the diffusion of extrasynaptic AMPARs is high at the surface of hippocampal neurons[35] (Figure 15.3A,B). Within the synaptic area, AMPAR lateral diffusion markedly decreases over development, indicating that the number of AMPARs that exchange between the extrasynaptic and synaptic membrane is higher early one. Such developmental change can be due to several factors, including the formation of actin skeleton beneath the plasma membrane and the development of spines[70]. Regarding the NR1-containing NMDARs, they also diffuse laterally although at a lower level than AMPARs. Moreover, their surface exchange rate between extrasynaptic and synaptic membrane is consistently high in immature neurons[36]. To what extent these *in vitro* results apply to the *in vivo* situation remains unknown. However, there is now evidence that lateral diffusion of glutamate receptors plays a role *in vivo* in the Drosophila neuromuscular formation. Indeed, new glutamate synapses were found to form *de novo*, and the recruitment of GluR-IIA (related to the mammalian non-NMDAR type) to the newly formed PSDs occurs preferentially via surface diffusion of extrasynaptic receptors[71]. Thus, experimental evidences point toward a critical role of glutamate receptor lateral diffusion in synapse formation, although as for the neuromuscular junction, direct evidence for such a role is still lacking.

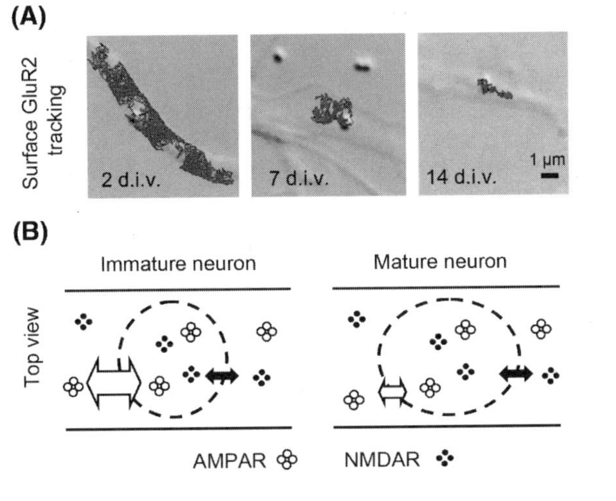


**Figure 15.3.** Lateral Diffusion of AMPARs and NMDARs During Synaptogenesis. (A) Lateral diffusion of extrasynaptic GluR2-containing AMPARs at the surface of developing hippocampal neurons (from 2 to 14 days *in vitro*, d.i.v.). Adapted, with permission, from ref. 35. (B) Schematic representation of the lateral diffusion (double head arrows) of surface glutamate receptors in immature (left panel) and mature (right panel) neurons. The synaptic area is represented by the circle (broken line).

Over the last decade, there have been some controversies about the insertion timing of AMPARs and NMDARs within developing synapses[72,73]. It emerges that, as mentioned above, already from the initial steps of synapse formation both NMDARs and AMPARs are present in the membrane, but the AMPA- and not NMDA-signalling is highly unstable. Indeed, AMPA signalling depends on whether spontaneous or evoked (test frequencies: 0.05–1 Hz) AMPAR currents are assessed[74–76]. In spontaneous conditions AMPAR as much as NMDAR contribute to synaptic activity whereas after evoked synaptic activity only NMDAR synaptic activity is recorded[76]. This stimulation-induced AMPA "silencing" was found in approximately half of the synapses, and was only observed in the neonatal hippocampus[76], consistent with the previously reported age dependent decrease of AMPA silent synapses in the developing hippocampus[77]. Thus, although both receptors are present in developing synapses, AMPARs are prone to be quickly removed from the postsynaptic membrane by synaptic stimulation. It is interesting to note that such effect is similar to the ACh-induced destabilization of AChRs in the neuromuscular junction[58], as if this process of neurotransmitter-induced receptor removal is a common feature of developing synapses. By which pathway are AMPARs removed from the postsynaptic membrane after synaptic stimulation? As already mentioned, AMPARs and NMDARs can undergo endocytosis from the plasma membrane to an intracellular pool. Since the endocytotic rate of glutamate receptors is high at early stages of development[45,46,78], the labile behaviour of AMPARs in immature synapses may be partly due to its high cycling rate. Consistently, AMPARs are more internalized over time than NMDARs[17,55,79,80]. When the lateral diffusion of AMPARs and NMDARs was compared during synaptogenesis, AMPAR lateral diffusion was higher than that of NMDARs[17], indicating that AMPARs are also less stable than NMDARs within the plasma membrane. Furthermore, changes in neuronal activity affected only AMPAR surface diffusion[17]. Thus, AMPAR signalling is highly unstable in immature synapse and this behaviour is possibly due to high lateral diffusion of AMPARs. Future experiments in which electrophysiological and SPT recordings will be coupled will certainly help to better understand this process.

## 8. CONCLUSIONS

Lateral diffusion of surface neurotransmitter receptors has emerged as a key pathway to regulate receptor trafficking and surface distribution, in addition to their cycling between intracellular and plasma membrane pools. The various diffusion patterns that a receptor display over time indicate that its lateral diffusion is modulated by the receptor environment, such as interactions with other proteins (e.g. scaffold proteins) as well as intracellular (e.g. actin skeleton) and extracellular (e.g. proteoglycan matrix) protein networks. During synaptogenesis, receptors aggregate and cluster within developing synapses. As presented in this chapter, lateral diffusion of receptors is likely to play an important role in such process since receptor lateral diffusion is high during synaptogenesis, providing favourable conditions for receptor trapping within synaptic contact. The "diffusion-trap" model for receptor accumulation in developing synapses, initially established for developing neuromuscular junctions, has then gain experimental supports from excitatory and inhibitory synapses. Direct evidence to test the role of receptor lateral diffusion in synapse maturation is however still lacking due to the absence of adequate tools to precisely control extrasynaptic receptor lateral diffusion. The development of several approaches to tackle this issue will certainly unravel receptor trafficking mechanisms involved in synaptic maturation and plasticity. Within the developing synapse, it also emerges that neurotransmitter release generates instability for AChRs and AMPARs, and their stabilization requires additional factors (e.g. agrin for AChR and NARP for AMPAR) or the accumulation of scaffold proteins. Finally, it has been proposed that receptors of the immune synapses reach the contact zone by undirected lateral diffusion, and interactions with other synaptic proteins are required for their stabilization[81]. Altogether we propose that the lateral diffusion of plasma membrane proteins represents a common pathway to build intercellular contacts.

## 9. REFERENCES


1. Axelrod, D., Ravdin, P., Koppel, D.E., Schlessinger, J., Webb, W.W., Elson, E.L., and Podleski, T.R. (1976) *Proc Natl Acad Sci U S A* **73**, 4594–4598.
2. Axelrod, D., Koppel, D.E., Schlessinger, J., Elson, E., and Webb, W.W. (1976) *Biophys J* **16**, 1055–1069.
3. Meier, J., Vannier, C., Sergé, A., Triller, A., and Choquet, D. (2001) *Nature Neurosci* **4**, 253–260.
4. Sergé, A., Fourgeaud, L., Hémar, A., and Choquet, D. (2002) *J Neurosci* **22**, 3910–3920.
5. Borgdorff, A., and Choquet, D. (2002) *Nature* **417**, 649–653.
6. Tovar, K.R., and Westbrook, G.L. (2002) *Neuron* **34**, 255–264.
7. Singer, S.J., and Nicolson, G.L. (1972) *Science* **175**, 720–731.
8. Kusumi, A., Ike, H., Nakada, C., Murase, K., and Fujiwara, T. (2005) *Semin Immunol* **17**, 3–21.
9. Murase, K., Fujiwara, T., Umemura, Y., Suzuki, K., Iino, R., Yamashita, H., Saito, M., Murakoshi, H., Ritchie, K., and Kusumi, A. (2004) *Biophys J* **86**, 4075–4093.
10. Saffman, P.G., and Delbruck, M. (1975) *Proc Natl Acad Sci U S A* **72**, 3111–3113.
11. Kucik, D.F., Elson, E.L., and Sheetz, M.P. (1999) *Biophys J* **76**, 314–322.
12. Axelrod, D., Ravdin, P., Koppel, D.E., Schlessinger, J., Webb, W.W., Elson, E.L., and Podleski, T.R. (1976) *Proc Natl Acad Sci U S A* **73**, 4594–4598.
13. Tardin, C., Cognet, L., Bats, C., Lounis, B., and Choquet, D. (2003) *EMBO J* **22**, 4656–4665.
14. Saxton, M.J., and Jacobson, K. (1997) *Annu Rev Biophys Biomol Struct* **26**, 373–399.
15. Schnapp, B.J., Gelles, J., and Sheetz, M.P. (1988) *Cell Motility Cytoskeleton* **10**, 47–53.
16. Geerts, H., De Brabander, M., Nuydens, R., Geuens, S., Moeremans, M., De Mey, J., and Hollenbeck, P. (1987) *Biophys J* **52**, 775–782.
17. Groc, L., Heine, M., Cognet, L., Brickley, K., Stephenson, F.A., Lounis, B., and Choquet, D. (2004) *Nat Neurosci* **7**, 695–696.
18. Boyer, D., Tamarat, P., Maali, A., Lounis, B., and Orrit, M. (2002) *Science* **297**, 1160–1163.
19. Dickson R.M., Norris D.J., Tzeng Y.L., and Moerner W.E., (1996) *Science* **274**, 966–969.
20. Schmidt, T., Schutz, G.J., Baumgartner, W., Gruber, H.J., and Schindler, H. (1996) *Proc Natl Acad Sci U S A* **93**, 2926–2929.
21. Harms G.S., Cognet L., Blab G.A., Lommerse P.H.M., Spaink H.P., and Schmidt, T. (2001) *Biophys J* **80**, 2396–2408.
22. Choquet, D., and Triller, A. (2003) *Nat Rev Neurosci* **4**, 251–265.
23. Jacobson, K., Sheets, E.D., and Simson, R. (1995) *Science* **268**, 1441–1442.
24. Simson, R., Yang, B., Moore, S.E., Doherty, P., Walsh, F.S., and Jacobson, K.A. (1998) *Biophys J* **74**, 297–308.
25. Edidin, M., Kuo, S.C., and Sheetz, M.P. (1991) *Science* **254**, 1379–1382.
26. Sako, Y., and Kusumi, A. (1995) *J Cell Biol* **129**, 1559–1574.
27. Salome, L., Cazeils, J.L., Lopez, A., and Tocanne, J.F. (1998) *Eur Biophys J* **27**, 391–402.
28. Edidin, M., Zuniga, M.C., and Sheetz, M.P. (1994) *Proc Natl Acad Sci U S A* **91**, 3378–3382.
29. Thoumine, O., Saint-Michel, E., Dequidt, C., Falk, J., Rudge, R., Galli, T., Faivre-Sarrailh, C., and Choquet, D. (2005) *Biophys J,* **89**, L40–42.
30. Kusumi, A., and Sako, Y. (1996) *Curr Opin Cell Biol* **8**, 566–574.
31. Edidin, M. (2001) *Trends Cell Biol* **11**, 492–496.
32. Fujiwara, T., Ritchie, K., Murakoshi, H., Jacobson, K., and Kusumi, A. (2002) *J Cell Biol* **157**, 1071–1081.
33. Sheets, E.D., Lee, G.M., Simson, R., and Jacobson, K. (1997) *Biochemistry* **36**, 12449–12458.
34. Winckler, B., Forscher, P., and Mellman, I. (1999) *Nature* **397**, 698–701.
35. Borgdorff, A.J., and Choquet, D. (2002) *Nature* **417**, 649–653.

36. Tovar, K.R., and Westbrook, G.L. (2002) *Neuron* **34**, 255–264.
37. Serge, A., Fourgeaud, L., Hemar, A., and Choquet, D. (2002) *J Neurosci* **22**, 3910–3920.
38. Meier, J., Vannier, C., Serge, A., Triller, A., and Choquet, D. (2001) *Nat Neurosci* **4**, 253–260.
39. Dahan, M., Lévi, S., and Triller, A. (2003) *Science* **302,** 442–445.
40. Thomas, P., Mortensen, M., Hosie, A.M., and Smart, T.G. (2005) *Nat Neurosci* **8**, 889–897.
41. Meissner, O., and Haberlein, H. (2003) *Biochemistry* **42**, 1667–1672.
42. Hegener, O., Prenner, L., Runkel, F., Baader, S.L., Kappler, J., and Haberlein, H. (2004) *Biochemistry* **43**, 6190–6199.
43. Carroll, R.C., Beattie, E.C., von Zastrow, M., and Malenka, R.C. (2001) *Nat Rev Neurosci* **2**, 315–324.
44. Noel, J., Ralph, G.S., Pickard, L., Williams, J., Molnar, E., Uney, J.B., Collingridge, G.L., and Henley, J.M. (1999) *Neuron* **23**, 365–376.
45. Blanpied, T.A., Scott, D.B., and Ehlers, M.D. (2002) *Neuron* **36**, 435–449.
46. Petralia, R.S., Wang, Y.X., and Wenthold, R.J. (2003) *Eur J Neurosci* **18**, 3207–3217.
47. Passafaro, M., Piech, V., and Sheng, M. (2001) *Nat Neurosci* **4**, 917–926.
48. Nusser, Z. (2000) *Curr Opin Neurobiol* **10**, 337–341.
49. Pin, J.P., and Duvoisin, R. (1995) *Neuropharmacology* **34**, 1–26.
50. Kennedy, M.B. (2000) *Science* **290**, 750–754.
51. Walikonis, R.S., Jensen, O.N., Mann, M., Provance, D.W., Jr., Mercer, J.A., and Kennedy, M.B. (2000) *J Neurosci* **20**, 4069–4080.
52. Husi, H., Ward, M.A., Choudhary, J.S., Blackstock, W.P., and Grant, S.G. (2000) *Nat Neurosci* **3**, 661–669.
53. Bredt, D.S., and Nicoll, R.A. (2003) *Neuron* **40**, 361–379.
54. Schnell, E., Sizemore, M., Karimzadegan, S., Chen, L., Bredt, D.S., and Nicoll, R.A. (2002) *Proc Natl Acad Sci U S A* **99**, 13902–13907.
55. Lin, J.W., Ju, W., Foster, K., Lee, S.H., Ahmadian, G., Wyszynski, M., Wang, Y.T., and Sheng, M. (2000) *Nat Neurosci* **3**, 1282–1290.
56. Zhou, Q., Xiao, M., and Nicoll, R.A. (2001) *Proc Natl Acad Sci U S A* **98**, 1261–1266.
57. Garner, C.C., Zhai, R.G., Gundelfinger, E.D., and Ziv, N.E. (2002) *Trends Neurosci* **25**, 243–251.
58. Misgeld, T., Kummer, T.T., Lichtman, J.W., and Sanes, J.R. (2005) *Proc Natl Acad Sci U S A* **102**, 11088–11093.
59. Young, S.H., and Poo, M.M. (1983) *J Neurosci* **3**, 225–231.
60. Poo, M. (1982) *Nature* **295**, 332–334.
61. Schlessinger, J., Koppel, D.E., Axelrod, D., Jacobson, K., Webb, W.W., and Elson, E.L. (1976) *Proc Natl Acad Sci U S A* **73**, 2409–2413.
62. Akaaboune, M., Culican, S.M., Turney, S.G., and Lichtman, J.W. (1999) *Science* **286**, 503–507.
63. Gally, C., Eimer, S., Richmond, J.E., and Bessereau, J.L. (2004) *Nature* **431**, 578–582.
64. Maric, D., Liu, Q.Y., Grant, G.M., Andreadis, J.D., Hu, Q., Chang, Y.H., Barker, J.L., Joseph, J., Stenger, D.A., and Ma, W. (2000) *J Neurosci Res* **61**, 652–662.
65. Sah, D.W., Ray, J., and Gage, F.H. (1997) *J Neurobiol* **32**, 95–110.
66. Hagimura, N., Tsuzuki, K., Iino, M., Takatsuru, Y., Yoshida, Y., Kishi, S., and Ozawa, S. (2004) *Brain Res Dev Brain Res* **152**, 213–223.
67. Tyzio, R., Represa, A., Jorquera, I., Ben-Ari, Y., Gozlan, H., and Aniksztejn, L. (1999) *J Neurosci* **19**, 10372–10382.
68. Matsuzaki, M., Ellis-Davies, G.C., Nemoto, T., Miyashita, Y., Iino, M., and Kasai, H. (2001) *Nat Neurosci* **4**, 1086–1092.
69. Bresler, T., Shapira, M., Boeckers, T., Dresbach, T., Futter, M., Garner, C.C., Rosenblum, K., Gundelfinger, E.D., and Ziv, N.E. (2004) *J Neurosci* **24**, 1507–1520.
70. Fiala, J.C., Feinberg, M., Popov, V., and Harris, K.M. (1998) *J Neurosci* **18**, 8900–8911.
71. Rasse, T.M. et al. (2005) *Nat Neurosci* **8**, 898–905.
72. Isaac, J.T. (2003) *Neuropharmacology* **45**, 450–460.
73. Voronin, L.L., and Cherubini, E. (2004) *J Physiol* **557**, 3–12.
74. Myme, C.I., Sugino, K., Turrigiano, G.G., and Nelson, S.B. (2003) *J Neurophysiol* **90**, 771–779.
75. Groc, L., Gustafsson, B., and Hanse, E. (2002) *J Neurosci* **22**, 5552–5562.
76. Xiao, M.Y., Wasling, P., Hanse, E., and Gustafsson, B. (2004) *Nat Neurosci* **7**, 236–243.
77. Durand, G.M., Kovalchuk, Y., and Konnerth, A. (1996) *Nature* **381**, 71–75.
78. Blanpied, T.A., Scott, D.B., and Ehlers, M.D. (2003) *Neurobiol Aging* **24**, 1095–1104.
79. Man, H.Y., Lin, J.W., Ju, W.H., Ahmadian, G., Liu, L., Becker, L.E., Sheng, M., and Wang, Y.T. (2000) *Neuron* **25**, 649–662.
80. Ehlers, M.D. (2000) *Neuron* **28**, 511–525.
81. Iber, D. (2005) *Cell Mol Life Sci* **62**, 206–213.

# Part IV

## SYNAPTIC CYTOSKELETON AND MORPHOGENIC SIGNALING

# 16

# ASSEMBLY OF PRESYNAPTIC ACTIVE ZONES

Thomas Dresbach[†], Anna Fejtová[*], and Eckart D. Gundelfinger[*]

## 1. SUMMARY

The active zone defines the area of the presynaptic plasma membrane where synaptic vesicles (SV) dock, mature, and fuse with the cell membrane in a regulated manner. The active zone is defined by an electron-dense cytomatrix of specialized proteins, the cytomatrix assembled at the active zone (CAZ). Some of these proteins, e.g., RIMs, Munc13s, ERC/CASTs as well as Bassoon and Piccolo are exclusive components of the CAZ, and are thought to mediate specific functions, including the localization and regulation of the SV cycle. Recently, major progress has been achieved in studying the assembly of the active zone and the underlying cytoskeletal matrix, suggesting it is not assembled 'brick-by-brick' as originally assumed. Rather, mounting evidence suggests that active zone components are preassembled within the neuronal soma, presumably at a trans-Golgi compartment, and then transported on distinct precursor vesicles to primordial presynaptic sites where they fuse with the cell membrane. As Piccolo and Bassoon are unique marker proteins of these vesicles they are called Piccolo–Bassoon transport vesicles (PTVs). The present review summarizes evidence for this active zone precursor vesicle hypothesis, which predicts that presynaptic active zones are assembled in a more or less quantal manner.

[†]Institute for Anatomy and Cell Biology, University of Heidelberg, Im Neuenheimer Feld 307, D-69120 Heidelberg, Germany; thomas.dresbach@urz.uni-heidelberg.de
[*]Leibniz Institute for Neurobiology, Brenneckestr. 6, D-39118 Magdeburg, Germany; gundelfi@ifn-magdeburg.de

## 2. INTRODUCTION

Synaptic transmission relies on spatially restricted and mechanistically regulated release of neurotransmitter. What seems like an obvious statement to make translates into cell biological questions of great importance: How is it brought about that SVs, the 50 nm-diameter lipid-bounded organelles that store neurotransmitter, are accumulated selectively at sites of neurotransmitter release, the so-called active zones? What mechanisms restrict their exocytic fusion, the event underlying neurotransmitter release, to these sites? Which molecular mechanisms account for the regulation of kinetics, amount, mode, and probability of neurotransmitter release at a given synapse, i.e. events that are thought to underlie conversion of synaptic transmission into information? Current notions hold that a network of cytoplasmic proteins that we have termed the cytomatrix assembled at the active zones (CAZ)[1], which is a characteristic and specific feature of neurotransmitter release sites, plays a central role in the functional organization of synapses and may in fact mediate most of the above events.

Other chapters in this book deal with questions as to how the timing and the site of active zone formation are determined, and what extracellular and intracellular signals are involved in these events. In this chapter, we focus on questions as to how the components of active zones are assembled to generate the molecular machinery of neurotransmitter release. To this end, we define the individual components and highlight their characteristics. Moreover, we describe recent advances suggesting that active zone components are transported to nascent active zones on the surface of specialized vesicles.

## 3. NERVE TERMINALS

Active zones may be considered as central components of a larger machinery designed to guarantee controlled neurotransmitter release and local maintenance of the contributing components. These components are (a) SVs, (b) active zones, and (c) periactive zones. Clusters of SVs are selectively accumulated close to active zones. Functionally, these SVs can be divided into several pools, depending on their responsiveness to stimulation of transmitter release[2–4]. Exocytic release of neurotransmitter occurs from SVs fusing with the active zone plasma membrane and involves formation of a complex between the SNARE-proteins VAMP, a transmembrane protein of SVs, and syntaxin and SNAP25 in the plasma membrane[2,5]. Not all SVs fuse with the same probability upon nerve stimulation. Acquisition of fusion competence is thought to be controlled critically by proteins of the CAZ.

At some synapses, the active zone does not only release neurotransmitter, but also regenerates fused or fusing SVs locally by clathrin-independent mechanisms termed kiss-and-run and kiss-and-stay[2,3,6]. Usually, synapses are capable of retrieving fused SVs by clathrin-mediated endocytosis, which occurs at the regions peripheral to the active zone, termed periactive zones[2,7]. Such SVs are locally routed back into the SV cloud. CAZ proteins are likely to be involved in linking the exocytic and endocytic zones in presynaptic boutons[8]. As discussed below a major question regarding active zone assembly turns out to be: To what extent are the components of the release machinery recruited simultaneously to nascent synapses?

## 4. ACTIVE ZONES

Active zones, a term coined by Couteaux and Pecot-Dechavassine (1970), are unanimously defined as the sites located along an axon or in a nerve terminal where neurotransmitter release occurs. Such sites can be readily detected morphologically because clusters of dozens to hundreds of SVs are located close to them. Deep-etch electron microscopic studies have revealed that at central nervous system synapses these SV clusters are embedded in a filamentous meshwork of proteins, which has been termed the presynaptic cytomatrix. Key proteins of the presynaptic cytomatrix are synapsins, peripheral membrane proteins of SVs that are thought to reversibly crosslink SVs and to bind them to actin in a phosphorylation-dependent manner[7,9,10]. Those SVs of the cluster that appear to touch the plasma membrane at the release site, generally referred to as "docked vesicles," are embedded in an even thicker meshwork of filaments, which is immediately obvious in electron micrographs as an electron-dense thickening of the cytoplasm and the presynaptic plasma membrane. End face views of this area suggest that the electron-dense material is organized like an egg carton with pyramidal projections surrounding depressions that have just about the size to each harbor an SV. As these structures display a more or less regular hexagonal array they were termed "presynaptic grid"[7,11–13]. Based on ultrastructural observations at the Mauthner cell of teleosts it was hypothesized that the presynaptic grid may undergo activity-dependent deformations and that the release of individual vesicles may lead to distortions in the grid, which in turn limit the release probability for other vesicles[14].

Despite agreement on the concept of active zones there are subtle variances in the terminology. For example, active zones are sometimes defined as the "electron-dense, biochemically insoluble material located at the presynaptic plasma membrane precisely opposite the synaptic cleft" and as "the specialized region of the cortical cytoplasm of the presynaptic nerve terminal that directly faces the synaptic cleft"[2,3]. On the other hand, active zones may be considered as subcellular compartments, including the region of the plasma membrane facing the synaptic cleft plus the underlying dense material and the SVs embedded in it (e.g., ref. 15). For the purpose of this article we employ the latter nomenclature and refer to the specialized dense material specifically as the CAZ[7]. This notion emphasizes that active zones are functional units promoting regulated and spatially restricted exocytosis of neurotransmitter, and it takes into account that transmembrane proteins and lipids of the active zone plasma membrane, a set of SVs, and the molecules of the CAZ interact with each other to orchestrate regulated neurotransmitter release. Using this notion, the "CAZ" is synonymous with "the presynaptic particle web"[16], and "active zones" consist of an area of plasma membrane as well as underlying CAZ material.

## 5. THE PLASMA MEMBRANE AT ACTIVE ZONES

The active zone plasma membrane must be designed (i) to mediate exocytic release of neurotransmitter from SVs, (ii) to allow the local entry of calcium ions from the extracellular space to trigger transmitter release, and (iii) to organize *trans*-synaptic cell adhesion via membrane proteins. In fact, it does contain the target membrane SNARE proteins syntaxin and SNAP25, which by binding to the

SV transmembrane protein VAMP mediate SV exocytosis[5]. Of note, syntaxin and SNAP25 are located all over the neuronal plasma membrane, indicating that these proteins cannot account for the restriction of transmitter release to active zones[17]. The active zone plasma membrane also contains voltage-gated $Ca^{2+}$ channels, although the degree of accumulation with respect to nonactive zone regions is not entirely clear and may vary between synapse types. Theoretical considerations suggest that docked SVs may be randomly distributed at an active zone while calcium channels may be clustered in subregions of the active zone plasma membrane. Distinct distances between SVs and calcium channels might account for distinct neurotransmitter release probabilities. Release competent SVs must be located close to calcium channels (less than 300 nm, presumably less than 50 nm) to account for the short time delay of 0.2 ms between $Ca^{2+}$ entry and exocytosis[3,18,19]. The emerging importance of active zone transmembrane proteins for cell adhesion, synapse formation, and *trans*-synaptic signaling is discussed in other chapters of this book (Chapters 4–10). Of note, synaptic junctions can be purified biochemically, suggesting tight *trans*-synaptic binding interactions. Electron microscopy shows that such purified junctions contain both the CAZ and its postsynaptic equivalent, the postsynaptic density (PSD), as well as fibrous structures spanning the synaptic cleft[7,16]. This suggests that a protein scaffold composed of transmembrane proteins as well as the proteins of the CAZ and the PSD may serve as the backbone of synaptic junctions.

## 6. THE CYTOMATRIX AT THE ACTIVE ZONE (CAZ)

Decades after the CAZ had first been observed by electron microscopy its protein constituents are being identified. Five molecules have meanwhile been identified that are exclusively localized to the CAZ and may be considered as core CAZ components. These are Munc13s, RIM, CAST/ERC, Bassoon, and Piccolo[7,20,21]. These proteins are large multidomain proteins lacking a transmembrane region, which directly or indirectly bind to each other and to several other synaptic proteins, thus potentially generating a supramolecular structure of unusual size. Intriguingly, CAZ proteins also bear crucial importance for several steps of neurotransmitter release.

Munc13 isoforms are essential for evoked neurotransmitter release, presumably by conferring fusion competence to docked SVs[21,22]. In addition, these proteins play a role in synaptic plasticity[23,24]. Likewise, isoforms of RIM are required in normal SV exocytosis and in synaptic plasticity[25,26]. RIM binds a large number of presynaptic proteins, including the CAZ molecules Munc13 and Liprins[27–29]. Liprins in turn interact with receptor tyrosine phosphatases called LAR, which are involved in active zone assembly[30]. CAST/ERC is thought to act as scaffolding protein; it interacts with RIM, Liprin, Bassoon, and Piccolo[31,32]. Bassoon and Piccolo are the largest CAZ proteins with 420 and 550 kDa, respectively. At conventional synapses, Bassoon is required for the function of a subset of synapses[33]. At photoreceptor and inner ear ribbon synapses Bassoon has additional roles in the formation and function of synaptic ribbons, i.e., specialized cytoskeletal structures involved in SV dynamics[15, 34–36].

Thus, the CAZ represents an exceptional accumulation of proteins controlling neurotransmitter release. These proteins directly or indirectly interact with each other and may in fact represent one giant supramolecular complex ultrastructurally visible as the presynaptic grid (see above). Via various interactions they are directly linked to proteins of the exocytic machinery. For example, Munc13 interacts with syntaxin, RIM interacts with the SV-associated protein Rab3a and with the putative calcium sensor protein of SVs, synaptotagmin[7,37] (Figure 16.1; Colorplate 10). The current notion holds that the CAZ confers the spatial and temporal coordination to the exocytosis machinery that is a signature of neurotransmitter release.

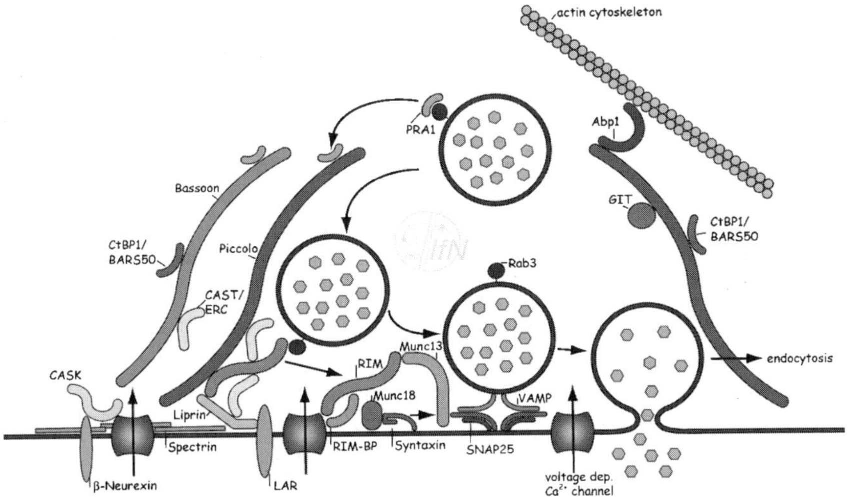


**Figure 16.1.** Molecular Organization of the CAZ. Few CAZ-specific proteins, including the Rab3-interacting molecules (RIMs), the SV vesicle priming factor Munc13, the two related scaffolding proteins Bassoon and Piccolo, and the CAZ-associated structural protein CAST/ERC, are specifically localized at the active zone. They are thought to localize and organize membrane trafficking events of the SV cycle and connect it to the active zone membrane proteins, including voltage-gated $Ca^{2+}$ channels and cell adhesion molecules such as neurexins. Further components that are not exclusive CAZ components include the $Ca^{2+}$/calmodulin kinase domain-containing membrane-associated guanylate kinase CASK, the transcriptional co-repressor CtBP1/BARS50, the RIM-binding proteins (RIM-BP), the prenylated Rab3 acceptor protein PRA1, the ARF-GTPase-activating protein GIT, the receptor tyrosine phosphatase LAR and its interacting protein Liprin, components of the SNARE complex and its control elements (e.g., Munc18). The interaction between Piccolo and the actin-binding protein Abp1 is thought to link the active zone to the neighboring endocytic zone. For further details see Gundelfinger, Altrock and Fejtova, Active zone in: *Encyclopedic Reference of Neuroscience* (Field editor: M. Takahashi, Springer Berlin, in press). The picture was taken from the website of the Institute for Neurobiology (http://www.ifn-magdeburg.de/en/departments/neurochemistry_and_molecular_biology). See Colorplate 10.

## 7. ACTIVE ZONE ASSEMBLY AS STUDIED IN MAMMALS: IDENTIFICATION OF PRECURSOR VESICLES

Functional active zones can form within 30–60 min of initial contact between axon and dendrite[38,39]. Piccolo and Bassoon are among the earliest proteins to appear at nascent synapses, and arrive simultaneously with or prior to functional SVs, consistent with a role of these proteins in synapse assembly[40,41]. How are active zone proteins transported to nascent synaptic sites? In particular, how are CAZ molecules transported and what is their contribution to synapse assembly?

Recent studies have addressed this question in detail and identified a novel type of vesicular organelles that appear to represent mobile active zone precursors. Using antibodies against either Piccolo or Bassoon a previously unknown type of vesicular organelles could be affinity-purified from embryonic rat brain (day E18)[41,42]. Ultrastructurally, the purified organelles resembled vesicles detected in electron microscopy studies in axons of developing spinal chord[43], and were later also detected in axons of cultured neurons[41]. These vesicles are 80 nm in diameter, have an electron-dense lumen, and are coated with electron-dense projections highly reminiscent of CAZ material of active zones[41,43]. This morphology raised the possibility that the vesicles carry CAZ material on their cytoplasmic surface to the presynapse. Indeed, the projections emanating from these vesicles are immunopositive for Bassoon and Piccolo in electron microscopy[41], and were therefore termed Piccolo–Bassoon transport vesicles (PTVs). Initial biochemical analysis of purified PTVs revealed that they carry not only Piccolo and Bassoon, but also the plasma membrane SNAREs Syntaxin and SNAP25, and *N*-cadherin, a cell adhesion protein found at mature synapses. By contrast, SV proteins, such as synaptophysin, synaptotagmin, synaptobrevin/VAMP2, or vesicular neurotransmitter transporter GAT1, were not detected on PTVs. Further biochemical analysis revealed that PTVs indeed contain a comprehensive set of active zone proteins, including RIM1, Munc13, N-type voltage-gated calcium channels, and the syntaxin binding protein Munc18 (Figure 16.2)[41,42].

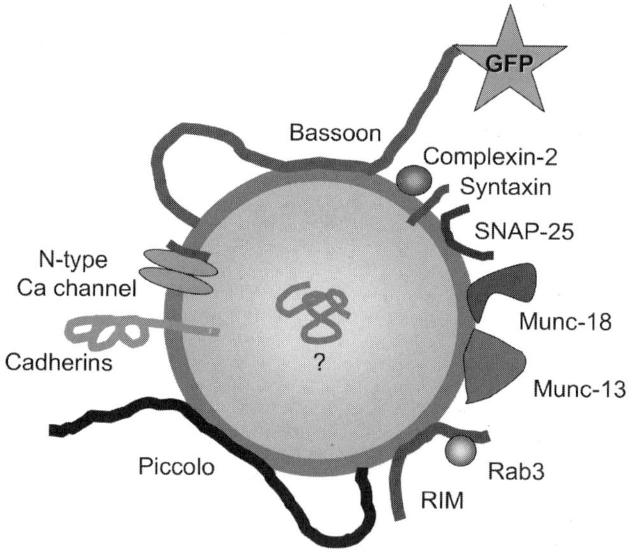


**Figure 16.2.** Proposed Molecular Composition of a Piccolo–Bassoon Transport Vesicle (PTV). In addition to the exclusive CAZ proteins Bassoon, Piccolo, RIMs and Munc-13s, the postulated active zone precursor vesicle incorporates integral membrane proteins, such as N-type calcium channels and *N*-cadherin, and proteins that control priming and fusion of SVs[41,42]. EGFP-tagged Bassoon has been very instrumental for the analysis of PTV trafficking.

Based on this striking composition the active zone precursor vesicle hypothesis emerged, which predicts that (i) PTVs may allow for transport of the entire complement of active zones proteins, and (ii) exocytosis of very few such vesicles could generate primordial active zone, suggesting a quantal and rapid assembly of presynaptic sites for neurotransmitter release[41,42] (Figures 16.3 and 16.4).

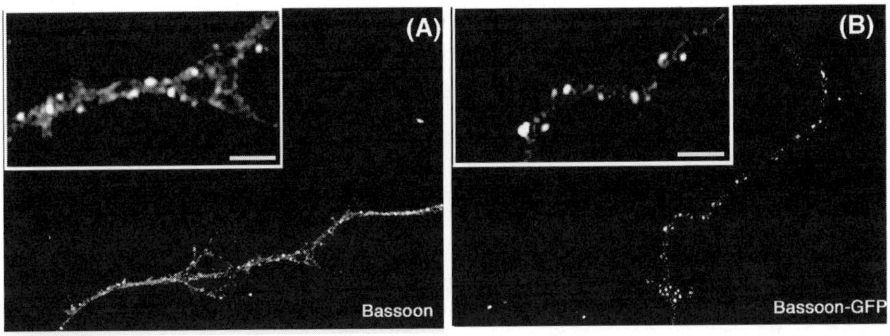


**Figure 16.3.** Visualization of PTVs in Immature Primary Neurons. (A) Immunofluorescence localization of Bassoon in an axon. The punctate distribution most likely represents Bassoon bound to PTVs. The inset shows a three-fold magnification, the scale bar represents 500 nm. (B) Fluorescence of GFP-Bassoon in a transfected neuron. The recombinant protein is distributed similar to the endogenous protein. The size and distribution of the puncta are reminiscent of the endogenous protein. Such GFP-Bassoon puncta are highly mobile[42] and can be recruited to nascent synapses[38]. The neurons were maintained for 4 days in culture.

The appearance of cadherins on PTVs is of particular interest as a recent study has shown that cadherins in conjunction with β-catenin is involved in presynaptic assembly and localization of SVs[44]. Deletion of β-catenin in hippocampal primary neurons results in a reduced size of reserve pool SVs and in an impaired response to repetitive synaptic stimulation. This cadherin/β-catenin function seems to involve the PDZ binding motif of β-catenin and its interaction with a downstream PDZ domain protein.

## 8. QUANTAL TRANSPORT OF PRIMORDIAL ACTIVE ZONES VIA PTVS

The results of several studies strongly support the idea that active zone components are transported in packages. Immunofluorescence localization revealed a punctate distribution and colocalization of Bassoon and Piccolo in axons of immature primary neurons in culture, consistent with an early association of these proteins with vesicles[41,45]. Likewise RIM showed a punctate distribution, but only an about 50% colocalization with Bassoon and Piccolo, suggesting that RIM might be associated with distinct organelles in addition[42]. The functional characterization of PTVs has been promoted by the use GFP-tagged versions of Bassoon, which has revealed to be a reliable marker for PTVs in live imaging studies performed on young neurons[38,42], and for synapses in mature neurons[46]. For example, using GFP-Bassoon as a tool, PTVs were shown to be highly mobile in immature neurons, consistent with the notion that they serve as transport entities[38,42]. Combining these live imaging studies with immunofluoresence localization studies of endogenous proteins in quantitative analyses[38,42] has generated the following emerging picture: There are small, mobile as well as larger, stable puncta positive for CAZ markers. The smaller puncta are typically negative, the larger puncta are typically positive for markers of mature synapses. The larger, synaptic puncta are on average two-fold brighter in fluorescence as compared to the smaller, mobile puncta. This would be consistent with mature active zones being built from material delivered by transport vesicles, and that typically more than one PTV generates an active zone.

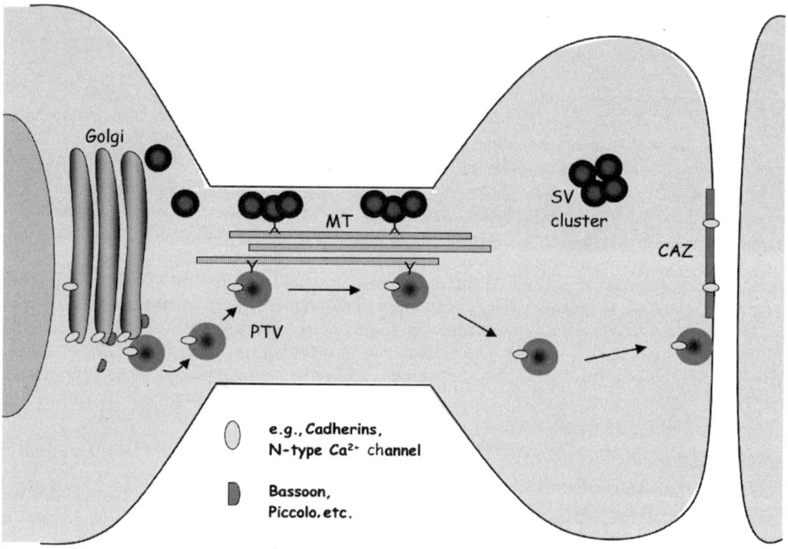


**Figure 16.4**. The Active Zone Transport Vesicle Hypothesis: PTVs are Thought to Bud from the Trans-Golgi-Complex. At least Bassoon and Piccolo are packaged onto the cytoplasmic surface of PTVs already at this site, most likely at a sub-compartment of the trans-Golgi-network, and this event seems obligatory for presynaptic targeting of these proteins[47]. At the same site, PTVs may receive the entire complement of proteins (Figure 16.2). PTVs are supposed to travel along the axon via microtubule-based transport and are directed to nascent active zones by unknown signaling mechanisms. Fusion of PTVs (usually 2–3; refs. 38,42) with the axonal plasma membrane results in formation of an active zone. The topology of membrane fusion predicts that the CAZ proteins traveling on PTVs become located immediately underneath the plasma membrane, their contents are released, and transmembrane proteins become incorporated into the plasma membrane. SV precursors are also generated at the Golgi complex (left) and move along the axon along microtubules using independent pathways. (adapted from Dresbach, Altrock, Gundelfinger, Neuroforum 03/2003).

But does an important postulate of the PTV hypothesis hold true, i.e., that a PTV delivers an entire set of active zone molecules? The answer appears to be yes based on further quantitative studies and modeling approaches: While there is a 1:2 ratio between CAZ-protein content of PTVs and synapses on average, both PTVs and mature synapses have a broad spectrum of individual fluorescence intensity levels. Taking into account the relative occurrence of individual fluorescence intensities of Bassoon associated with PTVs and synapses, reveals a striking correlation: The synaptic content of Bassoon can be accounted for by incorporation of integer numbers of fluorescence units into synapses. Specifically, two, three, and more rarely four fluorescence units of Bassoon appear to account for the Bassoon content of mature synapses in hippocampal primary neurons. The same applied to Piccolo and RIM. Moreover, such integer multiples calculated for Bassoon correctly predicted the synaptic content of RIM in the same experiment. These quantitative data strongly support the hypothesis that active zones may be assembled from unitary insertion of a small number of PTVs into the plasma membrane.

Further time-lapse imaging studies have provided additional support for this notion based on analysis of individual synapse formation events, as opposed to analysis of synapse populations. In these experiments, the occurrence of GFP-tagged Bassoon was imaged during formation of novel synapses. Synapse formation was monitored live by repeated application of the styryl dye FM4-64, which in the paradigms applied is taken up into and released from SVs at active synapses only. Mobile puncta of GFP-tagged Bassoon, presumably PTVs, were shown to arrive at sites where a novel synapse appeared in the course of the time lapse. Strikingly, when mobile puncta became stable at a new synapse, the fluorescence of the tagged protein increased stepwise at this site. In particular, this stepwise increase occurred from incorporation of two, three, four, or five puncta. These studies corroborate the active zone precursor hypothesis based on analysis of individual synapse formation events. Notably, several postsynaptic proteins were shown to be incorporated into newly forming synapses in a more gradual manner suggesting that postsynaptic assembly may differ from active zone assembly[38]. However, as discussed in detail by P.E. Washbourne (Chapter 14) various mechanisms may contribute to postsynaptic assembly. It should be noted that, to date, Bassoon and Piccolo seem to be the only CAZ components that are exclusively transported via PTVs, while others might be transported to the presynapse on various pathways. Moreover, CAZ formation via PTVs seems to be a major if not exclusive process during the major period of synaptogenesis during brain development. An important question for the future is: How is synapse formation and turnover mediated in the differentiated nervous system?

## 9. THE SITE OF PTV GENERATION AND CAZ PRECURSOR FORMATION

If PTVs carry unitary sets of CAZ molecules on their cytoplasmic surface, and possibly also of active zone transmembrane proteins, where are these units generated and how are they packaged onto nascent PTVs? Recent data suggest that formation of CAZ-precursor complexes may already occur at the level of the trans-Golgi-network (TGN): Bassoon and Piccolo, bona fide CAZ markers, are associated with a subcompartment of the TGN. Blocking vesicle biogenesis at the level of the Golgi apparatus prevents axonal trafficking of Bassoon and Piccolo, and causes retention of these CAZ proteins in the neuronal soma. Likewise, perturbing Golgi association of Bassoon specifically by blocking its Golgi association capacity prevents trafficking of Bassoon to synapses[47]. The apparent importance of Golgi association and vesicle-based transport for axonal trans-transport and for synaptic delivery of Bassoon and Piccolo is intriguing in view of the nature of CAZ proteins as nontransmembrane proteins, for which a necessity for vesicle association is not obvious per se. One reason for this pathway could be that active zone generation demanded an event of simultaneous packaging of CAZ molecules during vesicle biogenesis, e.g., to ensure a more or less fixed ratio of distinct molecules per vesicle.

## 10. ADDITIONAL PATHWAYS OF ACTIVE ZONE ASSEMBLY

While some CAZ molecules, such as Bassoon and Piccolo, seem to be specifically carried by PTVs, other sets of proteins may be transported by distinct vesicles. A study using the SV protein VAMP fused to GFP for time-lapse imaging in cultured hippocampal neurons revealed a distinct type of mobile, vesicular packet that appeared to contain SV proteins, voltage-gated calcium channels and proteins related to SV endocytosis. Such packets are stabilized at young synaptic contacts[48]. Electron microscopy revealed a heterogeneous population of vesicular and tubulovesicular structures at such sites. Therefore, various presynaptic components may be transported on such vesicles or packets of vesicles. Membrane remodeling may subsequently produce both the uniform SVs typical of mature synapses, and contribute to formation of active zone and periactive zone plasma membrane domains.

## 11. CONCLUSIONS AND FUTURE DIRECTIONS

In conclusion, it is an emerging scenario that active zone components are transported by vesicular organelles which may be used as modular units in synapse formation in developing neuronal networks. In the future, the following questions may move into the center of this field of research: what exactly is the subcellular compartment where active zone components meet initially? Do all active zone components meet there? Which ones are incorporated later? How many distinct pathways for trafficking of active zone components exist? Moreover, how are CAZ-proteins transported in invertebrates and in specialized synapses such as ribbon synapses? How are PTVs transported in general and in distinct systems, how are they targeted to nascent synaptic sites? Furthermore, do PTVs really fuse with the plasma membrane during active zone formation? Which molecules are involved, how is this regulated? Finally, how are synaptogenesis and synaptic turnover mediated in the adult nervous system? Do PTVs contribute to plasticity of mature synapses?[*]

## 12. REFERENCES

1. Garner, C.C., Kindler, S., and Gundelfinger, E.D. (2000) *Curr Opin Neurobiol* **10,** 321–327.
2. Murthy, V.N., and De Camilli, P. (2003) *Annu Rev Neurosci* **26,** 701–728.
3. Sudhof, T.C. (2004) *Annu Rev Neurosci* **27,** 509–547.
4. Sara, Y., Virmani, T., Deak, F., Liu, X., and Kavalali, E.T. (2005) *Neuron* **45,** 563–573.
5. Rizo, J., and Sudhof, T.C. (2002) *Nat Rev Neurosci* **3,** 641–653.
6. An, S., and Zenisek, D. (2004) *Curr Opin Neurobiol* **14,** 522–530.
7. Dresbach, T., Qualmann, B., Kessels, M.M., Garner, C.C., and Gundelfinger, E.D. (2001) *Cell Mol Life Sci* **58,** 94–116.
8. Gundelfinger, E.D., Kessels, M.M., and Qualmann, B. (2003) *Nat Rev Mol Cell Biol* **4,** 127–139.
9. Landis, D.M., Hall, A.K., Weinstein, L.A., and Reese, T.S. (1988) *Neuron* **1,** 201–209.
10. Hirokawa, N., Sobue, K., Kanda, K., Harada, A., and Yorifuji, H. (1989) *J Cell Biol* **108,** 111–126.
11. Bloom, F.E., and Aghajanian, G.K. (1968) *J Ultrastruct Res* **22,** 361–375.

---

[*]The work was supported by DFG (DR373/3-1, DR373/3-2, SFB 426/A1, GRK 1167), European Commission (SynScaff), the Land Saxony-Anhalt (Schwerpunktprogramm N2), the Fonds der Chemischen Industrie, Max-Planck-Award by the Max Planck Society, and the Alexander von Humboldt Society to EDG. We thank Werner Zuschratter for providing picture material.

12. Pfenninger, K., Akert, K., Moor, H., and Sandri, C. (1972) *J Neurocytol* **1,** 129–149.
13. Sur, C., Triller, A., and Korn, H. (1995) *J Comp Neurol* **351,** 247–260.
14. Triller, A., and Korn, H. (1985) *J Neurocytol* **14,** 177–192.
15. Zhai, R.G., and Bellen, H.J. (2004) *Physiology (Bethesda)* **19,** 262–270.
16. Phillips, G.R., Huang, J.K., Wang, Y., Tanaka, H., Shapiro, L., Zhang, W., Shan, W.S., Arndt, K., Frank, M., Gordon, R.E., Gawinowicz, M.A., Zhao, Y., and Colman, D.R. (2001) *Neuron* **32,** 63–77.
17. Garcia, E.P., McPherson, P.S., Chilcote, T.J., Takei, K., and De Camilli, P. (1995) *J Cell Biol* **129,** 105–120.
18. Meinrenken, C.J., Borst, J.G., and Sakmann, B. (2002) *J Neurosci* **22,** 1648–1667.
19. Meinrenken, C.J., Borst, J.G., and Sakmann, B. (2003) *J Physiol* **547,** 665–689.
20. Garner, C.C., Zhai, R.G., Gundelfinger, E.D., and Ziv, N.E. (2002) *Trends Neurosci* **25,** 243–251.
21. Rosenmund, C., Rettig, J., and Brose, N. (2003) *Curr Opin Neurobiol* **13,** 509–519.
22. Varoqueaux, F., Sigler, A., Rhee, J.S., Brose, N., Enk, C., Reim, K., and Rosenmund, C. (2002) *Proc Natl Acad Sci U S A* **99,** 9037–9042.
23. Rhee, J.S., Betz, A., Pyott, S., Reim, K., Varoqueaux, F., Augustin, I., Hesse, D., Sudhof, T.C., Takahashi, M., Rosenmund, C., and Brose, N. (2002) *Cell* **108,** 121–133.
24. Junge, H.J., Rhee, J.S., Jahn, O., Varoqueaux, F., Spiess, J., Waxham, M.N., Rosenmund, C., and Brose, N. (2004) *Cell* **118,** 389–401.
25. Kaeser, P.S., and Sudhof, T.C. (2005) *Biochem Soc Trans* **33,** 1345–1349.
26. Calakos, N., Schoch, S., Sudhof, T.C., and Malenka, R.C. (2004) *Neuron* **42,** 889–896.
27. Betz, A., Thakur, P., Junge, H.J., Ashery, U., Rhee, J.S., Scheuss, V., Rosenmund, C., Rettig, J., and Brose, N. (2001) *Neuron* **30,** 183–196.
28. Dulubova, I., Lou, X., Lu, J., Huryeva, I., Alam, A., Schneggenburger, R., Sudhof, T.C., and Rizo, J. (2005) *EMBO J* **24,** 2839–2850.
29. Schoch, S., Castillo, P.E., Jo, T., Mukherjee, K., Geppert, M., Wang, Y., Schmitz, F., Malenka, R.C., and Sudhof, T.C. (2002) *Nature* **415,** 321–326.
30. Serra-Pages, C., Medley, Q.G., Tang, M., Hart, A., and Streuli, M. (1998) *J Biol Chem* **273,** 15611–15620.
31. Ko, J., Na, M., Kim, S., Lee, J.R., and Kim, E. (2003) *J Biol Chem* **278,** 42377–42385.
32. Takao-Rikitsu, E., Mochida, S., Inoue, E., Deguchi-Tawarada, M., Inoue, M., Ohtsuka, T., and Takai, Y. (2004) *J Cell Biol* **164,** 301–311.
33. Altrock, W.D., Dieck, S.T., Sokolov, M., Meyer, A.C., Sigler, A., Brakebusch, C., Fassler, R., Richter, K., Boeckers, T.M., Potschka, H., Brandt, C., Loscher, W., Grimberg, D., Dresbach, T., Hempelmann, A., Hassan, H., Balschun, D., Frey, J.U., Brandstatter, J.H., Garner, C.C., Rosenmund, C., and Gundelfinger, E.D. (2003) *Neuron* **37,** 787–800.
34. Dick, O., Dieck, S.T., Altrock, W.D., Ammermuller, J., Weiler, R., Garner, C.C., Gundelfinger, E.D., and Brandstatter, J.H. (2003) *Neuron* **37,** 775–786.
35. Khimich, D., Nouvian, R., Pujol, R., Dieck, S.T., Egner, A., Gundelfinger, E.D., and Moser, T. (2005) *Nature* **434,** 889–894.
36. Dieck, S.T., Altrock, W.D., Kessels, M.M., Qualmann, B., Regus, H., Brauner, D., Fejtova, A., Bracko, O., Gundelfinger, E.D., and Brandstatter, J.H. (2005) *J Cell Biol* **168,** 825–836.
37. Zhen, M., and Jin, Y. (2004) *Curr Opin Neurobiol* **14,** 280–287.
38. Bresler, T., Shapira, M., Boeckers, T., Dresbach, T., Futter, M., Garner, C.C., Rosenblum, K., Gundelfinger, E.D., and Ziv, N.E. (2004) *J Neurosci* **24,** 1507–1520.
39. Ziv, N.E., and Garner, C.C. (2001) *Curr Opin Neurobiol* **11,** 536–543.
40. Friedman, H.V., Bresler, T., Garner, C.C., and Ziv, N.E. (2000) *Neuron* **27,** 57–69.
41. Zhai, R.G., Vardinon-Friedman, H., Cases-Langhoff, C., Becker, B., Gundelfinger, E.D., Ziv, N.E., and Garner, C.C. (2001) *Neuron* **29,** 131–143.
42. Shapira, M., Zhai, R.G., Dresbach, T., Bresler, T., Torres, V.I., Gundelfinger, E.D., Ziv, N.E., and Garner, C.C. (2003) *Neuron* **38,** 237–252.
43. Vaughn, J.E. (1989) *Synapse* **3,** 255–285.
44. Bamji, S.X., Shimazu, K., Kimes, N., Huelsken, J., Birchmeier, W., Lu, B., and Reichardt, L.F. (2003) *Neuron* **40,** 719–731.
45. Zhai, R., Olias, G., Chung, W.J., Lester, R.A., Dieck, S.T., Langnaese, K., Kreutz, M.R., Kindler, S., Gundelfinger, E.D., and Garner, C.C. (2000) *Mol Cell Neurosci* **15,** 417–428.
46. Dresbach, T., Hempelmann, A., Spilker, C., Dieck, S.T., Altrock, W.D., Zuschratter, W., Garner, C.C., and Gundelfinger, E.D. (2003) *Mol Cell Neurosci* **23,** 279–291.
47. Dresbach, T., Torres, V., Wittenmayer, N., Altrock, W.D., Zamorano, P., Zuschratter, W., Nawrotzki, R., Ziv, N.E., Garner, C.C., and Gundelfinger, E.D. (2006) *J Biol Chem* **281,** 6038–6047.
48. Ahmari, S.E., Buchanan, J., and Smith, S.J. (2000) *Nat Neurosci* **3,** 445–451.

# Color Plate 1

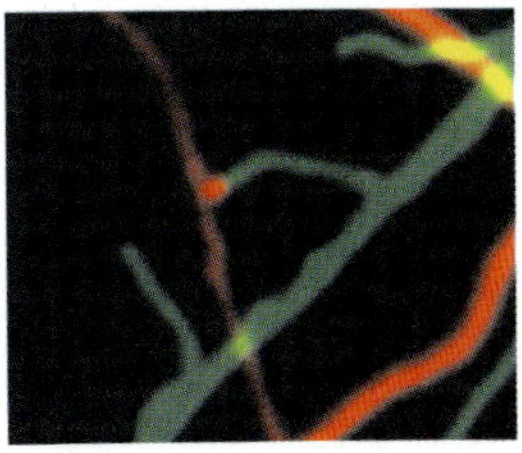

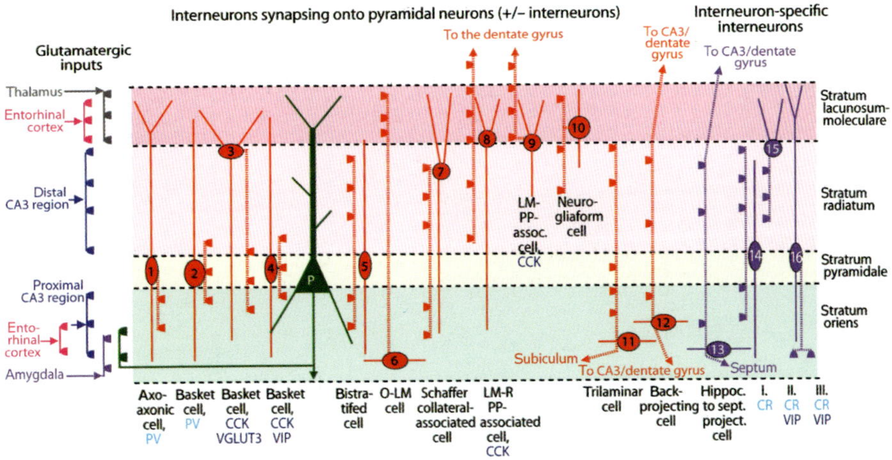


**Figure 0.1.** Schematic Diagram of the Innervation Patterns of a Pyramidal Cell in the CA1 Region of the Hippocampus by 12 Types of GABAergic Interneurons. The main laminar-specific glutamatergic inputs are indicated on the left. The somata and dendrites of interneurons innervating pyramidal cells (green) are shown in orange, those innervating mainly or exclusively other interneurons are shown in lilac. The main termination zones of GABAergic synapses are shown by trapeziform symbols. The proposed names of neurons, some of them abbreviated, are under each schematic cell and a minimal list of molecular cell markers is given, which in combination with the axonal patterns help the recognition and characterization of each class. Note that one molecular cell marker may be expressed by several distinct cell types. The number of interneurons shown is not exhaustive or complete. Note the association of the output synapses of different sets of cell types with the perisomatic region, and either the Schaffer collateral, commissural, or the entorhinal pathway termination zones, respectively. CB, calbindin; CR, calretinin; LM-PP, lacunosum-moleculare–perforant path; LM-R-PP; lacunosum-moleculare–radiatum–perforant path; m2, muscarinic-receptor type 2; NPY, neuropeptide tyrosine; PV, parvalbumin; SM, somatostatin; VGLUT3, vesicular glutamate transporter 3. Modified with permission from ref. 1.

# Color Plate 2

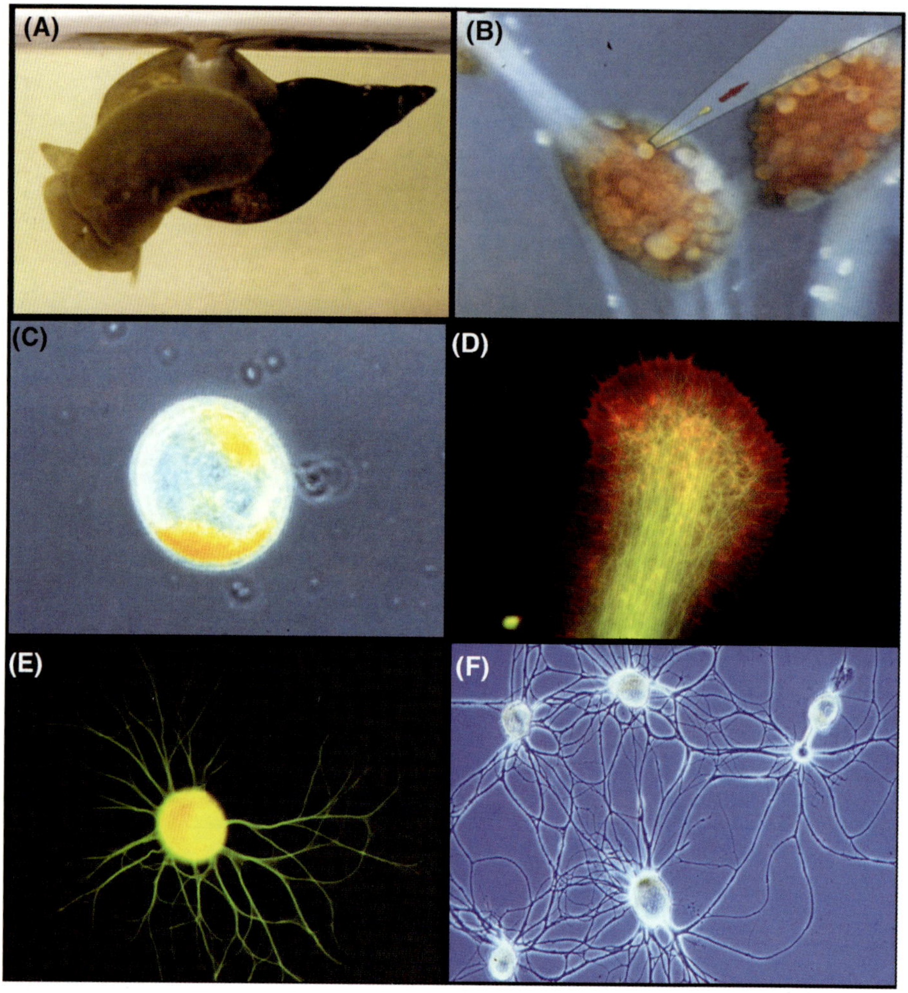


**Figure 2.1.** Cell Culture Techniques. (A) The pond snail *Lymnaea stagnalis*. (B) A cell extraction pipette is positioned to isolate individual neuronal somata. The arrow depicts neuronal extraction in progress. (C) An acutely isolated *Lymnaea* neuron in culture. (D) A *Lymnaea* growth cone fluorescently labeled with actin (red) and tubulin (green) antibodies. (E) An isolated neuron exhibiting neurite outgrowth overnight. (F) Interspecies synapse formation between *Lymnaea* and *Helisom*a neurons. A neuronal "hybrid" network reconstructed from two different snail species. Original copyright notice: Figure 2.1A reproduces Figure 3A on page 543 in *J Comp Physiology A*, volume 169, from the article by Syed, N.I., Harrison, D., and Winlow, W. (1991) Respiratory behavior in the pond snail *Lymnaea stagnalis*. Figures 2.1B, C and E reproduce Figures 2B, C (page 363), and 4C (page 372), respectively, of the book on *Modern Techniques in Neuroscience Research* (1999) Eds: Windhorst, U. and Johansson, H. Authors: Syed, N.I., Zaidi, H., and Lovell, P. Chapter 12: *In vitro* reconstruction of neuronal networks: a simple model system approach. All figures are reproduced with kind permission of Springer Science and Business Media.

# Color Plate 9

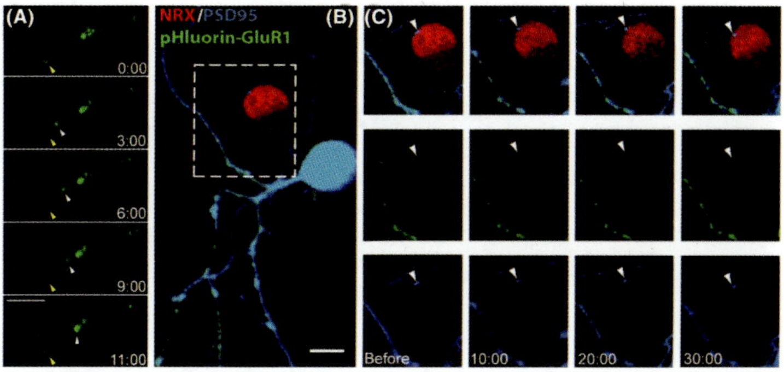


**Figure 14.2.** AMPA Receptor Transport and Recruitment. (A) Time-lapse images of AMPA receptor clusters *in vitro*. Clusters of GFP-tagged AMPA receptor subunit GluR1 are mobile (white arrow) in the dendrites of cortical neurons cultured for 4 days. Yellow arrow shows original location; time in minutes and seconds; scale bar = 6 μm. (B) Insertion of pH-dependent GFP-tagged GluR1 (pHluorin-GluR1) at new sites of synapse formation. Contact of a β-neurexin expressing PC12 cell (red) indues the accumulation of CFP-PSD-95 (blue). Scale bar = 10 μm. (C) The appearance of pHluorin-GluR1 at the PSD-95 cluster (arrowheads) is induced by the bath application of glutamate and glycine in brief pulses. Time in minutes and seconds. Reproduced with permission from ref. 88. Copyright (2005) National Academy of Sciences, U.S.A.

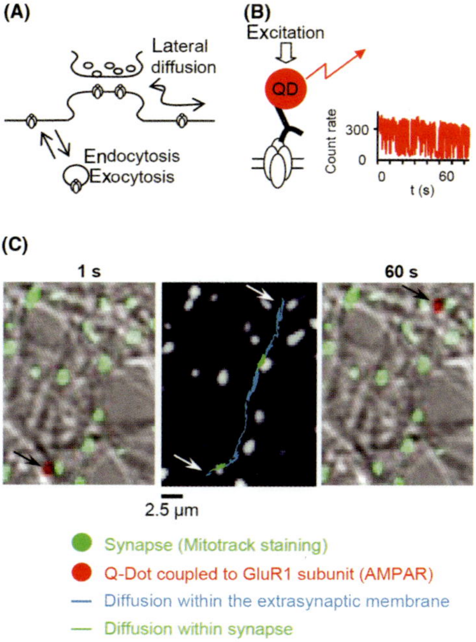


**Figure 15.2.** Receptor Lateral Diffusion at the Surface of Live Neurons Revealed by Single Particle Tracking. (A) Schematic representation of the two main pathways for neurotransmitter receptor trafficking. First, neurotransmitter receptors traffic to and from the plasma membrane by exocytosis and endocytosis, respectively. Second, neurotransmitter receptors diffuse laterally within the plasma membrane and they can even exchange between the extrasynaptic and synaptic membranes. (B) Quantum dot-based single particle tracking is based on the coupling between a single quantum dot and the receptor of interest via an antibody directed against the extracellular domain of the receptor. Note that the fluorescence emission of the quantum dot is stable over time, although blinking occurs (right panel). (C) Trajectories of a GluR1 containing AMPAR (60 s duration). Neurites are shown on right and left panels (differential interference contrast images). Note that the AMPAR diffused along a neurite and crossed two synaptic areas over this time period.

# Color Plate 10

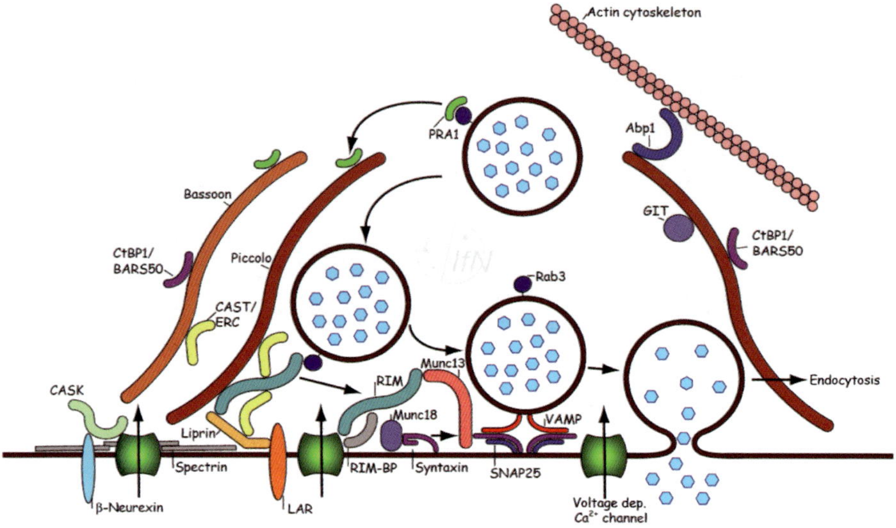


**Figure 16.1.** Molecular Organization of the CAZ. Few CAZ-specific proteins, including the Rab3-interacting molecules (RIMs), the SV vesicle priming factor Munc13, the two related scaffolding proteins Bassoon and Piccolo, and the CAZ-associated structural protein CAST/ERC, are specifically localized at the active zone. They are thought to localize and organize membrane trafficking events of the SV cycle and connect it to the active zone membrane proteins, including voltage-gated $Ca^{2+}$ channels and cell adhesion molecules such as neurexins. Further components that are not exclusive CAZ components include the $Ca^{2+}$/calmodulin kinase domain-containing membrane-associated guanylate kinase CASK, the transcriptional co-repressor CtBP1/BARS50, the RIM-binding proteins (RIM-BP), the prenylated Rab3 acceptor protein PRA1, the ARF-GTPase-activating protein GIT, the receptor tyrosine phosphatase LAR and its interacting protein Liprin, components of the SNARE complex and its control elements (e.g., Munc18). The interaction between Piccolo and the actin-binding protein Abp1 is thought to link the active zone to the neighboring endocytic zone. For further details see Gundelfinger, Altrock and Fejtova, Active zone in: *Encyclopedic Reference of Neuroscience* (Field editor: M. Takahashi, Springer Berlin, in press). The picture was taken from the website of the Institute for Neurobiology (http://www.ifn-magdeburg.de/en/departments/neurochemistry_and_molecular_biology).

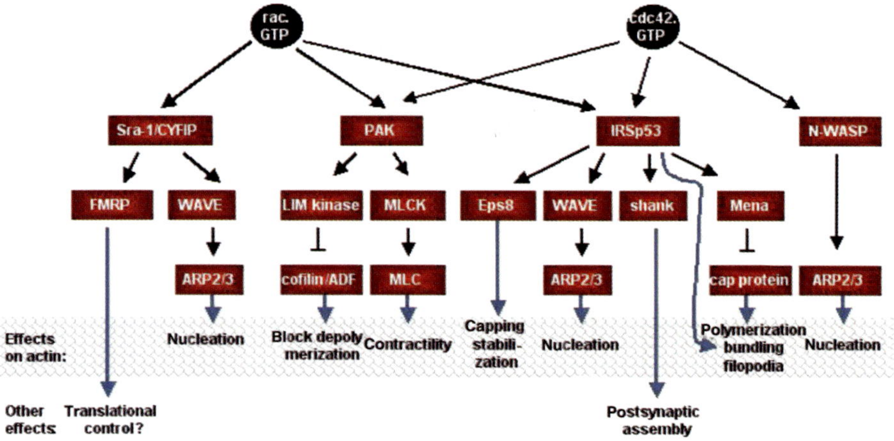


**Figure 17.2.** Possible Signaling Pathways Used by the Small GTPases rac and cdc42 in Neuronal Dendrites. Activation is denoted by a black arrow, inhibition by an interrupted black line. The final downstream effect of each pathway is indicated by a blue arrow. Note that for IRSp53, activation of Eps8 and WAVE has been described only in connection with rac signaling, whereas linkage to shank or Mena has been described after activation by cdc42. Several known effectors were omitted for clarity.

# Color Plate 11

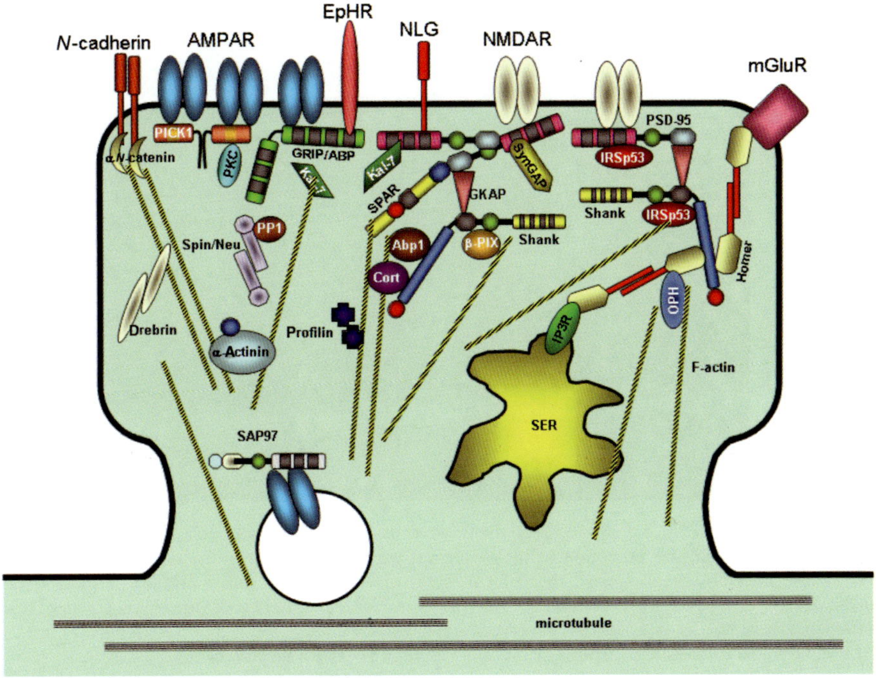


**Figure 18.2.** The Synaptic Scaffold Proteins Involved in Spine Morphogenesis. NLG, neuroligin; Kal-7, Kalirin-7; Spin/Neu, Spinophilin and Neurabin I; Cort, Cortactin; OPH, Oligophrenin.

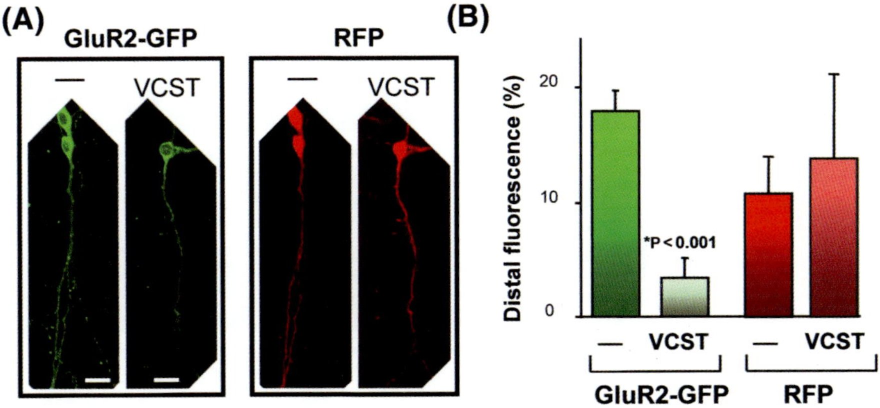


**Figure 24.1.** Dendritic Trafficking of AMPA Receptors is Impaired by Destabilization of the Microtubular Cytoskeleton. (A) The GFP-tagged GluR2 subunit of the AMPA receptor and a cytosolic red fluorescence protein (RFP) were co-expressed in organotypic hippocampal slice cultures[1] with ("VCST") or without ("–") 5 nM vincristine. Confocal fluorescence images were acquired 36 h after transfection. Scale bar: 20 μm. (B) Quantification of GluR2–GFP and RFP fluorescence 170–200 μm away from the cell soma. Fluorescence intensity values are normalized to those found at the cell soma. Plotted are average values and standard error of the mean from four experiments as the one shown in (A). GluR2–GFP presence in distal dendrites was drastically reduced in neurons incubated with vincristine.

# Color Plate 12

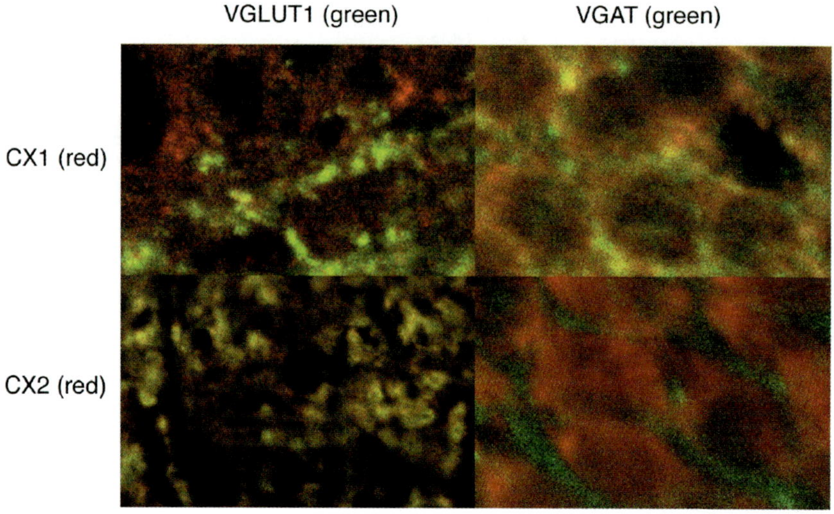


**Figure 26.1.** Confocal Microscopy of Presynaptic Proteins in Rat Hippocampus. All images from the granule cell layer with the exception of the lower left panel obtained from the mossy fiber terminal zone. These images largely support the colocalization of complexin I with inhibitory terminals and complexin II with excitatory terminals. However, in other subfields of the hippocampus this relationship was not as consistently observed. *Abbreviations:* Cx1 and Cx2, complexins 1 and 2; vGLUT, vesicular glutamate transporter 1; vGAT, vesicular GABA transporter.

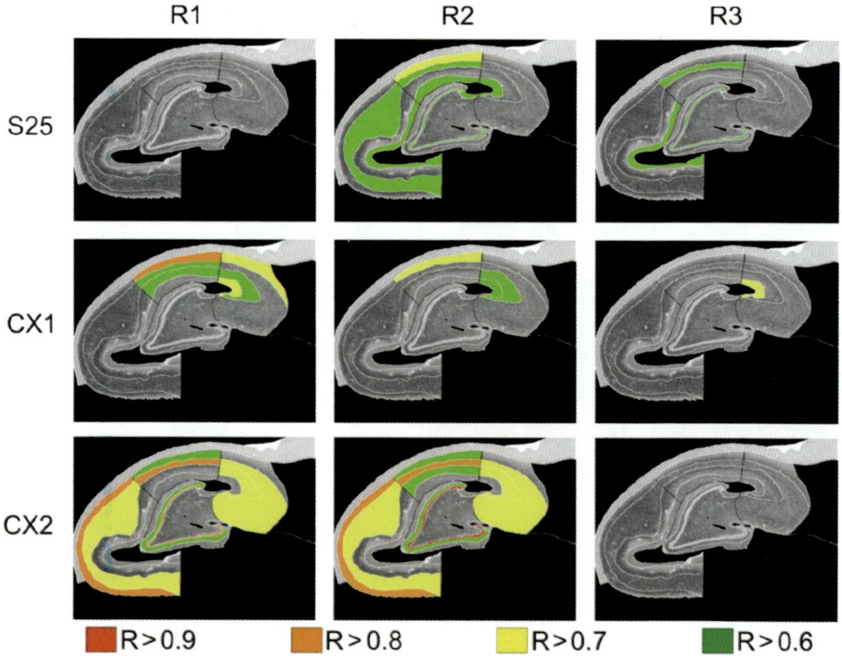


**Figure 26.2.** Maps of the Distribution of Hippocampal Subfields with Statistically Significant ($P \leq 0.05$) Correlations Between Presynaptic Protein Levels and Specific Domains of Cognitive Dysfunction in Schizophrenia. *Abbreviations:* S25, SNAP-25; Cx1, complexin I; Cx2, complexin II; *r*, Spearman correlation coefficient; R1, R2, and R3, clinical dementia ratings for memory, orientation, and judgment, respectively.

# Color Plate 3

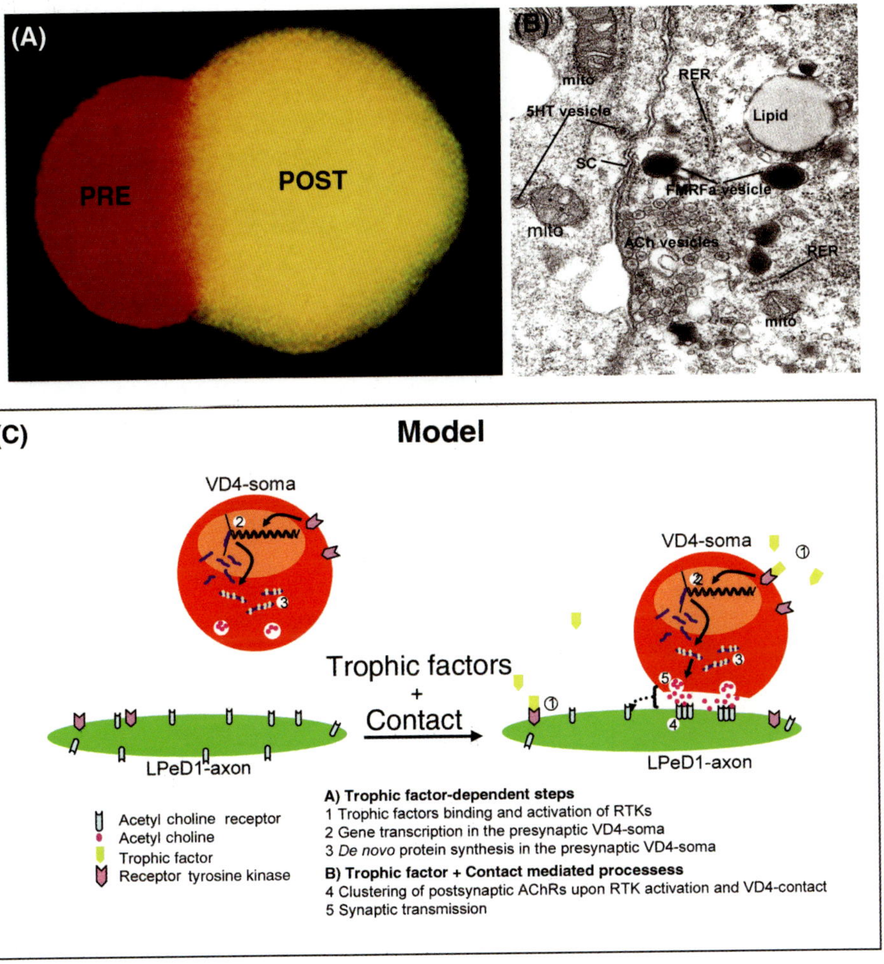


**Figure 2.2.** Mechanisms Regulating Synapse Formation in Various Model Preparations. (A) *Lymnaea* presynaptic neuron (visceral dorsal 4 – VD4 – fluorescently labeled with red dye, sulforhodamine) and postsynaptic neuron (left pedal dorsa 1 – LPeD1 – injected with Lucifer yellow) were soma–soma paired. In this configuration, most molluscan neurons develop appropriate excitatory and inhibitory synapses similar to those seen *in vivo*. (B) An electron micrograph showing the nature of synaptic contacts between soma–soma paired neurons. Vesicles dock at presynaptic site juxtaposed against the postsynaptic cell (Figure courtesy of Dr. Matthias Amrein, University of Calgary). (C) Model depicting steps and mechanisms underlying trophic factor and target cell contact-induced synapse formation between *Lymnaea* neurons in a soma–axon configuration. The model predicts that both target cell contact and extrinsic trophic support are required for appropriate, excitatory synapse formation.

# Color Plate 4

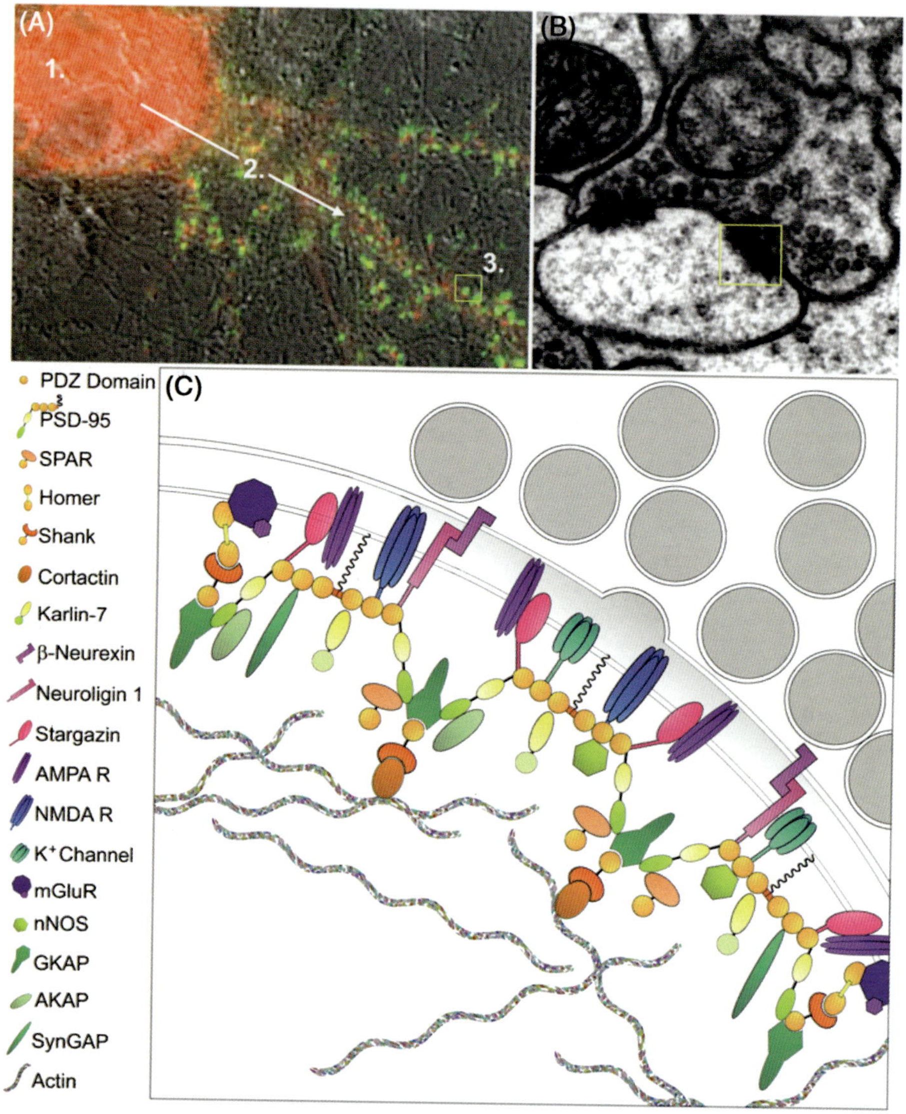


**Figure 4.1.** Protein Complexes at Excitatory Synapses. (A) An image of a neuron stained with the presynaptic marker synaptophysin (green), to identify synaptic contacts. This panel illustrates steps involved in the assembly of proteins at contact sites. Synapse formation is generally thought to involve three basic steps which include production of proteins in the cell soma (A-1), transport of these proteins to early sites of contact between axons and dendrites (A-2), and assembly of protein complexes at synapses (A-3). (B) The intense clustering of proteins seen at the PSD of excitatory synapses is highlighted in the electron micrograph shown in (B). A schematic diagram of this region is blown up in, illustrating the role of scaffolding molecules such as PSD-95 in assembly of large protein complexes. PSD-95 forms the core of the protein network, which is associated with the membrane through palmitoylation, and anchored within the postsynaptic compartment by several proteins that associate with actin. Coupling of PSD-95 to adhesion molecules such as neuroligins allows for *trans*-synaptic signaling.

# Color Plate 5

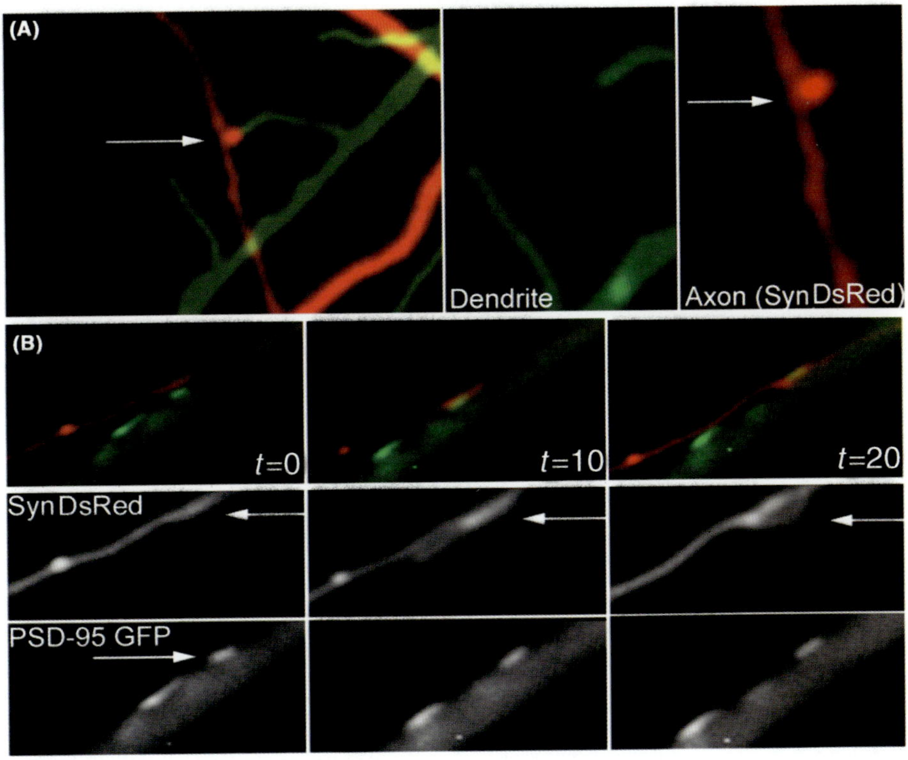


**Figure 4.5.** Recruitment of Clusters of Presynaptic and Postsynaptic Proteins at Early Sites of Contact Between Axons and Dendrites. (A) Shows an accumulation of a synaptophysin cluster at a contact site between dendritic filopodia of a cell transfected with a membrane targeted GFP and an axon from a neuron transfected with synaptophysin tagged with DsRed (SYN DsRed). (B) Time-lapse images showing accumulation of SYN DsRed at a site apposed to an existing PSD-95 GFP cluster, occurred over a time (*t*) period of 20 min.

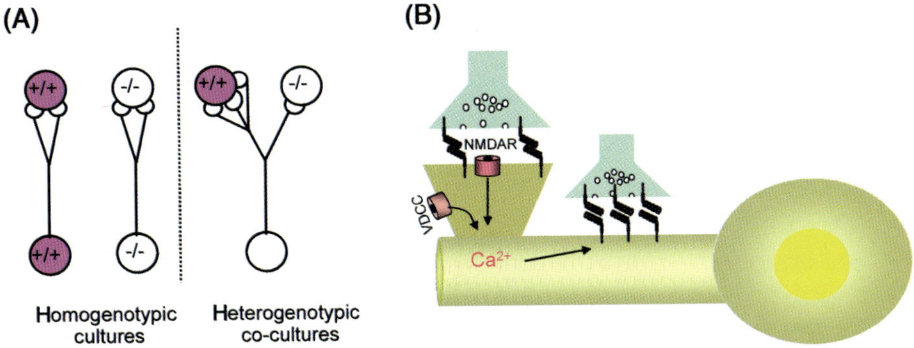


**Figure 6.2.** NCAM Promotes Synaptogenesis in a Choice Situation. (A) A schematic diagram depicting normal synaptic coverage of NCAM-deficient (–/–) neurons, as compared to wild-type neurons (+/+) in homogenotypic cultures, and of reduced synaptic coverage in heterogenotypic co-cultures. (B) A hypothetical model according to which synaptic activity and activation of NMDA receptors (NMDAR) and voltage-dependent $Ca^{2+}$ channels (VDCC) during induction of LTP may lead to increases in NCAM expression in stimulated neurons or in neighborhood of stimulated synapses, which may promote synaptogenesis in these cells/subcellular domains.

# Color Plate 6

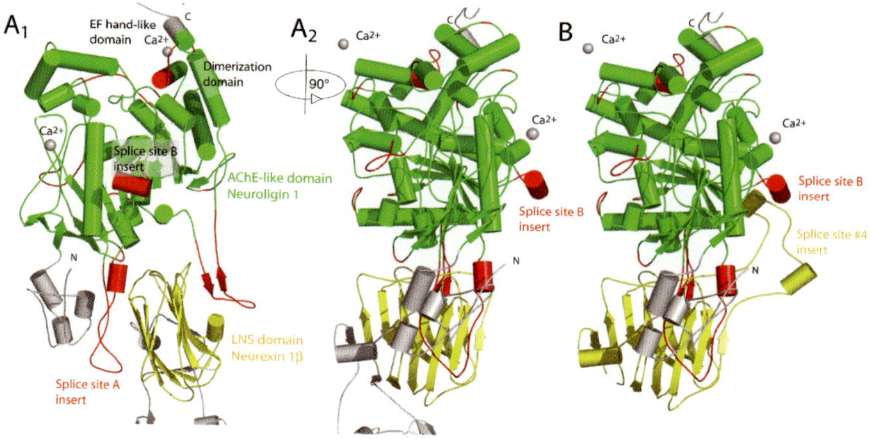


**Figure 7.1.** Model Structure of the Extracellular Domains of Neuroligin 1 and Neurexin 1β. An interaction in a head-to-head *trans*-complex is proposed because it best fits the available biochemical data. The AChE-like domain of neuroligin (green) interacts with the laminin (LNS)-like domain of neurexin (yellow) when the latter contains no insert in splice site #4 (panels $A_1$ and $A_2$, the latter showing the structure at a different angle as indicated). Several loops (red) in neurologin 1 are much longer than in other proteins with an AChE-like fold, including insertions in the two splice sites A and B. This interaction is severely hindered when β-neurexin contains an insert at splice site #4 (panel B), presumably by sterical interfering with the position of the insert B in neuroligin. This model is supported by a recent finding that lack of insert B allows binding to all neurexins irrespective of their splice combination[3]. While this interaction should be less efficient when neuroligin is glycosylated at the splice site insert B, their $Ca^{2+}$-dependence is more difficult to explain based on this model structure: the proposed major calcium binding site at a degenerate EF-hand-like motif is located at the C-terminus, and a second potential site is close to splice site B (best seen in panel $A_2$). The structures have been built using coordinates from protein data bank (PDB, www.rcsb.org) entries 1c4r for neurexin, and 1mah & 1fss for neuroligin. The structures of the N-terminal sequences as well as those regions linking the domains have been predicted using software programs SPDBViewer and Threader, and the web service Hmmstr/Rosetta.

# Color Plate 7

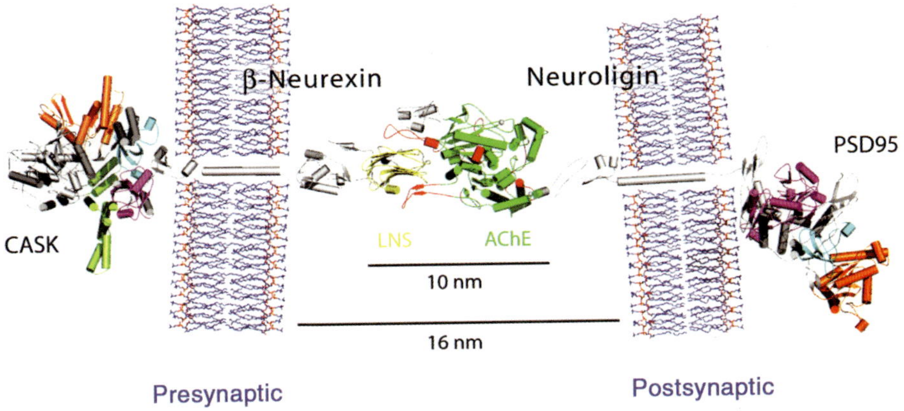


**Figure 7.2.** Model Structure of the Neuroligin/β-Neurexin *Trans*-Synaptic Complex Together with Their Putative Intracellular Binding Partners PSD-95 and CASK. The size of the complex formed by the laminin (LNS)-like domain (yellow) of β-neurexin with the AChE-like domain (green) of neuroligin is about 10 nm (compare also Figure 7.1) and for itself not sufficient to span the synaptic cleft. Each of these domains is linked to the adjacent membrane by a sequence that may build structurally flexible domains (gray) and would have to expand to the required lengths, building a bridge of at least 16 nm over the synaptic cleft. The model is based on the structures of the LNS domain from β-neurexin, the PDZ and the guanylate kinase domain from CASK, and the three PDZ domains from PSD-95 that have been solved by X-ray crystallography. The structures of the other domains have been obtained by homology modeling using coordinates from the protein data bank (PDB, www.rcsb.org), i.e., entries 1kgd, 1kwa, 1jxm, 1y74, and 1rso for CASK; 1c4r for neurexin; and 1mah & 1fss for neuroligin; 1be9, 1iu0, 1jxm, 1qlc, and 1v1t for PSD-95. Predictions were made using programs SPDBViewer and Threader, and the web service Hmmstr/Rosetta, and represent most compact forms[25]. The bio-informatical work for this review was supported by grant SFB406-C9 (to MM).

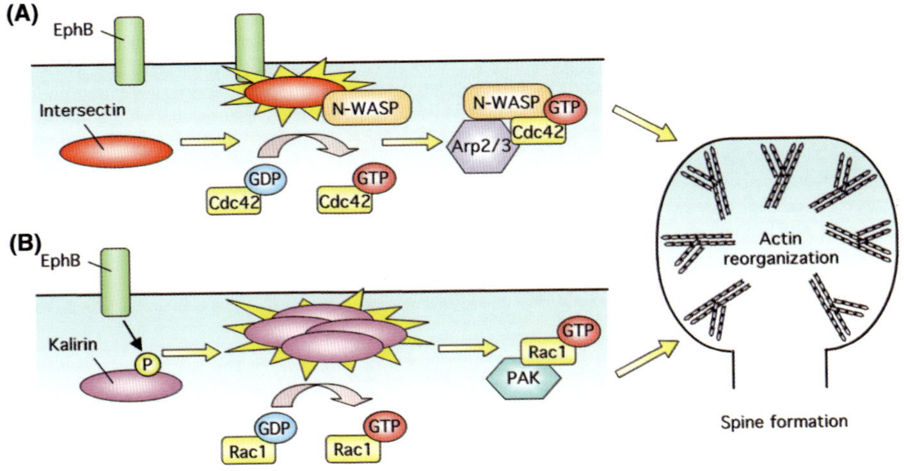


**Figure 10.2.** EphB Downstream Signaling Pathways Leading to Actin Polymerization and Spine Morphogenesis. (A) The Cdc42 GEF intersectin-l is recruited to the site of ephrinB–EphB signaling through its interaction with EphB receptors. Association of EphB2 and N-WASP has a synergistic effect to promote the GEF activity of intersectin-l, which in turn causes local activation of Cdc42. Together with N-WASP, activated Cdc42 triggers Arp2/3-mediated actin polymerization into a branched network[9]. (B) EphB receptors phosphorylate the Rac1 GEF Kalirin and induce its translocation to synapses and dustering. Activated Kalirin then stimulates the Rac1-PAK cascade, resulting in actin rearrangement[21].

# Color Plate 8

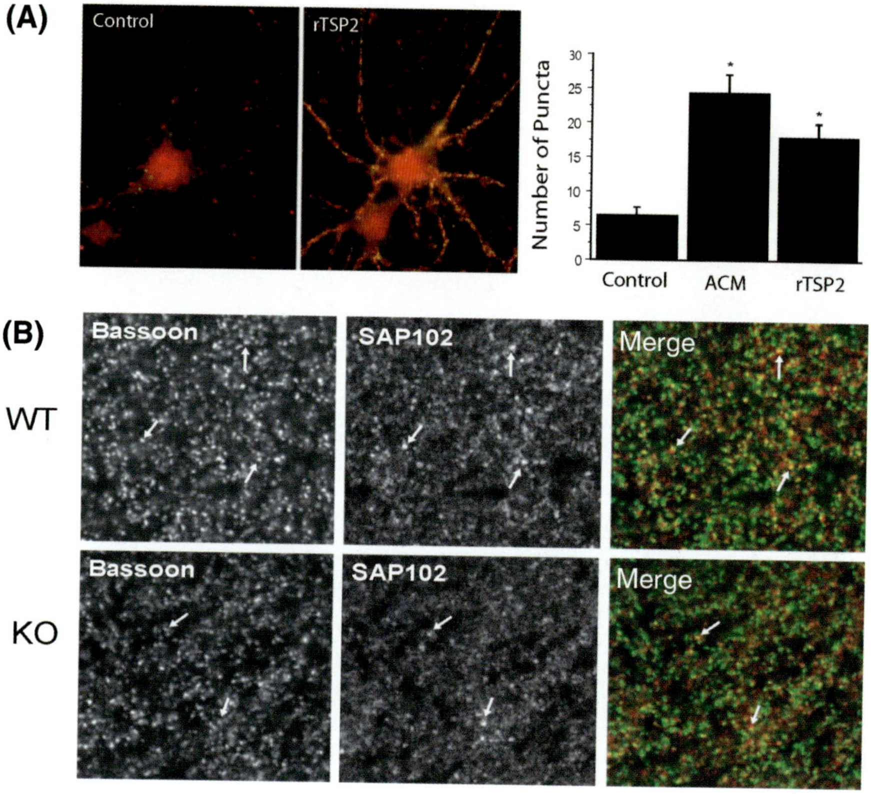


**Figure 11.2.** TSP is Necessary and Sufficient for Synapse Formation. (A) Purified CNS neurons do not form synapses under control condition (Control) as seen by the absence of synaptic puncta after staining for presynaptic synaptotagmin and postsynaptic PSD-95. Addition of recombinant TSP2 (rTSP2) induces a dramatic increase in the number of colocalized presynaptic puncta and postsynaptic puncta (yellow puncta in Colorplate 8). Graph shows a nearly 4-fold increase in synapse number after addition of rTSP2 that is comparable to the increase elicited by the astrocyte-conditioned medium (ACM). (B) Post-natal day 21 wild-type (WT) and TSP1/2 double knockout (KO) mice brains were sectioned and stained for presynaptic marker Bassoon and postsynaptic marker SAP102. KO animals show a reduction in the number of colocalized synaptic puncta (arrows). Reprinted from ref. 30, Copyright (2005) with permission of Elsevier.

# Color Plate 14

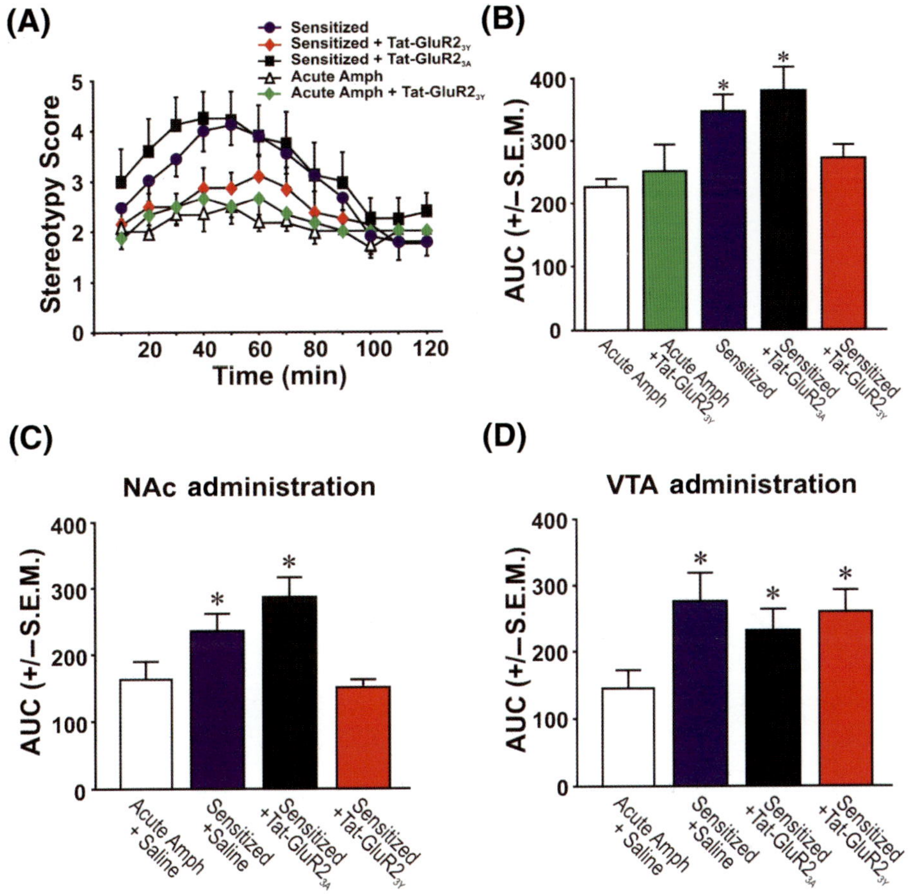


**Figure 32.5**. Interference Tat-GluR2$_{3Y}$ Peptide Blocks *D*-Amphetamine-Induced Behavioral Sensitization. (A) D-amphetamine-sensitized rats ($n$=6–8/group) were pretreated with Tat-GluR2$_{3Y}$, Tat-GluR2$_{3A}$, or saline by i.v. injection (1.5 nM/gr) 90 min prior to a challenge dose of *D*-amphetamine (2 mg/kg, i.p). Stereotypy scores were measured at 10-min intervals over a 2-hour session. Pretreating D-amphetamine-sensitized rats with Tat-GluR2$_{3Y}$ (red diamond), but not Tat-GluR2$_{3A}$ (black square), prevented the sensitized behavioral response to a challenge dose of *D*-amphetamine (blue circle). Tat-GluR2$_{3Y}$ abolished only the expression of behavioral sensitization, as it did not affect the response of naïve rats to D-amphetamine (green diamond) when compared with saline treated control (white triangle). (B) Histogram shows the summary of changes in stereotypy scores across the 2-hour test session converted to Area Under the Curve (AUC) for individual groups depicted in graph (A). (C) The blockade of behavioral sensitization was reproduced when Tat-GluR2$_{3Y}$ (15 pmol; in 0.3 μl) was microinjected directly into the NAc. Injecting peptides into VTA (D) did not affect behavioral sensitization (* $p<0.05$ versus acute D-amphetamine group. $n$=6–7 in each group). Redrawn from ref. 41.

# 17

# ASSEMBLY OF POSTSYNAPTIC PROTEIN COMPLEXES IN GLUTAMATERGIC SYNAPSES

Hans-Jürgen Kreienkamp*

## 1. SUMMARY

Formation of postsynaptic specializations requires the positioning of a huge macromolecular complex, the postsynaptic density, at the tip of the dendritic spine. Assembly of this protein complex is enabled by tight control of intra- and intermolecular interactions between PSD scaffold proteins. In parallel, rearrangements of the actin cytoskeleton are necessary to generate a spine and bring it into its mature form. Actin treadmilling in spines is controlled by small GTPases of the rho family via a range of downstream effectors, including PAK, WAVE, and IRSp53 proteins. Aberrant spine formation is a hallmark of the pathology of inherited forms of mental retardation (MR), and proteins of the rho G-protein cycle are mutated in specific forms of MR. In addition, the loss of the RNA-binding protein FMRP in patients suffering from fragile X MR syndrome suggests that translational control at or near synapses may also contribute to the correct assembly of the postsynaptic apparatus.

## 2. INTRODUCTION

The postsynaptic density of excitatory synapses is a huge protein complex, with an estimated molecular weight of about 1 GDa[1,2]. It is composed of neurotransmitter receptors, scaffold and signaling proteins and is localized at the tip of dendritic spines. In contrast to the presynaptic cytomatrix of the active zone, which is transported to axons in a preassembled form, assembly of the PSD may occur locally concomitant with the maturation of the spine, possibly from filopodial precursors[3]. This process requires the coordination of several cellular processes, including reorganization of the local actin cytoskeleton, regulated interaction between PSD molecules, transport of membrane proteins into dendrites, and possibly also local synthesis of some components from dendritically localized

*Institut für Humangenetik, Universitätsklinikum Hamburg-Eppendorf, Martinistrasse 52, 20246 Hamburg, Germany; kreienkamp@uke.uni-hamburg.de

mRNAs. In addition, recent evidence shows that selected postsynaptic proteins undergo regulated degradation by the ubiquitin proteasome system[4]. I discuss here the contributions of these mechanisms to synaptogenesis, as well as their pathological relevance, as discovered in inherited forms of MRs.

## 3. INTRA- AND INTERMOLECULAR INTERACTIONS OF POSTSYNAPTIC SCAFFOLD PROTEINS

Several postsynaptic proteins are present in large copy numbers within the PSD, including members of the PSD-95, GKAP/SAPAP, shank and homer families, which have been estimated to contribute several 100 molecules each[1,2]. Given the propensity of these proteins to engage in multiple intermolecular contacts, and to form large clusters when co-expressed together in heterologous systems[5], one may ask why PSD proteins do not aggregate at any random place within the dendrite. In fact, shank1 multimerizes in heterologous cells in a filamentous form associated with keratin filaments, due to an interaction between the ankyrin and SH3 domains of shank1. Presence of the interaction partner GKAP/SAPAP leads to an aggregation of both proteins in aggresomes. Only when a third protein of the complex, PSD-95 is present, regular clusters of all three proteins are observed in overexpressing cells as well as in neurons. This indicates that PSD proteins influence each other during complex formation, and precise control of intermolecular interaction is required to avoid unproductive aggregation of these proteins as observed in heterologous systems[5].

So, where does assembly take place? A transport package model, where postsynaptic complexes might be assembled within the cell body and then be transported in a vesicular form to axo-dendritic contact sites, would be analogous to the situation at the presynapse, where Shapira and co-workers[6] have identified a population of transport vesicles containing multiple critical elements of the cytomatrix of the active zone. Within dendrites, membrane proteins obviously need assistance in terms of motor–protein dependent vesicular transport. This was reported, e.g., by Washbourne and colleagues[7] who observed large fluorescent NMDA-receptor particles to be transported to the postsynaptic site (see also Chapters 6 and 13). The role of postsynaptic scaffold molecules, such as PSD-95 or GRIP in linking postsynaptic cargoes to their respective motor proteins, is reviewed in Chapter 12.

Despite these and some other studies demonstrating existence of large postsynaptic transporting complexes, the dominant view is that components of the PSD are largely assembled locally in dendrites. Evidence for local assembly comes from the observation that postsynaptic clusters of GFP-labeled PSD-95 accumulate in dendrites, apparently from diffuse cytoplasmic stores of the molecule which are present locally[8]. Extending this approach to other postsynaptic components, Bresler and co-workers recently demonstrated that several postsynaptic proteins exhibit similar kinetics in cluster formation, i.e., single exponentials with rate constants in the range of 12–14 min[9]. This is consistent with a mechanism of co-clustering of these molecules, whereby addition of one postsynaptic protein to the complex requires the previous addition of another one, and so on. Thus PSD-95, shank proteins, and the NMDA-receptor subunits analyzed in this study do not form clusters by themselves but require certain signals such as the previous association with other postsynaptic proteins in order to form clusters. As all these proteins are apparently also present within dendrites in a nonclustered form, this implies that individual PSD proteins are somehow prevented from association with each other; one possibility would be that molecules such as PSD-95 or shank are in a "closed" conformation, in which individual domains are not accessible for binding to their respective ligands. An important question would then be which signals would lead to a conformational change, allowing binding of protein ligands and triggering complex assembly.

Several molecular mechanisms have been discovered recently which converge on the PSD-95 family of proteins. PSD-95 (and other MAGuK family members) exhibit an intramolecular interaction in their C-terminal region, which involves the SH3 und GK domains[10,11]. Analysis of the 3-dimensional structure of this part of the molecule has revealed that the GK domain is actually inserted into a noncontiguous SH3 domain, in a way that sequences N- and C-terminal to the GK domain link up to form a β-sheet that is part of the SH3 fold[12,13]. This interaction may occur in an intramolecular or intermolecular fashion[10], potentially allowing for the regulated assembly of a multimeric array of PSD-95 molecules[14]. The importance of this arraying capacity is demonstrated by mutations which interfere with the SH3/GK interaction and which disrupt channel clustering by MAGuK proteins. In *Drosophila* similar mutations lead to a loss of MAGuK function, suggesting that the ability of the PSD-95 homolog *discs large* to form membrane associated complexes critically depends on the ability of the SH3/GK domain to multimerize. Further evidence for intramolecular interactions in the related SAP97 protein was presented by Wu et al.[15], who identified a regulatory effect of the N-terminus of this protein on the ability of the C-terminal GK domain to bind GKAP.

Regulation of PSD-95 appears to occur also at the N-terminal end of the protein via different mechanisms. El-Husseini et al.[16,17] demonstrated in an extensive series of experiments that reversible palmitoylation events lead to modification of two Cys residues in the N-terminus; palmitoylation is necessary for the synaptic targeting of PSD-95. The N-terminal 64 residues of PSD-95 are also required for multimerization[18]. This multimerization can be prevented through phosphorylation of this region by cdk5, leading to a reduced size of clusters in recombinant systems as well as *in vivo* in a mouse carrying a deletion in the cdk5 gene[19]. Taken together, these findings show that the ability of PSD-95 and related proteins to multimerize and cluster their binding partners is tightly controlled by intramolecular interactions and various post-translational modifications.

## 4. ACTIN REGULATORY PROTEINS IN DENDRITIC SPINES

Actin filaments are the predominant form of cytoskeleton in dendritic spines, which determines their morphology as well as functional characteristics. Recent FRAP studies have shown that actin in spines of cultured neurons undergoes rapid turnover, with a half-life of less than a minute[20]. Activity-dependent changes in spine size are associated with changes in the ratio of monomeric G-actin to filamentous F-actin, such that the increase in spine size observed during LTP shifts the balance toward F-actin[21]. In addition, several studies have indicated that mature dendritic spines are generated from filopodial precursors on dendrites. These long, thin, and highly motile protrusions may function as a "scout" to detect incoming axon terminals. Filopodia are characterized by actin bundles, whereas mature spines are likely to contain a more branched actin network[22]. Taken together these observations indicate that signals regulating the shape of the actin cytoskeleton are of fundamental importance for any type of structural change occurring in dendrites, either in a developmental phase or during plasticity. Knowledge about actin regulatory processes arises mainly from non-neuronal systems, such as fibroblasts. Here, nucleation of actin filaments, as well as the generation of branches on existing filaments, requires the Arp2/3 complex which is recruited to specific cellular sites by the VCA (verprolin homology, cofilin homology, acidic) domain of WASP (Wiskott–Aldrich-syndrome protein) family members (WASP, N-WASP, WAVE1–3; ref. 23). Activation of WASP proteins in different cellular pathways is achieved by exposure of the VCA region, which otherwise is locked in an intramolecular interaction in the resting state of the protein. In addition, the actin-binding protein cortactin contains an acidic region and may recruit the Arp2/3 complex[24]. Growth of filaments is stimulated by profilin, which aids the addition of monomeric actin to the growing or (+) end. Actin filaments are stabilized by capping proteins, including CapZ, gelsolin, and Eps8, whereas members of the Ena/VASP family may compete with capping proteins for free ends, allowing further polymerization. Cofilin/actin deploymerizing factor disassembles F-actin at the (–) end; it is thus required for actin treadmilling as it eventually provides the actin monomers that can be reinserted into filaments by profilin. Many of these actin regulatory proteins are present in neuronal dendrites and contribute to the constant turnover of F-actin indicated by the FRAP measurements reported in ref. 20. The equilibrium between the different forms of actin filaments is dictated by various signaling molecules, among which members of the rho GTPase family are most notorious.

## 5. RHO GTPASES: NATURALLY BORN TRIGGERS OF POSTSYNAPTIC ASSEMBLY

Small GTPases from the rho, ras, rab, and other families are switched "on" by specific guanine nucleotide exchange factors (GEFs) which catalyze the exchange of bound GDP by GTP. GTP binding induces a conformational change which enables the now active protein to bind to and regulate the activity of numerous effector molecules. This occurs usually by changing the conformation of target proteins, thus enabling additional protein/protein interactions. Small GTPases may eventually be switched "off" by their namesake intrinsic GTPase activity. As this activity is rather slow, it may be enhanced by GTPase activating proteins (GAPs), which therefore serve to terminate G-protein signaling. Thus, the G-protein returns to its inactive, GDP-bound state which in general does not bind to any effector molecules. Members of several different GTPase families affect form or function of the postsynaptic specialization, notable examples being rap[25] and ras[26]. For the sake of simplicity, I focus here on the Rho family of small GTPases, which contains among others RhoA, rac, and cdc42. Rho proteins are considered as regulators of dynamic reorganizations of the actin cytoskeleton since individual Rho proteins have been shown to induce rather specific morphological changes in the fibroblast model system[27]. Given the dependence of both spine development and synaptic plasticity on changes in actin filament assembly, rho family GTPases may be considered as natural born molecular triggers in these processes.

Attempts to establish signaling pathways involving rho proteins which might be initiated by axonal contact have identified cell adhesion molecules from the ephrin family (see Chapter 10). EphB2 is a transmembrane tyrosine kinase which is present at the postsynaptic membrane; it may be activated by membrane-associated ephrinB ligands, or experimentally by clustered ephrinB-Fc fusions.

Several GEFs, in particular kalirin, intersectin, and βPIX are present in dendritic spines or developing dendrites and might transduce the signal to rho GTPases. Activation of rho itself has been implicated in the inhibitory control of dendrite branching, most likely through activation of rho kinase and phosphorylation of myosin light chain[28,29]. Both rac and cdc42 are likely to play an active role in spine formation and synapse maturation. EphB2 activation has been suggested to locally activate either rac (via kalirin[30] ) or cdc42 (via intersectin[31]). Kalirin is attached to members of the PSD-95 family via a PDZ-type interaction, and this form of attachment to postsynaptic molecules is required for its effect on spine growth[32]. Similarly, βPIX can be anchored at postsynaptic sites via a PDZ-type interaction with shank, potentially allowing for localized activation of cdc42 or rac[33]. The nature of signaling downstream of the GTPases is then determined by the potential effector molecules present, i.e., molecules which can bind the GTP-bound form through cdc42/rac interactive-binding (CRIB) motifs, or other target sequences. Specificity in small GTPase signaling for particular pathways can be achieved when downstream targets are physically associated with the GEFs that initially generate the active GTPase (Figure 17.1).

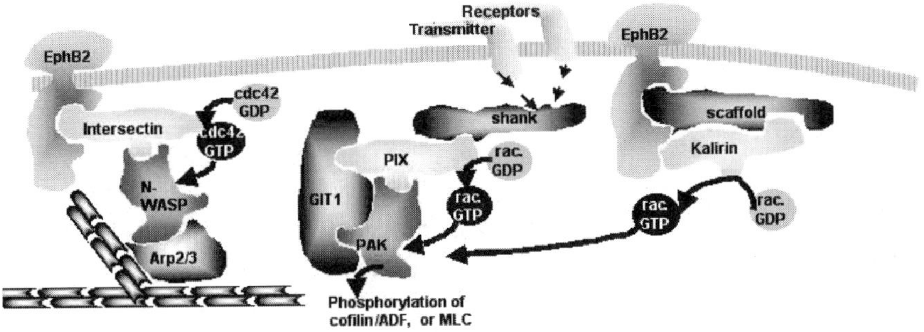


**Figure 17.1.** Signaling Modules Controlling the Activity of Small GTPases in Dendrites. In case of the GIT1/PAK/PIX module, shank is not required for functional activity but has been described to target this protein complex to postsynaptic sites. Note that EphB2 apparently acts via different exchange factors. Kalirin is translocated to the synapse upon EphB2 activation; however it is not clear if synaptic targeting involves a scaffold such as PSD-95, or if kalirin can interact directly with (Tyr-phosphorylated) active EphB2.

A remarkable example for this is the so-called PIX/PAK/GIT1 signaling module. p21-activated kinases (PAK1-3) are typically involved in cytoskeletal regulation in other cell systems and have also been implicated in synaptogenesis. PAK is physically associated with one of the aforementioned GEFs, PIX, and both proteins are additionally linked by further interactions with the adaptor protein GIT1. This trimolecular complex is involved in the establishment and regulation of focal adhesions in non-neuronal cells[34]. PAK, and presumably also GIT1, are targeted to the PSD due to the interaction of the βPIX C-terminal PDZ ligand motif with the PDZ domain of shank proteins[33]. The role of this signaling module for the establishment of dendritic spines was recently demonstrated in two studies by Zhang et al.[35,36] who showed that spine and synapse formation was reduced by interference with either GIT1 or PAK activity. The authors elegantly demonstrated local activation of rac by the use of fluorescence resonance energy transfer (FRET) probes which combine rac with a G-protein binding motif flanked by different fluorophores[36]. The FRET signal obtained from this probe, indicative of local exchange activity, was found in dendritic protrusions making contact to presynaptic terminals, while the main shaft of the dendrite exhibited little or no signal.

The downstream activities of locally activated PAK might involve the phosphorylation and activation of either the regulatory light chain of myosin II[36], or, as inferred from non-neuronal systems, LIM kinase. LIM kinase would then in turn phosphorylate and thus inhibit cofilin/actin depolymerizing factor. The road from an extracellular stimulus to the activation of LIM kinase and inhibition of cofilin may also be more direct, as a receptor for bone morphogenetic protein, BMPRII, can interact directly with LIM kinase at the cell membrane. By allowing actin polymerization, BMP thus exerts a rather direct influence on dendritic growth[37].

Activated cdc42 activates PAK isoforms, and in addition it can interact with N-WASP. Activation of N-WASP leads to recruitment of the Arp2/3 complex, which is required for actin nucleation and the formation of a branched actin network by providing attachment points for new actin filaments alongside preexisting ones. Intersectin (which is activated by EphB2, see above) also interacts with N-WASP and may constitute another signaling module which stimulates Arp2/3 mediated actin branching. Taken together, several of these signaling modules have been identified which are present in neuronal dendrites or spines, and have the ability to transduce an extracellular signal to changes in F-actin (and therefore spine) structure.

## 6. DIVERGENCE OF SMALL GTPASE PATHWAYS IN DENDRITES

Actin nucleation may also be achieved by rac, through homologs of the WASP family termed WAVE/Scar proteins. WAVE proteins do not possess a rac-binding motif, and therefore require intermediate factors. One such intermediate is the formation of a pentameric complex consisting of the rac-target Sra1/CYFIP1, together with Nap, Abi, HSPC300 and WAVE[38]. The relevance of this complex in dendrites is not known yet (however, see below and Figure 17.2; Colorplate 10). The other possibility for rac to activate WAVE is via the insulin receptor substrate of 53 kDa (IRSp53) which was shown by Miki et al.[39] to bind and control of WAVE, thereby initiating actin nucleation *in vitro*. IRSp53 is prominently expressed in neurons where it is a major constituent of the postsynaptic density[40], indicating that the IRSp53/WAVE pathway is likely to be involved in rac-dependent control of spine morphology. IRSp53 is targeted to synapses by interaction with PSD-95[41–43]. By siRNA and expression of dominant negative constructs, Choi et al. demonstrated that IRSp53 contributes to spine formation by increasing linear spine density on dendrites, and rac/IRSp53/WAVE2 signaling may provide the necessary amount of actin nucleation to achieve this. However, the example of IRSp53 also illustrates the complexity of molecular events, as its SH3 domain binds not only to WAVE2, but also to Mena (implicated in formation of filopodia[44]), Eps8 (a barbed end capping protein[45] ), and shank (a postsynaptic scaffold promoting spine formation and recruitment of postsynaptic components; refs. 46,47; see Figure 17.2; Colorplate

10). Interaction with Mena and shank does not depend on activation by rac, but instead requires GTP-bound cdc42 which binds to a different site on IRSp53. This promiscuity in terms of upstream activators as well as downstream effectors makes it difficult to clearly identify a single signaling pathway. Instead, it is likely that IRSp53 fulfils multiple functions depending on the local and temporal availability of activators as well as effectors. Thus, Mena might be a signal for filopodia formation on dendrites in early phases of development, as it is expressed only transiently during neuronal development. The N-terminal actin bundling domain of IRSp53 would support this process[48]. In later phases, shank might replace Mena as ligand for the SH3 domain, leading to increased assembly of PSD proteins. Consistently, co-expression of shank1 with IRSp53 in heterologous cells interferes with the formation of filopodia induced by the expression of IRSp53 alone[46].

Several studies which highlight the effect of rho family proteins on spine formation (presumably via regulation of the actin cytoskeleton) demonstrate that these GTPases should also be involved in postsynaptic complex formation, which is usually visualized as clustering of PSD-95. Thus kalirin expression is not only required for the maintenance of spines, but also for the maintenance of the pre- and postsynaptic protein complexes, visualized as PSD-95 and bassoon clusters[49]. So far it is unclear whether formation of PSD-95 clusters is secondary to achievement of a certain status of actin polymerization and branching in the spine, or whether the activity of small GTPases also has a direct effect on clustering or association of PSD scaffolds. The cdc42-dependent association of IRSp53 with shank indicates that small GTPases can have a direct effect on postsynaptic protein assembly. However, it remains to be shown whether other postsynaptic scaffolds are influenced by GTPases in a similar manner.

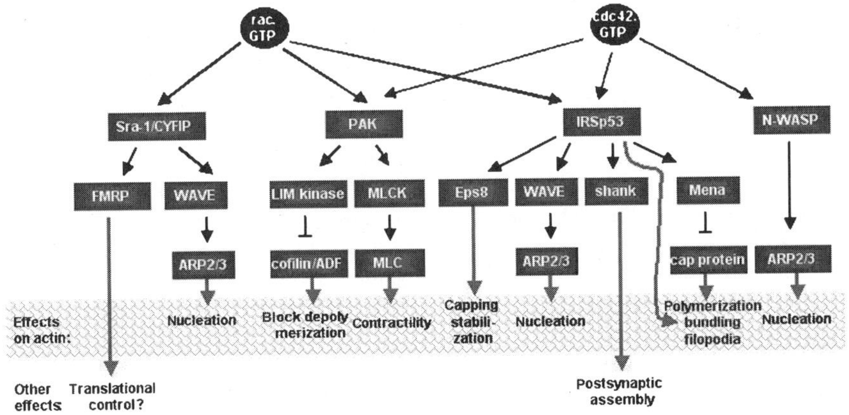


**Figure 17.2.** Possible Signaling Pathways Used by the Small GTPases rac and cdc42 in Neuronal Dendrites. Activation is denoted by a black arrow, inhibition by an interrupted black line. The final downstream effect of each pathway is indicated by a blue arrow. Note that for IRSp53, activation of Eps8 and WAVE has been described only in connection with rac signaling, whereas linkage to shank or Mena has been described after activation by cdc42. Several known effectors were omitted for clarity. See Colorplate 10.

## 7. SHANK PROTEINS: A CASE FOR LOCAL TRANSLATION OF POSTSYNAPTIC PROTEINS

Shank proteins, like other multidomain postsynaptic scaffolds, link receptors and the cytoskeleton. The role of shank proteins for the generation of dendritic spines and as important scaffolds within the PSD is described in Chapter 18. Remarkably, overexpression of shank induces enhanced structural and functional maturation of the postsynaptic receptor complex, concomitant with an earlier morphological maturation of dendritic spines[50,51]. In aspiny neurons, overexpression of shank is sufficient to induce spine-like structures[52], suggesting that shank is required in the transition from immature dendritic filopodia, which do not carry postsynaptic specializations, to proper, mushroom-shaped, postsynaptic spines. Thus the regulation of shank should be of crucial importance during synapse formation. Whereas so far no signaling mechanisms have been identified which act on the shank proteins themselves, an interesting feature of all three shank family members is the prominent dendritic localization of their coding mRNAs[53]. This became evident in *in situ* hybridization experiments by a strong labeling obtained in the so-called molecular layers of the hippocampus and the cerebellum, i.e., regions which contain mostly dendrites, axons, and synapses, but few cell bodies. Labeling was observed in dendritic fields originating from neurons in the CA1–CA3 regions, as well as the dentate gyrus in the hippocampus, and from Purkinje cells in the cerebellum. Dendritic targeting of mRNAs could be reproduced in an expression model in cultured neurons, leading to the conclusion that a 200 bp targeting sequence within the shank1 3' untranslated region is sufficient for targeting of the mRNAs. It remains to be shown if local translation of shank mRNAs contributes to maturation of the postsynapse; intriguingly, during postnatal development of the cerebellum high levels of shank2/ProSAP1 mRNA in the molecular layer were observed at a phase of intense synaptogenesis, i.e., during the second week *post partum*[53].

The number of other mRNAs coding for PSD proteins which are localized in dendrites is rather limited; the most notable example is the α-subunit of the $Ca^{2+}$/calmodulin-dependent protein kinase (αCaMKII); dendritic presence of this mRNA is required for synaptic plasticity, as detected in long-term potentiation paradigms[54]. In addition, mRNA coding for the arg3.1/arc gene product, a protein of unknown function, is rapidly induced and transported into distal dendrites upon neuronal activity[55]. Out of the multidomain structural proteins of the PSD, mRNAs coding for homer2[56] and the SAPAP/GKAP isoform 3 (SAPAP3[57,58]) have been found in hippocampal dendrites besides the shank mRNAs mentioned above. Based on sophisticated PCR approaches and molecular imaging techniques, claims have been made for the dendritic localization of many more mRNAs. However, confirmation of their dendritic localization by *in situ* hybridization either on brain sections or cultivated neurons is missing, suggesting that only a rather small (but growing) subset of mRNAs may be translated in dendrites in significant quantities.

The concept of dendritically localized mRNAs, which may be translated "on demand" requires strong translational control mechanisms, so that translation should be suppressed during transport, and only be allowed upon a certain stimulus such as activation of cell surface receptors or Ca influx. Several imaging studies on dendritic RNA transport particles, as well as biochemical purification of these particles, now suggest that mRNAs are transported within huge RNP particles[59]. These include putative translational regulators such as the fragile X mental retardation protein (FMRP) and its homologs FXR1 and FXR2, transacting factors

# Color Plate 13

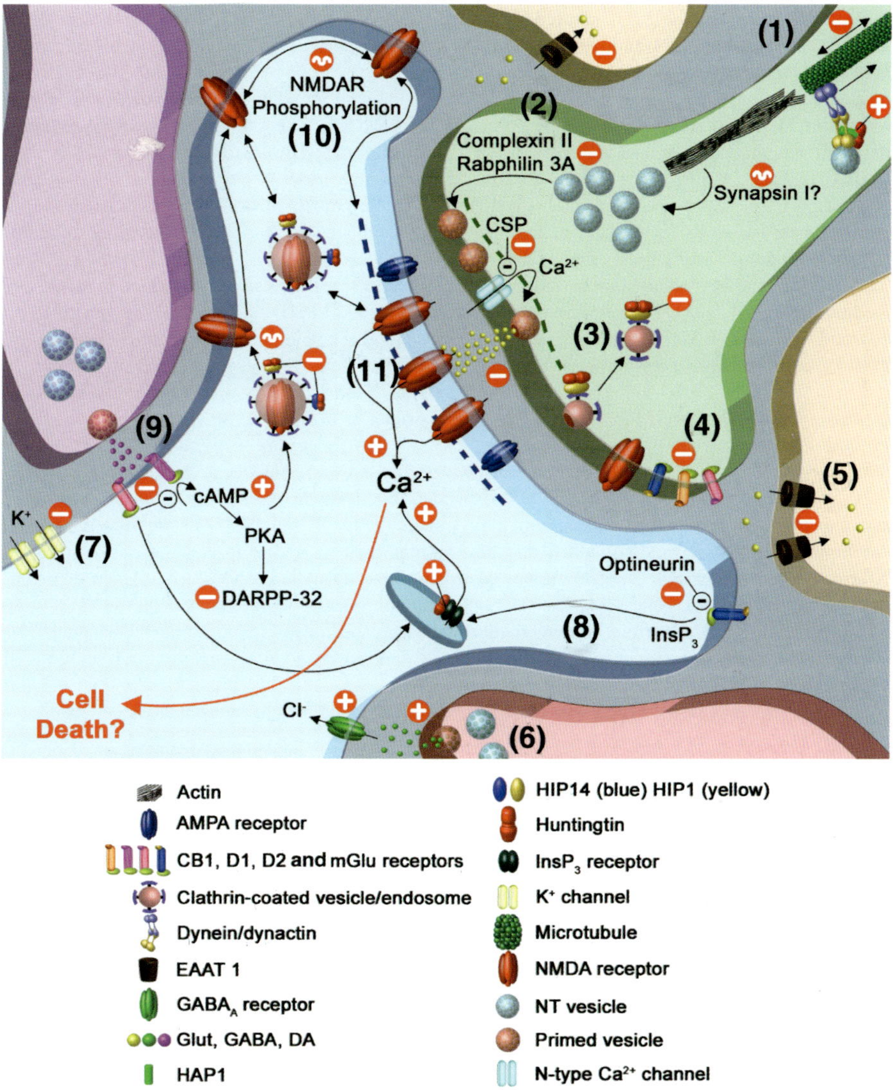


**Figure 31.1.** Summary of Alterations in Cortico-Striatal Synaptic Function Implicated in HD. Simplified figure of a cortico-striatal synapse; a cortical afferent (green) synapses with a spine of a striatal MSN (blue), astrocytes (brown) surround the synaptic cleft, and GABAergic (red) and dopaminergic (purple) terminals are also shown. Red circles indicate where positive, negative, and nonlinear alterations have been reported in HD and HD models. (1) Axonal transport along microtubules is reduced by polyQ expansion[9–12], e.g., disrupted BDNF transport is associated with the detachment of a dynactin p150/HAP1/mhtt complex from microtubules and may be due to increased HAP1/huntingtin binding produced by polyQ expansion[13–15]. (2) Synapsin 1 hyperphosphorylation[20] may alter synaptic vesicle reserve pool regulation; this and decreased levels of the vesicular proteins complexin II[23,25,26,31] and rabphilin 3A[22] may perturb docking and priming of NT vesicles within the active zone (broken green line). $Ca^{2+}$ influx, through N-type $Ca^{2+}$ channels, triggers vesicle fusion and NT release; this influx may be increased in HD by the loss of cysteine string protein (CSP)-mediated inhibition of N-type channels[32]. Earlier increases and later decreases in glutamate release are observed

# Color Plate 13 (Contd.)

in some HD mice[18]. (3) HIP1 is involved in clathrin-mediated endocytosis[35], a process which may be disrupted by reduced HIP1/huntingtin binding by polyQ expansion[34]. In addition, htt binding of the PACSIN1 regulator of vesicle recovery increases proportionally to polyQ, and PACSIN1 is abnormally located away from the synapse in HD[36]. (4) Receptors, known to modulate release presynaptically, are reduced in HD[48]. (5) Astrocytic excitatory amino acid transporter 1 (EAAT1/GLT1) mRNA levels are reduced[37] and glutamate uptake is altered in HD[20,38]. (6) Inhibitory GABA release is increased[39,44], as is GABA receptor labeling[44], possibly due to perturbations in HAP1-regulated GABA receptor recycling[81]. (7) $K^+$ channel currents and subunits are decreased in HD models[47]. (8) Activation of group I mGluRs results in the generation of $InsP_3$ and mhtt facilitates the action of the mGluR inhibitory protein optineurin[54] leading to reduced $InsP_3$ signaling; however polyQ-enhanced binding between $InsP_3$ receptor and htt results in increased $Ca^{2+}$ release from intracellular stores (reviewed in ref. 53). (9) Reductions in D1 and D2 dopamine receptors are observed in HD[21,49,60,75,76–78] (reviewed in ref. 74) and increased cAMP labeling[70], decreased DARPP-32 labeling, and a loss of dopaminergic modulation of other NT receptor systems is also observed[21,79]. D1 receptor agonism induces the redistribution of htt, HIP1 and clathrin to the membrane[71] and may play a role in regulating surface expression of NMDARs. (10) Increases in NR1[43] and decreases in NR2A/B protein (43, but see ref. 60), in addition to altered PSD-95 interactions[66,67] and differences in reported NMDAR subunit phosphorylation state[49,60,67,70] together suggest that trafficking of NMDARs is altered in HD. In addition, NMDARs are internalized by clathrin-dependent endocytosis; reduced mhtt/HIP-1 binding[33,34] and a breakdown in clathrin interaction may contribute to altered NMDAR surface expression. Additionally, HIP14/htt binding, which may also be important for endocytosis and protein trafficking, is also reduced by expanded polyQ (reviewed in ref. 68). (11) Altered NMDAR surface expression likely accounts for early increases in NMDAR function seen across HD models[40,43,46,52,56–59]. The resultant increase in $Ca^{2+}$ influx, in combination with increased release from intracellular stores[53], has severe implications for cortico-striatal $Ca^{2+}$ signaling, phosphatase/kinase activity, and consequently synaptic plasticity and the regulation of gene transcription. Moreover, such disruptions in $Ca^{2+}$ signaling and homeostasis may be the triggers for more severe neuronal dysfunction and ultimately cell death.

for RNA localization such as the RNA-binding proteins staufen, PUR, and elongation factor EF1α. One study implied that ribosomes are co-transported in these RNPs and that disassembly of the particle on a depolarizing stimulus would allow for translation of the transported mRNAs by these ribosomes[60].

## 8. REGULATED DEGRADATION OF PSD PROTEINS BY THE UBIQUITIN/PROTEASOME SYSTEM

As protein synthesis is tightly regulated in dendrites, it probably does not come as a surprise that the same holds true for protein degradation. Several studies have now begun to illuminate how neurons creatively use the forces of protein destruction in order to restructure their synaptic connections. Depending on previous neuronal activity, individual components of the PSD are selectively targeted for destruction by ubiquitinylation, followed by proteolysis in the 26S proteasome. One example is Spar, a GAP for Rap GTPases (i.e., a RapGAP). Spar is an actin-binding protein which is present in spines and contributes significantly to spine morphology[25]. Strong neuronal activity induces the expression of the protein kinase SNK which phosphorylates Spar and thus marks it for proteasome-dependent degradation[61]. Ehlers[4] systematically analyzed changes in the protein composition of the PSD from cultivated neurons which were induced by electrical activity. By pharmacological inhibition of proteasome activity and pull-down assays with polyubiquitin-binding proteins, he could identify a subgroup of PSD scaffold proteins which were targeted for degradation. Among these are the A-kinase anchoring protein AKAP, as well as shank and GKAP/SAPAP proteins. GKAP/SAPAP proteins are physically associated with shank, and both are believed to constitute some core fraction of PSD scaffold proteins. Selective removal of both proteins may allow for restructuring of the PSD. Interestingly, both shank and SAPAP3 mRNAs are localized to neuronal dendrites, suggesting that the expression or the turnover of these interacting proteins may be co-regulated. Further work is required to sort out which signaling pathways contribute to selection of PSD proteins for ubiquitinylation. In particular, it will be interesting to find out which types of E3 ubiquitin ligases confer substrate specificity to the process of polyubiquitinylation at postsynaptic sites.

## 9. X-LINKED MENTAL RETARDATION

Nonsyndromic MR in human patients is frequently associated with abnormal formation of dendritic spines. The characterization of inherited forms of MR, especially in patients suffering from X-chromosome linked MR, has supported the prominent role of rho family GTPases in establishing neuronal circuitry. Thus one putative GEF for rac and/or cdc42 (αPIX[62]), one GAP specific for rho (oligophrenin[63]), and one downstream target of cdc42 and rac (PAK3[64]) have been shown to be mutated in affected patients. This led to the proposal that the dysregulation of small GTPase cycles, and therefore dysregulation of spine actin might be a common denominator in the pathogenesis of MR[65]. Indeed knockdown of oligophrenin or PAK3 expression in hippocampal neurons changes the structure of dendritic spines. Reduction of oligophrenin levels leads to a decrease in spine length, which can also be mimicked experimentally by constitutively active RhoA, suggesting that derepression of the rho/Rho kinase pathway may contribute to MR in patients[66]. Suppression of PAK3 expression in neurons blocks spine maturation and leads to production of filopodial protrusions instead[67]. However, things may not be so straightforward, as the deletion of the PAK3 gene in mice does not lead to changes in the structure of the dendritic tree of hippocampal neurons, or changes in spine morphology[68]. PAK3 deficient mice do exhibit synaptic deficits, evident as a decrease in the late phase of LTP. However, these deficits are probably not due to a different regulation of the actin cytoskeleton, as the key step in this process, phosphorylation of cofilin, is not affected by loss of PAK3. Instead, the authors showed a decreased phosphorylation of the transcription factor CREB (cAMP responsive element-binding protein), which is known to contribute significantly to synaptic plasticity. The example of the PAK3 knockout mice demonstrates that there are probably many ways to generate deficits in synaptic plasticity and the cognitive abilities of mice or men, and interference with normal spine morphology may only be one of several aspects in the pathogenesis of MR.

Deficiencies in translational control at or near synapses have been implicated as a cause for the cognitive deficits associated with the fragile X mental retardation syndrome, which is the most frequent inherited form of MR. Patients exhibit loss of expression of FMRP due to a triplet repeat expansion in the promoter region of the gene. Consistent with studies on other molecules implicated in the pathogenesis of MR, the brains of mice lacking FMRP expression do exhibit irregular thin and elongated spines which are also the hallmark of MR patients[69,70]. FMRP is an RNA-binding protein which is implicated in translational repression of responsive mRNAs. One current view on FMRP is that its loss in MR patients or knockout mice leads to a translational derepression of certain target mRNAs. As the FMRP protein is present in dendrites, it is likely that especially local translational control in the dendrite is absent in patients. Recent work suggests that this might involve small nontranslated RNAs such as the BC1 (in rodents) or BC200 (in humans) as translational corepressors, both of which are also present in dendrites[71]. However, it has been a matter of debate which mRNAs are exactly controlled in their translational efficiency by the FMRP protein. A large group of mRNAs was found to be specifically associated with the FMRP protein in the brains of mice *in vivo*[72]. RNA binding by FMRP is favored by specific G-quartet sequence motifs within target mRNAs; association of these mRNAs with FMRP changed their translational status[73]. Other studies have provided additional lists of mRNAs which are associated with FMRP in neurons, with only limited overlap to the first study[74]. Studies on the related *Drosophila* protein (dFMR1) were helpful in this respect as it became clear that both the fly and the mammalian protein may associate with and regulate translation of the message coding for microtubule associated protein MAP1b (*futsch* in flies). MAP1b/*futsch* protein is required for development of neuronal morphology[75]. Nevertheless it remains unclear at present how the large number of mRNAs/proteins regulated by FMRP actually contributes to aberrant neuronal development or synapse formation.

An interesting link between the fragile X syndrome and those MR forms caused by mutations in rho GTPase regulators is constituted by the cytosolic FMRP interacting protein CYFIP1, also known as Sra-1 (specifically rac1 associated protein). As mentioned above, CYFIP1 is involved in a hetero-pentameric complex, including WAVE. Upon rac binding to CYFIP1, this complex dissociates, leading to activation of WAVE. Similarly, the interaction of

*Drosophila* CYFIP with dFMRP is disrupted by rac binding, potentially establishing control of the translational apparatus via a rac-CYFIP-FMRP pathway[76]. Thus, there may be a strong interdependence of MR gene products controlling the cytoskeleton, and FMRP having a translational control function. However, the role of CYFIP and its various associated proteins for the morphological and functional development of the mammalian nervous system has not been analyzed yet.

## 10. CONCLUSIONS AND FUTURE DIRECTIONS

The current catalog of PSD proteins, and the diversity of signaling pathways that contribute to spine formation and PSD assembly, present a rather diverse picture of cellular activities which shape formation of the postsynaptic specialization. If one takes into account the promiscuity of the protein/protein interactions engaged by the various signaling proteins, it is at present rather difficult to assess the contribution of individual proteins. Thus rac or cdc42 can bind to quite a number of downstream target proteins, and it is not known which of these targets are present during dendrite development and spine formation at significant quantities. The same holds true for signaling intermediates such as IRSp53, which may control a substantial variety of dendritic actin regulatory proteins as well as postsynaptic scaffolds. The case of the rac target Sra-1/CYFIP is particularly intriguing: does it control the actin nucleation activity of WAVE in dendrites, or the translational inhibitory activity of FMRP? If it does both, does the lack of FMRP in MR patients lead to a loss of translation control, or does it simply mean that more Sra-1/CYFIP is available to inhibit WAVE? Methods providing subcellular quantification of possible binding partners should help in this respect. At the microscopic level fluorescence methods enabling temporal and spatial resolution are required to understand when and where a particular signaling event might occur. In this respect, the detection of local activation of rac by the FRET probes employed by Zhang and colleagues[36] is an interesting development which should be applicable to other signaling pathways.

## 11. REFERENCES


1. Chen, X., Vinade, L., Leapman, R.D., Petersen, J.D., Nakagawa, T., Phillips, T.M., Sheng, M., and Reese, T.S. (2005) *Proc Natl Acad Sci U S A* **102**, 11551–11556.
2. Sugiyama, Y., Kawabata, I., Sobue, K., and Okabe, S. (2005) *Nat Methods* **2**, 677–684.
3. Okabe, S., Miwa, A., and Okado, H. (2001) *J Neurosci* **21**, 6105–6114.
4. Ehlers, M.D. (2003) *Nat Neurosci* **6**, 231–242.
5. Romorini, S., Piccoli, G., Jiang, M., Grossano, P., Tonna, N., Passafaro, M., Zhang, M., and Sala, C. (2004) *J Neurosci* **24**, 9391–9404.
6. Shapira, M., Zhai, R.G., Dresbach, T., Bresler, T., Torres, V.I., Gundelfinger, E.D., Ziv, N.E., and Garner, C.C. (2003) *Neuron* **38**, 237–252.
7. Washbourne, P., Bennett, J.E., and McAllister, A.K. (2002) *Nat Neurosci* **5**, 751–759.
8. Bresler, T., Ramati, Y., Zamorano, P.L., Zhai, R., Garner, C.C., and Ziv, N.E. (2001) *Mol Cell Neurosci* **18**, 149–167.
9. Bresler, T., Shapira, M., Böckers, T., Dresbach, T., Futter, M., Garner, C.C., Rosenblum, K., Gundelfinger, E.D., and Ziv, N.E. (2004) *J Neurosci* **24**, 1507–1520.
10. Shin, H., Hsueh, Y.P., Yang, F.C., Kim, E., and Sheng, M. (2000) *J Neurosci* **20**, 3580–3587.
11. McGee, A.W., and Bredt, D.S. (1999) *J Biol Chem* **274**, 17431–17436.
12. McGee, A.W., Dakoji, S.R., Olsen, O., Bredt, D.S., Lim, W.A., and Prehoda, K.E. (2001) *Mol Cell* **8**, 1291–1301.
13. Tavares, G.A., Panepucci, E.H., and Brunger, A.T. (2001) *Mol Cell* **8**, 1313–1325.
14. Mayer, B.J. (2001) *J Cell Sci* **114**, 1253–1263.

15. Wu, H., Reissner, C., Kuhlendahl, S., Coblentz, B., Reuver, S., Kindler, S., Gundelfinger, E.D., and Garner, C.C. (2000) *EMBO J* **19**, 5740–5751.
16. El-Husseini, A.E., Craven, S.E., Chetkovich, D.M., Firestein, B.L., Schnell, E., Aoki, C., and Bredt, D.S. (2000) *J Cell Biol* **148**, 159–172.
17. El-Husseini Ael, D., Schnell, E., Dakoji, S., Sweeney, N., Zhou, Q., Prange, O., Gauthier-Campbell, C., Aguilera-Moreno, A., Nicoll, R.A., and Bredt, D.S. (2002) *Cell* **108**, 849–863.
18. Hsueh, Y. P., Kim, E., and Sheng, M. (1997) *Neuron* **18**, 803–814.
19. Morabito, M.A., Sheng, M., and Tsai, L.H. (2004) *J Neurosci* **24**, 865–876.
20. Star, E.N., Kwiatkowski, D.J., and Murthy, V.N. (2002) *Nat Neurosci* **5**, 239–246.
21. Okamoto, K., Nagai, T., Miyawaki, A., and Hayashi, Y. (2004) *Nat Neurosci* **7**, 1104–1112.
22. Fifkova, E., and Delay, R.J. (1982) *J Cell Biol* **95**, 345–350.
23. Takenawa, T., and Miki, H. (2001) *J Cell Sci* **114**, 1801–1809.
24. Weaver, A.M., Karginov, A.V., Kinley, A.W., Weed, S.A., Li, Y., Parsons, J.T., and Cooper, J.A. (2001) *Curr Biol* **11**, 370–374.
25. Pak, D.T., Yang, S., Rudolph-Correia, S., Kim, E., and Sheng, M. (2001) *Neuron* **31**, 289–303.
26. Vazquez, L.E., Chen, H.J., Sokolova, I., Knuesel, I., and Kennedy, M.B. (2004) *J Neurosci* **24**, 8862–8872.
27. Etienne-Manneville, S., and Hall, A. (2002) *Nature* **420**, 629–635.
28. Jones, S.B., Lu, H.Y., and Lu, Q. (2004) *J Neurosci* **24**, 8510–8521.
29. Martinez, M.C., Ochiishi, T., Majewski, M., and Kosik, K.S. (2003) *J Cell Biol* **162**, 99–111.
30. Penzes, P., Beeser, A., Chernoff, J., Schiller, M.R., Eipper, B.A., Mains, R.E., and Huganir, R.L. (2003) *Neuron* **37**, 263–274.
31. Irie, F., and Yamaguchi, Y. (2002) *Nat Neurosci* **5**, 1117–1118.
32. Penzes, P., Johnson, R.C., Sattler, R., Zhang, X., Huganir, R.L., Kambampati, V., Mains, R.E., and Eipper, B.A. (2001) *Neuron* **29**, 229–242.
33. Park, E., Na, M., Choi, J., Kim, S., Lee, J.R., Yoon, J., Park, D., Sheng, M., and Kim, E. (2003) *J Biol Chem* **278**, 19220–19229.
34. Zhao, Z.S., Manser, E., Loo, T.H., and Lim, L. (2000) *Mol Cell Biol* **20**, 6354–6363.
35. Zhang, H., Webb, D.J., Asmussen, H., and Horwitz, A.F. (2003) *J Cell Biol* **161**, 131–142.
36. Zhang, H., Webb, D.J., Asmussen, H., Niu, S., and Horwitz, A.F. (2005) *J Neurosci* **25**, 3379–3388.
37. Lee-Hoeflich, S.T., Causing, C.G., Podkowa, M., Zhao, X., Wrana, J.L., and Attisano, L. (2004) *EMBO J* **23**, 4792–4801.
38. Eden, S., Rohatgi, R., Podtelejnikov, A.V., Mann, M., and Kirschner, M.W. (2002) *Nature* **418**, 790–793.
39. Miki, H., Yamaguchi, H., Suetsugu, S., and Takenawa, T. (2000) *Nature* **408**, 732–735.
40. Abbott, M.A., Wells, D.G., and Fallon, J.R. (1999) *J Neurosci* **19**, 7300–7308.
41. Soltau, M., Berhörster, K., Kindler, S., Buck, F., Richter, D., and Kreienkamp, H.J. (2004) *J Neurochem* **90**, 659–665.
42. Choi, J., Ko, J., Racz, B., Burette, A., Lee, J.R., Kim, S., Na, M., Lee, H.W., Kim, K., Weinberg, R.J., and Kim, E. (2005) *J Neurosci* **25**, 869–879.
43. Hori, K., Yasuda, H., Konno, D., Maruoka, H., Tsumoto, T., and Sobue, K. (2005) *J Neurosci* **25**, 2670–2681.
44. Krugmann, S., Jordens, I., Gevaert, K., Driessens, M., Vandekerckhove, J., and Hall, A. (2001) *Curr Biol* **11**, 1645–1655.
45. Funato, Y., Terabayashi, T., Suenaga, N., Seiki, M., Takenawa, T., and Miki, H. (2004) *Cancer Res* **64**, 5237–5244.
46. Soltau, M., Richter, D., and Kreienkamp, H.J. (2002) *Mol Cell Neurosci* **21**, 575–583.
47. Bockmann, J., Kreutz, M.R., Gundelfinger, E.D., and Böckers, T.M. (2002) *J Neurochem* **83**, 1013–1017.
48. Yamagishi, A., Masuda, M., Ohki, T., Onishi, H., and Mochizuki, N. (2004) *J Biol Chem* **279**, 14929–14936.
49. Ma, X.M., Huang, J., Wang, Y., Eipper, B.A., and Mains, R.E. (2003) *J Neurosci* **23**, 10593–10603.
50. Sala, C., Piech, V., Wilson, N.R., Passafaro, M., Liu, G., and Sheng, M. (2001) *Neuron* **31**, 115–130.
51. Quitsch, A., Berhörster, K., Liew, C.W., Richter, D., and Kreienkamp, H.J. (2005) *J Neurosci* **25**, 479–487.
52. Roussignol, G., Ango, F., Romorini, S., Tu, J.C., Sala, C., Worley, P.F., Bockaert, J., and Fagni, L. (2005) *J Neurosci* **25**, 3560–3570.
53. Bockers, T.M., Segger-Junius, M., Iglauer, P., Bockmann, J., Gundelfinger, E.D., Kreutz, M.R., Richter, D., Kindler, S., and Kreienkamp, H.J. (2004) *Mol Cell Neurosci* **26**, 182–190.

54. Miller, S., Yasuda, M., Coats, J.K., Jones, Y., Martone, M.E., and Mayford, M. (2002) *Neuron* **36**, 507–519.
55. Link, W., Konietzko, U., Kauselmann, G., Krug, M., Schwanke, B., Frey, U., and Kuhl, D. (1995) *Proc Natl Acad Sci U S A* **92**, 5734–5738.
56. Schratt, G.M., Nigh, E.A., Chen, W.G., Hu, L., and Greenberg, M.E. (2004) *J Neurosci* **24**, 7366–7377.
57. Kindler, S., Rehbein, M., Classen, B., Richter, D., and Bockers, T.M. (2004) *Brain Res Mol Brain Res* **126**, 14–21.
58. Welch, J.M., Wang, D., and Feng, G. (2004) *J Comp Neurol* **472**, 24–39.
59. Kanai, Y., Dohmae, N., and Hirokawa, N. (2004) *Neuron* **43**, 513–525.
60. Krichevsky, A.M., and Kosik, K.S. (2001) *Neuron* **32**, 683–696.
61. Pak, D.T., and Sheng, M. (2003) *Science* **302**, 1368–1373.
62. Kutsche, K., Yntema, H., Brandt, A., Jantke, I., Nothwang, H.G., Orth, U., Boavida, M.G., David, D., Chelly, J., Fryns, J.P., Moraine, C., Ropers, H.H., Hamel, B.C., van Bokhoven, H., and Gal, A. (2000) *Nat Genet* **26**, 247–250.
63. Billuart, P., Bienvenu, T., Ronce, N., des Portes, V., Vinet, M.C., Zemni, R., Roest Crollius, H., Carrie, A., Fauchereau, F., Cherry, M., Briault, S., Hamel, B., Fryns, J.P., Beldjord, C., Kahn, A., Moraine, C., and Chelly, J. (1998) *Nature* **392**, 923–926.
64. Allen, K.M., Gleeson, J.G., Bagrodia, S., Partington, M.W., MacMillan, J.C., Cerione, R.A., Mulley, J.C., and Walsh, C.A. (1998) *Nat Genet* **20**, 25–30.
65. Ramakers, G.J. (2002) *Trends Neurosci* **25**, 191–199.
66. Govek, E.E., Newey, S.E., Akerman, C.J., Cross, J.R., Van der Veken, L., and Van Aelst, L. (2004) *Nat Neurosci* **7**, 364–372.
67. Boda, B., Alberi, S., Nikonenko, I., Node-Langlois, R., Jourdain, P., Moosmayer, M., Parisi-Jourdain, L., and Muller, D. (2004) *J Neurosci* **24**, 10816–10825.
68. Meng, J., Meng, Y., Hanna, A., Janus, C., and Jia, Z. (2005) *J Neurosci* **25**, 6641–6650.
69. Irwin, S.A., Patel, B., Idupulapati, M., Harris, J.B., Crisostomo, R.A., Larsen, B.P., Kooy, F., Willems, P.J., Cras, P., Kozlowski, P.B., Swain, R.A., Weiler, I.J., and Greenough, W.T. (2001) *Am J Med Genet* **98**, 161–167.
70. Comery, T.A., Harris, J.B., Willems, P.J., Oostra, B.A., Irwin, S.A., Weiler, I.J., and Greenough, W.T. (1997) *Proc Natl Acad Sci U S A* **94**, 5401–5404.
71. Zalfa, F., Giorgi, M., Primerano, B., Moro, A., Di Penta, A., Reis, S., Oostra, B., and Bagni, C. (2003) *Cell* **112**, 317–327.
72. Brown, V., Jin, P., Ceman, S., Darnell, J.C., O'Donnell, W.T., Tenenbaum, S.A., Jin, X., Feng, Y., Wilkinson, K.D., Keene, J.D., Darnell, R.B., and Warren, S.T. (2001) *Cell* **107**, 477–487.
73. Darnell, J.C., Jensen, K.B., Jin, P., Brown, V., Warren, S.T., and Darnell, R.B. (2001) *Cell* **107**, 489–499.
74. Miyashiro, K.Y., Beckel-Mitchener, A., Purk, T.P., Becker, K.G., Barret, T., Liu, L., Carbonetto, S., Weiler, I.J., Greenough, W.T., and Eberwine, J. (2003) *Neuron* **37**, 417–431.
75. Zhang, Y.Q., Bailey, A.M., Matthies, H.J., Renden, R.B., Smith, M.A., Speese, S.D., Rubin, G.M., and Broadie, K. (2001) *Cell* **107**, 591–603.
76. Schenck, A., Bardoni, B., Langmann, C., Harden, N., Mandel, J.L., and Giangrande, A. (2003) *Neuron* **38**, 887–898.

18

# REGULATION OF DENDRITIC SPINE MORPHOLOGY AND SYNAPTIC FUNCTION BY SCAFFOLDING PROTEINS

Stefano Romorini, Giovanni Piccoli, and Carlo Sala*

## 1. SUMMARY

Glutamate receptors on central nervous system (CNS) synapses are directly or indirectly associated with various scaffold proteins, most of which are localized at the postsynaptic density (PSD). Although the function of a number of these proteins is still unknown, it is now generally accepted that some regulate dendritic spine morphology and synapse function. Scaffold proteins bind and recruit proteins that regulate actin cytoskeleton remodeling and signaling transduction, thus linking neurotransmitter receptor activation to intracellular cytoskeletal and signaling modifications. Interestingly, mutations of some of these scaffold proteins have been implicated in the manifestation of severe forms of mental retardation and autism, which suggests that they are fundamental elements for synapse and dendritic spine structure and function. The aim of this review is to summarize recent findings concerning the scaffold protein functions involved in regulating the morphology of dendritic spines and, consequently, the function of excitatory synapses.

## 2. INTRODUCTION

Neurons communicate with each other using specialized cell junctions known as synapses. An individual neuron receives two major types of synaptic input, excitatory and inhibitory, which enormously differ in their molecular composition, morphology and function. Excitatory synapses are localized mostly at dendritic spines and are asymmetric synapses at the electron microscopic images due to the presence of a distinct structure known as the PSD. In contrast, inhibitory synapses

*CNR Institute of Neuroscience, Cellular and Molecular Pharmacology, Department of Pharmacology, University of Milan, Via Vanvitelli 32, Milan 20129, Italy; c.sala@in.cnr.it

are essentially formed on the dendritic shaft and are mostly void of a structurally defined PSD (symmetric synapses). In this chapter we focus on discussing the recent discoveries on the machinery involved in the differentiation of excitatory synapses. In particular we focus on scaffold proteins involved in regulating excitatory synapses function and morphology.

Over the last 10 years, a number of scaffold proteins have been identified and cloned as major components of the PSD in excitatory synapses. Some of these proteins were identified biochemically and genetically because of their specific synapse localization. The first scaffold proteins to be identified were members of the PSD-95 family, which are currently considered to be among the most abundant proteins in the PSD. PSD-95 is by far the most widely studied and probably one of the most important PSD scaffold proteins. The PSD-95 family of proteins associates with a number of receptors and transmembrane proteins, and is also coupled to various scaffold and signaling proteins. For example coupling of PSD-95 to GKAP/SAPAP proteins links it to Shank, which in turn binds to another important PSD-resident protein family called Homer.

The complex formed by these four proteins is apparently a major component of the PSD (see Chapter 17 for a description of how the complex is assembled at postsynaptic sites during synapse formation) and thus is thought to play a critical role in the development and stability of excitatory contacts.

Although their association with the PSD is regulated by synaptic activity, these scaffold proteins can be considered among key resident molecules in the PSD, but there are also various other scaffold proteins and actin-binding proteins that probably shuttle in and out of the PSD and synapse, thus regulating actin polymerization upon glutamate receptor stimulation. The PSD is therefore a highly complex structure in which resident and shuttling proteins are involved in organizing signaling at synapses. A number of scaffold proteins are formed by combining modules of protein domains whose function is protein–protein interactions. One of the most abundant interaction domains used by synaptic proteins is the PSD-95/Dlg/ZO-1 homology (PDZ) domain. PSDs are localized at the tip of small dendritic protrusions known as dendritic spines. Most of the dendritic spine cytoskeleton consists of F-actin, which is why various actin-binding scaffold proteins are concentrated in the spines. The development of these spines will be mainly discussed in Chapter 10, whereas we discuss how a number of the PSD scaffold proteins regulate dendritic spine morphology. The importance of studying spine morphogenesis is related to the fact that various genetic diseases have been identified in which defects in dendritic spine morphology are associated with cognitive disorders such as mental retardation and autism. Future challenges will be to analyze the function of these scaffold proteins in detail in order to discover which of the molecular mechanisms regulating spine formation and synapse function are deregulated in such a way as to cause mental disabilities.

## 3. DENDRITIC SPINE STRUCTURE

The functions and properties of a synapse localized on a dendritic spine are probably largely dependent and directly controlled by the shape, structure, and molecular composition of the spine itself, and also the number and shape of dendritic spines are controlled by synaptic activity. One finding that supports the importance of spine shape and number in cognitive functions is the fact that both parameters are altered in various genetic diseases involving mental retardation, although it is not clear whether this is a consequence or the cause of the diseases themselves[1]. It is therefore fundamental to understand how spine shape and structure are regulated and controlled.

Dendritic spines essentially consist of a head and neck attached to the dendritic membrane. Extensive electron and light microscopy studies of brain tissue have shown that dendritic spines have multiform shapes that are classified as thin, stubby, or mushroom shape (Figure 18.1), all of which can be found simultaneously on the same dendrites[2,3]. The classic and most representative mushroom-shaped spines have a large head and narrow neck, whereas thin spines have a smaller head; stubby spines show no obvious constriction between the head and its attachment to the dendritic shaft (Figure 18.1). However, the static microscopic view of spine shapes only partially reflects the real *in vivo* situation because, at least in developing neurons, the shape of about 50% of the spines changes over periods of minutes or hours[4]. Spine motility and modifications are developmentally regulated and, in mature neurons, there are fewer transitions between categories and the number of stable spines increases[5]. Mature spines usually make contact with one or two excitatory presynaptic terminals located at the head and, essentially, a dendritic spine represents the main unitary postsynaptic compartment for excitatory input.

Immature spines are often very long and have filopodia-like shape. At later stages of neuronal contact development, their mean length decreases and the number of filopodia is greatly reduced. During the maturation process there is an increase in spine density, a decrease in overall length, and a decrease in the number of dendritic filopodia. Recent studies showed that filopodia rapidly protrude and retract from dendrites, especially during the early stages of synaptogenesis[6], and thus, it is widely believed that dendritic filopodia are the precursors of dendritic spines, although how the transition from filopodia to spines occur has not been completely clarified.

Remarkable differences have been observed in the intracellular composition of each spine, which consists of the PSD organelle facing the presynaptic button and the cytoskeletal structure consisting mainly of F-actin. About 50% of the spines on hippocampal CA1 cells and virtually all spines on Purkinje cells also have a smooth endoplasmic reticulum (SER)[7]. Some of the dendritic spines of pyramidal neurons contain the spine apparatus, an organelle consisting of two or more disks of SER separated by an electron-dense material. Large spines usually have a morphologically and functionally large synapse, and often contain an SER and spine apparatus, organelles that are formally absent in small spines[7]. As the SER plays a role in $Ca^{2+}$ handling, the regulation of spine size may be critical for controlling calcium homeostasis. It has recently been demonstrated that the NMDA receptor-dependent increase in spine $Ca^{2+}$ is less in large spines (which have a large neck that permits a greater efflux of $Ca^{2+}$ into the dendritic shaft) and more in small spines that have a smaller neck. Spine–neck geometry may govern the ability of small spines to be preferential sites for the isolated induction of long-term potentiation (LTP)[8]. Other important components are the polyribosomes, which are in a close relationship with each spine and in which proteins can be specifically synthesized. They are frequently found in the spines of different neuron subtypes, where they relocate after LTP[9].

The spine PSD occupies about 10% of the surface area exactly opposite the presynaptic active zone. The PSD and spine volume are also proportional to the area of the active zone, which is proportional to the number of docked vesicles[10], and the number of docked vesicles in turn correlates with the amount of neurotransmitter release per action potential. All of these data suggest that large spines represent stronger synapses for both pre- and postsynaptic properties, and that the growth of the spine head during development probably correlates with reinforced synaptic transmission[10].

The shape of the PSD is not fixed, but may change with alterations in the strength of synaptic activity, such as those occurring during LTP, and this may reflect enhanced AMPA receptor insertion into the postsynaptic membrane (as an early phase of synapse duplication and spine division)[11]. Overexpression of the AMPA GluR2 subunit induces spine formation in hippocampal neurons, an effect that depends on the subunit's N-terminal domain (NTD) and not on receptor channel properties. These data suggest that GluR2 plays a structural role in spine formation[12].

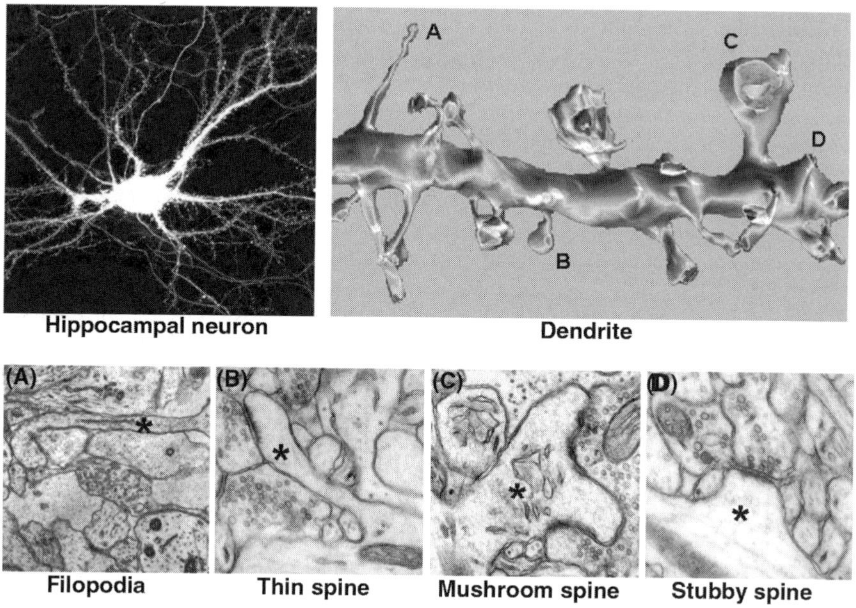


**Figure 18.1.** The Different Dendritic Spine Structures and Morphologies. Dendritic spine morphology can be observed in a hippocampal neuron transfected with GFP as shown in the top left panel. A more detailed analysis of spine shape has been observed by a 3D reconstruction of serial electron microscopy (EM) images as shown in the top right panel. In panels A–C are represented examples of different dendritic spine shapes viewed at EM level (spines labeled with asterisks). Similar images have been used for the 3D reconstruction. EM and 3D-recontruction images were adapted from: *Synapse Web, Atlas of Ultrastructural Neurocytology Josef Spacek MUDr DrSc, John C. Fiala PhD, Kristen M. Harris PhD, Funded by the Human Brain Project, NIH grant R01 MH/DA 57351,* http://synapses.mcg.edu; and from Fiala, J. C., Spacek, J., and Harris, K. M. (2002) *Brain Res Brain Res Rev* **39**, 29–54, with permission from Elsevier.

## 4. THE FUNCTION OF SCAFFOLD PROTEINS AT SYNAPSES

By definition, scaffolds have the function of assembling a complex of proteins that bind to their concatenated protein–protein interaction domains. At CNS excitatory synapses, each glutamate receptor subclass is associated with a complex of different and often interconnected scaffold proteins. The first scaffold proteins identified as being associated with glutamate receptors were PSD-95 for the NMDA receptor, GRIP for AMPA receptors, and Homer for metabotropic GluRs (mGluRs) (Figure 18.2; Colorplate 11)[13]. One of the first observations made in those early days was that most of these proteins have a PDZ domain. We will discuss PSD-95 and Homer function in detail later, and here concentrate on the function of the PDZ domain and the complex of scaffold proteins interacting with AMPA-type receptors.

PDZ domain-containing proteins play a central role at the synapses, as is shown by the fact that several synaptic scaffold proteins contain at least one PDZ domain (for a review, see 14). The major function of these PDZ-containing proteins is to act as scaffolds for the assembly of large protein complexes at specific subcellular locations, particularly on the cell surface. The PDZ functions as a modular domain that is well suited to be inserted into multi-PDZ proteins, or combined with other modular protein interaction domains in order to generate more complex scaffolds. Most PDZ domains recognize a few amino acids at the C-termini of proteins, and thus interact with the large majority of transmembrane proteins whose C-termini face the cytoplasm, as in the case of all glutamate receptors. It is therefore not surprising that PDZ proteins are well suited for postsynaptic membrane functions. It is also important to note that PDZ domains have different binding specificities, which means that the combination of different PDZ domains within a scaffold protein determines the composition of the protein complex assembled around the scaffold. Finally, most PDZ-containing proteins can multimerize to increase the size and, potentially, the heterogeneity of the PDZ-based complex[15].

One clear example of the synaptic function of PDZ-containing scaffold proteins came from studies of the mechanisms that modulate the synaptic trafficking of AMPA-type glutamate receptors. The number of synaptic AMPA receptors directly regulates synapse potentiation and depression, which are respectively associated with LTP and long-term depression (LTD). AMPA receptors are linked to two major scaffold proteins, the glutamate-receptor-interacting protein/AMPAR-binding protein (GRIP/ABP, encoded by the two distinct genes *GRIP1* and *ABP/GRIP2*), and the protein interacting with C kinase 1 (PICK1); these interactions may account for the dynamic cell-biological behavior of AMPARs at synapses.

The C-terminal of the GluR2/3 subunit specifically binds to the PDZ5 domain of GRIP, but the PDZ4 domain is also required for a strong interaction by stabilizing the PDZ5 structure through interdomain interactions[16]. GRIP is believed to be involved in the synaptic trafficking and/or stabilization of AMPARs and other interacting proteins (also see Chapter 24). The widespread cellular distribution of GRIP and its interaction with motor proteins [directly with conventional kinesin KIF5[17] or indirectly with KIF1A via liprin-α[18]] suggest that multiple motor proteins may contribute to the transport of AMPARs.

GRIP has up to seven PDZ domains through which it can interact with many proteins, including Eph receptors and their ephrin ligands[19], a RAS guanine nucleotide exchange factor (RasGEF)[20], liprin- α[21], the transmembrane protein Fraser syndrome 1 (FRAS1)[22], and the metabotropic and kainate-type glutamate receptors[23]. Thus GRIP can participate in synaptic function not only by interacting with AMPARs, but also by associating with Eph receptors and their ephrin ligands, which have been found to be involved in dendrites and dendritic spine morphogenesis and hippocampal synaptic plasticity[24,25].

PICK1 is located at synaptic and nonsynaptic sites in neurons. Its PDZ domain has relatively promiscuous binding properties, and both pre- and postsynaptic partners have been found: PKCα, GluR2/3, the netrin receptor UNC5H[26], various metabotropic glutamate receptor subtypes[27,28], the dopamine plasma-membrane transporter[29], and the erythroblastic leukemia viral oncogene homolog 2 (ErbB2) receptor tyrosine kinase[30]. In most of these cases, the subcellular localization and/or surface expression of these partners seems to be regulated by interactions with PICK1.

## 5. THE PSD-95 FAMILY

The PSD-95 family is encoded by four genes, PSD-95/SAP90 (synapse-associated protein 90), PSD-93/chapsyn-110, SAP102, and SAP97. The structure of these proteins is characterized by the assembly of three PDZ domains, an SRC homology (SH3) domain, and a guanylate kinase-like (GK) domain. SAP97 also has a LIN2/LIN7 (L27) domain at the N-terminal. The PSD-95 family belongs to a protein superfamily called membrane-associated guanylate kinase (MAGUK), which is characterized by the presence of at least one PDZ and one GK domain.

Immuno-EM and tomography studies have indicated that PSD-95 is localized very close to the postsynaptic membrane (a mean distance of 12 nm), a good position for interacting with postsynaptic membrane proteins such as receptors, ion channels, and cell-adhesion molecules, as well as with cytoplasmic proteins (Figure 18.2; Colorplate 11). It has been suggested that these interactions are important for PSD-95 clustering and targeting of postsynaptic membrane proteins. It is known that PSD-95 can cluster NMDARs and Shaker-type $K^+$ channels on the surface of heterologous cells[31], but the best *in vivo* demonstration that it also clusters postsynaptic proteins has been found in *Drosophila melanogaster*, in which mutations in discs large (Dlg), the *Drosophila* homolog of PSD-95, abolish the synaptic clustering of Shaker $K^+$ channels that bind to the PDZ domains of Dlg[32]. However, there is still disagreement as to whether PSD-95 is essential for the synaptic clustering of NMDA receptors in mammals, which is certainly not eliminated by the genetic disruption of PSD-95[33].

PSD-95 probably plays an important role in synapse formation by means of its interaction with neuroligin, a postsynaptic membrane protein that interacts *trans*-synaptically with β-neurexins, which in turn bind to the PDZ domain of CASK/LIN2 (another scaffold of the MAGUK superfamily of proteins), which is enriched on both sides of the synapse and interacts with other synaptic membrane proteins, such as syndecan and SynCAM[34,35]. The *trans*-synaptic neuroligin-β–neurexin interaction seems to be important in inducing pre- and postsynaptic synaptic differentiation[36–38], and the amount of PSD-95 regulates the balance between the number of inhibitory and excitatory synapses[36].

It now seems clear that the most important biochemical function of PSD-95 is to organize the signaling complexes at the postsynaptic membrane. PSD-95 interacts with a wide variety of cytoplasmic signaling molecules and thus, by physically bringing together cytoplasmic signal-transducing enzymes and surface receptors, may facilitate signal coupling in the PSD. For example, the over-expression of PSD-95 in hippocampal neurons increases the number of dendritic spines and the maturation of excitatory synapses[39], which may be related to the recruitment of transmembrane proteins and intracellular signaling proteins. Some of these proteins are directly involved in spine formation, including kalirin-7, a guanine nucleotide exchange factor (GEF) for RAC1 that promotes spine formation possibly as a downstream effector of the EphB receptor (also see Chapter 10)[40], SPAR, an inhibitory GAP for RAP that promotes the growth of dendritic spines[41], and IRSp53, a downstream effector of Cdc42 small GTPase[42]. However, whether the effect of PSD-95 on spine depends on its interactions with kalirin, SPAR or IRSp53 is not yet known (Figure 18.2; Colorplate 11).

A number of other signaling proteins are associated with PSD-95. One of these is the nitric oxide synthase (nNOS), a $Ca^{2+}$/calmodulin-activated enzyme that produces the nitric oxide involved in regulating neurotransmission and excitotoxicity. Interestingly, the ternary NMDAR–PSD-95–nNOS complex may functionally couple NMDAR gating to nNOS activation, as is suggested by the observation that disrupting the NMDAR–PSD-95 interaction with a synthetic peptide that mimics the last nine residues of NR2B reduces NMDAR-induced excitotoxicity without affecting NMDAR function[43]. Another abundant PSD signaling molecule that binds to PSD-95 is the synaptic Ras GTPase-activating protein (SynGAP), a GTPase-activating protein (GAP) for the Ras small GTPase which, after activation by CaMKII, suppresses the ERK pathway regulating synaptic plasticity. Heterozygous mice mutated for SynGAP show not only reduced LTP in the CA1 region of the hippocampus and impaired spatial learning, but also accelerated dendritic spine and synapse maturation[44,45]. PSD-95 also associates with the nonreceptor tyrosine kinases of the Src family[46] and their upstream activator, proline-rich tyrosine kinase 2 (Pyk2)[47], both of which are thought to be important for synaptic plasticity. It is therefore possible that it may localize the Pyk2–Src signaling cascade close to NMDARs, although the importance of PSD-95 scaffolds in synaptic regulation by tyrosine phosphorylation has not been directly investigated.

PSD-95 also interacts with other scaffold proteins such as GKAP that binds Shank (see below), which in turn binds to additional signaling and scaffold proteins. PSD-95 is in any case more abundant (in molar terms) than NMDARs, and several times more abundant than GKAP and Shank[48], thus suggesting its importance in integrating a large network of signaling and adaptor proteins, many of which also contain PDZ domains, although most of them do not interact directly with PSD-95.

Importantly, PSD-95 greatly influences synaptic transmission and plasticity, mainly because it recruits the tetraspanning membrane protein stargazin to synapses by binding to its C-terminus. Stargazin and its relatives are associated with AMPA receptors and are essential for their surface expression and synaptic accumulation and function[49,50] (also see Chapter 24). These data may explain why PSD-95 overexpression potentiates AMPAR-mediated excitatory postsynaptic currents (EPSCs), but not the currents of the directly linked NMDA receptor.

The synaptic location of PSD-95 depends on the palmitoylation of two N-terminal cysteines (Cys3 and Cys5)[51], and synaptic activity induces the removal of PSD-95 by depalmitoylating the two Cys residues[52]. A set of enzymes capable of inducing PSD-95 palmitoylation has recently been identified, but some controversy remains as to which of them is specific for PSD-95 and whether their function is regulated by synaptic activity[53,54]. PSD-95 can be degraded through the ubiquitin–proteasome pathway by means of direct ubiquitination of PSD-95[55] or, indirectly, by the ubiquitination and degradation of the interacting protein SPAR[56].

In general, the activity-dependent dispersal or degradation of PSD-95 is often associated with a loss of synaptic AMPARs and weakened synapses, or changes in glutamate receptor-induced intracellular signaling such as CREB and MAPK phosphorylation[57].

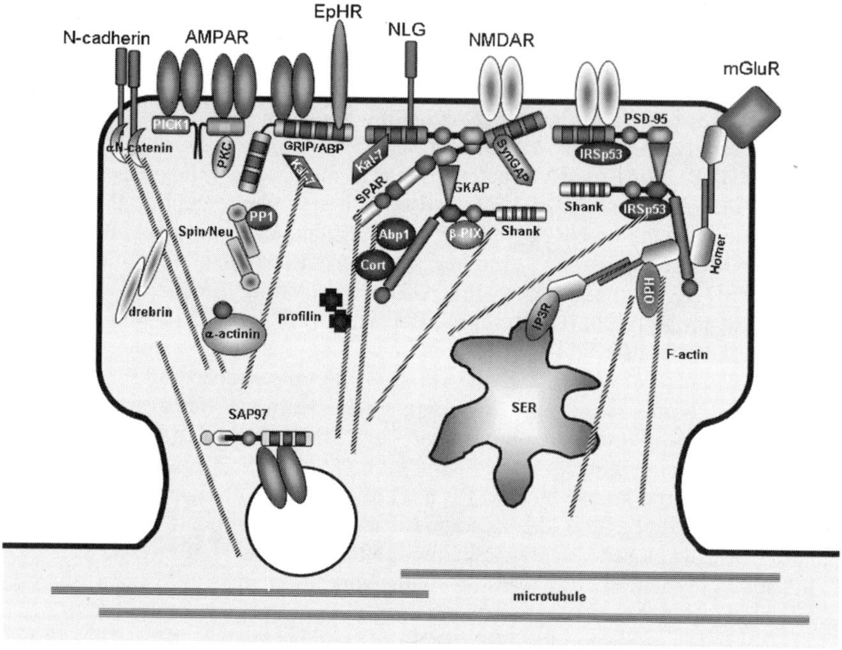


**Figure 18.2.** The Synaptic Scaffold Proteins Involved in Spine Morphogenesis. NLG, neuroligin; Kal-7, Kalirin-7; Spin/Neu, Spinophilin and Neurabin I; Cort, Cortactin; OPH, Oligophrenin. See Colorplate 11.

Finally, it is important to underline the different properties of each member of the PSD-95 family. The proteins of the PSD-95 family have clearly different distributions in both brain and neurons. PSD-95 and PSD-93 are mainly enriched in the PSD, whereas SAP102 and SAP97 are abundant in both cytoplasm and synapses, and found in dendrites and axons. The expression of SAP102 begins early in postnatal development, whereas PSD-95 and PSD-93 are more abundant in later stages[58]. *In vivo*, PSD-95 family members apparently interact with different but overlapping sets of proteins, with PSD-95 being preferentially associated with the NR2A subunit of the NMDA receptor, whereas SAP102 is preferentially associated with the NR2B subunit[58]. This suggests that the properties of the NR2B–SAP102 complex may be different from those of the NR2A–PSD-95/PSD-93 complex, and that the functional properties of synaptic NMDARs may depend on the prevalence of one or the other[59].

SAP102 and SAP97 are involved in the trafficking of NMDA and AMPA receptors, respectively. By interacting with the PDZ-binding domain of Sec8, SAP102 can associate with the exocyst complex and regulate the delivery of NMDARs to the surface of neuronal cells[60]. SAP97 interacts directly with the AMPAR GluR1 subunit[61], and the fact that the SAP97–GluR1 complex has been found early in the secretory pathway indicates that SAP97 can regulate the trafficking of GluR1[62]. CaMKII phosphorylation of SAP97 in the N-terminal L27 domain promotes the synaptic targeting of SAP97 and GluR1[63]. To some extent like PSD-95, the overexpression of SAP97 increases the number of synaptic AMPA receptors, induces spine enlargement, and increases the frequency of miniature EPSCs[64,65].

## 6. THE SHANK AND HOMER FAMILIES

The Shank and Homer protein families are two major components of the PSD that directly interact with each other.

Shank, also known as proline-rich synapse-associated protein (ProSAP), somatostatin receptor interacting protein (SSTRIP), cortactin-binding protein (cortBP), Synamon and Spank, is a large scaffold protein whose multidomain organization consists of an ankyrin repeat near the N-terminal, followed by an SH3, a PDZ domain, a long proline-rich region, and a sterile alpha motif domain (SAM) at the C-terminus[66]. Shank proteins (codified by three genes, *Shank1–3*) molecularly link two glutamate receptor subtypes: NMDA receptors and type-I mGluRs. The Shank PDZ domain binds to the C-terminal of GKAP. The Homer interaction with the proline-rich domain ensures the association of Shank with type I mGluRs.

Homer proteins are encoded by three genes (*Homer 1–3*), and typically consist of an N-terminal EVH1 domain followed by a coiled-coil domain that mediates dimerization with other Homer proteins. The Ena/VASP homology 1 (EVH1) domain of Homer1 binds to a PPXXF or a very similar sequence motif present in Shank, mGluR1/5, inositol-1,4,5-trisphosphate ($IP_3$) receptor, ryanodine receptor, and to different members of the TRPC family of ion channels[67,68]. Through their ability to self-associate, Homer isoforms containing the coiled-coil domain (termed "CC-Homer" or Homer1b for the Homer1 gene) can physically and functionally link the proteins and receptors that bind to the EVH1 domain. Homer1a is a short-splice variant of Homer1 that contains the EVH1 domain but lacks the coiled-coil domain. Importantly, Homer1a expression is induced at mRNA level by synaptic activity.

There is consistent experimental evidence suggesting that Homer1a functions as a natural dominant negative protein, because it cannot dimerize. For example, Homer1a overexpression attenuates mGluR-evoked intracellular calcium release, probably by interfering with Homer-mediated coupling between type I mGluRs and $IP_3$ receptors; similarly, Homer1a inhibits the Homer1b control of mGluR1/5 constitutive activity[69] and disassembles the TRPC1-Homer-IP3R complex by inducing TRPC1 channel activation[68].

The overexpression of Shank1 and Homer1b in hippocampal neurons accelerates the maturation of filopodial-like protrusions in mature spines, and promotes the enlargement of mature spines (which acquire the classical mushroom shape) without increasing their number. Shank and Homer also cooperate to promote the accumulation of PSD proteins in dendritic spines such as GKAP and NR1, and increase the F-actin content of spines[70].

Roussignol et al. (2005) have more recently shown that the overexpression of Shank3 in cerebellum granule cells induces dendritic spine and synapse formation by recruiting different subtypes of glutamate receptors, whereas the inhibition of Shank3 expression in hippocampal neurons reduces the number of dendritic spines. One feature of the global effect of Shank on synapse maturation is that its over-expression also induces the maturation of the presynaptic compartment[70,71].

It is therefore not surprising that, like PSD-95, Shank and Homer interact with a number of actin-binding proteins. Shank binds to cortactin, Abp1[72], fodrin[73], the Rac1 and Cdc42 exchange factor βPIX[74], and Cdc42-binding protein IRSp53[75]; and Homer binds to Rho GTPase-activating protein oligophrenin-1[76]. Interestingly, the interaction of Shank1 with Homer seems to be essential for inducing spine maturation, and the interaction with cortactin seems to be equally important for Shank3[71].

Shank1 and Homer1b can also recruit the entire ER compartment to dendritic spines, which may contribute to their enlargement effect[77]. The ability of Shank1 and Homer1 to promote spine morphogenesis depends on their ability to form a complex with each other, and correlates with their accumulation in spines[70,78]. Interestingly, Homer 1a disrupts the interaction between full-length Homer1b and Shank, and inhibits the synaptic targeting of both proteins[78]. As a consequence, the overexpression of Homer1a destabilizes synapses and decreases the number and size of dendritic spines, also reducing the synapse number of both AMPA and NMDA receptors. In this case, the actions of Homer1a may contribute to the global activity-dependent loss of spines in a neuron, and the negative regulation of unstimulated synapses[78] (Figure 18.3).

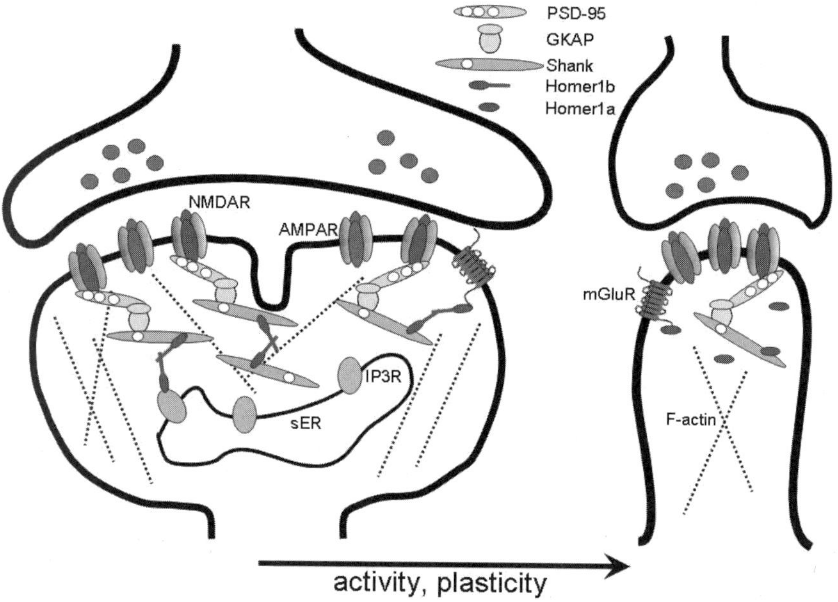


**Figure 18.3.** Shank Synaptic Targeting Regulation by Homer Proteins. Shank targeting and PSD structure are regulated by competitive Shank and Homer interactions: Shank and the constitutively expressed Homer1b protein increase dendritic spine size, whereas the activity-dependent expression of Homer1a reduces Shank localization to the spine and reduces the size of the spine and PSD.

All of the effects of Shank on synapses and spines strictly depend on its PSD localization, and so the mechanism that regulates Shank association at synapses also influences dendritic spine morphology and synapse function. At least in the case of Shank1, the interaction with GKAP and PSD-95 seems to be essential for Shank stability and targeting to synapses, also because the dissociation of PSD-95 from this complex induces the degradation of Shank and GKAP[79]. Synaptic activity can also induce the ubiquitination and degradation of Shank and GKAP[57]. The synaptic targeting of Shank2 and Shank3 seems instead to depend on their C-terminal, and not on their interaction with GKAP in the PDZ domain[80].

The function of GKAP family of proteins is less characterized. The four members of the family were originally identified as proteins interacting with the GK domain of PSD-95. GKAP has five repeats of 14 amino acids involved in the interaction of PSD-95, and binds to *S*-SCAM, nArgBP2, Dynein light chain, and Shank, and may therefore function as a scaffolding protein that link PSD protein complexes to motor proteins[81].

Other important Shank partners are proteins that regulate dendritic morphology and formation, such as Dasm1[82] and Densin-180[83], whose activities also seem to be regulated by Shank. Finally, Shank proteins can potentially be produced locally due to the presence of its mRNA in the dendrites[84].

## 7. ACTIN-BINDING PROTEINS

It has been demonstrated that a number of actin-binding proteins (usually effectors of Rho and Rac GTPases regulated by $Ca^{2+}$) control the actin cytoskeleton in dendritic spines. Most of them are formed by the concatenation of different protein interaction domains, and thus function as scaffolds assembled in multimolecular complexes.

We here describe the most important actin-binding scaffold proteins regulating dendritic spine morphology (Figure 18.2; Colorplate 11).

One of these is drebrin, an F-actin-binding protein, which is mainly expressed in neurons and highly concentrated in dendritic spines. Drebrin overexpression in cortical neurons increases the length of dendritic spines[85], and its activity may be mediated by its binding to F-actin and other acting-regulatory proteins, such as profilin, myosin, and gelsolin. Also, it has more recently been shown[86] that, in hippocampal neurons, drebrin promotes actin assembly and the synaptic clustering of PSD-95 in the PSD. Importantly, drebrin is redistributed to dendritic spines, together with an increase in F-actin content, after the induction of LTP in the dentate gyrus[87], which suggests it is important in inducing F-actin polymerization and/or stabilization.

Like drebrin, profilin II and α*N*-catenin are recruited to spines by general synaptic activation and by NMDA receptors activation or LTP induction, and can induce actin polymerization and stabilization. Profilin is a small actin-binding protein that promotes actin polymerization by positioning the actin monomers at the barbed end of the growing F-actin. The accumulation of profilin II in dendritic spines persists for hours beyond the initiating stimulus and this depends on an increase in postsynaptic $Ca^{2+}$ levels and, probably, on the association of profilin II with Ena/VASP family proteins. Profilin II stabilizes spine morphology in a mature state and suppresses dendritic spine motility by reducing actin dynamics[88]. These results suggest that profilin II plays a role in linking the activation of glutamate receptors with the actin-based stabilization of synapse morphology.

α*N*-catenin is a cadherin-associated protein and, together with β-catenin, links the adhesion molecules to the cytoskeleton and actin. Abe et al. (2004) have recently showed that, in the absence of α*N*-catenin, dendritic spines are more motile and their filopodia rapidly protrude and retract from the spine heads, a sign of unstable synaptic contacts. Conversely, the overexpression of α*N*-catenin accelerates dendritic spine maturation and decreases spine motility, thus suggesting that it promotes spine morphogenesis and stabilization. As it appears to accumulate in activated synapses, α*N*-catenin may mediate neural activity-dependent signals that stabilize synapses by suppressing motility and turnover[89]. One possible bidirectional interaction between the profilin/actin and cadherin/α*N*-catenin systems has been proposed in which neural activity contribute to changes in spine shape by modifying any of these molecular interactions[89].

Spinophilin (or neurabin II) and neurabin I are two related F-actin-binding proteins with similar domain structures containing a PDZ and a coiled-coil domain that form homo- and heterodimers. Via its actin-binding domain, spinophilin (as its name suggests) is predominantly localized on the dendritic spines of pyramidal neurons; it seems to be required for the correct maturation of dendritic spines because knockout mice have more filopodia and immature spines and altered glutamatergic transmission[90]. Its activity may be related to its ability to bind PP1 phosphatase, thus regulating the phosphorylation state of PP1 substrates such as AMPA and NMDA glutamate receptors and the myosin regulatory light chain. In one case, spinophilin can modulate the stability of the actin cytoskeleton in spines by modulating the ion currents of glutamate receptors[91]; it may also regulate dendritic spine morphology by promoting the dephosphorylation of the myosin light chain, thus inhibiting the assembly and contractility of actin filaments[92]. On the contrary, neurabin I overexpression induces the formation of dendritic filopodia in immature cultured hippocampal neurons and promotes the enlargement of dendritic spines in older neurons[93]. Moreover, it has been recently shown that the neurabin I amino-terminal fragment, which contains the actin-binding domain, increases the density and length of dendritic spines also by increasing actin polymerization and spine motility[94]. This suggests that the neurabin I carboxy-terminal portion plays a negative regulatory role. It is not clear how spinophilin and neurabin I have opposite effect on spines, although they both bind a similar set of proteins in synapses. Finally, the coiled-coil domains of neurabin I and spinophilin have more recently been found to interact with Lfc, a Rho GEF that regulates the Rho-dependent organization of F-actin in spines[95].

α-actinin-2, one of the α-actinin isoforms, is a member of the spectrin/dystrophin family of actin-binding proteins, and is enriched in dendritic spines of pyramidal neurons. Its overexpression increases the length and density of dendritic protrusions in cultured hippocampal neurons[96], and it binds to the NR1 and NR2B subunits of NMDA receptors, thus providing a cytoskeletal bridge between NMDA receptors and actin. However, its association with NR1 can be competitively displaced from by $Ca^{2+}$/calmodulin, and it may thus dissociate from the NMDA receptor as a result of $Ca^{2+}$ influx into the postsynaptic terminal[97]. Finally α-actinin is also linked to AMPA receptors via the reversion-induced LIM protein RIL, and participates in the regulation of AMPA receptor trafficking within spines[98].

IRSp53 not only interacts with Shank, but also with PSD-95 and PSD-93, and these interactions are required for its spine localization: its overexpression in cultured neurons increases the density of dendritic spines, while its siRNA-mediated knockdown reduces spine density, length and width. The overexpression of IRSp53 with a point mutation in the SH3 domain (where WAVE2 binds) reduces spine density and size, thus suggesting that IRSp53 plays a role in linking PSD-95 to activated Rac1/Cdc42 and downstream effectors of actin regulation in spines[42].

## 8. LESSONS FROM MUTANT ANIMALS AND GENETIC DISEASES

The functions of a number of scaffold proteins have been analyzed in knockout mice and, in the case of some actin-binding proteins (such as α*N*-catenin and spinophilin), a clear phenotype has been identified as a change in dendritic spine morphology and synapse function. Many others will probably soon be available, thus clarifying the relative roles of these proteins in spine structure.

Among the PSD-95 family, the most interesting phenotype observed so far is that of PSD-95 mutant mice, whose spatial learning is impaired although LTP is enhanced and LTD is defective[33]. The behavioral sensitization induced by chronic cocaine administration is also eliminated in these mice[99] and the maturation of orientation preference in the visual cortex fails to mature[100]. These *in vivo* results demonstrate that PSD-95 is important for the regulation of synaptic plasticity.

Even more interesting is the observation that PSD-95 may be important for the correct balance of excitatory and inhibitory synapses in brain. A possible mutation of PSD-95 has been found in autistic patients[101], and it was proposed that an incorrect balance between excitatory and inhibitory synapses is associated with autism[102] (also see Chapter 27).

As mentioned above, a number of genetic diseases involving mental retardation are often correlated with defects in dendritic spine morphology. One form of mental retardation associated with severe expressive language delay and minor facial dysmorphisms has been found in several patients with a 22q13.3 distal deletion that causes Shank3 haploinsufficiency[103,104]. It is still not known whether these patients have altered synapses and spine morphology, but the presence of this genetic disease in humans strongly suggests that scaffold proteins play a major role in brain function.

## 9. CONCLUSION AND FUTURE DIRECTIONS

Over the last few years, our knowledge of the molecules and molecular mechanisms involved in the formation and function of synapses and dendritic spines has considerably increased, and a number of protein complexes and signaling pathways have been independently identified by different laboratories. The next step is to put all of this information together. The ability of some scaffold proteins to assemble different protein complexes may enable them to link different signaling pathways in synapses and spines, thus making it important to identify all of the scaffold protein-binding partners and how their interactions are regulated. It is even more important to understand how the synthesis, post-translational modification and degradation of scaffold proteins are regulated. All of these studies will be more informative if carried out at the level of single neurons and single spine because it is now clear that the mechanisms regulating spine and synapse formation and function may not only differ in various types of neurons, but also in different spines and synapses belonging to the same neuron.

It is therefore necessary to combine biochemical, morphological, and genetic techniques in order to define better how scaffold and associated proteins are specifically involved in regulating each single synapse and spine. The final aim of this complex work is to increase our knowledge of the origin of various cognitive disorders associated with dendritic spine and synapse abnormalities, which will make it possible to interfere with and repair the damaged molecular mechanism, thus improving the treatment of patients with severe cognitive dysfunctions.

## 10. REFERENCES

1. Fiala, J.C., Spacek, J., and Harris, K.M. (2002) *Brain Res Brain Res Rev* **39**, 29–54.
2. Hering, H., and Sheng, M. (2001) *Nat Rev Neurosci* **2**, 880–888.
3. Harris, K., and Kater, S. (1994) *Annu Rev Neurosci* **17**, 341–371.
4. Parnass, Z., Tashiro, A., and Yuste, R. (2000) *Hippocampus* **10**, 561–568.
5. Holtmaat, A.J., Trachtenberg, J.T., Wilbrecht, L., Shepherd, G.M., Zhang, X., Knott, G.W., and Svoboda, K. (2005) *Neuron* **45**, 279–291.
6. Ziv, N.E., and Smith, S.J. (1996) *Neuron* **17**, 91–102.
7. Spacek, J., and Harris, K. (1997) *J Neurosci* **17**, 190–203.
8. Noguchi, J., Matsuzaki, M., Ellis-Davies, G.C., and Kasai, H. (2005) *Neuron* **46**, 609–622.
9. Ostroff, L.E., Fiala, J.C., Allwardt, B., and Harris, K.M. (2002) *Neuron* **35**, 535–545.
10. Schikorski, T., and Stevens, C.F. (1997) *J Neurosci* **17**, 5858–5867.
11. Toni, N., Buchs, P.A., Nikonenko, I., Bron, C.R., and Muller, D. (1999) *Nature* **402**, 421–425.
12. Passafaro, M., Nakagawa, T., Sala, C., and Sheng, M. (2003) *Nature* **424**, 677–681.
13. Kennedy, M.B. (2000) *Science* **290**, 750–754.
14. Kim, E., and Sheng, M. (2004) *Nat Rev Neurosci* **5**, 771–781.
15. Sheng, M., and Sala, C. (2001) *Annu Rev Neurosci* **24**, 1–29.
16. Feng, W., Shi, Y., Li, M., and Zhang, M. (2003) *Nat Struct Biol* **10**, 972–978.
17. Setou, M., Seog, D.H., Tanaka, Y., Kanai, Y., Takei, Y., Kawagishi, M., and Hirokawa, N. (2002) *Nature* **417**, 83–87.
18. Shin, H. et al. (2003) *J Biol Chem* **278**, 11393–11401.
19. Bruckner, K., Pablo Labrador, J., Scheiffele, P., Herb, A., Seeburg, P.H., and Klein, R. (1999) *Neuron* **22**, 511–524.
20. Ye, B., Liao, D., Zhang, X., Zhang, P., Dong, H., and Huganir, R. (2000) *Neuron* **26**, 603–617.
21. Wyszynski, M., Kim, E., Dunah, A.W., Passafaro, M., Valtschanoff, J.G., Serra-Pages, C., Streuli, M., Weinberg, R.J., and Sheng, M. (2002) *Neuron* **34**, 39–52.
22. Takamiya, K., Kostourou, V., Adams, S., Jadeja, S., Chalepakis, G., Scambler, P. J., Huganir, R.L., and Adams, R.H. (2004) *Nat Genet* **36**, 172–177.
23. Hirbec, H. et al. (2003) *Neuron* **37**, 625–638.
24. Henkemeyer, M., Itkis, O.S., Ngo, M., Hickmott, P.W., and Ethell, I.M. (2003) *J Cell Biol* **163**, 1313–1326.
25. Hoogenraad, C.C., Milstein, A.D., Ethell, I.M., Henkemeyer, M., and Sheng, M. (2005) *Nat Neurosci*.
26. Williams, M.E., Wu, S.C., McKenna, W.L., and Hinck, L. (2003) *J Neurosci* **23**, 11279–11288.
27. Perroy, J., El Far, O., Bertaso, F., Pin, J.P., Betz, H., Bockaert, J., and Fagni, L. (2002) *EMBO J* **21**, 2990–2999.
28. Hirbec, H., Perestenko, O., Nishimune, A., Meyer, G., Nakanishi, S., Henley, J.M., and Dev, K.K. (2002) *J Biol Chem* **277**, 15221–15224.
29. Torres, G.E., Yao, W.D., Mohn, A.R., Quan, H., Kim, K.M., Levey, A.I., Staudinger, J., and Caron, M.G. (2001) *Neuron* **30**, 121–134.
30. Jaulin-Bastard, F., Saito, H., Le Bivic, A., Ollendorff, V., Marchetto, S., Birnbaum, D., and Borg, J.P. (2001) *J Biol Chem* **276**, 15256–15263.
31. Kim, E., Niethammer, M., Rothschild, A., Jan, Y.N., and Sheng, M. (1995) *Nature* **378**, 85–88.
32. Tejedor, F.J., Bokhari, A., Rogero, O., Gorczyca, M., Zhang, J., Kim, E., Sheng, M., and Budnik, V. (1997) *J Neurosci* **17**, 152–159.
33. Migaud, M., Charlesworth, P., Dempster, M., Webster, L.C., Watabe, A.M., Makhinson, M., He, Y., Ramsay, M.F., Morris, R.G., Morrison, J.H., O'Dell, T.J., and Grant, S.G. (1998) *Nature* **396**, 433–439.
34. Biederer, T., Sara, Y., Mozhayeva, M., Atasoy, D., Liu, X., Kavalali, E.T., and Sudhof, T.C. (2002) *Science* **297**, 1525–1531.
35. Hsueh, Y.-P., Yang, F.-C., Kharazia, V., Naisbitt, S., Cohen, A.R., Weinberg, R.J., and Sheng, M. (1998) *J Cell Biol* **142**, 139–151.
36. Prange, O., Wong, T.P., Gerrow, K., Wang, Y.T., and El-Husseini, A. (2004) *Proc Natl Acad Sci USA* **101**, 13915–13920.
37. Chih, B., Engelman, H., and Scheiffele, P. (2005) *Science* **307**, 1324–1328.
38. Graf, E.R., Zhang, X., Jin, S.X., Linhoff, M.W., and Craig, A.M. (2004) *Cell* **119**, 1013–1026.

39. El-Husseini, A.E., Schnell, E., Chetkovich, D.M., Nicoll, R.A., and Bredt, D.S. (2000) *Science* **290**, 1364–1368.
40. Penzes, P., Beeser, A., Chernoff, J., Schiller, M.R., Eipper, B.A., Mains, R.E., and Huganir, R.L. (2003) *Neuron* **37**, 263–274.
41. Pak, D.T., Yang, S., Rudolph-Correia, S., Kim, E., and Sheng, M. (2001) *Neuron* **31**, 289–303.
42. Choi, J., Ko, J., Racz, B., Burette, A., Lee, J.R., Kim, S., Na, M., Lee, H.W., Kim, K., Weinberg, R.J., and Kim, E. (2005) *J Neurosci* **25**, 869–879.
43. Aarts, M., Liu, Y., Liu, L., Besshoh, S., Arundine, M., Gurd, J.W., Wang, Y.T., Salter, M.W., and Tymianski, M. (2002) *Science* **298**, 846–850.
44. Kim, J.H., Lee, H.K., Takamiya, K., and Huganir, R.L. (2003) *J Neurosci* **23**, 1119–1124.
45. Vazquez, L.E., Chen, H.J., Sokolova, I., Knuesel, I., and Kennedy, M.B. (2004) *J Neurosci* **24**, 8862–8872.
46. Kalia, L.V., and Salter, M.W. (2003) *Neuropharmacology* **45**, 720–728.
47. Huang, Y. et al. (2001) *Neuron* **29**, 485–496.
48. Peng, J., Kim, M.J., Cheng, D., Duong, D.M., Gygi, S.P., and Sheng, M. (2004) *J Biol Chem* **279**, 21003–21011.
49. Tomita, S., Adesnik, H., Sekiguchi, M., Zhang, W., Wada, K., Howe, J.R., Nicoll, R.A., and Bredt, D.S. (2005) *Nature* **435**, 1052–1058.
50. Chen, L., Chetkovich, D.M., Petralia, R.S., Sweeney, N.T., Kawasaki, Y., Wenthold, R.J., Bredt, D.S., and Nicoll, R.A. (2000) *Nature* **408**, 936–943.
51. Craven, S.E., El-Husseini, A.E., and Bredt, D.S. (1999) *Neuron* **22**, 497–509.
52. El-Husseini Ael, D., Schnell, E., Dakoji, S., Sweeney, N., Zhou, Q., Prange, O., Gauthier-Campbell, C., Aguilera-Moreno, A., Nicoll, R.A., and Bredt, D.S. (2002) *Cell* **108**, 849–863.
53. Fukata, M., Fukata, Y., Adesnik, H., Nicoll, R.A., and Bredt, D.S. (2004) *Neuron* **44**, 987–996.
54. Huang, K. et al. (2004) *Neuron* **44**, 977–986.
55. Colledge, M., Snyder, E.M., Crozier, R.A., Soderling, J.A., Jin, Y., Langeberg, L.K., Lu, H., Bear, M.F., and Scott, J.D. (2003) *Neuron* **40**, 595–607.
56. Pak, D.T., and Sheng, M. (2003) *Science* **302**, 1368–1373.
57. Ehlers, M.D. (2003) *Nat Neurosci* **6**, 231–242.
58. Sans, N., Petralia, R.S., Wang, Y.X., Blahos, J., 2nd, Hell, J.W., and Wenthold, R.J. (2000) *J Neurosci* **20**, 1260–1271.
59. Kim, M.J., Dunah, A.W., Wang, Y.T., and Sheng, M. (2005) *Neuron* **46**, 745–760.
60. Sans, N., Prybylowski, K., Petralia, R.S., Chang, K., Wang, Y.X., Racca, C., Vicini, S., and Wenthold, R.J. (2003) *Nat Cell Biol* **5**, 520–530.
61. Leonard, A.S., Davare, M.A., Horne, M.C., Garner, C.C., and Hell, J.W. (1998) *J Biol Chem* **273**, 19518–19524.
62. Sans, N., Racca, C., Petralia, R.S., Wang, Y.X., McCallum, J., and Wenthold, R.J. (2001) *J Neurosci* **21**, 7506–7516.
63. Mauceri, D., Cattabeni, F., Di Luca, M., and Gardoni, F. (2004) *J Biol Chem* **279**, 23813–23821.
64. Rumbaugh, G., Sia, G.M., Garner, C.C., and Huganir, R.L. (2003) *J Neurosci* **23**, 4567–4576.
65. Nakagawa, T., Futai, K., Lashuel, H.A., Lo, I., Okamoto, K., Walz, T., Hayashi, Y., and Sheng, M. (2004) *Neuron* **44**, 453–467.
66. Sheng, M., and Kim, E. (2000) *J Cell Sci* **113**, 1851–1856.
67. Xiao, B., Tu, J.C., and Worley, P.F. (2000) *Curr Opin Neurobiol* **10**, 370–374.
68. Yuan, J.P., Kiselyov, K., Shin, D.M., Chen, J., Shcheynikov, N., Kang, S.H., Dehoff, M.H., Schwarz, M.K., Seeburg, P.H., Muallem, S., and Worley, P.F. (2003) *Cell* **114**, 777–789.
69. Ango, F., Prezeau, L., Muller, T., Tu, J.C., Xiao, B., Worley, P.F., Pin, J.P., Bockaert, J., and Fagni, L. (2001) *Nature* **411**, 962–965.
70. Sala, C., Piech, V., Wilson, N.R., Passafaro, M., Liu, G., and Sheng, M. (2001) *Neuron* **31**, 115–130.
71. Roussignol, G., Ango, F., Romorini, S., Tu, J.C., Sala, C., Worley, P.F., Bockaert, J., and Fagni, L. (2005) *J Neurosci* **25**, 3560–3570.
72. Qualmann, B., Boeckers, T.M., Jeromin, M., Gundelfinger, E.D., and Kessels, M.M. (2004) *J Neurosci* **24**, 2481–2495.
73. Bockers, T.M., Mameza, M.G., Kreutz, M.R., Bockmann, J., Weise, C., Buck, F., Richter, D., Gundelfinger, E.D., and Kreienkamp, H.J. (2001) *J Biol Chem* **276**, 40104–40112.
74. Park, E., Na, M., Choi, J., Kim, S., Lee, J.R., Yoon, J., Park, D., Sheng, M., and Kim, E. (2003) *J Biol Chem* **278**, 19220–19229.
75. Soltau, M., Richter, D., and Kreienkamp, H.J. (2002) *Mol Cell Neurosci* **21**, 575–583.
76. Govek, E.E., Newey, S.E., Akerman, C.J., Cross, J.R., Van der Veken, L., and Van Aelst, L. (2004) *Nat Neurosci* **7**, 364–372.
77. Sala, C., Roussignol, C., Meldolesi, J., and Fagni, L. (2005) *J Neurosci* **25**, 4587–4592.
78. Sala, C., Futai, K., Yamamoto, K., Worley, P.F., Hayashi, Y., and Sheng, M. (2003) *J Neurosci* **23**, 6327–6337.

79. Romorini, S., Piccoli, G., Jiang, M., Grossano, P., Tonna, N., Passafaro, M., Zhang, M., and Sala, C. (2004) *J Neurosci* **24**, 9391–9404.
80. Boeckers, T.M., Liedtke, T., Spilker, C., Dresbach, T., Bockmann, J., Kreutz, M.R., and Gundelfinger, E.D. (2005) *J Neurochem* **92**, 519–524.
81. Naisbitt, S., Valtschanoff, J., Allison, D., Sala, C., Kim, E., Craig, A., Weinberg, R., and M, Sheng. (2000) *J Neurosci.* **20**, 4524–4534.
82. Shi, S.H., Cheng, T., Jan, L.Y., and Jan, Y.N. (2004) *Proc Natl Acad Sci U S A* **101**, 13346–13351.
83. Quitsch, A., Berhorster, K., Liew, C.W., Richter, D., and Kreienkamp, H.J. (2005) *J Neurosci* **25**, 479–487.
84. Bockers, T.M., Segger-Junius, M., Iglauer, P., Bockmann, J., Gundelfinger, E.D., Kreutz, M.R., Richter, D., Kindler, S., and Kreienkamp, H.J. (2004) *Mol Cell Neurosci* **26**, 182–190.
85. Hayashi, K., and Shirao, T. (1999) *J Neurosci* **19**, 3918–3925.
86. Takahashi, H., Sekino, Y., Tanaka, S., Mizui, T., Kishi, S., and Shirao, T. (2003) *J Neurosci* **23**, 6586–6595.
87. Fukazawa, Y., Saitoh, Y., Ozawa, F., Ohta, Y., Mizuno, K., and Inokuchi, K. (2003) *Neuron* **38**, 447–460.
88. Ackermann, M., and Matus, A. (2003) *Nat Neurosci* **6**, 1194–1200.
89. Abe, K., Chisaka, O., Van Roy, F., and Takeichi, M. (2004) *Nat Neurosci* **7**, 357–363.
90. Feng, J., Yan, Z., Ferreira, A., Tomizawa, K., Liauw, J.A., Zhuo, M., Allen, P.B., Ouimet, C.C., and Greengard, P. (2000) *Proc Natl Acad Sci USA* **97**, 9287–9292.
91. Yan, Z., Hsieh-Wilson, L., Feng, J., Tomizawa, K., Allen, P.B., Fienberg, A.A., Nairn, A.C., and Greengard, P. (1999) *Nat Neurosci* **2**, 13–17.
92. Fernandez, A., Brautigan, D.L., Mumby, M., and Lamb, N.J. (1990) *J Cell Biol* **111**, 103–112.
93. Oliver, C.J., Terry-Lorenzo, R.T., Elliott, E., Bloomer, W.A., Li, S., Brautigan, D.L., Colbran, R.J., and Shenolikar, S. (2002) *Mol Cell Biol* **22**, 4690–4701.
94. Zito, K., Knott, G., Shepherd, G.M., Shenolikar, S., and Svoboda, K. (2004) *Neuron* **44**, 321–334.
95. Ryan, X.P., Alldritt, J., Svenningsson, P., Allen, P.B., Wu, G.Y., Nairn, A.C., and Greengard, P. (2005) *Neuron* **47**, 85–100.
96. Nakagawa, T., Engler, J.A., and Sheng, M. (2004) *Neuropharmacology* **47**, 734–745.
97. Wyszynski, M., Lin, J., Rao, A., Nigh, E., Beggs, A.H., Craig, A.M., and Sheng, M. (1997) *Nature* **385**, 439–442.
98. Schulz, T.W. et al. (2004) *J Neurosci* **24**, 8584–8594.
99. Yao, W.D., Gainetdinov, R.R., Arbuckle, M.I., Sotnikova, T.D., Cyr, M., Beaulieu, J.M., Torres, G.E., Grant, S.G., and Caron, M.G. (2004) *Neuron* **41**, 625–638.
100. Fagiolini, M., Katagiri, H., Miyamoto, H., Mori, H., Grant, S.G., Mishina, M., and Hensch, T.K. (2003) *Proc Natl Acad Sci U S A* **100**, 2854–2859.
101. Rubenstein, J.L., and Merzenich, M.M. (2003) *Genes Brain Behav* **2**, 255–267.
102. Zoghbi, H.Y. (2003) *Science* **302**, 826–830.
103. Bonaglia, M.C., Giorda, R., Borgatti, R., Felisari, G., Gagliardi, C., Selicorni, A., and Zuffardi, O. (2001) *Am J Hum Genet* **69**, 261–268.
104. Wilson, H.L., Wong, A.C., Shaw, S.R., Tse, W.Y., Stapleton, G.A., Phelan, M.C., Hu, S., Marshall, J., and McDermid, H.E. (2003) *J Med Genet* **40**, 575–584.

# 19

# COMPOSITION AND ASSEMBLY OF GABAERGIC POSTSYNAPTIC SPECIALIZATIONS

Yunhee Kang[*#] and Ann Marie Craig[*]

## 1. SUMMARY

GABAergic synapses are the major sites of inhibitory transmission in the brain. The key functional components, $GABA_A$ receptors, are GABA-gated chloride channels sensitive to benzodiazepines and barbiturates. Among a wide diversity of subunit compositions, most common is a pentamer of two α1, two β2, and one γ2 subunit. The γ2 subunit contributes to synaptic localization of the complex. Point mutant knock-in mice of individual α subunits revealed specific roles for α1 in sedation and amnesia and α2 in anxiety. Trafficking of $GABA_A$ receptors is regulated by interacting proteins and by extracellular signals such as epileptogenic activity. Other key components of GABAergic synapses include cell adhesion complexes of cadherins, and presynaptic neurexins with postsynaptic dystroglycan or neuroligin-2. Recent evidence suggests that neurexins and neuroligins function to recruit components to GABA synapses and to control the balance of GABA and glutamate synapses. A major intracellular postsynaptic component is gephyrin, which anchors some $GABA_A$ receptors and may recruit other molecules to the postsynaptic site. Despite recent progress, the molecular linkages among key components such as $GABA_A$ receptors, gephyrin, and neuroligins have not been identified, indicating that much is yet to be discovered about the composition and assembly of GABAergic synapses.

## 2. INTRODUCTION

GABAergic synapses are the major sites of inhibitory transmission in the CNS. GABA (γ-aminobutyric acid) is synthesized from glutamate by glutamic acid

[*]Brain Research Centre and Department of Psychiatry, University of British Columbia, Vancouver, BC, Canada V6T 2B5; kimp1117@dreamwiz.com, amcraig@interchange.ubc.ca
[#] Department of Anatomy and Division of Brain Korea 21 Biomedical Science, Korea University College of Medicine, Seoul, South Korea

decarboxylase (GAD) and loaded into synaptic vesicles by vesicular GABA/glycine transporter (VGAT). GABAergic terminals use molecular machinery in common with other terminal types to control calcium-dependent synaptic vesicle fusion and recycling. There are some important differences, for example, the absence of SNAP-25 at GABAergic terminals mediates differences in calcium dynamics and release rates in comparison with SNAP-25-dependent glutamatergic terminals[1]. But for the most part, the transmitter release apparatus is conserved between GABAergic and glutamatergic terminals. In contrast, the postsynaptic apparatus of GABAergic synapses is distinct from that of glutamatergic synapses. GABAergic postsynaptic elements lack the glutamatergic scaffolding and signaling proteins such as PSD-95 and CaMKII, but contain their own set of scaffolding and signaling proteins. Major components of GABAergic postsynaptic elements are $GABA_A$ receptor subunits, gephyrin, cadherins–catenins, the dystrophin glycoprotein complex (DGC), and neuroligin-2 (Figure 19.1). We review here these major elements of GABAergic postsynaptic sites, with emphasis on the primary functional element, $GABA_A$ receptors. Furthermore, as discussed below, none of these five major components have been found to directly interact, one of the indications that the study of GABAergic postsynaptic elements is still in its infancy, with much still to be discovered.

A particularly interesting aspect of GABAergic synapses is their distribution on postsynaptic neurons. Whereas glutamatergic synapses occur primarily on dendritic spines of mature pyramidal neurons, GABAergic synapses occur on dendrite shafts, soma, and the axon initial segment (Figure 19.2). Thus some inhibitory synapses are close to the site of action potential initiation and may have a strong influence on cell firing. In cortex *in vivo*, innervation of different postsynaptic domains is mediated by different classes of GABAergic neurons[2]. Double bouquet cells innervate distal dendrites, dendrite targeting cells innervate proximal dendrites, basket cells innervate somata and proximal dendrites, and axo-axonic cells selectively innervate the axon initial segment. This circuitry contributes to shape integrative properties at multiple levels, from locally shunting excitatory currents and regulating excitability within dendritic segments to defining oscillatory activity patterns of ensembles of cortical cells.

## 3. $GABA_A$ RECEPTORS: STRUCTURE AND FUNCTION

$GABA_A$ receptors mediate the majority of fast synaptic inhibition in the mammalian CNS. The larger family of GABA receptors is dividend into two classes, ionotropic ($GABA_A$ and $GABA_C$) and metabotropic ($GABA_B$). $GABA_B$ receptors are G-protein coupled receptors formed by heterodimerization of the 7-transmembrane proteins $GABA_{B1}$ and $GABA_{B2}$[3]. Through G-protein signaling, $GABA_B$ receptors regulate voltage-gated $K^+$ and $Ca^{++}$ channels to mediate slow postsynaptic inhibition or to reduce transmitter release. $GABA_C$ receptors are structurally similar to $GABA_A$ receptors but are generally composed of distinct subunits ρ1-3. Expression of $GABA_C$ receptors is the highest in retinal bipolar terminals where they inhibit glutamate release[4]. Thus unlike $GABA_A$ receptors which function as the main signal transducers at GABAergic postsynaptic sites, $GABA_B$ and $GABA_C$ receptors primarily modulate transmission, often from the presynaptic side. A question still under investigation is whether in some instances $GABA_A$ and $GABA_C$ receptor subunits may co-assemble to mediate fast inhibition[5].

$GABA_A$ receptors are members of the ligand-gated ion channel family which includes glycine receptors, nicotinic acetylcholine receptors, serotonin type 3 receptors, and the 5-hydroxytryptamine type 3 (5-HT3) receptors, and are composed of pentamers (Figure 19.1; ref. 6). By binding two molecules of GABA, a conformational change occurs and an integral chloride channel is opened to flow $Cl^-$ ions down a concentration gradient. $GABA_A$ receptors are considered inhibitory because in mature systems they mediate $Cl^-$ influx and hyperpolarization. However, early in development, when the intracellular concentration of $Cl^-$ is higher than the extracellular concentration, $GABA_A$ receptor activation leads to neuronal excitation[7]. Benzodiazepines, appreciated clinically for their sedative, anticonvulsive, and anxiolytic effects, function by binding to $GABA_A$ receptors and allosterically potentiating the effects of GABA, effectively enhancing currents at submaximal GABA concentrations.

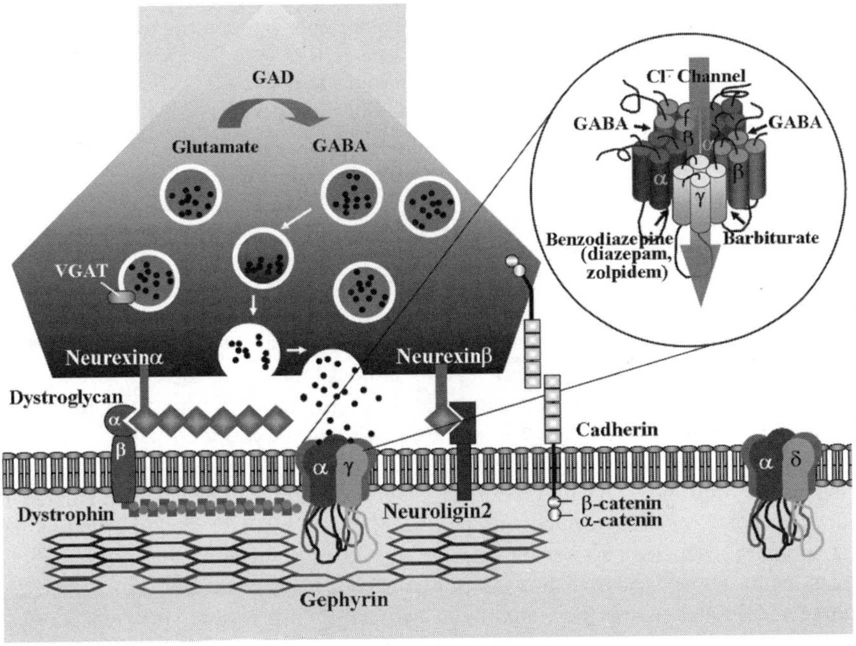


**Figure 19.1.** Schematic Representation of the Major Postsynaptic Proteins at GABAergic Synapses. The subunit interfaces where GABA and benzodiazepines bind to $GABA_A$ receptors is indicated. Notably, no direct interactions among $GABA_A$ receptors, gephyrin, cadherins–catenin, the dystroglycan complex, and neuroligin-2 have been reported, indicating the presence of additional as yet unidentifed components.

Electron microscopic imaging of GABA receptors revealed a channel with a diameter of about 8 nm in the plane of the membrane, around which five subunits are arranged pseudosymmetrically[8]. The subunits constituting a pentameric receptor have a common structure made of a large extracellular amino terminus which contains a signature cysteine–cysteine loop prior to the first of four transmembrane domains (TM1–4) and a very short extracellular carboxyterminus.

TM2 is thought to form a major lining of the ion channel. A large intracellular loop between TM3 and TM4 is the most divergent and includes multiple binding sites for trafficking and postsynaptic scaffolding proteins and phosphorylation sites for diverse serine/threonine and tyrosine kinases. We further review here $GABA_A$ receptor distributions and trafficking; for reviews of functional modulation by phosphorylation see refs. 9 and 10.

A huge diversity of mammalian $GABA_A$ receptors is generated by combinations of subunits α1-6, β1-3, γ1-3, δ, ε, π, and θ[11]. Spatiotemporal patterns of expression and certain rules for generating a functional receptor limit the number of possible combinations. Nonetheless, heterogeneity of $GABA_A$ receptor subtypes confers differential localization and functional and pharmacological properties, making possible fine tuning of GABAergic synaptic transmission. It is generally believed that most receptors consist of two α subunits, two β subunits, and one γ subunit. Almost 60% of all mammalian $GABA_A$ receptors are thought to be α1β2γ2 type, followed in abundance by α2β3γ2 (15–20%) and α3βxγ2 (10–15%)[12]. Receptors containing α4–6 or β1 form a minor population, and γ1, γ3, δ, ε, π, or θ are thought to replace the γ2 subunit in less abundant $GABA_A$ receptor subtypes. In the rare receptors containing nonidentical α subunits, their arrangement in the pentamer confers distinct properties. Finally, additional diversity is generated by alternative splicing of the γ2 subunit. The short (γ2S) and long (γ2L) forms are distinguished by the absence or presence of eight amino acids within the cytoplasmic loop between TM3 and TM4[13].

The physiological significance of individual $GABA_A$ receptor subunits has been studied intensively with genetically targeted mice (Table 1; reviewed in ref. 14). The γ2 subunit is essential for mouse survival, presumably due to its presence in the majority of $GABA_A$ receptors. Heterozygous γ2 +/– mice exhibit increased reactivity toward aversive stimuli and enhanced responsiveness in trace fear conditioning, and thus may provide an animal model of anxiety disorders[33]. Deletion of the β3 subunit also results in a severe phenotype, poor coordination, hyperactivity, hyper-responsiveness, increased seizures, and cleft palate, the latter due to non-neuronal expression of β3[29,38]. In contrast, deletion of β2 results in little phenotype; β3 may be more essential due to its expression earlier in development. A series of mice with point mutations (H101R) rendering individual α subunits insensitive to benzodiazepine site ligands revealed that α1 is largely responsible for the sedative and amnesic effects of benzodiazepines, whereas α2 is largely responsible for their anxiolytic actions[12]. Furthermore, in visual cortex, GABAergic circuitry involving α1 but not α2 subunit mediates critical period ocular dominance plasticity, as also demonstrated using the H101R mutant mice[39].

## 4. $GABA_A$ RECEPTORS: SYNAPTIC AND EXTRASYNAPTIC DISTRIBUTIONS

The remarkable heterogeneity of $GABA_A$ receptor subunits appears in part related to different functions mediated by specific subunit combinations at different subcellular locations[10,40]. $GABA_A$ receptors α1,2,3,6βxγ2 are concentrated at postsynaptic sites in hippocampus, cerebral cortex, and cerebellum, where they mediate phasic inhibition initiated by discrete vesicular GABA release. Receptor combinations α5βxγ2 and α4,6βxδ are predominantly or exclusively

**Table 1.** Summary of $GABA_A$ Receptor Functional Analysis by Targeted Mutagenesis in Mice.

| Subtype (Reference) | Animal phenotype | Cellular phenotype |
|---|---|---|
| α1 KO (15–17) | Viable, fertile, body weight reduction, tremor on handling<br>No spontaneous seizures, but increased bicuculine-induced seizures | Loss of more than 50% of all $GABA_A$ receptors<br>Reduced α6,β2/3 and γ2<br>Increased α2, α3 |
| α1 H101R* (18,19) | Normal appearance<br>Decreased DZ-induced sedation and amnesia and anticonvulsant actions | |
| α2 H101R* (20,21) | Normal appearance<br>Strongly impaired DZ-induced anxiolysis and myorelaxation at low dose | |
| α3 H126R* (20,21) | Normal appearance<br>Reduction of DZ-induced myorelaxation at high dose | |
| α5 KO (22,23) | No overt neurological phenotype<br>Enhanced hippocampus-dependent spatial learning and trace fear conditioning<br>No alteration in non-hippocampal-dependent learning and anxiety tasks | Reduced β2/3 and γ2 Lack of tonic conductance in cultured hippocampal cells |
| α5 H105R* (24) | Normal locomotor function<br>Reduction of DZ-induced myorelaxation<br>Facilitated trace fear conditioning | Reduced α5 in hippocampus<br>Reduced β2/3 and γ2 |
| α6 KO (25–28) | Normal appearance<br>Enhanced DZ-induced motor impairment | Post-translational loss of δ in cerebellum<br>Reduced α1,β2, β3, and γ2<br>Compensatory increase of TASK-1 $K^+$ leak channels |
| β2 KO (16) | Viable, fertile, increased locomotor activity<br>No spontaneous seizures | Loss of more than 50% of all $GABA_A$ receptors<br>Reduced α1–α6 |
| β3 KO (29,30) | 90% die within 24 h, 57% cleft palate<br>Survivors fertile, but not nursing<br>Animal model for Angelman syndrome<br>spontaneous seizures, hyperactive, hyper-responsive<br>Heterozygous mice show almost normal phenotype | Loss of ~50% of all $GABA_A$ receptors |
| γ2 KO (31–33) | Normal embryonic development<br>Early postnatal lethality<br>γ2+/– mice are animal model for chronic anxiety with enhanced response in trace fear conditioning and ambiguous cue discrimination | Reduced postsynaptic clustering of $GABA_A$ receptors and gephyrin, which can be partially rescued by overexpression of γ3 |
| γ2 floxed X CaMKIIα-Cre (34) | Conditional knockout in 3-week-old mouse forebrain<br>Lethal epilepsy | Rapid loss of postsynaptic $GABA_A$ receptors |
| δ KO (35–37) | Viable, fertile, slightly smaller litters, occasional seizures | Increased γ2<br>Reduced α4 |

KO, knockout; *these point mutations render the receptor insensitive to diazepam; DZ, diazepam.

extrasynaptic, mediating tonic inhibition in response to GABA spill over from synaptic release sites. Thus δ tends to specify extrasynaptic localization, and γ2 synaptic localization, but targeting is also influenced by the α subunit since α5 can direct γ2 to an extrasynaptic localization. Selective synaptic versus extrasynaptic distributions, for example of α1 and α2 versus α5, can occur in dissociated hippocampal cell culture and thus are not dependent on specific innervation patterns[41]. However, *in vivo* there is selective enrichment of different synaptic type subunits opposite different inputs. In pyramidal cell somata and proximal dendrites, whereas α1 subunits are enriched opposite parvalbumin-positive basket cells, α2 subunits are enriched opposite parvalbumin-negative cholecystokinin-positive basket cells[42,43]. These selective distributions may relate to the different physiological roles of α1 more in vigilance and memory and α2 more in anxiety related behavior as revealed by the H101R mutant mouse studies. Developmental and cell-type specific expression patterns may also be important. In adult hippocampus, α2 is high in pyramidal neurons and α1 highest in a subset of interneurons, whereas in neocortex α1 replaces α2,3 as the major α subunit during postnatal development[44]. The molecular basis for this richness in subcellular targeting patterns of $GABA_A$ receptor subunit combinations is not yet known. To generate selective targeting of α1 versus α2 opposite different types of basket cell terminals clearly requires input-specific signals, perhaps differential activity or perhaps axon-specific molecular cues.

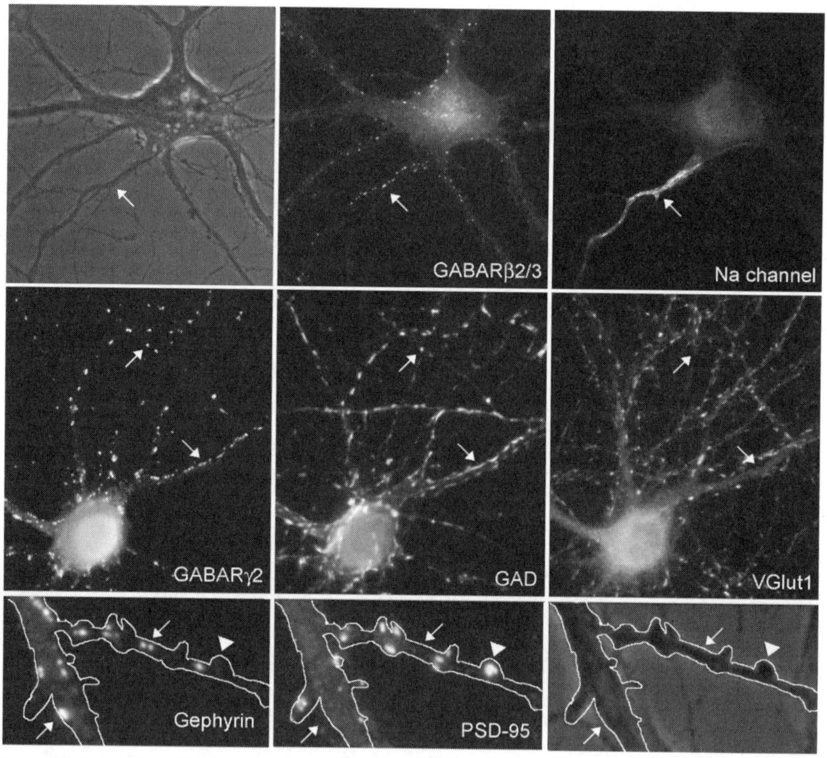


**Figure 19.2.** Cellular Distribution of GABAergic Synapses as Revealed by Immunofluorescence of Hippocampal Neurons in Dissociated Culture. Upper panels: $GABA_A$ receptors form postsynaptic clusters on the soma, dendrite shafts, and axon initial segment; the axon initial segment is identified by immunoreactivity for $Na^+$ channels (arrow). Middle panels: $GABA_A$ receptors cluster selectively opposite GABAergic terminals immunoreactive for glutamic acid decarboxylase (GAD, arrows) and not opposite glutamatergic terminals immunoreactive for vesicular glutamate transporter 1 (VGlut1). Lower panels: GABAergic synapses identified by clusters of the scaffolding protein gephyrin form on dendrite shafts (arrows), whereas glutamatergic synapses identified by clusters of the scaffolding protein PSD-95 form mainly on dendritic spines (arrowhead) protruding from the shafts. The dendrite outline is roughly traced. Upper panel images were generated by Dr. Anuradha Rao.

More generally, a molecular signal(s) from GABAergic axons, and not GABA itself, is the major cue for aligning postsynaptic receptors with the presynaptic release apparatus. Neuron culture studies revealed the surprising finding that, when GABAergic input is lacking or very limited, $GABA_A$ receptors and gephyrin do not remain diffusely localized but rather cluster opposite glutamatergic terminals[45,46]. The presence of such mismatched appositions was interpreted to indicate the existence of a shared family of proteins that participate in aligning pre- and post-synaptic elements at both GABA and glutamate synapses[45]. Indeed, neurexins and neuroligins were later found to fit this prediction. Neurexins alone presented to dendrites induce clustering of $GABA_A$ receptors and gephyrin via binding to neuroligin-2, and induce clustering of NMDA receptors and PSD-95 via binding to neuroligin-1[47]. Further information about neuroligins and neurexins, as well as other components of GABAergic synapses and their potential roles in localizing $GABA_A$ receptors, is discussed in separate sections below on gephyrin, cadherins–catenins, the DGC, and neuroligin-2.

Excellent progress has been made in defining the role of the $\gamma 2$ subunit in synaptic targeting of $GABA_A$ receptors, and mechanisms of $\gamma 2$ localization. Deletion of $\gamma 2$ in mice results in a profound decrease in postsynaptic clustering of $\alpha 1$ and $\alpha 2$ and a parallel reduction in mIPSC frequency[32]. Deletion of $\gamma 2$ also results in dispersal of the inhibitory postsynaptic scaffolding protein gephyrin, indicating that multiple postsynaptic components are dependent on $\gamma 2$ for synaptic localization. Deletion of $\gamma 2$ by a Cre-loxP strategy in hippocampus during the third postnatal week also results in loss of punctate immunoreactivity for $\alpha 2$ and gephyrin[34]. Thus $\gamma 2$ is required for development and ongoing maintenance of synaptic $GABA_A$ receptor localization. Loss of $\gamma 2$ did not affect presynaptic VGAT puncta, suggesting that postsynaptic receptor clustering is not important for maintaining contact by presynaptic terminals. Subunit $\gamma 3$ or the individual $\gamma 2S$ or $\gamma 2L$ splice variants could all partially compensate for $\gamma 2$ depletion and rescue the clustering of $\alpha 1$ and $\alpha 2$ subunit containing receptors[48,49]. To further dissect mechanisms of $\gamma 2$ synaptic targeting, Alldred et al.[50] tested chimeric $\alpha 2/\gamma 2$ constructs for rescue of synaptic receptor localization and function in hippocampal neuron cultures from $\gamma 2$ knockout mice. The surprising finding is that TM4, not the major cytoplasmic loop, is necessary and sufficient for postsynaptic clustering of $GABA_A$ receptors, but both TM4 and the cytoplasmic loop are required for synaptic gephyrin clustering and restoration of mIPSCs. Thus transmembrane proteins as well as cytoplasmic components will likely feature in future studies of mechanisms underlying synaptic targeting of $GABA_A$ receptors.

Like other neurotransmitter receptors[51], $GABA_A$ receptors are presumably localized to synapses via dynamic protein–protein interactions, and thus in flux with receptor pools on the extrasynaptic plasma membrane and in intracellular compartments. Even for $GABA_A$ receptors that are concentrated at synaptic sites, there are detectable extrasynaptic levels. Quantitative immunogold labeling estimates the relative enrichment of $GABA_A$ receptors at postsynaptic sites to be ~200 fold, for $\alpha 1$ and $\beta 2/3$ subunits in cerebellar granule cells[52]. Functional tagging of $\alpha 1$ and activity-dependent block was used to demonstrate mobility of functional $GABA_A$ receptors in cultured hippocampal neurons[53]. It was estimated that the entire cohort of synaptic receptors may turn over within ~14 min. Labeling $GABA_A$ receptors with antibodies against extracellular epitopes and monitoring redistribution in live cells also reveals considerable dynamics. Van Rijnsoever et al.[54] report that surface-labeled receptors accumulate in an intracellular pool beneath the postsynaptic membrane, although perhaps limited accessibility of synaptic receptors to extracellular antibodies combined with lateral mobility may also contribute to the dynamics observed. Clearly, mechanisms that regulate trafficking of $GABA_A$ receptors, from synthesis in the endoplasmic reticulum through exocytotic transport and surface membrane insertion, lateral diffusion, endocytosis and recycling, and finally to degradation, will all be important in determining the steady-state numbers of synaptic $GABA_A$ receptors.

## 5. $GABA_A$ RECEPTORS: TRAFFICKING

Many studies suggest that receptor assembly occurs by defined pathways and there exist specific mechanisms that limit the combinations and arrangements of GABA receptor subunits contributing to functional surface $GABA_A$ receptors[55]. Most single $GABA_A$ receptor subunits expressed alone in HEK cells or Xenopus oocytes are retained in the endoplasmic reticulum. Exceptionally, while $\beta 3$ subunits can form pentobarbitol-sensitive homomeric surface receptors[56], $\beta 3$ preferentially oligomerizes with other subunits, and $\beta 3$ homomers are not thought to exist in neurons. When co-expressed, $\alpha 1$, $\beta 2$, and $\gamma 2$ form heteromeric receptors, with $\alpha 1\gamma 2$ and $\beta 2\gamma 2$ combinations retained in the endoplasmic reticulum, and only $\alpha 1\beta 2$ and $\alpha 1\beta 2\gamma 2$ making functional surface receptor[57]. Additional studies have defined specific residues in the N-terminal regions that control subunit assembly[55]. Recombinant dimeric and trimeric concatenated subunits linking the C-terminus of one subunit to the N-terminus of the next subunit have been useful in defining the arrangement of subunits as $\gamma\beta\alpha\beta\alpha$ and in defining positional effects of different $\alpha$ subunits in the complex[58].

Trafficking of $GABA_A$ receptors to and from the plasma membrane is regulated by extracellular signals and by intracellular interacting proteins (Figure 19.3). While yeast two-hybrid screens with the large intracellular loop of $GABA_A$ receptor subunits have not been so successful in identifying new synaptic interacting partners, many receptor binding proteins that appear to function in regulating receptor trafficking have been found. Perhaps the best known of these is $GABA_A$ receptor-associated protein (GABARAP). GABARAP can bind the $\gamma 2$ intracellular domain, gephyrin, tubulin, and NSF, and is homologous to GATE-16, an intra-Golgi transport factor[59–61]. GABARAP also binds GRIP1, a protein initially named as a glutamate receptor-interacting protein but later found to bind the kinesin molecular motor and steer this motor to dendrites[62,63]. GABARAP is enriched in the Golgi apparatus and intracellular membranous compartments[61]. These data all suggest that GABARAP may regulate trafficking of $GABA_A$ receptors to the cell surface. The homology of GABARAP and GATE-16 to a yeast protein involved in

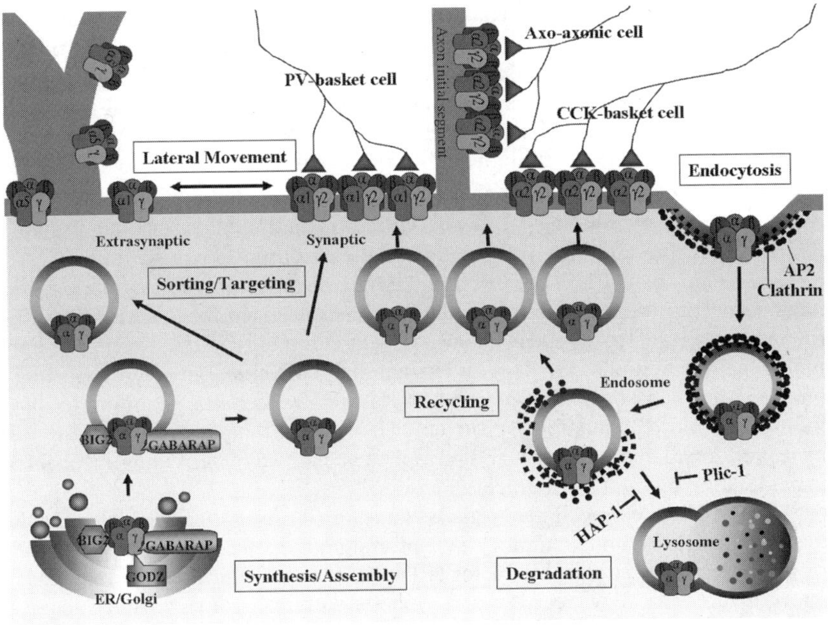


**Figure 19.3.** Schematic Representation of the Trafficking Pathways of $GABA_A$ Receptors. Receptor subunits are synthesized and assembled in the endoplasmic reticulum and Golgi complex, sorted, and targeted to the extrasynaptic or synaptic plasma membrane opposite appropriate inputs. $GABA_A$ receptor-interacting proteins GABARAP, BIG2, and GODZ are thought to regulate the exocytotic pathway. Receptors undergo constitutive endocytosis and recycling or degradation, regulated by Plic-1 and HAP-1.

ubiquitination-like modifications during autophagy, and participation of recombinant GABARAP in this pathway, suggest another potential role in post-translational modification and protein degradation[64]. Another Golgi-associated protein that may regulate trafficking of GABA receptors to the surface is brefeldin A-inhibited GDP/GTP exchange factor 2 (BIG2), via direct binding to β subunits[65].

A particularly interesting finding to come from the yeast two-hybrid screens is the association and direct palmitoylation of $GABA_A$ receptor γ2 by the Golgi-specific DHHC zinc finger palmitoyltransferase GODZ[66]. Palmitoylation, or thioacylation, of cysteine residues is a reversible post-translational modification appreciated as a key regulator of plasma membrane association of cytosolic proteins. A number of neuronal transmembrane proteins are also palmitoylated, including synaptotagmin I, Kv1.1 voltage-gated channel, nicotinic α7 acetylcholine receptor, AMPA receptor subunits GluR1-4, and $GABA_A$ receptor[67,68]. Palmitolyation can regulate trafficking of membrane proteins and modulate function independent of trafficking. Initial experiments suggest that palmitoylation is necessary to attain normal surface levels and synaptic clustering of $GABA_A$ receptors in hippocampal culture[69]. Regulation by palmitoylation can be quite complex; palmitoylation of AMPA receptors at one site by GODZ reduces surface levels while palmitoylation at another site regulates agonist-induced internalization[68].

$GABA_A$ receptors are subject to constitutive clathrin-mediated endocytosis and recycling, in a regulated manner (reviewed in refs. 9,10). For example, in cultured cortical neurons, within 30 min about one-quarter of surface β3-containing receptors are internalized, and about half of the internalized pool is recycled back to the plasma membrane over the same time period[70]. Efficient endocytosis occurs by binding of the clathrin adaptor AP2 to the intracellular loop of $GABA_A$ receptor β subunits[71]. $GABA_A$ receptor degradation is inhibited and thus surface levels increased by binding of β subunits to Huntingtin associated protein 1 (HAP1) and by binding of α or β subunits to the ubiquitin-like protein Plic-1[70,72]. It was suggested that Plic-1 reduces $GABA_A$ receptor polyubiquitination, thus reducing receptor targeting to the proteasome, as suggested for other ubiquitin-like proteins. For HAP1, the mechanisms of regulation of $GABA_A$ receptor might involve the interaction of HAP1 with Hrs and its involvement in receptor sorting or with the motor associated protein dynactin p150Glued. In addition to direct binding to these regulatory proteins, $GABA_A$ receptor trafficking is also extensively regulated by kinase and phosphatase pathways, providing mechanisms for modulation by multiple extracellular signals.

Three major extracellular signals have been found to regulate $GABA_A$ receptor trafficking and inhibitory synaptic signaling: insulin, epileptic activity, and brain-derived neurotrophic factor (BDNF). Insulin induces a rapid increase in surface levels of $GABA_A$ receptor and synaptic currents by a mechanism requiring phosphorylation of the β subunit by the downstream serine/threonine kinase Akt[73,74]. Perhaps of greatest physiological relevance is the regulation of $GABA_A$ receptors by epileptic activity. Kindling-induced epilepsy increases the synaptic content of $GABA_A$ receptors by ~75%, resulting in a corresponding increase in synaptic currents in hippocampal dentate granule neurons[75]. Increases in expression of multiple GABA receptor subunits have been found in dentate neurons in several experimental models of epilepsy[40]. Remarkably, these neurons also upregulate GAD67 and VGAT and are thought to co-release GABA with glutamate under epileptic conditions[76]. BDNF, released by pyramidal neurons in an activity-regulated manner, promotes the development of GABAergic interneurons. Several studies also suggest that BDNF regulates GABAergic signaling postsynaptically, by regulating cell surface levels of $GABA_A$ receptors, but the precise findings vary greatly[77–80]. Perhaps the magnitude and even direction of regulation of $GABA_A$ receptor trafficking by BDNF may depend on the cell type, stage of development, and duration of treatment.

## 6. GEPHYRIN

Gephyrin is a major scaffolding protein concentrated at most if not all GABAergic postsynaptic elements (reviewed in refs. 81, 82). Gephyrin is also concentrated at inhibitory glycinergic postsynaptic sites, and in fact was first identified as a protein that copurified with the glycine receptor. While gephyrin directly binds to the large intracellular loop of the glycine receptor β subunit[83], direct binding to $GABA_A$ receptor subunits has not been found. Gephyrin contains multiple sites of alternative splicing, and different splice variants exhibit different subcellular localizations in transfected cell lines[84,85]. Some of the gephyrin variants aggregate in a filamentous-type pattern. Gephyrin binds tubulin, probably via a short internal region with 60–80% identity with the repeat motifs of MAP2 and tau mediating tubulin polymerization and microtubule binding[85]. Surprisingly, the primary sequence of gephyrin suggests that it evolved by insertion of a novel sequence into genes encoding molybdenum cofactor biosynthetic enzymes. Indeed, gephyrin retains molybdenum cofactor biosynthetic activity[86], and mice lacking gephyrin are defective in molybdoenzyme activity in non-neural tissues[87]. Thus gephyrin appears to have dual unrelated functions.

The role of gephyrin in localizing $GABA_A$ receptors to the postsynaptic site has been probed with antisense oligonucleotide and mouse knockout experiments[32,88–90]. Both types of experiments revealed a partial dependence of GABA receptor synaptic clustering on gephyrin, to varying degrees. While synaptic clustering of $GABA_A$ receptor subunits α2, α3, β2/3, and γ2 was partially dependent on gephyrin, clustering of α1 and α5 was completely gephyrin independent[89,90]. For α2 and γ2, dendrite surface levels were unaffected by loss of gephyrin, indicating a role for gephyrin in synaptic trapping or anchoring but not in trafficking receptors to the cell surface. Functionally, there was a decrease in amplitude but no change in frequency of mIPSCs in hippocampal neurons lacking gephyrin compared with wild type[90]. Developmental compensation in the knockout is always a possibility, but the complete absence of glycine receptor clustering in both spinal cord and hippocampal neurons from the gephyrin knockout indicates that other proteins cannot compensate for this aspect of gephyrin's function[87,90]. Indeed, the knockout mice die within one day of birth due to an apparent failure of glycinergic transmission[87]. Taken together, these results indicate the existence of a major gephyrin-independent pathway for synaptic clustering of $GABA_A$ receptors.

In contrast, the localization of gephyrin itself to the postsynaptic domain is highly dependent on $GABA_A$ receptors. Knockout of the γ2 subunit, present in >80% of $GABA_A$ receptors, results in a dramatic reduction in synaptic clustering of α subunits and gephyrin and in mIPSC frequency[32]. Rescue experiments in primary neuron cultures from these mice indicate that both the major cytoplasmic loop and TM4 of γ2 are necessary for recruiting gephyrin to the synapse[50].

An idea emerging from this evidence and the growing list of gephyrin interacting proteins is that gephyrin may function as a scaffolding protein to recruit components in addition to or other than $GABA_A$ receptors to the postsynaptic domain. In addition to the glycine receptor β subunit and tubulin, gephyrin can bind to the GDP-GTP exchange factor collybistin[91], the actin regulatory protein profilin[92], the translation regulator rapamycin and FKBP12 target protein RAFT1[93], the motor associated protein dynein light chain[94], and the $GABA_A$ receptor binding protein GABARAP[60]. Most of these proteins are not specifically concentrated at GABAergic postsynaptic elements, but they may be recruited for signaling purposes and/or be involved in trafficking of gephyrin (e.g., ref. 95).

## 7. CADHERINS AND CATENINS

A family of proteins that functions at both GABA and glutamate synapses is the cadherin cell adhesion molecules and catenin binding partners. Cadherins are a large family of calcium-dependent homophilic cell adhesion molecules that link to F-actin and to signaling pathways via catenins. In hippocampal cultures, *N*-cadherin and β-catenin are among the earliest components of developing GABAergic and glutamatergic synapses[96]. While *N*-cadherin is lost from GABA synapses as they mature, β-catenin is retained, suggesting that some other cadherin may be present. A dominant-negative form of *N*-cadherin had a strong but transient effect on disrupting both GABA and glutamate synapse assembly early but not late in development of hippocampal cultures[97,98]. These results suggest that *N*-cadherin normally contributes to early stage GABA synapse development but it is not absolutely required. The cadherin-related γ-protocadherins may also contribute to inhibitory synapse formation in some brain regions, in spinal cord but not hippocampus (refs. 99,100 and Chapter 10).

Unfortunately, cadherins and catenins as yet have been little studied specifically in the context of GABAergic synapses. Characterization of the cadherins present in GABAergic pathways *in vivo* would be a good first step. Detailed analysis of GABAergic synapses in the catenin knockout mice would also be interesting. An α-*N*-catenin knockout mouse survives but exhibits defects in spine maturation and stability[101]; whether GABA synapses are less stable is not known. Deletion of β-catenin specifically in pyramidal neurons resulted in dispersal of synaptic vesicles and a reduced reserve pool[102]; whether β-catenin is required in a similar way for presynaptic organization at GABAergic synapses is not known. The dominant-negative *N*-cadherin did not disrupt the peri-somatic proximo-distal density gradient of GABAergic synapses along dendrites in hippocampal culture[98], and other studies suggest the involvement of neurofascin adhesion molecules in the peri-somatic targeting of basket cell GABAergic axons[103]. Analyses of knockout mice indicate a contribution of other Ig domain family cell adhesion molecules such as L1, and extracellular matrix molecules such as tenascin-R, to proper development of GABAergic synapses[104,105]. A key question related to these other cell adhesion protein families is whether they function directly at the GABAergic synapse or perhaps indirectly via a role in earlier developmental events.

## 8. THE DYSTROPHIN GLYCOPROTEIN COMPLEX (DGC)

Originally identified in relation to muscular dystrophy, the DGC has been found as a major component of mature inhibitory GABAergic synapses. Mutations in dystrophin are a major etiology of Duchenne and Becker muscular dystrophy[106]. In addition, the finding that some patients with Duchenne muscular dystrophy display severe mental retardation suggests that dystrophin may have an important role in the CNS. Dystrophin is a large actin-binding protein primarily expressed in muscle and is also present in several isoforms in brain. Dystrophin binds the transmembrane protein dystroglycan, which binds laminin, agrin, and perlecan, thus forming a link from the cytoskeleton to the extracellular matrix[107]. Synthesized from a single precursor, dystroglycan is proteolytically cleaved to produce α and β subunits. α-Dystroglycan exhibits extensive and tissue-specific glycosylation, and defects in its glycosylation are thought to underlie Fukuyama congenital muscular dystrophy[108]. Links to additional intracellular signaling molecules are provided via binding of β-dystroglycan to Grb-2, and dystrophin to syntrophins and dystrobrevin.

A role for the DGC at GABAergic synapses was first suggested by Kneusel et al.[109] who found dystrophin immunoreactivity in brain extensively colocalized with gephyrin and $GABA_A$ receptor subunits α1 and α2. Furthermore, the number and the size of $GABA_A$ receptor clusters was significantly reduced in mdx mice deficient in long dystrophin isoforms, although gephyrin clustering was unaltered. Further studies in hippocampal neuron cultures showed that dystrophin, α- and β-dystroglycan all concentrate at a subset of GABAergic synapses late in development[110]. Since targeted deletion of dystroglycan is early embryonic lethal in mice, due to its role in basement membrane formation, a conditional knockout was generated. Cultured neurons lacking dystroglycan still clustered gephyrin and $GABA_A$ receptors opposite GABAergic terminals, with no apparent defects. Moreover, brain-specific deletion of dystroglycan *in vivo* leads to major defects in cell migration and hippocampal long-term potentiation, but the mice survive and hippocampal slices exhibit normal field potentials in response to stimulation[111]. These studies indicate that the DGC is not essential for GABAergic synapse formation, including synaptic recruitment of gephyrin and GABA receptors, but suggest that the DGC may be involved in stabilization or other signaling at the mature GABAergic synapse. The finding that dystroglycan in brain is largely complexed with α-neurexin[112,113] further supports some role in *trans*-synaptic maintenance or signaling.

While β-neurexin binding to neuroligin-2 can recruit gephyrin and $GABA_A$ receptors and may represent an early step in the formation of functional GABAergic synapses (see below), these interactions do not recruit the DGC[47]. The DGC is recruited to GABA synapses by a mechanism independent of gephyrin and of $GABA_A$ receptor γ2 via a signal unique to GABA but not glutamate axons[110,114]. How the observed interaction of dystroglycan with α-neurexins[113] fits into the recruitment or function of the DGC at GABA synapses is not yet clear.

## 9. NEUROLIGIN-2

Neuroligin-2 is a recently discovered component of GABA synapses. Neuroligins and their presynaptic binding partners, neurexins, can mediate calcium-dependent cell adhesion[112,115]. By virtue of binding the glutamate postsynaptic scaffolding molecule PSD-95[116], neuroligins are thought to be critical for glutamate synapse development. However, of the 4–5 neuroligins in mammals, neuroligin-2 does not localize to glutamate synapses but rather localizes to GABA synapses, as reported in two independent studies[47,117]. Two other key studies also indicated a role of neuroligins in GABA synapse development. Prange et al.[118] first showed that altering the levels of neuroligin-1 or PSD-95 can alter the number of GABAergic as well as glutamatergic synapses. Finally, Chih et al.[119] reported that RNAi knock-down of neuroligins-1, -2, and -3, or each individually, disrupted both glutamatergic and GABAergic synapses, GABAergic more severely. Thus, neuroligin-2 may have a critical function at GABA synapses.

Structurally, neuroligins have an extracellular inactive acetylcholinesterase homologous domain that binds to the LNS domain of neurexins. Both neuroligins and neurexins terminate intracellularly in PDZ domain binding sites. Whether the PDZ domain binding site on neuroligin-2 functions at GABA synapses, or whether it is there primarily to allow cross-talk with key PDZ domain proteins at glutamate synapses is not yet clear. Neuroligins are also multimeric[120]. Reconciling the multimeric nature with the selective localization of neuroligin-2 would suggest that neuroligin-2 oligomerizes with itself in preference to neuroligins-1, -3, or -4. Finally, a site of alternative splicing in neuroligin-2, the "A" site in the acetylcholinesterase homologous domain[121], adds further complexity, of as yet unknown functional significance.

Four functions for neuroligin-2 have been suggested, all mutually compatible: mediating adhesion between the correct pre- and post-synaptic partners, inducing or maintaining presynaptic differentiation, balancing glutamate and GABA synaptic inputs, and mediating GABAergic postsynaptic differentiation. While testing of the first hypothesis, selective synaptic adhesion, awaits *in vivo* knockout analyses, significant *in vitro* evidence exists to support the other functions.

As first shown by Scheiffele et al.[122], neuroligins expressed on non-neuronal cells are able to induce presynaptic differentiation in contacting axons. These induced 'hemi-presynapses' consist of clusters of synaptic vesicles able to undergo activity-dependent recycling with properties similar to bona fide synapses, and thus must contain a fairly complete presynaptic apparatus[123]. Multiple neuroligins are able to induce such presynaptic differentiation in GABAergic or glutamatergic axons[47]. Neuroligin-induced presynaptic differentiation is mediated by aggregation of neurexin on the axon surface[120]. One obvious interpretation of these results is that neuroligin-2 at GABA postsynaptic sites may function to induce and/or maintain synaptic vesicle clusters and associated release apparatus in the apposing axon.

The ability of neuroligins on the dendrite to maintain apposing presynaptic specializations combined with the binding of neuroligins to PSD-95 appears to underlie their role in regulating the balance of excitatory versus inhibitory synaptic inputs[124], as discussed more extensively in Chapter 5. The loss of GABA synapses observed upon overexpression of PSD-95 may occur by translocation of neuroligin-2 from GABA to glutamate synapses, thus destabilizing GABA synapses[47,118,125]. Whether there are conditions that induce movement of the other neuroligins to GABA synapses, or whether they can partially compensate for the function of neuroligin-2 is not yet clear.

Finally, neuroligin-2 may function within the postsynaptic side to mediate clustering of $GABA_A$ receptors and gephyrin. Graf et al.[47] showed that neurexin presented to dendrites induces focal postsynaptic differentiation, forming separate clusters of $GABA_A$ receptors with gephyrin or NMDA glutamate receptors with PSD-95. Furthermore, direct aggregation of neuroligins on the dendrite surface indicated a link from all neuroligins to PSD-95 but only neuroligin-2 to gephyrin. Thus, aggregation of neuroligin-2 on the dendrite surface by neurexins on GABAergic axons may function to recruit and/or stabilize $GABA_A$ receptors and gephyrin at a developing GABAergic synapse. Supporting this idea, mislocalized expression of neuroligin or RNAi knockdown resulted in a reduction in morphological and functional GABAergic synapses[47,119]. Defining the molecular link between neuroligin-2, $GABA_A$ receptors, and gephyrin is an area ripe for investigation, as none of these components are reported to bind directly. The dystroglycan complex, another major known component of mature GABAergic postsynaptic sites, also does not bind any of these components directly but does bind neurexins[113], thus linking it to neuroligin.

## 10. CONCLUSIONS

Considering the equal importance of GABAergic and glutamatergic synapses, and the arguably greater potential of GABA synapses as therapeutic targets given the variety of $GABA_A$ receptors, surprisingly little is known about how GABAergic synapses are assembled. It seems that the GABAergic synaptic components are not as amenable to assay by yeast two-hybrid methods, and there has yet to be developed a selective inhibitory postsynaptic density biochemical fractionation method. Thus the next major step in the field will be to identify more of the constituent molecules of GABAergic postsynaptic elements. Of the components that have been studied, the cadherin–catenin and neuroligin–neurexin pairs may function early to mediate adhesion between the pre- and post-synaptic partners and to recruit synaptic components. The $GABA_A$ receptor $\gamma 2$ subunit is important for targeting $\alpha$ and $\beta$ subunits as well as gephyrin to the developing or mature synapse, by interactions of TM4 and the large intracellular loop with as yet unidentified partners. The scaffolding protein gephyrin contributes in part to recruiting some $GABA_A$ receptor subunits and to recruiting other components, including glycine receptors. Finally, the DGC acts late in development, perhaps in stabilization of the mature GABAergic synapse. The recent discovery of neuroligin-2 as a component of GABAergic but not glutamatergic postsynaptic sites, and the ability of neurexins to trigger GABAergic postsynaptic differentiation via neuroligin-2, is likely to lead to rapid further progress. Such progress will necessarily involve identifying the missing links between neuroligin, cadherin, gephyrin, the DGC, and $GABA_A$ receptors. *In vivo* mutagenesis of neuroligins may prove fertile, as have the *in vivo* point mutants altering benzodiazepine sensitivity in individual $GABA_A$ receptor subunits. Eventually, we look forward to molecular analyses to understand the basis of the complex differential cellular domain targeting of GABAergic input types and differential postsynaptic trafficking of $GABA_A$ receptor subunit combinations.

## 11. REFERENCES


1. Verderio, C., Pozzi, D., Pravettoni, E., Inverardi, F., Schenk, U., Coco, S., Proux-Gillardeaux, V., Galli, T., Rossetto, O., Frassoni, C., and Matteoli, M. (2004) *Neuron* **41**, 599–610.
2. Somogyi, P., Tamas, G., Lujan, R., and Buhl, E.H. (1998) *Brain Res Brain Res Rev* **26**, 113–135.
3. Bettler, B., Kaupmann, K., Mosbacher, J., and Gassmann, M. (2004) *Physiol Rev* **84**, 835–867.
4. McCall, M.A., Lukasiewicz, P.D., Gregg, R.G., and Peachey, N.S. (2002) *J Neurosci.* **22**, 4163–4174.
5. Milligan, C.J., Buckley, N.J., Garret, M., Deuchars, J., and Deuchars, S.A. (2004) *J. Neurosci.* **24**, 7241–7250.
6. Connolly, C.N., and Wafford, K.A. (2004) *Biochem Soc Trans* **32**, 529–534.
7. Ben-Ari, Y. (2002) *Nat Rev Neurosci* **3**, 728–739.
8. Nayeem, N., Green, T.P., Martin, I.L., and Barnard, E.A. (1994) *J Neurochem* **62**, 815–818.
9. Kittler, J.T., and Moss, S.J. (2003) *Curr Opin Neurobiol* **13**, 341–347.

10. Luscher, B., and Keller, C.A. (2004) *Pharmacol Ther* **102**, 195–221.
11. Whiting, P.J., Bonnert, T.P., McKernan, R.M., Farrar, S., Le Bourdelles, B., Heavens, R.P., Smith, D.W., Hewson, L., Rigby, M.R., Sirinathsinghji, D.J., Thompson, S.A., and Wafford, K.A. (1999) *Ann N Y Acad Sci* **868**, 645–653.
12. Mohler, H., Fritschy, J.M., and Rudolph, U. (2002) *J Pharmacol Exp Ther* **300**, 2–8.
13. Whiting, P., McKernan, R.M., and Iversen, L.L. (1990) *Proc Natl Acad Sci U S A* **87**, 9966–9970.
14. Korpi, E.R., Grunder, G., and Luddens, H. (2002) *Prog Neurobiol* **67**, 113–159.
15. Vicini, S., Ferguson, C., Prybylowski, K., Kralic, J., Morrow, A.L., and Homanics, G.E. (2001) *J Neurosci* **21**, 3009–3016.
16. Sur, C., Wafford, K.A., Reynolds, D.S., Hadingham, K.L., Bromidge, F., Macaulay, A., Collinson, N., O'Meara, G., Howell, O., Newman, R., Myers, J., Atack, J.R., Dawson, G.R., McKernan, R.M., Whiting, P.J., and Rosahl, T.W. (2001) *J Neurosci* **21**, 3409–3418.
17. Kralic, J.E., Korpi, E.R., O'Buckley, T.K., Homanics, G.E., and Morrow, A.L. (2002) *J Pharmacol Exp Ther* **302**, 1037–1045.
18. Rudolph, U., Crestani, F., Benke, D., Brunig, I., Benson, J.A., Fritschy, J.M., Martin, J.R., Bluethmann, H., and Mohler, H. (1999) *Nature* **401**, 796–800.
19. McKernan, R.M., Rosahl, T.W., Reynolds, D.S., Sur, C., Wafford, K.A., Atack, J.R., Farrar, S., Myers, J., Cook, G., Ferris, P., Garrett, L., Bristow, L., Marshall, G., Macaulay, A., Brown, N., Howell, O., Moore, K.W., Carling, R.W., Street, L.J., Castro, J.L., Ragan, C.I., Dawson, G.R., and Whiting, P.J. (2000) *Nat Neurosci* **3**, 587–592.
20. Low, K., Crestani, F., Keist, R., Benke, D., Brunig, I., Benson, J.A., Fritschy, J.M., Rulicke, T., Bluethmann, H., Mohler, H., and Rudolph, U. (2000) *Science* **290**, 131–134.
21. Crestani, F., Low, K., Keist, R., Mandelli, M., Mohler, H., and Rudolph, U. (2001) *Mol Pharmacol* **59**, 442–445.
22. Fritschy, J.M., Benke, D., Johnson, D.K., Mohler, H., and Rudolph, U. (1997) *Neuroscience* **81**, 1043–1053.
23. Collinson, N., Kuenzi, F.M., Jarolimek, W., Maubach, K.A., Cothliff, R., Sur, C., Smith, A., Otu, F.M., Howell, O., Atack, J.R., McKernan, R.M., Seabrook, G.R., Dawson, G.R., Whiting, P.J., and Rosahl, T.W. (2002) *J Neurosci* **22**, 5572–5580.
24. Crestani, F., Keist, R., Fritschy, J.M., Benke, D., Vogt, K., Prut, L., Bluthmann, H., Mohler, H., and Rudolph, U. (2002) *Proc Natl Acad Sci U S A* **99**, 8980–8985.
25. Jones, A., Korpi, E.R., McKernan, R.M., Pelz, R., Nusser, Z., Makela, R., Mellor, J.R., Pollard, S., Bahn, S., Stephenson, F.A., Randall, A.D., Sieghart, W., Somogyi, P., Smith, A.J., and Wisden, W. (1997) *J Neurosci* **17**, 1350–1362.
26. Korpi, E.R., Koikkalainen, P., Vekovischeva, O.Y., Makela, R., Kleinz, R., Uusi-Oukari, M., and Wisden, W. (1999) *Eur J Neurosci* **11**, 233–240.
27. Nusser, Z., Ahmad, Z., Tretter, V., Fuchs, K., Wisden, W., Sieghart, W., and Somogyi, P. (1999) *Eur J Neurosci* **11**, 1685–1697.
28. Brickley, S.G., Revilla, V., Cull-Candy, S.G., Wisden, W., and Farrant, M. (2001) *Nature* **409**, 88–92.
29. Homanics, G.E., DeLorey, T.M., Firestone, L.L., Quinlan, J.J., Handforth, A., Harrison, N.L., Krasowski, M.D., Rick, C.E.M., Korpi, E.R., Makela, R., Brilliant, M.H., Hagiwara, N., Ferguson, C., Snyder, K., and Olsen, R.W. (1997) *Proc Natl Acad Sci U S A* **94**, 4143–4148.
30. DeLorey, T.M., Handforth, A., Anagnostaras, S.G., Homanics, G.E., Minassian, B.A., Asatourian, A., Fanselow, M.S., Delgado-Escueta, A., Ellison, G.D., and Olsen, R.W. (1998) *J Neurosci* **18**, 8505–8514.
31. Gunther, U., Benson, J., Benke, D., Fritschy, J.M., Reyes, G., Knoflach, F., Crestani, F., Aguzzi, A., Arigoni, M., Lang, Y. et al. (1995) *Proc Natl Acad Sci U S A* **92**, 7749–7753.
32. Essrich, C., Lorez, M., Benson, J.A., Fritschy, J.-M., and Luscher, B. (1998) **1**, 563–571.
33. Crestani, F., Lorez, M., Baer, K., Essrich, C., Benke, D., Laurent, J.P., Belzung, C., Fritschy, J.M., Luscher, B., and Mohler, H. (1999) *Nat Neurosci* **2**, 833–839.
34. Schweizer, C., Balsiger, S., Bluethmann, H., Mansuy, I.M., Fritschy, J.M., Mohler, H., and Luscher, B. (2003) *Mol Cell Neurosci* **24**, 442–450.
35. Mihalek, R.M., Banerjee, P.K., Korpi, E.R., Quinlan, J.J., Firestone, L.L., Mi, Z.P., Lagenaur, C., Tretter, V., Sieghart, W., Anagnostaras, S.G., Sage, J.R., Fanselow, M.S., Guidotti, A., Spigelman, I., Li, Z., DeLorey, T.M., Olsen, R.W., and Homanics, G.E. (1999) *Proc Natl Acad Sci U S A* **96**, 12905–12910.
36. Tretter, V., Hauer, B., Nusser, Z., Mihalek, R.M., Hoger, H., Homanics, G.E., Somogyi, P., and Sieghart, W. (2001) *J Biol Chem* **276**, 10532–10538.

37. Korpi, E.R., Mihalek, R.M., Sinkkonen, S.T., Hauer, B., Hevers, W., Homanics, G.E., Sieghart, W., and Luddens, H. (2002) *Neuroscience* **109**, 733–743.
38. Hagiwara, N., Katarova, Z., Siracusa, L.D., and Brilliant, M.H. (2003) *Dev Biology* **254**, 93–101.
39. Fagiolini, M., Fritschy, J.-M., Low, K., Mohler, H., Rudolph, U., and Hensch, T.K. (2004) *Science* **303**, 1681–1683.
40. Fritschy, J.M., and Brunig, I. (2003) *Pharmacol Ther* **98**, 299–323.
41. Brunig, I., Scotti, E., Sidler, C., and Fritschy, J.M. (2002) *J Comp Neurol* **443**, 43–55.
42. Nyiri, G., Freund, T.F., and Somogyi, P. (2001) *Eur J Neurosci* **13**, 428–442.
43. Klausberger, T., Roberts, J.D.B., and Somogyi, P. (2002) *J Neurosci* **22**, 2513–2521.
44. Fritschy, J., Paysan, J., Enna, A., and Mohler, H. (1994) *J Neurosci* **14**, 5302–5324.
45. Rao, A., Cha, E.M., and Craig, A.M. (2000) *J Neurosci* **20**, 8344–8353.
46. Christie, S.B., Miralles, C.P., and De Blas, A.L. (2002) *J Neurosci* **22**, 684–697.
47. Graf, E.R., Zhang, X., Jin, S.X., Linhoff, M.W., and Craig, A.M. (2004) *Cell* **119**, 1013–1026.
48. Baer, K., Essrich, C., Benson, J.A., Benke, D., Bluethmann, H., Fritschy, J.M., and Luscher, B. (1999) *Proc Natl Acad Sci U S A* **96**, 12860–12865.
49. Baer, K., Essrich, C., Balsiger, S., Wick, M.J., Harris, R.A., Fritschy, J.M., and Luscher, B. (2000) *Eur J Neurosci* **12**, 2639–2643.
50. Alldred, M.J., Mulder-Rosi, J., Lingenfelter, S.E., Chen, G., and Luscher, B. (2005) *J Neurosci* **25**, 594–603.
51. Choquet, D., and Triller, A. (2003) *Nat Rev Neurosci* **4**, 251–265.
52. Nusser, Z., Roberts, J.D., Baude, A., Richards, J.G., and Somogyi, P. (1995) *J Neurosci* **15**, 2948–2960.
53. Thomas, P., Mortensen, M., Hosie, A.M., and Smart, T.G. (2005) *Nat Neurosci* **8**, 889–897.
54. van Rijnsoever, C., Tauber, M., Choulli, M.K., Keist, R., Rudolph, U., Mohler, H., Fritschy, J.M., and Crestani, F. (2004) *J Neurosci* **24**, 6785–6790.
55. Bollan, K., Robertson, L.A., Tang, H., and Connolly, C.N. (2003) *Biochem Soc Trans* **31**, 875–879.
56. Taylor, P.M., Thomas, P., Gorrie, G.H., Connolly, C.N., Smart, T.G., and Moss, S.J. (1999) *J Neurosci* **19**, 6360–6371.
57. Connolly, C.N., Krishek, B.J., McDonald, B.J., Smart, T.G., and Moss, S.J. (1996) *J Biol Chem* **271**, 89–96.
58. Minier, F., and Sigel, E. (2004) *Proc Natl Acad Sci U S A* **101**, 7769–7774.
59. Wang, H., Bedford, F.K., Brandon, N.J., Moss, S.J., and Olsen, R.W. (1999) *Nature* **397**, 69–72.
60. Kneussel, M., Haverkamp, S., Fuhrmann, J.C., Wang, H., Wassle, H., Olsen, R.W., and Betz, H. (2000) *Proc Natl Acad Sci U S A* **97**, 8594–8599.
61. Kittler, J.T., Rostaing, P., Schiavo, G., Fritschy, J.M., Olsen, R., Triller, A., and Moss, S.J. (2001) *Mol Cell Neurosci* **18**, 13–25.
62. Setou, M., Seog, D.-H., Tanaka, Y., Kanai, Y., Takei, Y., Kawagishi, M., and Hirokawa, N. (2002) *Nature* **417**, 83–87.
63. Kittler, J.T., Arancibia-Carcamo, I.L., and Moss, S.J. (2004) *Biochem Pharmacol* **68**, 1649–1654.
64. Tanida, I., Sou, Y.-s., Ezaki, J., Minematsu-Ikeguchi, N., Ueno, T., and Kominami, E. (2004) *J Biol Chem* **279**, 36268–36276.
65. Charych, E.I., Yu, W., Miralles, C.P., Serwanski, D.R., Li, X., Rubio, M., and De Blas, A.L. (2004) *J Neurochem* **90**, 173–189.
66. Keller, C.A., Yuan, X., Panzanelli, P., Martin, M.L., Alldred, M., Sassoe-Pognetto, M., and Luscher, B. (2004) *J Neurosci* **24**, 5881–5891.
67. Huang, K., and El-Husseini, A. (2005) *Curr Opin Neurobiol* **15**, 527–535.
68. Hayashi, T., Rumbaugh, G., and Huganir, R.L. (2005) *Neuron* **47**, 709–723.
69. Rathenberg, J., Kittler, J.T., and Moss, S.J. (2004) *Mol Cell Neurosci* **26**, 251–257.
70. Kittler, J.T., Thomas, P., Tretter, V., Bogdanov, Y.D., Haucke, V., Smart, T.G., and Moss, S.J. (2004) *Proc Natl Acad Sci U S A* **101**, 12736–12741.
71. Kittler, J.T., Delmas, P., Jovanovic, J.N., Brown, D.A., Smart, T.G., and Moss, S.J. (2000) *J Neurosci* **20**, 7972–7977.
72. Bedford, F.K., Kittler, J.T., Muller, E., Thomas, P., Uren, J.M., Merlo, D., Wisden, W., Triller, A., Smart, T.G., and Moss, S.J. (2001) *Nat Neurosci* **4**, 908–916.
73. Wan, Q., Xiong, Z.G., Man, H.Y., Ackerley, C.A., Braunton, J., Lu, W.Y., Becker, L.E., MacDonald, J.F., and Wang, Y.T. (1997) *Nature* **388**, 686–690.
74. Wang, Q., Liu, L., Pei, L., Ju, W., Ahmadian, G., Lu, J., Wang, Y., Liu, F., and Wang, Y.T. (2003) *Neuron* **38**, 915–928.

75. Nusser, Z., Hajos, N., Somogyi, P., and Mody, I. (1998) *Nature* **395**, 172–177.
76. Gutierrez, R. (2005) *Trends Neurosci* **28**, 297–303.
77. Brunig, I., Penschuck, S., Berninger, B., Benson, J., and Fritschy, J.-M. (2001) *Eur J Neurosci* **13**, 1320–1328.
78. Mizoguchi, Y., Kanematsu, T., Hirata, M., and Nabekura, J. (2003) *J Biol Chem* **278**, 44097–44102.
79. Elmariah, S.B., Crumling, M.A., Parsons, T.D., and Balice-Gordon, R.J. (2004) *J Neurosci* **24**, 2380–2393.
80. Jovanovic, J.N., Thomas, P., Kittler, J.T., Smart, T.G., and Moss, S.J. (2004) *J Neurosci* **24**, 522–530.
81. Kneussel, M., and Betz, H. (2000) *J Physiol* **525 Pt 1**, 1–9.
82. Sassoe-Pognetto, M., and Fritschy, J.M. (2000) *Eur J Neurosci* **12**, 2205–2210.
83. Meyer, G., Kirsch, J., Betz, H., and Langosch, D. (1995) *Neuron* **15**, 563–572.
84. Meier, J., De Chaldee, M., Triller, A., and Vannier, C. (2000) *Mol Cell Neurosci* **16**, 566–577.
85. Ramming, M., Kins, S., Werner, N., Hermann, A., Betz, H., and Kirsch, J. (2000) *Proc Natl Acad Sci U S A* **97**, 10266–10271.
86. Stallmeyer, B., Schwarz, G., Schulze, J., Nerlich, A., Reiss, J., Kirsch, J., and Mendel, R.R. (1999) *Proc Natl Acad Sci U S A* **96**, 1333–1338.
87. Feng, G., Tintrup, H., Kirsch, J., Nichol, M.C., Kuhse, J., Betz, H., and Sanes, J.R. (1998) *Science* **282**, 1321–1324.
88. Kneussel, M., Brandstatter, J.H., Laube, B., Stahl, S., Muller, U., and Betz, H. (1999) *J Neurosci* **19**, 9289–9297.
89. Fischer, F., Kneussel, M., Tintrup, H., Haverkamp, S., Rauen, T., Betz, H., and Wassle, H. (2000) *J Comp Neurol* **427**, 634–648.
90. Levi, S., Logan, S.M., Tovar, K.R., and Craig, A.M. (2004) *J Neurosci* **24**, 207–217.
91. Kins, S., Betz, H., and Kirsch, J. (2000) *Nat Neurosci* **3**, 22–29.
92. Mammoto, A., Sasaki, T., Asakura, T., Hotta, I., Imamura, H., Takahashi, K., Matsuura, Y., Shirao, T., and Takai, Y. (1998) *Biochem Biophys Res Commun* **243**, 86–89.
93. Sabatini, D.M., Barrow, R.K., Blackshaw, S., Burnett, P.E., Lai, M.M., Field, M.E., Bahr, B.A., Kirsch, J., Betz, H., and Snyder, S.H. (1999) *Science* **284**, 1161–1164.
94. Fuhrmann, J.C., Kins, S., Rostaing, P., El Far, O., Kirsch, J., Sheng, M., Triller, A., Betz, H., and Kneussel, M. (2002) *J Neurosci* **22**, 5393–5402.
95. Harvey, K., Duguid, I.C., Alldred, M.J., Beatty, S.E., Ward, H., Keep, N.H., Lingenfelter, S.E., Pearce, B.R., Lundgren, J., Owen, M.J., Smart, T.G., Luscher, B., Rees, M.I., and Harvey, R.J. (2004) *J Neurosci* **24**, 5816–5826.
96. Benson, D.L., Yoshihara, Y., and Mori, K. (1998) *J Neurosci Res* **52**, 43–53.
97. Togashi, H., Abe, K., Mizoguchi, A., Takaoka, K., Chisaka, O., and Takeichi, M. (2002) *Neuron* **35**, 77–89.
98. Bozdagi, O., Valcin, M., Poskanzer, K., Tanaka, H., and Benson, D.L. (2004) *Mol Cell Neurosci* **27**, 509–521.
99. Wang, X., Weiner, J.A., Levi, S., Craig, A.M., Bradley, A., and Sanes, J.R. (2002) *Neuron* **36**, 843–854.
100. Weiner, J.A., Wang, X., Tapia, J.C., and Sanes, J.R. (2005) *Proc Natl Acad Sci U S A* **102**, 8–14.
101. Abe, K., Chisaka, O., Van Roy, F., and Takeichi, M. (2004) *Nat Neurosci* **7**, 357–363.
102. Bamji, S.X., Shimazu, K., Kimes, N., Huelsken, J., Birchmeier, W., Lu, B., and Reichardt, L.F. (2003) *Neuron* **40**, 719–731.
103. Ango, F., di Cristo, G., Higashiyama, H., Bennett, V., Wu, P., and Huang, Z.J. (2004) *Cell* **119**, 257–272.
104. Saghatelyan, A.K., Dityatev, A., Schmidt, S., Schuster, T., Bartsch, U., and Schachner, M. (2001) *Mol Cell Neurosci* **17**, 226–240.
105. Saghatelyan, A.K., Nikonenko, A.G., Sun, M., Rolf, B., Putthoff, P., Kutsche, M., Bartsch, U., Dityatev, A., and Schachner, M. (2004) *Mol Cell Neurosci* **26**, 191–203.
106. Hoffman, E.P., and Kunkel, L.M. (1989) *Neuron* **2**, 1019–1029.
107. Henry, M.D., and Campbell, K.P. (1999) *Curr Opin Cell Biol* **11**, 602–607.
108. Michele, D.E., Barresi, R., Kanagawa, M., Saito, F., Cohn, R.D., Satz, J.S., Dollar, J., Nishino, I., Kelley, R.I., Somer, H., Straub, V., Mathews, K.D., Moore, S.A., and Campbell, K.P. *Nature* (2002) **418**, 417–421.

109. Knuesel, I., Mastrocola, M., Zuellig, R.A., Bornhauser, B., Schaub, M.C., and Fritschy, J.M. (1999) *Eur J Neurosci* **11**, 4457–4462.
110. Levi, S., Grady, R.M., Henry, M.D., Campbell, K.P., Sanes, J.R., and Craig, A.M. (2002) *J Neurosci* **22**, 4274–4285.
111. Moore, S.A., Saito, F., Chen, J., Michele, D.E., Henry, M.D., Messing, A., Cohn, R.D., Ross-Barta, S.E., Westra, S., Williamson, R.A., Hoshi, T., and Campbell, K.P. (2002) *Nature* **418**, 422–425.
112. Boucard, A.A., Chubykin, A.A., Comoletti, D., Taylor, P., and Sudhof, T.C. (2005) *Neuron* **48**, 229–236.
113. Sugita, S., Saito, F., Tang, J., Satz, J., Campbell, K., and Sudhof, T.C. (2001) *J Cell Biol* **154**, 435–445.
114. Brunig, I., Suter, A., Knuesel, I., Luscher, B., and Fritschy, J.M. (2002) *J Neurosci* **22**, 4805–4813.
115. Nguyen, T., and Sudhof, T.C. (1997) *J Biol Chem* **272**, 26032–26039.
116. Irie, M., Hata, Y., Takeuchi, M., Ichtchenko, K., Toyoda, A., Hirao, K., Takai, Y., Rosahl, T.W., and Sudhof, T.C. (1997) *Science* **277**, 1511–1515.
117. Varoqueaux, F., Jamain, S., and Brose, N. (2004) *Eur J Cell Biol* **83**, 449–456.
118. Prange, O., Wong, T.P., Gerrow, K., Wang, Y.T., and El-Husseini, A. (2004) *Proc Natl Acad Sci U S A* **101**, 13915–13920.
119. Chih, B., Engelman, H., and Scheiffele, P. (2005) *Science* **307**, 1324–1328.
120. Dean, C., Scholl, F.G., Choih, J., DeMaria, S., Berger, J., Isacoff, E., and Scheiffele, P. (2003) *Nat Neurosci* **6**, 708–716.
121. Ichtchenko, K., Nguyen, T., and Sudhof, T.C. (1996) *J Biol Chem* **271**, 2676–2682.
122. Scheiffele, P., Fan, J., Choih, J., Fetter, R., and Serafini, T. (2000) *Cell* **101**, 657–669.
123. Sara, Y., Biederer, T., Atasoy, D., Chubykin, A., Mozhayeva, M.G., Sudhof, T.C., and Kavalali, E.T. (2005) *J Neurosci* **25**, 260–270.
124. Levinson, J.N., and El-Husseini, A. (2005) *Neuron* **48**, 171–174.
125. Levinson, J.N., Chery, N., Huang, K., Wong, T.P., Gerrow, K., Kang, R., Prange, O., Wang, Y.T., and El-Husseini, A. (2005) *J Biol Chem* **280**, 17312–17319.

# 20

# ROLE OF SYNAPTOGENESIS IN MORPHOLOGIC STABILIZATION OF DEVELOPING DENDRITES

Kurt Haas*

## 1. SUMMARY

Recent studies employing time-lapse imaging of growing neurons within intact brains, brain slices and retinal explants have discovered a remarkable amount of morphological remodeling as dendritic arbors grow. Most of this dynamic growth involves rapid turnover of dendritic filopodia, a small fraction of which stabilize and grow to become persistent branches. Why do growing neurons expend such extensive resources extending processes that are mostly retracted, and what influences or confers stabilization on the few filopodia that persist? One hypothesis is that filopodia sample their environment for appropriate presynaptic contacts. This 'synaptotropic' model is supported by findings that afferent input and synaptogenesis are significant contributors to filopodial stabilization and dendrite growth. Remarkably, increased dendrite growth in response to brief periods of enhanced sensory stimuli, mediated by glutamatergic transmission, has been directly observed in the intact brain of developing animals. *In vivo* imaging of PSD-95 puncta in growing dendrites demonstrates that synapses occur on filopodia and that the presence of synapses confers morphologic stabilization. Further *in vitro* studies have identified components of activity-regulated molecular cascades including $Ca^{2+}$ influx, and the activation of CaMKII, CaMKIV, CREB, CREST, and small GTPases that may translate afferent input into both synaptogenesis and cytoskeletal plasticity. Innervation that triggers both synapse

*Brain Research Centre, University of British Columbia, Vancouver, BC, Canada V6T2B5; kurt.haas@ubc.ca

formation and local dendrite growth promotes arborization in regions with appropriate afferent input.

## 2. INTRODUCTION

Dendrites are the scaffolding onto which the majority of synapses occur. The size and shape of dendritic arbors limits the number and type of synapses that a neuron can form, and therefore, determines the complexity and function of the neuronal network to which it belongs. Since dendrite shape also influences spread and convergence of transmembrane potentials associated with synaptic input and cellular excitability, dendritic arbor morphology is integrally involved in functional properties of signal processing[1,2]. The diversity of patterns of dendritic arbor structures and the high conservation of these patterns within specific cell types also suggest an important link between form and function. How do developing neurons achieve elaborate dendritic arbor morphologies? The ability of neurons within a specific class to faithfully adhere to a general arbor pattern, and the recapitulation, to some degree, of this pattern when neurons are dissociated and regrown in cultures largely devoid of normal environmental cues strongly suggests a genetic basis for general arbor structure. However, the unique characteristics of individual neuronal arbors in specific dendrite characteristics such as exact positions of branch points, branch lengths, and branch number supports influence from factors extrinsic to the cell. A model involving the interplay between genetic programming of general dendritic arbor form and influence from multiple extrinsic environmental factors is supported by the recent findings of transcription factors associated with dendritic arbor shapes[3,4], and a growing list of growth-modifying extracellular molecules, including neurotrophic growth factors[5–9], and chemoaffinity guidance molecules[10,11]. Another critical component of the growing neuron's environment is neurotransmission from afferent innervation. Here we review evidence implicating neurotransmission in regulating dendrite growth and focus on the potential involvement of activity-dependent synaptogenesis in this process.

## 3. GROWTH OF DENDRITES IN INTACT TISSUES

Recent advances in imaging early stages of neuronal development within intact tissues have led to new understandings of how dendrites grow and which factors direct formation of morphologies required for correct connectivity and function[12–20]. Brain slices and retinal explants have allowed examination of growth in semi-intact environments while offering increased experimental accessibility. Imaging neurons within intact preparations, however, has provided observation of growth within 3-dimensional environments complete with the full complement of external growth influencing molecules and neuronal activity. Much of our knowledge of how dendrites grow within the intact brain comes from studies of optic tectal neurons in the albino *Xenopus laevis* tadpole and zebrafish. The transparency of both organisms allows direct imaging of labeled brain neurons at early stages of development during extensive neurogenesis and dendritogenesis. Both preparations benefit from and easily targeted population of newly differentiated neurons produced from a proliferative zone in the caudal optic tectum, and zebrafish offers methods for transgenic manipulation. Central to this work has been the development of powerful and less damaging *in vivo* imaging systems[21] and techniques to fluorescently label individual neurons to discern dendritic processes from the surrounding tissue[13,16,22–24]. Repeated time-lapse confocal or two-photon imaging with brief intervals ranging from seconds to hours allows direct examination of the motility associated with dendrite growth. Imaging the same neurons over days provide insight as to how such short-term growth events culminate to form persistent arbor structures.

## 4. LONG-INTERVAL IMAGING OF DENDRITIC ARBOR GROWTH

*In vivo* time-lapse imaging of developing tectal neurons reveals that dendritic arbor growth involves more that than straightforward extension to presynaptic targets[25]. Rather, there is a highly dynamic process involving considerable remodeling. Imaging the same developing neuron over multiple days within the intact brain reveals that neurons pass through distinct growth phases. Neurons are initially round cells devoid of processes. Short neurites extend, and axons differentiate and extend toward targets. During this initial axonal extension phase, the dendritic arbor is in a quiescent state with few dendritic branches or length added. Dendritic growth then switches to a period of rapid arbor elaboration. In *Xenopus* tectal neurons, a dynamic period of growth lasting 4 days ensues (Figure 20.1). New branches are added both by extension of dendritic growth cones and addition of interstitial branches, with the majority of growth associated with extension of interstitial branches. New branch length then supports the emergence of further new interstitial branches, thereby continuously increasing arbor complexity. Comparing dendrites from the same neuron from one day to the next reveals that even relatively large portions of the arbor can be partially or completely retracted. After 4 days, growth rates plateau and no further net increase in arbor length or branch number occurs.

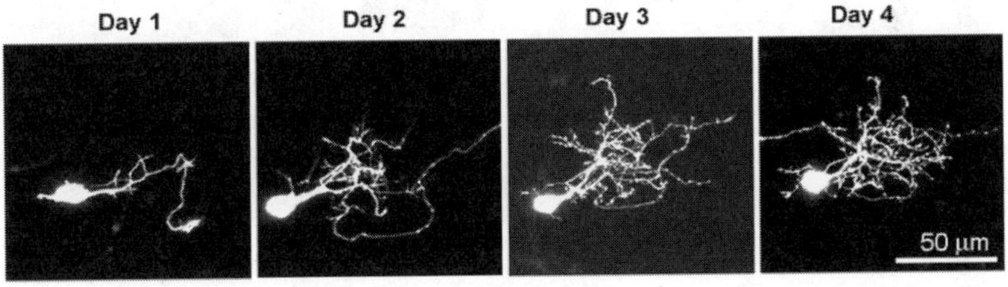


**Figure 20.1.** *In Vivo* Imaging of Dendritic Arbor Growth in the Optic Tectum of the *Xenopus laevis* Tadpole. A single neuron was transfected for EGFP expression using single-cell electroporation and imaged by two-photon time-lapse microscopy with 24 h intervals to capture dramatic changes in arbor size and shape over 4 days.

## 5. SHORT-INTERVAL IMAGING OF FILOPODIAL MOTILITY

Imaging growing dendrites using short intervals ranging from seconds to hours reveals a remarkable level of motility present in immature dendritic arbors. The majority of this motility is associated with short protrusions that rapidly extend and retract from growing dendrites[12–15,19,20,26]. Over short periods of seconds to minutes, these short protrusions can emerge *de novo* from existing dendrites, rapidly extend, retract, and disappear (Figure 20.2). These short (usually less than 10 μm), highly motile processes are actin-rich and are therefore referred to as dendritic filopodia[13,19,27,28]. There appear to be at least three distinct populations of dendritic filopodia with different growth behaviors: filopodia on developing dendrites extending from growth cones and shafts, and filopodia on mature dendrites. Filopodia of mature neurons may be associated with synaptogenesis and spine formation[29–31], but not with dendrite growth since they do not give rise to new branches. In early periods of neuronal development during the most extensive period of dendrite growth, filopodia are not associated with spines. The height of expression of filopodia associated with dendrite growth occurs at developmental stages well before the appearance of spines in spiny neurons, and they are similarly found expressed in nonspiny neurons[14,32–36]. Portera-Cailliau and colleagues imaged dendritic filopodial motility in dye-filled pyramidal neurons within acute cortical slices derived from rat pups 0 to 12 days old[13]. Using rapid, 30 s-interval time-lapse imaging, they found a maturational change in occurrence of filopodial, with the height of expression around postnatal day 5 (one week before spines appear in these neurons). They observed that filopodia crowning dendritic growth cones behaved differently than shaft filopodia in that they are longer, more motile, have shorter life-times, and are insensitive to activity.

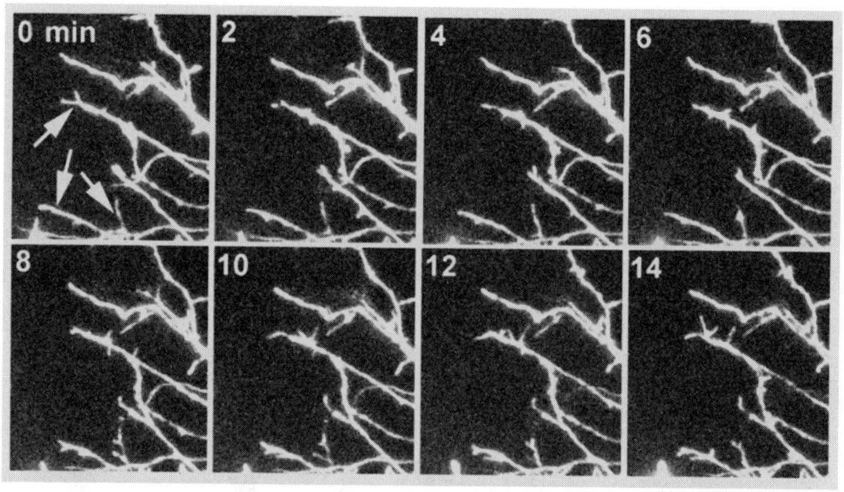


**Figure 20.2.** Short-Interval *In Vivo* Time-Lapse Imaging of a Portion of the Dendritic Arbor of a Single Developing *Xenopus* Tectal Neuron Demonstrates the Rapid Motility and Turnover of Dendritic Filopodia. Pictures were taken at 2-min intervals and arrows point to three sites of dramatic filopodial motility.

The surprising finding from short-interval imaging of dendrite growth is that the great majority of newly extended filopodia retract over seconds or minutes. Cortical pyramidal neurons imaged in acute slices from postnatal day 2 to 3 showed a remarkable turnover rate calculated to result in extension and retraction of approximately 50,000 shaft filopodia in a single dendritic arbor in 1 day. Even with this high turnover, dendritic arbors rapidly grow in total branch length and branch number through both elongation of existing branches and addition and maintaining new branches[12,25]. A small fraction of the pool of newly added filopodia stabilize and grow to become persistent components of the dendritic arbor, and subsequently support further extension of additional filopodia. The purpose of such high turnover of dendritic filopodia remains unclear, but it may involve sampling the environment for growth-promoting factors, to guide growth, and/or to establish correct synapses. Recent work, however, sheds some light on factors that promote stabilization of the few filopodia that contribute to persistent dendrite growth.

## 6. SENSORY STIMULATION INFLUENCES NEURONAL GROWTH

The potential involvement of afferent innervation in regulating dendrite growth originates from a rich history of studies in which animals are reared under conditions of altered sensory stimulation. Animals raised in environments enriched with complex sensory stimuli and mentally challenging activities develop cortical neurons with more complex dendritic arbors compared to neurons of animals reared in less stimulating environments[37–39]. More drastic interventions created by sensory occlusion demonstrate that loss of normal activity during development leads to lasting alterations in dendritic morphology. Dark rearing during critical periods of visual system development alters the pattern and distribution of dendrites of layer 4 dendrites in rat visual cortex[40]. Monocular deprivation alters dendrite development in lateral geniculate nucleus (LGN) and visual cortex[41,42], with dendrites of denervated layer 4 cortical spiny stellate neurons abnormally extending into neighboring nondeprived ocular dominance columns[43]. Further evidence of afferent input directing dendrite growth comes from the three-eyed frog in which the additional retinal input creates abnormal ocular dominance bands in the tectum, and dendritic arbors of tectal neurons are restricted to regions innervated by one eye[44]. Afferent input-regulated growth is not limited to the visual system since occluding auditory input by plugging one ear results in shortening of dendrites of laminar nucleus neurons innervated by that ear[45]. However, visual deprivation does not significantly alter dendritic growth of cortical layer 4 stellate cells[46] or layer 3 pyramidal cells[47], suggesting that factors besides neuronal activity associated with external sensory stimulation may promote correct arborization. Findings that de-afferentation produces a profound reduction in dendrite length, exceeding effects of occluding sensory input, imply that spontaneous activity or activity-independent mechanisms substantially contributes to growth[48]. The co-existence of multiple mediators of growth in addition to correlated afferent input, including spontaneous activity and activity-independent mechanisms that remain following sensory occlusion likely contributes to these varying results.

Important direct evidence for sensory stimuli regulating growth of brain neurons was shown using *in vivo* repeated time-lapse imaging of developing *Xenopus* tectal neurons[15]. Sin and colleagues took advantage of the ability to transfect newly formed neurons in the tectal proliferative zone and the subsequent growth of these neurons into a functional visual processing circuit. The primary afferent input to tectal neurons at this stage is glutamatergic synaptic input from retinal ganglion cells (RGCs), and tectal neurons respond to visual stimuli. Sin and colleagues controlled afferent glutamatergic innervation to tectal neurons by altering visual stimulation for short periods while directly imaging effects on growth of developing tectal neuron dendrites using *in vivo* two-photon microscopy. They find that exposure to brief periods (4 h) of darkness or visual stimulation induced decreased and increased growth, respectively. The increased dendrite growth during visual stimuli was mediated by an increase in new branch additions and filopodial stabilization. Further experiments found that when a brief period of visual stimulation precedes exposure to darkness, the growth-promoting effects of patterned visual stimulation persists even after the stimulus is removed. This suggests that sensory stimulation triggers long-lasting increased morphological plasticity that becomes independent of persistent activity.

## 7. EFFECTS OF TETRODOTOXIN ON DENDRITE GROWTH

Important evidence for involvement of neuronal activity in dendrite growth comes from pharmacological blockade of activity. Exposing developing neurons to tetrodotoxin (TTX), which blocks action potential-mediated transmission, has yielded varying effects on dendrite growth. In acute cortical slices, TTX exposure increases density and length of interstitial shaft filopodia, without affecting their motility, half-life or turnover, and filopodia associated with growth cones are insensitive to TTX[13]. TTX did not prevent afferent innervation-induced dendrite growth in hippocampal co-cultures[43]. Exposure to TTX in the eye or LGN in mammals produced no detectable effects on arborization of LGN dendrites[49,50]. Dendritic growth of axolotl Mauthner cells has been shown to be driven by afferent innervation that is TTX insensitive[51]. TTX also failed to alter dendrite growth *in vivo* in *Xenopus laevis* optic tectum[17] or filopodial motility of developing RGCs in retinal explants[26]. However, in ferret cortical brain slices, TTX blocks neurotrophin-induced dendritic growth[52]. The action of TTX to eliminate action potential-mediated synaptic transmission, but leaving spontaneous vesicular neurotransmitter release, induces a switch from strong correlated input to spontaneous weak innervation. Thus activity-driven events normally regulated by relative amounts of innervation from each synaptic contact would be removed, but activity-dependent mechanisms that can be driven by weak spontaneous input, as well as activity-independent mechanisms remain.

## 8. GLUTAMATERGIC TRANSMISSION AND DENDRITE GROWTH

Sensory stimuli and afferent innervation likely regulate dendrite growth through the primary excitatory neurotransmitter glutamate. The effects of glutamatergic synaptic transmission on dendrite growth have been studied using selective antagonists to the glutamatergic receptors subtypes NMDA and AMPA. Similar to studies of altered afferent innervation and TTX, various and, at times, contradictory results have been found. NMDA and AMPA receptor blockade reduces density and turnover of dendritic shaft filopodia in acute neonatal cortical brain slices[13], and decreases the motility and length of filopodial on developing RGC dendrites in retinal explants[26]. In both of these studies, TTX produced different effects than glutamate receptor antagonists in the same preparations, highlighting the potential importance of low levels of spontaneous activity in promoting growth that is completely blocked by receptor antagonists. Studies examining long-term effects of antagonists on growth find that NMDA and AMPA receptor blockade decreases dendrite growth in pyramidal neurons of neonatal rat cortex organotypic cultures[53] and tectal neurons in *Xenopus* tadpoles[17,18], and NMDA receptor antagonists suppress dendrite growth *in vivo* in rat spinal motor neurons[54] and supraoptic nucleus[55]. A developmental shift in glutamate receptor function in dendritogenesis has been shown through *in vivo* imaging of *Xenopus* tectum[17,18]. As developing tectal neurons elaborate dendritic arbors, the glutamatergic synapses on these dendrites undergo concomitant maturation[56,57]. Immature tectal neurons with small dendritic arbors express predominantly NMDA receptors at retinotectal synapses and express proportionally high levels of 'silent' synapses containing only NMDA receptors. Not surprisingly, at this stage, immature tectal neurons are susceptible to NMDA receptor antagonists, but not AMPA receptor blockade. As tectal neurons grow, AMPA receptors are incorporated into maturing glutamatergic synapses, and AMPA receptor antagonist-induced reduction in dendrite growth is seen. These results demonstrate that basal glutamatergic transmission contributes to dendrite growth and that different glutamate receptor subtypes mediate these influences at different stages of neuronal maturation. This developmental shift in glutamate receptor subtype expression may contribute to contradictory results from studies of receptor blockade on dendrite growth.

Glutamatergic transmission does not only underlie basal growth, but has also been shown to mediate activity-dependent dendritic growth. Both NMDA and AMPA receptor antagonists block filopodial stabilization and dendrite growth of *Xenopus* tectal neurons elicited by visual stimulation[15]. A direct effect of glutamate receptor stimulation on filopodia behavior was detected by time-lapse imaging of growing cortical pyramidal neurons in acute neonatal brain slices in response to local exogenous application of glutamate[13]. Glutamate released from a pipette in close proximity to dendritic shaft filopodia induced filopodial elongation. These filopodia may be detecting and growing toward the increasing glutamate gradient, and therefore may direct dendrite growth to presynaptic release sites.

## 9. DENDRITOGENESIS AND SYNAPTOGENESIS

Evidence for a relationship between glutamatergic afferent innervation and dendrite growth, and coincidence of developmental periods of maximal dendritogenesis and activity-dependent synaptogenesis[14] leaves the question of what is the relationship between synaptogenesis and dendritic growth. In rat cortex, the majority of dendrite growth occurs in the first 3 postnatal weeks and closely parallels the time course of afferent innervation. Likewise, in the cerebellum, Purkinje neuron dendritic growth coincides with innervation from granule cells, and interfering with this afferent innervation causes abnormal Purkinje dendrite formation[58,59]. In the *Xenopus* tadpole tectum, neurons undergo extensive dendritic arborization while RGC axons are entering the tectum and forming new synapses to establish the retinotopic map[56]. Although filopodia are highly dynamic and turnover rapidly, they can possess synapses[19,60]. Expression of fluorescently tagged components of the pre- and postsynaptic assemblies has proven useful for detecting location of synapses, the time course of synapse formation and elimination, and the relationship between synaptogenesis and dendrite growth. In dissociated hippocampal cultures, presynaptic proteins cluster within 30 min, and postsynaptic proteins cluster within 45 min of axonal contact with dendritic filopodia[61]. Clusters of the postsynaptic density protein PSD-95 fused to GFP are found to rapidly coalesce, move, and dissociate over a time period of minutes in rat hippocampal organotypic cultures from postnatal day 4 to 7[62]. PSD-95:GFP typically localized in close proximity to clustering of the presynaptic protein synapsin-1, suggesting that PSD-95: GFP clusters are indeed markers of functional synapses. Highly motile filopodia with short lifetimes tend not express PSD-95:GFP clusters, but PSD-95:GFP clusters were found in filopodia that persist[62,63]. A recent study by Niell and colleagues has extended these studies by using *in vivo* two-photon time-lapse microscopy to directly image filopodia behavior and clustering of synaptic proteins in developing neurons within the larval zebrafish optic tectum[19]. In this study, tectal neurons were labeled with the space-filling fluorophore DsRed for imaging filopodial morphology along with PSD-95:GFP to detect synapse locations. Zebrafish tectal neurons were repeatedly imaged using short intervals to capture filopodial motility and PSD-95:GFP clustering over minutes, as well as long-interval imaging to observe cumulative changes in arbor structure. They found that the number of PSD-95:GFP puncta increased in direct relation to dendritic arbor size over 5 days, demonstrating coincidence of morphogenesis and synaptogenesis. Using 20 min-interval imaging for up to 24 h allowed correlation of formation of PSD-95:GFP puncta with filopodial motility. Similar to that described in the *Xenopus* tadpole tectum[25] and acute rat cortical slice[13], zebrafish tectal neurons demonstrate highly dynamic dendritic filopodia that continuously extend and retract over a time course of minutes, leading to dendritic arbor remodeling over hours[14,20,64]. Niell and colleagues found both transient and persistent PSD-95:GFP puncta. PSD-95:GFP puncta appear through *de novo* coalescence at a specific site and not from the transport of large preassembled puncta as seen in rat hippocampus[62,65]. These puncta are lost in a similar fashion, by fading away, not moving *en masse* to new sites. As dendritic arbors mature, the proportion of stable to transient PSD-95 puncta increases. The majority (94%) of new stable PSD-95:GFP puncta appear on filopodia, with few occurring on dendritic shafts, and often more than one puncta will appear on a single filopodia. A functional relationship between synapses and cytoskeletal plasticity is evident from the finding that new filopodia extend from dendritic shafts at sites of stable PSD-95:GFP puncta. How do PSD-95:GFP puncta relate to filopodial motility, stabilization, and dendrite growth? Niell and colleagues observed that newly extending filopodia are devoid of puncta, but puncta appear approximately 30 min following extension if the new process had not retracted in that time. New puncta are dim, but increase in brightness and size with time. Filopodia containing PSD-95:GFP clusters exhibit continued dynamic extension and retraction, but typically do not retract beyond a stabilized PSD-95:GFP puncta. When filopodia with puncta do retract completely, the PSD-95:GFP cluster first disassembles over 20 min prior to retraction. Importantly, they found a direct relationship between puncta and filopodial stabilization. Filopodia stable for more than 1 h invariably possessed at least one puncta. In the developing tectum, filopodia puncta become shaft filopodia when the stabilized filopodia becomes a shaft with additional new filopodia extending from it. These findings demonstrate that dendritic filopodia are involved in both establishing synaptic contact and promoting growth of the dendritic arbor[26,31,60]. In the synaptotropic model, synaptogenesis and synapse strengthening induce local dendrite stabilization and promote further local growth, while weakened and lost synapses result in morphological retraction[66]. Since further growth provides additional potential sites for new synapses, synaptotropically driven growth represents a positive feed-forward mechanism to promote dendritic arborization in regions with appropriate innervation.

## 10. MECHANISMS OF NEUROTRANSMISSION-MEDIATED DENDRITE GROWTH

It is not yet clear how neuronal transmission translates into the activity-dependent changes in dendrite growth and synaptogenesis described above. One certain component of this pathway, however, is elevation of intracellular $Ca^{2+}$ [67,68]. At glutamatergic synapses, the primary source for extracellular $Ca^{2+}$ entry into dendrites is through NMDA receptors and L-type voltage-sensitive $Ca^{2+}$ channels (VSCCs)[30,67,69]. $Ca^{2+}$ can also be released from intracellular stores, including inositol-1,4,5-triphosphate (Ins(1,4,5)P3)-ryanodine-sensitive stores, and $Ca^{2+}$-induced release from the endoplasmic reticulum. Imaging $Ca^{2+}$ signals in developing neurons has detected spontaneous transient elevations that are either spatially restricted to short regions of dendrites, or globally distributed throughout the arbor. Attempts to correlate these spontaneous events to filopodial motility have yielded varying results that may be due to the neuronal maturational stage, cell type, or the precise spatial and temporal characteristics of the $Ca^{2+}$ transients observed. Exposing acute neonatal rat cortical slices to $Ca^{2+}$-free media increased filopodia length and density, suggesting that entry of extracellular $Ca^{2+}$ contributes to stabilization[13]. In developing RGCs, $Ca^{2+}$ transients attributed to local release from intracellular stores have been implicated in filopodia stabilization, while blocking release from these stores promotes branch retraction[70,71]. In immature hippocampal neurons, local $Ca^{2+}$ transients in filopodia and shafts could be well correlated to filopodial motility[71]. $Ca^{2+}$ levels are low in dendritic shafts when filopodia first extend, local $Ca^{2+}$ transient become more frequent as filopodia begin growing, and when the occurrence of $Ca^{2+}$ transient is high filopodia become immobile. Reducing local $Ca^{2+}$ transients increases filopodial growth, while uncaging $Ca^{2+}$ within dendrites reduced filopodial motility. These findings suggest that low levels of $Ca^{2+}$ promote outgrowth and high levels stabilize filopodial. $Ca^{2+}$ elevation associated with NMDA receptor activation and depolarization increases dendrite growth and has been shown to require activation and further $Ca^{2+}$ through VSCCs[55]. $Ca^{2+}$ entry through VSCCs is also necessary in developing ferret cortical slices for neurotrophin-induced dendrite growth of layer 4 pyramidal neurons[52]. One model to emerge from these studies involves glutamatergic innervation-induced entry of $Ca^{2+}$ through NMDA receptors and local depolarization-induced activation and $Ca^{2+}$ flux through L-type VSCCs. Entry of extracellular $Ca^{2+}$ triggers further local release from intracellular stores, and this spatially restricted $Ca^{2+}$ transients leads to stabilization of filopodia receiving appropriate innervation. Local filopodia stabilization is a necessary step in dendrite growth because it prevents retraction, but allows further subsequent extension to create a longer and persistent dendritic branch.

$Ca^{2+}$ appears to affect growth through two mechanisms: rapid action affecting local motility, and slower global effects on growth[72]. Little is known of the specific effectors of $Ca^{2+}$ that mediate these different effects, but downstream targets of $Ca^{2+}$ influx effecting dendrite plasticity are emerging, including the $Ca^{2+}$/calmodulin-dependent protein kinases CaMKII and CaMKIV, and mitogen-activated protein kinase (MAPK)[67,68,73]. Both CaMKII and CaMKIV are developmentally regulated in neurons with maximal levels during the maximal period of dendrite growth[68]. *In vivo* studies in *Xenopus* tectum find that inhibition of CaMKII increases dendrite growth[74,75], and expression of constitutively active CaMKII resticts growth[74], suggesting that CaMKII may play a role in dendritic branch stabilization and potentially contribute to rapid $Ca^{2+}$ effects. Supporting the connection between synaptogenesis and filopodia stabilization, expression of constitutively active CaMKII in immature tectal neurons promotes both glutamatergic synapse maturation (increased AMPA receptor expression) and dendrite stabilization. Overexpression of CaMKIV by itself did not promote growth, but greatly enhanced the growth-promoting effects of depolarization-induced $Ca^{2+}$ influx[68]. The effects of CaMKIV on dendrite growth are likely mediated by changes in gene expression since CaMKIV is targeted to the nucleus and has been shown to interact with the transcription factor cyclic-AMP-responsive-element binding protein (CREB). Growth induced by influx of $Ca^{2+}$ through VSCCs and activation of CaMKIV is blocked by expression of dominant negative CREB, demonstrating that CREB activation is a required downstream mediator of $Ca^{2+}$-dependent dendritic growth[68]. It remains unclear, however, how CREB activation leads to changes in dendrite growth. One transcriptional target of CREB is BDNF, which has been shown to regulate dendritic growth in cortical and cerebellar neurons[5,9,52]. Recently, another member of this pathway has been discovered called CREST[76]. CREST is a $Ca^{2+}$-responsive transactivator that is developmentally expressed in cortex during periods of dendrite growth. CREST binds to the CREB-binding protein (CBP), and $Ca^{2+}$ influx through NMDA receptors and VSCCs activate CREST-mediated transcription. Trangenic mice expressing mutated CREST express reduced dendrite growth and branching in hippocampal neurons, and expression of mutated CREST in cultured hippocampal neurons blocks depolarization-induced dendrite growth. Transcriptional regulation mediated through CaMKIV, CREB, and CREST may mediate the slow activating long-lasting effects of activity on growth.

## 11. CONCLUSIONS

Time-lapse imaging of individual developing neurons within intact tissues has revealed the dynamic nature of dendritic arbor growth during early brain development. Imaging within the intact developing brain has allowed direct observations of enhanced dendrite growth in response to visual stimulation which supports classical studies of the effects of environmental enrichment and sensory occlusion on neuronal morphology[15]. The molecular pathways responsible for such afferent innervation-directed neuronal growth are beginning to emerge, with initial activation of glutamate receptors and VSCCs as critical initiators of $Ca^{2+}$ influx and activated release of additional $Ca^{2+}$ from intracellular stores (see Figure 20.3).

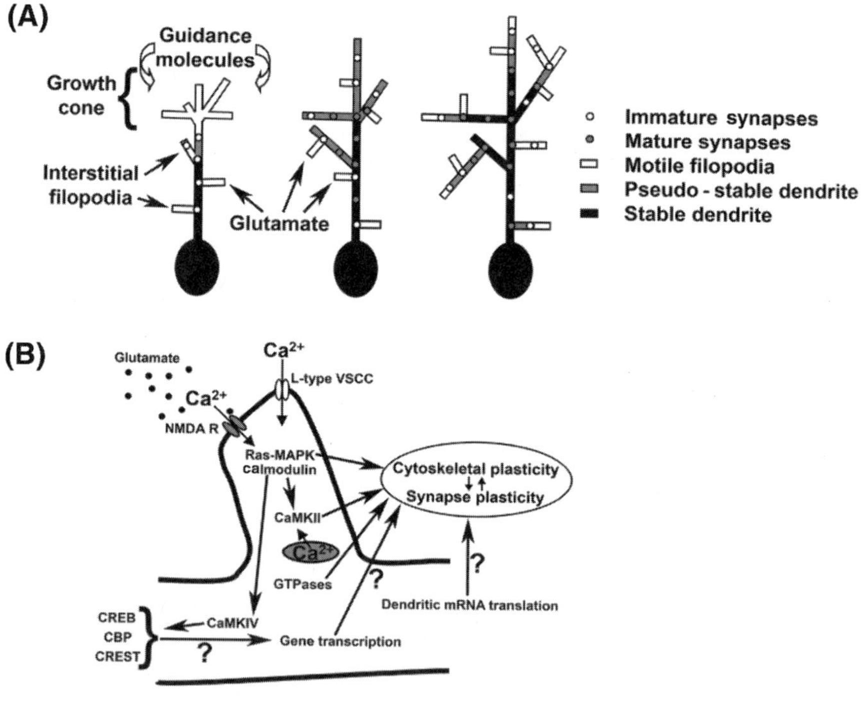


**Figure 20.3.** Synaptotropic Model for Dendrite Growth. (A) While growth cone filopodia are insensitive to activity, interstitial dendritic filopodia grow toward glutamate. Nascent synapses form on filopodia and confer structural stabilization. (B) $Ca^{2+}$ entry through NMDA receptors and VSCCs on dynamic interstitial dendritic filopodia, and subsequent evoked release of $Ca^{2+}$ from intracellular stores activate molecular cascades implicated in both synaptic and cytoskeletal plasticity. Direct and indirect interactions between synaptic and cytoskeletal proteins provide the possibility that synapse and morphologic stabilization/destabilization are coordinated events.

Alteration in cytoskeletal components that give rise to dynamic filopodial motility (actin) and stability of mature dendrites (microtubules) are certainly endpoints for these pathways, but the intervening components have not been completely elucidated. Rapid effects on growth spatially restricted to the area around active synapses are likely mediated by local kinases/phosphatases and activation of small GTPases (see Chapter 21). Transient increases in $Ca^{2+}$ may also regulate local translation of dendritic mRNA[77]. Slower effects that may induce larger changes in growth may involve transcriptional control mediated by CaMKIV, CREB, and CREST. Further work is required to determine which genes are activated by transcription regulators and how this expression translates into altered growth. Activation of the same pathways has been implicated in synaptic plasticity and imaging of the clustering of synaptic proteins shows a clear relationship between dendritic filopodial stabilization and the appearance of synapses. The synaptotropic model in which neuronal transmission and synapse formation contributes to dendritogenesis is strongly supported by these findings. Filopodial stabilization appears to involve a multistep process initiated by filopodial contact with presynaptic axons. Cell adhesion molecular interactions between pre- and postsynaptic membrane may provide early signals for cessation of filopodial growth since contact rapidly alters motility. The detectable coalescence of pre- and postsynaptic proteins is delayed following contact and may provide a distinct mechanism for increased stability through the accumulation of proteins with binding domains for cytoskeletal elements. Finally, activity-induced transcriptional and translational regulation may promote further stabilization of synapses and filopodia. Cytoskeletal stabilization and plasticity, however, must be well balanced to allow dendrite growth. Dynamic filopodia must be stabilized to prevent retraction, but retain plasticity for further extension to form longer mature dendritic branches from which additional motile filopodia can extend. Future investigations of mechanisms underlying activity-dependent growth during development will hopefully clarify the progressive changes in the sensitivity of dendrite growth to different external inputs as neurons progress through maturational states, each with its own complement of receptors, signaling cascades, and effectors.

## 12. REFERENCES


1. Yuste, R., and Tank, D.W. (1996) *Neuron* **16**, 701–716.
2. Connors, B.W., and Regehr, W.G. (1996) *Curr Biol* **6**, 1560–1562.
3. Grueber, W.B., Jan, L.Y., and Jan, Y.N. (2003) *Cell* **112**, 805–818.
4. Komiyama, T., Johnson, W.A., Luo, L., and Jefferis, G.S. (2003) *Cell* **112**, 157–167.
5. McAllister, A.K., Lo, D.C., and Katz, L.C. (1995) *Neuron* **15**, 791–803.
6. Baker, R.E., Dijkhuizen, P.A., Van Pelt, J., and Verhaagen, J. (1998) *Eur J Neurosci* **10**, 1037–1044.
7. Niblock, M.M., Brunso-Bechtold, J.K., and Riddle, D.R. (2000) *J Neurosci* **20**, 4165–4176.
8. Huang, E.J., and Reichardt, L.F. (2001) *Annu Rev Neurosci* **24**, 677–736.
9. Horch, H.W. (2004) *Rev Neurosci* **15**, 117–129.
10. Furrer, M.P., Kim, S., Wolf, B., and Chiba, A. (2003) *Nat Neurosci* **6**, 223–230.
11. Whitford, K.L., Marillat, V., Stein, E., Goodman, C.S., Tessier-Lavigne, M., Chedotal, A., and Ghosh, A. (2002) *Neuron* **33**, 47–61.
12. Wong, W.T., and Wong, R.O. (2001) *Nat Neurosci* **4**, 351–352.
13. Portera-Cailliau, C., Pan, D.T., and Yuste, R. (2003) *J Neurosci* **23**, 7129–7142.
14. Wu, G.Y., Zou, D.J., Rajan, I., and Cline, H. (1999) *J Neurosci* **19**, 4472–4483.
15. Sin, W.C., Haas, K., Ruthazer, E.S., and Cline, H.T. (2002) *Nature* **419**, 475–480.
16. Niell, C.M., and Smith, S.J. (2004) *Annu Rev Physiol* **66**, 771–798.

17. Rajan, I., and Cline, H.T. (1998) *J Neurosci* **18**, 7836–7846.
18. Rajan, I., Witte, S., and Cline, H.T. (1999) *J Neurobiol* **38**, 357–368.
19. Niell, C.M., Meyer, M.P., and Smith, S.J. (2004) *Nat Neurosci* **7**, 254–260.
20. Kaethner, R.J., and Stuermer, C.A. (1997) *J Neurobiol* **32**, 627–639.
21. Denk, W., and Svoboda, K. (1997) *Neuron* **18**, 351–357.
22. McAllister, A.K. (2000) *Sci STKE* **2000**, PL1.
23. Gan, W.B., Grutzendler, J., Wong, W.T., Wong, R.O., and Lichtman, J.W. (2000) *Neuron* **27**, 219–225.
24. Haas, K., Sin, W.C., Javaherian, A., Li, Z., and Cline, H.T. (2001) *Neuron* **29**, 583–591.
25. Cline, H.T. (2001) *Curr Opin Neurobiol* **11**, 118–126.
26. Wong, W.T., Faulkner-Jones, B.E., Sanes, J.R., and Wong, R.O. (2000) *J Neurosci* **20**, 5024–5036.
27. Matus, A. (2000) *Science* **290**, 754–758.
28. Lendvai, B., Stern, E.A., Chen, B., and Svoboda, K. (2000) *Nature* **404**, 876–881.
29. Trachtenberg, J.T., Chen, B.E., Knott, G.W., Feng, G., Sanes, J.R., Welker, E., and Svoboda, K. (2002) *Nature* **420**, 788–794.
30. Dailey, M.E., and Smith, S.J. (1996) *J Neurosci* **16**, 2983–2994.
31. Ziv, N.E., and Smith, S.J. (1996) *Neuron* **17**, 91–102.
32. Lund, J.S., Boothe, R.G., and Lund, R.D. (1977) *J Comp Neurol* **176**, 149–188.
33. Mason, C.A. (1983) *J Comp Neurol* **217**, 458–469.
34. Wong, R.O., Yamawaki, R.M., and Shatz, C.J. (1992) *Eur J Neurosci* **4**, 1387–1397.
35. Ulfhake, B., and Cullheim, S. (1988) *J Comp Neurol* **278**, 88–102.
36. Linke, R., Soriano, E., and Frotscher, M. (1994) *Brain Res Dev Brain Res* **83**, 67–78.
37. Holloway, R.L., Jr. (1966) *Brain Res* **2**, 393–396.
38. Greenough, W.T., and Volkmar, F.R. (1973) *Exp Neurol* **40**, 491–504.
39. Volkmar, F.R., and Greenough, W.T. (1972) *Science* **176**, 1445–1447.
40. Borges, S., and Berry, M. (1978) *J Comp Neurol* **180**, 277–300.
41. Coleman, P.D., and Riesen, A.H. (1968) *J Anat* **102**, 363–374.
42. Wiesel, T.N., and Hubel, D.H. (1963) *J Neurophysiol* **26**, 978–993.
43. Kossel, A.H., Williams, C.V., Schweizer, M., and Kater, S.B. (1997) *J Neurosci* **17**, 6314–6324.
44. Katz, L.C., and Constantine-Paton, M. (1988) *J Neurosci* **8**, 3160–3180.
45. Gray, L., Smith, Z., and Rubel, E.W. (1982) *Brain Res* **244**, 360–364.
46. Lund, J.S., Holbach, S.M., and Chung, W.W. (1991) *J Comp Neurol* **309**, 129–140.
47. Tieman, S.B., Zec, N., and Tieman, D.G. (1995) *Brain Res Dev Brain Res* **84**, 39–45.
48. Rubel, E.W., Smith, Z.D., and Steward, O. (1981) *J Comp Neurol* **202**, 397–414.
49. Riccio, R.V., and Matthews, M.A. (1987) *J Neurosci Res* **17**, 440–451.
50. Dalva, M.B., Ghosh, A., and Shatz, C.J. (1994) *J Neurosci* **14**, 3588–3602.
51. Goodman, L.A., and Model, P.G. (1990) *J Neurobiol* **21**, 283–294.
52. McAllister, A.K., Katz, L.C., and Lo, D.C. (1996) *Neuron* **17**, 1057–1064.
53. Baker, R.E., Wolters, P., and van Pelt, J. (1997) *Brain Res Dev Brain Res* **104**, 31–39.
54. Kalb, R.G. (1994) *Development* **120**, 3063–3071.
55. Chevaleyre, V., Moos, F.C., and Desarmenien, M.G. (2002) *J Neurosci* **22**, 265–273.
56. Wu, G., Malinow, R., and Cline, H.T. (1996) *Science* **274**, 972–976.
57. Cantallops, I., Haas, K., and Cline, H.T. (2000) *Nat Neurosci* **3**, 1004–1011.
58. Rakic, P. (1975) *Adv Neurol* **12**, 117–134.
59. Altman, J., and Anderson, W.J. (1972) *J Comp Neurol* **146**, 355–406.
60. Fiala, J.C., Feinberg, M., Popov, V., and Harris, K.M. (1998) *J Neurosci* **18**, 8900–8911.
61. Friedman, H.V., Bresler, T., Garner, C.C., and Ziv, N.E. (2000) *Neuron* **27**, 57–69.
62. Marrs, G.S., Green, S.H., and Dailey, M.E. (2001) *Nat Neurosci* **4**, 1006–1013.
63. Prange, O., and Murphy, T.H. (2001) *J Neurosci* **21**, 9325–9333.
64. Jontes, J.D., Buchanan, J., and Smith, S.J. (2000) *Nat Neurosci* **3**, 231–237.
65. Bresler, T., Ramati, Y., Zamorano, P.L., Zhai, R., Garner, C.C., and Ziv, N.E. (2001) *Mol Cell Neurosci* **18**, 149–167.
66. Vaughn, J.E. (1989) *Synapse* **3**, 255–285.
67. Ghosh, A., and Greenberg, M.E. (1995) *Science* **268**, 239–247.
68. Redmond, L., Kashani, A.H., and Ghosh, A. (2002) *Neuron* **34**, 999–1010.
69. Chen, N., Murphy, T.H., and Raymond, L.A. (2000) *J Neurophysiol* **84**, 693–697.
70. Lohmann, C., Myhr, K.L., and Wong, R.O. (2002) *Nature* **418**, 177–181.
71. Lohmann, C., Finski, A., and Bonhoeffer, T. (2005) *Nat Neurosci* **8**, 305–312.
72. Wong, R.O., and Ghosh, A. (2002) *Nat Rev Neurosci* **3**, 803–812.
73. Vaillant, A.R., Zanassi, P., Walsh, G.S., Aumont, A., Alonso, A., and Miller, F.D. (2002) *Neuron* **34**, 985–998.
74. Wu, G.Y., and Cline, H.T. (1998) *Science* **279**, 222–226.
75. Zou, D.J., and Cline, H.T. (1999) *J Neurosci* **19**, 8909–8918.

76. Aizawa, H., Hu, S.C., Bobb, K., Balakrishnan, K., Ince, G., Gurevich, I., Cowan, M., and Ghosh, A. (2004) *Science* **303**, 197–202.
77. Sutton, M.A., Wall, N.R., Aakalu, G.N., and Schuman, E.M. (2004) *Science* **304**, 1979–1983.
78. Furuyashiki, T., Arakawa, Y., Takemoto-Kimura, S., Bito, H., and Narumiya, S. (2002) *Proc Natl Acad Sci U S A* **99**, 14458–14463.
79. Oertner, T.G., and Matus, A. (2005) *Cell Calcium* **37**, 477–482.

# 21

# PROTEIN KINASES AND SYNAPTOGENESIS

Jochen C. Meier*

## 1. SUMMARY

Our brain is the interface for communication between the inside of the body and the outer world. Appropriate function of the central nervous system (CNS) relies on a number of intra- and intercellular communication events during and after CNS development, and synaptic connectivity of neurons provides a basis for adequate intercellular communication. Synapse formation and removal are highly dynamic processes, which occur on a time scale of minutes. Disturbance of a single parameter of the intra- and intercellular communication system can already provoke CNS-associated diseases. For example, the appearance of neurofibrillary tangles, which arise from aberrant phosphorylation of a single microtubule-associated protein (MAP) tau, is associated with a number of neurodegenerative diseases, such as Alzheimer's disease. Protein kinases, thus, play a crucial role in regulating intra- and intercellular communication and thereby orchestrate the functionally appropriate assembly and maintenance of our brain. They evolved therefore as valuable therapeutic targets in the treatment of brain disorders. However, prior to the development of novel therapeutic agents, the molecular mechanisms of kinase actions and their integration into signaling pathways should be fully understood. This chapter aims at providing an overview over the molecular mechanisms underlying protein kinase signaling in the context of synapse formation.

## 2. INTRODUCTION

What is the purpose of dendrites and why do they exhibit such an overwhelming variety and complexity of shapes? In his *Histology of the Nervous System* Ramon y Cajal was the first to establish that neuronal processes divide into two types, namely axons and dendrites[1], and that the complexity of the dendrite might indeed reflect the number of synaptic contacts a neuron receives. Apart from

*Charité – Universitätsmedizin Berlin, Tucholskystraße 2, 10117 Berlin, Germany; jochen.meier@charite.de

being the major integrative elements involved in the further downstream processing of synaptic signals, which is not discussed here, dendrites provide the receptive surface area for synapses. With respect to the mechanisms underlying synapse formation, dendrites were therefore considered as somatic protrusions aimed at increasing the synapse-receptive surface area by simultaneously economizing volume. For example, 97% of the 370,000 $\mu m^2$ of somatodendritic surface of motor neurons is dendritic and engages as little as 300,000 $\mu m^3$ [2]. Considering that 80% of the surface area of proximal motor neuron dendrites is covered with synapses suggests that the number of synapses might indeed correlate with the size of the synapse-receptive area[3]. In contrast to this, the axon often extends far away from the neuron soma, arborizes at its termination zone, and covers only 1–2% of the neurons total surface. Axons thus ensure the propagation of the result of dendritic integrative processes to target neurons, which may actually be localized at considerable distance to the former neuron. Therefore, by modulating axon path finding or axon and dendrite geometry the synaptic connectivity may also be affected. Indeed, the synaptotrophic hypothesis[4], according to which new dendritic branches grow out at points of synaptic contacts that in turn stabilize the newly formed branches was recently supported by the elegant experiments of Smith and colleagues[5]; see also Chapter 20. With respect to the potential role of dendrite arborization in the assembly of new synapses they reported that the formation of a glutamatergic synapse occurs at newly extended dendritic filopodia and that the concomitant recruitment of the postsynaptic neurotransmitter receptor anchoring protein postsynaptic density 95 kDa (PSD-95) stabilizes the corresponding dendritic filopodium, which subsequently matures into a dendritic branch[5].

To date, a multitude of protein kinases and phosphatase are well known to participate in the determination of neuron shape by modulating neurite elongation and arborization, and to influence synapse formation by regulating the recruitment of synaptic proteins as well as their functional properties. This chapter provides an abridged overview of the complex intracellular protein kinase network and aims at discussing some principles and rules of kinase action with respect to neuron geometry and synapse formation. Firstly, however, some general considerations on the mechanisms by which phosphorylation of target residues modulates the functional properties of phosphoproteins have to be made. As a rule, phosphorylation of target residues provokes alterations in the protein conformation, which consequently leads to exposure or masking of protein domains relevant for protein–protein interaction and/or affects the integral functional properties of the target phosphoproteins. Protein kinases are intracellular, intermediate computational elements located between a physiological cue from the extracellular milieu and the corresponding intracellular effector molecules (Figure 21.1A). At this intermediate stage the information content of a given extracellular signal is determined by complex enzymatic interplay between different protein kinases and/or between protein kinases and phosphatases, which regulates the relative phosphorylation status of a given residue within the target phosphoprotein. Moreover, protein kinases act on different hierarchy levels, and the closer the hierarchy level to the effector protein the more direct its impact on the functional output will be. Thus, this phosphoprotein network generates a considerable computational protein capacity within the cell, which is far from being behind on

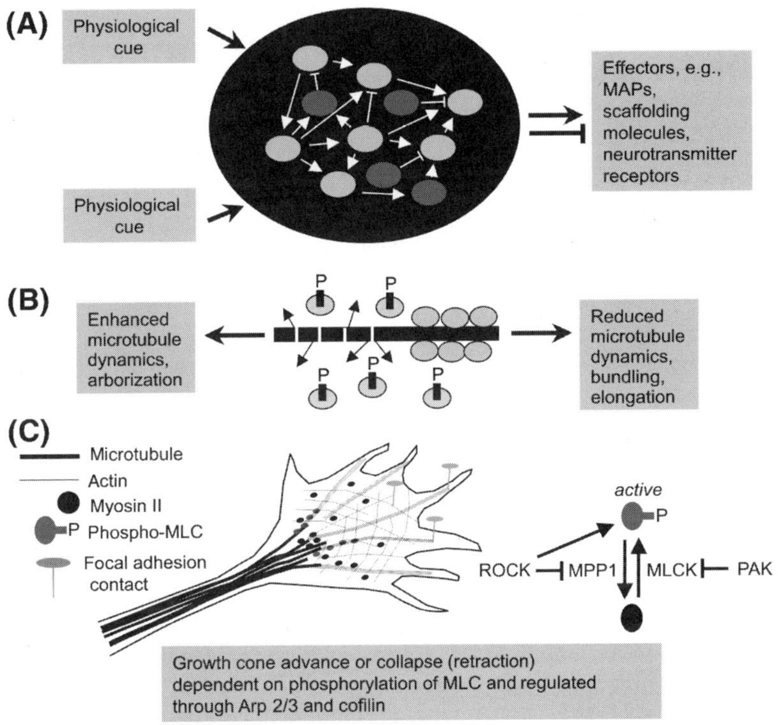


**Figure 21.1.** A Scheme Exemplifying the Intracellular Signal Processing Machinery Composed of Protein Kinases and Phosphatases and Their Impact on Cytoskeletal Dynamics. (A) The information content of a physiological cue from the extracellular milieu is determined by the functional interplay between protein kinases (white rings) and phosphatases (gray circles) and is finally communicated to the respective effector proteins, such as MAPs, scaffolding proteins, and neurotransmitter receptors. As kinases and phosphatases are themselves phosphoproteins, their enzymatic activity influences each other, either in a stimulatory (→) or inhibitory way (⊥). (B) Principle of phosphorylation-dependent regulation of cytoskeletal dynamics. Dephosphorylated MAPs bind to microtubules and thereby contribute to their stabilization, bundling into parallel filaments and elongation. Phosphorylation of MAPs provokes conformational alterations responsible for the observed lower binding affinity to microtubules. Phosphorylated MAPs thus dissociate from microtubules, destabilize them, and increase microtubule dynamics, which favors arborization. (C) Actin cytoskeleton and regulation of growth cone advance or collapse through phosphorylation of myosin light chain (MLC). Parallel actin filament bundles and branched filamentous actin co-exist in the growth cone. The appearance of parallel actin filaments is promoted by actin-binding proteins, such as α-actinin, and active, phosphorylated MLC (phospho-MLC), which exerts traction forces on actin bundles. Branched actin filaments occur in case of cross-linking through larger actin-binding proteins, such as filamin. Besides this, actin-related protein 2/3 (Arp 2/3) binding to actin nucleates the assembly of new filaments on already existing ones and produces a network of short and branched actin filaments, which is more effective than unbranched filaments in pushing the leading edge forward (see also Figure 21.2). Thereby, the relative MLC phosphorylation status determines the net direction of the growth cone movement (either forward powered by dephosphorylation of MLC or backward powered by phospho-MLC). MLC is a substrate for myosin light chain kinase (MLCK) and the phosphorylation status of MLC is modulated by several protein kinases and phosphatases. Rho-associated coiled-coil containing kinase (ROCK) inhibits myosin protein phosphatase 1 (MPP1) and phosphorylates MLC directly, both leading to myosin activation and promoting growth cone retraction. The p21-activated kinase (PAK) phosphorylates and inhibits MLCK, which promotes the appearance of dephosphorylated MLC, branched actin filaments, and growth cone forward movement.

the information processing capacity of, for example, neuronal networks within the CNS. This chapter is therefore focused only on a few selected protein kinases and their effector proteins such as MAPs, scaffolding molecules, and neurotransmitter receptors (Figure 21.1A).

Secondly, a few definitions with respect to neuron geometry and synapse formation have to be given here. In this chapter, it will be distinguished between neurites, dendrites, and axons, and between elongation and arborization. Neurites will be addressed as undifferentiated somatic protrusions while differentiated somatic protrusions will be designated dendrites (somatodendritic or basolateral compartment) and axons (axonal or apical compartment), according to the compartmentalization originally found in polarized epithelial cells (see Chapter 13). The terminus "elongation" will be used in its proper meaning, i.e., the extension in length of an already existing somatic protrusion. The use of the terminus "arborization" will be restricted to the process of ramification of already existing protrusions. We will not discuss here initiation, i.e., the outgrowth of new protrusions directly from the soma (primary dendrites).

Thirdly, I would like to exemplify here the impact of protein phosphorylation on neurite arborization using a previously established model[6–8]. To date, it is believed that phosphorylation of MAPs (e.g., MAP2 and tau) decreases their binding capacity to microtubules and thereby favors microtubule destabilization and dynamics. On the other hand, dephosphorylation of MAPs favors their interaction with microtubules, which promotes their stabilization and reduces microtubule dynamics (Figure 21.1B). Consistent with the idea that destabilization of cytoskeletal structures is required for dynamic changes in neurite shape, phosphorylation of MAPs was found to promote neurite arborization[9]. The elongation of microtubules and, thus, of neurites (arborized or bundled) is mediated by addition of $\alpha/\beta$ tubulin heterodimers preferentially to the (+) ends and occurs only when a critical tubulin concentration is exceeded. When the addition of tubulin heterodimers is faster than guanosine triphosphate (GTP) hydrolysis (of the GTP cap at the growing end) the microtubule gains in length. Once GTP hydrolysis occurs, however, microtubules begin to shrink. Interestingly, the GTP binding site of tubulins is structurally similar to the nucleotide-binding site of GTP-binding proteins of the Ras superfamiliy. Dephosphorylated MAPs speed up polymerization of tubulin heterodimers as they bind to both tubulin polymers and tubulin monomers (type I MAPs, e.g., MAP1A/B). Dephosphorylated type II MAPs (e.g., MAP2, tau) cross-link microtubules and thus contribute to their bundling.

Within the growth cone (Figure 21.1C), which edges the tips of growing neurites, the assembly of actin into filamentous actin carries the aforementioned MAP-dependent processes forward and critically depends on the relative amount of phospho-MLC. The phosphorylation status of MLC is determined by the enzymatic activities of MLCK and MPP1 (Figure 21.1C). PAK phosphorylates and inhibits MLCK, and ROCK phosphorylates and inhibits MPP1 or directly phosphorylates MLC[10]. According to the treadmill hypothesis, phospho-MLC is associated with retrograde actin flow and assembly of myosin into bipolar filaments and stress fibers (Figure 21.1C), which opposes growth cone forward movement, and, if predominating over dephosphorylated MLC, which powers growth cone arborization and forward movement, stimulates contractibility of actin filaments and withdraws protrusions toward the growth cone center[11]. Therefore, the net direction (forward or backward) of growth cone movements is determined by the amount of phospho-MLC relative to dephosphorylated MLC (Figures 21.1C, 21.2).

## 3. NEURON GEOMETRY AND SYNAPSE FORMATION

### 3.1. Protein Kinases and Axon Geometry

The ephrin-ligand (ephrin) ephrin-receptor (Eph) cell adhesion system is one of the most intensively studied intercellular signaling systems. The identification of the molecular processes underlying ephrin-mediated signaling led to prominent and widely accepted models of axon guidance and neuronal map formation. Ephs are members of the tyrosine kinase receptor family, which become activated upon ligand binding and heterotetramer formation. The intrinsic tyrosine kinase activity of Ephs contributes to this by auto- and transphosphorylation of Ephs, which provides binding sites for the 60 kDa nonreceptor tyrosine kinase (Src) homology domain-2 (SH2) containing scaffold or adaptor proteins (forward signaling). In addition, the formation of the Eph–ephrin heterotetramers repositions ephrin transmembrane and cytoplasmic domains, converting them into a signaling configuration. This is followed by phosphorylation of the ephrin cytoplasmic tail on tyrosine residues initiating reverse signaling through, among others, regulation of G-proteins. The complexity of ephrin-associated signaling might actually be interpreted as a spatial and temporal signaling event network, where ephrinA-dependent processes represent repulsive cues that lead to growth cone collapse and neurite retraction and ephrinB-related processes represent rather attractive cues, which promote arborization at the axon termination zone. However, it is important to note that ephrin signaling may exert divergent effects on axon geometry in different brain regions (for enhanced reading see refs. 12,13 and Chapter 10).

Studies on the formation of the retino-topic representation map within the superior colliculus (SC) provided valuable insights into the spatial aspects of ephrin-associated signaling, which represents a cellular mechanism underlying the establishment of the spatially highly ordered retinal projection system. It was shown that axon arborization of retinal ganglion cells (RGCs) within the SC occurs along two sets of orthogonally oriented axes: (1) the nasal–temporal retinal axis corresponds to the anterior-posterior SC axis, and (2) the dorsal–ventral retinal axis corresponds to the lateral–medial SC axis. EphrinA receptor (EphA) and ephrinA ligand expression increases and decreases, respectively, along the nasal–temporal retinal axis, while ephrinA ligand and EphA expression increases and decreases, respectively, along the anterior–posterior SC axis. EphrinB receptor (EphB) and ephrinB ligand expression increases and decreases, respectively, along the dorsal–ventral retinal axis while ephrinB ligand and EphB expression increases and decreases, respectively, along the lateral–medial SC axis. These orthogonally oriented counter gradients precisely determine spatial positions within the SC, where retinal axon arborization is required, i.e., they define the retino-topic representation of retinal afferents within the SC[14].

The temporal aspect of ephrin-mediated signaling involves protein tyrosine phosphatase (PTP)B1 activation[15]. By tyrosine dephosphorylation, PTPB1 allows for a switch from an initially Src-kinase-related signaling, which elicits axon repulsive mechanisms by modulating the activity of Rho GTPases (see below), to a PSD-95/discs large/zonula occludens (PDZ) domain-mediated signaling[16], which implies a means for the recruitment of synaptic proteins to newly formed synapses (see below, 4.1). This also makes clear that the degree of ephrin-ligand tyrosine phosphorylation, which changes along the respective Eph expression gradients, determines the relative contribution of Src- and PDZ-related signaling events to processes involved in axon geometry and synapse formation.

The small GTPases RhoA, Rac and cell division cycle 42 (cdc42) are among the effector molecules of ephrin-associated signaling with respect to cytoskeletal dynamics (Figure 21.2A). RhoA and Rac/cdc42 signaling is inversely arranged. RhoA activation, which is negatively controlled by protein kinase A (PKA) phosphorylation, and concomitant Rac inactivation appears to mediate axon repulsion.

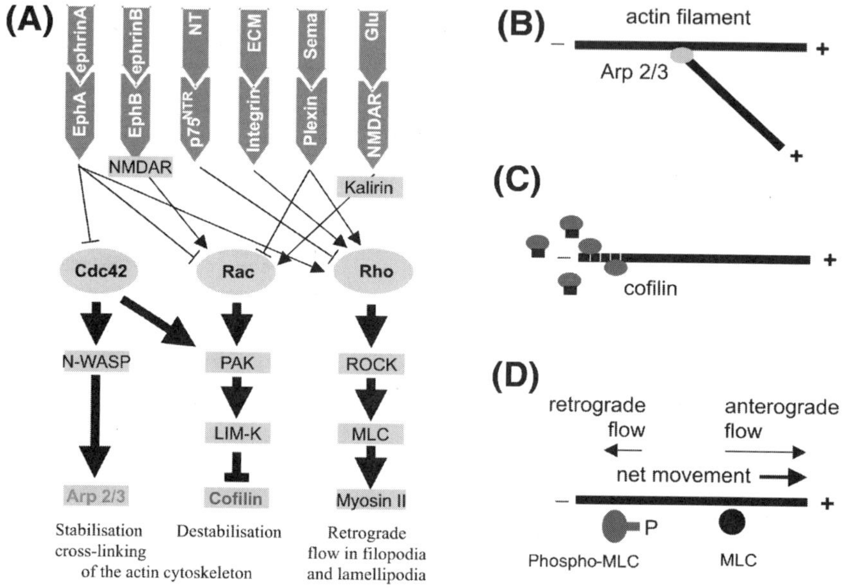


**Figure 21.2.** Compendium of the Roles of RhoA, Rac, and cdc42 Signaling in Actin Cytoskeleton Dynamics. (A) Inverse relationship between activation of RhoA and Rac/cdc42 pathways. Physiological cues, signaling network, and effectors are illustrated. Activation of RhoA leads to growth cone retraction through MLC phosphorylation (see Figure 21.1C). Activation of Rac/cdc42 promotes the formation of branched actin filaments and, if dominating over ROCK-dependent mechanisms, leads to growth cone forward movement and exploration of the extracellular milieu. Please note that EphA-dependent downstream signaling to Rho and Rac/cdc42 engages the GDP–GTP exchange factor ephexin as an intermediate step (not shown). (B,C) Effector molecules regulating actin dynamics within the growth cone. (B) Arp 2/3 nucleates *de novo* actin polymerization from existing actin filaments by binding actin. This contributes to the formation of branched actin filament networks and outgrowth of protrusions. (C) Cofilin is an actin depolymerization factor that destabilizes actin filaments. By inhibition of cofilin through Rac/cdc42, PAK, and LIM domain-containing kinase (LIM-K) the outgrowth of protrusions is favored. (D) As shown in Figure 21.1C and discussed in the text, the phosphorylation status of MLC relative to dephosphorylated MLC decides about the net direction of growth cone movements. Plus (+) and minus (–) ends of actin filaments are indicated. NT: neurotrophin, Sema: semaphorins, Glu: glutamate, N-Wasp: neuronal related to Wiskott–Aldrich syndrome protein, ECM: extracellular matrix, ROCK: Rho-associated coiled-coil containing kinase, LIM-K: LIM domain containing kinase.

The mechanism by which RhoA activation prevents axon outgrowth involves a ROCK-mediated phosphorylation and inhibition of the myosin phosphatase that leads to increased myosin II activity through phospho-MLC[17]. In fact, it is believed that enhancing myosin II activity increases the speed of retrograde actin flow in filopodia and lamellipodia, a process that results in a net retrograde movement and retraction (Figures 21.1C, 21.2D). The phosphorylation status of MLC relative to dephosphorylated MLC, thus, decides about the net direction of growth cone movements. ROCK activation indeed must play a crucial role in regulating cytoskeletal dynamics since its experimental activation *per se* appears to be sufficient for preventing axon outgrowth[18] and for eliciting dendritic[19] or axonal[20] retraction.

Rac/cdc42 and RhoA have opposite effects on growth cone dynamics, and further downstream Rac/cdc42 engage signaling pathways, which involve, among others, the p21-activated kinase (PAK) and a neuronal protein related to the Wiskott–Aldrich syndrome protein (N-WASP), respectively (Figure 21.2A). N-WASP is relieved from autoinhibition by cdc42 or by phosphatidylinositol bisphosphate ($PIP_2$) binding (see below, protein kinase C [PKC] signaling) and triggers *de novo* actin polymerization through Arp 2/3 activation (see Chapter 17). In addition, actin depolymerization is inhibited by decreasing cofilin activity as a consequence of LIM-K-mediated serine phosphorylation, which is under control of PAK activity[21]. Phosphorylation of cofilin causes it to dissociate from actin, which stabilizes the (–) ends of filamentous actin. However, actin release from phosphorylated cofilin increases its availability for assembly with the (+) ends of filamentous actin. These functionally antagonistic effects of cofilin phosphorylation with respect to actin filament dynamics may, however, be reconciled when considering that the temporal aspects of cofilin phosphorylation by LIM-K outline a dual mode of LIM-K function, where initial cofilin phosphorylation is followed by cofilin dephosphorylation presumably via the protein phosphatases (PP)1 and PP2A[21].

The focal adhesion kinase (FAK) was reported to control RhoA activation. FAK positively regulates the activity Rho GTPases by accelerating the GDP–GTP exchange rate through activation in consequence of tyrosine phosphorylation of the GDP–GTP exchange factor p190RhoGEF[22] and inhibition of the GTPase-activating protein p190RhoGAP, the latter mechanism including Src-kinase activation as intermediate step[20]. Upon integrin-mediated cell adhesion, FAK autophosphorylation and consecutive phosphorylation of additional tyrosine residues by Src kinase orchestrate the communication of the adhesive signal further downstream to the interior of the cell. Considering the effect that FAK activation exerts on RhoA activity one could expect that FAK is a negative regulator of axon arborization. Indeed, FAK inactivation was found to promote the arborization of axons and, as a consequence of their disability to retract due to reduced turnover of focal adhesion contacts, to increase the total length of axonal arbors[23]. Thus, FAK shifts the set point of the equilibrium between axon growth cone forward movement and retraction by selectively engaging intracellular signaling events responsible for axonal growth cone collapse and retraction.

The list of protein kinases that regulate axon elongation and arborization is ever growing. As mentioned above, PKA exerts a modulatory effect on neuron shape. In fact, the involvement of PKA in cytoskeletal dynamics is obvious because PKA activation was shown to phosphorylate and consequently inhibit RhoA activity, which implies a stimulatory effect on neurite outgrowth and arborization. PKA is activated in a cyclic adenosine monophosphate (cAMP) dependent way, and cAMP was recently shown to have a stimulatory effect on the mitogen-activated protein kinase (MAPK) pathway. MAPK activation occurs in response to growth factor binding to, and activation of, tyrosine kinase receptors (e.g., tropomyosin-related kinase [Trk]A) and Ras GTPase, which are well known to regulate neuronal differentiation and survival[24,25]. Activated MAPK phosphorylates MLC and affects growth cone actin dynamics through the formerly discussed myosin II-dependent mechanism. However, MAPK activation also promotes phosphorylation of MAPs through phosphorylation-dependent activation of glycogen synthase kinase 3β (GSK3β), which results in enhanced neurite arborization[26]. Inhibition of GSK3β activity by phosphatidylinositol-3-kinase (PI3K) activation results in enhanced binding of MAPs to microtubules and microtubule stabilization, which favors elongation. Neurite arborization and elongation, thus, also depends on the appropriate balance between unstable and stable microtubules, regulated through GSK3β and PI3K, respectively[9].

A multitude of other signals, including activation of protein kinase C (PKC) upon fibroblast growth factor-2 (FGF-2) binding to its tyrosine kinase receptor (FGFR), converge into the Ras signaling cascade and culminate in the activation of the MAPK pathway. A growing number of studies identified PKC as a target of various signaling components known to be involved in regulating cytoskeletal dynamics. Among these components are neurotrophins, such as brain-derived neurotrophic factor (BDNF), integrins, arachidonic acid, and intracellular $Ca^{2+}$. PKC is activated in a phospholipase C (PLC)-dependent way. Metabotropic, G-protein coupled neurotransmitter receptor activation leads to PLC activation, which processes $PIP_2$ to generate two second messengers, $IP_3$ and diacylglycerol (DAG). DAG in conjunction with $IP_3$-mediated raise in intracellular $Ca^{2+}$ activates conventional, $Ca^{2+}$ sensitive PKC isoforms[27]. Indeed, many studies identified PKC as a positive regulator of neurite arborization and growth. As mentioned above, MAPs, such as MAP2 (preferentially in dendrites) and tau (preferentially in axons), are abundant substrates for serine/threonine kinases. Thus, the list of kinases affecting the phosphorylation status of MAPs includes GSK3β, PKA, PKC, the microtubule affinity regulating kinase (MARK), cyclin-dependent kinase 5 (cdk5), and $Ca^{2+}$/calmodulin-dependent protein kinases (CaMK). The degree of neurite branching is correlated with the phosphorylation status of MAP2 and tau. Tau hyperphosphorylation, however, has been associated with severe brain disorders, such as Alzheimer's disease. On the other hand, tau hypophosphorylation as a consequence of kinase inhibition, e.g., PKA, PKC, or CaMK, is associated with inhibition of axon branching[8] and might consequently provoke deficits in brain development and function. Again, the relative phosphorylation status of the target proteins, which results from a complex pattern of distinct kinase activities, plays a crucial role in regulating cytoskeletal dynamics[6].

### 3.2. Protein Kinases, Dendrite Geometry, and Synapse Formation

Principally, the intracellular signaling networks, which were discussed above in light of axon growth and arborization, also account for dendrite growth and arborization. There are, however, some exceptions to be made. For example, semaphorin 3A signaling exerts opposite effects with respect to axon and dendrite growth, respectively. How can this difference with respect to semaphorin 3A signaling be explained? As shown in Figure 21.2, semaphorin 3A activates RhoA and inhibits Rac pathways, respectively, thereby negatively modulating axon growth and providing a 'STOP' signal at the axon termination zone. Regarding dendrite growth, semaphorin 3A additionally engages other signaling pathways, which serve to convert repulsive into attractive cues. For example, Ghosh and colleagues found that the relative availability of guanylyl cyclase and in consequence of cyclic guanosine monophosphate (cGMP), and further downstream of serine/ threonine protein kinase G (PKG), might be critical for the conversion of repulsive into attractive semaphorin 3A cues[28,29]. In addition to this, nerve growth factor (NGF) binding to TrkA and concomitant MAPK activation was reported to counteract repulsive semaphorin 3-associated cues[30]; for advanced reading consult[31].

As for axonal growth cones, the outgrowth of dendritic filopodia is positively modulated by PKC phosphorylation of MAPs. PKC activation can result from arachidonic acid liberation as a consequence of integrin-mediated cell adhesion and phospholipase $A_2$ ($PLA_2$) processing of membrane precursors[32]. The intracellular signal cascade implicated in the activation of PKC by arachidonic acid involves the generation of second messengers upon lipoxygenase-catalyzed oxidation of arachidonic acid, which are processed further downstream by PLC to DAG[33]. This recent discovery showed that astrocytes participate in the processes underlying synapse formation, and it is consistent with the requirement of the PKC-ε isoform for arachidonic acid to translocate to the plasma membrane[34]. Thus, local integrin-mediated contact of astrocytes with neurons leads to a global enhancement of PKC activity, which in turn promotes neuronal differentiation including dendrite arborization and synapse formation.

In line with this, an increasing body of evidence identifies intracellular $Ca^{2+}$ fluctuations as a major regulatory element in the control of dendrite arborization and growth. Indeed, almost a decade ago, the role of voltage-activated $Ca^{2+}$ channels (VACC) and $Ca^{2+}$-induced $Ca^{2+}$ release was already realized[35]. Increasing intracellular $Ca^{2+}$ levels provides a direct link to the activation of CaMK and furthermore to the translocation of classical PKC isoforms to the plasma membrane, where they exert their function. CaMKIIβ activation stimulates the formation of glutamatergic synapses and, in addition, enhances dendrite complexity[36]. This effect is preferentially associated with CaMKIIβ, and not with CaMKIIα, because CaMKIIβ possesses an additional actin-binding sequence, which allows binding polymerized actin and regulation of actin polymerization[36]. However, $Ca^{2+}$/calmodulin binding to CaMKIIβ leads to a dissociation of the kinase from actin filaments, which is consistent with the idea that glutamatergic synaptic activity stabilizes dendritic branches and synapses by reducing actin-related cytoskeletal dynamics[37].

CaMKIIα behaves in the opposite way, that is synaptic activity and concomitant $Ca^{2+}$/calmodulin activation of CaMKIIα promotes association with α-actinin, which localizes the kinase to postsynaptic glutamatergic sites where it regulates the trafficking and stabilization of α-amino-3-hydroxy-5-methylisoxazole-4-propionate receptors (AMPARs) and *N*-methyl-D-aspartate receptors (NMDARs) at postsynaptic domains (Section 4.1)[38]. These considerations emphasize the fact that physiological conditions determine the functional output of kinases by tightly and dynamically regulating the balance between active and inactive kinases, and by the recruiting distinct active kinase isoforms to confined enzymatic compartments where phosphorylation of target proteins is required for appropriate neuron physiology.

To engross these thoughts, it should be considered here that glutamatergic synaptic transmission per se, which provides local sources for intracellular $Ca^{2+}$ oscillations, was identified as a potent regulatory element with respect to dendrite growth, arborization, and synapse formation[39]. In fact, it is the $Ca^{2+}$ permeability of NMDARs, which is a critical determinant in the development of dendritic arbors[40]. Interestingly, the $Ca^{2+}$ permeability of NMDARs can even be increased in response to binding of the extracellular EphB2 domain. As a consequence of this, the GDP–GTP exchange factors kalirin and tiam are activated, which leads to activation of the Rac pathway (see Section 3.1, Figure 21.2). Another modulator of intracellular $Ca^{2+}$ levels is the neurotrophin BDNF, which uses several pathways including the Src-family 59 kDa nonreceptor tyrosine kinase Fyn and PLCγ. For example,

BDNF enhances excitatory postsynaptic currents through phosphorylation-dependent mechanisms and this involves an interaction through Fyn between the full-length high affinity neurotrophin receptor tyrosine kinase TrkB and the NMDAR 2B subunit[41], whereby the TrkB-mediated tyrosine phosphorylation of the NMDAR increases the channel open probability[42]. Besides this, TrkB receptor activation provokes a depolarizing cation current through the transient receptor potential channel 3 (TrpC3), which consequently increases intracellular $Ca^{2+}$ levels by VACC activation[43].

An ever-growing number of kinases are associated with regulation of synapse formation, and to assemble some more pieces of the kinase – synaptogenesis puzzle, we also have to consider the modulatory effects PKA, MAPK, TrkB, and EphB exert on synapse formation (summarized in Table 1)[44–48]. However, rather than giving a detailed overview of the attributions of experimental manipulations of kinase activities and their effects on synapse numbers, I would like to discuss now some selected molecular signaling events situated between the experimental trigger and the determined synapse numbers, which orchestrate the results of protein phosphorylation up to the formation of new synapses.

## 4. CONTRIBUTION OF PROTEIN KINASES TO SYNAPSE FORMATION BY RECRUITMENT OF SYNAPTIC COMPONENTS

Recently, the prominent role of the cell adhesion molecules neurexin and neuroligin in the assembly of synaptic proteomes at newly formed synapses was realized (see Chapter 7). Neuroligins are transmembrane proteins found in the somatodendritic neuron compartment. They were demonstrated to be capable of inducing the recruitment of neurexin to future presynaptic terminals, even when they are heterogeneously expressed in, and presented by, non-neuronal cells[49]. Neuroligin 1 is preferentially associated with glutamatergic postsynaptic domains[50] while neuroligin 2 exerts its function mostly in γ-aminobutyric acid (GABA)ergic synapses[51]. Unfortunately, little is known about the potential contribution of protein kinases to neuroligin–neurexin-dependent signaling, i.e., the direct intermediaries of synapse assembly. The formation of synapses, however, not only requires the respective elements to meet and to interact via cell adhesion molecules, but also involves the recruitment of a number of proteins to both pre- and postsynaptic loci, which consequently enables the future synapse to acquire a functional status. In general, protein kinases add to this in a modulatory way.

At presynaptic sites for example, the synaptic recruitment of the vesicle associated proteins synaptophysin and synapsin I is enhanced in response to FGF-2 signaling and concomitant MAPK activation[45]. Further, PKC activation was clearly shown to be sufficient for formation of glutamatergic synapses and for recruitment of synaptophysin and synaptotagmin I, a presynaptic vesicle protein required for $Ca^{2+}$-regulation of neurotransmitter release probability, to newly formed presynaptic terminals[32]. Using *in vivo* fluorescence microscopy, Cohen-Cory and colleagues could nicely show that BDNF stimulates the formation of glutamatergic synapses and enhances recruitment of synaptobrevin II, which is also part of the presynaptic vesicle fusion machinery[52]. Presumably via TrkB receptor signaling, BDNF further stimulates the recruitment of synapsin I to future presynaptic glutamatergic terminals[53]. In this case, however, the overall amount of synapsin I expression remained unchanged, indicating that the BDNF stimulates the presynaptic recruitment of synapsin I rather than its overall expression level.

Thus, protein kinases contribute to the functional adjustment of presynaptic terminals. Another example of this was given by the discovery that noradrenergic activation of presynaptic β-adrenoreceptors engages PKA activity, which leads to increased loading of GABA into presynaptic vesicles and to enhancement of synchronous and asynchronous GABA release from presynaptic terminals of cerebellar interneurons[54]. On the other hand, adenosine $A_1$ receptor-mediated signaling was shown to converge into, and to downregulate the cAMP/PKA-signaling pathway and, thus, to contribute to presynaptic inhibition of spontaneous GABA release[55]. Therefore, once again, the relative activation status of protein kinases, such as PKA, is controlled by many different factors, and it is the relative amount of active kinases within a highly intermingled kinase network that determines the net kinase function[56].

The assembly of a functional synapse, thus, not only involves the assembly of presynaptic proteomes, but also relies on a number phosphorylation events, which are implicated in the functional adjustment of the presynaptic release machinery to physiological cues from the extracellular milieu. Of course, on the other side, i.e., the postsynaptic membrane, protein kinases also modulate the recruitment and functional adjustment of postsynaptic proteome constituents (summarized in Table 1).

## 4.1. Protein Kinases and Protein Recruitment at Glutamatergic Synapses

### *4.1.1. Receptor Diffusion*

To date it is well established that neurotransmitter receptors are mobile entities, which can bind and unbind to postsynaptic anchor proteins. Therefore, the classical static view of postsynaptic receptor distributions was challenged by the discovery that postsynaptic receptor numbers fluctuate as a consequence of receptor flux into and out of synapses. In fact, Choquet and colleagues provided the first insights into the dynamics underlying postsynaptic neurotransmitter receptor anchoring. They could nicely show that the lateral mobility of glycine receptors (GlyRs) is characterized by ultrafast alternations between two states, diffusive and confined[57]. The following studies confirmed this dynamic view of the glycinergic postsynaptic apparatus, and GlyR movements were characterized more in detail, at the single molecule level[58]. The perception of the synapse as a highly dynamic structure was further extended to other neurotransmitter receptor systems, such as AMPARs[59] and $GABA_A$Rs[60,61]. Therefore, by shifting this equilibrium between diffusive and confined receptor diffusion states protein kinases exert major effects on the postsynaptic responsiveness.

### *4.1.2. PDZ Domains*

The most common protein–protein interaction motive present in the postsynaptic proteome of glutamatergic synapses is the PDZ domain. PDZ domains can occur several times within the same protein and, thus, are well suited to cross-link synaptic proteome constituents. Therefore, they are considered to constitute the building blocks in the assembly of pre- and postsynaptic proteomes. As described elsewhere (Chapter 4), PSD-95, glutamate receptor interacting protein (GRIP), and Homer are among the so-called scaffolding molecules responsible for the postsynaptic anchoring of NMDARs and kainate receptors, AMPARs, and metabotropic glutamate receptors (mGluRs), respectively.

Intriguingly, the serine/threonine residues within the amino acid motives (S/TXV, S/T: serine or threonine, X: any amino acid, V: hydrophobic amino acid) capable of binding to PDZ domains are protein kinase substrates. So far, however, little is known about the potential impact that phosphorylation of the PDZ domain recognition motives may exert on protein–protein interaction. Some recent evidence suggests that phosphorylation of the threonine residue within PDZ recognition motives enhances its binding capacity[62]. Others have shown that phosphorylation of the PDZ recognition motive exerts a negative effect on protein–protein interaction. For example, the interaction of the potassium inward rectifying channel (Kir)2.3 with PSD-95 is reduced by PKA phosphorylation of the serine residue within the Kir2.3 PDZ domain recognition motive. By reducing the availability of postsynaptic Kir2.3 channels, which hyperpolarize the membrane potential, PKA might help to facilitate relief of the $Mg^{2+}$ block from NMDARs[63]. Furthermore, Kim and colleagues showed that PKA phosphorylation of the PDZ domain recognition sequence of stargazin, a protein that contributes to postsynaptic AMPAR targeting, reduces its interaction with PSD-95, which interferes with postsynaptic delivery of AMPARs[64]. Beyond this, PKA was found to contribute to postsynaptic AMPAR delivery through direct phosphorylation of AMPA receptor type 1 (GluR1)-containing AMPARs by facilitating its re-insertion into the plasma membrane after endocytosis. This, however, requires a concomitant $Ca^{2+}$ entry through NMDARs[65] and may therefore involve additional kinases, such as CaMKII (see below).

*4.1.3. PSD-95*

PSD-95 is a member of the membrane-associated guanylate kinase (GK) homologs (MAGUKs), which lack an ATP-binding site and consequently do not have kinase enzymatic activity. PSD-95 possesses three PDZ domains, which provide a platform for postsynaptic anchoring of NMDARs, kainate receptors and neuroligin, one Src-homology domain 3 (SH3), which provides a binding site for synaptic GTPase-activating protein (SynGAP), and finally one GK domain, which mediates interaction with the guanylate kinase-associated protein (GKAP) that enhances postsynaptic recruitment of group I mGluRs through the intermediates Shank and Homer. In line with this, overexpression of PSD-95 was shown to accelerate the development and size of glutamatergic synapses and furthermore to promote the maturation of contacting presynaptic terminals, probably due to a retrograde neuroligin signal[66]. The postsynaptic recruitment of PSD-95 is controlled by MAPK phosphorylation[45]. The Ras GTPase-activating protein SynGAP, which binds to the SH3 domain in PSD-95, negatively regulates MAPK activation, but is itself under negative control of CaMKII[67]. This implies that NMDAR activation and concomitant $Ca^{2+}$ influx leads to CaMKII-mediated inhibition of SynGAP and thereby enhances MAPK-dependent signaling, which enhances postsynaptic PSD-95 recruitment (Table 1).

It should be considered here that the functional output of CaMKII with respect to NMDAR efficacy renders CaMKII a key candidate for the treatment of chronic and in particular neuropathic pain disorders. In addition, postsynaptic recruitment of PSD-95 is controlled by tyrosine phosphorylation of β-catenin, which mediates interactions between the cell adhesion molecule cadherin and the actin cyto-skeleton. In that case, however, tyrosine phosphorylation of β-catenin negatively affects the differentiation of postsynaptic domains. Schuman and colleagues proposed that depolarization by glutamatergic synaptic activity drives a signaling

**Table 1.** Summary of the Effects of Protein Phosphorylation on the Composition and Functional Adjustment of Postsynaptic Proteomes at Glutamatergic and GABAergic Synapses. The mode of action (↑: enhancement, ↓: reduction) of a few selected protein kinases on neurotransmitter receptors, anchoring proteins, and cell adhesion molecules is shown. For details on the underlying mechanisms see Sections 4.1 and 4.2. References are given in brackets. (*) in this study, the number of glutamatergic synaptic profiles (dendritic spines) was unaffected by modulation of TrkB receptor signaling, however, TrkB receptor signaling promoted maturation or "activation" of excitatory synapses, which explains the increase in the number of functional glutamatergic synapses.

| Kinase | Phosphorylation Target | Consequences on | Effect | Synapse Number | |
|---|---|---|---|---|---|
| | | | | Glu | GABA |
| CaMKII | | | | ↑ (36) | |
| | AMPAR GluR1 | AMPAR GluR1 channel conductance | ↑ | | |
| | CaMKII | $Ca^{2+}$ requirement for CaMKII | ↓ | | |
| | | Synaptic AMPAR GluR1 recruitment through a protein–protein interaction cascade, which involves α-actinin, protein 4.1, and SAP-97 | ↑ | | |
| | NMDAR 2B | Aynaptic NMDAR recruitment; this is a reciprocal relationship where NMDAR 2B binding to CaMKII contributes to CaMKII stabilization in the "on-state" | ↑ | | |
| | SAP-97 (S39) | Synaptic AMPAR GluR1 recruitment | ↑ | | |
| | SAP-97 (S232) | Synaptic NMDAR 2A recruitment | ↓ | | |
| | $GABA_AR$ γ2L | Ethanol potentiation of $GABA_AR$ currents | ↑ | | |
| EphB | | | | ↑ (47) | |
| | EphB | Assembly of EphB signaling machinery | ↑ | | |
| | NMDAR | NMDAR $Ca^{2+}$ permeability | ↑ | | |
| MAPK | | | | ↑ (45) | |
| | PSD-95 | Synaptic PSD-95 recruitment | ↑ | | |
| | AMPAR GluR1 | Synaptic AMPAR GluR1 recruitment | ↑ | | |
| PKA | | | | ↑ (46) | |
| | Kir2.3 | Synaptic Kir2.3 recruitment | ↓ | | |
| | Stargazin | Synaptic Stargazin and AMPAR GluR2 recruitment | ↓ | | |
| | GluR1 | GluR1 surface re-insertion after endocytosis; requires concomitant $Ca^{2+}$ entry through NMDARs | ↑ | | |
| | $GABA_AR$ | Allosteric modulation of $GABA_AR$ currents by neurosteroids | ↑ | | |
| | GlyR α3 | GlyR α3-mediated currents, in response to $PGE_2$ signaling | ↓ | | |
| PKC | | | | ↑ (32) | ↑ (100) |
| | NMDAR | NMDAR surface expression | ↑ | | |
| | | NMDAR 2A-CaMKIIα association | ↓ | | |
| | | synaptic NMDAR recruitment | ↓ | | |
| | $GABA_AR$ γ2L | Synaptic $GABA_AR$ γ2L recruitment | ↑ | | |
| Src | | | | | |
| | β-catenin | Synaptic β-catenin, cadherin, PSD-95 recruitment; depolarization drives a signaling cascade that leads to tyrosine dephosphorylation (e.g., by PP1B) of β-catenin and enhanced synaptic recruitment of β-catenin, cadherin and PSD-95 | ↓ | | |
| | NMDAR 2B | NMDAR-mediated $Ca^{2+}$ influx | ↑ | | |
| TrkB | | | | ↑ (53) * | ↑(44,48) |
| | NMDAR 2B | NMDAR channel open probability | ↑ | | |
| | $GABA_AR$ | $GABA_AR$ current amplitudes | ↓ | | |

cascade that leads to phosphatase-mediated (e.g., PP1B) dephosphorylation of β-catenin and enhanced postsynaptic recruitment of β-catenin itself, cadherin, and PSD-95[68]. Finally, PSD-95 itself is under negative control of the cyclin-dependent kinase 5 (cdk5), which is a cell cycle-dependent kinase that downregulates postsynaptic PSD-95 accumulation[69]. Thus, cdk5-dependent serine/threonine phosphorylation of PSD-95 might constitute another negative regulatory element that restrains the development of glutamatergic synapses.

#### *4.1.4. CaMKII*

It is well recognized that the postsynaptic localization of neurotransmitter receptors is controlled by serine/threonine phosphorylation. For example, MAPK activation stimulates both the postsynaptic recruitment of PSD-95 (see above) and of AMPAR GluR1[45]. The recruitment of AMPARs to postsynaptic domains that previously contained solely NMDARs implies a means for the conversion of so-called silent synapses into functional ones and for long-term potentiation (LTP) of glutamatergic synaptic transmission. LTP is considered as the cellular basis for some types of learning and memory, and the induction of LTP requires the coincidence of synaptic activity and an adequately strong postsynaptic depolarization. The molecular basis for coincidence detection involves NMDAR-mediated $Ca^{2+}$ influx and CaMKIIα activation. CaMKII probably is the most intensively studied kinase and therefore merits a detailed discussion here (Figure 21.3). $Ca^{2+}$/calmodulin binding to CaMKIIα activates CaMKIIα by relief of the kinase domain from association with its autoinhibitory domain. As a consequence, activated CaMKIIα phosphorylates position T286, which renders the kinase domain to be less $Ca^{2+}$-dependent. This involves phosphorylation-dependent conformational alterations, which increase the affinity of $Ca^{2+}$/calmodulin for CaMKIIα in order to prevent the autoinhibitory CaMKIIα domain to associate with, and to inhibit, the kinase domain[70,71]. In this way, sufficient amounts of phospho-CaMKIIα can be maintained even under basal intracellular $Ca^{2+}$ levels, which guarantees a long-lasting CaMKIIα signaling beyond the induction of LTP (Figure 21.3A). CaMKIIα is therefore considered to be among the "memory molecules" that encode the intensity of synaptic usage[72]. However, considering the presence of protein phosphatases (PP1 or PP2A) and, of course, constitutive protein turnover, the question arises as to how this long-lasting CaMKIIα activity is actually maintained. The following considerations outline another principle of function of intracellular signaling networks. As for the establishment of synaptic contacts, intracellular signaling components must be held in close spatial proximity to each other in order to meet and exert their enzymatic activities. Within postsynaptic glutamatergic densities, this is the case for NMDARs, AMPARs, CaMKII, and also PP1. During the so-called "on state," the CaMKIIα phosphorylation rate dominates over PP1 enzymatic activity because PP1 is confined to postsynaptic densities where the enormous amounts of phospho-CaMKIIα, which reaches concentrations of up to 200 μM, saturate PP1. Before the CaMKIIα on state is attained, the enzymatic rate of PP1 predominates over the kinase rate so that CaMKIIα remains largely dephosphorylated (Figure 21.3A, off-state). If PP1 was not trapped within this separate enzymatic compartment of the postsynaptic density, as is the case for PP2A, the on–off-switch working mode of CaMKIIα would not be guaranteed[73]. Postsynaptic CaMKIIα activity might also be limited by restraining the access of $Ca^{2+}$/calmodulin, which leads to autophosphorylation of T305/306 (Figure 21.3A) and dislocation of CaMKIIα from postsynaptic sites[74]. This highlights once again that a tight regulation of the spatial and temporal aspects of kinase and phosphatase activities is indispensable for adequate neuron physiology.

### *4.1.5. AMPARs and NMDARs*

As mentioned above, AMPARs and NMDARs are integral parts of the glutamatergic postsynaptic density. CaMKII-mediated phosphorylation of AMPARs is known to increase the unitary AMPAR channel conductances[75], which strengthens glutamatergic synaptic transmission. Furthermore, the recruitment of AMPARs to postsynaptic densities is enhanced by CaMKII phosphorylation and this ensures a long-lasting enhancement of postsynaptic responses[76]. However, as AMPAR phosphorylation by CaMKII is not necessarily associated with enhanced AMPAR recruitment to postsynaptic densities[77], the question arises as to how CaMKII signaling and enhanced postsynaptic AMPAR delivery may be achieved. A likely answer to this question is given in a model proposed by Lisman[73], where active (i.e., phosphorylated) CaMKIIα binds α-actinin, which interacts with actin, which binds the actin-binding protein 4.1, which in its turn is known to bind to synapse-associated protein of 97 kDa (SAP-97), the postsynaptic anchor protein for GluR1 subunit-containing AMPARs (Figure 21.3B). The fact that phosphorylation of CaMKIIα enhances its association with the actin cytoskeleton ensures efficient postsynaptic CaMKIIα recruitment, which is required for saturation of PP1 (see above) and, hence, long-lasting CaMKIIα activity. In synergy with this, CaMKIIα phosphorylates SAP-97 (Figure 21.3B) and drives S39-phosphorylated SAP-97 into spines where it contributes to GluR1 anchoring[78].

Moreover, CaMKIIα can bind to two distinct residues within the NMDAR 2B subunit. By mimicking a segment of the CaMKIIα autoinhibitory domain, NMDAR 2B binding to CaMKIIα generates an autonomous and $Ca^{2+}$/calmodulin trapping state of CaMKIIα that cannot be reversed by phosphatases ($Ca^{2+}$/calmodulin-regulated NMDAR-CaMKIIα interaction)[79] and does no longer require $Ca^{2+}$/calmodulin binding (Figure 21.3A). This conformation also prevents CaMKIIα from inhibitory autophosphorylation at T305/306. The second CaMKIIα binding site is phosphorylation regulated and requires both the NMDAR and CaMKIIα to be phosphorylated[79]. These mechanisms add another dimension to the on–off-switch working mode for CaMKIIα because the postsynaptic recruitment of CaMKIIα activity relies on several distinct, but synergistic, mechanisms, including (1) the classical $Ca^{2+}$- and phosphorylation-dependent mechanism of CaMKIIα recruitment to postsynaptic sites, (2) confinement of enzymatic activities causing phosphatase saturation, and (3) complete autonomy of CaMKIIα by conformational means upon NMDAR 2B binding. Thus, an intimate linkage between active CaMKIIα, NMDARs, and AMPARs at postsynaptic sites is established during, and required for, LTP. CaMKIIβ, however, which contains an additional short actin-binding sequence as compared to CaMKIIα, is capable of binding polymerized actin in the unphosphorylated state (Section 3.2). As mentioned before (Section 3.2), synaptic activity favors dissociation of CaMKIIβ from actin. This implies that both CaMKIIα and CaMKIIβ signaling translate synaptic activity in cellular responses, which stabilize the concerned glutamatergic synapse by means of enhanced receptor function (e.g., enhanced postsynaptic AMPAR delivery and channel conductance) and reduced cytoskeletal dynamics.

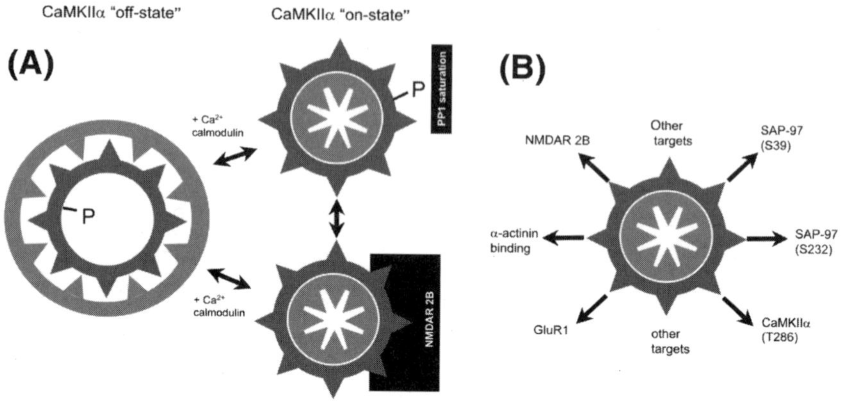


**Figure 21.3.** CaMKIIα Mode of Action and Effector Mechanisms. (A) Complex interplay between CaMKIIα activation and deactivation cues. $Ca^{2+}$/calmodulin binding to CaMKIIα leads to kinase activation by conformational means, i.e., the autoinhibitory domain ceases from the kinase domain, which autophosphorylates T286 residue ("on-state"). Saturation of PP1 activity due to sequestration of the kinase activity within the compartment of the postsynaptic density fraction favors a prolonged "on-state" working mode, which is relevant for learning and memory. Furthermore, T286 phosphorylation dramatically enhances the affinity of CaMKIIα to $Ca^{2+}$/calmodulin, which implies that less $Ca^{2+}$/calmodulin is required for prolonged kinase activity. Postsynaptic CaMKIIα activation can also be achieved in response to $Ca^{2+}$/calmodulin binding and subsequent NMDAR 2B association with CaMKIIα, which stabilizes the kinase in the "on-state" conformation without a further need for $Ca^{2+}$/calmodulin, comparable to T286 autophosphorylation. Limiting the amount of $Ca^{2+}$/calmodulin, e.g., due to reduced synaptic usage, leads to dissociation of $Ca^{2+}$/calmodulin from CaMKIIα and consequently autophosphorylation of T305/306 residues, which stabilizes the kinase in the "off state" conformation. (B) Consequences of CaMKIIα activation. Among other targets, which will not be discussed here, active CaMKIIα phosphorylates itself at T286, phosphorylates GluR1-containing AMPARs and SAP-97. SAP-97 phosphorylation at S39 promotes SAP-97 association with the postsynaptic density fraction and consequently provides more anchoring sites for NMDARs and AMPARs. In contrast, SAP-97 phosphorylation at S232 limits postsynaptic localization of NMDAR 2A subunits. GluR1 phosphorylation leads to increase in the AMPAR channel conductance. Enhancement of GluR1-containing AMPAR function is further achieved by sequential binding of phosphorylated CaMKIIα to α-actinin, actin, protein 4.1, and SAP-97.

It emerges that NMDAR trafficking is also strongly dependent on PKA and PKC phosphorylation. Surface NMDARs are S890-, but not S896-phosphorylated. S896 phosphorylation, however, is required for relief of NMDARs from retention within the endoplasmatic reticulum and, hence, increases their delivery to the cell surface. Once sorted into the exocytotic pathway, S896 is rapidly dephosphorylated. S890 and S896 are phosphorylation targets of PKC α and β, and PKC γ, respectively[80]. However, the serine residue S897 is selectively phosphorylated by PKA, and this enhances postsynaptic NMDAR recruitment[81]. PKA further contributes to strengthening of glutamatergic synapses by phosphorylation of AMPARs, which increases their channel peak open probability[82]. The fact that distinct protein kinases and furthermore distinct PKC isoforms mediate selective phosphorylations of AMPARs and NMDAR residues S890, S896, and S897 emphasizes once more the profound physiological relevance of individual phosphorylation target residues in the determination of kinase function. However, postsynaptic NMDAR recruitment is restrained by PKC phosphorylation, which probably serves to protect neurons from overexcitation and concomitant cell death by excitotoxicity[83]. Furthermore, this coincides in a functionally synergistic way with the reduced $Ca^{2+}$ requirement for CaMKII activity. Thus, PKC and CaMKIIα act in a synergistic way with respect to NMDARs: PKC activation, for example in response to group I mGluR activation, leads to phosphorylation of the NMDAR 2A subunit, which consequently promotes NMDAR dissociation from CaMKIIα in postsynaptic densities[84]. Furthermore, NMDAR 2A recruitment to synapses is negatively regulated through CaMKIIα phosphorylation of SAP-97 at S232[85]. Thus, with respect to SAP-97, CaMKIIα exerts a dual effect on postsynaptic AMPAR and NMDAR recruitment. It enhances postsynaptic AMPAR recruitment and reduces postsynaptic NMDAR 2A accumulation by using two different phosphorylation sites within SAP-97 (Figure 21.3B).

In addition to these serine/threonine phosphorylation-related mechanisms, tyrosine phosphorylation plays a crucial role in synapse formation. Activation of Src-kinase, which in its turn phosphorylates NMDAR at three tyrosine residues[86], can be achieved in several ways, including PKC activation and ephrin signaling. With respect to ephrin signaling we know that ephrin-ligand to EphB2 engages Src-kinase, which leads to enhanced NMDAR-mediated $Ca^{2+}$ influx. Thus, tyrosine phosphorylation of NMDARs renders neurons more sensitive to glutamate and might therefore represent an initial step in the assembly of the glutamatergic postsynaptic density. Moreover, by strengthening synaptic connections this mechanism might convert activity-independent cell adhesion cues into activity-dependent synaptic processes, such as LTP. However, as already mentioned (Section 3.2), binding *per se* of the extracellular EphB2 domain to NMDARs appears to be sufficient for the observed enhancement of NMDAR $Ca^{2+}$ permeability[47,87]. In addition, EphB2 was found to bind to GRIP, the postsynaptic anchor protein for AMPARs. Thus, GRIP interaction with EphB2 triggers EphB2 clustering at postsynaptic sites, which enhances postsynaptic AMPAR recruitment and simultaneously, at presynaptic terminals, initiates reverse ephrin signaling that involves presynaptic PKA activation and enhancement of glutamate release[88]. Altogether, this reveals that in addition to their contribution to postsynaptic protein recruitment and to synapse formation, protein kinases are selectively associated with the control of several determinants of channel function and presynaptic neurotransmitter release. These complex, spatially and temporally highly intermingled molecular mechanisms orchestrate the functionally adequate maturation of pre- and postsynaptic proteomes. Once again, these insights emphasize the critical importance of phosphorylation hierarchies, confinement of enzymatic activities, and, down to an intramolecular level, of positioning enzymatic kinase activities to selected target residues within the proteins of interest.

## 4.2. Protein Kinases and Protein Recruitment at Inhibitory Synapses

Although it is well known that the postsynaptic $GABA_AR$ and GlyR anchor protein gephyrin is a substrate for serine/threonine kinase, nothing is known about the functional consequences of gephyrin phosphorylation[89]. However, protein phosphorylation also influences postsynaptic recruitment and functional modulation of $GABA_ARs$. Compared with excitatory neurotransmitter receptor systems, however, much less is known. PKC phosphorylation of $GABA_AR$ β and γ subunits was repeatedly shown to depress $GABA_AR$-mediated currents[90–93]. In line with this, BDNF was identified to depress inhibitory postsynaptic currents through TrkB receptor activation and, further downstream, recruitment of PKC activity to inhibitory postsynaptic sites[94]. In hippocampal neurons, however, BDNF was found to exert a dual effect on postsynaptic $GABA_AR$ currents, which was characterized by a rapid (within 5–10 min) transitory increase in phosphorylation of the $GABA_AR$ β3 subunit, increasing the amplitudes of miniature inhibitory postsynaptic currents (mIPSC), followed by PP2A-mediated dephosphorylation, decreasing mIPSC amplitudes[95]. In this case, the BDNF-induced phosphorylation of the $GABA_AR$ β3 subunit not only involved the activation of PKC, but also recruited PP2A activity, which are both located downstream to the TrkB receptor-initiated signal. Apparently, phosphorylation of β subunit-containing $GABA_ARs$ not only modulates the functional properties of $GABA_AR$ currents, but also differentially affects association of PKC and PP2A with these subunits, that is reduction of, and increase in, PKC and PP2A recruitment, respectively, to phosphorylated $GABA_AR$ β subunits. PKC shares its phosphorylation target residues within the $GABA_AR$ β3 subunit with a number of other kinases, including PKA. PKA activation was found to differentially influence $GABA_AR$ function according to the use of either two adjacent serine residues (S408 and S409) as phosphorylation targets, which potentiates GABA-activated responses, or the use of S409 only, which reduces GABA-elicited responses[96]. This scenario significantly gains complexity when considering that the protein phosphatase PP1A dephosphorylates $GABA_AR$ β3 subunits. The PP1A anchoring protein phospholipase C-related inactive protein type 1 (PRIP1) traps PP1A close to $GABA_AR$ β3 subunits. PKA phosphorylation of PRIP1 releases active PP1A from PRIP1, which then allows for dephosphorylation of β3 subunit-$GABA_ARs$. PKA activation, for example in response to dopamine D1 receptor activation[97], thus contributes to the fine tuning of the $GABA_AR$ phosphorylation state, and it emerges once again that it is a complex spatial and temporal interplay between kinases and phosphatases, which determines the functional output of kinases.

In this context, it important to remind that it is the $GABA_AR$ γ2 subunit, which is made responsible for postsynaptic $GABA_AR$ stabilization and which also is a phosphorylation target of a number of protein kinases[98–100]. Alternative splicing of $GABA_AR$ γ2 transcripts results in the generation of two splice variants γ2S and γ2L, the latter being generated by insertion of an eight-amino acid long sequence into the large intracellular loop. The γ2S subunit contains several serine residues, which are substrates for protein kinases[101]. Interestingly, the insertion of the additional eight amino acids, however, adds another phosphorylation target residue to the $GABA_AR$ γ2 subunit. It is this additional phosphorylation target residue (S343), which emerges as functionally relevant with respect to postsynaptic $GABA_AR$ recruitment. Consistent with this, CaMKII and PKC were identified to specifically phosphorylate the γ2L subunit[102], and PKC activation by phorbol ester specifically enhanced postsynaptic accumulation of the $GABA_AR$ γ2L intracellular domain[103]. Thus, PKC was identified as positive modulator of GABAergic synapse formation[100]. As for kinase signaling at glutamatergic synapses, protein phosphatases also contribute to signaling at GABAergic synapses. For example, the protein phosphatase calcineurin (PP2B) leads to dephosphorylation of S327, present in both γ2S and γ2L loops. This is accompanied by depression of inhibitory synaptic responses[104], but, in this case, it is not clear if the depression is due to reduced postsynaptic $GABA_AR$ recruitment or due to modulation of $GABA_AR$ channel function.

Altogether, this is a challenging issue because the balance between γ2S and γ2L $GABA_AR$ isoforms seems to be related to schizophrenia. In fact, a predominance of $GABA_AR$ γ2L over γ2S was found in the prefrontal cortex of humans suffering from schizophrenia[105]. In addition, the expression levels of the GABA-transporter GAT1 and the GABA-synthesizing enzyme GAD-67 were found to be reduced[106]. Therefore, the upregulation of $GABA_AR$ γ2L may be considered as a compensatory mechanism to counteract the deficits in the GABA release machinery, or it might actually trigger schizophrenia since both $Ca^{2+}$-sensitive (isoform γ) and $Ca^{2+}$-insensitive PKCs (isoform ζ) were shown to be overactive in schizophrenics[107], presumably as a consequence of enhanced membrane-anchoring of these PKCs through binding of the receptor for activated C-kinase 1 (RACK1)[108]. Altogether, modulation of channel function and postsynaptic $GABA_AR$ recruitment depends on (1) the subunit that is phosphorylated, (2) the phosphorylation target residue within each of the subunits, and (3) the relative phosphorylation status of target proteins, resulting from a complex interplay between several protein kinases and phosphatases.

## 5. CONCLUSIONS

Protein kinases and phosphatases are intracellular signaling elements, which orchestrate the appropriate assembly of synapses and, thus, of the intercellular communication system in our brain. This chapter illuminated a number of molecular mechanisms governing protein kinase and phosphatase signaling with respect to synapse formation. In the introduction, the synaptotrophic hypothesis was forwarded, and consistent with this idea that neuron geometry involves synapse formation, we expected to find congruence of kinase signaling with respect to neuritic and synaptic growth. Indeed, activation of the kinases discussed here and summarized in Table 1 promotes both neuritic and synaptic growth, irrespective of the neurotransmitter phenotypes, GABA or glutamate. However, it should be considered that a functional diversity of kinase activation can be found when dissecting neuritic and synaptic growth with respect to the recruitment of postsynaptic constituents and their functional fine tuning, once a synaptic contact is physically established. For example, PKA and PKC activation promote synaptogenesis both at a neuritic and synaptic growth level, but both PKA and PKC restrain the postsynaptic recruitment of AMPARs and NMDARs, respectively. Therefore, synaptic growth, i.e., the assembly of scaffolding molecules to novel synaptic skeletons, might coincide with increased somatodendritic surface areas; however, only those skeletons that are equipped with all elements required for synaptic transmission would acquire a full functional status.

Derailment of kinase function underlies neurological disorders, such as Alzheimer's disease. Therefore, therapeutic possibilities emerge, as we understand more of the molecular world of protein kinases. However, as kinases are the elementary intracellular signaling units beyond the blood brain barrier it remains to be shown whether pharmacological manipulation of kinase enzymatic activity tuned out to effective in the medication of such diseases. One of the main take-to-home messages of this chapter is that appropriate protein kinase signaling is the result of tightly regulated shifts in the temporally and spatially highly ordered balance between kinase and phosphatase enzymatic activities. Considering the diversity of protein kinases and their highly intermingled signaling machinery it will certainly be a major task in future research to fully understand this intracellular world of protein phosphorylation.

## 6. REFERENCES


1. Ramon y Cajal, S. (1995) *Histology of the Nervous System of Man and Vertebrates*, Oxford Univ Press, NY; English translation by N.Swanson and L.W.Swanson; Originally published: Histologie du systeme nerveux de l'homme et des vertebres.(Trans. L.Azoulay), Paris, 1909–1911.
2. Ulfhake, B. and Kellerth, J.O. (1981) *J Comp Neurol* **202,** 571–583.
3. Kellerth, J.O., Berthold, C.H., and Conradi, S. (1979) *J Comp Neurol* **184,** 755–767.
4. Vaughn, J.E. (1989) *Synapse* **3,** 255–285.
5. Niell, C.M., Meyer, M.P., and Smith, S.J. (2004) *Nat Neurosci* **7,** 254–260.
6. Hely, T.A., Graham, B., and Ooyen, A.V. (2001) *J Theor Biol* **210,** 375–384.
7. Sanchez, C., Diaz-Nido, J., and Avila, J. (2000) *Prog Neurobiol* **61,** 133–168.
8. Audesirk, G., Cabell, L., and Kern, M. (1997) *Brain Res Dev Brain Res* **102,** 247–260.
9. Goold, R.G. and Gordon-Weeks, P.R. (2004) *Biochem Soc Trans* **32,** 809–811.
10. Sanders, L.C., Matsumura, F., Bokoch, G.M., and de Lanerolle, P. (1999) *Science* **283,** 2083–2085.
11. Schmidt, J.T., Morgan, P., Dowell, N., and Leu, B. (2002) *J Neurobiol* **52,** 175–188.
12. Wilkinson, D.G. (2003) *Nat Cell Biol* **5,** 851–853.
13. Pasquale, E.B. (2005) *Nat Rev Mol Cell Biol* **6,** 462–475.
14. Hindges, R., McLaughlin, T., Genoud, N., Henkemeyer, M., and O'Leary, D.D. (2002) *Neuron* **35,** 475–487.
15. Pathre, P., Arregui, C., Wampler, T., Kue, I., Leung, T.C., Lilien, J., and Balsamo, J. (2001) *J Neurosci Res* **63,** 143–150.
16. Palmer, A., Zimmer, M., Erdmann, K.S., Eulenburg, V., Porthin, A., Heumann, R., Deutsch, U., and Klein, R. (2002) *Mol Cell* **9,** 725–737.
17. Bito, H., Furuyashiki, T., Ishihara, H., Shibasaki, Y., Ohashi, K., Mizuno, K., Maekawa, M., Ishizaki, T., and Narumiya, S. (2000) *Neuron* **26,** 431–441.
18. Hirose, M., Ishizaki, T., Watanabe, N., Uehata, M., Kranenburg, O., Moolenaar, W.H., Matsumura, F., Maekawa, M., Bito, H., and Narumiya, S. (1998) *J Cell Biol* **141,** 1625–1636.
19. Nakayama, A.Y., Harms, M.B., and Luo, L. (2000) *J Neurosci* **20,** 5329–5338.
20. Billuart, P., Winter, C.G., Maresh, A., Zhao, X., and Luo, L. (2001) *Cell* **107,** 195–207.
21. Aizawa, H. et al. (2001) *Nat Neurosci* **4,** 367–373.
22. Zhai, J., Lin, H., Nie, Z., Wu, J., Canete-Soler, R., Schlaepfer, W.W., and Schlaepfer, D.D. (2003) *J Biol Chem* **278,** 24865–24873.
23. Rico, B., Beggs, H.E., Schahin-Reed, D., Kimes, N., Schmidt, A., and Reichardt, L.F. (2004) *Nat Neurosci* **7,** 1059–1069.
24. Atwal, J.K., Massie, B., Miller, F.D., and Kaplan, D.R. (2000) *Neuron* **27,** 265–277.
25. Kaplan, D.R. and Miller, F.D. (2000) *Curr Opin Neurobiol* **10,** 381–391.
26. Goold, R.G. and Gordon-Weeks, P.R. (2005) *Mol Cell Neurosci* **28,** 524–534.
27. Oancea, E. and Meyer, T. (1998) *Cell* **95,** 307–318.
28. Polleux, F., Giger, R.J., Ginty, D.D., Kolodkin, A.L., and Ghosh, A. (1998) *Science* **282,** 1904–1906.
29. Polleux, F., Morrow, T., and Ghosh, A. (2000) *Nature* **404,** 567–573.
30. Atwal, J.K., Singh, K.K., Tessier-Lavigne, M., Miller, F.D., and Kaplan, D.R. (2003) *J Neurosci* **23,** 7602–7609.
31. Miller, F.D. and Kaplan, D.R. (2003) *Curr Opin Neurobiol* **13,** 391–398.
32. Hama, H., Hara, C., Yamaguchi, K., and Miyawaki, A. (2004) *Neuron* **41,** 405–415.
33. Auer, K.L. and Jacobson, B.S. (1995) *Mol Biol Cell* **6,** 1305–1313.
34. Shirai, Y., Kashiwagi, K., Yagi, K., Sakai, N., and Saito, N. (1998) *J Cell Biol* **143,** 511–521.
35. McAllister, A.K., Katz, L.C., and Lo, D.C. (1996) *Neuron* **17,** 1057–1064.
36. Fink, C.C., Bayer, K.U., Myers, J.W., Ferrell, J.E., Jr., Schulman, H., and Meyer, T. (2003) *Neuron* **39,** 283–297.
37. Wu, G.Y. and Cline, H.T. (1998) *Science* **279,** 222–226.
38. Rongo, C. and Kaplan, J.M. (1999) *Nature* **402,** 195–199.
39. Metzger, F., Wiese, S., and Sendtner, M. (1998) *J Neurosci* **18,** 1735–1742.

40. Hoogenraad, C.C., Milstein, A.D., Ethell, I.M., Henkemeyer, M., and Sheng, M. (2005) *Nat Neurosci* **8,** 906–915.
41. Lin, S.Y., Wu, K., Levine, E.S., Mount, H.T., Suen, P.C., and Black, I.B. (1998) *Brain Res Mol Brain Res* **55,** 20–27.
42. Levine, E.S., Crozier, R.A., Black, I.B., and Plummer, M.R. (1998) *Proc Natl Acad Sci USA* **95,** 10235–10239.
43. Li, H.S., Xu, X.Z., and Montell, C. (1999) *Neuron* **24,** 261–273.
44. Rico, B., Xu, B., and Reichardt, L.F. (2002) *Nat Neurosci* **5,** 225–233.
45. Li, A.J., Suzuki, S., Suzuki, M., Mizukoshi, E., and Imamura, T. (2002) *Eur J Neurosci* **16,** 1313–1324.
46. Yamamoto, M., Urakubo, T., Tominaga-Yoshino, K., and Ogura, A. (2005) *Brain Res* **1042,** 6–16.
47. Dalva, M.B., Takasu, M.A., Lin, M.Z., Shamah, S.M., Hu, L., Gale, N.W., and Greenberg, M.E. (2000) *Cell* **103,** 945–956.
48. Seil, F.J. and Drake-Baumann, R. (2000) *J Neurosci* **20,** 5367–5373.
49. Scheiffele, P., Fan, J., Choih, J., Fetter, R., and Serafini, T. (2000) *Cell* **101,** 657–669.
50. Chih, B., Engelman, H., and Scheiffele, P. (2005) *Science* **307,** 1324–1328.
51. Varoqueaux, F., Jamain, S., and Brose, N. (2004) *Eur J Cell Biol* **83,** 449–456.
52. Alsina, B., Vu, T., and Cohen-Cory, S. (2001) *Nat Neurosci* **4,** 1093–1101.
53. Vicario-Abejon, C., Collin, C., McKay, R.D., and Segal, M. (1998) *J Neurosci* **18,** 7256–7271.
54. Saitow, F., Suzuki, H., and Konishi, S. (2005) *J Physiol* **565,** 487–502.
55. Jeong, H.J., Jang, I.S., Nabekura, J., and Akaike, N. (2003) *J Neurophysiol* **89,** 1214–1222.
56. Doi, A., Ishibashi, H., Jinno, S., Kosaka, T., and Akaike, N. (2002) *Neurosci* **109,** 299–311.
57. Meier, J., Vannier, C., Serge, A., Triller, A., and Choquet, D. (2001) *Nat Neurosci* **4,** 253–260.
58. Dahan, M., Levi, S., Luccardini, C., Rostaing, P., Riveau, B., and Triller, A. (2003) *Science* **302,** 442–445.
59. Tardin, C., Cognet, L., Bats, C., Lounis, B., and Choquet, D. (2003) *EMBO J* **22,** 4656–4665.
60. Meissner, O. and Haberlein, H. (2003) *Biochemistry* **42,** 1667–1672.
61. Thomas, P., Mortensen, M., Hosie, A.M., and Smart, T.G. (2005) *Nat Neurosci* **8,** 889–897.
62. Adey, N.B., Huang, L., Ormonde, P.A., Baumgard, M.L., Pero, R., Byreddy, D.V., Tavtigian, S.V., and Bartel, P.L. (2000) *Cancer Res* **60,** 35–37.
63. Cohen, N.A., Brenman, J.E., Snyder, S.H., and Bredt, D. S. (1996) *Neuron* **17,** 759–767.
64. Choi, J., Ko, J., Park, E., Lee, J.R., Yoon, J., Lim, S., and Kim, E. (2002) *J Biol Chem* **277,** 12359–12363.
65. Ehlers, M.D. (2000) *Neuron* **28,** 511–525.
66. El Husseini, A.E., Schnell, E., Chetkovich, D.M., Nicoll, R.A., and Bredt, D.S. (2000) *Science* **290,** 1364–1368.
67. Kim, J.H., Liao, D., Lau, L.F., and Huganir, R.L. (1998) *Neuron* **20,** 683–691.
68. Murase, S., Mosser, E., and Schuman, E.M. (2002) *Neuron* **35,** 91–105.
69. Morabito, M.A., Sheng, M., and Tsai, L.H. (2004) *J Neurosci* **24,** 865–876.
70. Fukunaga, K. and Miyamoto, E. (2000) *Neurosci Res* **38,** 3–17.
71. Meyer, T., Hanson, P.I., Stryer, L., and Schulman, H. (1992) *Science* **256,** 1199–1202.
72. Miller, S.G. and Kennedy, M.B. (1986) *Cell* **44,** 861–870.
73. Lisman, J.E. and Zhabotinsky, A.M. (2001) *Neuron* **31,** 191–201.
74. Elgersma, Y., Fedorov, N.B., Ikonen, S., Choi, E.S., Elgersma, M., Carvalho, O.M., Giese, K.P., and Silva, A.J. (2002) *Neuron* **36,** 493–505.
75. Derkach, V., Barria, A., and Soderling, T.R. (1999) *Proc Natl Acad Sci USA* **96,** 3269–3274.
76. Shi, S.H., Hayashi, Y., Petralia, R.S., Zaman, S.H., Wenthold, R.J., Svoboda, K., and Malinow, R. (1999) *Science* **284,** 1811–1816.
77. Hayashi, Y., Shi, S.H., Esteban, J.A., Piccini, A., Poncer, J.C., and Malinow, R. (2000) *Science* **287,** 2262–2267.
78. Mauceri, D., Cattabeni, F., Di Luca, M., and Gardoni, F. (2004) *J Biol Chem* **279,** 23813–23821.
79. Bayer, K.U., De Koninck, P., Leonard, A.S., Hell, J.W., and Schulman, H. (2001) *Nature* **411,** 801–805.
80. Sanchez-Perez, A.M. and Felipo, V. (2005) *Neurochem Int* **47,** 84–91.
81. Crump, F.T., Dillman, K.S., and Craig, A.M. (2001) *J Neurosci* **21,** 5079–5088.
82. Banke, T.G., Bowie, D., Lee, H., Huganir, R.L., Schousboe, A., and Traynelis, S.F. (2000) *J Neurosci* **20,** 89–102.
83. Fong, D.K., Rao, A., Crump, F.T., and Craig, A.M. (2002) *J Neurosci* **22,** 2153–2164.
84. Gardoni, F., Bellone, C., Cattabeni, F., and Di Luca, M. (2001) *J Biol Chem* **276,** 7609–7613.
85. Gardoni, F., Mauceri, D., Fiorentini, C., Bellone, C., Missale, C., Cattabeni, F., and Di Luca, M. (2003) *J Biol Chem* **278,** 44745–44752.
86. Takasu, M.A., Dalva, M.B., Zigmond, R.E., and Greenberg, M.E. (2002) *Science* **295,** 491–495.

87. Grunwald, I.C., Korte, M., Wolfer, D., Wilkinson, G.A., Unsicker, K., Lipp, H.P., Bonhoeffer, T., and Klein, R. (2001) *Neuron* **32,** 1027–1040.
88. Contractor, A., Rogers, C., Maron, C., Henkemeyer, M., Swanson, G.T., and Heinemann, S.F. (2002) *Science* **296,** 1864–1869.
89. Langosch, D., Hoch, W., and Betz, H. (1992) *FEBS Lett* **298,** 113–117.
90. Brandon, N.J., Delmas, P., Kittler, J.T., McDonald, B.J., Sieghart, W., Brown, D.A., Smart, T.G., and Moss, S.J. (2000) *J Biol Chem* **275,** 38856–38862.
91. Brandon, N.J., Jovanovic, J.N., Smart, T.G., and Moss, S.J. (2002) *J Neurosci* **22,** 6353–6361.
92. Krishek, B.J., Xie, X., Blackstone, C., Huganir, R.L., Moss, S.J., and Smart, T.G. (1994) *Neuron* **12,** 1081–1095.
93. Brunig, I., Penschuck, S., Berninger, B., Benson, J., and Fritschy, J.M. (2001) *Eur J Neurosci* **13,** 1320–1328.
94. Henneberger, C., Jüttner, R., Rothe, T., and Grantyn, R. (2002) *J Neurophysiol* **88,** 595–603.
95. Jovanovic, J.N., Thomas, P., Kittler, J.T., Smart, T.G., and Moss, S.J. (2004) *J Neurosci* **24,** 522–530.
96. McDonald, B.J., Amato, A., Connolly, C.N., Benke, D., Moss, S.J., and Smart, T.G. (1998) *Nat Neurosci* **1,** 23–28.
97. Terunuma, M., Jang, I.S., Ha, S.H., Kittler, J.T., Kanematsu, T., Jovanovic, J.N., Nakayama, K.I., Akaike, N., Ryu, S.H., Moss, S.J., and Hirata, M. (2004) *J Neurosci* **24,** 7074–7084.
98. Schweizer, C., Balsiger, S., Bluethmann, H., Mansuy, I.M., Fritschy, J.M., Mohler, H., and Luscher, B. (2003) *Mol Cell Neurosci* **24,** 442–450.
99. Krishek, B.J., Xie, X., Blackstone, C., Huganir, R.L., Moss, S.J., and Smart, T.G. (1994) *Neuron* **12,** 1081–1095.
100. Meier, J., Akyeli, J., Kirischuk, S., and Grantyn, R. (2003) *Mol Cell Neurosci* **23,** 600–613.
101. Moss, S.J., Doherty, C.A., and Huganir, R.L. (1992) *J Biol Chem* **267,** 14470–14476.
102. Machu, T.K., Firestone, J.A., and Browning, M.D. (1993) *J Neurochem* **61,** 375–377.
103. Meier, J. and Grantyn, R. (2004) *J Physiol* **559,** 355–365.
104. Wang, J., Liu, S., Haditsch, U., Tu, W., Cochrane, K., Ahmadian, G., Tran, L., Paw, J., Wang, Y., Mansuy, I., Salter, M.M., and Lu, Y. (2003) *J Neurosci* **23,** 826–836.
105. Huntsman, M.M., Tran, B.V., Potkin, S.G., Bunney, W.E.J., and Jones, E.G. (1998) *Proc Natl Acad Sci U S A* **95,** 15066–15071.
106. Volk, D.W. and Lewis, D.A. (2002) *Physiol Behav* **77,** 501–505.
107. Wang, H.Y. and Friedman, E. (1996) *Biol Psychiatry* **40,** 568–575.
108. Wang, H. and Friedman, E. (2001) *Biol Psychiatry* **50,** 364–370.

22

# SIGNALING FROM SYNAPSE TO NUCLEUS AND BACK

Imbritt König and Michael R. Kreutz*

## 1. SUMMARY

Long-lasting changes in neuronal excitability and synapto-dendritic cytoarchitecture require gene transcription and it is thought that synaptic activity by itself triggers signaling pathways to the nucleus that will eventually control transcriptional regulation. Signaling from the synapse to the nucleus might therefore directly control the making of proteins involved in processes broadly referred to as synaptic plasticity. At present relatively little is known about the exact role of plasticity related genes and how their expression is regulated at the cellular level. Moreover, it is largely unclear how this process affects synaptic transmission and whether there is any specific feedback between the nucleus and activated synapses that drive transcriptional regulation. These questions are of pivotal importance to understand how the wiring of the brain is brought about during development and how long-term use-dependent changes of synaptic efficacy are mediated at the molecular level.

## 2. INTRODUCTION

Activity-dependent gene transcription is known to be of vital importance for long-lasting neural plasticity. Major efforts have therefore been undertaken in recent years to elucidate how biochemical signals initiated at the synapse are conveyed to the nucleus. But why is signaling from the synapse to the nucleus and back a significant cell biological problem? Synaptic plasticity defined broadly as any form of use-dependent change in synapto-dendritic input covering a broad range of plasticity phenomena, including LTP and LTD, is known to be established at individual synapses. Intuitively, one might therefore assume that a privileged connection exists between synapses undergoing plastic events and activity-dependent

*AG 'Molecular Mechanisms of Plasticity', Department of Neurochemistry/Mol. Biology, Leibniz Institute for Neurobiology, Brenneckestr. 6, 39118 Magdeburg, Germany; kreutz@ifn-magdeburg.de

nuclear gene transcription that will eventually stabilize synaptic changes for long periods. Signaling from the individual synapse to the nucleus has been convincingly shown to exist in invertebrate model systems[1,2]. It has, however, been particularly hard to establish such signaling pathways in vertebrate neurons. One major problem has been to demonstrate that the translocation of molecules to the nucleus is initiated at a single synapse. Alternatively, it could result from the net activity of many synapses that is integrated in the somatic cytoplasm from where then transcriptional regulators will eventually enter the nucleus. Moreover, it is largely unclear why and more precisely which type of gene transcription is actually necessary for long-term plasticity processes. Finally, it remains to be resolved how altered gene expression feeds back to the activated synapses that originally induced it.

To address these questions we initially review the present knowledge how synaptic activity might regulate the major signaling pathways that have been shown to trigger nuclear gene transcription in response to synaptic activation in vertebrate neurons. We then summarize how these pathways are integrated at the level of the nucleus. Another part of the chapter is devoted to the question how plasticity-driven gene transcription might feed back to the synapse. The scope of the chapter is limited to signaling pathways from the excitatory postsynaptic compartment to the nucleus, excluding signaling pathways from the inhibitory synapse or retrograde presynaptic signaling, since less is known about such pathways.

## 3. CALCIUM IS THE PIVOTAL MESSENGER THAT TRIGGERS NEURONAL SIGNALING PATHWAYS TO THE NUCLEUS

Experimental work from the last decade has highlighted the importance of many signaling molecules that mediate the transduction of synaptic signals to transcriptional regulation in the nucleus at different levels. A pivotal role within the framework of synapse-to-nucleus communication has been attributed to the regulation of intracellular $Ca^{2+}$ levels. Several studies from recent years have shown that activity-dependent gene expression crucially depends upon neuronal $Ca^{2+}$-signaling and that the dynamics and spatio-temporal characteristics of $Ca^{2+}$ signals determine the type of transcriptional response. Albeit nuclear $Ca^{2+}$ concentrations seem to be one crucial transcriptional regulator, synaptic activity may elicit $Ca^{2+}$ transients that are confined near the site of entry but are capable of regulating gene expression. Within such $Ca^{2+}$ microdomains the rise of $Ca^{2+}$ levels downstream of synaptic activation can be conveyed to the nucleus by several different mechanisms and it has been shown that synapse-to-nucleus signaling from these domains propagates to the nucleus independent of global increases in $Ca^{2+}$-concentrations.

Several different mechanisms have been proposed to account for the prominent role of $Ca^{2+}$ signaling for transcriptional regulation. Such mechanisms include a $Ca^{2+}$ dependent translocation of the transcription factor itself, as it has been shown to occur for the nuclear factor of activated T-cells (NFAT) family of transcription factors[3]. Upon opening of voltage dependent L-type $Ca^{2+}$ channels NFAT rapidly translocates from the cytoplasm through the nuclear pore complex into the nucleus[3]. The nuclear import depends upon a calcineurin-mediated dephosphorylation of NFAT, which in turn unmasks a variety of nuclear localization signals (NLS) (Figure 22.1A).

Another well established $Ca^{2+}$ activated transcriptional event is the synaptic-activity dependent phosphorylation of the transcription factor $Ca^{2+}$-/cAMP responsive element binding protein (CREB). In this case, a slow and a fast $Ca^{2+}$ pathway has been postulated to converge in the nucleus at the same target molecule, leading to the phosphorylation of CREB at the serine residue-133 and thereby rendering CREB transcriptionally active (Figure 22.1A). In both pathways *N*-methyl-D-aspartate receptors (NMDARs) and L-type $Ca^{2+}$ channels constitute the upstream activators for the signaling pathway regulating nuclear CREB phosphorylation[4–7]. Interestingly, elevations of intracellular $Ca^{2+}$ levels from other sources like intracellular $Ca^{2+}$ stores or P/Q- and N-type $Ca^{2+}$-channels are ineffective to trigger the crucial phosphorylation of CREB at Ser-133[4]. Moreover, despite nuclear $Ca^{2+}$ elevations have been frequently observed in response to strong neuronal activation, elevation of nuclear $Ca^{2+}$ levels was found to be insufficient to trigger CREB phosphorylation at Ser-133 in some studies[4] but not in others[8]. Two pathways have been suggested to transduce the $Ca^{2+}$ signal (for review see ref. 5). A key event in both pathways is the binding of $Ca^{2+}$ to the $Ca^{2+}$-sensor protein Calmodulin (CaM) (Figure 22.1A). Following synaptic activation, the fast pathway is transduced via the CaM-dependent kinase (CaMK) cascade, whereas the slow pathway requires the activation of the mitogen activated protein kinase (MAPK)-cascade (Figure 22.1A). At present the status of the fast pathway is still under debate. Both pathways converge on the largely nuclear CaMKIV, which in turn seems to be responsible for the activity-dependent phosphorylation of CREB in neurons. In its $Ca^{2+}$-bound state CaM has been proposed to translocate rapidly to the nucleus in at least certain neuronal cell types (for review see ref. 5), where it subsequently binds to CaM-dependent protein kinase and activate CaMKIV (Figure 22.1A). This, however, only leads to a transient phosphorylation of CREB. For sustained transcriptional activation the slower MAPK pathway has to be activated by a yet to be identified mechanism (Figure 22.1A). The role of CaM in this process is still unclear.

A third type of $Ca^{2+}$-controlled transcription mechanism has been characterized in recent years where the free $Ca^{2+}$ transients after neuronal excitation are directly coupled to $Ca^{2+}$ binding of the nuclear EF-hand calcium binding protein downstream regulatory element antagonistic modulator (DREAM). DREAM is a transcription factor that under resting conditions is bound to a downstream regulatory element (DRE), which acts as a repressor or gene silencer in genes like prodynorphin[9]. Elevation of nuclear $Ca^{2+}$ and $Ca^{2+}$-binding of DREAM leads to a dissociation of the protein and subsequent activation of gene expression[9]. It was shown recently that $Mg^{2+}$ binding at the EF-hand 2 structurally bridges DREAM to DNA targets whereas the $Ca^{2+}$-induced protein dimerization via $Ca^{2+}$-binding to DREAM's EF-hands disrupts DNA binding[10] (Figure 22.1A). Although this mechanism appears to be surprisingly simple, a number of questions still remain to be answered. These include: how is nuclear and extranuclear localization of DREAM regulated? Are synaptic signaling events coupled to the transcriptional repressor role of DREAM? And by which mechanism this could feed back to the synaptic function, if at all?

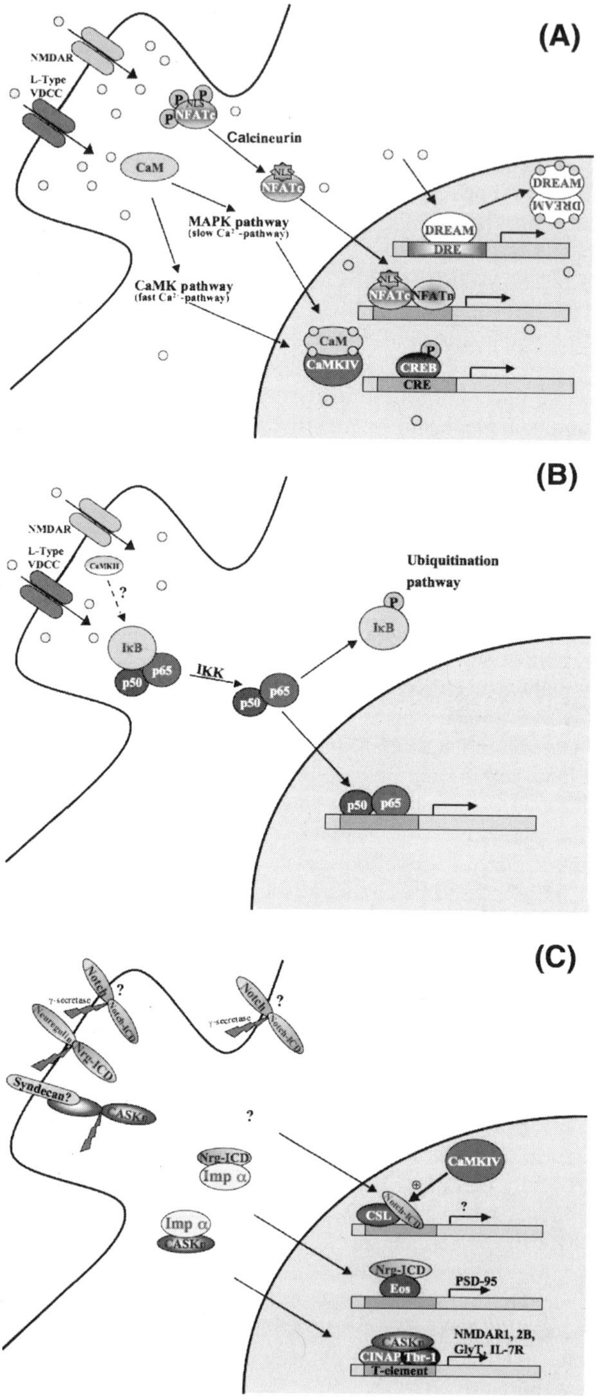


**Figure 22.1.** Neuronal Synapto-Nuclear Signaling Pathways. (A) $Ca^{2+}$ is an important messenger in neuronal signaling and regulates transcription in several ways. $Ca^{2+}$ binding to the repressor DREAM releases the transcription factor from its regulatory element, thereby permitting transcription. Transcription factors of the NFAT family translocate directly from the cyto-plasm into the nucleus upon the opening of VDCC. This process depends on the dephos-phorylation of NFAT by calcineurin and the unmasking of the NLS. Phosphorylation of CREB results from the activation of different pathways, which converge on CaMKIV. (B) Activation and nuclear translocation of NF-κB are also controlled by $Ca^{2+}$. Phos-phorylation of IκB leads to the dissociation of the inhibitory subunit from the NF-κB complex and the subsequent degradation of the subunit. (C) In response to synaptic activity, some synaptic transmembrane and scaffolding proteins are cleaved via a specific proteolytic activity. The resulting intracellular frag-ments are imported into the nucleus in part with the help of importinα, and regulate transcription of numerous synaptic proteins by binding to other trans-cription activators.

## 4. SYNAPTIC ACTIVITY AND THE NUCLEAR TRANSLOCATION OF NF-κ-B

A further $Ca^{2+}$-controlled synapse-to-nucleus signaling event is constituted by the activity-dependent nuclear translocation of nuclear factor κB (NF-κB). NF-κB is a transcription factor that consists of two DNA-binding subunits: a 50 kDa subunit (p50) and a 65 kDa subunit (p65, also known as RelA). In brain NF-κB usually forms a complex with the inhibitory subunit IκB-α that masks the NLS in NF-κB. Phosphorylation of IκB-α via IκB kinase leads to its degradation and the subsequent nuclear translocation of NF-κB (Figure 22.1B). The idea that synaptic stimuli might trigger the NF-κB pathway was originally based on the observation that NF-κB is present in synapses and many reports suggest now that activation of glutamate receptors will activate NF-κB[11,12]. Moreover, blockade of NMDA-receptors and L-type $Ca^{2+}$-channels has been shown to reduce this activation[11,13,14], indirectly suggesting that a $Ca^{2+}$-responsive pathway controls NF-κB activation. Part of this $Ca^{2+}$-responsive pathway upstream of NF-κB seems to be CaMKII dependent, since inhibiting CaMKII reduces the $Ca^{2+}$-induced activation of NF-κB, whereas a constitutively active CaMKII mutant enhances NF-κB activation[13]. Interestingly, stimulation of excitatory synapses also leads to a redistribution of NF-κB from neurites to the nucleus, indicating that the transcription factor can translocate from distal dendritic sites to the nucleus in response to excitatory stimuli[13,15]. The underlying mechanisms are not well understood but NF-κB trafficking appears to be exclusively retrograde in direction to the nucleus[13,15]. Although these findings make it rather plausible that synaptic activity will enhance NF-κB induced gene transcription in a spatio-temporal segregated manner, it is at present still unclear whether activity of single synapses or the integrated activity of larger dendritic segments drives this process and how this relates to plasticity processes at the synapse itself.

## 5. TRANSMEMBRANE AND SCAFFOLDING PROTEINS OF THE SYNAPSE WITH A POTENTIAL ROLE IN SYNAPTO-NUCLEAR SIGNALING

In addition to the pathways outlined above, some transmembrane and scaffolding proteins have been implicated in synapse-to-nucleus communication in recent years (Figure 22.1C). First, the C-terminal part of the synaptic scaffolding protein CASK has been shown to translocate to the nucleus[16]. Nuclear CASK binds to a specific DNA sequence (the T-element) in a complex with the transcription factor TBR1 and the histone-associated protein CINAP[17]. These three proteins are thought to form a complex involved in nucleosome assembly to allow for Tbr-1 induced transcription of T-element containing genes with CASK acting as a coactivator of Tbr-1[16–18]. More than 20 identified genes containing the nonpalindromic T-element in the 5' regulatory region were found to be expressed in brain[17,18]. Gene targets for this transcription-inducing complex include the NMDAR1 and NMDAR2B receptors, reelin, the glycine transporter, and the interleukin 7 receptor (Figure 22.1C). More recently also the intracellular domain (ICD) of Neuregulin-1 was shown to translocate to the nucleus and to regulate PSD-95 transcription in an activity-dependent manner in cochlear afferent cells[19]. The Neuregulin-1-ICD enhances the transcriptional activity of the PSD-95 gene promoter by binding to the zinc-finger transcription factor Eos[19] (Figure 22.1C). This enhancement is facilitated by synaptic activity[19]. Notably, nuclear Neuregulin-1-ICD-Eos signaling was also shown to operate in cortical and hippocampal primary neurons, which indicates that this signaling pathway is active in a broad range of neurons[19].

Interestingly, also the cell-surface receptor Notch, that is involved in cell–cell interaction that allow neighboring cells to adopt different fates during development, seems to be associated with synapse-to-nucleus communication (Figure 22.1C). Notch has been localized to synaptic membranes and it was suggested to contribute to synaptic function[20]. During the differentiation of cortical neurons the ICD of Notch is cleaved via γ-secretase activity and translocates to the nucleus, where it binds to the nuclear cofactor CSL (CBF1/RBP-Jκ/Su(H)/Lag-1)[20]. Moreover, the modulation of gene transcription mediated by the Notch ICD can be potentiated via activation of CaMKIV pathway[21]. Of particular interest is the finding that the nuclear accumulation of Notch's ICD promotes dendritic branching and inhibits dendritic growth during neuronal development[20]. Interestingly, mice with reduced Notch levels have significantly decreased basal and stimulation-induced NF-κB activity and show impaired LTP at hippocampal CA1 synapses[22]. Whether these two findings are causally related and whether the nuclear accumulation of Notch's ICD plays a role for these phenotypes is currently unknown. In addition, it remains to be established whether a direct link exists between plastic events at the synapse and a Notch synapse-to-nucleus pathway in mature neurons. In this context it should be mentioned that also the ICD of APP[23] and the ErbB-4 receptor tyrosine kinase[24] translocate to the cell nucleus after cleavage by γ-secretase. Both proteins have been localized to synapses, but it is not known at present whether a nuclear translocation occurs in neurons.

Most intriguing, also components of the classical nuclear import pathway, the α-importins, which bind the NLS that is essential for the nuclear import of larger proteins, have recently been shown to translocate from dendrites to the nucleus in an NMDA-receptor-dependent manner[25] (Figure 22.1C). Although these findings are at first glance isolated, the emerging picture suggests that a steadily growing number of molecules with a signaling function in the synapto-dendritic compartment also have a specific regulatory role in gene transcription. This role usually requires proteolytical cleavage of the protein and a translocation mechanism, which might be correlated to the activity-dependent nuclear import of α-importins[25]. In this scenario sustained synaptic activation would trigger nuclear gene transcription not only by classical intracellular messenger pathways but also by the import of protein fragments of synaptic molecules. Interestingly, numerous gene targets regulated by these translocation events encode for synaptic proteins[17–19].

## 6. MOLECULES AT THE SYNAPSE IN CONTROL OF SYNAPTO-NUCLEAR SIGNALING

Several studies in recent years have shown that it is not the increase of intracellular $Ca^{2+}$ levels *per se* that leads to activation of gene expression and that it is therefore an open question of what type of $Ca^{2+}$ signal is needed for activity-dependent transcriptional regulation. The local activation of signaling pathways to the nucleus by plasma membrane $Ca^{2+}$ channels requires the concerted action of many proteins that are supposed to be present in highly organized macromolecular complexes close to the synaptic membrane (Figure 22.2). Most models of synapto-nuclear signaling implicitly assume that upon synaptic release of glutamate $Ca^{2+}$ influx via NMDARs and L-type $Ca^{2+}$ channels is restricted to the microcompartment of dendritic spine synapses (Figure 22.2). These synaptic $Ca^{2+}$ signals are subsequently thought to be transduced to the nucleus via activation of the CaMKII, Ras/extracellular signal-regulated protein kinase (ERK), and stress-activated-protein-kinase (SAPK) pathways[5,7]. Interestingly, for all these pathways a potential role in coupling synaptic signals to nuclear CREB phosphorylation has been demonstrated[5,7].

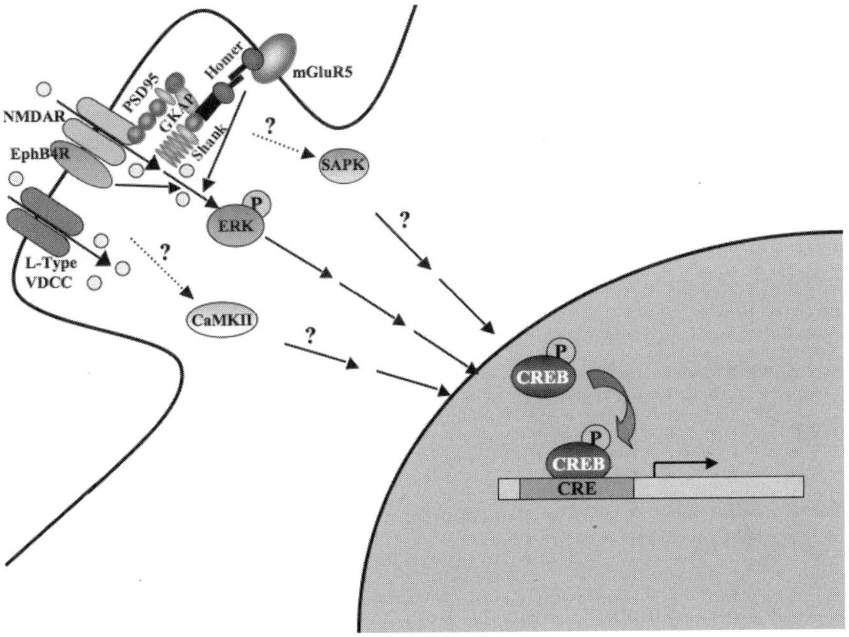


**Figure 22.2.** Molecules at the Synapse in Control of Synapto-Nuclear Signaling. Synaptic $Ca^{2+}$ signals are thought to be transduced to the nucleus via activation of the CaMKII, Ras/ERK, SAPK pathways, and CREB phosphorylation. The $Ca^{2+}$-dependent signaling initiated by the activation of NMDA receptors and L-type $Ca^{2+}$ channels is modulated by $Ca^{2+}$-independent mechanisms mediated by EphB receptors and mGluR5, and requires the concerted action of many proteins that are highly organized in macromolecular complexes at the synaptic membrane.

Specificity for the coupling of synaptic activity to nuclear gene expression seems to be brought about by the subcellular targeting and localization of the signaling molecules involved. It has been suggested that the activation of ERK1/2 by a spatially restricted pool of $Ca^{2+}$ is likely due to the association of ERK1/2 and their activator molecules with the NMDA receptor complex (Figure 22.2). Along these lines it was speculated that the NMDA-receptor activated $Ca^{2+}$ influx represents the most sensitive $Ca^{2+}$ pathway for synapse-to-nucleus signaling[26]. Of particular interest in this regard is the finding that $Ca^{2+}$ influx through synaptic NMDARs and L-type $Ca^{2+}$ channels as well as activity of their downstream effectors can be modulated by receptor tyrosine kinases and G-protein coupled receptors[27–29]. Moreover, the tyrosine kinase activity of EphB receptors as well as the ligand-induced activation of the metabotropic glutamate receptor 5 (mGluR5) also modulate NMDA-receptor controlled gene expression[27,28] (Figure 22.2). In both cases the ligand–receptor interactions induce a synergistic increase of ERK phosphorylation that is independent of the conventional $Ca^{2+}$ signaling derived from NMDA receptors ($Ca^{2+}$ influx) and mGluR5 (intracellular $Ca^{2+}$ release). Furthermore, the phosphorylation of ERK and CREB requires the interaction of adaptor proteins or the EphB4 with the NMDA-receptor complex[27–29]. Thus, $Ca^{2+}$-independent NMDA-receptor coupled signaling pathways to ERK exist that are not directly triggered by synaptic activity as a read-out of $Ca^{2+}$ signaling, and that will add another level of regulation to synapse-to-nucleus communication. Furthermore, an intriguing different outcome on CREB phosphorylation on Ser-133 was reported after stimulation of synaptic and extrasynaptic NMDA receptors. Whereas activation of synaptic NMDA receptors results in enhanced nuclear CREB phosphorylation on Ser-133, the opposite was found after stimulation of extrasynaptic NMDA receptors (Figure 22.2)[30]. In conclusion, it can be hypothesized that the spatio-temporal aspect of $Ca^{2+}$ signaling and thereby the localization of the $Ca^{2+}$ entry or release channel, might to a large extent determine the type of regulatory event triggered in the nucleus.

Additional specificity with regard to synaptic input is probably brought about by the spatial or temporal features of the signal that is generated within a neuron in response to synaptic activity. Thus, synaptic stimulation protocols that strengthen synaptic transmission often generate high frequency $Ca^{2+}$ transients that accumulate to produce relatively high elevations of intracellular $Ca^{2+}$ *in vitro*[31]. By contrast, protocols that weaken synaptic connections often evoke low-frequency $Ca^{2+}$ transients that moderately elevate intracellular $Ca^{2+}$ [31]. The precise underlying molecular mechanisms as well as their spatio-temporal dynamics are, however, largely unclear.

## 7. THE STRUCTURAL BASIS FOR INTEGRATING SYNAPTIC ACTIVITY IN THE NUCLEUS

From what is outlined above, it can already be deduced that a high degree of convergence and synergy must exist between synapse-to-nucleus signaling pathways. The integration of many different upstream signaling events into a coordinated nuclear response is a major challenge for synapse-to-nucleus communication. Thus, many divergent intracellular signaling cascades can regulate the same nuclear target providing a high degree of fine-tuning that might be of importance in controlling activity-dependent gene transcription. One prominent example for the integration of different $Ca^{2+}$ signals that control gene transcription is the brain-derived neurotrophic factor (BDNF) promoter III, which requires induction of three $Ca^{2+}$-responsive elements (Figure 22.3). The exon-III containing splice isoforms are particularly sensitive to regulation by synaptic activity and it was shown that a region 170bp upstream of the exon III initiation site is responsible for the activity-dependent induction of BDNF mRNA[32,33]. Within this region three $Ca^{2+}$-responsive elements were identified, one of which constituting a $Ca^{2+}$/cAMP response element that is bound to CREB[32,33]. Phosphorylation of CREB is necessary but not sufficient for the expression of exon III[32,33]. Accordingly, BDNF exon III expression is selectively activated by $Ca^{2+}$ influx but not by other stimuli that enhance the phosphorylation of CREB, like for instance, elevated cAMP levels[34]. The two other $Ca^{2+}$-responsive elements were recently characterized as binding sites for MeCP2[35–37] and a novel $Ca^{2+}$-responsive transcription factor termed CaRF[34]. MeCP2 functions as a transcriptional repressor of BDNF gene expression whereas CaRF is a transcriptional activator. Membrane depolarization triggers a $Ca^{2+}$-dependent phosphorylation and subsequent release of MeCP2 from the BDNF exon III promoter thereby facilitating transcription. The transcriptional activity of CaRF also requires elevated $Ca^{2+}$ levels but the mechanism of $Ca^{2+}$ activation is not yet known. Interestingly, coordination of the activity of all three transcription factors bound at the $Ca^{2+}$-responsive elements within the BDNF exon III promoter seems to be required for the expression of BDNF exon III containing transcripts in cortical primary cultures. Thus, although distinct $Ca^{2+}$-signaling mechanisms for CaRF, CREB, and MeCP2 have not been identified yet, it is tempting to speculate that differences in $Ca^{2+}$ regulation for these proteins might determine specificity for gene induction in response to $Ca^{2+}$ influx.

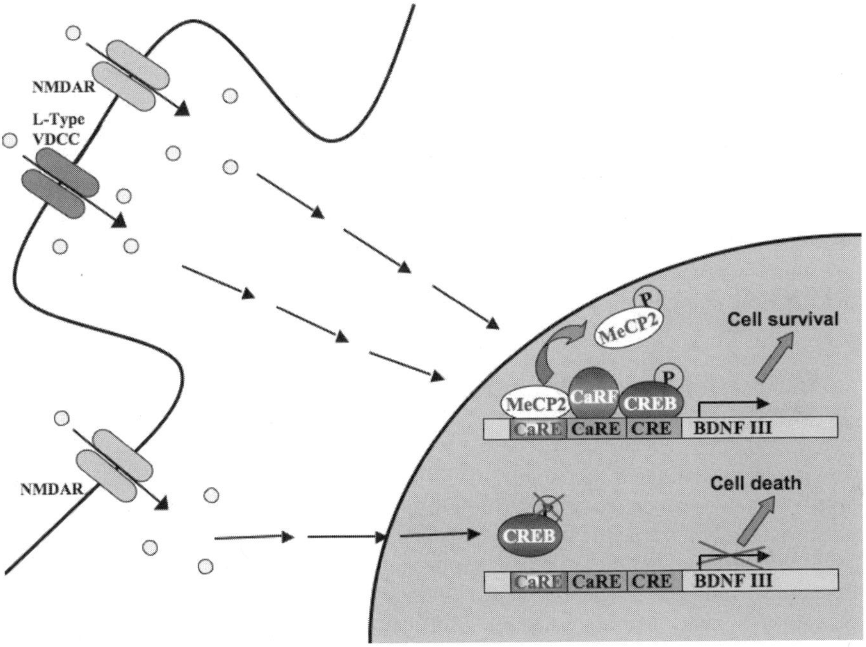


**Figure 22.3.** Integration of Synaptic Activity in the Nucleus. Transcription from the BDNF promoter III requires induction of three $Ca^{2+}$-responsive elements. Synaptic $Ca^{2+}$ influx induces CREB and MeCP2 phosphorylation, thereby binding CREB to the promoter and releasing MeCP2 from it. It also activates CaRF by a yet unknown mechanism. Transcription of BDNF then promotes cell survival. Extrasynaptic NMDA-receptor activation on the other hand suppresses CREB function, leading to cell death.

Finally, it will be interesting to determine which molecular mechanisms are responsible for the convergence of different synapse-to-nucleus signaling pathways. The prevailing idea in recent years is that nuclear chromatin will provide the structure that integrates potentially hundreds of signals derived from synaptic activity (for a recent review see ref. 38). Among the most modified nuclear proteins are histones, the major scaffolding proteins associated with nuclear DNA in chromatin. Histones interact with each other in the form of nucleosomes, an octamer of histones, from which the amino (N)-terminal tail protrudes beyond the genomic DNA. This tail can eventually serve as a signal integration platform and post-translational modifications of the amino terminus of histones are supposed to result in long-lasting changes in regulation of gene transcription. Within this region, lysines are the most modified residues that can be acetylated, methylated, ubiquitinated, or sumoylated (for a recent review see ref. 39).

In recent years also some evidence has accumulated that synaptic activity will lead to enduring modifications in the organization of nucleosomes. Several forms of synaptic plasticity that require gene transcription concomitantly require the activation of the NMDA receptor and ERK/MAPK signaling and it was therefore important to show that activation of NMDA-receptors leads to the acetylation of histone H3[40]. The NMDA-induced acetylation could be blocked by inhibition of the ERK/MAPK cascade, indirectly suggesting a link between the activation of excitatory synapses and post-translational modifications of the nucleosome. In support of this idea is the finding that also ERK-dependent increases in histone phosphorylation were reported after glutamate receptor stimulation[41]. Although the exact mechanisms that regulate nucleosome modification in response to synaptic activity are still elusive, it is conceivable that the regulated accumulation of macromolecular complexes consisting of transcription factors and their co-activators will subsequently govern plasticity-related gene expression in an integrated manner. The structural basis for the assembly of these complexes could be in principle stable for a long time, which in turn is a prerequisite for long-lasting forms of synaptic plasticity. Much more work, however, is needed to fill the gaps in our present knowledge about the molecular mechanisms that control activity-dependent gene transcription.

## 8. SYNAPTIC PLASTICITY-RELATED GENE EXPRESSION AFTER SYNAPTIC ACTIVATION

It can be estimated that about 30–350 genes expressed in neurons show some type of activity-dependent regulation. Of those, several belong to immediate early genes and only a subset of them like SNK[42], Narp[43,44], Arc[45], CPG15[46], or Homer-1a[47], seem to have a direct role in synaptic function. These functions can, however, be quite opposite. Whereas Narp induces synaptic clustering of AMPA-receptors[43,44] and increases synaptic efficacy, and CPG15 promotes synaptic maturation[46], SNK[42] and Homer-1a[47] destabilize and reduce the number of synaptic contacts. Other immediate early genes encode transcription factors like c-Fos that control the transcription of other genes and thereby provide an activity-dependent control mechanism for the production of proteins that might also have a specific role in synaptic signaling. Finally, increased transcription levels after synaptic stimulation have been reported for a number of genes like BDNF that might have direct effects on processes of synaptic plasticity (for review see ref. 45). BDNF signaling at glutamate synapses enhances the translation of newly transported Arc and locally stored (i.e., α-CaMKII) mRNA in dendrites. Thus, the emerging picture is rather complex and no clear-cut functional scheme for activity-driven gene expression can be deduced from the present data for synaptic plasticity.

This might change with a more elaborate picture that takes into account that different protocols to induce synaptic activity might also lead to different patterns of synaptic plasticity-related gene expression. A major obstacle for many years has been to decipher the molecular pathways that induce activity-dependent neuronal gene transcription and we are just at the beginning to understand these molecular mechanisms (see also above). Apart from the putative consequences of activity-dependent gene transcription for synaptic input there are, however, also a number of other intriguing questions that have not been addressed yet in much detail. These concern the putative cellular targets for *de novo* transcribed or upregulated gene products. It is thought that newly synthesized mRNAs or proteins are targeted specifically to activated synapses by means of a synaptic tag[48]. The 'tagging' hypothesis proposes that potentiated synapses are able to capture newly synthesized proteins by not yet clearly defined mechanisms. Synaptic tagging could explain how local events at the synapse might provide the basis for the realization of plastic events at individual synapses via such a capturing mechanism without the necessity for synapse-to-nucleus communication. This concept is particularly attractive because it is at present still essentially unclear whether a privileged connection between a single synapse undergoing plastic events and the nuclear gene transcription that is induced by synaptic activity exists and whether this is at all possible. A number of studies argue against this possibility and foster the idea that synaptic plasticity-related signaling to the nucleus can be achieved without direct transport of molecules from individual activated synapses. One of the strongest arguments against a role of synapse-to-nucleus signaling in establishing long-lasting plastic events at the synapse comes from studies of Dudek and co-workers who could show that the conversion of early-phase LTP to late-phase LTP can be induced by antidromic stimulation of CA1 neurons in the absence of excitatory synaptic activity[49]. Moreover, in the same study it was found that somatic action potentials are sufficient for the phosphorylation of ERK and CREB. In contrast, however, Deisseroth et al.[4] reported that synaptic stimuli but not electrically induced action potentials are a prerequisite for the phosphorylation of CREB. Further arguments against the necessity of synapse-to-nucleus communication for the maintenance of long-term synaptic plasticity were introduced in a recent review from Adams and Dudek[50]. The authors state that the amount of protein needed in the nucleus to be relevant for gene transcription is relatively high as compared to conceivable protein concentrations in single dendritic spines. Moreover, the translocation to the nucleus would dilute this concentration at least 2000-fold with respect to the much larger nuclear volume. It should be taken into account, however, that signaling molecules for synapse-to-nucleus communication are efficiently targeted to active sites of gene transcription of plasticity-relevant genes. A more stringent point is raised by the finding that some plasticity-relevant genes like Arc can be induced as early as 2 min following sustained neuronal activation via electroconvulsive shock[51]. This rapid nuclear response seems to be too fast for a synapse-to-nucleus transport process. But it should be emphasized that electroconvulsive shock does not reflect the *in vivo* situation of synaptic transmission even under pathophysiological conditions and that motor proteins like dynein exist which could transport signaling molecules within a few minutes from dendritic compartments to the nucleus. Thus, in light of the conflicting evidence provided so far it is still a matter of debate whether synapse-to-nucleus signaling is a prerequisite for the regulation of plasticity-related gene expression.

## 9. CONCLUSION AND FUTURE DIRECTIONS

The aim of this chapter was to provide a synopsis of the present knowledge and concepts of synapse-to-nucleus-and-back signaling mechanisms. Despite the progress in recent years a consistent theoretical framework for the role of activity-driven gene expression in the induction and maintenance of synaptic plasticity is still lacking. A unifying hypothesis of synapto-nuclear signaling will have to solve many problems that are currently not convincingly addressed. The view that molecules are transported from single synapses to the nucleus in response to synaptic activity is largely based on evidence observed in Aplysia motor neurons but has not been convincingly shown in vertebrate neurons like the pyramidal cells of the cortex and hippocampus. Moreover, it needs to be addressed whether dendritic transport of signaling molecules is fast enough to account for the rapid induction of plasticity-related gene expression. An integrated view of synapse-to-nucleus communication will also have to deal with the problem of input specificity. It has been convincingly shown that the spatial and temporal diversity of neuronal $Ca^{2+}$ transients has important consequences for gene transcription. Thus, activity-dependent gene expression depends on the $Ca^{2+}$ entry site, the amplitude and the temporal dynamics of the $Ca^{2+}$ signal, and the subcellular compartment it invades. Furthermore, the $Ca^{2+}$-binding proteins sensing these $Ca^{2+}$ signals and triggering the target interactions upon $Ca^{2+}$ binding will most likely not only include CaM and Calcineurin. The identification of other neuronal calcium sensor proteins involved in synapse-to-nucleus communication might therefore provide even more specificity for the transduction of $Ca^{2+}$ signals to the nucleus. Next, it will be important to elucidate how activity-driven gene expression feeds back to the activated synapses. Part of this problem is to understand why the expression of activity-regulated genes is necessary for a synapse to undergo long-term structural changes. Thus, it will be important to understand the molecular mechanisms by which the presence of activity-regulated genes render a potentiated synapse specific as compared to other synaptic input, and whether the expression of plasticity-related genes is instructive to induce certain forms of synaptic plasticity even in the absence of enhanced synaptic activity. Answers to these questions will probably help to learn more why neurons invented synapse-to-nucleus communication and activity-driven gene expression despite the possibility of local protein synthesis from dendritic mRNAs. A possible answer might be that dendritic spine synapses need a nuclear feedback mechanism to adjust synaptic weights during processes generally referred to as synaptic scaling. Synapto-nuclear signaling could be a mechanism to provide the neuron with proteins needed for input specificity by destabilizing less efficient synapses (i.e., proteins like SNK and Homer1a) and contributing protein components that transform pre-existing synaptic macromolecular complexes into less plastic structures, i.e., the mushroom-shaped mature spine that survives potentially for many years or even the lifespan.*

*The work of M.R.K. is supported by the BMBF, DAAD, DFG, the Land Sachsen-Anhalt, and the Schram Foundation.

## 10. REFERENCES

1. Martin, K.C., Casadio, A., Zhu, H., Yaping, E., Rose, J.C., Chen, M., Bailey, C.H., and Kandel, E.R. (1997) *Cell* **91**, 927–938.
2. Martin, K.C., Michael, D., Rose, J.C., Barad, M., Casadio, A., Zhu, H., and Kandel, E.R. (1997) *Neuron* **18**, 899–912.
3. Graef, I.A., Mermelstein, P.G., Stankunas, K., Neilson, J.R., Deisseroth, K., Tsien, R.W., and Crabtree, G.R. (1999) *Nature* **401**, 703–708.
4. Deisseroth, K., Bito, H., and Tsien, R.W. (1996) *Neuron* **16**, 89–101.
5. Deisseroth, K., Bito, H., and Tsien, R.W. (2003) *Curr Opin Neurobiol* **13**, 354–365.
6. Dolmetsch, R.E., Pajvani, U., Fife, K., Spotts, J.M., and Greenberg, M.E. (2001) *Science* **294**, 333–339.
7. West, A.E., Griffith, E.C., and Greenberg, M.E. (2002) *Nat Rev Neurosci* **3**, 921–931.
8. Hardingham, G.E., Arnold, F.J., and Bading, H. (2001) *Nat Neurosci* **4**, 261–267.
9. Carrion, A.M., Link, W.A., Ledo, F., Mellstrom, B., and Naranjo, J.R. (1999) *Nature* **398**, 80–84.
10. Osawa, M., Dace, A., Tong, K.I., Valiveti, A., Ikura, M., and Ames, J.B. (2005) *J Biol Chem* **280**, 18008–18014.
11. Guerrini, L., Blasi, F., and Denis-Donini, S. (1995) *Proc Natl Acad Sci U S A* **92**, 9077–9081.
12. Kaltschmidt, C., Kaltschmidt, B., and Baeuerle, P.A. (1995) *Proc Natl Acad Sci U S A* **92**, 9618–9622.
13. Meffert, M.K., Chang, J.M., Wiltgen, B.J., Fanselow, M.S., and Baltimore, D. (2003) *Nat Neurosci* **6**, 1072–1078.
14. Lilienbaum, A., and Israel, A. (2003) *Mol Cell Biol* **23**, 2680–2698.
15. Wellmann, H., Kaltschmidt, B., and Kaltschmidt, C. (2001) *J Biol Chem* **276**, 11821–11829.
16. Hsueh, Y.P., Wang, T.F., Yang, F.C., and Sheng, M. (2002) *Nature* **404**, 298–302.
17. Wang, G.S., Hong, C.J., Yen, T.Y., Huang, H.Y., Ou, Y., Huang, T.N., Jung, W.G., Kuo, T.Y., Sheng, M., Wang, T.F., and Hsueh, Y.P. (2004) *Neuron* **42**, 113–128.
18. Wang, T.F., Ding, C.N., Wang, G.S., Luo, S.C., Lin, Y.L., Ruan, Y., Hevner, R., Rubenstein, J.L., and Hsueh, Y.P. (2004) *J Neurochem* **91**, 1483–1492.
19. Bao, J., Lin, H., Ouyang, Y., Lei, D., Osman, A., Kim, T.W., Mei, L., Dai, P., Ohlemiller, K.K., and Ambron, R.T. (2004) *Nat Neurosci* **7**, 1250–1258.
20. Redmond, L., Oh, S.R., Hicks, C., Weinmaster, G., and Ghosh, A. (2000) *Nat Neurosci* **3**, 30–40.
21. McKenzie, G.J., Stevenson, P., Ward, G., Papadia, S., Bading, H., Chawla, S., Privalsky, M., and Hardingham, G.E. (2005) *J Neurochem* **93**, 171–185.
22. Wang, Y., Chan, S.L., Miele, L., Yao, P.J., Mackes, J., Ingram, D.K., Mattson, M.P., and Furukawa, K. (2004) *Proc Natl Acad Sci U S A* **101**, 9458–9462.
23. Cao, X., and Südhof, T.C. (2001) *Science* **293**, 115–120.
24. Ni, C.Y., Murphy, M.P., Golde, T.E., and Carpenter, G. (2001) *Science* **294**, 2179–2181.
25. Thompson, K.R., Otis, K.O., Chen, D.Y., Zhao, Y., O'Dell, T.J., and Martin, K.C. (2004) *Neuron* **44**, 997–1009.
26. Hardingham, G.E., Arnold, F.J., and Bading, H. (2001) *Nat Neurosci* **4**, 565–566.
27. Mao, L., Yang, L., Tang, Q., Samdani, S., Zhang, G., and Wang, J.Q. (2005) *J Neurosci* **25**, 2741–2752.
28. Takasu, M.A., Dalva, M.B., Zigmond, R.E., and Greenberg, M.E. (2002) *Science* **295**, 491–495.
29. Yang, L., Mao, L., Tang, Q., Samdani, S., Liu, Z., and Wang, J.Q. (2004) *J Neurosci* **24**, 10846–10857.
30. Hardingham, G.E., Fukunaga, Y., and Bading, H. (2002) *Nat Neurosci* **5**, 405–414.
31. Fields, R.D., Lee, P.R., and Cohen, J.E. (2005) *Cell Calcium* **37**, 433–442.
32. Shieh, P.B., Hu, S.C., Bobb, K., Timmusk, T., and Ghosh, A. (1998) *Neuron* **20**, 727–740.
33. Tao, X., Finkbeiner, S., Arnold, D.B., Shaywitz, A.J., and Greenberg, M.E. (1998) *Neuron* **20**, 709–726.
34. Tao, X., West, A.E., Chen, W.G., Corfas, G., and Greenberg, M.E. (2002) *Neuron* **3**, 383–395.
35. Chen, W.G., West, A.E., Tao, X., Corfas, G., Szentirmay, M.N., Sawadogo, M., Vinson, C., and Greenberg, M.E. (2003) *J Neurosci* **23**, 2572–2581.
36. Chen, W.G., Chang, Q., Lin, Y., Meissner, A., West, A.E., Griffith, E.C., Jaenisch, R., and Greenberg, M.E. (2003) *Science* **302**, 885–889.
37. Martinowich, K., Hattori, D., Wu, H., Fouse, S., He, F., Hu, Y., Fan, G., and Sun, Y.E. (2003) *Science* **302**, 890–893.
38. Levenson, J.M., and Sweatt, J.D. (2005) *Nat Rev Neurosci* **6**, 108–118.
39. Santos-Rosa, H.,and Caldas, C. (2005) *Eur J Cancer* **41**, 2381–2402.
40. Levenson, J.M., O'Riordan, K.J., Brown, K.D., Trinh, M.A., Molfese, D.L., and Sweatt, J.D. (2004) *J Biol Chem* **279**, 40545–40559.
41. Crosio, C., Heitz, E., Allis, C.D., Borrelli, E., and Sassone-Corsi, P. (2003) *J Cell Sci* **116**, 4905–4914.

42. Pak, D.T., and Sheng, M. (2003) *Science* **302**, 1368–1373.
43. O'Brien, R.J., Xu, D., Petralia, R.S., Steward, O., Huganir, R.L., and Worley, P. (1999) *Neuron* **23**, 309–323.
44. Xu, D., Hopf, C., Reddy, R., Cho, R.W., Guo, L., Lanahan, A., Petralia, R.S., Wenthold, R.J., O'Brien, R.J., and Worley, P. (2003) *Neuron* **39**, 513–528.
45. Bramham, C.R., and Messaoudi, E. (2005) *Prog Neurobiol* **76**, 99–125.
46. Cantallops, I, Haas, K., and Cline, H.T. (2000) *Nat Neurosci* **3**, 1004–1011.
47. Sala, C., Futai, K., Yamamoto, K., Worley, P.F., Hayashi, Y., and Sheng, M. (2003) *J Neurosci* **23**, 6327–6337.
48. Frey, U., and Morris, R.G. (1998) *Trends Neurosci* **21**, 181–188.
49. Dudek, S.M., and Fields, R.D. (2002) *Proc Nat Acad Sci U S A*, **99**, 3962–3967.
50. Adams, J.P., and Dudek, S.M. (2005) *Nat Rev Neurosci* **6**, 737–743.
51. Guzowski, J.F., McNaughton, B.L., Barnes, C.A., and Worley, P.F. (1999) *Nat Neurosci* **2**, 1120–1124.

# Part V

## SYNAPTIC PLASTICITY IN LEARNING AND MEMORY

# 23

# STRUCTURAL SYNAPTIC CORRELATES OF LEARNING AND MEMORY

Daniel A. Nicholson[†] and Yuri Geinisman[*]

## 1. SUMMARY

The majority of excitatory synapses in the brain are found on small dendritic protrusions called spines. Although the primary postsynaptic component of a synapse, the postsynaptic density (PSD), can only be resolved at the electron microscopic level, recent studies at the light microscopic level have shown that dendritic spines are extremely dynamic, capable of changing their shape and disappearing or emerging *de novo* from dendritic shafts. Despite the potential limitation of electron microscopic studies providing only a snapshot of the brain at a single time point, several groups have provided evidence of learning-related synaptic plasticity that corresponds well to what is known about cytoskeletal changes among dendritic spines. An emerging view from these electron microscopic studies is that learning involves the formation of new axospinous synapses and a remodeling of existing ones that includes alterations in their size, and perhaps strength. The application of modern methodologies to electron microscopy, such as unbiased stereological sampling and postembedding immunogold localization of synaptic proteins, will continue to elaborate our understanding of synaptic plasticity associated with learning, while also complementing research conducted at the electrophysiological and light microscopic levels.

[†]Department of Cell and Molecular Biology, Feinberg School of Medicine, Northwestern University, Chicago, IL 60611 USA; d-nicholson@northwestern.edu
[*]Department of Cell and Molecular Biology, Feinberg School of Medicine and Institute for Neuroscience, Northwestern University, Chicago, IL 60611 USA; yurig@northwestern.edu

## 2. INTRODUCTION

It is a general belief that the establishment of long-term memory is associated with its long-lasting storage in synapses, which is accomplished either by an addition of new synaptic contacts or an increase in the strength of existing ones. Numerous attempts have been made to define structural correlates of these processes with the aid of two basic approaches. One of these, electron microscopy, has a serious limitation because it provides static images of synapses fixed at a given moment after a new behavior has been learned. The other approach involves the use of two-photon laser scanning microscopy in combination with molecular probes for time-lapse imaging of dendritic spines on living neurons in slices and *in vivo*. The latter technical developments have produced a flurry of studies demonstrating that spines, which typically receive a single excitatory synapse, are highly dynamic structures (reviewed in refs. 1–3). Although some spines constantly outgrow from and are retracted into their parent dendrites, there are spine subpopulations that remain stable over a period of months and may represent loci of long-term information storage. In spite of its advantages, however, two-photon laser scanning microscopy lacks the resolution necessary to visualize synapses. Therefore, electron microscopy, with the aid of modern stereological techniques, remains the ultimate method for obtaining rigorous estimates of synapse number and for examining structural synaptic changes. The earlier studies on changes in synapse number and structure associated with learning and memory were extensively reviewed previously[4–8]. In this chapter, we focus our discussion on the data obtained over the last decade by quantitative electron microscopic analyses of the mammalian brain following acquisition of newly learned behaviors.

## 3. EVIDENCE FOR A LEARNING-INDUCED ADDITION OF SYNAPSES

Synapses are the primary link between neurons and have both presynaptic and postsynaptic components, separated by a small space called the synaptic cleft, all of which can be seen easily in osmium-fixed tissue at the electron microscopic level. The presynaptic component comprises a vesicle-filled axon terminal, with a small number of vesicles fused to or docked near the presynaptic membrane, and a very thin layer of electron-dense material immediately adjacent to the synaptic cleft. The postsynaptic component usually consists of an electron-dense plate called the PSD that is located on the cytosolic face of the postsynaptic membrane. The PSD is a protein-rich organelle containing neurotransmitter receptors, scaffolding molecules, actin-binding proteins, and a variety of other molecules involved in signal transduction. At the electron microscopic level, the PSD can be either approximately as thick as the presynaptic thickening, or it can be considerably thicker. The former type is a symmetric synapse involved in inhibitory synaptic transmission, whereas the latter is an asymmetric synapse mediating excitatory synaptic transmission. Most asymmetrical synapses in the brain are found on dendritic spines. A spine is a small protrusion that extends out from the parent dendrite and terminates as a head, which represents a postsynaptic element of a single excitatory synapse.

Axospinous excitatory synapses can be divided according to their PSD shape into perforated and nonperforated synaptic junctions[9,10]. When viewed in consecutive serial sections passing perpendicular or at an angle to the synaptic cleft, perforated synapses usually exhibit sectional PSD profiles showing a discontinuity or perforation that lacks the electron-dense material (Figure 23.1), hence the term "perforated." The PSD of perforated synapses can assume a fenestrated, horseshoe, or segmented shape. Serial sectioning of nonperforated synapses produces exclusively continuous PSD profiles, and the shape of a nonperforated PSD may be approximated by that of a flat circular or elliptical disc. Although a functional distinction between these two synaptic subtypes has not been

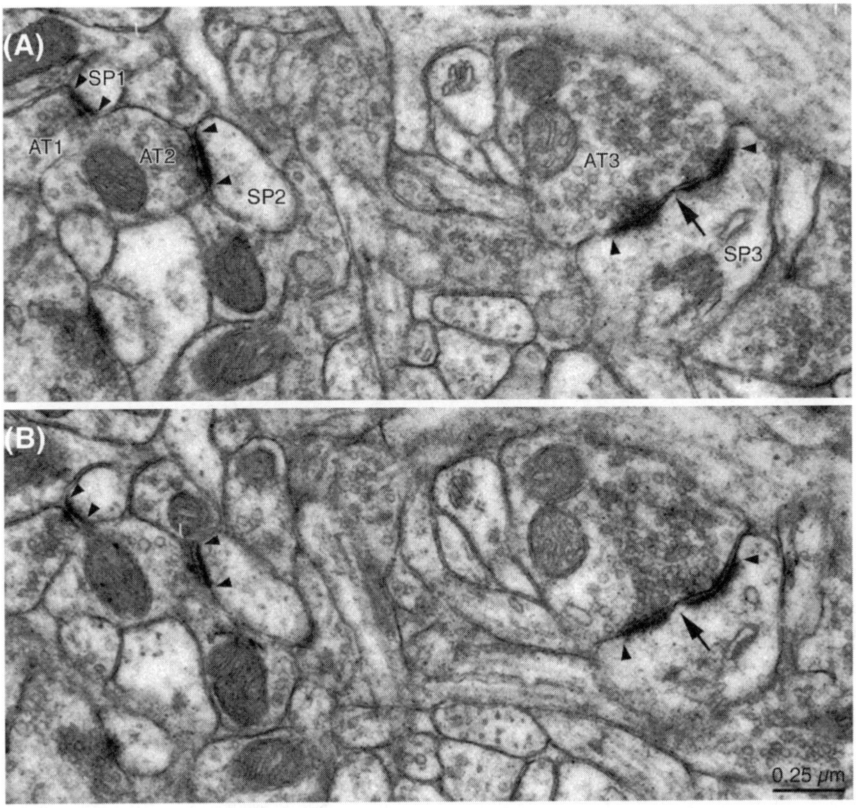


**Figure 23.1.** Electron Micrographs of Consecutive Ultrathin Sections (A,B) through the Rat CA1 Stratum Radiatum Demonstrating Axospinous Synapses. Sectional profiles of postsynaptic densities (PSDs) are marked by arrowheads. Two small synapses between presynaptic axon terminals (labeled AT1 and AT2 in (A)) and postsynaptic dendritic spines (labeled SP1 and SP2 in (A)) display continuous PSD profiles (all of which are shown) and belong therefore to the nonperforated synaptic category. The large synapse formed by an axon terminal (labeled in AT3 in (A)) and dendritic spine (labeled SP3 in (A)) exhibits PSD profiles with discontinuities or perforations (arrows) and belongs therefore to the perforated synaptic category.

shown directly, perforated synapses contain many more AMPA receptors (AMPARs) and NMDA receptors (NMDARs) than their nonperforated counterparts[11,12] and are therefore likely to generate much larger unitary synaptic potentials.

Many electron microscopic studies of learning-related synaptic plasticity have taken advantage of the diverse morphology among synaptic subtypes. By necessity, however, electron microscopic studies are limited to a very small portion of brain tissue and thus have inherent technical limitations and biases, especially when examining possible loci of synaptic plasticity. Modern stereological techniques that have been designed to minimize such biases include the physical disector[13] and systematic random sampling[14]. The physical disector is a three-dimensional stereological probe that consists of two adjacent ultrathin sections and has a volume limited by the sampling area and the thickness of the sampling section. This probe provides unbiased estimates of synaptic numerical density per unit volume by counting synapses present in one section (the sampling or reference section), but not in the other (the look-up section). Systematic random sampling is an unbiased method of choosing the area of tissue to be analyzed. The main advantage of systematic random sampling over other sampling schemes is that each area of tissue, along all three axes of the brain, has an equal probability of being chosen. Essentially, the first sampling field is chosen randomly, with subsequent sampling fields being systematically separated by a given distance throughout the extent of the brain region. Although this method requires multiple samples from each animal, the precision and accuracy afforded by this sampling technique is well worth the extra work[14].

### 3.1. Increases in Synapse Number Associated with Learning and Memory

#### *3.1.1. Effect of Acrobatic Task Acquisition*

A series of studies examining the effect of acrobatic training on synapse number comes from the work of Greenough and colleagues[15–19]. Adult rats were given acrobatic training during which they had to traverse elevated obstacles of an increasing difficulty for food reward over a period of 30–38 days. Three different groups of rats were used as controls. One of them received forced exercise on a treadmill, another was subjected to voluntary exercise on a freely accessible running wheel, and the third, inactive group was kept in standard laboratory cages. At the conclusion of training, the paramedian lobule of the cerebellar cortex, which is known to receive somatosensory and proprioceptive inputs from the forelimbs and hindlimbs, was examined. The estimates of synapse[18,19] and Purkinje cell[15,16,18,19] densities per unit volume of the molecular layer were obtained with the physical disector method. The number of synapses per Purkinje cell was calculated as the ratio of synaptic to cell numerical density. The major finding of these studies is that rats from the acrobatic group have 21–36% more synapses per Purkinje cell than those from the control groups. This learning-dependent increase in overall synapse number within the cerebellar cortex is accomplished primarily through the addition of parallel fiber synapses involving Purkinje cell spines[19]. A study of motor cortex layer II/III under the same experimental conditions reports similar results[17]. The authors additionally demonstrate that a significant learning-related increase in the number of synapses per neuron is more pronounced during the maintenance phase (days 5, 10, and 20 of training) than during the acquisition phase (days 1 and 2) of the learning curve. Taken together, these results led to the conclusion that motor learning required of the acrobatic animals, but not mere repetitive motor activity, generates new synapses in the cerebellar cortex and cerebral motor cortex. Importantly, the learning-dependent synaptic modifications observed in the cerebellar cortex of acrobatic rats persist for at least 4 weeks after completion of training[18] and may, therefore, represent a substrate of long-term memory storage.

### *3.1.2. Effect of Skilled Reaching Task Acquisition*

Skilled reaching paradigm of motor learning was used in two studies[20, 21]. In the earlier study[20], adult rats were trained for 10 consecutive days to grasp food pellets with a front paw from a table placed 1 cm away from a slotted opening in the front of a cage until a total of 400 reaches were made per day. Rats from a motor activity control group were placed in a cage that had a small lever located beneath the opening. Pressing the lever dispensed a food pellet into a nearby receptacle. These animals were given 400 pellets per day for 10 days, used their tongue and mouth to retrieve each pellet and, hence, did not develop skilled reaching movements. After the final training session, microelectrode stimulation techniques were applied to map the motor cortex contralateral to the trained paw, following which its medial and lateral boundaries were demarcated by injections of fluorescent microspheres of different colors. The area of the motor cortex was separately measured in electrophysiologically defined regions of rostral forelimb, caudal forelimb, and hindlimb representations. Within layer V of these three regions, the numerical density of neurons and synapses was estimated with the physical disector technique, and the number of synapses per layer V pyramidal cell was calculated. It was found that rats trained on the skilled reaching task exhibited a selective areal expansion of the caudal forelimb region (i.e., the region of digit and wrist movement representations) as compared to the controls. Correspondingly, trained animals also had significantly more synapses per neuron relative to controls only within the caudal forelimb region. The subsequent study[21] addressed the question of whether the observed synaptic alteration took place during the early (3 days of training) or late (7 or 10 days of training) phase of skilled reach learning. Although significant improvements in reaching accuracy occurred after 3 days, additional significant improvements were registered after 7 days, with no difference between 7- and 10-day animals. Significant expansion of distal movement representations in the caudal forelimb region was not detectable until after 10 days of training. The number of synapses per neuron was significantly higher in the caudal forelimb region of rats trained for 7 or 10 days than in the motor activity controls, indicating that the addition of synapses is characteristic of the late phase of learning when the consolidation of reaching motor skills occurs. It was hypothesized that the functional reorganization of the motor cortex is achieved through changes in cortical circuitry that involves new synapse formation. These data suggest that there may be two phases of plasticity in the motor cortex during skilled reaching training: one that is transient and does not involve an increase in synapse number; and another that is relatively long lasting and involves synaptogenesis. One untested possibility is that early-phase training increases the size and perhaps efficacy, but not the number, of synapses in the motor cortex.

### *3.1.3. Effect of Delay Eyeblink Conditioning*

Although acrobatic task acquisition and skilled reaching training involve behavioral plasticity and therefore learning, Kleim et al.[22] used the behavioral paradigm of delay eyeblink conditioning to examine whether increases in synapse number are associated with a more traditional, associative form of learning. Adult rats were given five daily training sessions, each one consisting of 100 paired presentations of a tone conditioned stimulus and periorbital shock unconditioned stimulus. During each presentation, the commencement of the conditioned stimulus preceded that of unconditioned one after which the stimuli were paired and then co-terminated. One group of control animals received explicitly unpaired presentations of the stimuli while the other was not stimulated. Synapses were quantified in the interpositus nucleus of the cerebellum, which is known to support long-term retention of the delayed eyeblink conditioned response. The numerical density of synapses and neurons was estimated in the anterior part of the interpositus nucleus with the physical disector method, and the number of synapses per neuron was calculated from synaptic and neuronal density values. Paired animals exhibited a significant increase in the percentage of trials with conditioned responses over the 5-day training period, whereas unpaired animals did not. Correspondingly, paired rats had a significantly greater number of synapses per neuron than both the unpaired and unstimulated controls. This conditioning-induced change involved asymmetrical, presumably excitatory, synapses, while the number of symmetrical, presumably inhibitory, synapses remained constant among the conditioned and control groups. It was proposed that the formation of additional excitatory synapses within the interpositus nucleus might represent a neural mechanism by which long-term memory is encoded in the cerebellum.

### *3.1.4. Effect of Spatial Memory Acquisition*

A quantitative electron microscopic study of the hippocampal dentate gyrus used the hippocampus-dependent Morris water maze task, which involves finding a hidden escape platform, to behaviorally characterize the spatial learning capacity of adult rats[23]. Rats were divided into three groups that received 4, 12, or 20 training trials. A probe trial was given 3 h after the end of training, and the animals were perfusion fixed for electron microscopy immediately following the probe trial. Two control groups were used: the "swim only" group time yoked to learners without any platform to locate and the naïve group with no water maze experience. The numerical densities of nonperforated axospinous synapses and granule cells were estimated per unit volume of the middle molecular layer and granule cell layer, respectively, using the physical disector technique. From these density values, the number of nonperforated synapses per granule cell was computed. Data from the probe trial showed that rats trained for 12 and 20 trials, but not four trials, displayed significant retention of platform location. In comparison to controls, a significant increase in the number of nonperforated synapses per neuron was detected in rats trained for 12 trials, but not in rats that received less (four trials) or more (20 trials) training. According to the authors, the transient nature of the observed increase in synapse number may mirror the notion that the hippocampal formation plays only an initial, time-limited role in memory acquisition and consolidation, but is not the final site of memory storage.

### *3.1.5. Negative Results and Their Interpretations*

An experiment similar to that described above examined the hippocampal dentate gyrus and CA1 in rats given Morris water maze training on 5 consecutive days[24]. In this study, however, the animals were perfused for electron microscopy 6 days after the last training session (as opposed to 3 hours after the probe trial). A control, nonspatial learning group was trained for 5 days to find a visible escape platform. The numerical density of dentate granule cells and CA1 pyramidal cells, as well as of synapses in the dentate middle molecular layer and proximal portions of CA1 basal dendrites, was assessed with the physical disector method. In both hippocampal regions studied, the numerical density of synapses and neurons was not changed. Therefore, the number of hippocampal synapses per neuron remained stable 6 days after spatial learning. Again, this negative finding could point to a transient role for hippocampal synaptic plasticity during learning.

Another quantitative electron microscopic study[25] examined whether the total number of synapses in CA1 stratum radiatum is altered by trace eyeblink conditioning. The latter is a hippocampus-dependent form of associative learning that is accompanied by an enhancement of synaptic responsiveness among CA1 pyramidal neurons. Adult rabbits were trained in pairs and received either trace eyeblink conditioning or pseudoconditioning. Conditioned rabbits were given daily 80 trial sessions to a criterion of 80% conditioned responses in a session. During each trial, the conditioned (tone) and unconditioned (corneal airpuff) stimuli were presented with a stimulus-free, or trace, interval of 500 ms. Control rabbits were pseudoconditioned with an equal number of random presentations of the same stimuli. Nine pairs of conditioned and pseudoconditioned animals were examined. Brain tissue was perfusion fixed for morphological analyses 24 h after the last training session. Synapses were sampled in a systematic random manner throughout the entire extent of the CA1 stratum radiatum, and their numerical density was estimated with the physical disector technique. The entire volume of the stratum radiatum was assessed according to the unbiased principle of Cavalieri. The total synapse number, including both the perforated and nonperforated subtypes, was calculated as the product of synaptic numerical density and stratum radiatum volume. It was virtually the same in conditioned and pseudoconditioned rabbits, the difference between the group means being less than 0.1%.

As indicated above, the dynamics of learning-induced increases in synapse number is different in various brain regions, which might be related to different roles of brain regions in the acquisition and maintenance of new behaviors. For example, overall gain in the number of synapses per neuron is observed in the motor and cerebellar cortices only during the late phase of learning a complex acrobatic task, and this change in cerebellar synapses lasts at least 4 weeks after the end of training[17,18]. Similarly, an increase in the number of synapses per neuron occurs in the motor cortex only during the late phase, but not the early phase, of motor learning of a skilled reaching task[21]. In the hippocampal formation, on the other hand, synapse number per dentate granule cell is increased after an intermediate amount of training, but then returns to the control level after further training on a hippocampus-dependent version of water maze task[23]. These data are consistent with the concepts that motor and cerebellar cortices may be involved in a long-term retention of memories for motor skills, whereas the hippocampal formation plays only a transient role in the acquisition and consolidation of spatial memory. In the two studies demonstrating a lack of change in the number of hippocampal synapses after learning of hippocampus-dependent tasks, synapses were quantified only at a single, relatively late time point following behavioral acquisition[24, 25]. If quantitative ultrastructural analyses were performed in these studies at earlier time points along the acquisition/consolidation curve, the presence of additional synaptic junctions could have been detected as well.

### 3.2. Increases in the Number of Multiple-Synapse Boutons Associated with Learning and Memory

Studies reviewed above indicate that learning of a new behavior is typically accompanied by synaptogenesis in a learning-relevant region of the mammalian brain, which is manifested by an increase in the number of synapses. Another ultrastructural synaptic alteration involving synaptogenesis is multiple-synapse bouton (MSB) formation, which is characteristic of various forms of plasticity (reviewed in 26). Each MSB establishes separate synaptic contacts with two or more discrete postsynaptic elements (Figure 23.2) instead of only one synapse with a single postsynaptic element as is the case for a single-synapse bouton. We next consider studies estimating MSB numbers after learning.

#### *3.2.1. Effect of Acrobatic Task Acquisition*

One such study[27] used the behavioral paradigm of acrobatic motor learning developed by Black et al.[15] and summarized above (see Section 3.1.1.). Adult rats were trained to traverse an elevated obstacle course for 30 days. Control animals were given a voluntary motor exercise on a wheel or were kept inactive in their home cages. The paramedian lobule of the cerebellar cortex was examined. The numerical density of parallel fiber varicosities establishing synaptic contacts with two Purkinje cell spines was assessed per unit volume of the molecular layer, and the numerical density of Purkinje cells was estimated per unit volume of cell body layer, using the method of physical disector. MSB number per Purkinje cell was calculated from the density values. MSBs were counted if they were observed contacting two postsynaptic spines in a single section plane. This approach to counting of MSBs greatly underestimates their occurrence[21], and the magnitude of the underestimation depends on MSB size, shape, and orientation. Nevertheless, the results of this experiment indicate that rats undergoing acrobatic motor learning have more MSBs per neuron than active or inactive controls. The formation of additional MSBs probably contributes to the increase in overall synapse number detected earlier in the molecular layer of adult rat cerebellar cortex following acrobatic learning[15,16,18,19]. Additionally, the establishment of a second synaptic contact between a preexisting parallel fiber varicosity and another Purkinje cell spine was interpreted as an indication of a selective, learning-related strengthening of this particular pathway that may represent a mechanism of neural encoding[27].

#### *3.2.2. Effect of Trace Eyeblink Conditioning*

The work from our laboratory has demonstrated that the total number of all synapses and of axospinous synaptic junctions in the rabbit CA1 stratum radiatum remains stable 24 h after trace eyeblink conditioning[25]. It was possible, however, that learning- induced synaptogenesis under such conditions might be confined to a specific subset of synaptic connections formed by MSBs. The validity of this supposition was tested in a study[26] using the behavioral protocol of trace eyeblink conditioning outlined above (see Section 3.1.5.). Modern stereological techniques and serial section analyses were employed for obtaining unbiased estimates of the total number of MSBs in the stratum radiatum of hippocampal subfield CA1. All MSBs entirely included in section series were identified as axonal swellings and with two or more separate spines. The

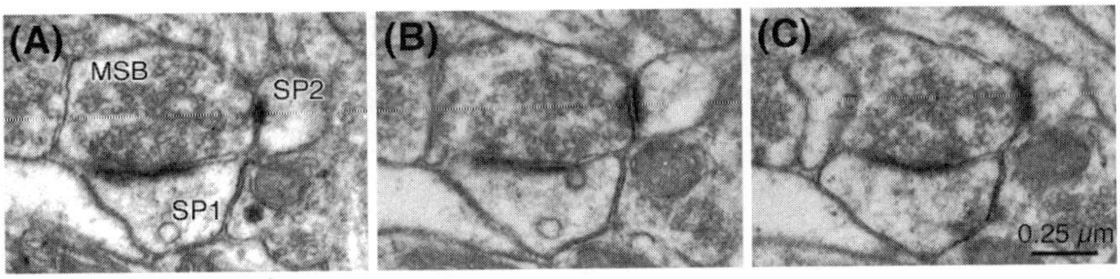


**Figure 23.2.** Electron Micrographs of Consecutive Serial Sections (A–C) Through the Rat CA1 Stratum Radiatum Demonstrating a Multiple-Synapse Bouton. The bouton (labeled MSB in (A)) makes synapses with two spines (labeled SP1 and SP2 in (A)).

numerical density of MSBs per unit volume was assessed with physical disectors, and the total MSB number was calculated as the product of MSB numerical density and CA1 stratum radiatum volume. It was found that conditioned rabbits had significantly more MSBs (18%) as compared with either pseudoconditioned or untrained controls, whereas the difference on this measure between the two control groups was small (2.8%) and not statistically significant. These data demonstrate that trace eyeblink conditioning is associated with the formation of MSBs. For MSBs to be formed, some axonal terminals must make new synaptic contacts with additional dendritic spines. Therefore, hippocampus-dependent associative learning promotes a specific synaptogenesis resulting in the formation of MSBs.

The latter structural synaptic change, which is not accompanied by a gain in total synapse number[25], can be explained based on the phenomena of spine protrusive motility and turnover that were initially observed in brain slices maintained *in vitro* (reviewed in 28) and then confirmed by *in vivo* experiments (reviewed in 2, 3). We will consider, as an example, the data reported by Svoboda and colleagues[29, 30] who imaged apical dendritic segments of mouse neocortical pyramidal neurons for periods of days to months. Under conditions of stable sensory input (i.e., visual and whisker-related stimuli), some spines were observed for only a few days or less before retracting into parent dendrites. Other spines formed *de novo* and grew toward axonal varicosities. Despite such a turnover of spines, their density per unit length of dendritic branch remained stable, whereas the retraction or elongation of dendritic and axonal processes did not occur. Subsequent serial section electron microscopy of the dendritic segments imaged *in vivo* revealed that spine retraction was associated with synapse elimination and spine outgrowth with synapse formation. Although the remarkably rapid turnover and associated motility of transient spines are most prominent during early postnatal development, they are also retained in adulthood, though to a lesser degree. In the somatosensory cortex, for example, the transient spine fraction reaches ~65% in 16–25-old-day mice but still amounts to ~25% in middle-aged (6-month-old) mice. Interestingly, plasticity of cortical synaptic fields following partial sensory deprivation was accompanied by a marked increase in spine turnover, but not by a change in spine density.

These data, taken together, suggest at least two models of structural plasticity that may underlie the learning-related addition of MSBs[26]. According to one model, newly formed spines establish synaptic contacts with single-synapse boutons activated by conditioning stimulation; simultaneously, some spines contacting nonactivated single-synapse boutons are retracted into parent dendrites. An alternative model postulates that some existing postsynaptic spines contacting nonactivated single-synapse boutons leave their presynaptic partners, relocate to boutons activated by conditioning stimulation, and synapse with them. In either case, some boutons would lack synaptic contacts, and such boutons without synapses are indeed encountered in the CA1 stratum radiatum[31].

The structural reorganization of synaptic connectivity following trace eyeblink conditioning may have different functional implications, depending on the origin of the postsynaptic spines contacting the newly formed MSBs. It has been documented with the aid of three-dimensional reconstruction from serial sections that MSBs in the CA1 stratum radiatum can synapse with spines arising from the same or different dendrites[32, 33]. In our study, however, we could not identify parent dendrites of many spines that were postsynaptic to MSBs. If these spines emerged from the same dendrite, the strength of the conditioned synaptic input to target CA1 neurons would be amplified. If, on the other hand, multiple spines synapsing with additional MSBs emanated from dendrites of different neighboring neurons, this might contribute to the synchronous activation of the latter and thus to the assembly of functional multineuronal units tuned to the synaptic input activated by conditioning stimulation. In either case, the synaptic plasticity that develops as a result of trace eyeblink conditioning would enhance synaptic inputs to and outputs from CA1 pyramidal neurons.

## 4. EVIDENCE FOR LEARNING-INDUCED REMODELING OF EXISTING SYNAPSES

Among various organelles of excitatory glutamatergic synapses involving dendritic spines, the PSD is most likely to undergo learning-related structural alterations. Although the PSD consists of many different components (reviewed in 34), the addition of AMPARs to the PSD emerges as a leading mechanism for enhancing the efficacy of synaptic transmission and enlarging PSD size. AMPARs mediate most of fast excitatory synaptic transmission, and their number is a major determinant of synaptic strength[35]. Being highly dynamic PSD components, AMPARs move into and out of the PSD on a timescale of minutes in an activity-dependent manner (reviewed in 36). There is a strong positive correlation between PSD dimensions and AMPAR content in axospinous synapses[11, 37–39]. This raises the question of whether PSD size is increased in existing axospinous synapses as a consequence of behavioral learning.

The results of earlier studies based on analysis of single ultrathin sections indicate that the length of PSD profiles is increased in relevant regions of the vertebrate brain following acquisition of new behaviors (reviewed in 8). These results, however, are biased by the size, shape, and orientation of synapses, as well as by section thickness and overprojection (i.e., apparent enlargement of opaque profile images because of the finite thickness of transparent sections). Recently, advanced stereological techniques have been employed to assess the size of PSD area. In our studies, the length of PSD profiles was measured in all consecutive serial sections through each sampled synapse, and the PSD area was estimated using appropriate corrections for variations in section thickness and overprojection[25]. The other serial section approach was to make three-dimensional reconstructtion of dendritic segments together with spines emanating from them and to measure PSD surface area on reconstructed images[40].

## 4.1. Enlargement of PSD Area in Axospinous Synapses After Learning

### *4.1.1. Effect of Trace Eyeblink Conditioning*

We did not detect any change in the total number of axospinous synapses or their morphological subtypes in the rabbit CA1 stratum radiatum 24 h after trace eyeblink conditioning[25]. Then we examined if an enlargement of PSD area in existing axospinous synapses is characteristic of this type of learning[25]. Synapses were sampled with disectors from the entire extent of the synaptic layer in a systematic random fashion, and the area of their PSDs was estimated. Comparison of conditioned and pesudoconditioned animals showed that the PSD area was significantly increased following conditioning in nonperforated, but not in perforated, axospinous synapses. To determine whether a PSD enlargement in the conditioned rabbits or PSD shrinkage in the pseudoconditioned controls caused the observed difference in size of nonperforated PSDs between the two groups of animals, unstimulated control rabbits were additionally investigated. It was found that conditioned animals had a significantly larger nonperforated PSD area than unstimulated controls, which were not different in this regard from pseudo-conditioned controls. Thus, the enlargement of nonperforated PSD area in conditioned animals relative to the pseudoconditioned controls represents a conditioning-induced increase in the PSD size. This change involves a decrease in the proportion of nonperforated synapses with PSDs that fell into the smallest size category (PSD area $< 20 \times 10^3$ $nm^2$) and a concomitant increase in the proportions of those nonperforated synaptic junctions that have larger PSD areas.

The discovery of postsynaptically silent synapses led to the suggestion that the learning-related enlargement of nonperforated PSDs might be associated with an addition of AMPARs[25]. Electrophysiological experiments have revealed that some synapses in the rat CA1 stratum radiatum exhibit functional NMDARs, but not functional AMPARs[41, 42]. This makes such synapses postsynaptically silent: they do not generate a postsynaptic response to a release of glutamate at normal resting membrane potentials because NMDAR channels are blocked by extracellular magnesium. Correspondingly, immunohistochemical studies have provided evidence for the existence of hippocampal axospinous synapses exhibiting only NMDAR, but not AMPAR, immunoreactivity associated with relatively small nonperforated PSDs (reviewed in 12). Silent synapses acquire the capability of evoking AMPA-type responses after LTP induction in the rat CA1 stratum radiatum[41,42], indicating that they may be transformed into functional synaptic junctions due to the insertion of AMPARs into their PSDs. These findings suggest the hypothesis that the same mechanism may underlie the conditioning-induced PSD enlargement, which would explain why the latter structural synaptic modification is selective for nonperforated axospinous synapses.

### *4.1.2. Effect of Spatial Memory Acquisition*

The hippocampus-dependent version of the Morris water maze task was used in a study aimed at determining whether PSD dimensions are changed in hippocampal axospinous synapses as a consequence of spatial learning[40]. During the acquisition training session (eight massed trials), adult rats had to find a hidden platform placed in the same pool quadrant. Progressive learning was detected in animals, as indicated by a decrease in the distance to reach the platform through these trails. The reversal training session (four trials) was administered on the next day, with the platform located in the quadrant opposite to the one used previously. Control rats were handled daily. The brain was fixed for electron microscopy 15 min after completing reversal training or handling. Digital reconstructive analyses of micrographs of serial sections through the CA3 stratum lucidum were performed to compute the PSD surface area in synapses formed by mossy fiber terminals and spine protrusions called thorny excrescences. A significant enlargement of PSD surface area was observed in water maze trained animals relative to controls. This change might reflect a pronounced (240%) increase in the proportion of perforated PSDs that also occurred following water maze training because perforated PSDs are known to be typically larger than nonperforated ones.

An earlier study from the same group[23] used a different protocol of water maze training (see Section 3.1.4.) and analyzed nonperforated axospinous synapses in the dentate middle molecular layer. The synaptic numerical density per unit volume was estimated. The total length of PSD profiles was measured in single sections, expressed per unit area, and converted into a PSD density estimate per unit volume. The PSD area per synapse was calculated as the ratio of synaptic density to PSD density. The results showed that the mean PSD area was significantly decreased in the water maze trained group as compared with untrained control group, but not the swim-only group. These results are difficult to interpret because they were derived from PSD measurements performed in single sections and might be therefore biased by synapse size, shape, and orientation.

## 4.2. Stability of Postsynaptic Density Size in Hippocampal Synapses of Aged Rats with Preserved, but not with Impaired, Spatial Learning

Another research strategy for defining structural synaptic correlates of learning is to compare patterns of synaptic ultrastructure in aged animals with preserved or impaired learning capacities. Although a cardinal feature of normal aging is a decline in learning and memory, it has long been noted that some aged individuals exhibit preserved cognitive functions even at advanced chronological ages. It has been demonstrated with unbiased stereological techniques that deficits in hippocampus-dependent spatial learning observed in a subpopulation of aged rats are not associated with a loss of principal hippocampal neurons[43,44] or synapses involving them[45]. In the absence of neuronal and synaptic loss, age-related spatial learning impairments may be caused by structural modifications of existing synapses. The validity of this supposition was tested in our study[46] evaluating whether age-related deficits in spatial learning are associated with a reduction of PSD area in hippocampal axospinous synapses because such a structural modification is likely to have a deleterious effect on excitatory synaptic transmission. A hippocampus-dependent version of the Morris water maze task was used to behaviorally characterize aged (27-month-old) rats and to separate them into learning-unimpaired and learning-impaired groups. Additionally, young adult (6-month-old) rats with good spatial learning were studied. Brain tissue was collected 4 weeks after the completion of the behavioral testing. A pilot experiment using different aged rats showed that their performance on the Morris water maze task was the same (either preserved or impaired) over a period of 4 weeks. Axospinous synapses were analyzed in the CA1 stratum radiatum, and unbiased estimates of PSD area in perforated and nonperforated synaptic junctions were obtained as described[25]. The major finding of this study is that a marked (~30%)

and significant decrease in PSD area occurs in perforated synapses of aged learning-impaired rats relative to both their aged learning-unimpaired counterparts and young adults. This change involves a substantial proportion of perforated synapses, regardless of whether they exhibit a fenestrated, horseshoe-shaped or segmented PSD, but it is not detected in nonperforated synapses.

The area of perforated PSDs positively correlates with the content of postsynaptic AMPARs in axospinous synapses from the rat CA1 stratum radiatum[11,39], and it is conceivable that the observed reduction of perforated PSD area in aged learning-impaired rats may result from a diminished AMPAR expression. Provided a significant proportion of perforated synapses indeed express only a few AMPARs, this would render such synaptic contacts less efficient and hence contribute to the age-related cognitive decline. Conversely, the size stability of perforated PSDs in aged learning-unimpaired rats, which reflects the retention of the full complement of postsynaptic AMPARs, appears to be a prerequisite for the maintenance of normal mnemonic functions during aging.

## 5. CONCLUSIONS

Quantitative electron microscopic studies of the mammalian brain have provided evidence that the cellular mechanisms of memory formation following behavioral learning include synaptogenesis manifested by an increase in the number of excitatory synapses involving dendritic spines in pertinent brain regions (Figure 23.3B). The structural synaptic modification of this kind has been reported most frequently and appears to be typical of different forms of behavioral learning. As opposed to the earlier electron microscopic work, synapse quantification during the last decade was performed with unbiased stereological methods for obtaining estimates of synapse number per postsynaptic neuron or per entire volume of a synapse-containing layer. In spite of this, some recent studies failed to detect a learning-induced increase in synapse number when a single time point along the learning curve was examined. This underscores the necessity of probing the time course of behavioral acquisition for the establishment of changes in synapse number associated with learning. Because the implementation of this approach at the electron microscopic level is impractical, a promising avenue of future research would be to combine electron microscopic quantification of axospinous synapses with pilot time-lapse imaging of dendritic spines on living neurons in slices or *in vivo*. Then quantitative electron microscopic analyses can be carried out only when additional new spines are found to be formed following learning. Another form of learning-induced restructuring of synaptic connectivity, which also involves synaptogenesis, is the addition of MSBs (Figure 23.3C). Although the existence of this structural synaptic alteration is well documented, its relationship to the *de novo* formation of single-synapse boutons at various phases of the learning process needs to be investigated.

Quantitative electron microscopic analyses have also revealed that behavioral learning promotes a structural remodeling of existing synapses. The most demonstrative example of this is an enlargement of the PSD (Figure 23.3D). Such a change was shown to selectively involve nonperforated axospinous synapses that had the smallest PSDs. The latter usually lack the AMPAR immunoreactivity, which probably makes them postsynaptically silent. The increase in nonperforated PSD area was postulated to reflect the insertion of AMPARs and to represent a structural correlate of the conversion of postsynaptically silent synapses into functional ones. An attractive line of future investigations would be to use quantitative immunogold electron microscopy for elucidating learning-induced alterations in both the area of PSDs and the AMPAR immunoreactivity. Interestingly, a recent study has demonstrated that associative auditory fear conditioning drives GluR1 subunit of AMPARs into thalamo-amygdala synapses

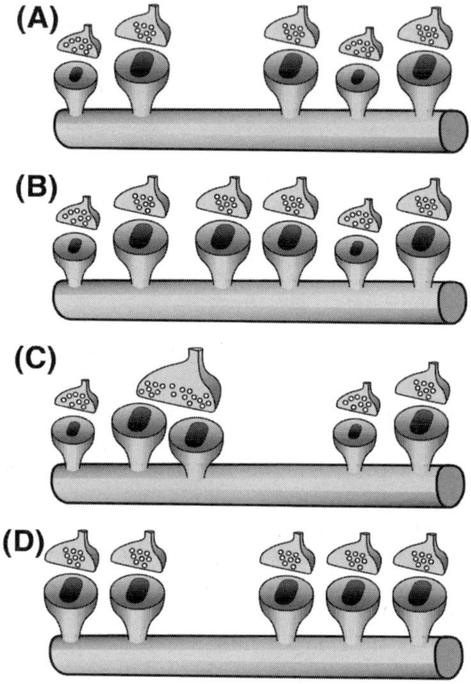


**Figure 23.3.** Schematic Demonstrating the Patterns of Changes in Axospinous Synapses that are most Frequently Observed at the Electron Microscopic Level in the Mammalian Brain Following Learning. As compared with controls (A), behavioral learning promotes synaptogenesis manifested by an addition of axospinous synapses established by single-synapse boutons (B) and by an increase in the number of multisynapse boutons (C). Additionally, behavioral learning is also associated with the restructuring of existing synapses that involves an enlargement of their PSD area (D).

while blockade of synaptic GluR1-receptor incorporation disrupts fear memory[47]. It would be advantageous, therefore, to use immunogold electron microscopy for quantifying the expression of the GluR1 subunit, and perhaps associated scaffolding and trafficking molecules like SAP-97 and stargazin in synapses, as well as possible changes in PSD area following various forms of learning.

To elucidate what kind of structural synaptic modifications are likely to account for synaptic plasticity associated with learning and memory, numerous studies have examined alterations in synapse number and structure following the induction of NMDA receptor-dependent hippocampal long-term potentiation (LTP), which is viewed as a synaptic model of memory. One of the most consistent observations of such studies is that LTP elicits the formation of additional perforated axospinous synapses in the hippocampal dentate gyrus[48] and CA1 region[49,50]. This process proceeds rapidly, but is transient. In CA1, for example, the proportion of perforated synapses gradually increases at 5, 15, and 30 min after cessation of potentiating stimulation; then it decreases at 45 min and returns to control values at 60 min[50]. It is not surprising, therefore, that no change in the proportion of perforated synapses is detected in CA1 at 2 h following LTP induction[51]. The LTP-related addition of perforated synapses is largely due to an increase in the number of their subtype that exhibits a segmented, completely partitioned PSD[52,53]. Such synapses are distinguished from other axospinous junctions by the highest content of AMPA receptors[11] and may contribute thereby to a marked enhancement of synaptic transmission soon after LTP induction. It has been proposed that LTP induction leads to the conversion of existing nonperforated synapses into segmented, completely partitioned ones[7]. This would explain why additional synapses of the latter subtype appear within a time frame that is unlikely to be long enough to allow the assembly of new excitatory hippocampal synapses[54]. In any case, the formation of segmented, completely partitioned synapses may represent an efficient way of rapidly and transiently augmenting synaptic transmission before other, perhaps permanent, forms of synaptic plasticity can emerge. Future studies will show if a similar pattern of synapse restructuring is also characteristic of early stages of the learning process.

## 6. REFERENCES


1. Nimchinsky, E.A., Sabatini, B.L., Svoboda, K. *Annu Rev Physiol* **64**, 313–333.
2. Segal, M. (2005) *Nat Rev Neurosci* **6**, 277–284.
3. Hayashi, Y., and Majewska, A.K. (2005) *Neuron* **46**, 529–532.
4. Greenough, W.T., and Bailey, C.H. (1988) *Trends Neurosci* **11**, 142–147.
5. Bailey, C.H., and Kandel, E.R. (1993) *Annu Rev Physiol* **55**, 397–426.
6. Klintsova, A.Y., and Greenough, W.T. (1999) *Curr Opin Neurobiol* **9**, 203–208.
7. Geinisman, Y. (2000) *Cereb Cortex* **10**, 952–962.
8. Geinisman, Y., Berry, R.W., and Ganeshina, O.T., in: In Riedel, G., and Platt, B., (Eds*), From Messengers to Molecules: Memories Are Made of These*, edited by Riedel, G., and Platt, B. (Eurekah.com and Kluwer Academic/Plenum Publishers, New York, 2004) pp. 544–563; http://www.eurekah.com.
9. Peters, A., and Kaiserman-Abramof, I.R. (1969) *Z Zellforsch* **100**, 487–506.
10. Geinisman, Y. (1993) *Hippocampus* **3**, 417–434.
11. Ganeshina, O., Berry, R.W., Petralia, R.S., Nicholson, D.A., and Geinisman, Y. (2004) *Neuroscience* **125**, 615–633.
12. Ganeshina, O., Berry, R.W., Petralia, R.S., Nicholson, D.A., and Geinisman, Y. (2004) *J Comp Neurol* **468**, 86–95.
13. Sterio, D.C. (1984) *J Microsc* **134**, 127–136.
14. Geinisman, Y., Gundersen, H.J., van der Zee, E.A., and West, M.J. (1996) *J Neurocytol* **25**, 805–819.
15. Black, J.E., Isaaks, K.R., Anderson B.J., Alcantara, A.A., and Greenough, W.T. (1990) *Proc Natl Acad Sci U S A* **87**, 5568–5572.
16. Anderson, B.J., Alcantra, A.A., and Greenough, W.T. (1996) *Neurobiol Learn Mem* **66**, 221–229.
17. Kleim, J.A., Lussing, E., Schwarz, E.R., Comery, T.A., and Greenough, W.T. (1996) *J Neurosci* **16**, 4529–4535.
18. Kleim, J.A., Vij, K., Ballard, D.H., and Greenough, W. T. (1997) *J Neurosci* **17**, 717-721
19. Kleim, J.A., Swain, R.A., Armstrong, K.A., Napper, R.M.A., Jones, T.A. and Greenough, W.T. (1998) *Neurobiol Learn Mem* **69**, 274–289.
20. Kleim, J.A., Barbay, S., Cooper, N.R., Hogg, T.M., Reidel, C.N., Remple, M.S., and Nudo, R.J. (2002) *Neurobiol Learn Mem* **77**, 63–77.
21. Keim, J.A., Hogg, T.M., VandenBerg, P.M., Cooper, N.R., Bruneau, R., and Remple, M. (2004) *J Neurosci* **24**, 628–633.
22. Kleim, J.A., Freeman, J.H., Bruneau, R., Nolan, B.C., Cooper, N.B., Zook, A., and Walters, D. (2002) *Proc Nat Acad Sci U S A* **99**, 13228–13231.
23. Eyre, M.D., Richter-Levin, G., Avital, A., and Stewart, M.G. (2003) *Eur J Neurosci* **17**, 1973–1980.

24. Rusakov, D.A., Davies, H.A., Harrison, E., Diana, G., Richter-Levin, G., Bliss, T.V.P., and Stewart, M.G. (1997) *Neuroscience* **80**, 69–77.
25. Geinisman, Y., Disterhoft, J.F., Gundersen, H.J.G., McEchron, M.D., Persina, I.S., Power, J.M., Van der Zee, E.A., and West, M.J. (2000) *J Comp Neurol* **417**, 49–59.
26. Geinisman, Y., Berry, R.W., Disterhoft, J.F., Power, J.M. and Van der Zee, E.A. (2001) *J Neurosci* **21**, 5560–5573.
27. Federmeier, K.D., Kleim, J.A., and Greenough, W.T. (2002). *Neurosci Lett* **332**, 180–184.
28. Matus, A. (2000) *Science* **290**, 734–738.
29. Trachtenberg, J.T., Chen, B.E., Knott, G.W., Feng, G., Sanes, J.R. Welker, E. and Svoboda, K. (2002) *Nature* **420**, 788–794.
30. Holtmaat, A.J.G.D., Trachtenberg, J.T., Wilbrecht, L., Shepherd, G.M., Zhang, X., Knott, G.W., and Svoboda, K. (2005) *Neuron* **45**, 279–291.
31. Shepherd, G.M.G., and Harris, K.M. (1998) *J Neurosci* **18**, 8300–8310.
32. Sorra, K.E. and Harris, K.M. (1993) *J Neurosci* **13**, 3736–3748.
33. Toni, N., Buchs, P.-A., Niconenko, I., Bron, C.R. and Muller, D. (1999) *Nature* **402**, 421–425.
34. Kennedy, M.B. (2000) *Science* **290**, 750–754.
35. Malenka, R.C. and Nicoll, R.A. (1999) *Science* **285**, 1870–1874.
36. Bredt, D.S. and Nicoll, R.A. (2003) *Neuron* **40**, 361–379.
37. Nusser, Z., Lujian, R., Laube, G., Roberts, D.S., Molnar, E., and Somogyi, P. (1998) *Neuron* **21**, 545–
559.
38. Takumi, Y., Ramirez-Leon, V., Laake, P., Rinvik, E., and Ottersen, O.P. (1999) *Nat Neurosci* **2**, 618–624.
39. Racca, C., Stephenson, P.A., Streit, P., Roberts, J.D., and Somogyi, P. (2000) *J Neurosci* **20**, 2512–
2522.
40. Stewart, M.G., Davies, H.A., Sandi, C., Kraev, I.V., Rogachevsky, V.V., Peddie, C.J., Rodriguez, J.J., Cordero, M.I., Donohue, H.S., Gabbott, P.L.A., and Popov, V.I. (2005) *Neuroscience* **131**, 43–54.
41. Isaac, J.T.R., Nicoll, R.A., and Malenka, R.C. (1995) *Neuron* **15**, 427–434.
42. Liao, D., Hessler, N.A., and Malinow, R. (1995) *Nature* **375**, 400–404.
43. Rapp, P.R., and Gallagher, M. (1996) *Proc Natl Acad Sci U S A* **93**, 9926–9930.
44. Rasmussen, T., Schliemann, T., Sorensen, J.C., Zimmer, J., and West, M.J. (1996) *Neurobiol Aging*
**17**, 143–147.
45. Geinisman Y., Ganeshina, O., Yoshida, R., Berry, R.W., Disterhoft, J.F., and Gallagher, M. (2004)
*Neurobiol Aging* **25,** 407–416.
46. Nicholson, D.A., Yoshida, R., Berry, R.W., Gallagher, M., and Geinisman, Y. (2004) *J Neurosci* **24**,
7648–7653.
47. Rumpel, S., LeDoux, J., Zador, A., and Malinow, R. (2005) *Science* **308**, 83–88.
48. Geinisman, Y., de Toledo-Morrell, L., and Morrell, F. (1991) *Brain Res* **566,** 77–88.
49. Buchs, P.-A., and Muller, D. (1996) *Proc Natl Acad Sci U S A* **93,** 8040–8045.
50. Toni, E., Buchs, P.-A., Nikonenko, I., Bron, C.R. and Muller, D. (1999) *Nature* **402,** 421–423.
51. Sorra, K.E., and Harris, K.M. (1998) *J Neurosci* **18,** 658–671.
52. Geinisman, Y., de Toledo-Morrell, L., Morrell, F., Heller, R.E., Rossi, M., and Parshall, R.F. (1993) *Hippocampus* **3,** 435–445.
53. Toni, E., Buchs, P.-A., Nikonenko, I., Povilaitite, P., Parisi, L., and Muller, D. (2001) *J Neurosci* **21,** 6245–6251.
54. Friedman, H.V., Bresler, T., Garner, C.C., and Ziv, E. (2000) *Neuron* **27,** 57–69.

# 24

# INTRACELLULAR TRAFFICKING OF AMPA-TYPE GLUTAMATE RECEPTORS

José A. Esteban*

## 1. SUMMARY

AMPA receptors are one of the most dynamic components of excitatory synapses. They are responsible for most excitatory transmission in the brain, and therefore, their presence at synapses is tightly controlled. We now know that AMPA receptors reach their synaptic targets after a complicated intracellular trafficking pathway, in which almost every step is subject to precise regulation. In particular, neuronal activity is able to trigger the addition or removal of AMPA receptors at synapses, leading to long-lasting forms of synaptic plasticity known as long-term potentiation (LTP) and long-term depression (LTD). This chapter summarizes our current knowledge of the intracellular trafficking of AMPA receptors, and its relation to synaptic function and plasticity.

## 2. INTRODUCTION

Most excitatory transmission in the brain is mediated by two types of glutamate receptors, namely, α-amino-3-hydroxy-5-methylisoxazole-4-proprionic acid (AMPA) and *N*-methyl-D-aspartate (NMDA) receptors. These two types of receptors play very different roles in synaptic function. AMPA receptors (AMPARs) are responsible for most excitatory responses in conditions of basal synaptic transmission. In contrast, NMDA receptors (NMDARs) remain silent at resting membrane potential, but they are crucial for the induction of specific forms of synaptic plasticity, such as LTP and LTD[2].

Although AMPARs and NMDARs reside in the same synapses in most brain regions, they reach their final synaptic targets following very different programs. In early postnatal development, most excitatory synapses contain only NMDARs, whereas the prevalence of AMPARs gradually increases as the brain develops[3].

*Department of Pharmacology, University of Michigan Medical School, 1150 W Medical Center Dr., Ann Arbor, MI 48109 USA; estebanj@umich.edu

Interestingly, the delivery of AMPARs into synapses is a regulated process that depends on NMDAR activation and underlies some forms of synaptic plasticity in early postnatal development and in mature neurons[4].

Although dendritic synthesis of AMPARs has been recently reported[5], most receptors are likely to be synthesized in the neuronal cell body, far away from their synaptic targets. Therefore, newly synthesized receptors have to engage in a long journey that starts at their points of biosynthesis, continues with their transport along dendrites, and ends with their translocation into dendritic spines and insertion into the postsynaptic membrane. This review summarizes our current knowledge of the intracellular trafficking pathways that lead to the synaptic delivery of AMPA receptors, with special emphasis on the late stages that contribute to synaptic plasticity. Mobile NMDA receptor clusters have also been observed, as described in Chapter 14 of this series. The intracellular transport of these mobile NMDARs is covered in other recent reviews[6].

## 3. AMPA RECEPTOR SYNTHESIS AND REGULATED EXIT FROM THE ENDOPLASMIC RETICULUM

AMPARs are hetero-tetrameric molecules[7] composed of different combinations of GluR1, GluR2, GluR3, and GluR4 subunits[8]. In the mature hippocampus, most AMPARs are composed of GluR1–GluR2 or GluR2–GluR3 combinations[9], whereas GluR4-containing AMPARs are expressed mainly in early postnatal development[10]. These oligomeric combinations are formed in the endoplasmic reticulum (ER) through mechanisms that are not well understood but that seem to depend on interactions between the luminal, N-terminal domains of the subunits[11] and the presence of an edited arginine residue ($R^{607}$) at the channel pore of the GluR2 subunit[12]. GluR1–GluR2 hetero-oligomers exit the ER rapidly, and traffic to the Golgi compartment where they become fully glycosylated[13]. In contrast, GluR2–GluR3 hetero-tetramers have a much longer residence time at the ER. In fact, a significant fraction of the GluR2 subunit is retained within the ER as an immature protein, in an active manner that depends on the presence of the positively charged $R^{607}$ at the channel pore[13]. GluR1, GluR3, and GluR4 mRNAs are not edited at this position, and therefore, these subunits contain a noncharged glutamine residue at the channel pore and do not undergo retention at the ER. The retention protein that prevents immature GluR2 from exiting the ER is not known; however, a fraction of AMPARs associates with the ER chaperones BiP and calnexin[14], and GluR2 colocalizes extensively with BiP in the ER[13]. Therefore, it is plausible that chaperones residing at the ER are related with the retention mechanism.

Additionally, export of AMPARs from the ER may require the interaction of the cytoplasmic, C-terminal domain of the AMPAR subunits with other proteins. The GluR2 C-terminus has a PDZ motif (-SVKI) that interacts with several PDZ domain-containing proteins, including PICK1[15], which is thought to be necessary for GluR2's exit from the ER[13]. The GluR1 C-terminus also contains a PDZ motif (-ATGL), which interacts with SAP97[16]. This interaction is known to occur early in the secretory pathway, probably while the receptor is still in the ER[17]. However, it is not known whether the SAP97–GluR1 interaction is necessary for this subunit to exit the ER.

Finally, AMPAR exit from the ER and acquisition of mature glycosylation at the Golgi complex is assisted by a family of transmembrane AMPA receptor regulatory proteins (TARPs)[18], of which stargazin is the founding member[19]. Indeed, TARPs are currently considered as auxiliary proteins for AMPARs[20,21], which may have chaperone-like functions during their early trafficking stages. The relevance of this family of proteins for the delivery of AMPARs into the cell surface and into synapses is discussed below.

## 4. AMPA RECEPTOR TRANSPORT ALONG THE CYTOSKELETON IN DENDRITES AND IN SPINES

The long-range dendritic trafficking of AMPARs depends on the microtubular cytoskeleton that runs along dendritic shafts. This can be demonstrated by monitoring AMPAR dendritic trafficking in neurons in which the tubulin cytoskeleton is destabilized pharmacologically. As shown in Figure 24.1 (Colorplate 11), incubation with a low concentration of vincristine (a microtubule destabilizer) impairs the dendritic transport of newly synthesized green fluorescence protein (GFP)-tagged AMPARs, to the point that their accumulation in distal dendrites is severely reduced. As a control, the same treatment does not affect the dendritic distribution of a co-expressed cytosolic protein, such as red fluorescence protein (RFP) (Figure 24.1; Colorplate 11).

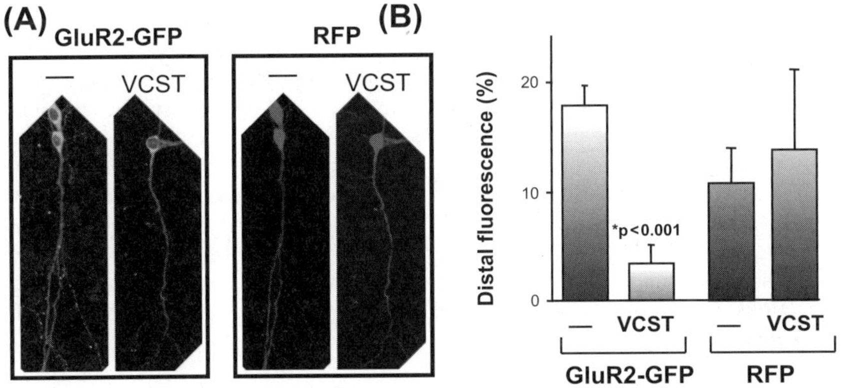


**Figure 24.1.** Dendritic Trafficking of AMPA Receptors is Impaired by Destabilization of the Microtubular Cytoskeleton. (A) The GFP-tagged GluR2 subunit of the AMPA receptor and a cytosolic red fluorescence protein (RFP) were co-expressed in organotypic hippocampal slice cultures[1] with ("VCST") or without ("–") 5 nM vincristine. Confocal fluorescence images were acquired 36 h after transfection. Scale bar: 20 μm. (B) Quantification of GluR2–GFP and RFP fluorescence 170–200 μm away from the cell soma. Fluorescence intensity values are normalized to those found at the cell soma. Plotted are average values and standard error of the mean from four experiments as the one shown in (A). GluR2–GFP presence in distal dendrites was drastically reduced in neurons incubated with vincristine. See Colorplate 11.

The transport of membrane organelles on microtubule tracks is an active process powered mainly by motor proteins of the kinesin and dynein superfamilies[22]. Therefore, membrane compartments bearing AMPARs are likely to be recognized and transported by some of these motor proteins. The molecular mechanisms underlying these processes are just being uncovered (the role of motor proteins in the transport of synaptic proteins is also discussed in Chapter 13 of this book).

The PDZ domain-containing protein GRIP1/ABP interacts directly with the heavy chain of conventional kinesin[23] and with the C-terminal PDZ motif of GluR2 and GluR3[24]. Therefore, GRIP1/ABP may serve as the link between AMPARs and microtubular motor proteins. In addition, liprin-α associates with both the GluR2–GRIP1/ABP complex[25] and with the kinesin family member KIF1[26], suggesting that the GRIP1–AMPAR complex can be transported along dendrites by multiple kinesin motor proteins. It seems likely that additional links between AMPARs and microtubular motor proteins will be discovered in the future, possibly mediated by adaptor scaffolding molecules.

Most excitatory synapses in the adult brain are formed on small dendritic protuberances called spines. Dendritic spines lack microtubular cytoskeleton, but they are rich in highly motile actin filaments. Therefore, at some point, AMPAR-containing organelles trafficking along microtubular tracks, must be transferred to the spine actin-based cytoskeleton for their final delivery into synapses. In fact, pharmacological depolymerization of actin filaments leads to the removal of AMPARs from dendritic spines[27]. So far, two molecular links between AMPARs and the actin cytoskeleton have been described: 4.1N and RIL. The different members of the protein 4.1 family are known to link the spectrin-actin cytoskeleton to different membrane-associated proteins[28]. The neuronal isoform 4.1N directly interacts with GluR1[29] and GluR4[30], and this interaction is important for the surface expression of these AMPAR subunits in heterologous systems. RIL is a PDZ domain-containing protein that links GluR1 with α-actinin, and it is important for the transport of AMPA receptors into dendritic spines[31].

The transport of AMPARs along the spine actin cytoskeleton is likely to be bidirectional, since AMPARs are known to move in and out of synapses in a very dynamic manner. This expectation has recently been confirmed by the identification of an actin-based motor protein, myosin VI, as a mediator of the endocytic removal of AMPARs from synapses[32]. Interestingly, myosin VI interacts with the GluR1-binding protein SAP97[33], providing a mechanistic link between AMPARs and a motor protein that drives their internalization. Undoubtedly, further studies will be required to unravel what is likely to be a network of interactions mediating the bidirectional transport of AMPARs along the actin cytoskeleton at dendritic spines.

## 5. TWO DISTINCT PATHWAYS FOR THE DELIVERY OF AMPA RECEPTORS INTO SYNAPSES

The last step in the long journey of AMPARs is their delivery into the specialized dendritic membrane that constitutes the postsynaptic terminal. AMPARs can reach synapses by two distinct pathways depending on their subunit composition. GluR2–GluR3 oligomers continuously cycle in and out of synapses in a manner largely independent from synaptic activity[1,34]. This process (constitutive pathway) preserves the total number of receptors at synapses, and therefore, it has been proposed to serve to maintain synaptic strength in the face of protein turnover[35]. This constitutive cycling is very fast (half-time of minutes) and it requires a direct interaction between GluR2 and NSF[36], which seems to be important for the recycling of AMPARs back into synapses[37]. The continuous synaptic cycling of AMPARs also requires the molecular chaperon Hsp90[38], although the mechanistic link between AMPARs and Hsp90 has not been elucidated yet.

There are multiple evidence that the constitutive cycling of AMPARs involves endocytic and exocytic trafficking; however, its connection with the known intracellular machinery that controls recycling endosomes is far from clear. For instance, AMPAR synaptic cycling requires the small GTPase Rab8[39], which is involved in trans-Golgi network (TGN) trafficking and some forms of endosomal recycling[40]. However, other Rab proteins typically associated with recycling endosomes, such as Rab4 and Rab11, are not involved[39]. As for the endocytic part of this cycle, the best characterized endocytic Rab protein, Rab5[40], does not participate in this constitutive cycling either[41]. Obviously, more work will be required to elucidate the cellular basis of this very dynamic aspect of the intracellular trafficking of AMPARs.

In contrast with their constitutive cycling, AMPARs containing GluR1[42], GluR2-long (a splice variant of GluR2)[43], or GluR4[10] subunits are added into synapses in an activity-dependent manner during synaptic plasticity. This regulated pathway acts transiently upon plasticity induction, leading to a net increase in the number of AMPARs present at synapses. This addition of new receptors results in the long-lasting enhancement of synaptic strength also known as LTP.

LTP is one of the best characterized paradigms of synaptic plasticity[4]. It is now well established that the opening of NMDARs with the concomitant rise in intracellular $Ca^{2+}$ at the postsynaptic terminal triggers the regulated insertion of new AMPARs into synapses. Multiple signaling cascades are thought to be activated downstream from $Ca^{2+}$ entry, and it is likely that complicated interactions between different signaling pathways control the regulated addition of receptors. There is abundant molecular and electrophysiological evidence supporting critical roles for the $Ca^{2+}$-calmodulin-dependent protein kinase II (CaMKII), PKA, PKC, and the mitogen-activated protein kinase (MAPK) in synaptic plasticity, and in particular, in LTP (see ref. 44 for a recent review). In a model recently proposed[45], LTP is mediated by the activation of CaMKII, which in turns activates Ras by inhibiting a synapse-localized Ras GTPase-activating protein (SynGap). The downstream effectors of Ras that mediate AMPAR synaptic delivery are still unclear, but they may involve well-known Ras effectors, such as p42-44 MAPK, and/or phosphatidylinositol 3-kinase (PI3K)[46]. Still, the precise mechanisms that trigger AMPAR synaptic delivery remain unknown.

The AMPAR subunits are known to be direct targets for some of these signaling cascades. For instance, PKA phosphorylates GluR1 at Ser845, and this event is required but not sufficient for AMPAR synaptic delivery and/or stabilization[47,48]. In addition, a new PKC phosphorylation site was very recently identified in GluR1 (Ser818), and this phosphorylation event seems to be directly involved in AMPAR synaptic delivery during LTP[49].

Interestingly, the signaling cascades controlling the delivery of AMPARs to synapses, as well as the AMPAR subunits involved, change during development. Thus, early in postnatal development of the hippocampus, the regulated delivery of AMPARs involves GluR4-containing receptors[10], and PKA-mediated phosphorylation of this subunit is necessary and sufficient for triggering delivery[47]. Around the second postnatal week, LTP is mostly mediated by the synaptic delivery of GluR2-long[43]. Then, later in development, the regulated addition of AMPARs involves GluR1, and its phosphorylation by PKA is required, but no longer sufficient for delivery[47]. Therefore, the number of signaling pathways that need to be activated for AMPAR synaptic delivery increases during development, in agreement with the empirical observation that synaptic plasticity is more difficult to trigger later in life.

## 6. ROLE OF TARPs IN AMPA RECEPTOR TRAFFICKING

Transmembrane AMPAR regulatory proteins, or TARPs, are the only known transmembrane proteins found to be associated with AMPARs. The first TARP to be identified was stargazin, which was found as a spontaneous mutation in the *stargazer* mouse[50] and is critically required for cell surface expression of AMPARs in cerebellar granule cells[19]. By sequence and structural homology, stargazin belongs to a large group of proteins that includes γ-subunits of $Ca^{2+}$ channels and the claudin family of cell-adhesion molecules. Nevertheless, only four of these proteins bind AMPARs and affect their trafficking: stargazin, γ-3, γ-4, and γ-8. Therefore, these are the four proteins that are designated as TARPs. Interestingly, different TARPS display specific patterns of expression in brain, which are to some extent, complementary[18].

TARPs associate with AMPARs early in their biosynthetic pathway and are able to combine with all AMPAR populations, irrespective of their subunit composition[18]. The most striking property of TARPs is their critical role for the expression of AMPARs at the extrasynaptic neuronal surface. Genetic ablation of stargazin, the TARP member most abundantly expressed in cerebellum, results in a virtual depletion of AMPARs from the extrasynaptic surface in granule cells[19]. A similar result is obtained in pyramidal neurons of the hippocampus after removal of γ-8, a TARP member that is almost exclusively expressed in hippocampus[51]. Interestingly, TARPs seem to be a limiting factor for AMPAR cell surface delivery, since overexpression of the appropriate TARP in different brain regions leads to a marked increase in the number of AMPARs expressed in the neuronal surface[19,51]. The role of these extrasynaptic surface receptors is still being debated, although morphological and electrophysiological evidence indicates that they are much more abundant than AMPARs present at the synaptic membrane. It has been observed that extrasynaptic surface receptors are highly mobile, and it has been demonstrated that they can reach the postsynaptic membrane through lateral diffusion from the extrasynaptic membrane[52]. This topic is specifically discussed in Chapter 15 of this book.

TARPs also participate in the trafficking of AMPARs into the synaptic membrane, although the mechanisms involved seem to be largely independent from the role of TARPs in extrasynaptic cell surface expression. TARPs contain a PDZ consensus sequence at the C-terminus, which can bind synaptic scaffolding molecules containing PDZ domains, such as PSD95 and PSD93[19]. The interaction between TARPs and PSD95 is particularly interesting, since PSD95 has been shown to be involved in the delivery of AMPARs into synapses. For instance, overexpression of PSD95 leads to an increase in the number of AMPARs at synapses[53]. Conversely, reduction of synaptic PSD95 levels through depalmitoylation[54] produces a concomitant decrease in synaptic AMPARs. Therefore, the synaptic levels of PSD95 and AMPARs seem to be controlled in parallel. The relevance of the stargazin-PSD95 interaction for AMPAR synaptic delivery was directly demonstrated in an elegant study in which compensatory mutations in the PDZ domain of PSD95 and in the C-terminus of stargazin rescued the ability of these proteins to drive AMPARs into synapses[55]. These results have led to the hypothesis that TARPs mediate the synaptic delivery of AMPARs in two steps. First, TARPs would assist in the insertion of AMPARs into the cell surface, in a process that does not require PDZ interactions. And second, the interaction between the PDZ sequence of TARPs and PSD95 would help in inserting and stabilizing AMPARs at synapses. These two separate roles of TARPs in AMPAR trafficking (cell surface expression and synaptic delivery through PSD95 interaction) are well established. However, it remains to be demonstrated whether they constitute necessary sequential steps for the delivery of AMPARs into synapses.

## 7. SUBCELLULAR ORGANIZATION OF AMPA RECEPTOR SYNAPTIC DELIVERY: ROLE OF RAB PROTEINS AND THE EXOCYST

AMPARs are integral membrane proteins. Therefore, all the trafficking processes described in this chapter necessarily involve the transport of the receptor along intracellular membrane organelles or receptor sliding through the cell surface (or a combination of both). In this sense, it may come as a surprise that so little is known about the subcellular organization of the membrane trafficking machinery that mediates AMPAR synaptic delivery. One plausible explanation for this gap in our understanding is that most of the trafficking events that are relevant for synaptic function may occur in the micron, or even submicron scale, within the environment of dendritic spines. This poses a significant challenge for the molecular and morphological identification of the membrane trafficking events that mediate AMPAR synaptic delivery. In this section, I will mention a few studies that have started to address this issue by investigating the components of the intracellular membrane sorting machinery that control AMPAR transport into synapses.

Most intracellular membrane sorting in eukaryotic cells is governed by small GTPases of the Rab family[40]. Therefore, the identification of specific Rab proteins involved in AMPAR trafficking may give us some clues as to how the intracellular sorting and synaptic targeting of AMPARs is organized in neurons. It was recently described that the small GTPase Rab8 mediates AMPAR synaptic delivery during both their constitutive cycling and regulated insertion upon LTP induction[39]. These results suggest that the TGN and recycling endosomes (both compartments in which Rab8 resides) are intracellular membrane sources for AMPAR synaptic delivery. However, the situation is likely to be more complicated, since Rab11, another small GTPase involved in the trafficking of recycling endosomes, was also shown to be absolutely required for AMPAR delivery during LTP[56]. Therefore, it is likely that the activity-dependent delivery of AMPARs involves a relay of at least two distinct endosomal compartments, whose sorting would be controlled by Rab11 and Rab8 in a sequential manner. Interestingly, as mentioned above, the constitutive cycling of AMPARs only requires Rab8, but not Rab11 or Rab4 (another Rab protein associated to recycling endosomes)[39]. This result emphasizes that the two pathways for AMPAR synaptic trafficking involve, to some extent, separate intracellular membrane compartments (Figure 24.2).

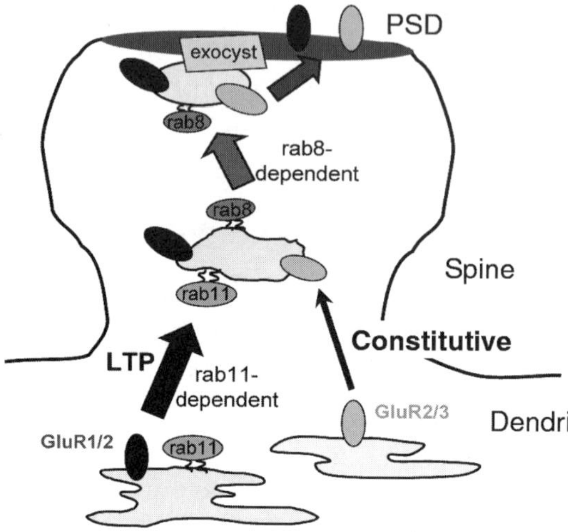


**Figure 24.2.** Subcellular Organization of the Membrane Trafficking Machinery that Mediates AMPA Receptor Synaptic Delivery. Schematic model for the regulated insertion of GluR1–GluR2 receptors across Rab11- and Rab8-controlled endosomal compartments (left). GluR2–GluR3 receptors constitutively reach Rab8-controlled endosomes within the spine. The last stages of receptor insertion into the synaptic membrane are mediated by Rab8 and the exocyst for both populations of AMPA receptors (see further explanation in the main text).

The functional studies mentioned above have not addressed the subcellular localization of the endosomal compartments that sort and deliver AMPARs into the synaptic membrane. However, the different requirements of the constitutive and regulated trafficking of AMPARs suggest a potential spatial organization of these membrane transport events. It is now well established that the regulated delivery of GluR1-containing AMPARs involves their translocation from the dendritic shaft into the adjacent spine, in a process that can be triggered by LTP induction[57]. In contrast, GluR2 AMPARs constitutively enter the spine compartment, as part of their continuous cycling[1]. As mentioned above, the regulated delivery of GluR1 requires both Rab11 and Rab8, whereas the constitutive delivery of GluR2 receptors is only mediated by Rab8, which is not required for receptor translocation into spines[39]. Since GluR1 delivery involves one extra regulated step (transport into the spine) and requires one additional Rab protein (Rab11), it is tempting to speculate that the translocation of GluR1 receptors from the dendritic shaft into the spine is governed by Rab11. According to this model, once inside the spine, the receptor would reach a different endosomal compartment, whose sorting would be controlled by Rab8. In this scenario, GluR2 would not traffic through the Rab11-controlled compartment, and therefore, would be free to enter the spine constitutively and reach the Rab8-containing compartment. Finally, the delivery of both GluR1 and GluR2 receptors into the synaptic membrane from within the spine would be controlled by Rab8 (see Figure 24.2 for a schematic representation of this scenario). Although this model has not been proved yet, it is compatible with all the electrophysiological and morphological evidence available.

Another important aspect of the intracellular trafficking of AMPARs is how the receptor is finally inserted into the postsynaptic membrane. Preliminary results from our laboratory point toward the exocyst complex as a critical component for this late step in AMPAR synaptic delivery. The exocyst is a multiprotein complex originally described in yeast to be required for exocytosis[58]. Our recent experiments suggest that the exocyst may act as a downstream effector of Rab8 for the targeted insertion of AMPARs at the postsynaptic density[59]. In particular, we have found that blockade of the exocyst subunit Exo70 does not prevent AMPAR translocation into spines, but drastically impairs AMPAR fusion with the postsynaptic membrane. In fact, interference with Exo70 function leads to the accumulation of AMPARs inside the spine, forming a complex physically associated, but not yet fused with the postsynaptic density. These recent results indicate that the exocyst is one of the latest factors that mediates AMPAR insertion at synapses, probably acting as a landmark for Rab8-driven exocytic trafficking of AMPAR-containing vesicles within the spine (see Figure 24.2).

## 8. AMPA RECEPTOR ENDOCYTOSIS AND REMOVAL FROM SYNAPSES

The controlled endocytic removal of AMPARs from synapses is as important for synaptic function and plasticity as the delivery of new AMPARs. Similar to the constitutive and regulated pathways for synaptic delivery, there are also constitutive and regulated pathways for the trafficking of AMPARs out of synapses. The continuous movement of receptors from the synaptic membrane into extrasynaptic compartments was originally hypothesized from the fast run-down of AMPAR-mediated synaptic transmission when interfering with the exocytic machinery or with the specific interaction between NSF and GluR2[36]. These experiments have now been corroborated and extended with molecular studies establishing that the GluR2–GluR3 population of AMPARs continuously cycle in and out of synapses, as mentioned earlier. Although the mechanisms for this cycling are unknown, NSF might play a critical role in this process by regulating the interactions between GluR2 and PICK1[37].

In addition to the continuous movement of AMPARs out of synapses, there is also ample evidence for the activity-dependent removal of receptors, which leads to long-lasting depression of synaptic strength (i.e., LTD). In contrast with the clear subunit-specific pathways for receptor delivery, it is still discussed which AMPAR populations are affected by this regulated removal. Hippocampal neurons lacking both GluR2 and GluR3 subunits can undergo normal LTD[60], suggesting that at least GluR1-containing receptors are subject to regulated removal. On the other hand, a switch of GluR2 binding partners from GRIP1 to PICK1 accompanies LTD[61,62], suggesting that the GluR2 subunits is critical for LTD. In fact, there is evidence that AMPAR removal during LTD may affect all AMPAR populations. This has been demonstrated in dissociated neuronal cultures, where the endocytosis of different receptor subunits can be directly monitored[63,64]. More recently, it has been found that Rab5 is the critical Rab protein that mediates AMPAR removal during LTD in hippocampal slices[41]. Interestingly, Rab5 is able to remove all AMPAR populations from synapses[41], suggesting that LTD affects all AMPAR subunits also in slices.

Still, subunit-specific internalization pathways do exist, as it has been described for GluR2-specific endocytosis and redistribution after internalization in hippocampal neurons[64–66]. Also, it is well established that PKC phosphorylation of GluR2 at Ser880 is correlated with LTD in the hippocampus and cerebellum[62,67]. The most accepted model for this regulated removal involves the preferential interaction of unphosphorylated GluR2 with GRIP1, which would favor the presence of the receptor at the synapse. After phosphorylation by PKC, GluR2 would dissociate from GRIP1 and bind PICK1, and this new interaction would retain GluR2 away from synapses[37]. This interplay of GluR2 interactions with GRIP1 and PICK1 has been reinforced with the finding that GRIP1 and PICK1 also interact directly[68].

It is also well established that LTD correlates with GluR1 dephosphorylation in the hippocampus[48,69], and that phosphatase activity is required for AMPAR removal and LTD expression[63,64,70,71]. These results suggest that phosphatase activity and GluR1 dephosphorylation are triggering events for AMPAR endocytic removal. However, it has also been proposed that GluR1 dephosphorylation occurs downstream of AMPAR endocytosis, and it is required to retain the receptor away from the synapse and prevent its reinsertion[41]. In fact, this interpretation is consistent with the original observation that phosphatase activity is required for LTD maintenance[71].

The subcellular organization of the endocytic machinery for the synaptic removal of AMPARs is not fully established yet, but some morphological and molecular studies have provided important clues. Multiple experimental observations indicate that there are endocytic "hotspots" at the extrasynaptic edges of the postsynaptic membrane[72], suggesting that these could be the gates for AMPAR endocytosis[73]. In support of this scenario, the endocytic protein Rab5 is also clustered at extrasynaptic sites laterally from the postsynaptic membrane[41]. As mentioned above, Rab5 drives AMPAR removal during LTD[41]. Rab5 is involved in clathrin-dependent endocytosis, and therefore, its role in AMPAR endocytosis during LTD is in good agreement with the established observation that AMPAR removal during LTD is mediated by clathrin-dependent mechanisms[66,74]. Interestingly, most Rab5 is present in its inactive configuration (GDP-bound) in basal conditions in the hippocampus. However, upon NMDAR activation, Rab5-GTP accumulates rapidly and transiently[41]. These results suggest a model in which LTD induction triggers the activation of Rab5 at pre-established endocytic hotspots, which in turn produces a transient endocytic wave, resulting in the synaptic removal of AMPARs. An important new aspect of this model is that the endocytic machinery does not act as a passive mediator of AMPAR removal, but as a regulated component in the signaling cascade that underlies LTD.

## 9. CONCLUSIONS

The field of intracellular trafficking of AMPARs is advancing at an incredible pace. New proteins interacting with AMPARs or with the AMPAR trafficking machinery are identified almost every week. In parallel, the number of neuronal systems that have been demonstrated to display plasticity because of AMPAR movement has also increased steadily. Importantly, we are now in a situation where specific mechanisms are being described for the new factors that participate in AMPAR intracellular trafficking. Undoubtedly, more studies will be required to unambiguously distinguish the initial steps in the dendritic transport of the receptor from the processes that orchestrate their dynamic behavior close to the synapse, where the regulated insertion and removal of receptors is used to control synaptic strength during plasticity. New investigations will also lead us to the identification of the core cellular machinery that directly transports AMPARs, as well as the regulatory molecules that gate or modulate specific trafficking steps. These are exciting times, when the field of AMPAR trafficking and synaptic plasticity is started to be integrated within the general framework of intracellular membrane transport and sorting.

## 10. REFERENCES


1. Shi, S., Hayashi, Y., Esteban, J.A., and Malinow, R. (2001) *Cell* **105**, 331–343.
2. Bear, M.F., and Malenka, R.C. (1994) *Curr Opin Neurobiol* **4**, 389–399.
3. Petralia, R.S., Esteban, J.A., Wang, Y.X., Partridge, J.G., Zhao, H.M., Wenthold, R.J., and Malinow, R. (1999) *Nat Neurosci* **2**, 31–36.
4. Malinow, R., and Malenka, R.C. (2002) *Annu Rev Neurosci* **25**, 103–126.
5. Ju, W., Morishita, W., Tsui, J., Gaietta, G., Deerinck, T.J., Adams, S.R., Garner, C.C., Tsien, R.Y., Ellisman, M.H., and Malenka, R.C. (2004) *Nat Neurosci* **7**, 244–253.
6. Wenthold, R.J., Prybylowski, K., Standley, S., Sans, N., and Petralia, R.S. (2003) *Annu Rev Pharmacol Toxicol* **43**, 335–358.
7. Rosenmund, C., Stern-Bach, Y., and Stevens, C.F. (1998) *Science* **280**, 1596–1599.
8. Hollmann, M., and Heinemann, S. (1994) *Annu Rev Neurosci* **17**, 31–108.
9. Wenthold, R.J., Petralia, R.S., Blahos, J., II, and Niedzielski, A.S. (1996) *J Neurosci* **16**, 1982–1989.
10. Zhu, J.J., Esteban, J.A., Hayashi, Y., and Malinow, R. (2000) *Nat Neurosci* **3**, 1098–1106.
11. Leuschner, W.D., and Hoch, W. (1999) *J Biol Chem* **274**, 16907–16916.
12. Greger, I.H., Khatri, L., Kong, X., and Ziff, E.B. (2003) *Neuron* **40**, 763–774.
13. Greger, I.H., Khatri, L., and Ziff, E.B. (2002) *Neuron* **34**, 759–772.
14. Rubio, M.E., and Wenthold, R.J. (1999) *J Neurochem* **73**, 942–948.
15. Xia, J., Zhang, X., Staudinger, J., and Huganir, R.L. (1999) *Neuron* **22**, 179–187.
16. Leonard, A.S., Davare, M.A., Horne, M.C., Garner, C.C., and Hell, J.W. (1998) *J Biol Chem* **273**, 19518–19524.
17. Sans, N., Racca, C., Petralia, R.S., Wang, Y.X., McCallum, J., and Wenthold, R.J. (2001) *J Neurosci* **21**, 7506–7516.
18. Tomita, S., Chen, L., Kawasaki, Y., Petralia, R.S., Wenthold, R.J., Nicoll, R.A., and Bredt, D.S. (2003) *J Cell Biol* **161**, 805–816.
19. Chen, L., Chetkovich, D.M., Petralia, R.S., Sweeney, N.T., Kawasaki, Y., Wenthold, R.J., Bredt, D.S., and Nicoll, R.A. (2000) *Nature* **408**, 936–943.
20. Vandenberghe, W., Nicoll, R.A., and Bredt, D.S. (2005) *Proc Natl Acad Sci U S A* **102**, 485–490
21. Fukata, Y., Tzingounis, A.V., Trinidad, J.C., Fukata, M., Burlingame, A.L., Nicoll, R.A., and Bredt, D.S. (2005) *J Cell Biol* **169**, 399–404.
22. Goldstein, L.S., and Yang, Z. (2000) *Annu Rev Neurosci* **23**, 39–71.
23. Setou, M., Seog, D.H., Tanaka, Y., Kanai, Y., Takei, Y., Kawagishi, M., and Hirokawa, N. (2002) *Nature* **417**, 83–87.
24. Dong, H., O'Brien, R.J., Fung, E.T., Lanahan, A.A., Worley, P.F., and Huganir, R.L. (1997) *Nature* **386**, 279–284.
25. Wyszynski, M., Kim, E., Dunah, A.W., Passafaro, M., Valtschanoff, J.G., Serra-Pages, C., Streuli, M., Weinberg, R.J., and Sheng, M. (2002) *Neuron* **34**, 39–52.
26. Shin, H., Wyszynski, M., Huh, K.H., Valtschanoff, J.G., Lee, J.R., Ko, J., Streuli, M., Weinberg, R.J., Sheng, M., and Kim, E. (2003) *J Biol Chem* **278**, 11393–11401.
27. Allison, D.W., Gelfand, V.I., Spector, I., and Craig, A.M. (1998) *J Neurosci* **18**, 2423–2436.
28. Hoover, K.B., and Bryant, P.J. (2000) *Curr Opin Cell Biol* **12**, 229–234.
29. Shen, L., Liang, F., Walensky, L.D., and Huganir, R.L. (2000) *J Neurosci* **20**, 7932–7940.
30. Coleman, S.K., Cai, C., Mottershead, D.G., Haapalahti, J.P., and Keinanen, K. (2003) *J Neurosci* **23**, 798–806.
31. Schulz, T.W., Nakagawa, T., Licznerski, P., Pawlak, V., Kolleker, A., Rozov, A., Kim, J., Dittgen, T., Kohr, G., Sheng, M., Seeburg, P.H., and Osten, P. (2004) *J Neurosci* **24**, 8584–8594.
32. Osterweil, E., Wells, D.G., and Mooseker, M.S. (2005) *J Cell Biol* **168**, 329–338.
33. Wu, H., Nash, J.E., Zamorano, P., and Garner, C.C. (2002) *J Biol Chem* **277**, 30928–30934.
34. Passafaro, M., Piech, V., and Sheng, M. (2001) *Nat Neurosci* **4**, 917–926.
35. Malinow, R., Mainen, Z.F., and Hayashi, Y. (2000) *Curr Opin Neurobiol* **10**, 352–357.
36. Nishimune, A., Isaac, J.T., Molnar, E., Noel, J., Nash, S.R., Tagaya, M., Collingridge, G.L., Nakanishi, S., and Henley, J.M. (1998) *Neuron* **21**, 87–97.
37. Hanley, J.G., Khatri, L., Hanson, P.I., and Ziff, E.B. (2002) *Neuron* **34**, 53–67.
38. Gerges, N.Z., Tran, I.C., Backos, D.S., Harrell, J.M., Chinkers, M., Pratt, W.B., and Esteban, J.A. (2004) *J Neurosci* **24**, 4758–4766.
39. Gerges, N.Z., Backos, D.S., and Esteban, J.A. (2004) *J Biol Chem* **279**, 43870–43878.
40. Zerial, M., and McBride, H. (2001) *Nat Rev Mol Cell Biol* **2**, 107–117.
41. Brown, T.C., Tran, I.C., Backos, D.S., and Esteban, J.A. (2005) *Neuron* **45**, 81–94.
42. Hayashi, Y., Shi, S.H., Esteban, J.A., Piccini, A., Poncer, J.C., and Malinow, R. (2000) *Science* **287**, 2262–2267.

43. Kolleker, A., Zhu, J.J., Schupp, B.J., Qin, Y., Mack, V., Borchardt, T., Kohr, G., Malinow, R., Seeburg, P.H., and Osten, P. (2003) *Neuron* **40**, 1199–1212.
44. Malenka, R.C., and Bear, M.F. (2004) *Neuron* **44**, 5–21.
45. Zhu, J.J., Qin, Y., Zhao, M., Van Aelst, L., and Malinow, R. (2002) *Cell* **110**, 443–455.
46. Qin, Y., Zhu, Y., Baumgart, J.P., Stornetta, R.L., Seidenman, K., Mack, V., van Aelst, L., and Zhu, J.J. (2005) *Genes Dev* **19**, 2000–2015.
47. Esteban, J.A., Shi, S.H., Wilson, C., Nuriya, M., Huganir, R.L., and Malinow, R. (2003) *Nat Neurosci* **6**, 136–143.
48. Lee, H.K., Barbarosie, M., Kameyama, K., Bear, M.F., and Huganir, R.L. (2000) *Nature* **405**, 955–959.
49. Boehm, J., Kang, M., Esteban, J.A., Huganir, R.L., and Malinow, R. (2005) in *Annual Meeting of the Society for Neuroscience*, Washington, DC.
50. Letts, V.A., Felix, R., Biddlecome, G.H., Arikkath, J., Mahaffey, C.L., Valenzuela, A., Bartlett, F.S., 2nd, Mori, Y., Campbell, K.P., and Frankel, W.N. (1998) *Nat Genet* **19**, 340–347.
51. Rouach, N., Byrd, K., Petralia, R.S., Elias, G.M., Adesnik, H., Tomita, S., Karimzadegan, S., Kealey, C., Bredt, D.S., and Nicoll, R.A. (2005) *Nat Neurosci* **8**, 1525–1533.
52. Borgdorff, A.J., and Choquet, D. (2002) *Nature* **417**, 649–653.
53. El-Husseini, A.E., Schnell, E., Chetkovich, D.M., Nicoll, R.A., and Bredt, D.S. (2000) *Science* **290**, 1364–1368.
54. El-Husseini Ael, D., Schnell, E., Dakoji, S., Sweeney, N., Zhou, Q., Prange, O., Gauthier-Campbell, C., Aguilera-Moreno, A., Nicoll, R.A., and Bredt, D.S. (2002) *Cell* **108**, 849–863.
55. Schnell, E., Sizemore, M., Karimzadegan, S., Chen, L., Bredt, D.S., and Nicoll, R.A. (2002) *Proc Natl Acad Sci U S A* **99**, 13902–13907.
56. Park, M., Penick, E.C., Edwards, J.G., Kauer, J.A., and Ehlers, M.D. (2004) *Science* **305**, 1972–1975.
57. Shi, S.H., Hayashi, Y., Petralia, R.S., Zaman, S.H., Wenthold, R.J., Svoboda, K., and Malinow, R. (1999) *Science* **284**, 1811–1816.
58. TerBush, D.R., Maurice, T., Roth, D., and Novick, P. (1996) *EMBO J* **15**, 6483–6494.
59. Gerges, N.Z., Backos, D.S., and Esteban, J.A. (2005) in *Annual Meeting of the Society for Neuroscience*, Washington, DC.
60. Meng, Y., Zhang, Y., and Jia, Z. (2003) *Neuron* **39**, 163–176.
61. Perez, J.L., Khatri, L., Chang, C., Srivastava, S., Osten, P., and Ziff, E.B. (2001) *J Neurosci* **21**, 5417–5428.
62. Kim, C.H., Chung, H.J., Lee, H.K., and Huganir, R.L. (2001) *Proc Natl Acad Sci U S A* **98**, 11725–11730.
63. Beattie, E.C., Carroll, R.C., Yu, X., Morishita, W., Yasuda, H., von Zastrow, M., and Malenka, R.C. (2000) *Nat Neurosci* **3**, 1291–1300.
64. Lin, J.W., Ju, W., Foster, K., Lee, S.H., Ahmadian, G., Wyszynski, M., Wang, Y.T., and Sheng, M. (2000) *Nat Neurosci* **3**, 1282–1290.
65. Lee, S.H., Simonetta, A., and Sheng, M. (2004) *Neuron* **43**, 221–236.
66. Man, H.Y., Lin, J.W., Ju, W.H., Ahmadian, G., Liu, L., Becker, L.E., Sheng, M., and Wang, Y.T. (2000) *Neuron* **25**, 649–662.
67. Chung, H.J., Steinberg, J.P., Huganir, R.L., and Linden, D.J. (2003) *Science* **300**, 1751–1755.
68. Lu, W., and Ziff, E.B. (2005) *Neuron* **47**, 407–421.
69. Lee, H.K., Kameyama, K., Huganir, R.L., and Bear, M.F. (1998) *Neuron* **21**, 1151–1162.
70. Ehlers, M.D. (2000) *Neuron* **28**, 511–525.
71. Mulkey, R.M., Herron, C.E., and Malenka, R.C. (1993) *Science* **261**, 1051–1055.
72. Blanpied, T.A., Scott, D.B., and Ehlers, M.D. (2002) *Neuron* **36**, 435–449.
73. Petralia, R.S., Wang, Y.X., and Wenthold, R.J. (2003) *Eur J Neurosci* **18**, 3207–3217.
74. Wang, Y.T., and Linden, D.J. (2000) *Neuron* **25**, 635–647.

# 25

# LEARNING-INDUCED CHANGES IN SENSORY SYNAPTIC TRANSMISSION

Min Zhuo*

## 1. SUMMARY

It is proposed that the central nervous system undergoes plastic changes in response to peripheral stimuli and new experiences. Consequently, alterations along central signaling processing system affect subsequent responses of animals and humans to the environment. Such long-term plastic changes not only play roles in important physiological functions such as learning and memory, but also are indispensable in unwanted pathological conditions such as chronic pain, drug addiction, and mental disorders. Investigation of such long-term plasticity using different animal models helps us to understand the molecular mechanism of brain functions. Here, I review recent work in the spinal cord dorsal horn, amygdale, and anterior cingulate cortex (ACC), and discuss the injury and learning-related changes in different regions of the brain and their implications in brain functions and mental disorders.

## 2. INTRODUCTION

Psychiatric and neurological illnesses are among the most common and serious health problems. The clinical treatment of neurological and psychiatric diseases requires advances in neuroscience for their elucidation, prevention, and treatment. In the past 20 years or so, we see successful applications of molecular genetics and cell biology to the central nervous system. With the complete mapping of both human and mouse genomes, new disease-related genes and proteins are being identified at a rapid speed. Discovery of diseases-related genes and proteins opens the door for the future understanding of the pathogenesis and lays the ground for their potential clinical treatment of patients with brain illnesses

*Department of Physiology, Faculty of Medicine, University of Toronto Center for the Study of Pain, University of Toronto, Ontario, Canada M5S 1A8; min.zhuo@utoronto.ca

such as schizophrenia, autism, mental retardation, chronic pain, depression, and Alzheimer and Parkinson diseases.

Knowing the sequence of a new gene or the structure of a new protein is not enough for understanding its role in brain functions. Neurons communicate with each other through a structure called the synapse. One major principle explaining how synapses work is that synaptic strength between two neurons should increase when the neurons exhibit coincident activity as proposed by Donald Hebb in his now-famous book, *The Organization of Behavior*. Subsequent findings of long-term potentiation (LTP) in the hippocampus provide a key synaptic model for studying brain functions. Recent advances in the ability to genetically modify mice made it possible to relate specific genes to both synaptic transmission/plasticity and intact animal behaviors, including memory, drug addiction, and persistent pain[1–4].

While we enjoy the success of connecting genes in synaptic function to behavioral phenotypic changes, we are confronted with a more difficult question: how changes in synaptic plasticity such as LTP may affect brain functions at systemic and behavioral levels? Recent progresses in neurobiology and genetics have significantly helped us to address it. First is the introduction of various genetically modified mice. This approach can overcome the lack of drugs for the interested proteins as well as their potential side effects. Due to the introduction of this approach in neuroscience of learning and memory by Susumu and Kandel, there are increasing numbers of publications relating molecules to behavioral phenotypes, learning, memory, pain, drug addiction, and various genetical diseases. Second is the cross marriage of chemistry and biology. Based on the understanding of the structures of proteins, many chemical inhibitors have been generated. In the near future, combination of gene-knockout mice with the use of pharmacological inhibitors will provide better studies of the mechanisms of brain functions. One major task for neuroscientists is to integrate recent findings in molecular and cellular research to the neuronal network and physiological functions of intact animals.

Here, I review recent progress in sensory-related central synapses, and their roles in sensory transmission, modulation, as well as long-term storage of fear-related information. I propose that these synaptic mechanisms provide the basis for future neuronal network studies as well as for design of better medicines to treat chronic pain and fear.

## 3. SPINAL DORSAL HORN: THE FIRST SENSORY SYNAPSE

### 3.1. Sensory Transmission

Neurons in the spinal cord dorsal horn and related areas receive sensory inputs, including noxious information, and convey them to supraspinal structures. Studies using pharmacological and behavioral approaches show that glutamate and neuropeptides, including substance P (SP), are excitatory transmitters for pain. Electrophysiological investigation of sensory synaptic responses between primary afferent fibers and dorsal horn neurons provided evidence that glutamate is the principal fast excitatory transmitter, and synaptic responses are mediated by postsynaptic glutamate receptors. While α-amino-3-hydroxy-5-methyl-4-isoxazole propionate (AMPA) receptors mediate the largest component of postsynaptic currents, kainate (KA) receptors preferentially contribute to synaptic responses induced by higher (noxious) stimulation intensities. Consistent with this, antagonism of both KA and AMPA receptor yields greater analgesic effects in adult animals than AMPA receptor antagonism alonc. These findings suggest that sensory modality may be coded in part by different postsynaptic neurotransmitter receptors[5].

In addition to glutamate, several neuropeptides, including SP, are thought to act as sensory transmitters. For many years, there has been a lack of electrophysiological evidence that SP can mediate monosynaptic responses, since SP-mediated responses had a very slow onset. Recent studies using whole-cell patch-clamp recordings reveal relatively fast SP- and neurokinin A-mediated synaptic currents in synapses between primary afferent fibers and dorsal horn neurons. Excitatory postsynaptic currents (EPSCs) in response to the burst activity of the primary afferent fibers may affect the excitability of spinal dorsal horn neurons. Together with glutamate-mediated synaptic responses, these neuropeptide-mediated EPSCs may cause dorsal horn neurons to fire action potentials at a high frequency for a long period of time. Therefore, the combination of glutamate- and neuropeptide-mediated EPSCs allow nociceptive information to be conveyed from the periphery to the central nervous system[5].

### 3.2. Serotonin (5-HT)-Induced Potentiation

Spinal dorsal neurons receive innervations from descending serotonin systems from the brainstem[6]. Application of 5-HT or 5-$HT_2$ subtype receptor agonist induced long-term facilitation of synaptic response[7]. One mechanism for the facilitation is the recruitment of silent synapses through interaction of glutamate AMPA receptors with proteins containing postsynaptic density-95/discs large/zona occludens-1 (PDZ) domains. To examine the functional significance of GluR2/3-PDZ interactions in sensory synaptic transmission, we made a synthetic peptide corresponding to the last 10 amino acids of GluR2 ("GluR2-SVKI": NVYGIESVKI) that disrupts binding of GluR2 to GRIP[7]. As expected, GluR2-SVKI peptide blocked the facilitatory effect of 5-HT. The effect of GluR2-SVKI on synaptic facilitation is rather selective because baseline EPSCs and currents evoked by glutamate application did not change over time in these neurons[7]. Furthermore, synaptic facilitation induced by phorbol 12,13-dibutyrate (PDBu) was also blocked by GluR2-SVKI, suggesting that synaptic facilitation mediated by protein kinase C activation is similar to that produced by 5-HT in its dependence on GluR2/3 C-terminal interactions[7] (Figure 25.1).

Activation of several receptors for sensory transmitters such as glutamate and calcitonin gene related peptide (CGRP) has been reported to raise cAMP levels. In a recent study, application of forskolin did not significantly affect synaptic responses induced by dorsal root stimulation in slices of adult mice. However, co-application of 5-HT and forskolin produced long-lasting facilitation of synaptic responses. Possible contributors to the increase in the cAMP levels are calcium-sensitive adenylyl cyclases (AC). We found that the facilitatory effect induced by 5-HT and forskolin was completely blocked in mice lacking AC1, indicating that calcium-sensitive AC1 is important[8]. The interaction between cAMP and 5-HT may provide an associative heterosynaptic form of central plasticity in the spinal dorsal horn to allow sensory inputs from the periphery to act synergistically with central modulatory influences descending from the brainstem rostroventral medulla (RVM).

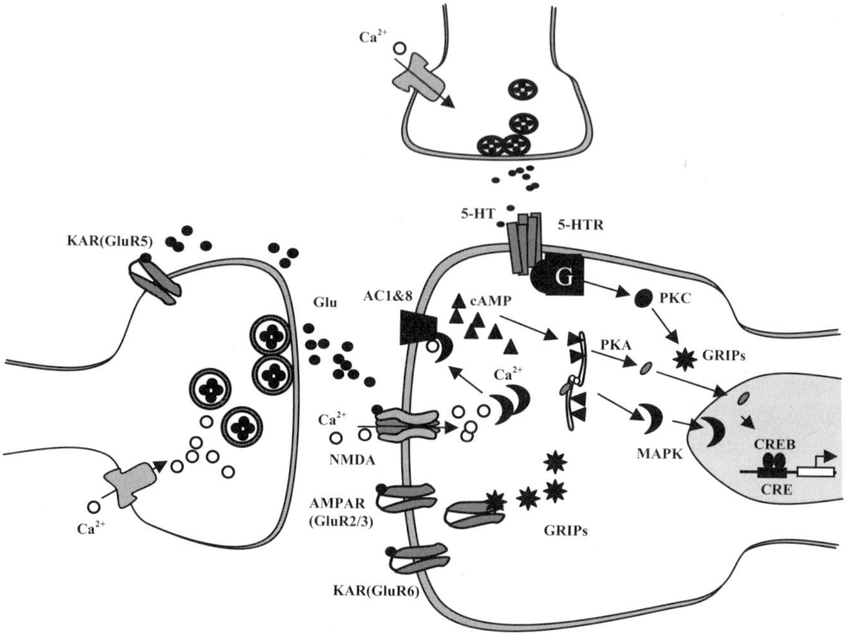


**Figure 25.1.** Signaling Pathways Contribute to Synaptic Potentiation in Spinal Sensory Synapses. Glutamate (Glu) is the major fast excitatory transmitter in the spinal cord. Both AMPA and KA GluR6 receptors contribute to synaptic responses in normal conditions. Activation of glutamate NMDA receptors leads to an increase in postsynaptic $Ca^{2+}$ in dendritic spines. $Ca^{2+}$ serves as an important intracellular signal for triggering a series of biochemical events that contribute to the potentiation of synaptic transmission. $Ca^{2+}$ binds to CaM and leads to activation of calcium-stimulated ACs, including AC1 and AC8 and $Ca^{2+}$/CaM dependent protein kinases (PKC, CaMKII, and CaMKIV) (not shown). The $Ca^{2+}$/CaM dependent protein kinases phosphorylate glutamate AMPA receptors, increase their sensitivity to glutamate and increase the number of functional synapses. In addition to the homosynaptic LTP, recruitment of postsynaptic AMPA receptors contributes to serotonin-induced facilitation. Neurons in the rostroventral medulla (RVM) project to the spinal dorsal horn and modulate sensory synaptic transmission in the spinal cord. Serotonin (5-HT) is the transmitter that mediates this facilitatory effect. The facilitation induced by 5-HT likely requires the activation of specific subtypes of serotonin receptors and coactivation of cAMP signaling pathways to induce facilitation in adult spinal dorsal horn neurons. 5-HT activates postsynaptic PKC through G protein receptors. PKC activation and subsequent AMPA receptor and GRIP interactions cause the recruitment of AMPA receptors to the synapse. Due to enhanced synaptic efficacy between primary afferent fibers and dorsal horn neurons, spike (action potential) responses to the stimulation of afferent fibers and behavioral nociceptive responses are enhanced (e.g., decrease in response latencies). Glutamatergic transmission is also under presynaptic regulation through the activation of auto-KA GluR5 receptors.

### 3.3. Homosynaptic LTP

Studies of LTP in spinal dorsal horn neurons draw much attention because it is believed that potentiation of sensory responses after injury may explain chronic pain[9]. While it has been demonstrated that spike responses of dorsal horn neurons to peripheral stimulation are enhanced after injury[9], it remains to be investigated if enhanced spike responses are simply due to enhanced synaptic transmission between the dorsal root ganglion (DRG) cells and dorsal horn neurons. Unlike synapses in other areas such as hippocampus, synaptic potentiation in the spinal dorsal horn neurons is not induced by strong tetanic stimulation. Recent studies further show that LTP only happens in some spinal projecting cells[10] (Figure 25.1). Activation of neurokinin subtype 1 (NK1) receptors or NMDA receptors is required for spinal LTP. Additionally, only spinal cord dorsal horn neurons that express SP receptors undergo potentiation, although in many other areas of the brain, there is no requirement of SP for the induction of NMDA receptor dependent LTP[10]. It will be important in the future to investigate why LTP cannot be induced in dorsal horn neurons that do not express SP receptors.

Trying to link changes in the spinal cord with behavioral responses has proved to be difficult. Not only do spinal sensory synapses receive inputs from the periphery, but they also receive biphasic modulation from supraspinal structures. In addition, many supraspinal structures also contribute to pain-related behavioral responses. Therefore, it is premature to directly connect what found in the spinal cord with behavioral changes.

## 4. AMYGDALA: FEAR AND ITS LONG-TERM STORAGE

### 4.1. Amygdala and Fear

Fear memory requires the involvement of higher brain structures including the hippocampus, the amygdala, and related cortical structures[2,11–13]. Studies from animals and humans consistently suggest that the amygdala is important for fear memory formation[14–16]. Two forms of fear memory are commonly studied in animals: auditory and contextual fear memory. For auditory fear memory, evidence from experiments using different approaches suggests that the amygdala plays an important role in the acquisition and retention of the memory[2]. Neurons in the amygdala, particularly in the lateral amygdala (LA), receive auditory (used as conditioning stimulus, CS, in fear conditioning) and somatosensory inputs (used as unconditioned stimuli, US) from the periphery[2]. LA neurons receive auditory inputs from both the auditory thalamus and the auditory cortex. Both pathways contribute to auditory fear memory, although they may play differential and complementary roles in different forms of auditory memory. For somatosensory inputs, including noxious inputs (used as US), neurons in the LA receive somatosensory inputs from thalamus. LA neurons are responsive to nociceptive stimulation, and some LA cells respond to both auditory and noxious stimuli[17]. For contextual fear memory, it is believed to require the involvement of both the amygdala and hippocampus[18,19]. In addition to the amygdala and hippocampus, cortical areas that process nociceptive stimuli project to the LA and other amygdala nuclei[20,21].

### 4.2. Fear and LTP

How is new information learned and stored in amygdala circuits? One leading hypothesis is that synaptic transmission undergoes long-term plastic changes after training in the amygdala. Studies using multiple approaches, including electrophysiological recordings in brain slices and single units or field synaptic responses in whole animals, consistently indicate that LTP is the most likely synaptic mechanism underlying fear memory in the amygdala. First, electrophysiological studies using *in vitro* amygdala slices or *in vivo* recordings showed that auditory afferent pathways, including auditory thalamus-amygdala and auditory cortex-amygdala pathways, undergo synaptic potentiation after LTP-inducing stimulation paradigm[22–24]. Second, the associative nature of LTP in the amygdala supports the idea that fear conditioning requires the convergence of auditory and nociceptive inputs onto single neurons in the LA[24]. Third, neural activity in the LA has been shown to be modified during auditory fear conditioning in a manner similar to that observed after artificial LTP induction[25,26]. Fourth, drugs that inhibit the induction and/or consolidation of LTP in the amygdala also inhibit fear memory[27–29] (see ref. 2 for reviews). Finally, a recent study shows that fear conditioning occluded LTP-induced presynaptic enhancement of synaptic transmission in the cortical pathway to the lateral amygdala[30].

### 4.3. Thalamic-Amygdala LTP

Synaptic mechanisms for LTP have been best investigated in the amygdala slices. It has been noted that the contribution of NMDA receptors and/or L-type voltage gated calcium channels (L-VGCCs) to the LTP induction are somewhat dependent on induction protocols[31] (Figure 25.2). Activation of NMDA receptors is required for the induction of the thalamic-amygdala LTP. Interestingly, bath application of a selective NR2B receptor antagonist blocked the induction of LTP[32,33]. The role of NR2A containing NMDA receptors has not yet been investigated. In addition to the NMDA receptors, KA GluR6 receptors are reported to be important for the LTP. Field potential recordings and whole-cell patch-clamp recordings revealed that LTP is blocked in mice lacking GluR6 subunit. In contrast, LTP in the GluR5 subunit knockout mice is normal. The exact synaptic mechanisms for the GluR6 contribution to the LTP remain to be investigated. Additionally, inhibition of metabotropic glutamate receptor (mGluR) subtype 5 has been reported to block LTP[32]. Thus, it is likely that multiple receptor systems are involved in synaptic potentiation in the thalamic-amygdala pathways.

The involvement of protein kinases in amygdala LTP has also been investigated. cAMP-dependent protein kinase (PKA) and CaMKII are reported to contribute to the induction of LTP. In addition, CaMKIV has been implicated in early LTP in the amygdala[34]. The expression of LTP in the thalamic-amygdala pathways contains both presynaptic and postsynaptic mechanisms. A recent study suggests that during LTP, paired-pulse facilitation (PPF) was altered in the thalamic-amygdala pathways, raising the possibility for presynaptic contribution to LTP. Furthermore, inhibition of the production of nitric oxide (NO), a retrograde messenger, blocked the LTP[35] (see Figure 25.2 for a model), suggesting a possible role of NO as a retrograde messenger in the expression of thalamic-amygdala LTP.

### 4.4. LTP in the Cortical-Amygdala Pathway

Induction of LTP in the cortical pathway to the LA is postsynaptic, whereas the expression of LTP is presynaptic. The induction of LTP is dependent on postsynaptic depolarization, on the influx of $Ca^{2+}$ into the postsynaptic cell either through NMDA receptors[23,30] or L-VGCCs[30]. In a well-executed experiment by Tsvekov et al., it was found that LTP was completely blocked when both NMDA receptor and L-VDCCs were blocked. Small conductance $Ca^{2+}$ activated $K^+$ channels (SK) have also been reported to contribute to the LTP in the cortical-amygdala pathways[36]. Blockade of SK channels greatly enhanced the LTP. These results suggest that $Ca^{2+}$ influx through NMDA receptor activates SK channels and shunts the resultant excitatory postsynaptic potential.

The expression of the cortical-amygdala LTP is mostly presynaptic[23,30]. LTP is associated with a decrease of PPF and is blocked by bath application but not by postsynaptic injection of inhibitors of PKA. Moreover, forskolin could induce LTP in this pathway, which occluded the tetanus-induced LTP. Many other signaling molecules/proteins have been reported to contribute to PI-3 kinase inhibitors blocked tetanus-induced LTP (field recordings) in cortico-amygdala pathway. Tetanus and forskolin-induced activation of MAPK was blocked by PI-3 kinase inhibitors, which also inhibited CREB phosphorylation. These results suggest PI-3 is upstream for MAPK and CREB activation, thereby contributing to synaptic plasticity in the amygdala.

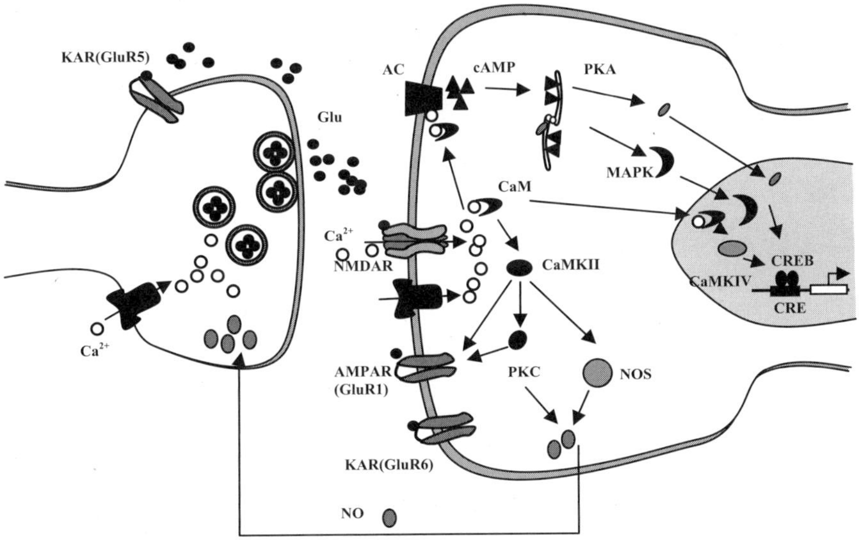


**Figure 25. 2.** Presynaptic and Postsynaptic Signaling Pathways that Contribute to LTP in the Amygdala. NMDA receptors are important for the induction of LTP. $Ca^{2+}$ serves as an important intracellular signal for triggering a series of biochemical events that contribute to the expression of LTP. The $Ca^{2+}$/CaM dependent protein kinases phosphorylate glutamate AMPA receptors, increase their sensitivity to glutamate and the number of functional AMPA receptors. Activation of CaMKIV, a kinase predominantly expressed in the nuclei, will trigger CaMKIV-dependent CREB that contribute to LTP. The involvement of L-VGCCs has also been implicated, depending on the LTP induction protocol. The production of retrograde messenger nitric oxide (NO) is thought to be important for the expression of LTP.

The late phase of LTP (L-LTP) in the cortico-amgdala is mediated by PKA and MAPK[37]. The L-LTP is associated with phosphorylation of CREB and protein synthesis. Most molecules studied are involved in the thalamic-amygdala LTP, including GluR1[38], NR2B[33], GluR6[39], NO[35], CaMKIV[34], and CaMKII[31], but their roles in cortical LTP has not been elucidated yet.

## 5. ANTERIOR CINGULATE CORTEX (ACC): AN INTEGRATIVE CENTER

### 5.1. Synaptic Transmission in the ACC

Glutamate is the major fast excitatory transmitter in the ACC[7]. Different types of glutamate receptors, including AMPA, KA, NMDA, and mGluRs, are found in the ACC. Whole-cell patch-clamp recordings from ACC pyramidal neurons of rats and mice showed that excitatory synaptic responses are mediated by glutamate AMPA/KA receptors[7,40]. While the majority of fast EPSCs are mediated by postsynaptic AMPA receptors, a recent study reported that some synaptic currents can be mediated by glutamate KA receptors in ACC neurons of adult mice[40]. Furthermore, genetic studies using a KA subtype receptor knockout mice showed that postsynaptic KA receptor-mediated currents require GluR5 and GluR6 subunits[40]. In addition to fast synaptic responses, NMDA receptor-mediated slow synaptic responses were also recorded from adult ACC slices at physiological temperatures or in vivo recordings from the ACC of freely moving animals[41,42]. These findings indicate that under physiological conditions, postsynaptic NMDA receptors can contribute to sensory synaptic transmission in the ACC.

### 5.2. LTP in the ACC

Glutamatergic synapses in the ACC can undergo long-lasting potentiation in response to theta-burst stimulation, a paradigm more closely related to the activity of ACC neurons by using field recording in ACC slices of adult rats or mice. The potentiation lasts for at least 40–120 min[34]. LTP can be recorded using whole-cell patch-clamp recordings in slices of adult mice. Long-term synaptic potentiation can be induced by using other two different induction protocols[43]. Activation of postsynaptic NMDA receptors and postsynaptic $Ca^{2+}$ signaling is critical for the induction of LTP. During LTP, no obvious changes in PPF are detected, suggesting that potentiation is unlikely due to pure presynaptic mechanisms. Activation of NMDA NR2A and NR2B subunits is critical for the induction of LTP[43]. Blockade of both NR2A and NR2B subunits is required for the complete inhibition of the induction of cingulate LTP. In addition to NMDA receptors, activation of KA GluR6 receptors was also reported to contribute to cingulate LTP[39].

Cyclic AMP is a key second messenger in neurons. Recent studies using gene-knockout mice and pharmacological activators/inhibitors found that calcium-stimulated AC1 and AC8 are important for the cingulate LTP induced by theta-burst stimulation or pairing training protocol in pyramidal cells. Activation of MAP kinases is also found to be important for cingulate LTP. In addition, CaMKIV is also required for the induction of LTP[34]. Thus, it is likely that multiple signaling pathways are critical for the induction of cingulate LTP. Future studies are needed to address the possible contribution of AMPA receptor subunits in cingulate LTP (Figure 25.3).

### 5.3. Behavioral Fear and the ACC

It has been reported that stimulation of ACC generates fear memory, and the neuronal activity in the ACC is required for retrieval of remote fear memory[44].

Recent studies using pharmacological and genetic approaches showed that inhibition of NMDA NR2B receptors in the ACC significantly reduced the formation of classic fear memory, indicating that ACC neuronal activity contributes to fear memory formation[43]. To support the roles of the ACC in fear, Tang and co-workers reported that stimulation of the ACC generates fear memory without classic foot shocks[42]. This ACC stimulation-induced fear memory requires activity in the amygdala, suggesting that the ACC and amygdala are likely to work together to memorize fearful cues.

## 6. CHRONIC PAIN

In order to investigate the molecular and cellular mechanisms for pain-related plasticity in the ACC, we decided to use genetic approaches together with integrative neuroscience techniques. First, we wanted to test if persistent pain may be enhanced by genetically enhanced NMDA receptor function, a key mechanism for triggering central plasticity in the brain[4]. Functional NMDA receptors contain heteromeric combinations of the NR1 subunit plus one or more of NR2A-D. While NR1 shows a widespread distribution in the brains, NR2 subunits exhibit regional distribution. In humans and rodents, NR2A and NR2B subunits predominate in forebrain structure. NR2A and NR2B subunits confer distinct properties to NMDA receptors; heteromers containing NR1 plus NR2B mediate a current that decays three to four times more slowly than receptors composed of NR1 plus NR2A. Unlike other ionotropic channels, NMDA receptors are 5–10 times more permeable to $Ca^{2+}$ than to $Na^{+}$ or $K^{+}$. NMDA receptor-mediated currents are long lasting compared with the rapidly desensitizing kinetics of AMPA and KA receptor channels. In transgenic mice with forebrain-targeted NR2B overexpression, the normal developmental change in NMDA receptor kinetics was reversed[45]. NR2B subunit expression was observed extensively throughout the cerebral cortex, striatum, amygdala, and hippocampus, but not in the thalamus, brainstem, or cerebellum. In both the ACC and insular cortex, NR2B expression was significantly increased, and NMDA receptor-mediated responses were enhanced[46]. However, NMDA receptor-mediated responses in the spinal cord were not affected. NR2B transgenic and wild-type mice were indistinguishable in tests of acute nociception, but NR2B transgenic mice exhibited enhanced behavioral responses after peripheral injection of formalin. Late phase nociceptive responses but not early responses were enhanced. Furthermore, mechanical allodynia measured in the complete Freund's adjuvant (CFA) model (i.e., behavioral withdrawal responses to a non-noxious stimulus) were significantly enhanced in NR2B transgenic mice. These findings provide the first genetic evidence that forebrain NMDA receptors play a critical role in chronic pain.

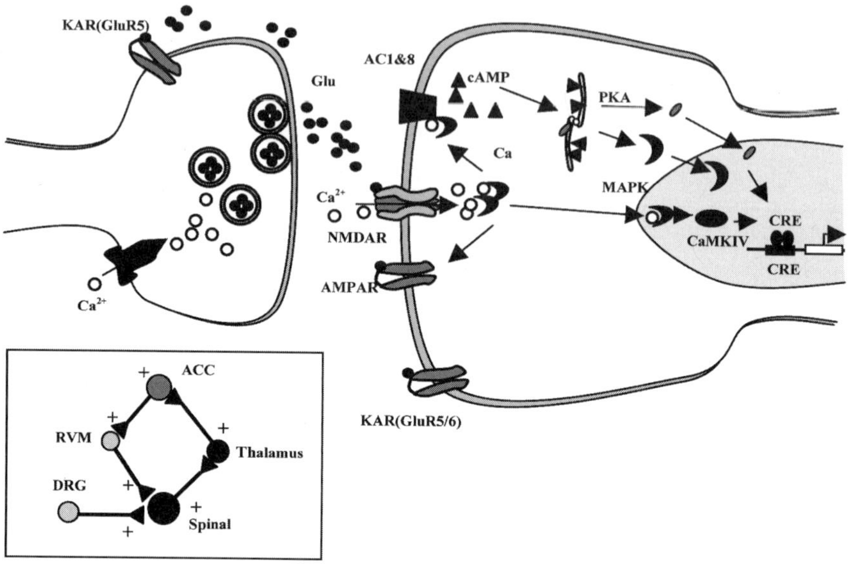


**Figure 25.3.** Signaling Pathways from the Postsynaptic Membrane to the Nuclei that Lead to LTP in the ACC. Neural activity triggered by injury releases glutamate in the ACC synapses. Subsequent to activation of glutamate NMDA receptors, $Ca^{2+}$ binds to CaM and leads to activation of calcium-stimulated ACs, including AC1 and AC8 and $Ca^{2+}$/CaM dependent protein kinases (PKC, CaMKII, and CaMKIV). Activation of CaMKIV, a kinase predominantly expressed in the nuclei, will trigger CaMKIV-dependent CREB. In addition, activation of AC1 and AC8 leads to activation of PKA, and subsequently CREB. CREB and other immediate early genes (e.g., Egr1) in turn activate targets that are thought to lead to more permanent structural changes. Inset: the simplified neuronal network for sensory synaptic transmission, plasticity, and regulation in the central nervous system. DRG: dorsal root ganglion.

Next, we wanted to know if inhibition of NMDA receptor dependent, calcium-stimulated signaling pathways in the ACC would help reduce chronic pain while keeping acute pain sensation intact (critical for animal or human self-protection). AC1 and AC8, the two major CaM-stimulated ACs in the brain, couple NMDA receptor activation to cAMP signaling pathways. In the ACC, strong and homogeneous patterns of AC1 and AC8 expression were observed in all cell layers[47]. Behavioral studies found that wild-type, AC1, AC8, or AC1&AC8 double knockout mice were indistinguishable in tests of acute pain including the tail-flick test, hot-plate test, the mechanical withdrawal responses. However, behavioral responses to a peripheral injection of two inflammatory stimuli, formalin and CFA, were reduced in AC1 or AC8 single knockout mice. Deletion of both AC1 and AC8 in AC1&AC8 double knockout mice produced the greatest reduction in persistent pain[47]. Importantly, microinjection of an AC activator, forskolin, can rescue defects in chronic pain in AC1 and AC8 double knockout mice. Consistently, pharmacological intervention at NMDA receptors as well as cAMP signaling pathways within the ACC also produced inhibitory effects on persistent pain in normal and wild-type animals, supporting the roles of ACC in persistent pain. Microinjection of NMDA receptor antagonists or PKA inhibitors reduced or blocked mechanical allodynia related to inflammation[47]. A recent study showed that persistent pain induced by tissue inflammation or nerve injury was significantly reduced in PSD-93 knockout mice, in part due to the lower level of NR2B expression at the spinal and cortical levels[48].

## 7 CONCLUSIONS

It becomes clear that peripheral sensory inputs can trigger long-term plasticity in the central nervous system, and that plasticity depends on both the intensity (non-noxious versus noxious) and the duration of the stimulus (acute or chronic). In the case of brief noxious stimulation, fear memory can be coded along the sensory transmission pathways (such as the amygdala and the ACC). By altering synaptic transmission through pre- and postsynaptic mechanisms, fearful and painful information is physiologically important for animals and humans to gain knowledge about dangerous information in the environment, and learn to avoid such stimuli and protect themselves in the future. In the case of injury, long-term plastic changes are likely to happen along sensory transmission pathways, including dorsal horn synapses and central synapses in the forebrains. These abnormal enhancements are not beneficial to health, and may contribute to chronic pain and its related fear, anxiety, and depression. There are many reports of possible structural changes after the injury in the central nervous system such as cortical organization. Future studies are clearly needed to understand the synaptic and molecular basis of these permanent structural changes. Identifying molecular signaling molecules and proteins that are involved in plasticity will help us to understand brain mechanisms for memory, and design better medicines to treat patients with different brain diseases.

## 8. REFERENCES


1. Nestler, E.J. (2001) *Nat Rev Neurosci* **2**, 119–128.
2. LeDoux, J.E. (2000) *Annu Rev Neurosci* **23**, 155–184.
3. Hyman, S.E., and Malenka, R.C. (2001) *Nat Rev Neurosci* **2**, 695–703.
4. Zhuo, M. (2002) *Drug Discov Today* **7**, 259–267.
5. Zhuo, M. (2005) *Can J Neurol Sci* **32**, 27–36.
6. Robinson, D.A., Calejesan, A.A., Wei, F., Gebhart, G.F., and Zhuo, M. (2004) *Curr Neurovas Res* **1**, 11–20.
7. Li, P., Kerchner, G.A., Sala, C., Wei, F., Huettner, J.E., Sheng, M., and Zhuo, M. (1999) *Nat Neurosci* **2**, 972–977.
8. Wang, G.D., and Zhuo, M. (2002) *J Neurophysiol* **87**, 732–739.
9. Willis, W.D. (2002) *Brain Res Brain Res Rev* **40**, 202–214.
10. Ikeda, H., Heinke, B., Ruscheweyh, R., and Sandkuhler, J. (2003) *Science* **299**, 1237–1240.
11. Medina, J.F., Christopher Repa, J., Mauk, M.D., and LeDoux, J.E. (2002) *Nat Rev Neurosci* **3**, 122–131.
12. Maren, S. (2001) *Annu Rev Neurosci* **24**, 897–931.
13. Davis, M. (1992) *Annu Rev Neurosci* **15**, 353–375.
14. LaBar, K.S., Gatenby, J.C., Gore, J.C., LeDoux, J.E., and Phelps, E.A. (1998) *Neuron* **20**, 937–945.
15. Phelps, E.A., O'Connor, K.J., Gatenby, J.C., Gore, J.C., Grillon, C., and Davis, M. (2001) *Nat Neurosci* **4**, 437–441.
16. Anderson, A.K., and Phelps, E.A. (2001) *Nature* **411**, 305–309.
17. Romanski, L.M., and LeDoux, J.E. (1993) *Cereb Cortex* **3**, 515–532.
18. Maren, S., Aharonov, G., and Fanselow, M.S. (1997) *Behav Brain Res* **88**, 261–274.
19. Rudy, J.W., Barrientos, R.M., and O'Reilly, R.C. (2002) *Behav Neurosci* **116**, 530–538.
20. Turner, B.H., and Zimmer, J. (1984) *J Comp Neurol* **227**, 540–557.
21. McDonald, A.J. (1998) *Prog Neurobiol* **55**, 257–332.
22. Rogan, M.T., and LeDoux, J.E. (1995) *Neuron* **15**, 127–136.

23. Huang, Y.Y., and Kandel, E.R. (1998) *Neuron* **21**, 169–178.
24. Weisskopf, M.G., and LeDoux, J.E. (1999) *J Neurophysiol* **81**, 930–934.
25. McKernan, M.G., and Shinnick-Gallagher, P. (1997) *Nature* **390**, 607–611.
26. Rogan, M.T., Staubli, U.V., and LeDoux, J.E. (1997) *Nature* **390**, 604–607.
27. Miserendino, M.J., Sananes, C.B., Melia, K.R., and Davis, M. (1990) *Nature* **345**, 716–718.
28. Schafe, G.E., Atkins, C.M., Swank, M.W., Bauer, E.P., Sweatt, J.D., and LeDoux, J.E. (2000) *J Neurosci* **20**, 8177–8187.
29. Schafe, G.E., and LeDoux, J.E. (2000) *J Neurosci* **20**, RC96.
30. Tsvetkov, E., Carlezon, W.A., Benes, F.M., Kandel, E.R., and Bolshakov, V.Y. (2002) *Neuron* **34**, 289–300.
31. Rodrigues, S.M., Schafe, G.E., and LeDoux, J.E. (2004) *Neuron* **44**, 75–91.
32. Rodrigues, S.M., Schafe, G.E., and LeDoux, J.E. (2001) *J Neurosci* **21**, 6889–6896.
33. Bauer, E.P., Schafe, G.E., and LeDoux, J.E. (2002) *J Neurosci* **22**, 5239–5249.
34. Wei, F., Qiu, C.S., Liauw, J., Robinson, D.A., Ho, N., Chatila, T., and Zhuo, M. (2002) *Nat Neurosci* **5**, 573–579.
35. Schafe, G.E., Bauer, E.P., Rosis, S., Farb, C.R., Rodrigues, S.M., and LeDoux, J.E. (2005) *Eur J Neurosci* **22**, 201–211.
36. Faber, E.S., Delaney, A.J., and Sah, P. (2005) *Nat Neurosci* **8**, 635–641.
37. Huang, Y.Y., Martin, K.C., and Kandel, E.R. (2000) *J Neurosci* **20**, 6317–6325.
38. Rumpel, S., Ledoux, J., Zador, A., and Malinow, R. (2005) *Science* **308**, 83–99.
39. Ko, S., Zhao, M.G., Toyoda, H., Qiu, C.S., and Zhuo, M. (2005) *J Neurosci* **25**, 977–984.
40. Wu, L.J., Zhao, M.G., Toyoda, H., Ko, S.W., and Zhuo, M. (2005) *J Neurophysiol* **94**, 1805–1813.
41. Liauw, J., Wang, G.D., and Zhuo, M. (2003) *Sheng Li Xue Bao* **55**, 373–380.
42. Tang, J., Ko, S., Ding, H.K., Qiu, C.S., Calejesan, A.A., and Zhuo, M. (2005) *Mol Pain* **1**, 6.
43. Zhao, M.G., Toyoda, H., Lee, Y.S., Wu, L.J., Ko, S.W., Zhang, X.H., Jia, Y., Shum, F., Xu, H., Li, B.M., Kaang, B.K., and Zhuo, M. (2005) *Neuron* **47**, 859–872.
44. Frankland, P.W., Bontempi, B., Talton, L.E., Kaczmarek, L., and Silva, A.J. (2004) *Science* **304**, 881–883.
45. Tang, Y.P., Shimizu, E., Dube, G.R., Rampon, C., Kerchner, G.A., Zhuo, M., Liu, G., and Tsien, J.Z. (1999) *Nature* **401**, 63–69.
46. Wei, F., Wang, G.D., Kerchner, G.A., Kim, S.J., Xu, H.M., Chen, Z.F., and Zhuo, M. (2001) *Nat Neurosci* **4**, 164–169.
47. Wei, F., Qiu, C.S., Kim, S.J., Muglia, L., Maas, J.W., Pineda, V.V., Xu, H.M., Chen, Z.F., Storm, D.R., Muglia, L.J., and Zhuo, M. (2002) *Neuron* **36**, 713–726.
48. Tao, Y.X., Rumbaugh, G., Wang, G.D., Petralia, R.S., Zhao, C., Kauer, F.W., Tao, F., Zhuo, M., Wenthold, R.J., Raja, S.N., Huganir, R.L., Bredt, D.S., and Johns, R.A. (2003) *J Neurosci* **23**, 6703–6712.

# Part VI

## SYNAPTOGENESIS AND BRAIN DISORDERS

# 26

# RELEVANCE OF PRESYNAPTIC PROTEINS TO NEUROPSYCHIATRIC DISORDERS

Alasdair M. Barr*, Clint E. Young*, Ken Sawada†, and William G. Honer*

## 1. SUMMARY

Schizophrenia is a psychiatric disorder that likely arises from a complex series of interactions between genetic and environmental factors, resulting in altered neural connectivity. Consistent with this position, an extensive body of research has demonstrated that the molecular machinery that regulates synaptic activity is altered in post mortem brain tissue in schizophrenia. In particular, the soluble *N*-ethylmaleimide-sensitive factor attachment protein receptor (SNARE) protein synaptosomal-associated protein-25 (SNAP-25) and the SNARE-interacting proteins complexin I and II are implicated in the pathophysiology of this disorder. These proteins regulate synaptic neurotransmission and are important for plasticity-associated changes in the brain. Recent evidence from our laboratory has demonstrated that levels of these proteins are decreased in the hippocampus in schizophrenia, and importantly, these decreases are significantly associated with cognitive impairment. These findings represent the first description significantly linking synaptic proteins to cognitive function in schizophrenia. Preclinical studies provide evidence that SNAP-25 and complexins may be implicated in the pathophysiology of additional psychiatric and neurological disorders, as well as in the therapeutic efficacy of drugs used in the treatment of these disorders.

*Department of Psychiatry, Center for Complex Disorders, University of British Columbia, Vancouver General Hospital Research Pavilion, 828 West 10th Avenue, Vancouver, BC, Canada V5Z 1L8; honer@interchange.ubc.ca
†Department of Neuropsychiatry, Kochi Medical School, Kochi, Japan

## 2. INTRODUCTION

Schizophrenia and related forms of psychosis are among the most severe, persistent, and debilitating illnesses affecting young people. The worldwide prevalence of schizophrenia (narrowly defined, i.e., with a nuclear syndrome consisting mainly of first-rank systems at onset) is almost 1%, and occurs with a similar incidence across most cultures and geographical regions. The World Health Organization has reported that schizophrenia is the fourth leading cause of worldwide disability. In 1996, the financial costs of schizophrenia in Canada were estimated at $2.35 billion, and in the United States the estimated financial burden exceeded that for all forms of cancer[1]. The costs to the individual and their family in schizophrenia are also among the most exacting of those in all mental illnesses. Schizophrenia is typically characterized by a lengthy prodromal phase that may disrupt the teenage years, while full onset of florid symptoms occurs in early adulthood and persists for the lifetime, rendering the individual unable to function within society without therapeutic intervention. Incidence rates of suicide are also significantly greater than for the general population.

Considerable progress has been made in recent years in identifying molecular substrates that may contribute to the etiology and expression of schizophrenia. The post mortem brain tissue in schizophrenia is typically bereft of neuronal loss and there is absence of astrogliosis; nor is there any substantial evidence for apoptotic processes, providing strong evidence that schizophrenia is not a neurodegenerative disorder. Current theories thus emphasize the role of neurodevelopmental factors in the etiology of this disorder, and significant risk factors for schizophrenia include early life events, such as history of prenatal and birth complications, developmental abnormalities, urban place of birth, exposure to viruses and childhood social interaction. Evidence for markers of altered neurodevelopment in post mortem brain tissue has been widely reported, although most neuropathological signs are relatively modest. A greater burden of evidence suggests that the physiological basis for the disorder may be one of altered neural connectivities, in which the underlying pathophysiology of the condition is manifested at a synaptic level in the brain by abnormal "miswiring."

Dissection of the cellular and molecular basis of abnormalities of neural connectivity in schizophrenia has been particularly fruitful at the level of the synapse, where presynaptic proteins involved in the regulation of neurotransmitter release have shown consistent changes. These molecules, many of which are described in detail by other authors in this book, are essential for neural plasticity and presumably underlie much of the brain's capacity for basic cognitive processes, such as learning and memory. In the present chapter, we discuss the evidence that presynaptic proteins are altered in the brain in schizophrenia. This discussion also evaluates in detail recent data indicating that three specific presynaptic proteins, SNAP-25, Complexin I (Cx I), and Complexin II (Cx II), are not only altered in schizophrenia, but also may be specifically related to the types of cognitive impairment that are a characteristic symptom of this disorder.

## 3. SCHIZOPHRENIA

### 3.1. Pathophysiology of Schizophrenia

Schizophrenia is a complex psychiatric disorder, which likely arises from an elaborate, and as yet unknown, series of interactions between genetic and environmental factors. Schizophrenia clearly displays a "progressive" component, in which clinical manifestations, including cognitive deficits[2], deteriorate with time, particularly during adolescence and early adulthood. Absence of neuro-degenerative pathology implicates neurodevelopmental factors in the etiology of psychotic disorders, and functional neuroimaging studies indicate abnormalities of connectivity, including altered patterns of correlation between regions for brain blood flow or metabolism[3], in addition to abnormal cerebral activation following cognitive challenges[4]. Unlike the more advanced stages of neurodegenerative diseases, schizophrenia has been associated with relatively subtle, and often inconsistent, changes in brain architecture or pathology. This has led to the hypothesis that schizophrenia is primarily a disorder of "neural connectivity" or altered neurotransmission. Post mortem studies have indicated that molecules at the level of the synapse are likely substrates for altered connectivity and transmission .We have recently reviewed in detail the nature of synaptic deficits in schizophrenia[5]. Briefly, 16 of 19 studies of presynaptic proteins demonstrated differences in schizophrenia, as did eight of 12 studies of mRNA. Most reports indicated decreased synaptic proteins. More recent studies suggest the importance of networks of interacting presynaptic proteins; for example, using a cDNA screening strategy, reduced mRNAs for the synaptic protein synapsin II was observed consistently in schizophrenia[6], while mRNAs coding for other presynaptic proteins were also affected, but less consistently. There is little information concerning functional implications of altered levels of presynaptic proteins. In cingulate and temporal cortex, increased synaptophysin immunoreactivity was correlated with increased severity of negative symptoms[7], while studies of Alzheimer's disease noted that hippocampal reductions in synaptophysin and syntaxin were correlated with cognitive impairment[8]. Current evidence thus suggests that presynaptic proteins are necessary for normal cognitive function, but the identity of which proteins, in which brain regions, and on which cognitive indices, remains at present poorly understood.

### 3.2. Cognitive Deficits in Schizophrenia

Recent studies using powerful statistical techniques indicate that the sequelae of schizophrenia can be categorized into multiple separate factors, typically including a cluster characterized by deficits in cognition. This latter group of deficits, which are strongly associated with neuropsychological impairment, has recently been highlighted by major funding groups in the USA, including the National Institute of Mental Health, by programs such as the Measurement and Treatment Research to Improve Cognition in Schizophrenia (MATRICS) cooperative initiative. Currently, efforts are underway to allow US government approval for the design of pharmacotherapies that treat specifically cognitive indications. Cognitive deficits reflect a diverse loss of function, so that it may be convenient to subcategorize them: it has been proposed that the three main types of cognitive deficit observed in schizophrenia occur in memory, attention and executive function[9].

Recent adoption of standardized, operational procedures, such as the tasks performed in the Cambridge Neuropsychological Test Automated Battery (CANTAB), has provided accurate data about the precise types of cognitive deficits that occur in psychotic disorders. CANTAB has become a popular standardized test procedure due to its ability to decompose a set of complex tasks commonly used in clinical assessment into their cognitive elements[10]. Furthermore, many CANTAB tests are based on data extrapolated from animal studies, allowing for closely homologous behavioral tests to be used between humans and rodents; rigorous evaluation of the tasks has confirmed that cognition in both humans and rodents involves similar brain regional activations and sensitivities to pharmacological manipulations. Cognitive deficits in schizophrenia include memory dysfunction, with problems in both working and long-term memory[11]. Attentional deficits in schizophrenia include those at both a preconscious level, in tasks such as prepulse inhibition (PPI) to a startling stimulus[12], as well as at a conscious level, in tasks such as sustained vigilance[13]. Tasks of executive function are performed especially poorly in schizophrenia, evident in tests such as the Wisconsin Card Sort Task and the "Tower of London" task in CANTAB[14]. Animal and human studies indicate that forebrain regions, including the frontal cortex, hippocampus, and basal ganglia, are particularly important for the performance of cognitive tasks of memory, attention, and executive function[15,16].

As part of the ongoing research in our laboratory, we are committed to understanding in greater detail the relationship between synaptic changes in post mortem schizophrenia and premorbid cognitive impairment. By measuring levels of presynaptic proteins in brain regions that are essential for normal cognitive function, we have recently demonstrated significant associations between altered levels of three presynaptic proteins, complexins Cxs I and II, and SNAP-25, and cognitive impairment. Furthermore, rodents trained in different cognitive tasks display plasticity-associated regional changes in levels of presynaptic proteins. We now discuss in detail the relevance of each of these proteins to schizophrenia.

## 4. COMPLEXINS AND SCHIZOPHRENIA

### 4.1. Presynaptic Localization of Complexins

The Cxs (also previously known as synaphins) are small, neuronal cytosolic proteins that are found at highest concentrations in the presynaptic nerve terminals, where they interact with the soluble NSF-attachment protein receptor (SNARE) complex[17]. Complexins interact directly with SNAREs, which form a thermodynamically stable complex that is hypothesized to drive fusion of vesicular and presynaptic membranes. Molecular studies have confirmed that Cxs are the only neuronal proteins known to bind tightly to the SNARE complex, in a 1:1 stoichiometry along the groove between the transmembrane regions of syntaxin and synaptobrevin (VAMP), possibly facilitating the interaction between these two proteins[18]. Complexins are necessary for $Ca^{2+}$-dependent neurotransmission, and in this manner act as positive regulators of synaptic vesicle exocytosis.

### 4.2. Expression of Complexins in the Brain

The precise regional localization of Cxs in the brain continues to be a matter of debate. A number of studies have published *in situ* hybridization data showing that Cx I is predominantly expressed in inhibitory interneurons, whereas Cx II is expressed mainly in glutamatergic neurons[19]. By contrast, other groups have reported that both glutamatergic and GABAergic neurons express high levels of Cx II, and that the differential expression of Cx isoforms is unrelated to the identity of the neurotransmitter; rather, it is hypothesized that isoform expression is more closely related to anatomical circuitry[20]. A possible resolution to this conundrum may be obtained by investigating the localization of Cxs at the level of the synapse. An initial electron microscopic study determined that Cx I was most prominent in the axon terminals of axosomatic synapses, whereas Cx II exhibited greater immunoreactivity in the axon terminals of axospinous and axodendritic synapses[21]. These results indicate that Cx I is more common at inhibitory synapses, while Cx II is more common at excitatory synapses, regardless of whether the different isoforms are expressed by different types of cell. We have recently provided additional evidence to support this generalization. Our quantitative immunohistochemistry studies of the hippocampus in both rats and humans clearly demonstrate that at a macroarchitectural level, Cx I is more strongly localized with markers of inhibitory terminals, such as the GABA transporter, while Cx II is more strongly localized with markers of excitatory terminals, such as the glutamate transporter[22].

Confocal microscopy of the rat hippocampus (Figure 26.1; Colorplate 12) also indicates that in general, Cx I is colocalized with the vesicular GABA transporter, while Cx II colocalizes with the vesicular glutamate transporter.

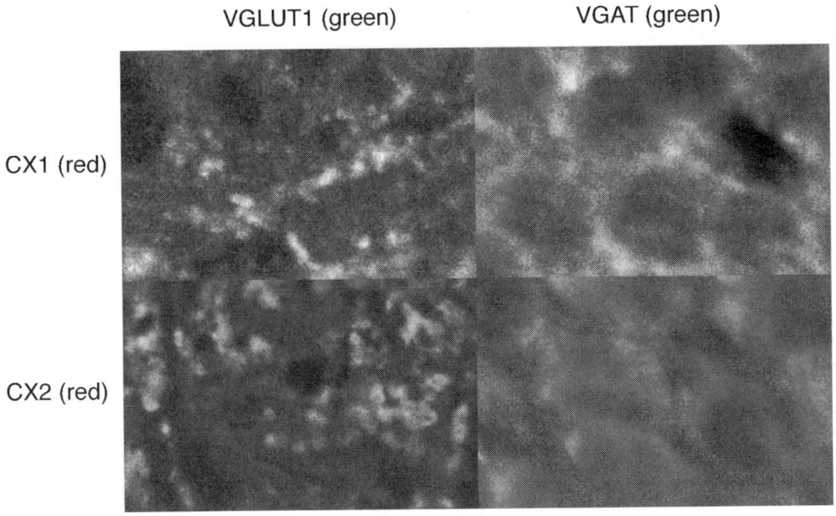


**Figure 26.1.** Confocal Microscopy of Presynaptic Proteins in Rat Hippocampus. All images from the granule cell layer with the exception of the lower left panel obtained from the mossy fiber terminal zone. These images largely support the colocalization of complexin I with inhibitory terminals and complexin II with excitatory terminals. However, in other subfields of the hippocampus this relationship was not as consistently observed. *Abbreviations:* Cx1 and Cx2, complexins 1 and 2; vGLUT, vesicular glutamate transporter 1; vGAT, vesicular GABA transporter. See Colorplate 12.

### 4.3. Physiological Functions of Complexins

At present, the physiological functions of Cxs are not well understood. While the presumed role of Cxs is to regulate neurotransmission, there is a surprising amount of conflicting evidence as to whether they are facilitatory or inhibitory to neurotransmission. One earlier study reported that injection of a peptide from the central, syntaxin-binding domain of Cxs into squid giant presynaptic terminals inhibited neurotransmitter release at a late stage of synaptic vesicle exocytosis[23]. Consistent with these findings, injection of anti-complexin II antibody into mature presynaptic neurons of Aplysia buccal ganglia caused a stimulation of neurotransmitter release, suggesting that Cx II is inhibitory to neurotransmitter release. This hypothesis was further supported by evidence that injection of recombinant Cx II caused depression of neurotransmitter release from nerve terminals[24]. However, these studies would appear to be in contrast to the data from other research groups, who have shown that there is a reduced $Ca^{2+}$ sensitivity of neurotransmitter release in mutant genetic mice lacking both Cxs I and II, although evoked EPSC amplitudes and spontaneous neurotransmitter release were normal in single knockouts[25]. Furthermore, Cx II mice display frequency-dependent deficits in long-term potentiation (LTP) of the hippocampus (see below), suggestive of a facilitatory role for this isoform. Insight into these contrasting sets of data may be provided by Morton and colleagues, who have demonstrated that PC12 cells overexpressing mutant huntingtin protein exhibit decreased levels of Cx II and concomitant decreases in neurotransmitter release. Increased expression of Cx II in PC12 cells that co-expressed the mutant huntingtin protein partially rescued the phenotype and increased neurotransmission, whereas increased Cx II expression in wild-type PC12 cells decreased neurotransmission[26]. These combined data suggest that Cx II may be required for neurotransmission when at optimal cytosolic concentrations, but along an inverted-∪ shaped gradient, so that decreases or increases in levels serve to inhibit neurotransmission. Substantially less research has been undertaken to define the physiological role of Cx I in regulation of neurotransmission.

Some degree of insight into the physiological role of Cxs has been provided by the generation of mice that have been genetically engineered for a deletion of the two isoforms. While the combined loss of both Cxs is perinatally lethal[25], the individual loss of either Cx results in viable offspring that can survive to puberty. Complexin I knockout mice display a more striking behavioral phenotype than Cx II mutant mice. In the first mention of the Cx I knockout animals, it was reported that these mutants exhibit a strong ataxia, were unable to reproduce, experience sporadic seizures, and die within 2–4 months after birth[25]. A more recent study, in which the behavioral phenotype of these mice was characterized extensively, revealed that Cx I knockout mice can survive into full adulthood if fed under special conditions. These mice display pronounced motor deficits, ataxia, reduced neuromuscular strength, and a resting tremor. In addition, there are also deficits in complex behaviors and increased reactivity in affective-related tasks[27]. By contrast, Cx II knockout mice display a relatively subtle behavioral phenotype. At present, there are two separate lines of Cx II mutant mice, both of which display seemingly normal development through to puberty[25,28]. However, behavioral and physiological deficits become more pronounced with age. The "Kochi" line of mice exhibit normal reproductive behavior, although hippocampal neurons from these mice display plasticity-associated deficits, indicated by a loss of LTP by high-frequency stimulation, recorded in the hippocampal CA1 area, whereas ordinary neural transmission remains normal[28]. In the most complete behavioral characterization of Cx II knockout mice by Morton and colleagues, a complex pattern of deficits was observed that were best described as relating to "higher function"[29]. These included age-dependent loss of social interaction, self-care, investigative activity and increased cognitive impairment, especially on spatially mediated tasks and procedures that required reversal learning. This discrete loss of complex behavior, in the absence of overt motor impairments, bears an interesting parallel to human neuropsychiatric disorders, such as schizophrenia and Huntington's disease.

We have recently demonstrated that Cxs are associated with cognitive-related synaptic plasticity in rodents. To confirm that Cxs in the hippocampus and frontal cortex were relevant to cognition, we conducted a series of experiments in rats to examine plasticity-associated changes in Cxs during learning and memory tasks. Animals were trained on either working or reference memory tasks and regional levels of Cxs were measured with quantitative immunohistochemistry. In the hippocampus, we observed that learning increased levels of Cx II and Cx II:I ratio throughout most of the hippocampus[22]. These findings imply that increases in Cx II and a relative increase in excitatory/inhibitory terminals in hippocampus may be important for working and reference memory.

### 4.4. Preclinical Data on Complexins

Converging data from preclinical animal studies implicate Cxs as molecules of relevance to neuropsychiatric disorders. We and others have shown previously that neurodevelopmental rodent models of schizophrenia, similar to the disorder itself, may be characterized by regional alterations in presynaptic proteins[30]. In a neurodevelopmental model of schizophrenia based on exposure to variable prenatal stress, it was noted that this manipulation decreased levels of Cx I in the frontal pole[31], although cognition was not assessed in these animals. We have recently observed decreased levels of both Cxs in the frontal cortex of rats in a model of prenatal ethanol exposure[32]. In this paradigm, *in utero* exposure to high blood alcohol levels of pregnant dams that voluntarily consume ethanol results in a wide range of neurohormonal and cognitive deficits in the offspring. By measuring levels of presynaptic proteins in the frontal cortex and hippocampus, where we had previously observed cognitive-related plastic changes in Cxs, we noted a selective loss of both Cx I and II in the frontal cortex. There was no effect of prenatal ethanol exposure on Cxs in the hippocampus, and decreased levels of Cxs in the frontal cortex were protein specific, as there was no effect of ethanol exposure on the presynaptic protein synaptophysin, which provides a global marker of presynaptic terminal density. The absence of changes in levels of hippocampal Cxs is somewhat surprising, given the well-characterized memory deficits in this paradigm and previous research associating Cxs with memory tasks. However, a possible explanation for the failure to observe decreases in hippocampal complexins is that while basal levels of the proteins remained unaltered, prenatal ethanol exposed rats may display specific plasticity-associated deficits, which would only be apparent after extended training in cognitive paradigms.

Interestingly, Cx II has also been implicated in Huntington's disease. As mentioned above, expression of mutant huntingtin protein – whose polyglutamine expansion within the N-terminus is a contributing factor to the disease – in PC12 cells decreases levels of Cx II and reduces exocytosis. Overexpression of Cx II in these cells reverses the decreases in neurotransmitter release, indicating a specific link between huntingtin and Cx II[26]. The progressive, neurocognitive deficits that are evident in Cx II knockout mice also exhibit a strong degree of homology with the neuropsychiatric symptoms of the earlier stages of Huntington's disease. In further support of this link, it was noted that there is a highly selective and consistent loss of Cx II in the R6/2 mouse model of Huntington's disease[33], which is correlated with progressive neuropathology. In this study, multiple presynaptic proteins were measured throughout the brain, and only the two presynaptic proteins Cx II and α-SNAP were reduced compared to wild-type mice. The loss of Cx II was significantly greater, and occurred significantly earlier, than with α-SNAP. In human postmortem brain, two independent research groups have reported a selective loss of Cx II. In one study, Cx II exhibited a marked reduction in the striatum of Huntington's disease[34], while another group reported a selective loss of Cx II in the frontal cortex of early pathological-grade Huntington's disease[35]. It is unknown why Cx II should be selectively decreased in this disorder, but if Cx II is predominantly localized to excitatory, glutamatergic terminals, there may be a link to neurotoxicity. There is less evidence of a role for Cx I in neurological disorders, although one recent study reported increased levels of Cx I in the substantia nigra of Parkinson's disease[36].

An alternate approach to understanding a possible link between Cxs and neuropsychiatric disorders is based on determining the effects of pharmacotherapies on levels of these proteins. For example, research from our laboratory has demonstrated that schizophrenia is associated with a loss of the presynaptic protein SNAP-25 (see below), while treatment of rats with the antipsychotic drugs haloperidol or chlorpromazine produces increases in levels of SNAP-25, opposite the direction of change in the disorder. With regard to Cxs, there are mixed data concerning the effects of antipsychotic drugs. We have previously reported that there is no effect in rats of treatment with the neuroleptic haloperidol for 21 days on levels of Cx proteins, measured by ELISA and quantitative immunohistochemistry, in the frontal cortex or hippocampus[22,37]. A study in which rats were administered haloperidol for 28 days according to a "depot" schedule determined that there was no effect of this drug on levels of Cx II mRNA in multiple brain regions. However, haloperidol significantly decreased levels of Cx I mRNA in the medial prefrontal cortex, nucleus accumbens and ventral tegmental area[38]. A more extensive study which evaluated the effects of both typical and atypical antipsychotics on Cxs reported diverse findings. In this study, levels of Cx II mRNA were increased in the frontoparietal cortex after treatment with the typical antipsychotic chlorpromazine, while levels of Cx I mRNA were elevated by the atypical antipsychotic olanzapine in the dorsolateral striatum and frontoparietal cortex; both olanzapine and haloperidol decreased the Cx II:I mRNA ratio in these latter two brain regions. These combined findings reveal apparently conflicting results, and further studies will be required, using more comparable doses and durations of drug administration between laboratories. Interestingly, a recent study determined the effects of different classes of antidepressant drugs on brain levels of Cxs[39]. The authors of this study noted that Cx I was induced only in habenular nuclei after treatment with fluoxetine. However, Cx II was significantly increased throughout multiple hippocampal subregions following treatment with the tricyclic antidepressant desipramine and the monoamine oxidase inhibitor tranylcypromine, but not with the selective serotonin reuptake inhibitor fluoxetine. These findings warrant further study of a possible link between Cxs and affective disorders.

## 4.5. Clinical Data on Complexins

Given the physiological role of Cxs in higher cognitive processes, and the convincing preclinical evidence for Cxs in multiple disease models, a number of studies have investigated a link between Cxs and mental illness. As described above, two independent groups have demonstrated a selective loss of Cx II in the frontal cortex and striatum in the earlier stages of Huntington's disease[34,35]. Our research group has primarily been interested in a role for the Cxs in psychiatric disorders. We have shown previously that Cxs are lower in both schizophrenia and MDD in frontal cortex. Using enzyme-linked immunoadsorbent assay (ELISA), we confirmed that Cx I immunoreactivity was significantly decreased in schizophrenia and MDD, and the ratio of Cx II:I was significantly increased by 32–34%[37]. This finding is particularly relevant to a putative loss of functional inhibition in frontal cortex in psychiatric disorders, as a selective loss of Cx I (inhibitory terminals) is consistent with reduced inhibitory synapses formed by GABAergic interneurons in prefrontal cortex, which are hypothesized to contribute to cognitive deficits.

More recently, we have investigated changes in levels of Cxs in the hippocampus in schizophrenia. Initial studies by other groups demonstrated that levels of both Cx I and Cx II mRNA were lower in hippocampus in schizophrenia, with a relatively greater loss of Cx II mRNA; protein studies found a selective loss of Cx II[40]. We confirmed these findings, and reported that Cx II was lower in hippocampal subregions of individuals with cognitive impairment than those not impaired, resulting in a significantly lower Cx II:I ratio in the hippocampus in cognitively impaired schizophrenics[22]. These findings are especially important for two reasons. Firstly, the confirmation of findings from a different laboratory, using different post mortem samples, indicates that changes in Cxs in the hippocampus are one of the most consistent molecular changes in schizophrenia. Even changes in excitatory and inhibitory markers, such as GABAergic or glutamatergic terminals, show inconsistent findings in the hippocampus (for example, two recent papers found contradictory changes of vGLUT1 in schizophrenia hippocampus[41,42]). The second important conclusion from our findings is that the Cxs, and the Cx II:I ratio, are the first synaptic changes to be confirmed as being associated specifically with cognitive deficits in schizophrenia. These links with cognitive impairment – including memory deficits – did not extend to other presynaptic proteins that we have measured in post mortem hippocampus from the same samples, including synaptophysin, suggesting a selective link between Cxs and cognitive dysfunction. These findings indicate that although Cxs are generally decreased in psychiatric disorders, the relative effects on Cx I versus Cx II depend on the specific brain region, as well as the type of disorder. It is therefore of interest that a recent study has confirmed a significant genetic link between a Cx II haplotype and schizophrenia in a Korean sample[43]. Currently, priorities for the study of Cxs and their relevance to psychiatric disorders include a better understanding of the functional role of Cxs in cognitive plasticity, which will help to provide insights into the nature of disease-related changes, as well as pharmacotherapy-induced changes.

## 5. SNAP-25 AND SCHIZOPHRENIA

### 5.1. Presynaptic Localization of SNAP-25

SNAP-25 is one of the three core presynaptic proteins that combine to form the SNARE complex. Structural members of the SNAP-25 protein family contribute two of the four alpha helices that compose the SNARE complex, which is necessary for $Ca^{2+}$-triggered exocytic transmitter release. The first characterized SNAP-25 proteins were the two isoforms SNAP-25A and SNAP-25B, which represent highly homologous proteins that differ by only nine amino acids[44]. SNAP-23 represents another member of this protein family, which shares an approximately 60% identity with SNAP-25, although the cellular distribution of the former protein is substantially more ubiquitous than the latter[45]. There is evidence that SNAP-23 may also be involved in the positive regulation of neurotransmitter release[46], unlike SNAP-29 – another member of the protein family – which appears to act as a negative modulator of neurotransmitter release, possibly by slowing recycling of the SNARE-based fusion machinery and synaptic vesicle turnover[47].

Current theories of vesicular membrane fusion emphasize the role of SNARE proteins. According to these models, syntaxin 1, synaptobrevin 2, and SNAP-25 regulate neuroexocytosis by forming a complex that forces vesicle and plasma membranes together, allowing the transmitter contained in the synaptic vesicle to enter the synaptic cleft (for review, see ref. 48). The SNARE proteins form a highly stable complex *in vitro* from helices in their cytoplasmic domains, with the formation of a 4-helix bundle - with 1 helix donated each by syntaxin 1 and synaptobrevin 2, and 2 helices donated by SNAP-25. The SNARE proteins are also specific substrates for the proteolytic action of the *Clostridial* neurotoxins, which inhibit neurotransmitter release.

### 5.2. Physiological Functions of SNAP-25

Data from *in situ* and immunohistochemical studies indicate that SNAP-25 displays a regional pattern of expression in the brain that changes developmentally. For example, some regions, such as the olfactory bulbs, exhibit high levels of SNAP-25 mRNA expression early during development, which decline to negligible levels in adulthood, whereas other brain regions, such as the neocortex, display developmental increases and persistent adult neuronal immunoreactivity[49]. The cellular distribution of SNAP-25 may also change during development, as in the striatum, where SNAP-25 in the caudate nucleus is initially concentrated in axons, but subsequently localized in presynaptic regions of these axons[49]. During development, SNAP-25 plays a key role in neurite elongation and axonal growth. Highest levels of SNAP-25 expression in adulthood are observed in neurons of the neocortex, hippocampus, piriform cortex, anterior thalamic nuclei, pontine nuclei, and granule cells of the cerebellum[50].

The primary function of SNAP-25 in the brain, in adulthood, is the regulation of calcium-dependent exocytosis of neurotransmitter, although it should be noted that SNAP-25 also regulates exocytosis in non-neuronal systems, such as insulin secretion by pancreatic beta cells[51], and histamine release by gastric enterochromaffin-like cells[52]. The exocytic activity of SNAP-25 is regulated by protein kinase-dependent phosphorylation at a number of different sites[53]. In addition to its role as part of the SNARE complex, SNAP-25 also displays a number of direct protein–protein interactions, including binding to voltage-gated $K^+$ and $Ca^{2+}$ channels[54,55], whose functional implications are continuing to be determined. While SNAP-25 serves as one of the three core proteins of the SNARE complex, details regarding its physiological activity during exocytosis indicate a more complex role than may be initially predicted. Recent evidence demonstrates that in the large, glutamatergic calyx of Held presynaptic terminal from rats, loss of SNAP-25 produces a graded loss of calcium sensitive transmitter release, unlike loss of syntaxin or synaptobrevin, which generates an "all-or-nothing" block of release[56]. Thus, while SNAP-25 remains essential for most calcium-dependent neurotransmitter release, it is likely, based on the findings described above, that our understanding of this molecule will continue to evolve.

Given the ubiquitous nature of SNAP-25, the protein is presumably of functional importance for numerous brain processes. Loss of SNAP-25, through genetic engineering of SNAP-25 null mutant mice, results in embryonic lethality[57]. However, heterozygous SNAP-25 mutant mice are robust and fertile. Insights into the role of SNAP-25 in behavioral processes have been provided by the heterozygous *Coloboma* mutant mouse (for review, see ref. 58). This mutant, which exhibits a contiguous chromosomal deletion encompassing 1.1 to 2.2 cM that includes the SNAP-25 gene, displays discrete physiological and behavioral deficits. The more obvious of these include spontaneous locomotor hyperactivity and head bobbing, and heterozygous *Coloboma* mice also exhibit delays in some tests of complex motor skills, such as righting reflex and bar holding. Locomotor hyperactivity in these mice has been hypothesized to result from regional alterations in monoamine neurotransmission, including decreased depolarization-evoked release of dopamine and serotonin in the dorsal striatum, and increased norepinephrine release in the striatum and nucleus accumbens[59]. Hippocampal-related deficits in heterozygous *Coloboma* mice include decreased induction of theta rhythmic activity after tail pinch, and reduced LTP in the dentate gyrus following high-frequency stimulation of the perforant path.

Evidence of a role for SNAP-25 in neural plasticity has been obtained from additional studies of LTP in nonmutant rats. Levels of both isoforms of SNAP-25 mRNA were increased 2 h after the induction of LTP in granule cells of the dentate gyrus following high frequency stimulation of the perforant path *in vivo*[60]. Hippocampal LTP was also associated with a significant increase in levels of SNAP-25 phosphorylation at multiple sites on the protein[61]. More recently, SNAP-25 was detected as one gene differentially expressed in the hippocampus in a learning and memory study in rats[62]. This same research group demonstrated that antisense oligonucleotides for SNAP-25 that were infused into the CA1 region of the hippocampus impaired long-term contextual fear memory and spatial memory, as well as interfered with the LTP of synaptic transmission in the CA1 region.

## 5.3. Preclinical Data on SNAP-25

Similar to findings regarding the Cxs, preclinical data linking SNAP-25 to psychiatric and neurological disorders are obtained from both pharmacological and *in vivo* studies. SNAP-25 is expressed at relatively high levels in the locus coeruleus, where it may play an important role in the regulation of norepinephrine release. Consistent with evidence that major depressive disorder may be due, in part, to overactivity of the locus coeruleus, we have shown previously that different modalities of antidepressant treatments in rats decrease SNAP-25 mRNA in this brain region[63]. More extensive investigation has measured the effects of antipsychotic drugs on levels of SNAP-25 in different brain regions in rodents. One study reported that there was no effect of chronic treatment with haloperidol on levels of SNAP-25 in multiple brain regions that are strongly innervated by dopamine neurons, including the prefrontal cortex, nucleus accumbens, striatum, substantia nigra, and ventral tegmental area[64]. Recently, we have assessed the effects of treatment for 21 days with the typical antipsychotic drugs haloperidol, chlorpromazine, and trifluoperazine on levels of SNAP-25 in the trisynaptic pathway of the hippocampus[65]. Levels of SNAP-25 were measured by quantitative immunohistochemistry. Our findings revealed that both haloperidol and chlorpromazine increased SNAP-25 throughout the hippocampus, with greatest effects for haloperidol-treated rats in the mossy fiber region, while chlorpromazine-treated rats exhibited largest increases of SNAP-25 in the mossy fiber and Schaffer collateral regions. However, these effects only remained significant for the haloperidol-treated rats following application of Bonferroni correction for multiple comparisons. These data, which are the first description of effects of treatment with antipsychotic drugs on levels of SNAP-25 within the hippocampus in rodents, suggest that SNAP-25 may be increased in the hippocampus following treatment with specific neuroleptics. This is particularly interesting, as schizophrenia has been associated with reduced levels of SNAP-25 in the hippocampus (see below).

Animal models have indicated a link between SNAP-25 and psychiatric disorders. We previously reported that levels of SNAP-25 were significantly greater in the amygdala of adult rats that were reared in isolation from weaning[66]. This experimental manipulation is known to induce a range of behavioral alterations, such as locomotor hyperactivity and deficits in PPI, that resemble homologous symptoms of psychotic disorders[30]. As isolation-reared rats also display increased anxiety, and the behavioral deficits in this paradigm have been associated with alterations in monoamine neurotransmission[67], our findings may indicate a novel substrate for anxiety in this model. The behavioral deficits evident in the heterozygous *Coloboma* mutant mouse have been hypothesized to model various symptoms of psychiatric disorders. The hyperactivity associated with the mutant is decreased by treatment with low doses of *d*-amphetamine that are sufficient to induce increased activity in wild-type mice[68]. Thus, it has been hypothesized that the heterozygous *Coloboma* mutant mouse may represent a valid animal model of attention deficit hyperactivity disorder (ADHD). However, many of the behavioral sequelae in this SNAP-25 haploinsufficient mutant are also of relevance to schizophrenia, and it is of interest that certain symptoms of ADHD, including hyperactivity and aggressiveness, may be treated effectively by low doses of both haloperidol and chlorpromazine[69] – the same drugs that increased levels of SNAP-25 in the hippocampus in our study.

## 5.4. Clinical Data on SNAP-25

Genetic studies have observed significant associations between SNAP-25 polymorphism haplotypes and ADHD[70]. These findings are consistent with the abnormal behavioral phenotype of the heterozygous *Coloboma* mutant mouse, and its differential response to low doses of *d*-amphetamine. Intriguingly, one group demonstrated that the association of SNAP-25 with ADHD is largely due to transmission of alleles from paternal chromosomes to affected probands, suggesting that this locus may be subject to genomic imprinting. By contrast, there has been no confirmed genetic link between SNAP-25 and schizophrenia, although it was recently noted that specific SNAP-25 polymorphisms were associated with a selective clinical response to antipsychotic drugs[71].

However, a number of studies by independent research groups have reported decreased levels of SNAP-25 in schizophrenia. Reduced levels of SNAP-25 protein were noted in the cerebellum, anterior frontal cortex, inferior temporal cortex, and prefrontal association areas[72–74], although CSF SNAP-25 is increased in schizophrenia[75]. We have previously used quantitative immunohistochemistry to measure SNAP-25 in the hippocampus. The multi-synaptic structure of the hippocampus, and its consequent cytoarchitectural complexity, allows for the measurement of subtle and regionalized changes. We have demonstrated that schizophrenia is associated with significant reductions in SNAP-25 immunoreactivity within the perforant path termination zones in the molecular layers of CA1 and CA2, as well as the outer molecular layer of the dentate gyrus, although no significant changes were observed in the presubiculum or CA3 region[76]. Similar, regionalized changes were observed by Fatemi and colleagues[77] who noted general reductions in hippocampal SNAP-25 immunoreactivity and significant decreases within the stratum granulosum. Additionally, reduced levels of SNAP-25 have also been observed in schizophrenia when entire hippocampal homogenates were quantified using immunoblotting[78].

### 5.5. Association Between Cognitive Function and Expression of SNAP-25 and Complexins in Schizophrenia

Previous work from our group examined presynaptic proteins in the hippocampus in schizophrenia, and we reported significantly lower levels of SNAP-25, Cx I, and Cx II in schizophrenia[22,76]. In the same series of cases compared with controls, we observed no significant differences in synaptophysin, a ubiquitously expressed presynaptic protein. Initial examination of the relationship between the Cxs and cognitive performance (in terms of memory, orientation, and judgment) revealed a number of significant associations. New analyses suggest there may be significant associations between SNAP-25 and cognitive function in these cases (Figure 26.2; Colorplate 12). Of note, there appeared to be some degree of specificity to the findings, for both individual proteins, and for individual hippocampal subfields.

## 6. CHANGES IN OTHER MOLECULAR MARKERS IN SCHIZOPHRENIA

While the focus of the present review is on the relationship between presynaptic proteins and neuropsychiatric disorders, this represents an analysis at just one level of these complex diseases. In the case of schizophrenia, there is also substantial interest at the neural "circuit" level. In particular, there has been recent interest in the nature of inhibitory circuits within the dorsolateral prefrontal cortex (DLPFC), and how alterations in these circuits may lead to cognitive deficits (reviewed in detail by Lewis and colleagues in ref. 79). Primate studies demonstrate that GABAergic interneurons in the DLPFC appear to play an essential role in working memory processes, by regulating both the spatial and temporal electrophysiological activity of surrounding pyramidal neurons. Multiple postmortem human studies have revealed that levels of glutamic acid decarboxylase (GAD67), an enzyme that synthesizes GABA and is specific to interneurons in the frontal cortex, is significantly decreased in the DLPFC in schizophrenia. This phenomenon is observed predominantly in DLPFC layers 3–5, wherein the number of GABAergic interneurons that express detectable levels of GAD67 is significantly reduced in schizophrenia, although absolute levels of GAD67 mRNA are similar to controls in those remaining neurons that do express detectable levels.

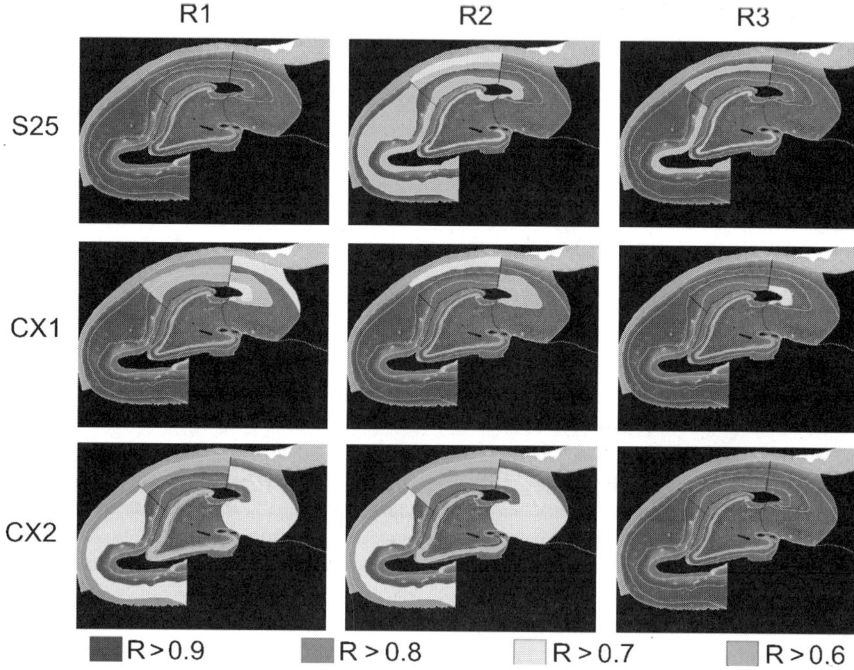


**Figure 26.2.** Maps of the Distribution of Hippocampal Subfields with Statistically Significant ($P \leq 0.05$) Correlations Between Presynaptic Protein Levels and Specific Domains of Cognitive Dysfunction in Schizophrenia. *Abbreviations:* S25, SNAP-25; Cx1, complexin I; Cx2, complexin II; *r*, Spearman correlation coefficient; R1, R2, and R3, clinical dementia ratings for memory, orientation, and judgment, respectively. See Colorplate 12.

Studies of frontal GABAergic interneurons revealed that there are multiple subtypes, which can be classified based on both neuroanatomical and electrophysiological properties, as well as the presence of the three calcium-binding proteins parvalbumin, calbindin, and calretinin. An analysis of the different subtypes of GABA interneurons, on a layer-by-layer basis in the DLPFC, indicates that parvalbumin expression is significantly decreased in layers 3 and 4 in schizophrenia, although the density of neurons expressing the gene was unaltered. However, of interest, it was noted that GAD67 mRNA was mostly reduced in neurons with lower levels of parvalbumin mRNA, suggesting that this class of neurons may be impaired by dysregulation of a number of related genes important for functional activity. These findings remain to be evaluated in alternate brain regions, such as the hippocampus, where evidence for global decreases in GABA markers is mixed, perhaps in part because of the greater macroarchitectural complexity of the region[80].

There is now converging evidence that functional alterations in GABAergic interneurons in schizophrenia may be related to additional molecular changes that have been reported post mortem in schizophrenia. For instance, one of the more consistent findings in schizophrenia has been decreased levels of the protein reelin in multiple brain regions, including frontal cortex, hippocampus, and cerebellum[81]. This protein plays an important role in the development of laminated brain structures, although its function in adulthood remains unknown. Loss of the protein, or its receptors, in rodent paradigms results in behavioral and anatomical deficits homologous to those of schizophrenia[82,83]. In the adult brain, reelin is expressed predominantly by GABAergic interneurons, and it was recently demonstrated that levels of reelin were decreased in GAD67-expressing interneurons in cortex. These interneurons also overexpress DNA-methyltransferase 1 (DNMT1), which contributes to promoter CpG island hypermethylation, causing a downregulation of promoter functions and possibly resulting in lower levels of GAD67 and reelin mRNA expression[84].

Genetic studies have also implicated a number of genes in the etiology of schizophrenia. Amongst these, one that has received much attention is the neuregulin 1 (NRG1) gene, which was identified by positional cloning as a susceptibility gene – an effect replicated in different ethnic populations, although sometimes with different haplotypes. The gene encodes a set of at least 15 known proteins, which act through the erbB family of membrane bound tyrosine kinase receptors. This class of receptors includes four genes, erbB1-4, although NRG1 only binds effectively to erbB3 and erbB4. Altered expression of erbB3 has been noted in the prefrontal cortex in schizophrenia. While the functional role of the NRG1/erbB complex remains under investigation, it is presently known to have diverse functions, including neural development and synaptic plasticity[85].

There is clearly opportunity to integrate the above observations in schizophrenia with our findings of regional alterations in presynaptic proteins. The reduced expression of GABAergic interneuron markers in the DLPFC in schizophrenia is consistent with our report of decreased levels of Cx I in the frontal cortex in schizophrenia. As Cx I is found mainly in inhibitory terminals, and is expressed in the same layers as parvalbumin, a possible connection between reduced levels of Cx I in parvalbumin/GAD67 expressing interneurons is worthy of further study. Similarly, reduced levels of Cx I in both the frontal cortex and hippocampus may be associated with a loss of reelin in GABAergic interneurons. While the physiological role of NRG1/erbB in synaptic plasticity continues to be determined, the presynaptic proteins SNAP-25 and Cx I/II represent promising candidates for co-investigation.

## 7. CONCLUSIONS

Work in animal models indicates that presynaptic proteins are markers for changes in synaptic function related to behavior and to medication treatments. Many of these studies examine individual proteins. Relationships between behavior or treatments and multiple, interacting proteins remain less clear. These possibilities are open for analysis, using combinations of transgenic animals, simultaneous assessment of families of proteins, and considering the interactions of genetic variation, environmental stressors, and medications. Research with human samples presents additional challenges in translating molecular findings into clinical implications. Careful assessment of genetic variants may yield insights, provided the consequences of such variants can be assayed directly in experimental systems. Prospective studies of cognition in subjects agreeing to participation in post mortem research may provide a link with animal studies. The availability of high-throughput strategies to assay multiple proteins in multiple brain regions will also provide a rich source of data for modeling the relationships between synaptic function, behavior, and illness. With such a foundation in place, truly novel therapeutic strategies may be developed.

## 8. REFERENCES


1. Thaker, G.K., and Carpenter, W.T., Jr. (2001) *Nat Med* **7**, 667–671.
2. Weickert, T.W., Goldberg, T.E., Gold, J.M., Bigelow, L.B., Egan, M.F., and Weinberger, D.R. (2000) *Arch Gen Psychiatry* **57**, 907–913.
3. Lahti, A.C., Holcomb, H.H., Medoff, D.R., Weiler, M.A., Tamminga, C.A., and Carpenter, W.T., Jr. (2001) *Am J Psychiatry* **158**, 1797–1808.
4. Andreasen, N.C., O'Leary, D.S., Cizadlo, T., Arndt, S., Rezai, K., Ponto, L.L., Watkins, G.L., and Hichwa, R.D. (1996) *Proc Natl Acad Sci U S A* **93**, 9985–9990.
5. Honer, W.G., and Young, C.E. (2004) *Int Rev Neurobiol* **59**, 175–199.
6. Mirnics, K., Middleton, F.A., Marquez, A., Lewis, D.A., and Levitt, P. (2000) *Neuron* **28**, 53–67.
7. Powchik, P., Davidson, M., Haroutunian, V., Gabriel, S.M., Purohit, D.P., Perl, D.P., Harvey, P.D., and Davis, K.L. (1998) *Schizophr Bull* **24**, 325–341.
8. Wakabayashi, K., Honer, W.G., and Masliah, E. (1994) *Brain Res* **667**, 24–32.
9. Marder, S.R., Fenton, W., and Youens, K. (2004) *Am J Psychiatry* **161**, 25.
10. Fray, P.J., and Robbins, T.W. (1996) *Neurotoxicol Teratol* **18**, 499–504.
11. Aleman, A., Hijman, R., de Haan, E.H., and Kahn, R.S. (1999) *Am J Psychiatry* **156**, 1358–1366.
12. Braff, D.L., Geyer, M.A., and Swerdlow, N.R. (2001) *Psychopharmacology (Berl)* **156**, 234–258.
13. Braff, D.L., and Light, G.A. (2004) *Psychopharmacology (Berl)* **174**, 75–85.
14. Pantelis, C., Barnes, T.R., Nelson, H.E., Tanner, S., Weatherley, L., Owen, A.M., and Robbins, T.W. (1997) *Brain* **120**, 1823–1843.
15. Stefani, M.R., Groth, K., and Moghaddam, B. (2003) *Behav Neurosci* **117**, 728–737.
16. Tunbridge, E.M., Bannerman, D.M., Sharp, T., and Harrison, P.J. (2004) *J Neurosci* **24**, 5331–5335.
17. McMahon, H.T., Missler, M., Li, C., and Sudhof, T.C. (1995) *Cell* **83**, 111–119.
18. Pabst, S., Margittai, M., Vainius, D., Langen, R., Jahn, R., and Fasshauer, D. (2002) *J Biol Chem* **277**, 7838–7848.
19. Eastwood, S.L., Cotter, D., and Harrison, P.J. (2001) *Neuroscience* **105**, 219–229.
20. Freeman, W., and Morton, A.J. (2004) *Brain Res Bull* **63**, 33–44.
21. Yamada, M., Saisu, H., Ishizuka, T., Takahashi, H., and Abe, T. (1999) *Neuroscience* **93**, 7–18.
22. Sawada, K., Barr, A.M., Nakamura, M., Arima, K., Young, C.E., Dwork, A.J., Falkai, P., Phillips, A.G., and Honer, W.G. (2005) *Arch Gen Psychiatry* **62**, 263–272.
23. Tokumaru, H., Umayahara, K., Pellegrini, L.L., Ishizuka, T., Saisu, H., Betz, H., Augustine, G.J., and Abe, T. (2001) *Cell* **104**, 421–432.
24. Ono, S., Baux, G., Sekiguchi, M., Fossier, P., Morel, N.F., Nihonmatsu, I., Hirata, K., Awaji, T., Takahashi, S., and Takahashi, M. (1998) *Eur J Neurosci* **10**, 2143–2152.
25. Reim, K., Mansour, M., Varoqueaux, F., McMahon, H.T., Sudhof, T.C., Brose, N., and Rosenmund, C. (2001) *Cell* **104**, 71–81.
26. Edwardson, J.M., Wang, C.T., Gong, B., Wyttenbach, A., Bai, J., Jackson, M.B., Chapman, E.R., and Morton, A.J. (2003) *J Biol Chem* **278**, 30849–30853.
27. Glynn, D., Drew, C.J., Reim, K., Brose, N., and Morton, A.J. (2005) *Hum Mol Genet* **14**, 2369–2385.
28. Takahashi, S., Ujihara, H., Huang, G.Z., Yagyu, K.I., Sanbo, M., Kaba, H., and Yagi, T. (1999) *Eur J Neurosci* **11**, 2359–2366.
29. Glynn, D., Bortnick, R.A., and Morton, A.J. (2003) *Hum Mol Genet* **12**, 2431–2448.
30. Barr, A.M., Young, C.E., Sawada, K., Trimble, W.S., Phillips, A.G., and Honer, W.G. (2004) *Eur J Neurosci* **20**, 303–307.
31. Kinnunen, A.K., Koenig, J.I., and Bilbe, G. (2003) *J Neurochem* **86**, 736–748.
32. Barr, A.M., Hofmann, C.E., Phillips, A.G., Weinberg, J., and Honer, W.G. (2005) *Alc Clin Exp Res* **29**, 1915–1920.
33. Freeman, W., and Morton, A.J. (2004) *Brain Res Bull* **63**, 45–55.
34. Morton, A.J., Faull, R.L., and Edwardson, J.M. (2001) *Brain Res Bull* **56**, 111–117.
35. Diprospero, N.A., Chen, E.Y., Charles, V., Plomann, M., Kordower, J.H., and Tagle, D.A. (2004) *J Neurocytol* **33**, 517–533.

36. Basso, M., Giraudo, S., Corpillo, D., Bergamasco, B., Lopiano, L., and Fasano, M. (2004) *Proteomics* **4**, 3943–3952.
37. Sawada, K., Young, C.E., Barr, A.M., Longworth, K., Takahashi, S., Arango, V., Mann, J.J., Dwork, A.J., Falkai, P., Phillips, A.G., and Honer, W.G. (2002) *Mol Psychiatry* **7**, 484 492.
38. Nakahara, T., Motomura, K., Hashimoto, K., Ueki, H., Gotoh, L., Hondo, H., Tsutsumi, T., Kuroki, T., Hirano, M., and Uchimura, H. (2000) *Neurosci Lett* **290**, 29–32.
39. Zink, M., Rapp, S., Gebicke-Haerter, P.J., Henn, F.A., and Thome, J. (2005) *Psychopharmacology (Berl)*, 1–6.
40. Harrison, P.J., and Eastwood, S.L. (1998) *Lancet* **352**, 1669–1673.
41. Harrison, P.J., Law, A.J., and Eastwood, S.L. (2003) *Ann N Y Acad Sci* **1003**, 94–101.
42. Talbot, K., Eidem, W.L., Tinsley, C.L., Benson, M.A., Thompson, E.W., Smith, R.J., Hahn, C.G., Siegel, S.J., Trojanowski, J.Q., Gur, R.E., Blake, D.J., and Arnold, S.E. (2004) *J Clin Invest* **113**, 1353–1363.
43. Lee, H.J., Song, J.Y., Kim, J.W., Jin, S.Y., Hong, M.S., Park, J.K., Chung, J.H., Shibata, H., and Fukumaki, Y. (2005) *Behav Brain Funct* **1**, 15.
44. Bark, I.C., and Wilson, M.C. (1994) *Gene* **139**, 291–292.
45. Ravichandran, V., Chawla, A., and Roche, P.A. (1996) *J Biol Chem* **271**, 13300–13303.
46. Chen, D., Minger, S.L., Honer, W.G., and Whiteheart, S.W. (1999) *Brain Res* **831**, 11–24.
47. Pan, P.Y., Cai, Q., Lin, L., Lu, P.H., Duan, S., and Sheng, Z.H. (2005) *J Biol Chem* **280**, 25769–25779.
48. Jahn, R., and Sudhof, T.C. (1999) *Annu Rev Biochem* **68**, 863–911.
49. Oyler, G.A., Polli, J.W., Higgins, G.A., Wilson, M.C., and Billingsley, M.L. (1992) *Brain Res Dev Brain Res* **65**, 133–146.
50. Oyler, G.A., Higgins, G.A., Hart, R.A., Battenberg, E., Billingsley, M., Bloom, F.E., and Wilson, M.C. (1989) *J Cell Biol* **109**, 3039–3052.
51. Gonelle-Gispert, C., Halban, P.A., Niemann, H., Palmer, M., Catsicas, S., and Sadoul, K. (1999) *Biochem J* **339**, 159–165.
52. Hohne-Zell, B., Galler, A., Schepp, W., Gratzl, M., and Prinz, C. (1997) *Endocrinology* **138**, 5518–5526.
53. Nagy, G., Reim, K., Matti, U., Brose, N., Binz, T., Rettig, J., Neher, E., and Sorensen, J.B. (2004) *Neuron* **41**, 417–429.
54. Michaelevski, I., Chikvashvili, D., Tsuk, S., Singer-Lahat, D., Kang, Y., Linial, M., Gaisano, H.Y., Fili, O., and Lotan, I. (2003) *J Biol Chem* **278**, 34320–34330.
55. Rettig, J., Sheng, Z.H., Kim, D.K., Hodson, C.D., Snutch, T.P., and Catterall, W.A. (1996) *Proc Natl Acad Sci U S A* **93**, 7363–7368.
56. Sakaba, T., Stein, A., Jahn, R., and Neher, E. (2005) *Science* **309**, 491–494.
57. Washbourne, P., Thompson, P.M., Carta, M., Costa, E.T., Mathews, J.R., Lopez-Bendito, G., Molnar, Z., Becher, M.W., Valenzuela, C.F., Partridge, L.D., and Wilson, M.C. (2002) *Nat Neurosci* **5**, 19–26.
58. Wilson, M.C. (2000) *Neurosci Biobehav Rev* **24**, 51–57.
59. Raber, J., Mehta, P.P., Kreifeldt, M., Parsons, L.H., Weiss, F., Bloom, F.E., and Wilson, M.C. (1997) *J Neurochem* **68**, 176–186.
60. Roberts, L.A., Morris, B.J., and O'Shaughnessy, C.T. (1998) *Neuroreport* **9**, 33–36.
61. Genoud, S., Pralong, W., Riederer, B.M., Eder, L., Catsicas, S., and Muller, D. (1999) *J Neurochem* **72**, 1699–1706.
62. Hou, Q., Gao, X., Zhang, X., Kong, L., Wang, X., Bian, W., Tu, Y., Jin, M., Zhao, G., Li, B., Jing, N., and Yu, L. (2004) *Eur J Neurosci* **20**, 1593–1603.
63. Barr, A.M., Conti, B., Wirz, S., Walker, J., Hayakawa, M., Sanna, P., Pemberton, J., Leonhard, S., Behrens, M., Fehlmann, D., Maier, R., Bilbe, G., Hoyer, D., and Bartfai, T. (2003) *Society for Neuroscience Abstracts* **28**, 957.957.
64. Nakahara, T., Nakamura, K., Tsutsumi, T., Hashimoto, K., Hondo, H., Hisatomi, S., Motomura, K., and Uchimura, H. (1998) *Brain Res Mol Brain Res* **61**, 238–242.
65. Barr, A.M., Young, C.E., Phillips, A.G., and Honer, W.G. *Int J Neuropsychopharmacol* (In press).
66. Barr, A.M., Young, C.E., Honer, W.G., and Phillips, A.G. (2000) *J Psychopharmacol* **14** Suppl, A64.
67. Lapiz, M.D., Fulford, A., Muchimapura, S., Mason, R., Parker, T., and Marsden, C.A. (2003) *Neurosci Behav Physiol* **33**, 13–29.
68. Hess, E.J., Collins, K.A., and Wilson, M.C. (1996) *J Neurosci* **16**, 3104–3111.
69. Bond, W.S. (1987) *Clin Pharm* **6**, 617–624.
70. Barr, C.L., Feng, Y., Wigg, K., Bloom, S., Roberts, W., Malone, M., Schachar, R., Tannock, R., and Kennedy, J.L. (2000) *Mol Psychiatry* **5**, 405–409.

71. Muller, D.J., Klempan, T.A., De Luca, V., Sicard, T., Volavka, J., Czobor, P., Sheitman, B.B., Lindenmayer, J.P., Citrome, L., McEvoy, J.P., Lieberman, J.A., Honer, W.G., and Kennedy, J.L. (2005) *Neurosci Lett* **379**, 81–89.
72. Thompson, P.M., Sower, A.C., and Perrone-Bizzozero, N.I. (1998) *Biol Psychiatry* **43**, 239–243.
73. Mukaetova-Ladinska, E.B., Hurt, J., Honer, W.G., Harrington, C.R., and Wischik, C.M. (2002) *Neurosci Lett* **317**, 161–165.
74. Honer, W.G., Falkai, P., Bayer, T.A., Xie, J., Hu, L., Li, H.Y., Arango, V., Mann, J.J., Dwork, A.J., and Trimble, W.S. (2002) *Cereb Cortex* **12**, 349–356.
75. Thompson, P.M., Kelley, M., Yao, J., Tsai, G., and van Kammen, D.P. (2003) *Biol Psychiatry* **53**, 1132–1137.
76. Young, C.E., Arima, K., Xie, J., Hu, L., Beach, T.G., Falkai, P., and Honer, W.G. (1998) *Cereb Cortex* **8**, 261–268.
77. Fatemi, S.H., Earle, J.A., Stary, J.M., Lee, S., and Sedgewick, J. (2001) *Neuroreport* **12**, 3257–3262.
78. Thompson, P.M., Egbufoama, S., and Vawter, M.P. (2003) *Prog Neuropsychopharmacol Biol Psychiatry* **27**, 411–417.
79. Lewis, D.A., Hashimoto, T., and Volk, D.W. (2005) *Nat Rev Neurosci* **6**, 312–324.
80. Benes, F.M. (1999) *Biol Psychiatry* **46**, 589–599.
81. Torrey, E.F., Barci, B.M., Webster, M.J., Bartko, J.J., Meador-Woodruff, J.H., and Knable, M.B. (2005) *Biol Psychiatry* **57**, 252–260.
82. Barr, A.M., Fish, K., Morgan, S., and Markou, A. (2004) *Int J Neuropsychopharmacol* **7**, S436.
83. Costa, E., Davis, J., Pesold, C., Tueting, P., and Guidotti, A. (2002) *Curr Opin Pharmacol* **2**, 56–62.
84. Veldic, M., Guidotti, A., Maloku, E., Davis, J.M., and Costa, E. (2005) *Proc Natl Acad Sci U S A* **102**, 2152–2157.
85. Stefansson, H., Steinthorsdottir, V., Thorgeirsson, T.E., Gulcher, J.R., and Stefansson, K. (2004) *Ann Med* **36**, 62–71.

# 27

# SYNAPTIC ABNORMALITIES AND CANDIDATE GENES IN AUTISM

Ridha Joober[†] and Alaa El-Husseini[*]

## 1. SUMMARY

Autism is one of the most genetically determined neuropsychiatric disorders. The risk for first-degree relatives is approximately 100-fold higher than the risk in the general population and the concordance in monozygotic (MZ) twins (60–90%) is much higher than the concordance in dizygotic (DZ) twins (~3%). Although specific genes implicated in autism have not been firmly identified, mutations in two members of the neuroligin family of cell adhesion molecules have been reported. Recent findings suggest that alterations in neuroligin function may result in an imbalance in excitatory to inhibitory synaptic transmission. These findings are consistent with a recent model for autism which proposes that enhanced neuronal excitability is the leading cause for autism. In this chapter we briefly discuss the difficulties encountered in identifying genes implicated in autism. We propose that advances in understanding the biological mechanism of synapse development will help to overcome these difficulties.

## 2. INTRODUCTION

Autistic disorder (AD)[1] is a relatively common condition, affecting approximately 0.1% of the population[2]. It is a chronic and debilitating syndrome characterized by complex behavioural and cognitive deficits, including abnormal social interaction and communication, repetitive behaviour, and atypical information processing. Onset is usually in the middle of the second year of life and always before age 3[3]. AD is part of a broad class of pervasive developmental disorders (PPD) together with Asperger's syndrome (AS) and pervasive

---

† McGill University, Departments of Psychiatry, Neurology and Human Genetics, McGill University, Montreal, QC, Canada; ridha.joober@douglas.mcgill.ca

* Department of Psychiarty and the Brain Research Centre, University of British Columbia, Vancouver, BC, V6T 1Z3, Canada; alaa@interchange.ubc.ca

development disorder not otherwise specified (PDDNOS). These three entities share most of their symptoms and may be genetically related[4].

Although the biological understanding of autism remains limited, the high concordance of the autism phenotype in MZ as compared to DZ twins strongly points to an important role of genes in the determination of autism. Also, the frequent co-morbidity of autism with markers of neurobiological impairment, like mental deficiency and epilepsy, is suggestive of an early neurodevelopmental origin. In recent years, several neuroimaging, neuropathological, and genetic studies have provided new insights into the putative pathogenesis of autism. However, the exact neuronal alterations underlying this disorder remain poorly understood and its treatment is still based on non-specific educational interventions and psychotropic medications. It is hoped that genetic research will identify specific genes increasing the risk for autism and subsequently lead to a better identification of the pathogenic mechanisms, which in turn will provide an early and reliable diagnosis and eventually specific treatments. However, in spite of intensive research in this field, the results of genetic studies remain non-conclusive and poorly replicated. This is particularly disappointing given that the human genome project has provided researchers with extraordinary tools to better investigate the genetic architecture of complex disorders. The purpose of this chapter is to provide an up-to-date view on the genetic results in autism. Subsequently, we briefly discuss the possible reasons for not identifying a greater number of genes implicated in autism and argue that a better knowledge of the biology of autism, particularly synaptic dynamics during development, will help in constructing complex models that may be tested genetically.

## 3. THE QUEST FOR AUTISM GENES

A number of studies have confirmed that autism has an important genetic component (for a review, see ref. 4). Twin studies show concordance rates of 36–91% for MZ twins and 1–23% for DZ twins[5–9]. The large difference between MZ and DZ concordance rates indicates that autism is a very strongly heritable condition (heritability > 90%). The risk of disease in siblings of autistic patients is estimated to be 3–7%, which is 30–70 times more than the risk of autism in the general population (recurrence risk). The recurrence risk of autism in second- and third-degree relatives of autistic subjects decreases very sharply (0.18% and 0.12%, respectively) compared to first-degree relatives. This dramatic fall-off in risk from first- to second- and third-degree relatives suggests that autism is a complex genetic trait involving more than one genetic locus and possibly between-loci interaction. This conclusion is consistent with family studies indicating that an oligogenic model of transmission is the most likely in autism[10].

A number of chromosomal alterations have been reported in autism. Two reviews of cytogenetics studies on autism found some interesting candidates regions. Gillberg et al. reported possible loci on chromosome (Chr.) 5q, 8, 13q, 17 and 18[11]. Lauritsen's review reported possible candidate regions on Chr. 7q21, 10q21.2, 15q11–13, 16q23 and 17p11.2[12]. In one study of chromosomal abnormalities in 100 autistic patients from the South Carolina Autism Project, abnormalities of the maternally inherited Chr. 15 emerged as the single most common cause[13]. The observation that the duplications of 15q11–13 are most frequently from maternal origin suggests the involvement of genomic imprinting. Also, the fact that increased head circumference was replicated in several studies[14,15] hints at possible imprinting effects. Indeed, it has been suggested that genetic imprinting anomalies may be associated with growth disturbances[16].

Numerous linkage genetic studies have been conducted to identify loci involved in autism. Because multiplex families with more than one child affected with autism are relatively rare, large international consortiums were formed to overcome this difficulty. These consortiums have reported genome scans using multiplex families (mostly having two affected children). Several loci have been identified but they were variable from one study to the other[17–20] and the results depended on the stringency of the clinical definition of autism[21] and its associated clinical features[22–24].

Association studies have also been used to test specific candidate genes in autism. Most studies focused on selected genes putatively implicated in development[25], growth[26], or neurotransmission, particularly of the serotonin synapse[27–32]. Results from these studies have not been conclusive as none of the candidate genes showed a reliable pattern of replication from one study to the other. For instance, Cook et al.[33] screened nine candidate genes in the 15q11–13 locus and identified linkage disequilibrium between autism and a polymorphism (1155CA-2) located in the gene coding for the subunit B3 of the gamma-aminobutyric acid receptor. Although the corrected $p$-value did not reach statistical significance, the convergence of functional and positional information implicating this gene in autism is of particular interest. More recently, some evidence for the involvement of this polymorphism[34] or other polymorphisms in the cluster of GABA-related genes located in this region was reported[35,36], though other studies failed to replicate these findings[37].

## 4. WHY IS IT SO DIFFICULT TO IDENTIFY GENES IN AUTISM: A TECHNOLOGICAL OR A BIOLOGICAL BOTTLENECK?

Are these conflicting results at all surprising in autism? Besides establishing the high heritability of autism, genetic epidemiology has helped to establish several other robust facts about autism. First, it is highly unlikely that one or few genes with a major effect produce autism. In a seminal paper published in 1990, Risch introduced the notion of the familial risk ratio ($\lambda R$) for a disorder, defined as the risk that a relative $R$ [$R$ = MZ twin, DZ twin, parent, sibling, second degree relative, etc.] of an affected individual has the disorder, divided by the population prevalence[38]. One major result of this study indicates that, for a disorder caused by a single gene or by many genes without gene–gene interaction (regardless of whether they interact with randomly distributed environmental factors or not), $\lambda R$–1 decreases by a factor of 2 with each degree of relatedness. For autism, the observed pattern of ($\lambda R$–1) from one degree of relatedness to the next departs very markedly from the linear decrease expected under an additive multi-locus model. In fact, the decrease is sharply curvilinear as $\lambda MZ - 1 \approx 90$, $\lambda DZ - 1 \approx 5$ and $\lambda$2nd-degree relatives $- 1 \approx 0.3$. This sharp decrease is compatible with a multiplicative model, that is, a model where 2–5 genes interact to increase the risk for the disease (epistasis). This places autism in the realm of complex oligogenic diseases rather than Mendelian genetics. Given these genetic epidemiological considerations, it is not at all surprising that linkage and candidate gene association studies testing one gene at a time failed to identify genes for autism even if the technological advances provided by the human genome project are delivering very detailed information about the structure and the sequence of the genome to improve the quality of within family linkage and population-based linkage disequilibrium mapping. What is really needed to improve our capacity of detecting genes in autism is a better biological knowledge about the pathogenesis of the disorder that will enable us to select a set of highly plausible candidate genes which will be tested for their main and interactive effects. Indeed, without having an a priori scheme for selecting candidate genes, testing for interactive effects of genes selected randomly will be very costly and practically not feasible[39].

## 5. SOME PROMISING LEADS INTO THE GENETICS OF AUTISM

Significant progress in the genetic and neurobiological understanding of PPDs related to autism, such as Rett syndrome and Fragile X, provides new models of pathogenesis that may shed some light on autism. For example, drawing a parallel between the clinical characteristics of Rett syndrome and autism, Zoghbi has proposed that autism may be due to a disruption in postnatal, or experience-dependent synaptic activity[40]. More specifically, she proposed that proteins involved in synapse formation and stability may be implicated in several PPDs. Similarly, Fragile X syndrome mutation 1 (FMR1), which causes a specific dysmorphic syndrome that is often associated with autistic features is more and more perceived as a disorder of the synapse (see Chapter 30). Indeed, it is now believed that FMR1 protein plays a key role in the regulation of dendritically localized mRNA where regulation of synaptic protein synthesis may influence synaptic structure, stability, and plasticity[41].

Given the clinical and developmental communalities between these two developmental disorders and autism, it is possible that autism is also a disease of the synapse. Some evidence implicating proteins playing a critical role in synapse fate has been reported in the last few years. In two brothers, one with X-linked autism and the other with X-linked Asperger syndrome, Jamain et al. (2003) identified a mutation in neuroligins 3 and 4 (NLG3 and NLG4), two proteins implicated in synapse development[42]. In another large French family, it was reported that a mutation in NLG4 that leads to a premature stop codon causes both mental retardation and autism[43]. Although several studies have failed to detect any major mutations in these two genes[44,45], it is possible that the few cases reported in the literature are rare examples of highly penetrant mutations which were revealed because of their co-existence with other genetic abnormalities that were not identified. In any case, these mutations underline the importance of synaptic integrity in autism and point to these two genes as potential players, in combination with other genes, in autism. A better understanding of the processes implicated in synapse development will help genetic research by constructing models for synapse development. These models will serve as a basis for choosing combinations of genes to be tested in an interactive genetic paradigm.

## 6. EXCITATORY/INHIBITORY IMBALANCE: A PLAUSIBLE MODEL FOR AUTISM

The complex clinical presentation of autism may be explained by a unifying model postulating an excess of glutamatergic excitatory synapses and/or a deficit of GABAergic inhibitory synapses[46]. This imbalance in favour of more excitable (more weakly inhibited) cortex is associated with less differentiated cortex, which is prone to epilepsy (observed in approximately 30% of autistic children), perceptual anomalies, atypical memory and cognitive style and motor stereotypy, and repetitive behaviour. This hypothesis, although difficult to prove directly in humans, is substantiated by several examples from animal models showing how the imbalance of the excitatory/inhibitory (E/I) inputs could lead to cognitive or perceptual anomalies reminiscent of autism manifestations. For example, recent investigations into the development of auditory systems in rat have shown that sound representation in the brain is determined during a 1-week critical period when the environmental auditory exposure will determine the "tonotopic map"[47,48]. Under some experimental conditions, such as continuous or modulated noise, the critical period will be either abruptly interrupted or indefinitely protracted leading in both cases to "noisy" cortical development, with proneness to epilepsy and possibly abnormal auditory experience. Other studies have shown that during development, the balance between excitation and inhibition governs the establishment of sensory system projections, including the onset of the critical period[49]. Another animal model, the GAD65 knockout mouse, shows how important the balance between glutamatergic to GABAergic inputs is to cortical development. Indeed, these animals do not develop binocular vision, an effect that is reversed by pharmacological enhancement of inhibitory activity[50]. Another example is provided by mice lacking the gastrin-related peptide receptor (GRPR). Normally, neurons with GRPR receptors enhance GABA activity, which keeps the glutamate activity under control in the amygdala. Contrary to wild-type mice that show extinction of fear reactivity to a conditioned stimulus that was previously paired to an unconditioned fearful stimulus, mice lacking the GRPR show no extinction of this conditioned fear[51]. This suggests that downregulation of GABA activity may be involved in the synaptic plasticity required for controlling adjustment to fear. More recently, it was determined that a mouse model characterized by a 50% reduction in GABAergic interneurons due to a knockout of the hepatocyte growth factor/scatter factor (HGF/SF) displays behaviours that are highly evocative of autism, with increased susceptibility to seizures, heightened anxiety, and diminished social interaction[52]. These examples illustrate the importance of experience-dependent and constitutive (E/I) inputs in the developing brain. It is possible that an imbalance in these inputs, due either to genetic anomalies or aberrant early environmental experiences, leads to highly excitable cortex and defects in the basic cognitive functions necessary to develop higher cognitive abilities.

In the autistic brain, altered neuronal connectivity may develop as a consequence of widespread alterations in synapse elimination and/or formation[53]. The timing of abnormal brain growth in autism in early postnatal development suggests defects in neuronal connectivity. The association with synaptic abnormalities is further strengthened by the fact that autistic behaviour is exhibited in fragile X syndrome, a disorder with a known genetic cause and substantial symptomatic overlap with autism. As discussed in more detail in Chapter 30, examinations of neuronal morphology in fragile X syndrome revealed specific structural alterations in the synapse (for review, see ref. 54). In particular, dendritic spines in specific cortical regions are abnormally long and thin. The decreased dendritic branching of hippocampal neurons of autistic patients further support a reduction in neuronal connectivity[55]. Studies on cerebral cortex in autism also indicate abnormalities in synaptic and columnar structure[56,57] and neuronal migration[58]. Further, analysis of changes in brain size of autistic patients revealed that much of the brain overgrowth occurs postnatally within the first 6–14 months[59], a period that coincides with rapid increase in synaptogenesis, dendritic arborization, and myelination[60]. Finally, anatomical analysis indicates that cortical areas most affected are those essential for complex cognitive functions such as attention, social behaviour, and language. However, additional work is necessary to characterize how defects in neuronal connectivity translate to abnormal behaviour associated with autism.

## 7. NEUROLIGINS: CANDIDATE SYNAPTIC PROTEINS AFFECTED IN AUTISM

As discussed earlier, abnormalities in neuronal connectivity and altered E/I synaptic input in early postnatal development are thought to be associated with autism[61]. To understand how excitation and inhibition is maintained it is important to reveal how specific number of excitatory and inhibitory synapses is achieved. Alteration of the E/I ratio may result from abnormal function/expression of molecules that regulate the number and activity of synaptic contacts in early neuronal development[62]. To gain in-depth knowledge about the biology of autism, it will be important to define molecular factors involved in synapse formation, differentiation, and stability that are affected in autism. Chromosomal aberrations associated with autism that harbour genes implicated in the regulation of the E/I synaptic balance include members of the neuroligin family and the postsynaptic density protein (PSD-95) family (see Table 1; see also Chapter 7). It is worth mentioning that the genes coding for PSD-95 and neuroligin 2 are located in very close proximity on chromosome 17p13.1[42]. This proximity may indicate that these two genes have common genetic mechanisms of regulation. These observations, combined with the identification of *de novo* mutations in genes coding for other members of the neuroligin family (neuroligins 3 and 4) in patients with autism[42,43], suggest that abnormalities in the function of neuroligins result in synaptic imbalance and predisposition toward autism. These results are consistent with the model discussed earlier which proposes an altered E/I ratio in autism.

How do these proteins control synaptic balance? At the postsynaptic density, an electron-dense cytoskeletal structure beneath the plasma membrane of mainly glutamatergic excitatory synapses, neurotransmitter receptors, and signalling molecules are clustered and thereby poised to respond to synaptic stimuli[63,64]. Specific members of the neuroligin family, enriched at glutamatergic synapses, have been recently implicated in the development of synaptic contacts. Neuroligins bind to neurexins, neuron-specific cell surface proteins present at the presynaptic terminal, which are proposed to act as a nucleation site for coupling cell adhesion molecules to synaptic vesicle exocytosis[62,65]. At excitatory postsynaptic sites, neuroligins associate with PSD-95 and this interaction is thought to modulate neuroligin clustering at the synapse[66]. PSD-95 is also implicated in clustering of specific neurotransmitter receptors and signalling proteins to excitatory synapses[62,67]. Recent studies revealed that PSD-95 has dual effects on synapse specificity: it enhances maturation of excitatory synapses, but reduces the number of inhibitory synaptic contacts. These effects correlated with enrichment of neuroligin 1 at excitatory synapses and loss of neuroligin 2 from inhibitory contacts (for details see Chapter 7). In general, these investigations indicate that PSD-95 modulates the E/I synaptic ratio by enhancing clustering of neuroligins at excitatory contacts at the expense of inhibitory synapses.

**Table 1.** List of Genes Coding for Neuroligins and Postsynaptic Density Proteins with Their Chromosomal Position and Relevant Positional Information in Relation to Genetic Studies in Autism.

| Name of the Protein, Official Symbol of the Gene | Location | Positional Information in Relation to Autism and Priority in Screening Genes | References |
|---|---|---|---|
| Neuroligin 1, NLG1 | 3q26–32 | 1. Linkage with autism was replicated at least once in a mega pedigree from Finland. Lod score is > 4.18<br>2. Strongest experimental evidence of involvement in synapse maturation and differentiation | (72–75) |
| Neuroligin 2, NLG2 | 17p13.2 | 1. Reports of chromosomal rearrangements in this region leading to autism phenotype<br>2. Report of linkage on chr 17p<br>3. 17p13.2 harbours a cluster of genes that represent good candidates for autism: NLGN2, DLG4 (coding for PSD-95), and GABA receptor-associated protein (GABARAP). GABARAP clusters GABA receptors by mediating interaction with the cytoskeleton<br>4. This cluster of genes may be submitted to a common developmental regulation | (20,76) |
| Neuroligin 3, NLG3 | Xq13.1 | Mutation already reported in autism. Need of further exploration | (40) |
| Neuroligin 4, NLG4 | Xp22.33 | Mutation already reported in autism. Need of further exploration | (40) |
| Postsynaptic density protein 95, DLG4 [discs, large homologue 4 (Drosophila)] | 17p13.1 | 1. Reports of chromosomal rearrangements in this region leading to autism phenotype<br>2. Report of linkage on chr 17p | (76)<br>(20) |
| Postsynaptic density protein 93, DLG2 [discs, large homologue 2, chapsyn-110 (Drosophila)] | 11q21 | 1. Linkage to this chromosomal region has been reported in autism | (77) |
| Neuroligin 4 Y linked, NLG4Y | Yq11.221 | Expressed in foetal and adult brain | (78) |

Genes are listed according to their priority of testing. All of these genes are expressed in the brain. DLG4 and DLG2 encode two members of the membrane-associated guanylate kinase (MAGUK) family. These two MAGUK proteins regulate clustering of receptors, ion channels, and associated signalling proteins.

Thus, mechanisms that govern assembly of neuroligins and their synaptic partners may be critical for controlling neuronal excitability. Worth mentioning, PSD-95 association with the PSD is dynamic and is regulated by synaptic activity and palmitate cycling on PSD-95[68]. Moreover, synaptic activity upregulates PSD-95 expression through a neuregulin mediated pathway[69]. In contrast, administration of cocaine, a drug known to cause hyperexcitability, results in downregulation of PSD-95 in the striatum[70]. Finally, mutation of FMR1 gene results in a loss of regulation of PSD-95 expression[71]. Thus, chromosomal aberrations or defects in pathways that control the expression of PSD-95 and neuroligins may induce profound effects on synaptic balance and disturb processes that influence memory and behaviour. However, further studies are required to address whether alteration in the expression of these proteins indeed results in altered E/I synaptic ratio and in the manifestation of abnormal behaviour associated with autism. Treatment with peptides that interfere with coupling of neuroligins with their synaptic partners that regulate their retention at the synapse may offer a new tool for manipulating the disturbed E/I ratio associated with brain disorders such as autism.

## 8. CONCLUSION

Recent data suggest that altering the balance between synaptic proteins such as PSD-95, neurexins, and neuroligins may result in altered E/I synaptic input. These results are intriguing since a recent model based on clinical observations proposes that autism may result from an abnormal increase in E/I synaptic ratio. Future studies are required to further examine whether alterations in the expression of specific genes implicated in autism may regulate the E/I synaptic balance and trigger abnormal neuronal excitability. However, considering the heterogeneity of factors involved in autism, it is possible that synaptic imbalance is triggered by many other genes unrelated to neuroligins but function similarly to regulate the E/I synaptic ratio. Thus, a better understanding in this area of research may clarify a fundamental mechanism responsible for triggering the abnormalities associated with autism.

## 9. REFERENCES

1. Kanner, L. (1943) *Nervous Child* **2**, 217–250.
2. Fombonne, E. (2002) *Mol Psychiatry* **7** Suppl 2, S4–S6.
3. American Psychiatric Association: (1994) Diagnostic and Statistical Manual of Mental Diseases, Fourth Edition. American Psychiatric Association., Washington, DC.
4. Rutter, M., Silberg, J., O'Connor, T., and Simonoff, E. (1999) *J Child Psychol Psychiatry* **40**, 19–55.
5. Bailey, A., Le Couteur, A., Gottesman, I., Bolton, P., Simonoff, E., Yuzda, E., and Rutter, M. (1995) *Psychol Med* **25**, 63–77.
6. Le Couteur, A., Bailey, A., Goode, S., Pickles, A., Robertson, S., Gottesman, I., and Rutter, M. (1996) *J Child Psychol Psychiatry Allied Disciplines* **37**, 785–801.
7. Ritvo, E.R., Freeman, B.J., Mason-Brothers, A., Mo, A., and Ritvo, A.M. (1985) *Am J Psychiatry* **142**, 74–77.
8. Steffenburg, S., Gillberg, C., Hellgren, L., Andersson, L., Gillberg, I.C., Jakobsson, G., and Bohman, M. (1989) *J Child Psychol Psychiatry* **30**, 405–416.
9. Szatmari, P., Jones, M.B., Zwaigenbaum, L., and MacLean, J.E. (1998) *J Autism Dev Disord* **28**, 351–368.
10. Pickles, A., Bolton, P., Macdonald, H., Bailey, A., Le Couteur, A., Sim, C.H., and Rutter, M. (1995) *A J Hum Genet* **57**, 717–726.
11. Gillberg, C. (1998) *J Autism Dev Disord* **28**, 415–425.
12. Lauritsen, M., Mors, O., Mortensen, P.B., and Ewald, H. (1999) *J Child Psychol Psychiatry* **40**, 335–345.
13. Schroer, R.J., Phelan, M.C., Michaelis, R.C., Crawford, E.C., Skinner, S.A., Cuccaro, M., Simensen, R.J., Bishop, J., Skinner, C., Fender, D., and Stevenson, R.E. (1998) *Am J Med Genet* **76**, 327–336.
14. Bailey, A., Luthert, P., Bolton, P., Le Couteur, A., Rutter, M., and Harding, B. (1993) *Lancet* **341**, 1225–1226.
15. Fombonne, E., Roge, B., Claverie, J., Courty, S., and Fremolle, J. (1999) *J Autism Dev Disord* **29**, 113–119.
16. Reik, W. (1996) *Exp Physiol* **81**, 161–172.
17. International Molecular Genetic Study of Autism Consortium (1998) *Hum Mol Genet* **7**, 571–578.
18. Barrett, S., Beck, J.C., Bernier, R., Bisson, E., Braun, T.A., Casavant, T.L., Childress, D., Folstein, S.E., Garcia, M., Gardiner, M.B., Gilman, S., Haines, J.L., Hopkins, K., Landa, R.,

Meyer, N.H., Mullane, J.A., Nishimura, D.Y., Palmer, P., Piven, J., Purdy, J., Santangelo, S.L., Searby, C., Sheffield, V., Singleton, J., and Slager, S. (1999) *Am J Med Genet* **88**, 609–615.

19. Philippe, A., Martinez, M., Guilloud-Bataille, M., Gillberg, C., Rastam, M., Sponheim, E., Coleman, M., Zappella, M., Aschauer, H., van Malldergerme, L., Penet, C., Feingold, J., Brice, A., and Leboyer, M. (1999) *Hum Mol Genet* **8**, 805–812.
20. Risch, N., Spiker, D., Lotspeich, L., Nouri, N., Hinds, D., Hallmayer, J., Kalaydjieva, L., McCague, P., Dimiceli, S., Pitts, T., Nguyen, L., Yang, J., Harper, C., Thorpe, D., Vermeer, S., Young, H., Hebert, J., Lin, A., Ferguson, J., Chiotti, C., Wiese-Slater, S., Rogers, T., Salmon, B., Nicholas, P., Petersen, P.B., Pingree, C., McMahon, W., Wong, D.L., Cavalli-Sforza, L.L., Kraemer, H.C., and Myers, R.M. (1999) *Am J Hum Genet* **65**, 493–507.
21. Liu, J., Nyholt, D.R., Magnussen, P., Parano, E., Pavone, P., Geschwind, D., Lord, C., Iversen, P., Hoh, J., Ott, J., and Gilliam, T.C. (2001) *Am J Hum Genet* 2001 **69**, 327–340.
22. Shao, Y., Wolpert, C.M., Raiford, K.L., Menold, M.M., Donnelly, S.L., Ravan, S.A., Bass, M.P., McClain, C., von Wendt, L., Vance, J.M., Abramson, R.H., Wright, H.H., Ashley-Koch, A., Gilbert, J.R., DeLong, R.G., Cuccaro, M.L., and Pericak-Vance, M.A. (2002) *Am J Med Genet* **114**, 99–105.
23. Shao, Y., Raiford, K.L., Wolpert, C.M., Cope, H.A., Ravan, S.A., Ashley-Koch, A.A., Abramson, R.K., Wright, H.H., DeLong, R.G., Gilbert, J.R., Cuccaro, M.L., and Pericak-Vance, M.A. (2002) *Am J Hum Genet* 70, 1058–1061.
24. Alarcon, M., Cantor, R.M., Liu, J., Gilliam, T.C., and Geschwind, D.H. (2002) *Am J Hum Genet* **70**, 60–71.
25. Petit, E., Herault, J., Martineau, J., Perrot, A., Barthelemy, C., Hameury, L., Sauvage, D., Lelord, G., and Muh, J.P. (1995) *J Med Genet* **32**, 269–274.
26. Herault, J., Martineau, J., Petit, E., Perrot, A., Sauvage, D., Barthelemy, C., Mallet, J., Muh, J.P., and Lelord, G. (1994) *J Autism Dev Disord* **24**, 233–236.
27. Cook, E.H., Jr., Courchesne, R., Lord, C., Cox, N.J., Yan, S., Lincoln, A., Haas, R., Courchesne, E., and Leventhal, B.L. (1997) *Mol Psychiatry* **2**, 247–250.
28. Yirmiya, N., Pilowsky, T., Nemanov, L., Arbelle, S., Feinsilver, T., Fried, I., and Ebstein, R.P. (2001) *Am J Med Genet* **105**, 381–386.
29. Klauck, S.M., Poustka, F., Benner, A., Lesch, K.P., and Poustka, A. (1997) *Hum Mol Genet* **6**, 2233–2238.
30. Tordjman, S., Gutknecht, L., Carlier, M., Spitz, E., Antoine, C., Slama, F., Carsalade, V., Cohen, D.J., Ferrari, P., Roubertoux, P.L., and Anderson, G.M. (2001) *Mol Psychiatry* **6**, 434–439.
31. Persico, A.M., Militerni, R., Bravaccio, C., Schneider, C., Melmed, R., Conciatori, M., Damiani, V., Baldi, A., and Keller, F. (2000) *Am J Med Genet* **96**, 123–127.
32. Betancur, C., Corbex, M., Spielewoy, C., Philippe, A., Laplanche, J.L., Launay, J.M., Gillberg, C., Mouren-Simeoni, M.C., Hamon, M., Giros, B., Nosten-Bertrand, M., and Leboyer, M. (2002) *Mol Psychiatry* **7**, 67–71.
33. Cook, E.H., Jr., Courchesne, R.Y., Cox, N.J., Lord, C., Gonen, D., Guter, S.J., Lincoln, A., Nix, K., Haas, R., Leventhal, B.L., and Courchesne, E. (1998) *Am J Hum Genet* **62**, 1077–1083.
34. Buxbaum, J.D., Silverman, J.M., Smith, C.J., Greenberg, D.A., Kilifarski, M., Reichert, J., Cook, E.H., Jr., Fang, Y., Song, C.Y., and Vitale, R. (2002) *Mol Psychiatry* **7**, 311–316.
35. Martin, E.R., Menold, M.M., Wolpert, C.M., Bass, M.P., Donnelly, S.L., Ravan, S.A., Zimmerman, A., Gilbert, J.R., Vance, J.M., Maddox, L.O., Wright, H.H., Abramson, R.K., DeLong, G.R., Cuccaro, M.L., and Pericak-Vance, M.A. (2000) *Am J Med Genet* **96**, 43–48.
36. Menold, M.M., Shao, Y., Wolpert, C.M., Donnelly, S.L., Raiford, K.L., Martin, E.R., Ravan, S.A., Abramson, R.K., Wright, H.H., DeLong, G.R., Cuccaro, M.L., Pericak-Vance, M.A., and Gilbert, J.R. (2001) *J Neurogenet* **15**, 245–259.
37. Salmon, B., Hallmayer, J., Rogers, T., Kalaydjieva, L., Petersen, P.B., Nicholas, P., Pingree, C., McMahon, W., Spiker, D., Lotspeich, L., Kraemer, H., McCague, P., Dimiceli, S., Nouri, N., Pitts, T., Yang, J., Hinds, D., Myers, R.M., and Risch, N. (1999) *Am J Med Genet* **88**, 551–556.
38. Risch, N. (1990) *Am J Hum Genet* **46**, 222–228.
39. Frankel, W.N., and Schork, N.J. (1996) *Nat Genet* **14**, 371–373.
40. Zoghbi, H.Y. (2003) *Science* **302**, 826–830.
41. Vanderklish, P.W., and Edelman, G.M. (2005) *Genes Brain Behav* **4**, 360–384.
42. Jamain, S., Quach, H., Betancur, C., Rastam, M., Colineaux, C., Gillberg, I.C., Soderstrom, H., Giros, B., Leboyer, M., Gillberg, C., and Bourgeron, T. (2003) *Nat Genet* **34**, 27–29.
43. Laumonnier, F., Bonnet-Brilhault, F., Gomot, M., Blanc, R., David, A., Moizard, M.P., Raynaud, M., Ronce, N., Lemonnier, E., Calvas, P., Laudier, B., Chelly, J., Fryns, J.P., Ropers, H.H., Hamel, B.C., Andres, C., Barthelemy, C., Moraine, C., and Briault, S. (2004) *Am J Hum Genet* **74**, 552–557.

44. Gauthier, J., Bonnel, A., St-Onge, J., Karemera, L., Laurent, S., Mottron, L., Fombonne, E., Joober, R., and Rouleau, G.A. (2005) *Am J Med Genet B Neuropsychiatr Genet* **132**, 74–75.
45. Vincent, J.B., Kolozsvari, D., Roberts, W.S., Bolton, P.F., Gurling, H.M., and Scherer, S.W. (2004) *Am J Med Genet B Neuropsychiatr Genet* **129**, 82–84.
46. Rubenstein, J.L., and Merzenich, M.M. (2003) *Genes Brain Behav* 2, 255–267.
47. Zhang, L.I., Bao, S., and Merzenich, M.M. (2002) *Proc Natl Acad Sci U S A* **99**, 2309–2314.
48. Zhang, L.I., Bao, S., and Merzenich, M.M. (2001) *Nat Neurosci* **4**, 1123–1130.
49. Ferster, D. (2004) *Science* **303**, 1619–1621.
50. Hensch, T.K., Fagiolini, M., Mataga, N., Stryker, M.P., Baekkeskov, S., and Kash, S.F. (1998) *Science* **282**, 1504–1508.
51. Shumyatsky, G.P., Tsvetkov, E., Malleret, G., Vronskaya, S., Hatton, M., Hampton, L., Battey, J.F., Dulac, C., Kandel, E.R., and Bolshakov, V.Y. (2002) *Cell* **111**, 905–918.
52. Levitt, P. (2005) *Epilepsia* 46 Suppl 7:22–28.
53. Sporns, O., Tononi, G., and Edelman, G.M. (2000) *Cereb Cortex* **10**, 127–141.
54. Beckel-Mitchener, A., and Greenough, W.T. (2004) *Ment Retard Dev Disabil Res Rev* **10**, 53–59.
55. Raymond, G.V., Bauman, M.L., and Kemper, T.L. (1996) *Acta Neuropathol* **91**, 117–119.
56. Williams, R.S., Hauser, S.L., Purpura, D.P., Delong, G.R., and Swisher, C.N. (1980) *Arch Neurol* **37**, 749–753.
57. Casanova, M.F., Buxhoeveden, D.P., Switala, A.E., and Roy, E. (2002) *Neurology* **58**, 428–432.
58. Bailey, A., Luthert, P., Dean, A., Harding, B., Janota, I., Montgomery, M., Rutter, M., and Lantos, P. (1998) *Brain* **121**, 889–905.
59. Courchesne, E., Carper, R., and Akshoomoff, N. (2003) *J Am Ment Assoc* **290**, 337–344.
60. Carper, R.A., Moses, P., Tigue, Z.D., and Courchesne, E. (2002) *Neuroimage* **16**, 1038–1051.
61. Waites, C.L., Craig, A.M., and Garner, C.C. (2005) *Annu Rev Neurosci* **28**, 251–274.
62. Levinson, J.N., and El-Husseini, A. (2005) *Neuron* **48**, 171–174.
63. Kim, E., and Sheng, M. (2004) *Nat Rev Neurosci* **5**, 771–781.
64. Rao, A., Kim, E., Sheng, M., and Craig, A.M. (1998) *J Neurosci* **18**, 1217–1229.
65. Scheiffele, P. (2003) *Annu Rev Neurosci* **26**, 485–508.
66. El-Husseini, A.E., Schnell, E., Chetkovich, D.M., Nicoll, R.A., and Bredt, D.S. (2000) *Science* **290**, 1364–1368.
67. Prange, O., Wong, T.P., Gerrow, K., Wang, Y.T., and El-Husseini, A. (2004) *Proc Natl Acad Sci U S A* **101**, 13915–13920.
68. Huang, K., and El-Husseini, A. (2005) *Curr Opin Neurobiol* **15**, 527–535.
69. Bao, J., Lin, H., Ouyang, Y., Lei, D., Osman, A., Kim, T.W., Mei, L., Dai, P., Ohlemiller, K.K., and Ambron, R.T. (2004) *Nat Neurosci* **7**, 1250–1258.
70. Yao, W.D., Gainetdinov, R.R., Arbuckle, M.I., Sotnikova, T.D., Cyr, M., Beaulieu, J.M., Torres, G.E., Grant, S.G., and Caron, M.G. (2004) *Neuron* **41**, 625–638.
71. Todd, P.K., Mack, K.J., and Malter, J.S. (2003) *Proc Natl Acad Sci U S A* **100**, 14374–14378.
72. Auranen, M., Varilo, T., Alen, R., Vanhala, R., Ayers, K., Kempas, E., Ylisaukko-Oja, T., Peltonen, L., and Jarvela, I. (2003) *Mol Psychiatry* **8**, 879–884.
73. Oliveira, G., Matoso, E., Vicente, A., Ribeiro, P., Marques, C., Ataide, A., Miguel, T., Saraiva, J., and Carreira, I. (2003) *J Autism Dev Disord* **33**, 177–185.
74. Auranen, M., Vanhala, R., Varilo, T., Ayers, K., Kempas, E., Ylisaukko-Oja, T., Sinsheimer, J.S., Peltonen, L., and Jarvela, I. (2002) *Am J Hum Genet* **71**, 777–790.
75. Konstantareas, M.M., and Homatidis, S. (1999) *J Autism Dev Disord* **29**, 275–285.
76. Mariner, R., Jackson, A.W., III, Levitas, A., Hagerman, R.J., Braden, M., McBogg, P.M., Smith, A.C., and Berry, R. (1986) *J Autism Dev Disord* **16**, 425–440.
77. Yonan, A.L., Alarcon, M., Cheng, R., Magnusson, P.K., Spence, S.J., Palmer, A.A., Grunn, A., Juo, S.H., Terwilliger, J.D., Liu, J., Cantor, R.M., Geschwind, D.H., and Gilliam, T.C. (2003) *Am J Hum Genet* **73**, 886–897.
78. Nagase, T., Ishikawa, K., Suyama, M., Kikuno, R., Hirosawa, M., Miyajima, N., Tanaka, A., Kotani, H., Nomura, N., and Ohara, O. (1999) *DNA Res* **6**, 63–70.

# 28

# SYNAPTIC PATHOLOGY IN DEPRESSION

Barbara Vollmayr, Fritz A. Henn, and Mathias Zink*

## 1. SUMMARY

Depression is an illness in which the reaction to stress and genetic vulnerability come together in such a way that the emotional tone and evaluation of information becomes biased in a negative direction. The pathophysiology lying behind these changes involves monoamine systems as well as the hypothalamus–pituitary–adrenal (HPA) axis and leads to changes in neuronal plasticity, as reflected by decreased levels of neurotrophins, decreased proliferation of neuronal precursors and, most importantly, changes in synaptic structure and function. These concepts are based on indirect evidence from human studies and have been confirmed by animal studies, allowing assessment of behavior as well as detailed functional and structural analyses. Several animal models, namely stress models and genetic models, have demonstrated a good validity for human depression and provided evidence for synaptic pathology in depression.

## 2. INTRODUCTION

Depression is a serious and widespread disorder which results from an interplay between genetic factors and environmental factors such as stress. Early life exposure to severe stress and acute exposure to stress appear to play a role in the development of the illness. The major theories concerning the etiology of depression involve alteration of the amines such as norepinephrine and serotonin or changes in the HPA axis that mediate stress. These theories reside at the receptor level on the surface of the cell membrane. More recent evidence suggests these are simply parts of a complex network altering second messenger systems and ultimately leading to changes in gene expression which may involve direct structural changes in the brain. These changes have been postulated to involve

*Zentralinstitut für Seelische Gesundheit, J5, D-68159 Mannheim, Germany; vollmayr@zi-mannheim.de

neurogenesis, or alternatively synaptogenesis. This chapter reviews the evidence to date implicating such structural change.

## 3. DEFINING DEPRESSION

Depression is one of the most common psychiatric disorders, it has a lifetime prevalence of up to 20%. Major depressive disorder (MDD) is associated with considerable suffering and impairment of the patient and consequently there is a high risk for committing suicide during a depressive episode. The WHO recently determined that depression may be the illness which contributes the most to total morbidity, and will clearly attain this position by 2020[1]. Diagnostic criteria for a depressive episode include depressed mood and the loss of interest or pleasure for more than 2 weeks as central features of the disorder. There is a wide variation of associated abnormalities: cognitive symptoms include inappropriate thoughts of guilt or worthlessness occasionally reaching delusional proportions; common are rumination, reduced concentration, and the impaired ability to make decisions. Neurovegetative symptoms include sleep and appetite disturbances, psychomotor changes like agitation or in contrast retardation, and symptoms like decreased energy, fatigue, and tiredness are very common.

Depression is an episodic and recurrent illness. Longitudinal studies suggest that only rarely will a person suffer from only a single episode of depression. The largest data set collected concerning longitudinal course (NIMH collaborative study on depression) showed that nearly 80% of all people suffering an episode of depression will have a reoccurrence within 8 years[2]. Thus, we see that the illness appears to be one with acute changes which are at least partially reversible and each episode somehow increases the risk for subsequent episodes.

The question which received the most research attention over the last three decades is what happens to cause an episode of depression. A host of epidemiological evidence has shown that stressful life events such as family or work-related conflicts, loss of close personal relationships, or major health problems play a role in the initiation of an episode[3].

Concurrently, there is an overwhelming evidence from epidemiologic studies that – apart from stress – early adverse events and genetic factors also play a role in increasing the risk for depression. With respect to early adverse events Heim and colleagues were able to show long-lasting effects of any kind of abuse in childhood on endocrine responses[4]. These persistent consequences may contribute to the diathesis for adulthood psychopathological conditions such as depression and have implication for therapeutic response and prognosis[5,6]. In addition, affective disorders share a significant genetic basis: Twin studies provide an estimate that roughly 33% of the risk for depression is genetically determined[7,8]. Despite this strong genetic influence, it has been difficult to identify the genes conferring the risk for depression with certainty. This is most likely explained by the fact that several genes are involved in mediating the risk and analysis is complicated by the strong genetic–environmental interaction. This means that genes conferring an increased risk for depression are thought to modulate the subject's ability to cope with environmental demands, and only if a certain environmental challenge – a stressful life event – occurs, the corresponding gene's property to contribute to the risk of depression can be detected. The polymorphism in the serotonin transporter (5HTT) gene regulatory region can serve as an example of this gene–environment interaction: the short variant of the 5HTT promoter gene results in decreased expression of 5HTT and is associated with increased anxiety[9]. Furthermore, individuals with two short alleles also are more sensitive to the depressiogenic effects of stressful life events[10,11].

## 4. ANIMAL MODELS OF DEPRESSION

No animal model is entirely comparable to the human disease, this is particularly true for psychiatric diseases which are mainly perceived subjectively and expressed verbally. However, it is possible to model many aspects of depression in rodents.

Screening paradigms like the Porsolt swim test or the tail suspension test are in a strict sense not animal models of depression because they do not induce long-lasting behavioral changes comparable to human depression in the animals and the pharmacologic effects are detectable without the time lag characteristic for antidepressant therapy. Nevertheless, screening tests are very suitable to predict the antidepressant effect of a pharmacological compound and they are routinely used to screen for potentially new compounds because they are much more easy to apply and much cheaper than the more complex animal models.

However, in order to understand the molecular changes underlying major depression true animal models are needed. The best animal model of depression simulates the etiology and replicates symptoms, course and treatment of human depression. In the case of depression stressful life events constitute a factor playing an etiological role. Acute and chronic uncontrollable stress has therefore been used to induce depressive-like behavior in rodents. All animal models have to define clear criteria which allow the validity of the model to be ascertained. This is somewhat difficult in psychiatric conditions, which are in part defined through subjective experience and cannot be assessed in animals. The face validity of a depressive model can be examined by looking at those vegetative symptoms that are altered clinically. These include concentration, appetite, sleep, and libido. One can look for physiological changes that are associated with the condition, in depression this involves altered HPA axis function. One central test is to determine if the behavioral changes induced are specifically reversible using clinically effective treatments. A clear and specific treatment response is central in developing a useful and valid animal model. Finally, one can utilize one symptom as a key marker and develop tests for this symptom. In stress models, anhedonia has been suggested by Willner as such a marker[12,13]. He has argued that anhedonia – as assessed as a decrease in the sensitivity to sweet palatable solutions – reflects the lack of pleasure and low mood essential for a diagnosis of depression.

Among the stress-induced depression models, learned helplessness – initially described by Seligman and Overmier – has excellent validity[14]. This model derives from a cognitive view of depression in which events are viewed negatively and interpreted as not controllable leading to feelings of anxiety and helplessness. In fact, animals exposed to uncontrollable and unpredictable stress such as inescapable shock develop helpless behavior when tested in an escape paradigm 24 h later. We established a reliable paradigm both for rats and mice, using shock of moderate intensity and carefully excluding possible artifacts[15,16]. When tested with this paradigm 20% of the rats and 30% of mice exposed to inescapable shocks show escape deficits that can be interpreted as persistent reduction in coping (helplessness, despair). Moreover, helpless animals demonstrate a variety of additional behavioral, vegetative, and endocrine symptoms analogous to human depression such as changes in body weight and food consumption, REM sleep disturbance, elevation of corticosterone, and nonsuppression in the dexamethasone (DEX) test, indicating a resistance to the feedback mechanisms of corticosteroids[17]. Finally, learned helpless rats have been found to be responsive to essentially every antidepressant treatment effective in human depression[18–20].

The fact that only a proportion of the animals exposed to stress develop helplessness suggested that this model has a genetic component and is suitable for studying the interaction of stress and genetic vulnerability. In an attempt to select for the genes predisposing to helpless or not helpless behavior, helpless and not helpless rats were mated selectively for more than 50 generations yielding two strains: congenitally learned helpless (cLH) rats exhibiting a helpless phenotype without exposure to uncontrollable shock, and a congenitally not learned helpless (cNLH) rats being resistant to the effects of inescapable shock[21]. In a series of experiments we verified that cLH rats have anhedonia and anergia which are seen in analogy to the hallmarks of depression: loss of interest and pleasure. The sensitivity to the rewarding properties of sucrose is regarded as an index of hedonia in rodents. We were able to show that cLH rats show a lower preference for sucrose in their home cage[22]. However, for conventional measurements of sucrose preference food deprivation is required and introduces several confounding factors of consummatory behavior like stress of deprivation and reduced body weight. We therefore established a procedure comparing several sucrose solutions over a range of concentrations against each other and detected a reduced sensitivity to sucrose in cLH rats which had no food deprivation[23]. Furthermore, applying a progressive ratio schedule we have demonstrated that cLH rats have less capacity to sustain bar pressing for sucrose reward which can be interpreted as anergia[24]. Altered expression of genes with functions in synaptic transmission is summarized below in Section 6.

In the chronic stress models developed initially by Katz[25] and further refined in several studies[12,13,26] an anhedonic state is induced by the repeated application of mild-to-moderate stressors over a prolonged time period. Rats were exposed to stressors like soiled cages, restricted food access, alterations of the light dark cycle, cage tilt, change of cage mate, and introduction of novel objects into the cage for up to 3 months. The model mirrors most of the findings seen in depressive episodes that can be examined in animals. Hedonic measures are the primary focus of the model and the effects of stress are assessed with repeated tests for preference for a palatable weak sucrose solution or saccharin solution. Chronic mild stress decreases preference for sweet solutions which is interpreted as a sign of anhedonia. Higher brain stimulation thresholds as well as decreased place preference conditioning also suggest decreased response to rewarding stimuli after chronic mild stress. After chronic mild stress animals show a wide variety of symptoms that parallel extensive and comprehensive features of human depression: vegetative changes such as decreases in locomotor activity, weight loss, and a decrease of sexual behavior. The animals also show altered diurnal rhythms and sleep disturbances with decreased REM latency and increased number of REM episodes. The model is pharmacologically sensitive and a variety of antidepressant treatments, including electroconvulsive therapy (ECT), are effective in reversing anhedonia after chronic mild stress[13].

There is an urgent need for mouse models of depression to study behavioral consequences of genetically altered mice. A recent study described a murine model of chronic stress, in which mice were exposed to a 4-week-long chronic stress procedure, which consisted of rat exposure, restrained stress, and tail suspension.

The mice showed reduced preference to sucrose solution versus drinking water, that was regarded as anhedonia. The following behavioral analyses of mice aimed to dissect the specific correlates of anhedonic status versus those related merely to the chronic stress. Since the employed chronic stress procedure resulted in hedonic deficit in the majority, but not in all animals, the individuals that did not develop anhedonic status were taken as an internal control of the effects of chronic stress alone. It was found that anhedonia, but not chronic stress per se correlates with key analogs of depressive symptoms, such as drastically increased floating behavior and decreased exploration of novelty. In contrast, increased anxiety and locomotor disturbances (hyperlocomotion triggered by light and hypolocomotion in stress-free situation) were found to be the consequences of chronic stress alone. In addition, individual vulnerability to the stress-induced anhedonia is related to passive coping in resident–intruder test[27].

Another way to validate or reject hypotheses regarding the biochemical or molecular mechanisms underlying depression is to alter the expression of genes related to stress reactions, namely the monoamine- and the HPA system[28]. In looking at the HPA axis the glucocorticoid receptors (GRs) are central in the feedback loop[29]. Mice with a general genetic downregulation of GRs (GR-heterozygote) show significantly increased helplessness[30]. Furthermore, these animals have a pathological DEX/CRH test (nonsuppression), similar to severely depressed patients. In line with these findings, mice that overexpress GRs via a yeast artificial chromosome (YGR mice) resulting in a twofold gene dose elevation are stress-resistant with reduced helplessness, and are "oversuppressors" in the DEX/CRH test. An interesting phenotype is also observed in mice with a brain-specific knockout of the GRs. In terms of the endocrine system these animals show a disinhibition of the HPA system with hypercortisolemia very similar to depressed patients. Because these animals still express GRs outside the nervous system, they exhibit a Cushing-like phenotype with body fat redistribution, osteoporosis, etc. However, they do not have behavioral signs of depression, because they do not express any GR in the brain, so the hypercortisolism cannot be detected in the brain areas mediating depressive-like behavior. In fact, on the behavioral level, these mice seem to represent a stress-resistant strain. Interestingly, GR-heterozygous mice show a downregulation of BDNF in the hippocampus, while GR-overexpressing mice exhibit an upregulation, both in accordance with the neurotrophin hypothesis of depression. Thus, GR-heterozygous mice represent a transgenic model with depression-like features on the molecular, neuro-endocrinological, and behavioral level.

## 5. PATHOPHYSIOLOGY OF DEPRESSION

The first coherent biological theory of depression was the catecholamine hypothesis[31] which suggested that when too little NE and/or 5HT are present in brains, this may lead to a depressive reaction. This theory was based the observation that reserpine which depletes catecholamines can cause depression and that anti-tubercular drugs which had monoamine oxidase (MAO) properties led to a manic-like state. With a better understanding of receptors this theory has been modified and still forms the basis of all currently approved medications for depression. Problems with this theory can be recognized by looking at the effectiveness of current antidepressants, all of which lay around 60–65% with a placebo response rate at over 40% on average. This suggests amine systems play a role in initiating depression but cannot be the sole explanation.

Subsequently it was found that the majority of people had a disturbance of HPA axis function during a depressive episode[32]. Since the HPA axis is a major pathway in mediating stress it was logical to consider that alteration in HPA axis function might be the underlying mechanism giving rise to depression. Here again a problem is that there are people with totally typical depressive illness who do not have alterations in HPA axis function, and some people with increased HPA axis function do not develop depression. Thus, it appears that HPA axis dysfunction cannot provide us with a necessary and sufficient explanation of depression. However, it is clear that the HPA axis does play a role in depression and considering this together with the catecholamine theory, it was postulated that there might be multiple routes which could lead to a depressive illness and there might be a general common final pathway leading to the expression of a depression.

With increasing knowledge of signal transduction pathways it was soon speculated that depression was the result of changes in gene expression mediated by transcription factors such as CREB. Nestler and Duman developed this idea and suggested that the activated genes might be neurotrophins[33–35]. They subsequently focused on the role of BDNF as a final common mediator of depression. It was found that a variety of factors such as stress could suppress BDNF and lead to a depressive state, antidepressants would activate BDNF transcription and elicit plastic changes which returned the organism to a balanced state relieving the depression[36–38]. The antidepressant effect of BDNF infusions into the dorsal raphe nucleus and the differentially regulated expression of BDNF mRNA after a stress challenge in the learned helplessness model support the importance of BDNF in depression[39,40]. This theory has the advantage that it suggests there are inputs from the catecholamine systems, HPA axis, and other peptide systems such as substance P and NPY which are integrated in a final common pathway. Importantly, there is some evidence for an association of MDD and polymorphisms in the BDNF gene[41].

The result of the activation of transcription factors and neurotrophins must involve changes in the neuronal network. Two major possibilities have been postulated as the end stage of a final common pathway for depression. One, a decrease in new neurons, i.e., decreased neurogenesis, and second, a decrease in synaptic activity which could result from elimination of synapses or decreased efficacy of synaptic connections.

The first possibility that neurogenesis plays a role in the etiology of depression was put forward by Jacobs and co-workers[42] and Duman and co-workers in 2000[43,44]. The primary evidence for this hypothesis is that antidepressants appear to increase neurogenesis and that 5HT appears to play a central role in the control of neurogenesis. However, neurogenesis is known to occur only in two places in the brain, the dentate gyrus of the hippocampus and the subventricular zone, where the cells migrate to the olfactory bulb[45]. Hippocampal neurogenesis is sensitive to stress, consistent with the role of stress in depression. Clinical studies have suggested that depressed patients have a smaller hippocampus than carefully matched controls, supporting the idea that neurogenesis could be impaired in this condition. This is also consistent with a greater secretion of cortisol in depression, which suppresses neurogenesis. Futhermore, when X-irradiation of a restricted region of mouse brain containing the hippocampus was used to inhibit neurogenesis, it also abrogated the behavioral effects of both 5HT specific and NE specific antidepressants[46]. The drugs failed to alter the latency to feeding in an open field feeding paradigm in which both antidepressants and antianxiety medications are normally effective in reducing the latency to feeding. These data taken together suggested that altered neurogenesis could be the driving factor in the etiology of depression.

The model of learned helplessness is suitable for separating unspecific stress effects from effects leading to pathologic behavioral and neurobiological changes in analogy to human depression. Using this model, we were able to show that inescapable stress leads to decreased dentate gyrus cell proliferation not only in helpless animals but also in not helpless animals, a finding arguing against a simple causal relation between reduced neurogenesis and depressive behavior[47,48]. We found that during the training phase of learned helplessness, when the animals are exposed to an unpredictable and uncontrollable stressor, neurogenesis decreased significantly and similarly for all animals exposed to the stressor, however only a small fraction developed helpless behavior. This suggested to us that helpless behavior and by analogy depression in humans may not result from decreases in the rate of neurogenesis, but from other structural changes, such as a decrease in synaptogenesis. Turning the experiment around and decreasing the rate of neurogenesis in naïve animals did not result in a greater proportion of animals becoming helpless, regardless if we used stress to decrease neurogenesis or an alkylating agent, "methylazoxymethanol acetate" (MAM ), again suggesting that neurogenesis had little to do with the behavioral changes which result in depression or helplessness. On this basis we have decided to investigate the changes in synaptogenesis in helpless state.

## 6. EVIDENCE FOR ALTERED SYNAPTOGENESIS IN DEPRESSION

The initial evidence that synaptic changes could play a central role in depression came from looking at the effects of stress on synaptogenesis. Over the last 15 years there have been more than a 100 articles documenting the role of stress in causing atrophy and a reduction in branch points in CA3 pyramidal neurons in the hippocampus[49,50]. Watanabe et al.[51] demonstrated atrophy of apical dendrites in the CA3 cells following chronic restraint stress. Impairments of long-term potentiation (LTP) were induced in the hippocampus and the amygdala-prefrontal cortex pathway by acute and chronic stress[52–54]. In parallel, differential expression of genes associated with synaptic functions was documented after stress exposure[55–57]. Stewart and colleagues[58] have recently shown that synaptic plasticity while reduced by stress is increased by learning tasks. In a similar experimental design, Sandi and co-workers showed a rapid reversal of stress induced loss of synapses in CA3 of rat hippocampus following water maze training[59]. This body of literature suggests that synaptic plasticity is a rapid mechanism involved in response to both negative and positive stimuli. The mechanism of such remodeling appears to involve glutamate, and excessive glutamate release may be the driving force in the remodeling of CA3 dendrites following chronic stress. This hypothesis is supported by the recent finding that chronic restraint stress induces the upregulation of glial glutamate transporter-1 (GLT-1), in an effort to reduce this stimulus, while the antidepressant tianeptine blocks this increase in glutamate release and thus protects against remodeling and limits the induction of GLT-1[60]. Not all antidepressants share this action. It has been shown that some SSRI's fail to block dendritic atrophy following stress exposure. This could lead to the suggestion that dendritic atrophy cannot be the etiological factor which is the final step in developing a depression. However the SSRI's appear to cause a restoration of synaptic plasticity even if they do not block the effects of chronic stress. In this respect they show effects analogous to ECT[61].

If synaptogenesis plays a role in depression it would be interesting to see how individual protein components involved specifically in synaptic function respond to chronic stress. As parts of the synaptic release mechanism, several gene families of synaptic vesicle proteins (SVP) regulate tethering, docking, priming, triggering, and fusion of synaptic vesicles in the presynaptic terminals[62,63]. A set of membrane-associated proteins participate in the fusion of internal membranes by assembling the so-called SNARE-complex in the cytosol (SNARE: soluble *N*-ethylmaleimide-sensitive-factor (NSF) attachment protein receptor)[64]. We examined the regulation of SVP by stress and found widespread effects throughout the hippocampus. In general synaptophysin was rapidly downregulated by acute stress, by about 50% while synaptotagmin was upregulated by both and acute and chronic stress, with acute stress again producing a somewhat larger effect[65]. These results are in good agreement with human post mortem data of Honer et al., showing abnormal SNARE-protein-expression in severe mental illness including depressed patients who committed suicide[66].

Since antidepressant agents are known to affect gene transcription[35] for example via activation of the cyclic AMP response element binding protein (CREB)[67,68], we were interested in describing transcriptional effects on SVP expression. In looking at the effects of antidepressant treatment on synaptic proteins, we found that both an SSRI, fluoxetine, and an MAO inhibitor, tranylcypromine, resulted in decreases in synaptotagmin 3 and VAMP 5 in both cerebral cortex and hippocampus, while these compounds increase the expression of synaptophysin[69]. This suggests that this reverses the stress-induced changes which we found earlier. All these findings and the report by Yamada and colleagues on antidepressive effects on VAMP 2 expression[70] indicate that SVP are responsive to both stress and antidepressant treatments, suggesting changes in synaptic function could underlie many of the behavioral changes seen in depression.

The most significant evidence of the involvement of synaptic alterations may be provided by studies on complexins (see Chapter 26). Complexins I and II, also known as synaphins, are highly conserved hydrophilic proteins and interact with the SNARE complex along the groove between the syntaxin and synaptobrevin coils[71]. In functional connection with synaptotagmin 1, the main $Ca^{2+}$ sensor, they are crucial for late steps of $Ca^{2+}$-dependent neurotransmitter release[72,73]. Expression studies have revealed differential localization at axospinal and axodendritic (complexin II) versus axosomatic (complexin I) synapses[74], implicating a link between complexin II and excitatory neurotransmission. Complexin I knockout mice exhibit severe neurological symptoms, whereas complexin II was found to be essential for cognitive function and learning behavior[75], and for the establishment of hippocampal LTP[76].

Post mortem studies suggested that complexin I is decreased in both schizophrenia and major depression in the prefrontal cortex[77]. Complexin II was shown to be decreased in the anterior cingulate cortex and the hippocampus in major depression and bipolar disorder, implicating a specific vulnerability of excitatory synapses[78,79]. The genomic localization of complexin I (4p13.3) and complexin II (5q35.2) has been associated with the risk to develop affective disorders in several independent investigations[80–82].

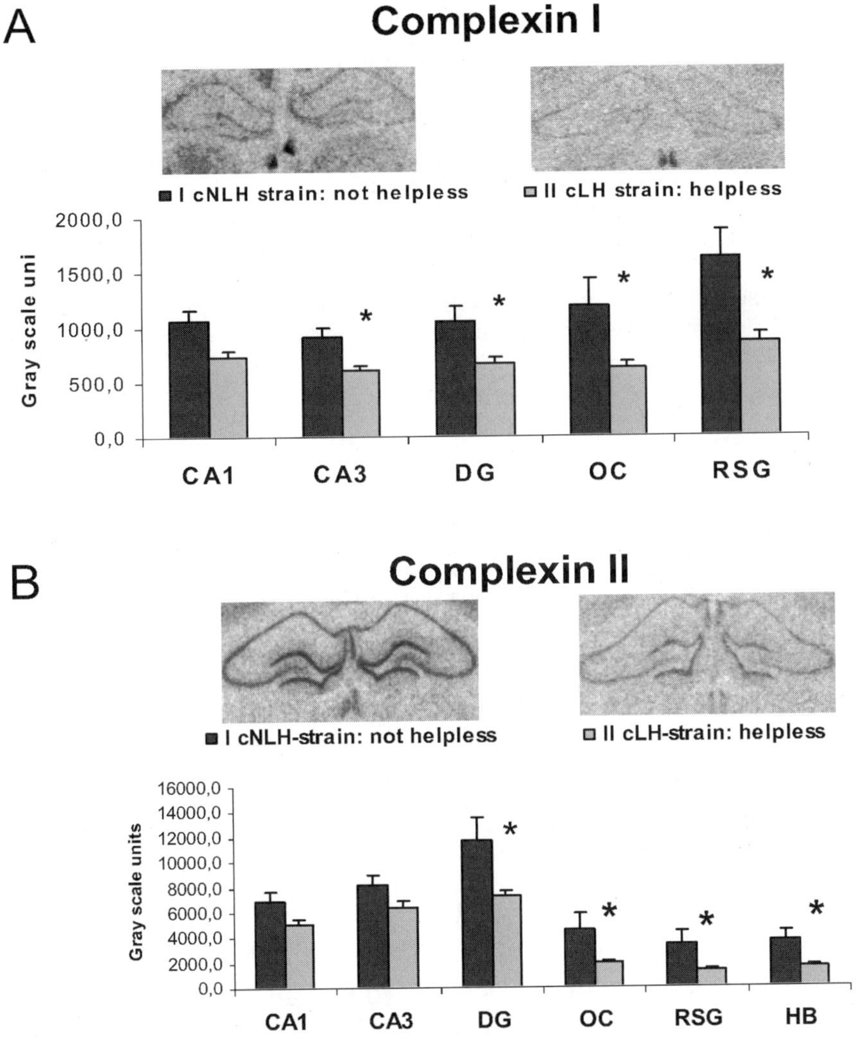


**Figure 28.1.** Expression of Complexin I and II is Reduced in Helpless Rats. Cohorts of adult male rats from inbred lines with congenital helpless or nonhelpless behavior (cLH or cNLH) were analyzed. Rats from the cNLH line exhibiting not helpless behavior ($n = 6$, 0.83 failures in 15 trials, cumulated mean time period of current exposure: (338.6 ± 85.6 s) in contrast to helpless animal of the cLH line ($n = 6$, 12.3 failures, mean current exposure 700 ± 96.2 s). The cohorts did not differ with respect to age and body weight. On brain sections of the dorsal hippocampus, semiquantitative *in situ* hybridizations were performed with cRNA probes corresponding to complexins I (A) and II (B), synaptophysin (SPH), and synaptotagmin 1 (ST1). No significant differences were observed regarding SPH and ST1. In contrast, the helpless animals showed reduced expression of complexins I (A) and II (B). Bars show means of cohorts ($n = 6$) ± S.E.M in different brain regions: Hippocampal subregions CA1, CA3, dentate gyrus (DG), cortical regions such as occipital cortex (OC), retrosplenial granular cortex (RSG), and medial habenula (HB). Levels of significance (Mann–Whitney U-test, Bonferroni correction for multiple testing) are given by asterisks (*$p < 0.05$).

This has prompted us to examine these proteins under antidepressant treatment in order to determine if these post mortem changes could be due to medication effects. Interestingly complexin I was induced only in the habenular nuclei after treatment with fluctine, while a tricyclic antidepressant (TCA), desimpramine, and an MAO inhibitor, tranylcypromine, induced only complexin II in several brain regions. This demonstrated that the post mortem effects were unlikely to be due to medication effects[83]. We have gone on to look at the changes in complexins in strains of animals bred for helplessness compared to animals resistant to the induction of helplessness. The former can be thought of as a genetic model of depression and we wished to see if these strains exhibited different levels of expression in the complexins. We found that both complexins I and II are decreased in hippocampus, habenula, and cortex of helpless animals (Figure 28.1) and observed an inverse correlation between mRNA abundance of complexins I and II and number of failures in the testing paradigm in several brain regions. This finding suggests that changes in synaptic proteins are associated with depression[84].

## 7. REFERENCES

1. Lopez, A.D., and Murray, C.C. (1998) *Nat Med* **4,** 1241–1243.
2. Mueller, T.I., Leon, A.C., Keller, M.B., Solomon, D.A., Endicott, J., Coryell, W., Warshaw, M., and Maser, J.D. (1999) *Am J Psychiatry* **156,** 1000–1006.
3. Kendler, K.S., Karkowski, L.M., and Prescott, C.A. (1999) *Am J Psychiatry* **156,** 837–841.
4. Heim, C., Newport, D.J., Heit, S., Graham, Y.P., Wilcox, M., Bonsall, R., Miller, A.H., and Nemeroff, C.B. (2000) *J Am Mental Assoc* **284,** 592–597.
5. Heim, C., Plotsky, P.M., and Nemeroff, C.B. (2004) *Neuropsychopharmacology* **29,** 641–648.
6. Nemeroff, C.B., and Vale, W.W. (2005) *J Clin Psychiatry* **66,** 5–13.
7. Fava, M., and Kendler, K.S. (2000) *Neuron* **28,** 335–341.
8. Sullivan, P.F., Neale, M.C., and Kendler, K.S. (2000) *Am J Psychiatry* **157,** 1552–1562.
9. Lesch, K.P., Bengel, D., Heils, A., Sabol, S.Z., Greenberg, B.D., Petri, S., Benjamin, J., Muller, C.R., Hamer, D.H., and Murphy, D.L. (1996) *Science* **274,** 1527–1531.
10. Caspi, A., Sugden, K., Moffitt, T.E., Taylor, A., Craig, I.W., Harrington, H., McClay, J., Mill, J., Martin, J., Braithwaite, A., and Poulton, R. (2003) *Science* **301,** 386–389.
11. Kendler, K.S. (2005) *Am J Psychiatry* **162,** 1243–1252.
12. Willner, P. (1991) *Trends Pharmacol Sci* **12,** 131–136.
13. Willner, P., and Mitchell, P.J. (2002) *Behav Pharmacol* **13,** 169–188.
14. Overmier, J.B., and Seligman, M.E. (1967) *J Comp Physiol Psychol* **63,** 28–33.
15. Vollmayr, B., and Henn, F.A. (2001) *Brain Res Prot* **8,** 1 7.
16. Chourbaji, S., Zacher, C., Sanchis-Segura, C., Dormann, C., Vollmayr, B., and Gass, P. (2005) *Brain Res Prot* **16,** 70–78.
17. Vollmayr, B., and Henn, F.A. (2003) *Clin Neurosci Res* **3,** 245–251.
18. Henn, F.A., Edwards, E., Anderson, D., and Vollmayr, B. (2002) *World Psychiatry* **1,** 115–117.
19. Sartorius, A., Vollmayr, B., Neumann-Haefelin, C., Ende, G., Hoehn, M., and Henn, F.A. (2003) *Neuroreport* **14,** 2199–2201.
20. Sherman, A.D., Sacquitne, J.L., and Petty, F. (1982) *Pharmacol Biochem Behav* **16,** 449–454.
21. Henn, F.A., and Vollmayr, B. (2005) *Neurosci Biobehav Rev* **29,** 799–804.
22. Vengeliene, V., Vollmayr, B., Henn, F.A., and Spanagel, R. (2005) *Psychopharmacology* **178,** 125–132.
23. Sanchis-Segura, C., Spanagel, R., Henn, F.A., and Vollmayr, B. (2005) *Behav Pharmacol* **16,** 267–270.
24. Vollmayr, B., Bachteler, D., Vengeliene, V., Gass, P., Spanagel, R., and Henn, F. (2004) *Behav Brain Res* **150,** 217–221.
25. Katz, R.J., Roth, K.A., and Carroll, B.J. (1981) *Neurosci Biobehav Rev* **5,** 247–251.
26. Willner, P. (1995) *Adv Biochem Psychopharm* **49,** 19–41.
27. Strekalova, T., Spanagel, R., Bartsch, D., Henn, F.A., and Gass, P. (2004) *Neuro-psychopharmacology* **29,** 2007–2017.
28. Urani, A., Chourbaji, S., and Gass, P. (2005) *Neurosci Biobehav Rev* **29,** 805–828.
29. Urani, A., and Gass, P. (2003) *Ann New York Acad Sci* **1007,** 379–393.
30. Ridder, S., Chourbaji, S., Hellweg, R., Urani, A., Zacher, C., Schmid, W., Zink, M., Hörtnagl, H., Flor, H., Henn, F.A., Schütz, G., and Gass, P. (2005) *J Neurosci* **25,** 6243–6250.

31. Heninger, G.R., Delgado, P.L., and Charney, D.S. (1996) *Pharmacopsychiatry* **29,** 2–11.
32. Plotsky, P.M., Owens, M.J., and Nemeroff, C.B. (1998) *Psychiatric Clin North Am* **21,** 293–307.
33. Duman, R.S., Heninger, G.R., and Nestler, E.J. (1997) *Arch Gen Psychiatry* **54,** 597–606.
34. Nestler, E.J., Barrot, M., Dileone, R.J., Eisch, A.J., Gold, S.J., and Monteggia, L.M. (2002) *Neuron* **34,** 13–28.
35. Coyle, J.T., and Duman, R.S. (2003) *Neuron* **38,** 157–160.
36. Altar, C.A. (1999) *Trends Pharmacol Sci* **20,** 59–61.
37. Rasmusson, A.M., Shi, L., and Duman, R. (2002) *Neuropsychopharmacology* **27,** 133–142.
38. Duman, R.S. (2002) *Mol Psychiatry* **7,** S29–S34.
39. Vollmayr, B., Faust, H., Lewicka, S., and Henn, F.A. (2001) *Mol Psychiatry* **6,** 471–474.
40. Siuciak, J.A., Lewis, D.R., Wiegand, S.J., and Lindsay, R.M. (1997) *Pharmacol Biochem Behav* **56,** 131–137.
41. Schumacher, J., Jamra, R.A., Becker, T., Ohlraun, S., Klopp, N., Binder, E.B., Schulze, T.G., Deschner, M., Schmal, C., Hofels, S., Zobel, A., Illig, T., Propping, P., Holsboer, F., Rietschel, M., Nothen, M.M., and Cichon, S. (2005) *Biol Psychiatry* **58,** 307–314.
42. Jacobs, B.L., Praag, H., and Gage, F.H. (2000) *Mol Psychiatry* **5,** 262–269.
43. Duman, R.S., Malberg, J., Nakagawa, S., and D'Sa, C. (2000) *Biol Psychiatry* **48,** 732–739.
44. Duman, R.S. (2004) *Biol Psychiatry* **56,** 140–145.
45. Ming, G.L., and Song, H. (2005) *Ann Rev Neurosci* **28,** 223–250.
46. Santarelli, L., Saxe, M., Gross, C., Surget, A., Battaglia, F., Dulawa, S., Weisstaub, N., Lee, J., Duman, R., Arancio, O., Belzung, C., and Hen, R. (2003) *Science* **301,** 805–809.
47. Henn, F.A., and Vollmayr, B. (2004) *Biol Psychiatry* **56,** 146–150.
48. Vollmayr, B., Simonis, C., Weber, S., Gass, P., and Henn, F. (2003) *Biol. Psychiatry* **54,** 1035–1040.
49. Magarinos, A.M., Verdugo, J.M., and McEwen, B.S. (1997) *Proc Natl Acad Sci U S A* **94,** 14002–14008.
50. McEwen, B.S. (2001) *Ann New York Acad Sci* **933,** 265–277.
51. Watanabe, Y., Gould, E., and McEwen, B.S. (1992) *Brain Res* **588,** 341–345.
52. Karst, H., and Joels, M. (2003) *J Neurophysiol* **89,** 625–633.
53. Alfarez, D.N., Joels, M., and Krugers, H.J. (2003) *Eur J Neurosci* **17,** 1928–1934.
54. Maroun, M., and Richter-Levin, G. (2003) *J Neurosci* **23,** 4406–4409.
55. Jay, T.M., Rocher, C., Hotte, M., Naudon, L., Gurden, H., and Spedding, M. (2004) *Neurotoxicity Res* **6,** 233–244.
56. Qin, Y., Karst, H., and Joels, M. (2004) *J Neurochem* **89,** 364–374.
57. Kinnunen, A.K., Koenig, J.I., and Bilbe, G. (2003) *J Neurochem* **86,** 736–748.
58. Stewart, M.G., Davies, H.A., Sandi, C., Kraev, I.V., Rogachevsky, V.V., Peddie, C.J., Rodriguez, J.J., Cordero, M.I., Donohue, H.S., Gabbott, P.L.A., and Popov, V.I. (2005) *Neuroscience* **131,** 43–54.
59. Sandi, C., Davies, H.A., Cordero, M.I., Rodriguez, J.J., Popov, V.I., and Stewart, M.G. (2003) *Eur J Neurosci* **17,** 2447–2456.
60. Reagan, L.P., Rosell, D.R., Wood, G.E., Spedding, M., Munoz, C., Rothstein, J., and McEwen, B.S. (2004) *Proc Natl Acad Sci U S A* **101,** 2179–2184.
61. Stewart, C.A., and Reid, I.C. (2000) *Psychopharmacology* **148,** 217–223.
62. Sudhof, T.C. (2004) *Ann Rev Neurosci* **27,** 509–547.
63. Jahn, R., Lang, T., and Sudhof, T.C. (2003) *Cell* **112,** 519–533.
64. Gerst, J.E. (2003) *Biochim Biophys Acta* **1641,** 99–110.
65. Thome, J., Pesold, B., Baader, M., Hu, M., Gewirtz, J.C., Duman, R.S., and Henn, F.A. (2001) *Biol Psychiatry* **50,** 809–812.
66. Honer, W.G., Falkai, P., Bayer, T.A., Xie, J., Hu, L., Li, H.Y., Arango, V., Mann, J.J., Dwork, A.J., and Trimble, W.S. (2002) *Cerebral Cortex* **12,** 349–356.
67. D'Sa, C., and Duman, R.S. (2002) *Bipolar Disorders* **4,** 183–194.
68. Thome, J., Sakai, N., Shin, K.-H., Steffen, C., Zhang, Y.-J., Impey, S., Storm, D., and Duman, R.S. (2000) *J Neurosci* **20,** 4030–4036.
69. Rapp, S., Baader, M., Hu, M., Jennen-Steinmetz, C., Henn, F.A., and Thome, J. (2004) *Pharmacogenomics J* **4,** 110–113.
70. Yamada, M., Takahashi, K., Tsunoda, M., Nishioka, G., Kudo, K., Ohata, H., Kamijima, K., Higuchi, T., and Momose, K. (2002) *Pharmacogenomics J* **2,** 377–382.
71. Marz, K.E., and Hanson, P.I. (2002) *Trends Neurosci* **25,** 381–383.
72. Reim, K., Mansour, M., Varoqueaux, F., McMahon, H.T., Sudhof, T.C., Brose, N., and Rosenmund, C. (2001) *Cell* **104,** 71–81.
73. Tokumaru, H., Umayahara, K., Pellegrini, L.L., Ishizuka, T., Saisu, H., Betz, H., Augustine, G.J., and Abe, T. (2001) *Cell* **104,** 421–432.
74. Yamada, M., Saisu, H., Ishizuka, T., Takahashi, H., and Abe, T. (1999) *Neurosci* **93,** 7–18.
75. Glynn, D., Bortnick, R.A., and Morton, J. (2003) *Hum Mol Genetics* **12,** 2431–2448.

76. Takahashi, S., Ujihara, H., Huang, G.Z., Yagyu, K.I., Sanbo, M., Kaba, H., and Yagi, T. (1999) *Eur J Neurosci* **11,** 2359–2366.
77. Sawada, K., Young, C.E., Barr, A.M., Longworth, K., Takahashi, S., Arongo, V., Mann, J.J., Dwork, A.J., Falkai, P., Phillips, A.G., and Honer, W.G. (2002) *Mol Psychiatry* **7,** 484–492.
78. Eastwood, S.L., and Harrison, P.J. (2001) *Brain Res Bull* **55,** 569–578.
79. Eastwood, S.L., and Harrison, P.J. (2000) *Mol Psychiatry* **5,** 425–432.
80. Curtis, D., Sherrington, R., Brett, P., Holmes, D.S., Kalsi, G., Brynjolfsson, J., Petursson, H., Rifkin, L., Murphy, P., Moloney, E., et al. (1993) *J Royal Soc Med* **86,** 506–510.
81. Zubenko, G.S., Maher, B., Hughes, H.B. 3rd, Zubenko, W.N., Stiffler, J.S., Kaplan, B.B., and Marazita, M.L. (2003) *Am J Med Genet B Neuropsychiatr Genet* **123,** 1–18.
82. Zubenko, G.S., Maher, B.S., Hughes, H.B. 3rd, Zubenko, W.N., Stiffler, J.S., and Marazita, M.L. (2004) *Am J Med Genet B Neuropsychiatr Genet* **129,** 47–54.
83. Zink, M., Rapp, S., Gebicke-Härter, P., Henn, F.A., and Thome, J. (2005) *Psychopharmacology* **181,** 560–565.
84. Zink, M., Vollmayr, B., Gebicke-Härter, P., Henn, F.A., and Thome, J. (2006) (submitted).

# 29

# SYNAPTIC PATHOLOGY IN DEMENTIA

Stephen W. Scheff*

## 1. SUMMARY

Alzheimer's disease (AD) is a progressive cognitive disorder characterized by the accumulation of structural and biochemical changes that occur in association areas of the neocortex and the hippocampus. Recent work has demonstrated that synaptic loss provides an excellent correlation with cognitive ability and may provide the best correlate of dementia. The loss of synaptic connectivity in the brain of individuals with AD appears to occur early in the disease process and may represent a loss of brain plasticity. Decline in synaptic plasticity does not appear to be an inevitable consequence of the aging process but may be disease related. This chapter reviews and summarizes some of the morphological evidence for Alzheimer-related synaptic loss in the neocortex and hippocampus.

## 2. INTRODUCTION

Individuals with AD have a progressive decline in cognitive function. Although it is unclear what mechanisms are responsible for this cognitive demise, structural and biochemical changes in cortical and subcortical regions have been closely scrutinized. Prospective clinicopathological studies have centered on the accumulation of amyloid plaques (SP) and neurofibrillary tangles (NFT) as possible structural correlates. The original study by Blessed and colleagues[1] and the excellent works by Braak and Braak[2–4] and others have demonstrated a strong relationship between dementia severity and the accumulation of SP and NFT. However, numerous neuropathological studies have now documented the presence of both NFT and SP in the cerebral cortex of nondemented elderly individuals, suggesting that perhaps the accumulation of these "AD hallmarks" may be a feature of normal CNS aging[5–8]. There is now accumulating evidence that one of the underlying mechanisms for AD dementia is a disruption of corticocortical connectivity resulting from a loss of synapses in key association areas of the cortex. Although there may be some cognitive decline in normal aging, what

*Center on Aging, University of Kentucky, Lexington, KY 40536-0230, USA; sscheff@email.uky.edu

differentiates AD cognitive decline from that associated with normal aging may be the failure to replace lost synaptic contacts.

## 3. NORMAL AGING AND SYNAPSE LOSS

It has been suggested in the literature that AD-associated cognitive decline may be an inevitable consequence of the aging process. Age-associated changes in neuron number are often cited as the primary reason for age-related cognitive change, yet careful cell counting studies have failed to support this idea[9]. Recent studies of the entorhinal cortex, an area believed to be an early site of AD-related pathology, shows no significant change in total neuron number as a function of aging[10,11]. Most notably there does appear to be a significant age-related loss of white matter, suggesting a loss of brain connectivity indicative of a loss of synapses[12–15]. The loss of such brain connectivity is important since recent work has demonstrated that synaptic loss is strongly associated with cognitive ability and provides a strong correlation with dementia[16–23].

Assessing inevitable age-related changes in synaptic connectivity are complicated since the methods for identifying and quantifying synaptic numbers in human tissue are either direct, using labor-intensive electron microscopy (EM), or indirect using immunohistochemical methods or an enzyme-linked immunoassay (ELISA). Relatively few studies have investigated normal age-related changes in the neocortex. Because of its involvement in AD, the frontal region (Brodmann areas 9, 10, 46) was one of the initial areas investigated and reported to undergo some age-related loss of synapses[24–27]. Other cortical regions, such as the inferior parietal (Brodmann area 39,40), inferior temporal (Brodmann area 20), and posterior cingulate cortex (Brodmann area 23), all with ties to AD, have also been implicated with age-related synaptic loss[26]. Not all regions of the cortex show such age-related changes. In a rather interesting ultrastructural study, Adams[28] reported an age-related loss in the precentral motor cortex (Brodmann area 4) but failed to observe an age-related change in the postcentral gyrus (Brodmann area 3), an area that is primary sensory cortex. An ELISA and immunoblotting study with synaptophysin[29] failed to detect any age-related differences in Brodmann area 17, the occipital cortex, the superior temporal cortex (Brodmann area 22), or hippocampus, supporting the idea that different regions of the neocortex age differently.

Most of the human studies assessing possible age-related changes in synaptic numbers have limited number of subjects in part because of restricted access to tissue from individuals without pathology. With the exception of two studies[27,30], all of the age-related synapse studies use subject pools with less than 40% of the subjects under the age of 60. One highly cited ultrastructural developmental study of the frontal cortex by Huttenlocher[31] used infants and very young children in the assessment. That study reported no age-related changes. When these data are re-evaluated after removing individuals below the age of 10, there is a clear trend toward an age-related loss in synaptic numbers. It is also interesting to note that the post mortem interval (PMI) for this tissue ranged from 0 to 36 h and could have had a significant effect on synaptic recognition. In a rather unprecedented study, we used an ultrastructural approach and sampled two different laminae (III and V) of Brodmann area 9[30]. The reason for examining laminae III and V stemmed from previous work involving cholinergic circuitry in the frontal cortex. Stereological methods were employed in an attempt to obtain an unbiased estimate of the synaptic packing density. Constraints on tissue acquisition at the time of autopsy precluded estimating total synaptic numbers. This study is unique in that at least five short post mortem cases were obtained for each decade of life. All cases were believed to be from individuals who were considered neurologically normal and without cognitive impairment. The subjects ranged in age from 20–89 years. Individuals 65 years of age and older had been cognitively tested within 1 year prior to death. A previous study has shown that sex of the individual is an important factor in predicting neocortical neuron number and can account for as much as 21% of the variance[32]. Each age group in our synapse study had almost equal numbers of males and females and almost all of the subjects were well educated.

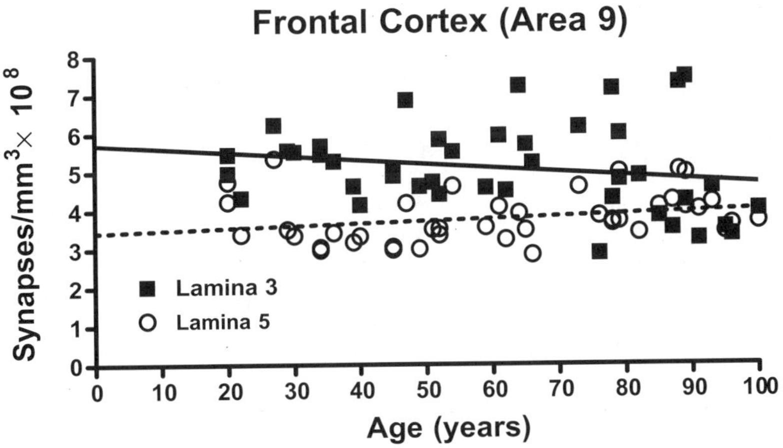


**Figure 29.1.** Scatter Plots with Linear Regression for Numerical Synaptic Density versus Subject's Age at the Time of Death for Both Lamina 3 and 5 of the Frontal Cortex (Brodmann Area 9). A minimum of five subjects were available for each decade of life and all subjects were considered cognitively normal at the time of death. There were no significant correlations ($p > 0.1$) for either lamina suggesting that in normal aging the frontal cortex manifests a plasticity response that maintains normal synaptic numbers.

We have also obtained additional cases from individuals older than 89 who also met the short PMI, education and neurologically intact, criterion. These additional five cases might represent individuals who could be considered "survivors" since they were outside of the normal human life expectancy[33]. The results of the entire group are shown in Figure 29.1. In this very objective study, we failed to observe any significant association between age and packing density of synapses in this Brodmann area, suggesting that the number of synapses remains stable in cognitively normal individuals. In addition there were no significant sex effects indicating that both males and females had equivalent packing densities. When comparing individuals in the ninth decade of life with those of the second and third decades, the very old group had significantly lower synaptic densities, suggesting that mechanism(s) that maintain synaptic numbers begin to wane in the oldest old, similar to other findings[34,35]. These results support the idea that cognitively normal individuals maintain synaptic connectivity throughout the normal life expectancy. Although individuals greater than 89 demonstrated a decline in synaptic numbers, they still were functioning at a fairly high cognitive level. It is tempting to categorize this as a case of cognitive reserve[36,37].

## 4. AD-RELATED SYNAPTIC ALTERATIONS IN NEOCORTEX

Early theories concerning structural mechanisms responsible for AD dementia considered synaptic changes as an important component[38,39]. This was based in part on prior Golgi studies reporting changes in dendritic arborization and dendritic spines, suggestive of AD-related synaptic pathology[40]. Two very early studies utilized conventional transmission EM to examine the frontal and temporal cortex in subjects with AD. In the earliest study[41], biopsy tissue was obtained from three patients who presented with signs of AD. These subjects were later confirmed to have a substantial amount of AD pathology. Numerous electron-dense dendritic spines were observed in close approximation to normal appearing presynaptic boutons. The second EM study by Gibson in 1983[24] reported no loss of synaptic numbers in the cortex in three AD subjects. Only one AD subject was included in the temporal lobe analysis and although the synaptic values were very low compared to almost all of the other individuals examined, two of the other 15 subjects were lower. A paper by Paula-Barbosa and colleagues appeared in 1986 and was the first study designed specifically to test for AD-related changes in neocortical synaptic numbers. Biopsy tissue from the superior frontal cortex (Brodmann areas 9 and 10) of severely demented subjects with AD was compared to "control" subjects undergoing biopsy during routine perfusion ventriculographies. The ultrastructural assessment showed no significant differences in the volume density of synapses between the two groups. The authors concluded that the cognitive changes in the AD subjects were probably not related to synaptic loss. It was speculated, that even though there was considerable neuronal loss in AD cortex[42–45], AD-related synaptic plasticity may maintain normal synaptic numbers[46].

Despite these negative findings, Davies and co-workers[47] used EM in an attempt to estimate whether the number of cortical synapses was altered in AD. Biopsy tissue from the middle temporal lobe (Brodmann area 21) was obtained from 12 early-stage AD subjects. The authors reported a significant loss (25–36%) of synaptic volume densities in laminae III and V. Why did this group find a loss of synapses and previous studies did not? A couple of variables may have been important. First, the cortical region analyzed was different. Brodmann area 9 was used in prior studies and area 21 for this analysis. As part of a side investigation, this group does report on 4 AD and 3 control cases with biopsy material for the superior frontal cortex. This limited analysis also found a 27% loss of synapses in AD but only for lamina V. Second, the study of Davies and co-workers only had three control subjects. Third, it is unclear whether or not the control cases in any of these studies were from cognitively normal individuals, and one might add that cognitively normal individuals would probably not be undergoing a brain biopsy. This study is significant in that it is the first quantitative study to report AD-related synapse loss in the cortex. Although all the AD cases were histologically confirmed, no attempt was made to correlate synaptic numbers with the accumulation of senile plaques or NFT.

In the early 1980s immunohistochemistry was beginning to become in vogue within the neuroscience communities and antibodies were developed against synaptic proteins. One of the earliest was the synapse-associated phosphoprotein synapsin I[48], which was used in a radioimmunoassay (RIA) to study possible AD-related changes[49]. Although five different brain regions, including the cingulate gyrus, were studied in autopsy material, only the hippocampus showed a significant change. Synaptophysin, a molecular marker for a presynaptic vesicle membrane protein[50], was also developed and gaining considerable popularity. In a seminal paper, Masliah and colleagues exploited the immunohistochemical technique by demonstrating significant declines in reactivity in the parietal, temporal, and frontal cortex of AD subjects[51]. Densitometric analysis of stained tissue demonstrated a 50% decrease in synaptophysin staining throughout these three cortical regions. The major advantage to this technique, compared to the labor-intensive EM, was the ease of staining and quantification, thus providing a very rapid evaluation of a large number of tissue samples. Subsequent studies demonstrated that this synaptic marker was stable at long PMIs[25]. One of the drawbacks is that alterations in staining might cause disease-related loss of antigenicity rather than an AD-related loss in the density of synapses. Currently, the synaptophysin antibody is the most commonly used marker to identify synaptic connectivity. It has been used for immunohistochemistry, immunoblotting, and ELISA quantification of synaptic numbers.

### 4.1. Frontal Cortex (Brodmann Areas 9, 10, 46)

Numerous different regions of the neocortex have been assessed for possible changes in synaptic numbers. Because of its association with executive functions known to be altered in AD, the frontal cortex has been the subject of several investigations. Every study designed specifically to assess synaptic numbers in the frontal cortex (Brodmann areas 9, 10, 46) has reported a disease-related decline. The percentage decline varies widely (11–48%) regardless of whether ultrastructural assessment[17,47,52,53] or immunohistochemical methods[16,20,26,51,54–59] were employed. The average observed AD-related decline in synaptic numbers within the frontal cortex is 36%. Reasons for the observed disparity between studies could relate to the subject's stage in the disease progression (early versus "end-stage") or to the limited region of analysis (entire cortical depth versus specific cortical lamina). Exactly what tissue is used as appropriate control and whether or not the groups are age and post mortem matched could also be important factors (Figure 29.2).

### 4.2. Temporal Cortex (Brodmann Areas 20, 21, 22)

The temporal lobe, consisting of the superior (area 22), middle (area 21), and inferior (area 20) regions, is well known for the accumulation of SP and NFT hallmarks of AD[60]. Several studies have evaluated possible synaptic change in these temporal lobe subregions. Markers for the synaptic protein synaptophysin decline approximately 50% in the superior temporal cortex[16,20,29,51] while ultrastructural studies report a 31% decline in AD[61]. Two different EM studies report a 32–36% loss of synapses in the middle temporal gyrus[47,61], while the inferior portion of the temporal lobe demonstrates a 31–49% AD-related decline[26,62]. These secondary and tertiary association regions of the cortex have reciprocal connections with numerous other cortical regions that are directly linked with limbic structures. The temporal lobe plays an important role in both language and visual spatial relationships, behaviors that are known to decline in AD.

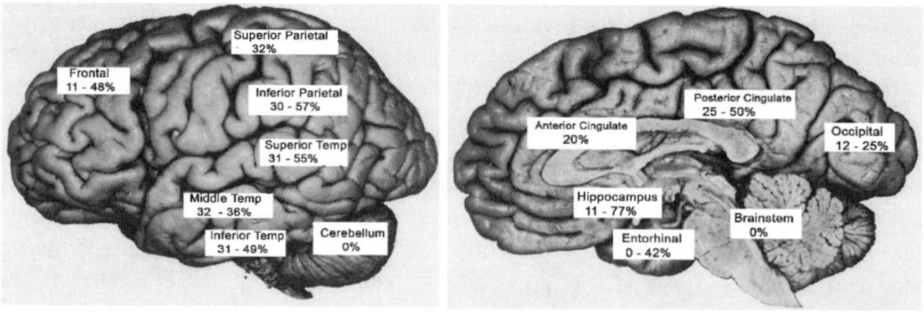


**Figure 29.2.** Lateral and Medial View of the Human Neocortex, Cerebellum, and Brainstem. Labels indicate regions of the brain that have been investigated for possible AD-related changes in synaptic numbers. The numbers indicate the percentage synaptic decline observed in the AD subjects compared to control subjects. For some regions there is a range depending upon which study reports the loss. For some regions there was no significant change in synaptic numbers.

### 4.3. Inferior Parietal Cortex (Brodmann Areas 39, 40)

Brodmann areas 39/40 (inferior parietal) are also heavily invested with AD pathology[63]. This cortical region is important in language and synaptic alterations could account for language dysfunction associated with AD[64]. Immunohistochemical studies using synaptophysin have reported 50% reductions in synaptic numbers in the inferior parietal cortex[16,26,51,54,56] with a single ultrastructural report of 31% in end-stage AD subjects[62]. A single synaptophysin study reported a 32% decline in synaptophysin in the superior parietal cortex (Brodmann area 7), a primary sensory association area. It is interesting that no studies have assessed possible synaptic changes in either primary motor or primary sensory areas, probably because these regions contain little if any SP and NFT[65].

### 4.4. Other Cortical and Subcortical Areas

Several studies have assessed the occipital region of the neocortex and while reporting loss of synaptophysin staining as much as 25%[56,59], some of the findings are not significantly different from age-matched controls[23,29]. Other regions of the neocortex that have synaptic loss appear to be areas known to be involved early in the disease process[3,7,66]. The cingulate gyrus has been identified as one of these cortical regions that may play an important role early in the progression of the disease. Two studies report a significant loss of synapses in the cingulate gyrus[26,67]. One study reported synaptic loss in both anterior and posterior cingulate regions with the caveat that lamina 3 of the anterior cingulate declines in synaptic numbers while lamina V does not[67]. The entorhinal cortex (Brodmann area 28) that is known for its early AD pathology has also been investigated by several groups for possible changes in synaptic numbers. Surprisingly, only two groups have reported significant declines in a synaptic markers[68,69] while the others have failed to detect any significant changes in this important limbic structures[23,56,70,71]. It is interesting that one study found no change in synaptophysin staining in the entorhinal cortex but did observe a decline in the synaptic protein SP6[69].

Several noncortical structures have been assessed, including the cerebellum, caudate, putamen, substantia nigra, pontine nuclei/mesencephalon, and oculomotor nucleus, all of which show no significant change[23,49,56,72,73]. The only subcortical structure that did show a noticeable decrease was the nucleus basalis but it is unclear if the loss is significant since it was not quantified[56].

## 5. SYNAPTIC ALTERATIONS IN THE HIPPOCAMPUS

Many of the studies investigating morphological substrates underlying the progressive AD dementia involved the hippocampal formation and specifically the molecular layer of the hippocampal dentate gyrus. The granule cell dendrites in the outer molecular layer (OML) of the dentate gyrus receive a very significant input from the ipsilateral entorhinal cortex, an area that is noted for its abundance of NFT-containing cells in AD. Immunohistochemical studies have concentrated on this region and reported significant declines in synaptophysin immunostaining ranging from 11% to 77%[23,56,59,68–70,74–76]. In two ultrastructural investigations, the actual number of synapse was counted in the molecular layer in both AD and age-matched controls. Ten short PMI (<8 h) nondemented control subjects were compared to age and PMI-matched AD subjects who had illness duration of approximately 9 years, indicative of end stage in the disease. A stereological sampling scheme was used to estimate the packing density of synapses in both the OML and inner molecular layer (IML). A significant 21% loss of synapses in the OML and a 15% decline in the IML was observed[77,78]. Along with these findings was a significant (26%) decline in the width of the dentate gyrus molecular layer in the AD group that could impact upon the synaptic packing density resulting in an underestimation of the total synaptic loss. The decline in the IML is particularly interesting since it supports the idea that the entire hippocampal formation is significantly affected[29,68,69]. This region of the dentate gyrus does not receive any direct input from the ERC but rather intrahippocampal connectivity.

The synaptic decline in the OML might be viewed as a reduction in hippocampal plasticity that occurs in the end- stage of the disease[46,79]. A very recent paper indicates that this region of the hippocampus fails to replace lost synaptic contacts even in the very early stages of the disease process[80]. If this region is vulnerable and fails to maintain a normal complement of synapses, it might explain some of the short-term memory changes observed in individuals with mild cognitive impairment. Two of the previously mentioned ultrastructural studies of the frontal cortex were carried out in biopsy samples from individuals in the early stages of the disease[17,47]. Both studies reported significant synapse loss, again supporting the idea that a decline in synaptic plasticity may be an early event and account for some of the dementia.

## 6. CHANGES IN APPOSITION SIZE

Early work on synaptic decline noted that the size of the synaptic contacts in the AD tissue was substantially larger compared to the nondemented controls[53,72]. Studies in both human and animal work had reported similar findings and suggested it might be part of a compensatory mechanism. As shown in Figure 29.3, there is a significant inverse correlation, such that as the number of synapses declined in a given region, the size of the residual synapses increases. Initially it was thought that this change in synaptic size may be a response found only in late-stage AD subjects. Studies clearly show that this compensatory response not only occurs in control subjects but also can be observed early in the course of the disease[17].

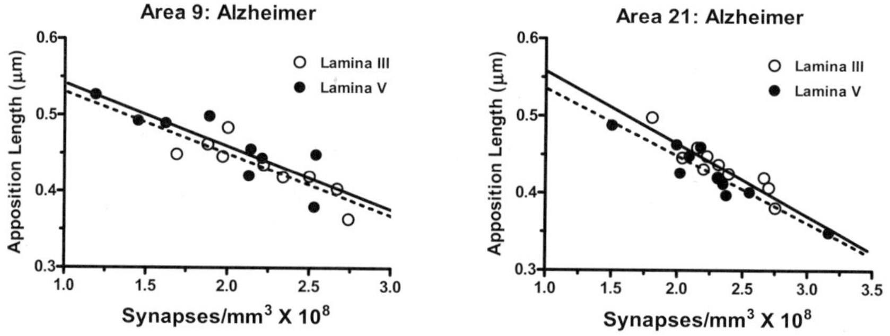


**Figure 29.3.** Relationship Between the Numerical Density of Synapses and Synaptic Apposition Size in the Neocortex of Individuals with Alzheimer's Disease. These graphs are representative of other regions of association cortex. Both laminae III and V show the same relationship. As the number of synapses declines, the size of the residual synapses increases. This may represent a compensatory mechanism used by the disease stricken cortex to maintain total synaptic contact area. The lines represent a simple regression analysis.

To test the hypothesis that the change in synapse size in the AD material might be the result of a preferential loss of small synapses from the neuropil, frequency histograms were constructed of synapse size in both the middle and superior temporal gyrus[61]. The shape of the histograms was very similar between both the AD and control subjects, with the exception that in AD the central location of synapse size was skewed to the right, suggesting that the change occurred throughout the entire distribution of synapses and that the loss was not targeted toward smaller synapses. Since the AD and nondemented controls were both age and PMI matched, the overall increase in apposition size in the AD group is most likely disease related. Other than a change in the size of the apposition length, the synapses in both the AD and non-demented control tissue appeared identical with approximately the same packing density of synaptic vesicles. Possible functional importance of this change in synapse size has been a topic of speculation and may relate to the brain's attempt at maintaining a certain amount of total synaptic contact area[81].

## 7. REASONS FOR SYNAPTIC DECLINE IN AD

The amyloid cascade hypothesis states that amyloid precursor protein (APP) and presenilin (PS) mutations combine to promote the formation of a highly insoluble amyloid beta (Aβ) and that the progressive accumulation of Aβ triggers neuronal and synaptic abnormalities associated with AD. Normally, APP is transported within cells, such as neurons, in a rapid anterograde fashion that accumulates in presynaptic terminals, where it most likely plays a role in cell adhesion. It is primarily processed through a nonamyloidogenic pathway in which cleavage occurs at the α-secretase site and the fragments are degraded. Cells that have mutant APP show increased Aβ as a result of cleavage at the β or γ secretase site. It is unclear if amyloid load correlates with changes in synaptic numbers. There is now evidence that a soluble form of amyloid-β (sAβ) is also found in brain tissue of normal and AD patients[82]. sAβ demonstrates a high correlation with synaptic loss in both a transgenic (tg) animal model of AD and in AD patients[83,84].

Oxidative stress and damage have also been implicated as a fundamental process involved in the pathogenesis of AD[85,86]. Oxidative damage is one of the earliest events in AD with some evidence that levels diminish with duration of the disease. It might be the case that sAβ accumulates early in the disease process and accelerates the formation of oxidative damage in synaptic terminals, ultimately impairing their function by affecting synaptic mitochondria and synaptic proteins eventually leading to synaptic loss.

The identification of the APP and PS mutations has allowed researchers to create tg animals with these mutant genes to gain a better understanding of cellular events that may underlie AD. Besides the overexpression of Aβ and amyloid deposits, tg mice also have many other histopathologic features, including activated microglia, reactive astrocytes, dystrophic neuritis, oxidative damage, and vascular amyloid. While most of the tg mouse models to date have not successfully demonstrated NFT or paired helical filaments, there are several animals that may have evidence of phosphorylated tau.

Transgenic mouse models that purport to be models of AD should demonstrate significant declines in synaptic numbers within regions that are known to be affected in the disease. In reviewing the current literature concerning tg models, it is surprising that while some studies report mild-to-moderate decreases[84,87–91], others report no change[92–99], or even significant increases[100–102] in markers of presynaptic boutons. These controversial results cannot be attributed to differences in presynaptic markers or the techniques employed since the majority of the studies use immunohistochemistry to stain for synaptophysin. These results question exactly what role amyloid may play in AD-related synaptic loss. It is also interesting that many of the tg mouse models also do not demonstrate significant changes in learning and memory.

One of the more current theories for the loss of synapses revolves around a preferential decline in axoplasmic flow. The accumulation of tau can block the transport of organelles in neurons by inhibiting the kinesin-dependent transport system. With the progressive buildup of paired helical filaments interacting with straight filaments the cytoskeleton of the neuron is disrupted, resulting in inhibited axoplasmic flow. The result would be a “starvation” of the synaptic terminals and eventually lead to not only synaptic loss but death of the neuron[103,104]. It has recently been demonstrated that with AD there is a significant reduction in vesicle trafficking-related genes[105]. Failure to dock specific synaptic vesicles would interfere with neurotransmitter release and synaptic connectivity resulting in a decline in cognitive function.

## 8. CONCLUSIONS AND FUTURE DIRECTIONS

Synapses are one of the basic building blocks of the nervous system and essential for cortical communication. These building blocks are not only modifiable but also replaceable. Changes in cortical connectivity most likely underlie our ability to learn and integrate new information. We know that normal aging in the nervous system does not necessary result in a net loss of neuronal connectivity but that such loss is most likely the result of a disease process. In AD there appears to be a significant loss in many regions of the cortex and also in the hippocampus. This loss occurs even in the early stages of the disease process and may reflect the systems inability to initiate and sustain a plasticity process leading to cognitive decline[106–109]. The ability to attend to a stimulus and subsequently determine the merit of that stimulus requires that multiple associational areas of the cortex work in concert. The ability to store important information and to recall that information also requires that multiple regions of the brain work together. The progressive cognitive decline observed in AD most likely reflects an anatomical disruption in the cooperation of various regions of the CNS. Synapse loss in multiple association regions of the cortex is a sign of a corticocortical disconnection. The challenge is to isolate the mechanism(s) underlying this loss of connectivity early in the disease process and to mount a therapeutic regime that halts the synaptic loss and encourages synaptogenesis.

## 9. REFERENCES


1. Blessed, G., Tomlinson, B., and Roth, M. (1968) *Br J Psychiatry* **114**, 797–811.
2. Braak, H., and Braak, E. (1990) *Acta Neuropathol* **80**, 479–486.
3. Braak, H., and Braak, E. (1991) *Acta Neuropathol* **82**, 239–259.
4. Braak, H., and Braak, E. (1994) Pathology of Alzheimer's disease. In: Calne, D.B. (ed). *Neurodegenerative Diseases*, W.B. Saunders, Co., Philadelphia.
5. Arriagada, P.V., Marzloff, K., and Hyman, B.T. (1992) *Neurology* **42**, 1681–1688.
6. Goldman, W.P., Price, J.L., Storandt, M., Grant, E.A., McKeel, D.W., Jr., Rubin, E.H., and Morris, J.C. (2001) *Neurology* **56**, 361–367.
7. Price, L.J., Davis, P.B., Morris, J.C., and White, D.L. (1991) *Neurobiol Aging* **12**, 295–312.
8. Schmitt, F.A., Davis, D.G., Wekstein, D.R., Smith, C.D., Ashford, J.W., and Markesbery, W.R. (2000) *Neurology* **55**, 370–376.
9. Pakkenberg, B., Pelvig, D., Marner, L., Bundgaard, M.J., Gundersen, H.J., Nyengaard, J.R., and Regeur, L. (2003) *Exp Gerontol* **38**, 95–99.
10. Kordower, J.H., Chu, Y., Stebbins, G.T., DeKosky, S.T., Cochran, E.J., Bennett, D., and Mufson, E.J. (2001) *Ann Neurol* **49**, 202–213.
11. Price, J.L., Ko, A.I., Wade, M.J., Tsou, S.K., McKeel, D.W., and Morris, J.C. (2001) *Arch Neurol* **58**, 1395–1402.
12. Salat, D.H., Tuch, D.S., Greve, D.N., van der Kouwe, A.J., Hevelone, N.D., Zaleta, A.K., Rosen, B.R., Fischl, B., Corkin, S., Rosas, H.D., and Dale, A.M. (2005) *Neurobiol Aging* **26**, 1215–1227.
13. Bartzokis, G., Beckson, M., Lu, P.H., Nuechterlein, K.H., Edwards, N., and Mintz, J. (2001) *Arch Gen Psychiatry* **58**, 461–465.

14. Guttmann, C.R., Jolesz, F.A., Kikinis, R., Killiany, R.J., Moss, M.B., Sandor, T., and Albert, M.S. (1998) *Neurology* **50**, 972–978.
15. Jernigan, T.L., Archibald, S.L., Fennema-Notestine, C., Gamst, A.C., Stout, J.C., Bonner, J., and Hesselink, J.R. (2001) *Neurobiol Aging* **22**, 581–594.
16. Brown, D.F., Risser, R.C., Bigio, E.H., Tripp, P., Stiegler, A., Welch, E., Eagan, K.P., Hladik, C.L., and White, C.L., 3rd. (1998) *J Neuropathol Exp Neurol* **57**, 955–960.
17. DeKosky, S.T., and Scheff, S.W. (1990) *Ann Neurol* **27**, 457–464.
18. Lipton, A.M., Cullum, C.M., Satumtira, S., Sontag, E., Hynan, L.S., White, C.L., 3rd, and Bigio, E.H. (2001) *Arch Neurol* **58**, 1233–1239.
19. Samuel, W., Masliah, E., Hill, L.R., Butters, N., and Terry, R. (1994) *Neurology* **44**, 2081–2088.
20. Terry, R.D., Masliah, E., Salmon, D.P., Butters, N., DeTeresa, R., Hill, R., Hansen, L.A., and Katzman, R. (1991) *Ann Neurol* **30**, 572–580.
21. Tigges, J., Herndon, J.G., and Rosene, D.L. (1996) *Acta Anat* **157**, 63–72.
22. Uylings, H.B., and de Brabander, J.M. (2002) *Brain Cogn* **49**, 268–276.
23. Sze, C.I., Troncoso, J.C., Kawas, C., Mouton, P., Price, D.L., and Martin, L.J. (1997) *J Neuropathol Exp Neurol* **56**, 933–944.
24. Gibson, P.H. (1983) *Acta Neuropathol* **62**, 127–133.
25. Liu, X., and Brun, A. (1995) *Dementia* **6**, 211–217.
26. Liu, X., Erikson, C., and Brun, A. (1996) *Dementia* **7**, 128–134.
27. Masliah, E., Mallory, M., Hansen, L., DeTeresa, R., and Terry, R.D. (1993) *Neurology* **43**, 192–197.
28. Adams, I. (1987) *Neurobiol Aging* **8**, 203–212.
29. Honer, W.G., Dickson, D.W., Gleeson, J., and Davies, P. (1992) *Neurobiol Aging* **13**, 375–382.
30. Scheff, S.W., Price, D.A., and Sparks, D.L. (2001) *Neurobiol Aging* **22**, 355–365.
31. Huttenlocher, P.R. (1979) *Brain Res* **163**, 195–205.
32. Pakkenberg, B., and Gundersen, J.G. (1997) *J Comp Neurol* **384**, 312–320.
33. Arias, E. (2004) *Natl Vital Stat Rep* **53**, 1–38.
34. Flood, D.G., Buell, S.J., Defiore, C.H., Horwitz, G.J., and Coleman, P.D. (1985) *Brain Res* **345**, 366–368.
35. Flood, D.G., Buell, S.J., Horwitz, G.J., and Coleman, P.D. (1987) *Brain Res* **402**, 205–216.
36. Stern, Y. (2002) *J Int Neuropsychol Soc* **8**, 448–460.
37. Stern, Y., Habeck, C., Moeller, J., Scarmeas, N., Anderson, K.E., Hilton, H.J., Flynn, J., Sackeim, H., and van Heertum, R. (2005) *Cereb. Cortex* **15**, 394–402.
38. Reisine, T.D., Yamamura, H.I., Bird, E.D., Spokes, E., and Enna, S.J. (1978) *Brain Res* **159**, 477–481.
39. Winblad, B., Hardy, J., Backman, L., and Nilsson, L.G. (1985) *Ann N Y Acad Sci* **444**, 255–268.
40. Scheibel, A.B., and Tomiyasu, U. (1978) *Exp Neurol* **60**, 1–8.
41. Paula-Barbosa, M.M., Cardoso, R.M., Guimaraes, M.L., and Cruz, C. (1980) *J Neuro Sci* **45**, 129–134.
42. Hubbard, B.M., and Anderson, J.M. (1985) *J Neuropath Appl Neurobiol* **11**, 369–382.
43. Mann, D.M.A., Yates, P.O., and Marcyniuk, B. (1985) *J Neurol Sci* **69**, 139–159.
44. Mountjoy, C.Q., Roth, M., Evans, N.J., and Evans, H.M. (1983) *Neurobiol Aging* **4**, 1–11.
45. Terry, R.D., Peck, A., DeTeresa, R., Schechter, R., and Horoupian, D.S. (1981) *Ann Neurol* **10**, 184–192.
46. Geddes, J., Monahan, D., Cotman, C., Lott, I.T., Kim, R., and Chui, H. (1985) *Science* **230**, 1179–1181.
47. Davies, C.A., Mann, D.M.A., Sumpter, P.W., and Yates, P.O. (1987) *J Neuro Sci* **78**, 151–164.
48. Goelz, S.E., Nestler, E.J., Chehrazi, B., and Greengard, P. (1981) *Proc Natl Acad Sci U S A* **78**, 2130–2134.
49. Perdahl, E., Adolfsson, R., Alafuzoff, I., Albert, K.A., Nestler, E.J., Greengard, P., and Winblad, B. (1984) *J Neural Transm* **60**, 133–141.
50. Wiedenmann, B., and Franke, W.W. (1985) *Cell* **41**, 1017–1028.
51. Masliah, E., Terry, R., DeTeresa, R., and Hansen, L. (1989) *Neurosci Lett* **103**, 234–239.
52. Paula-Barbosa, M.M., Saraiva, A., Tavares, M.A., Borges, M.M., and Verwer, R.W.H. (1986) *Acta Neurol Scand* **74**, 404–408.
53. Scheff, S.W., DeKosky, S.T., and Price, D.A. (1990) *Neurobiol Aging* **11**, 29–37.
54. Brun, A., Liu, X., and Erikson, C. (1995) *Neurodegeneration* **4**, 171–177.
55. Hansen, L.A., Daniel, S.E., Wilcock, G.K., and Love, S. (1998) *J Neurol Neurosurg Psychiatry* **64**, 653–656.
56. Masliah, E., Terry, R.D., Alford, M., DeTeresa, R., and Hansen, L.A. (1991) *Amer J Pathol* **138**, 235–246.
57. Masliah, E., Terry, R.D., Mallory, M., Alford, M., and Hansen, L.A. (1990) *Amer J Pathol* **137**, 1293–1297.

58. Reddy, P.H., Mani, G., Park, B.S., Jacques, J., Murdoch, G., Whetsell, W., Jr., Kaye, J., and Manczak, M. (2005) *J Alzheimers Dis* **7**, 103–117; discussion 173–180.
59. Zhan, S.-S., Beyreuther, K., and Schmitt, H.P. (1993) *Dementia* **4**, 66–74.
60. Mirra, S.S., Heyman, A., McKeel, D., Sumi, S.M., Crain, B.J., Brownlee, L.M., Vogel, F.S., Hughes, J.P., van Belle, G., and Berg, L. (1991) *Neurology* **41**, 479–486.
61. Scheff, S.W., and Price, D. (1993) *Ann Neurol* **33**, 190–199.
62. Scheff, S.W., and Price, D. (2006) *J Alzheimers Dis* (in press).
63. Arnold, S.E., Hyman, B.T., Flory, J., Damasio, A.R., and Van Hoesen, G.W. (1991) *Cereb Cortex* **1**, 103–116.
64. Lindeboom, J., and Weinstein, H. (2004) *Eur J Pharmacol* **490**, 83–86.
65. Pearson, R.C., Esiri, M.M., Hiorns, R.W., Wilcock, G.K., and Powell, T.P. (1985) *Proc Natl Acad Sci U S A* **82**, 4531–4534.
66. Hyman, B.T., Van Hoesen, G.W., and Damasio, A.R. (1990) *Neurology* **40**, 1721–1730.
67. Scheff, S.W., and Price, D.A. (2001) *J Alzheimers Dis* **3**, 495–505.
68. Heinonen, O., Soininen, H., Sorvari, H., Kosunen, O., Paljarvi, L., Koivisto, E., and Riekkinen, P.J. (1995) *Neuroscience* **64**, 375–384.
69. Wakabayaski, K., Honer, W.G., and Masliah, E. (1994) *Brain Res* **667**, 24–32.
70. Hamos, J., DeGennaro, L., and Drachman, D. (1989) *Neurology* **39**, 355–361.
71. Scheff, S.W., Sparks, D.L., and Price, D.A. (1993) *Ann Neurol* **34**, 356–361.
72. Bertoni-Freddari, C., and Fattoretti, P. (1989) *Path Res Pract* **185**, 799–802.
73. Davidsson, P., and Blennow, K. (1998) *Int Psychogeriatr* **10**, 11–23.
74. Cabalka, L.M., Hyman, B.T., Goodlett, C.R., Ritchie, T.C., and Van Hoesen, G.W. (1992) *Neurobiol Aging* **13**, 283–291.
75. Goto, S., and Hirano, A. (1990) *Acta Neuropathol* **79**, 545–550.
76. Lippa, C.F., Hamos, J.E., Pulaski-Salo, D., Degennaro, L.J., and Drachman, D.A. (1992) *Neurobiol Aging* **13**, 405–411.
77. Scheff, S.W., and Price, D.A. (1998) *J Neuropathol Exp Neurol* **57**, 1146–1153.
78. Scheff, S.W., Sparks, D.L., and Price, D.A. (1996) *Dementia* **7**, 213–220.
79. Represa, A., Duychaerts, C., Tremblay, E., Hauw, J.J., and Ben-Ari, Y. (1988) *Brain Res* **457**, 355–359.
80. Scheff, S.W., Price, D., Schmitt, F.A., and Mufson, E.J. (2005) *Neurobiol Aging* (in press).
81. Scheff, S.W., and Price, D.A. (2003) *Neurobiol Aging* **24**, 1029–1046.
82. Koudinov, A.R., Koudinova, N.V., Kumar, A., Beavis, R.C., and Ghiso, J. (1996) *Biochem Biophys Res Commun* **223**, 592–597.
83. Lue, L.F., Kuo, Y.M., Roher, A.E., Brachova, L., Shen, Y., Sue, L., Beach, T., Kurth, J.H., Rydel, R.E., and Rogers, J. (1999) *Amer J Pathol* **155**, 853–862.
84. Mucke, L., Masliah, E., Yu, G.Q., Mallory, M., Rockenstein, E.M., Tatsuno, G., Hu, K., Kholodenko, D., Johnson-Wood, K., and McConlogue, L. (2000) *J Neurosci* **20**, 4050–4058.
85. Perry, G., Nunomura, A., Hirai, K., Zhu, X., Perez, M., Avila, J., Castellani, R.J., Atwood, C.S., Aliev, G., Sayre, L.M., Takeda, A., and Smith, M.A. (2002) *Free Radic Biol Med* **33**, 1475–1479.
86. Perry, G., Taddeo, M.A., Nunomura, A., Zhu, X., Zenteno-Savin, T., Drew, K.L., Shimohama, S., Avila, J., Castellani, R.J., and Smith, M.A. (2002) *Comp Biochem Physiol C Toxicol Pharmacol* **133**, 507–513.
87. Buttini, M., Yu, G.Q., Shockley, K., Huang, Y., Jones, B., Masliah, E., Mallory, M., Yeo, T., Longo, F.M., and Mucke, L. (2002) *J Neurosci* **22**, 10539–10548.
88. Games, D., et al. (1995) *Nature* **373**, 523–526.
89. Hsia, A.Y., Masliah, E., McConlogue, L., Yu, G.Q., Tatsuno, G., Hu, K., Kholodenko, D., Malenka, R.C., Nicoll, R.A., and Mucke, L. (1999) *Proc Natl Acad Sci U S A* **96**, 3228–3233.
90. Rockenstein, E., Mallory, M., Mante, M., Sisk, A., and Masliaha, E. (2001) *J Neurosci Res* **66**, 573–582.
91. Wong, T.P., Debeir, T., Duff, K., and Cuello, A.C. (1999) *J Neurosci* **19**, 2706–2716.
92. Ashe, K.H. (2000) *Ann N Y Acad Sci* **924**, 39–41.
93. Boncristiano, S., Calhoun, M.E., Howard, V., Bondolfi, L., Kaeser, S.A., Wiederhold, K.H., Staufenbiel, M., and Jucker, M. (2005) *Neurobiol Aging* **26**, 607–613.
94. Cha, J.H., Farrell, L.A., Ahmed, S.F., Frey, A., Hsiao-Ashe, K.K., Young, A.B., Penney, J.B., Locascio, J.J., Hyman, B.T., and Irizarry, M.C. (2001) *Neurobiol Dis* **8**, 90–102.
95. Hu, L., Wong, T.P., Cote, S.L., Bell, K.F., and Cuello, A.C. (2003) *Neuroscience* **121**, 421–432.
96. Irizarry, M.C., McNamara, M., Fedorchak, K., Hsiao, K., and Hyman, B.T. (1997) *J Neuropathol Exp Neurol* **56**, 965–973.
97. Mucke, L., Masliah, E., Joihnson, W.B., Ruppe, M.D., Alford, M., Rockenstein, E.M., Forss-Petter, S., Pietropaolo, M., Mallory, M., and Abraham, C.R. (1994) *Brain Res* **666**, 151–167.

98. Rutten, B.P., Wirths, O., Van de Berg, W.D., Lichtenthaler, S.F., Vehoff, J., Steinbusch, H.W., Korr, H., Beyreuther, K., Multhaup, G., Bayer, T.A., and Schmitz, C. (2003) *Neurobiol Dis* **12**, 110–120.
99. Takeuchi, A., Irizarry, M.C., Duff, K., Saido, T.C., Hsiao Ashe, K., Hasegawa, M., Mann, D.M., Hyman, B.T., and Iwatsubo, T. (2000) *Amer J Pathol* **157**, 331–339.
100. Bell, K.F., de Kort, G.J., Steggerda, S., Shigemoto, R., Ribeiro-da-Silva, A., and Cuello, A.C. (2003) *Neurosci Lett* **353**, 143–147.
101. King, D.L., and Arendash, G.W. (2002) *Brain Res* **926**, 58–68.
102. Richardson, J.C., Kendal, C.E., Anderson, R., Priest, F., Gower, E., Soden, P., Gray, R., Topps, S., Howlett, D.R., Lavender, D., Clarke, N.J., Barnes, J.C., Haworth, R., Stewart, M.G., and Rupniak, H.T. (2003) *Neuroscience* **122**, 213–228.
103. Mandelkow, E.M., Stamer, K., Vogel, R., Thies, E., and Mandelkow, E. (2003) *Neurobiol Aging* **24**, 1079–1085.
104. Stamer, K., Vogel, R., Thies, E., Mandelkow, E., and Mandelkow, E.M. (2002) *J Cell Biol* **156**, 1051–1063.
105. Yao, P.J., Zhu, M., Pyun, E.I., Brooks, A.I., Therianos, S., Meyers, V.E., and Coleman, P.D. (2003) *Neurobiol Dis* **12**, 97–109.
106. Rowan, M.J., Klyubin, I., Cullen, W.K., and Anwyl, R. (2003) *Philos Trans R Soc Lond B Biol Sci* **358**, 821–828.
107. Rowan, M.J., Klyubin, I., Wang, Q., and Anwyl, R. (2004) *Exp Gerontol* **39**, 1661–1667.
108. Rowan, M.J., Klyubin, I., Wang, Q., and Anwyl, R. (2005) *Biochem Soc Trans* **33**, 563–567.
109. Trommer, B.L., Shah, C., Yun, S.H., Gamkrelidze, G., Pasternak, E.S., Blaine Stine, W., Manelli, A., Sullivan, P., Pasternak, J.F., and LaDu, M.J. (2005) *Neurobiol Dis* **18**, 75–82.

# 30

# A FRAGILE SYNAPSE: CHANGES AT THE SYNAPSE IN FRAGILE X SYNDROME

Alina J. Webber and Brian R. Christie*

## 1. SUMMARY

Researchers have long known that Fragile-X Syndrome is caused by a single gene defect; however it is only recently we have begun to understand how neuronal processing is affected by this disorder. Fragile-X syndrome is caused by the loss of the Fragile X Mental Retardation Protein (FMRP), the protein product of the FMR1 gene. This protein has a transient and activity-dependant expression and appears to play a number of roles in the neuron. Functioning as a shuttle and a repressor for mRNA, FMRP seems to have dramatic effects on both synaptic plasticity and synaptic structure. It is hoped that by restoring these functions in individuals affected by FXS, that new treatments for the disorder can be created. This chapter introduces the reader to the nature of Fragile X Syndrome (FXS) and describes how the loss of FMRP can have profound effects for both synaptic architecture and function.

## 2. INTRODUCTION

FXS is the most common form of inherited mental impairment, with symptoms ranging from mild learning disabilities to severe mental retardation. This X chromosome-linked disorder most severely affects males, leaving them often nonverbal and with IQs in the range of 20–70. Other features of the disorder include increased anxiety, attention deficits, autistic behaviors, and altered physical features that include connective tissue irregularities, macro-orchidism (enlarged testes) and long ears. Approximately 25% of individuals with FXS have epilepsy[1,2]. Female heterozygotes with FXS exhibit much milder symptoms, and

*Department of Psychology, University of British Columbia, 2136 West Mall, Vancouver, BC, Canada, V6T 1Z4; bchristie@psych.ubc.ca

only about one-third of females will show significant intellectual impairment. This disorder affects approximately 1 in every 4,000 males and 1 in every 6,000 females, with a more astonishing 1 in 800 men and 1 in 259 women carrying a premutation (see Section 3) that could cause this disease in their children[3]. Males who carry a premutation normally have no symptoms until late adulthood, when they may show Fragile X-Associated Tremor/Ataxia Syndrome (FXTAS). These symptoms involve parkinsonism, peripheral neuropathy, autonomic dysfunction, and cognitive impairment[4]. Females with a premutation are generally considered normal, with only a small portion developing anxiety and mood swings, and approximately 20% exhibiting premature ovarian failure[5].

## 3. GENETICS

Herbert Lubs' work led to the first chromosomal test for FXS when he noticed a chromosomal constriction, or fragile site, could be observed on the X chromosome of affected males during metaphase[6]. Later, the loss of the Fragile X Mental Retardation 1 (FMR1) gene, and the subsequent lack of its product, FMRP were determined to be the cause of FXS[7].

When looking at pedigrees for FXS, it became apparent that the disease struck individuals with an increasingly greater frequency in the more recent generations, a phenomenon that has become known as the Sherman paradox[8,9]. The Sherman paradox has to do with the nature of the FXS gene mutation in most cases. In over 95% of FXS cases, there is an expansion of a polymorphic CGG repeat in the 5′ UTR of exon 1[7,1]). Normally only 5–50 CGG repeats occur in this region. Persons are said to carry a premutation if they have around 50–200 CGG repeats in this region. The full mutation occurs with >200 CGG repeats. The CGG trinucleotide repeat is remarkably unstable, and individuals with 50–200 of these repeats are classified as having a "premutation" because the FMR1 gene is still translated. The CGG repeat is prone to repeat expansion in humans through a mechanism of abnormal processing during replication. Thus, the more CGG repeats there are, the greater the chances are that offspring will carry the full mutation.

When examining the pedigrees from more than 1,500 females with premutation alleles, the smallest premutation that ever expanded to a full mutation in one generation was 59 CGG repeats[11]. The chances that the number of CGG repeat will change from parent to offspring can be measured, regardless of whether or not the gene becomes a full mutation. Larger trinucleotide repeats were associated with a greater instability in transmitting the same allele from parent to offspring. In mothers with only 49–54 CGG repeats there were further genetic mutations in 19% of the cases, whereas 80% of mothers with 60–65 CGG repeats had some form of unstable transmission of the trinucleotide repeat[11].

Individuals with the full mutation can have thousands of CGG repeats, although any number of repeats in the full mutation category causes hypermethylation of CpG islands upstream in the gene. This recruits histone deacetlyases that condense the chromosomes, and effectively prevent transcription of FMR1 gene[12]. As FXS is caused by a hypermethylation sequence of DNA, blocking methyl-binding proteins, such as Mecp2, has been suggested as a potential treatment for FXS[13]. Downregulating methyl binding proteins has consequences, however, as the loss of Mecp2 has been implicated as a causal factor in Rett's syndrome, a neurological disorder that is also X linked[14].

## 4. MODELS OF FXS

A successful unstable triplet repeat has never been made in a mouse model as mouse genes do not have the same susceptibility to this particular repeat expansion as human genes do. However it is possible to delete the FMR1 gene in mice. The FMR1 gene is highly conserved in the mouse, with an amino acid and nucleotide sequences of 95% and 97%, respectively[15]. The mouse model has subtle behavioral symptoms that resemble some of those in the human disease, including mild spatial learning deficits[16–18]. The startle response in these animals has been reported to be both increased and decreased; however the background strain of the mouse can complicate the expression of subtle phenotypes[19]. One of the more robust phenotypes of the mice is macro-orchidism, with an increase in testicular size by about 30% in 6-month-old knockouts when compared to wild-type controls[20,21]. Macro-orchidism, seen in human patients, is caused from an increase increase in Sertoli cell proliferation[22]. Like their human counterparts, mice with FXS are more prone to seizures and exhibit subtle gross anatomical brain abnormalities[23].

A homologous gene to FMR1, dFMR1 has been found in Drosophila[24], and flies deficient in this gene exhibit phenotypic characteristics that include deficiencies in learning courtship behavior, and abnormal neurite extension, guidance, and branching[25,26].

## 5. CHANGES IN BRAIN ANATOMY AND THE SYNAPSE

Humans with FXS do not show gross cortical abnormalities and have normal brain weights. In some patients there is ventricular enlargement and an increase in hippocampal volumes. MRI studies indicate that the posterior vermis of the cerebellum is decreased and that the size of the caudate nucleus and hippocampus is increased. A diminished white-to-gray matter ratio was also observed in the same study[27]. The volumes of the cerebellum, caudate nucleus, and the size of the ventricles correlates positively with cognitive functioning[28,29]. Looking at these anatomical differences, it might seem easy to suggest that the anatomical differences are the cause of the cognitive deficits seen. Since the caudate volume correlates with the methylation status of the FMR1 gene[29], there is a possibility that gene expression is somehow linked to the size of these structures.

The first hints of changes in the synapse were seen in humans in 1985. Adult FXS males were shown to have longer thin spines, and less short mature spines[30]. Post mortem examination of FXS patient brain tissue revealed their cells to have fewer spines with mushroom or stubby (mature) morphologies. Paradoxically, there was also an increase in overall spine density in these FXS patients[30]. Although these results are indicative of dendritic changes in FXS, it should be noted that the dendritic spines of only six FXS patients of various ages have been examined, many of whom have been on medication for a significant part of their lives. Thus it is not clear if synaptic changes are central to the disorder, and turning to an animal model of FXS was an important step in validating any dendritic abnormalities.

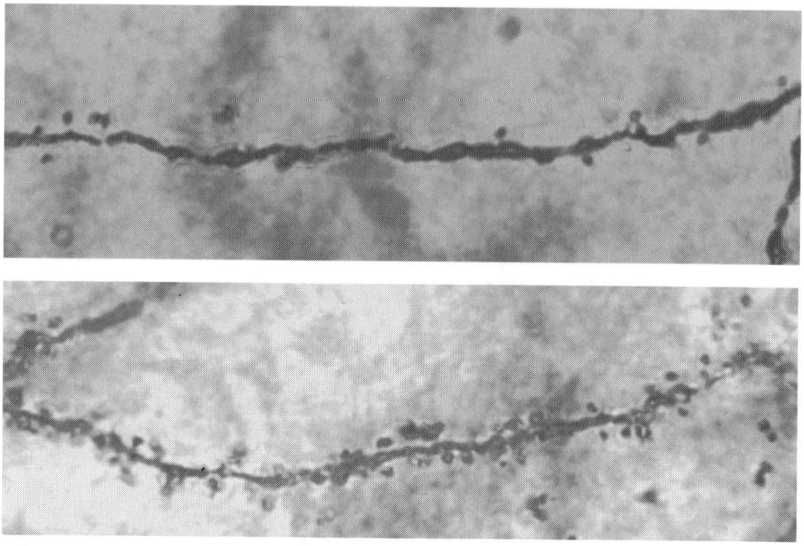

**Figure 30.1.** Abnormal Spine Numbers in FMR1 Knockout Mice. The photomicrograph depicts cortical dendrites in normal wild-type animals (top picture) and FMR1 knockout animals (bottom picture). Note the increase in spine numbers apparent in the dendrite from the FMR1 knockout animal.

The FMR1 knockout mouse on an FVB background did not give a clear indication of any changes in spine density. This appears to be due to to complications specific to this mouse strain, however overall their dendrites do appear similar to those reported in humans[31]. Recently, the C57 mouse background strain has shown that the FMR1 knockout mouse has an increased spine density, like humans (see Figure 30.1). In both background strains, the dendrites were characterized by long spines with a predominantly immature phenotype. In this study spine maturity was assessed independent of length[31–33]. It is important to note that some studies have pointed to a more transient expression of the spine phenotype in FMR1 knockout mice at different ages. Younger slice culture neurons (7–10 days) show immature spine morphologies, but these differences are less apparent at 2 or 4 weeks of age[34]. In cultured neurons at 3 weeks of age, short, low density spines are predominant[35]. Conversely, in FMR1 knockout mice, no changes in spine morphology are apparent at around 28 days of age, but there are changes later on (73–76 days of age)[36]. Interestingly, in human FXS patients, cognitive deficits seem to increase with age.

The work in slice culture neurons[34] has led to the hypothesis that a lack of FMRP leads to improper synapse pruning and maturation. FMRP appears to have an involvement in synapse formation, as evidenced by the paradoxically shorter dendrites, and reduced spine numbers observed in cultures. These cells also have fewer functional connections, with weaker excitatory currents that develop slower[35]. Another study in the Drosophila model showed that dFMRP is a negative regulator of neuronal elaboration and synaptic differentiation. This research group focused on the learning and memory center of the drosophila brain, the mushroom bodies[37]. It still remains to be shown whether or not these spine abnormalities are responsible for a limited capacity to process information.

## 6. SUMMARY OF FMRP PROPOSED MECHANISM OF ACTION

FMRP seems to have a number of functions in the adult nervous system, and it is highly expressed in neurons where one of its functions is to transport and translate mRNAs[38,39]. Quite possibly the most important feature of FMRP is it's role in mRNP complexes. In the nucleus, FMRP can bind to mRNA, forming a messenger ribonucleoprotein, or mRNP. Via a nuclear export signal (NES), FMRP can leave the nucleus and attach to an anterograde motor protein, with its mRNA still bound. The mRNA and FMRP are then transported down the dendrite to spines or filopodia where one of three events can occur. First, the FMRP can go to filopodia and play a role in synaptogenesis. Second, the FMRP can go to spines, where it can regulate some of the protein synthesis that is required for their structure and maintenance. Third, FMRP can also repress protein synthesis at spines. In addition, FMRP in the dendrites could also return to the nucleus via its Nuclear Localization Signal (NLS)[3,40,41]. A summary of these possibilities is shown in Figure 30.2.

## 7. FMRP, MRNA, AND PROTEIN INTERACTIONS

FMRP is in a family of RNA-binding proteins known as heterogeneous nuclear ribonucleoproteins (hnRNPs). FMRP interacts with other proteins and forms part of a larger messenger ribonuclearprotein (mRNP) complex in the nucleus. FMRP normally associates with actively translating polyribosomes, and this association is affected by I304N missense mutation, which alters its ability to bind mRNA *in vitro*[42–44]. Possibly the most convincing evidence that mRNA binding plays a critical role in FXS comes from a single human FXS patient with severe symptoms, but no chromosomal constriction point[45]. This particular case of FXS was the result of a missense mutation, I304N, altering the folding of FMRP's second KH domain[45]. This KH domain likely plays an important role in the normal function of the FMRP protein, judging from the patient's symptoms. The KH domains are RNA-binding domains. The second KH domain in particular binds to a tertiary RNA structure called a kissing complex[46]. In addition to having two KH domains, FMRP has an RGG-type RNA-binding domain (an RGG box), and an amino terminus with a strong affinity for mRNA. In addition to FMRP binding to its own mRNA it also binds with varying specificity to an impressive estimate of 4% of all fetal mRNA's[15]. Further adding to the notion of the importance of FMRP's function as an RNA-binding protein are the extensive summaries of mouse mRNA FMRP is known to bind to. In 2001, a growing list of 432 mRNAs was available, and 251 of these have already been demonstrated to show abnormal polyribosomal profile in the absence of FMRP[47]. During development, polyribosomal aggregates increase in response during experience-dependant synaptogenesis in the spines, suggesting their importance in normal development.

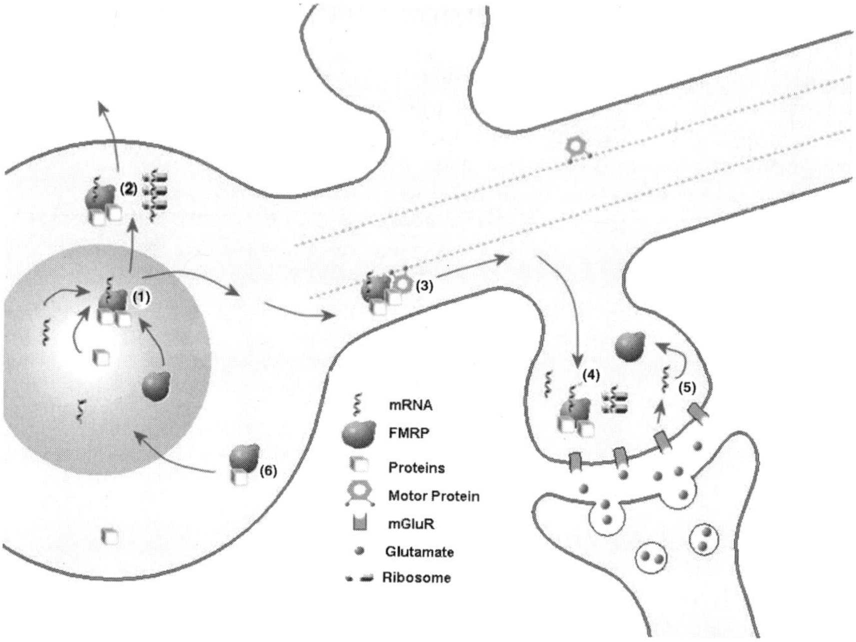


**Figure 30.2.** Summary of Putative FMRP Actions in Dendrites and the Nucleus. As depicted in the figure, FMRP can associate with mRNA and other proteins in the nucleus to form mRNP. (1) mRNP is transported out of the nucleus into the cytoplasm. (2) where the FMRP/mRNA complex can affect translation through its associations with actively translating polyribosomes. (3) The FMRP/mRNA complex can also associate with motor proteins such as kinesin, and be transported down the dendrites to dendritic spines. (4) Near the synapse, FMRP is thought to regulate the translation of mRNA and respond to synaptic messages, including mGluR signaling. (5) In turn, mGluR5 signaling can also cause the production of FMRP at the synapse. (6) FMRP can also enter the nucleus, as it contains a NLS (figure courtesy of Colleen Webber).

FMRP expression can also be modulated by synaptic activity, and it is rapidly synthesized in response to the addition of glutamate and group 1 metabotropic glutamate receptor agonists[48]. Thus, this also suggests a role for FMRP in synapse formation and development, and it has been proposed that FMRP may be linked to dendritic regression[48]. FMRP can also bind to BC1, a dendritic nontranslatable brain mRNA, and the human analog BC200. The binding of BC1 mRNA is strong and takes place at the N-terminus of FMRP. This is a novel binding motif as it leaves the other three RNA-binding domains of the FMRP protein open. This may explain how FMRP regulates the translation of specific mRNAs at the synapse when FMRP is bound to BC1 RNA[49]. In this case, FMRP, in concert with BC1, can then repress the translation of specific mRNAs such as microtubule-associated protein 1B (MAP1b). In normal brain development, active synaptogenesis requires the decline of MAP1b, and in the FXS neuron, this protein may not be negatively regulated.

Further evidence that FMRP is linked to the cytoskeleton comes from FMRP's involvement in the murine Rac 1 pathway[50,51]. This pathway is involved in actin remodeling and is altered in fibroblasts that either lack FMRP, or have a point mutation involving the KH1 or KH2 RNA-binding domain. The MAP1b and the

Rac1 pathways could therefore be an important link to the control of the cytoskeleton during development, and a link between FMRP and the abnormal spine phenotype seen in men and mice missing this protein (i.e., see Figure 30.1).

FMRP has also been suggested to be a general repressor of translation, based on the evidence that the FMRP 1304N mutation fails to suppress translation *in vivo*[52]. Indeed, FMRP seems to be capable of inhibiting the formation of 80S ribosomal complexes[52].

At the synapse, mRNA transport is important for transmission of signals. While it is difficult to identify the location of mRNA in the cell with many techniques, the development of "Antibody Positioned RNA Amplification" has shown for the first time that a loss of FMRP not only reduces certain mRNA's *in vivo*, but also results in a loss of their ability to localize to the dendrites[53].

FMRP also has a role in the nucleus where it is associated with both a functional NES, and an NLS[54]. FMRP has also been shown to bind to single-stranded and double-stranded DNA *in vitro*, binding to double stranded DNA with a lower affinity.

As with most other proteins FMRP also has a host of interacting proteins that have been isolated and characterized. FXR1 and FXR2 are protein homologs of FMRP. FXR1 shares an approximate 90% identity with FMR1, both having almost identical KH domains[55]. FXR2 is very similar to FMRP, sharing 60% of their identities, also containing 2 KH domains, and abilities to bind to mRNA. FMR1, FXR1, and FXR2 can all form heteromers with each other, and all can form homomers, as well. They all interact tightly in the cytoplasm and nucleus[56].

Two homologous cytoplasmic FMRP-interacting proteins, CYFIP1 and 2, also interact with FMRP, and CYFIP2 interacts with FXR1 and FXR2. Both are found localized at the synapses[51]. Although Drosophila only has dFMRP, lacking the orthologs for FXR1 and FXR2, there is a change in neural connectivity in this animal model. In the Drosophila model, Schenck et al. showed that expression of CYFIP, the Drosophila ortholog of CYFIP 1 and 2, was specific to the nervous system, and more importantly it links FMRP to the Rac1 GTPase pathway[51]. This essentially couples signal-dependant cytoskeleton remodeling and translation. Perhaps the most exciting link to FMRP's role in translation comes from recent electrophysiological data.

## 8. SYNAPTIC ELECTROPHYSIOLOGY AND THE mGLUR THEORY OF FXS

The loss of FMRP has been linked to several functional deficits in synaptic physiology. Alterations in synaptic plasticity have been observed in cortex[57], the anterior cingulate cortex, lateral amygdala[58], and hippocampus[59]. These alterations in synaptic physiology are not universal, and NMDA receptor-dependent forms of synaptic plasticity in the hippocampus are not altered in animals that lack FMRP[18,59,60]. Synaptic plasticity that involves group 1 metabotropic glutamate receptor activation is most dramatically affected by the loss of FMRP. Surprisingly, this does not produce a deficit, and instead mGluR-dependant LTD is significantly increased in the CA1 region of the hippocampus of FMR1 knockout mice. Moreover, this has been shown using two different methods for inducing mGluR LTD *in vitro*, while NMDAR-dependant LTD was found to be normal in these same experiments[59].

With over 400 mRNA-binding partners of FMRP, and only a subtle behavioral phenotype for FMR1 knockout mouse models, identifying the main problems of these mice is a daunting task. If mGluR-dependant LTD is increased, then the loss of FMRP seems to result in an increase in the activity of the mGluR signaling pathway. Keeping in mind that mGluR requires rapid translation of pre-existing mRNA, and the fact that FMRP has been implicated in repressing translation, it was thought that FMRP could function to inhibit further protein synthesis caused by mGluR activation. Conversely, a lack of FMRP, as in FXS, would then theoretically lead to exaggerated mGluR signaling.

In addition to the electrophysiological data, it was pointed out that a number of other observations also support a role for altered mGluR signaling in FXS[61]. First, prepulse inhibition of an auditory startle reflex is normally decreased in mGluR1 and mGluR5 knockout mice, while it is increased in the FMR1 knockout mouse. Second, FXS is often associated with loose bowels, and mGluR5 is involved in the innervation of the ileum and mGluR agonists increase intestinal motility. Third, FXS is also characterized by a hypersensitivity to tactile stimulation, while mGluR5 receptors are involved in nocioception. In addition, there are a number of other anecdotal findings that give credence to the mGluR hypothesis for FXS[61].

In summary, the mGluR theory predicts that the symptoms of FXS can be alleviated with specific mGluR antagonists. Theoretically, FMRP regulates a small portion of mRNAs, leaving other mRNAs to compete for resources for translation. Taking away FMRP upsets this delicate balance, and the potential treatment would involve restoring this balance with mGluR antagonists. Of course if the concentration of mGluR antagonists administered is too strong, the treatment could be equivocal to the disease itself as an imbalance would again be induced. For these reasons, and the fact that mGluR1 antagonists can disrupt cerebellar functioning, and cause ataxia, the mGluR5 antagonist 2-methyl-6-(phenylethynyl)pyridine (MPEP) has been more of a focus in current research. The effects of blocking mGluR5 are not likely to be beneficial however, as while mGluR5 knockout mice show decreased anxiety[62], they also show cognitive deficits[63].

The selective mGluR5 antagonist MTEP (3-[(2-methyl-1,3-thiazol-4-yl) ethynyl]-pyridine) also seems to have some antidepressant effect on rats and mice[64]. With so many studies on FXS focusing on the relation between a few proteins, the attraction of the mGluR theory is that it gives a link between a single molecular mechanism and a number of behavioral and cognitive symptoms. In addition, the mGluR theory links one mechanism to the misregulation of many proteins. Although it is just a theory, it has already inspired some exciting discoveries, some of which are outlined below.

## 9. REPAIRING THE FRAGILE SYNAPSE

### 9.1. MPEP and Other mGluR Antagonists

MPEP and other mGluR antagonists seem to be effective in reversing some of the deficits observed in mGluR5 knockout mice. Administration of MPEP resulted in reversals for both the increased susceptibility to audiogenic seizures, and the decrease in open field activity (a measure of anxiety) normally common to these knockout animals[65–67]. The knockout mice developed a tolerance to MPEP, but this could be overcome by using higher doses.

### 9.2. LiCl

Lithium chloride has long been in use in human patients for the treatment of mania, after being approved by the FDA in the 1970s. There are currently clinical trials ongoing to test the efficacy of lithium in FXS patients, and some evidence exists that there may be some benefits stabilizing mood swings in FXS patients[19]. The potential benefits of lithium have been brought to light by a research group that has looked at the effects of LiCl and MPEP in the Drosophila model of FXS. Both treatments restored some phenotypes of the animals missing dFMRP[65]. Only two mGluRs are present in the Drosophila genome, DmGluRA and DmGluRB, more closely resembling vertebrate group 2 mGluRs. Possibly because of its effects on these mGluRs LiCl seemingly restored memory for courtship behavior in knockout flies, along with mushroom body deficits caused by the lack of dFMRP. Lithium is linked to the mGluR pathway through it's effects on CREB binding and inositol triphosphate receptor mediated calcium release. This has been outlined previously in relation to mGluR signalling in a study by McBride and colleagues[65,68,70].

### 9.3. Gene Therapy

Gene therapy has the potential to cure FXS by simply restoring FMRP production. Admittedly, there are a number of obstacles to overcome, including finding the best structure of the gene, expressing this in cells, and the packaging and delivery of the gene. The use of an adenovirus has been suggested as one method of delivery, and it is hoped that the FMR1 knockout mouse will be an preclinical test of this therapy. The potential for the success of gene theraphy is illustrated in one large study, which has had success in both phase 1 and phase 2 clinical trials. Reduced lung function is the fatal symptom of cystic fibrosis, and using adenoviral infection as a method of delivery can improve human lung function in these patients[71]. Gene therapy does share some of the same problems as the pharmacological treatments however, as the ultimate goal remains to restore a delicate balance in the biochemical pathways involved with FMRP regulation. It remains to be determined how delicate the balance actually is, however, as male premutation carriers actually have elevated levels of FMRP and can exhibit symptoms associated with the premutation, FXTAS, later in life.

### 9.4. Other Options

FXS in humans is caused by hypermethylation and the recruitment of histone deacetylases which cause chromosome condensation that effectively shuts off this gene. Stopping this sequence of events would help greater than 95% of FXS patients who show a lack of FMRP as most individuals have a gene that can produce functional FMRP, rather than a point mutation. One study showed that in lymphoblastoid cells that contained 270–710 trinucleotide repeats (the full FXS mutation) FMRP production could still be seen[13]. In this case, the FMRP gene was reactivated modestly with the use of histone hyperacetylases. Using these in concert with DNA-demethylating agents, FMR1 mRNA levels were increased even more, often doubling the results seen with the use either one alone[13]. One recent review suggests that histone deacetylase inhibitors are potentially effective neuroprotective agents, and that further research in this area could help to treat diseases like FXS, leukemia, and various other cancers[72]. There is some evidence that Huntington's disease, Alzheimer's disease, amytrophic lateral sclerosis, and stroke could possibly benefit from this area of research, as well[72].

Although there are complex biochemical changes in mice that lack FMRP, the knockout mouse phenotype deficiencies can be largely rescued by environmental enrichment, as well as increasing alpha-amino-3-hydroxy-5-methyl-4-isoxazolepropionic acid (AMPA) glutamate receptor subunit 1 (GluR1) levels. Paradoxically, environmental enrichment does not increase FMRP levels in wild type mice[73]. In some cases however, experience can change the expression and localization of FMRP[74]. Visual experience for example has been shown to regulate this phenomena in the visual cortex neurons[74]. Whisker stimulation can also cause neurons in the barrel cortex to translate FMRP (but this is stimulation dependant), and requires the activation of Group 1 mGluRs[75].

## 10. CONCLUSIONS

It seems clear that FMRP can play several roles in neuronal functioning, and one of the major thrusts of current research will be to define the RNA and protein partners of FMRP so that we can understand more clearly how it is involved in mRNA translation, export and localization. The role of FMRP in synapse formation is also intriguing, and current efforts to understand how it interacts with mGluR receptors, and the role FMRP plays in the pruning of spines in the developing nervous system, will be exciting developments in FXS research. Clearly these recent efforts have improved our understanding of the effects of FXS on synaptic functioning and have given hope for the development of pharmacological therapies for FXS.[*]

## 11. REFERENCES

1. Fisch, G.S., Carpenter, N.J., Holden, J.J., Simensen, R., Howard-Peebles, P.N., Maddalena, A., Pandya, A., and Nance, W. (1999) *Am J Med Genet* **83**, 257–263.
2. O'Donnell, W.T., and Warren, S.T. (2002) *Annu Rev Neurosci* **25**, 315–338.
3. Oostra, B.A., and Willemsen, R. (2003) *Hum Mol Genet* **12**, R249–257.
4. Hagerman, R.J., Leehey, M., Heinrichs, W., Tassone, F., Wilson, R., Hills, J., Grigsby, J., Gage, B., and Hagerman, P.J. (2001) *Neurology* **57**, 127–130.
5. Hagerman, R.J., and Hagerman, P.J. (2002) *Curr Opin Genet Dev* **12**, 278–283.
6. Lubs, H.A. (1969) *Am J Hum Genet* **21**, 231–244.
7. Verkerk, A.J., Pieretti, M., Sutcliffe, J.S., Fu, Y.H., Kuhl, D.P., Pizzuti, A., Reiner, O., Richards, S., Victoria, M.F., Zhang, F.P., et al. (1991) *Cell* **65**, 905–914.
8. Sherman, S.L., Jacobs, P.A., Morton, N.E., Froster-Iskenius, U., Howard-Peebles, P.N., Nielsen, K.B., Partington, M.W., Sutherland, G.R., Turner, G., and Watson, M. (1985) *Hum Genet* **69**, 289–299.
9. Sherman, S.L., Morton, N.E., Jacobs, P.A., and Turner, G. (1984) *Ann Hum Genet* **48**, 21–37.
10. Crawford, D.C., Acuna, J.M., and Sherman, S.L. (2001) *Genet Med* **3**, 359–371.
11. Nolin, S.L., Brown, W.T., Glicksman, A., Houck, G.E., Jr., Gargano, A.D., Sullivan, A., Biancalana, V., Brondum-Nielsen, K., Hjalgrim, H., Holinski-Feder, E., Kooy, F., Longshore, J., Macpherson, J., Mandel, J.L., Matthijs, G., Rousseau, F., Steinbach, P., Vaisanen, M.L., von Koskull, H., and Sherman, S.L. (2003) *Am J Hum Genet* **72**, 454–464.
12. Coffee, B., Zhang, F., Warren, S.T., and Reines, D. (1999) *Nat Genet* **22**, 98–101.
13. Chiurazzi, P., Pomponi, M.G., Pietrobono, R., Bakker, C.E., Neri, G., and Oostra, B.A. (1999) *Hum Mol Genet* **8**, 2317–2323.
14. Amir, R.E., Van den Veyver, I.B., Wan, M., Tran, C.Q., Francke, U., and Zoghbi, H.Y. (1999) *Nat Genet* **23**, 185–188.

[*]This work was supported by grants from HELP (BC Ministry of Children and Development), The Fragile X Research Foundation of Canada, and the Scottish Rite Charitable Foundation of Canada to BRC. BRC is the BMO Young Investigator at UBC Hospital.

15. Ashley, C.T., Jr., Wilkinson, K.D., Reines, D., and Warren, S.T. (1993) *Science* **262**, 563–566.
16. D'Hooge, R., Nagels, G., Franck, F., Bakker, C.E., Reyniers, E., Storm, K., Kooy, R.F., Oostra, B.A., Willems, P.J., and De Deyn, P.P. (1997) *Neuroscience* **76**, 367–376.
17. Dobkin, C., Rabe, A., Dumas, R., El Idrissi, A., Haubenstock, H., and Brown, W.T. (2000) *Neuroscience* **100**, 423–429.
18. Godfraind, J.M., Reyniers, E., De Boulle, K., D'Hooge, R., De Deyn, P.P., Bakker, C.E., Oostra, B.A., Kooy, R.F., and Willems, P.J. (1996) *Am J Med Genet* **64**, 246–251.
19. Hagerman, R.J., and Hagerman, P.J. (2002) *Fragile X Syndrome: Diagnosis, Treatment and Research*, The John Hopkins University Press, Baltimore, MD.
20. Bakker, C.E., and Oostra, B.A. (2003) *Cytogenet Genome Res* **100**, 111–123.
21. Kooy, R.F., Reyniers, E., Verhoye, M., Sijbers, J., Bakker, C.E., Oostra, B.A., Willems, P.J., and Van Der Linden, A. (1999) *Eur J Hum Genet* **7**, 526–532.
22. Slegtenhorst-Eegdeman, K.E., de Rooij, D.G., Verhoef-Post, M., van de Kant, H.J., Bakker, C.E., Oostra, B.A., Grootegoed, J.A., and Themmen, A.P. (1998) *Endocrinology* **139**(1), 156–162.
23. Kooy, R.F. (2003) *Trends Genet* **19**, 148–154.
24. Wan, L., Dockendorff, T.C., Jongens, T.A., and Dreyfuss, G. (2000) *Mol Cell Biol* **20**, 8536–8547.
25. Dockendorff, T.C., Su, H.S., McBride, S.M., Yang, Z., Choi, C.H., Siwicki, K.K., Sehgal, A., and Jongens, T.A. (2002) *Neuron* **34**, 973–984.
26. Morales, J., Hiesinger, P.R., Schroeder, A.J., Kume, K., Verstreken, P., Jackson, F.R., Nelson, D.L., and Hassan, B.A. (2002) *Neuron* **34**, 961–972.
27. Hessl, D., Rivera, S.M., and Reiss, A.L. (2004) *Ment Retard Dev Disabil Res Rev* **10**, 17–24.
28. Mostofsky, S.H., Mazzocco, M.M., Aakalu, G., Warsofsky, I.S., Denckla, M.B., and Reiss, A.L. (1998) *Neurology* **50**, 121–130.
29. Reiss, A.L., Abrams, M.T., Greenlaw, R., Freund, L., and Denckla, M.B. (1995) *Nat Med* **1**(2), 159–167.
30. Rudelli, R.D., Brown, W.T., Wisniewski, K., Jenkins, E.C., Laure-Kamionowska, M., Connell, F., and Wisniewski, H.M. (1985) *Acta Neuropathol (Berl)* **67**, 289–295.
31. McKinney, B.C., Grossman, A.W., Elisseou, N.M., and Greenough, W.T. (2005) *Am J Med Genet B Neuropsychiatr Genet* **136**, 98–102.
32. Irwin, S.A., Galvez, R., and Greenough, W.T. (2000) *Cereb Cortex* **10**, 1038–1044.
33. Irwin, S.A., Idupulapati, M., Gilbert, M.E., Harris, J.B., Chakravarti, A.B., Rogers, E.J., Crisostomo, R.A., Larsen, B.P., Mehta, A., Alcantara, C.J., Patel, B., Swain, R.A., Weiler, I.J., Oostra, B.A., and Greenough, W.T. (2002) *Am J Med Genet* **111**, 140–146.
34. Nimchinsky, E.A., Oberlander, A.M., and Svoboda, K. (2001) *J Neurosci* **21**, 5139–5146.
35. Braun, K., and Segal, M. (2000) *Cereb Cortex* **10**, 1045–1052.
36. Galvez, R., and Greenough, W.T. (2005) *Am J Med Genet A* **135**, 155–160.
37. Pan, L., Zhang, Y.Q., Woodruff, E., and Broadie, K. (2004) *Curr Biol* **14**, 1863–1870.
38. Martin, I., Obradovic, B., Treppo, S., Grodzinsky, A.J., Langer, R., Freed, L.E., and Vunjak-Novakovic, G. (2000) *Biorheology* **37**, 141–147.
39. Steward, O., Bakker, C.E., Willems, P.J., and Oostra, B.A. (1998) *Neuroreport* **9**, 477–481.
40. Bagni, C., and Greenough, W.T. (2005) *Nat Rev Neurosci* **6**, 376–387.
41. Antar, L.N., Afroz, R., Dictenberg, J.B., Carroll, R.C., and Bassell, G.J. (2004) *J Neurosci* **24**, 2648–2655.
42. Feng, Y., Absher, D., Eberhart, D.E., Brown, V., Malter, H.E., and Warren, S.T. (1997) *Mol Cell* **1**, 109–118.
43. Feng, Y., Gutekunst, C.A., Eberhart, D.E., Yi, H., Warren, S.T., and Hersch, S.M. (1997) *J Neurosci* **17**, 1539–1547.
44. Stefani, G., Fraser, C.E., Darnell, J.C., and Darnell, R.B. (2004) *J Neurosci* **24**, 7272–7276.
45. De Boulle, K., Verkerk, A.J., Reyniers, E., Vits, L., Hendrickx, J., Van Roy, B., Van den Bos, F., de Graaff, E., Oostra, B.A., and Willems, P.J. (1993) *Nat Genet* **3**, 31–35.
46. Darnell, J.C., Fraser, C.E., Mostovetsky, O., Stefani, G., Jones, T.A., Eddy, S.R., and Darnell, R.B. (2005) *Genes Dev* **19**, 903–918.
47. Brown, V., Jin, P., Ceman, S., Darnell, J.C., O'Donnell, W.T., Tenenbaum, S.A., Jin, X., Feng, Y., Wilkinson, K.D., Keene, J.D., Darnell, R.B., and Warren, S.T. (2001) *Cell* **107**, 477–487.
48. Greenough, W.T., Klintsova, A.Y., Irwin, S.A., Galvez, R., Bates, K.E., and Weiler, I.J. (2001) *Proc Natl Acad Sci U S A* **98**, 7101–7106.
49. Zalfa, F., Giorgi, M., Primerano, B., Moro, A., Di Penta, A., Reis, S., Oostra, B., and Bagni, C. (2003) *Cell* **112**, 317–327.
50. Billuart, P., and Chelly, J. (2003) *Neuron* **38**, 843–845.
51. Schenck, A., Bardoni, B., Langmann, C., Harden, N., Mandel, J.L., and Giangrande, A. (2003) *Neuron* **38**, 887–898.
52. Laggerbauer, B., Ostareck, D., Keidel, E.M., Ostareck-Lederer, A., and Fischer, U. (2001) *Hum Mol Genet* **10**, 329–338.

53. Miyashiro, K.Y., Beckel-Mitchener, A., Purk, T.P., Becker, K.G., Barret, T., Liu, L., Carbonetto, S., Weiler, I.J., Greenough, W.T., and Eberwine, J. (2003) *Neuron* **37**, 417–431.
54. Eberhart, D.E., Malter, H.E., Feng, Y., and Warren, S.T. (1996) *Hum Mol Genet* **5**, 1083–1091.
55. Siomi, M.C., Siomi, H., Sauer, W.H., Srinivasan, S., Nussbaum, R.L., and Dreyfuss, G. (1995) *EMBO J* **14**, 2401–2408.
56. Zhang, Y., O'Connor, J.P., Siomi, M.C., Srinivasan, S., Dutra, A., Nussbaum, R.L., and Dreyfuss, G. (1995) *EMBO J* **14**, 5358–5366.
57. Li, J., Pelletier, M.R., Perez Velazquez, J.L., and Carlen, P.L. (2002) *Mol Cell Neurosci* **19**, 138–151.
58. Zhao, M.G., Toyoda, H., Ko, S.W., Ding, H.K., Wu, L.J., and Zhuo, M. (2005) *J Neurosci* **25**, 7385–7392.
59. Huber, K.M., Gallagher, S.M., Warren, S.T., and Bear, M.F. (2002) *Proc Natl Acad Sci U S A* **99**, 7746–7750.
60. Paradee, W., Melikian, H.E., Rasmussen, D.L., Kenneson, A., Conn, P.J., and Warren, S.T. (1999) *Neuroscience* **94**, 185–192.
61. Bear, M.F., Huber, K.M., and Warren, S.T. (2004) *Trends Neurosci* **27**, 370–377.
62. Brodkin, J., Bradbury, M., Busse, C., Warren, N., Bristow, L.J., and Varney, M.A. (2002) *Eur J Neurosci* **16**, 2241–2244.
63. Lu, Y.M., Jia, Z., Janus, C., Henderson, J.T., Gerlai, R., Wojtowicz, J.M., and Roder, J.C. (1997) *J Neurosci* **17**, 5196–5205.
64. Palucha, A., Branski, P., Szewczyk, B., Wieronska, J.M., Klak, K., and Pilc, A. (2005) *Pharmacol Biochem Behav* **81**, 901–906.
65. McBride, S.M., Choi, C.H., Wang, Y., Liebelt, D., Braunstein, E., Ferreiro, D., Sehgal, A., Siwicki, K.K., Dockendorff, T.C., Nguyen, H.T., McDonald, T.V., and Jongens, T.A. (2005) *Neuron* **45**, 753–764.
66. Yan, Q.J., Rammal, M., Tranfaglia, M., and Bauchwitz, R.P. (2005) *Neuropharmacology,* **49**, 1053–1066.
67. Zhang, L.Q., Wang, H., Lai, B.T., Yue, W.T., Zhan, X.P., Yang, X.H., Zhang, C.Y., and Zhang, T.M. (2005) *Zhonghua Jie He He Hu Xi Za Zhi* **28**, 333–336.
68. Berridge, M.J., and Irvine, R.F. (1989) *Nature* **341**, 197–205.
69. Berridge, M.J. (1993) *Nature* **361**, 315–325.
70. Grimes, C.A., and Jope, R.S. (2001) *J Neurochem* **78**, 1219–1232.
71. Liu, X., Luo, M., Zhang, L.N., Yan, Z., Zak, R., Ding, W., Mansfield, S.G., Mitchell, L.G., and Engelhardt, J.F. (2005) *Hum Gene Ther* **16**, 1116–1123.
72. Langley, B., Gensert, J.M., Beal, M.F., and Ratan, R.R. (2005) *Curr Drug Targets CNS Neurol Disord* **4**, 41–50.
73. Restivo, L., Ferrari, F., Passino, E., Sgobio, C., Bock, J., Oostra, B.A., Bagni, C., and Ammassari-Teule, M. (2005) *Proc Natl Acad Sci U S A* **102**, 11557–11562.
74. Gabel, L.A., Won, S., Kawai, H., McKinney, M., Tartakoff, A.M., and Fallon, J.R. (2004) *J Neurosci* **24**, 10579–10583.
75. Todd, P.K., Malter, J.S., and Mack, K.J. (2003) *Brain Res Mol Brain Res* **110**, 267–278.

# 31

# SYNAPTIC ABNORMALITIES ASSOCIATED WITH HUNTINGTON'S DISEASE

Austen J. Milnerwood and Lynn A. Raymond*

## 1. SUMMARY

In the little over a decade since the causal mutation in Huntington's disease (HD) was first identified, over a dozen genetic mouse models, as well as a variety of cell and invertebrate models, of HD have been produced. These models are continually enhancing our understanding of the varied roles of normal and mutated huntingtin proteins. This ubiquitously expressed and multi-functional protein has been found to be involved in a growing number of protein–protein interactions that, when dysfunctional, can have dire consequences for neuronal function and survival. The clinical features of HD have been regarded traditionally as resulting from neurodegeneration. Recent advances suggest that synaptic dysfunction may be the first detectable effect of the HD mutation and correlate with early clinical signs and symptoms of the disease. Moreover, mechanisms underlying synaptic dysfunction may also contribute to selective neuronal vulnerability in HD. Further investigation in this area will likely lead to novel therapies to treat symptoms, as well as to potentially delay onset and slow progression, of this devastating disorder.

## 2. INTRODUCTION TO HD

HD is an autosomal dominant, progressive and fatal neurodegenerative disease. The classic clinical features of adult onset HD include involuntary choreic movements and motor incoordination that progress to severe bradykinesia and rigidity. Emotional disturbance, personality changes, and dementia are prevalent and may be the first indications of illness, often many years prior to motor

*Department of Psychiatry and Brain Research Centre, University of British Columbia, 2255 Wesbrook Mall, Vancouver, Canada, V6T 1X7; lynnr@interchange.ubc.ca

symptoms. Presently there is no cure and no effective treatment, and death occurs following ~20 years of inexorable decline.

The incidence of HD is 5–10 in every 100,000 people; however, HD is one of an increasing number of similar neurodegenerative diseases caused by triplet repeat expansions. HD follows Mendelian autosomal dominant inheritance. Exon 1 of the HD gene includes a polymorphic CAG repeat[1], which translates to a tract of glutamine residues (polyglutamine repeat; polyQ) in the protein huntingtin (htt). HD gene carriers have larger repeats than normal individuals, and there is a negative correlation between repeat length and age of onset: adult onset (20–80 years) is attributed to >36 polyQ, while juvenile onset (2-20 years) generally occurs with >70 repeats. The neuropathology of HD is most evident in the corpus striatum of the basal ganglia, a group of interconnected nuclei involved in motor coordination, cognitive function, and subcortical memory processes. The most vulnerable striatal cells are GABAergic medium-sized spiny projection neurons (MSNs), which receive extensive glutamatergic input from all cortical areas and represent the main striatal output. HD pathology is classified grade 0–4 post mortem, ranging from no microscopic changes at grade 0 (despite substantial clinical evidence) to 95% loss of MSNs in grade 4[2]. HD is a severe neurodegenerative disease; however, recent advances suggest that many clinical features of the disease correlate best with cellular dysfunction prior to cell death.

This chapter reviews leading theories current to our understanding of synaptic dysfunction in HD. The chapter is constructed around a simplified distinction between pre- and postsynaptic function at the cortico-striatal synapse. Due to the synaptic focus, much evidence reviewed here was derived from mouse models of HD. Each model recapitulates different characteristics of human HD, and the severity of phenotype is inversely proportional to HD gene transcript length and directly proportional to polyQ repeat load and expression levels (reviewed in ref. 3); a comparison of their basic characteristics is outlined in Table 1.

**Table 1.** Summary of Referenced HD Mouse Models.

| Model | polyQ | Construct | Age at onset (months) Cognitive/behavioral | Motor | Death |
|---|---|---|---|---|---|
| R6/2, R6/1 | 150, 120 | N-terminal tg | <1, 2.5 | 1, 4.5 | 4, >12 |
| HD100 | 100 | N-terminal tg | NR | 3 | Normal |
| N171 | 82 | N-terminal tg | NR | 3 | 5 |
| YAC46, YAC72 | 46, 72 | Full length tg | Not seen, 3 | Not seen, 7 | Normal |
| CAG Knock-in | 94 | Tg knock-in | 2 | 4 | Normal |
| *HdhQ* | 80 | *Hdh* knock-in | <3 | Not seen | Normal |

NR = Not reported; Tg = transgenic; *Hdh* = Huntington's disease homolog (murine). For details see www.hdfoundation.org/PDF/hdmicetable.pdf

## 3. SELECTIVE NEURONAL VULNERABILITY IN HD

Normal htt is expressed throughout the brain. The protein associates with microtubules, vesicular organelles, and mitochondria and is believed to be involved in organelle transport[4]. Mutant huntingtin (mhtt) is also widely expressed and appears to localize similarly to the normal protein[5,6]. Selective neuronal vulnerability cannot be explained by regional variation in htt expression, since the highest levels are found in cerebellum, which is relatively preserved in HD, whereas levels of htt in the striatum are less than in many other brain regions[6]. It was proposed that the mutation confers a "gain of function", producing an aberrant protein that is widely expressed, yet harmful only to some neurons[5]. In support of this, the phenotype of htt deficient mice, which is embryonic lethality, is unlike that of HD; during early development htt appears to have a role in nutrient supply, hematopoiesis, rRNA and mRNA biogenesis, perinuclear membrane trafficking, and iron metabolism (reviewed in ref. 7) – all findings that support the proposed role for htt in microtubule transport in mature animals. However, such data do not preclude a loss of other unrelated htt functions in mature animals. For example, conditional (postnatal day 5) htt knockout mice mature, but later develop a neurological phenotype[7]. Recent research demonstrates that htt interacts with a wide variety of neuronal proteins, many of which are enriched in the synapse. Investigations of novel and altered protein interactions within striatal neurons may help explain the selective vulnerability of these cells to a mutated protein that is ubiquitously expressed and crucial to the normal function of a variety of cell types.

## 4. PRESYNAPTIC DYSFUNCTION

### 4.1. Axonal Transport

The discovery that htt microaggregates form preferentially in the neuropil of MSNs, and especially in their projection target areas, in knock-in HD mice (*HdhQ*) that do not develop an overt HD phenotype suggested that MSN axonal dysfunction is one of the earliest pathological changes in HD[8]. These aggregates also formed in mhtt-expressing cultured striatal cells where they blocked protein transport and caused neuritic degeneration[8]. Following this report, two studies added support to the axonal defect theory for HD pathogenesis, convincingly demonstrating that htt is required for normal axonal transport, and that expanded polyQ-containing peptides produce severe axonal transport deficits in *Drosophila*[9] and squid axoplasm[10]. Recent work has produced similar findings in mammalian neurons *in vivo* and *in vitro*[11]. Data from earlier subcellular localization studies also support the idea that axonal dysfunction may contribute to HD pathogenesis[4].

The first htt-interacting protein described, Huntingtin-Associated Protein 1 (HAP1), is highly expressed in the brain and binds htt in a manner that is enhanced by expanded polyQ[12]. Like htt, HAP1 localizes with membrane-bound organelles (including vesicles), dendritic and axonal microtubules[13], and the vesicular fractions of nerve terminals, where it interacts with the p150 subunit of dynactin[14,15], a protein required by the vesicular microtubule transport protein dynein. Perturbation of htt/HAP1-dependent vesicular trafficking by polyQ expansion may underlie the decreased synaptic vesicle number and HAP1–vesicle interactions, as well as impaired neurotransmitter (NT) release, observed after phenotypic onset in two HD mouse models[16]. Axonal transport of the MSN prosurvival factor, brain-derived neurotrophic factor (BDNF), is quickened by htt via its association with HAP1 and is much reduced by mhtt, resulting in neurotoxicity[15]. These data and others suggest that htt–HAP1 interactions are disrupted by polyQ expansion, causing a deficit in BDNF trophic support from the cortex to the striatum and the development of striatal neurodegeneration[14].

### 4.2. Vesicle Fusion and Neurotransmitter Release

Electrophysiological investigation of hippocampal synaptic plasticity in HD knock-in mice (*HdhQ*) demonstrated impaired NT release during periods of sustained activity[17]. Subsequent investigations have shown, at the cortico-striatal synapse of R6/2 mice, that unusually frequent large-amplitude synaptic events occur, with a concomitant and progressive reduction in the frequency of miniature excitatory postsynaptic currents (mEPSCs), beginning at the same time as the behavioral phenotype[18]. Together such observations suggest that mhtt interferes directly with the vesicle release cycle.

The synapsin family of proteins associates with synaptic vesicles and elements of the cytoskeleton and, although the exact function of synapsins remains unclear, this binding may regulate the size of the synaptic vesicle reserve pool (and concomitantly the readily releasable pool; RRP) at axon terminals (reviewed in ref. 19). Phosphorylation of synapsin 1 reduces its affinity for binding both vesicles and F-actin; and although synapsin 1 expression is unaltered in R6/2 HD mice[20,21], the levels of phosphorylated synapsin 1 are increased after onset of motor symptoms[20]. This suggests that, at this age, R6/2 mice may exhibit alterations in the size of the RRP. These data may help explain the unusually large spontaneous EPSCs observed at the cortico-striatal synapse around this time[18]. Synapsin 1 also interacts with the synaptic vesicle-associated small G protein Rab3A, which appears to negatively regulate a late step of synaptic vesicle fusion, controlling quantal release (reviewed in ref. 19). Thus, synapsin hyperphosphorylation[20] may also disrupt vesicular release and contribute to the reduced mEPSC frequency in R6/2 mice after the onset of symptoms[18]. More recent analysis of mRNA and protein levels of a number of synaptic proteins in the less severe phenotype of the R6/1 mouse has demonstrated that a Rab3A-interacting protein, Rabphilin 3A, is selectively reduced in cortex at the age at which symptoms appear[22].

Levels of the synaptic vesicle release protein complexin II (CPLXII) are selectively and progressively depleted in R6/2 mice[23] and human HD tissue[24]. Furthermore, PC12 cells expressing mhtt also show reduced CPLXII with concomitantly reduced exocytosis that was reversed by CPLXII overexpression[25]. Importantly, CPLXII mRNA expression is also reduced in individual striatal cells of R6/2 mice, prior to the onset of overt phenotype[26] and temporally associated with the appearance of cognitive deficits[27]. Analysis of CPLXI/II knock-out neurons suggests that complexins positively regulate synaptic vesicle fusion[28], consistent with a reduction of complexin II underlying reduced exocytosis in PC12 cells[25].

Mice lacking complexin II have near-normal basal hippocampal transmission but reduced short-term potentiation and long-term potentiation (LTP)[29], suggesting that NT release may be impaired during periods of sustained activity. These observations are strikingly similar to those from R6/2 mice[30]. Together, these studies suggest that reduced LTP and cognitive deficits in the R6/2 may be caused, at least in part, by altered CPLXII function. However, in the presence of the NMDA receptor antagonist AP5, short-term potentiation in R6/2 mice is similar to controls, suggesting that altered NMDAR activity, and not presynaptic NT release per se, is causal to differences in short-term potentiation[30]. Moreover, while complexin II knock-out mice do exhibit progressive cognitive and motor deficits[31], which are similar to those seen in R6/2 mice, the phenotype is mild in comparison to R6/2 and occurs at a later age. Thus, R6/2 cognitive and motor deficits may be only partially attributable to reduced CPLXII expression.

Other proteins that regulate NT release at the presynaptic terminal are also implicated in HD. Expanded mhtt, but not htt, binds the Cysteine String Protein (CSP), a molecular chaperone expressed on synaptic vesicles, and blocks its tonic inhibition of N-type $Ca^{2+}$ channels[32]. NT release is also modulated by presynaptic metabotropic glutamate (mGluRs), NMDA, cannabinoid and dopamine receptors (DARs), all of which are present on cortical terminals and altered in HD (see Section 5).

### 4.3. Vesicle Recovery

The huntingtin-interacting protein 1 (HIP1) binds and colocalizes with the N-terminal portion of huntingtin[33,34], and the efficacy of huntingtin/HIP1 binding is diminished by polyglutamine expansion[34]. HIP1 also binds clathrin and is thus implicated in clathrin-mediated endocytosis, a process that facilitates the rapid retrieval of synaptic vesicles at the presynaptic terminal[35].

The disruption of this and other interactions (e.g., see Section 4.1) by polyglutamine expansion is evidence for "loss of function" effects relating to NT release. In addition, "gain of function" effects may also be important in the HD vesicle cycle. For example, htt binding of the protein kinase C and casein kinase substrate in neurons protein 1 (PACSIN1: a regulator of vesicle recovery) is increased proportionally to polyQ repeat length[36], and PACSIN1 appears to be abnormally distributed away from synaptic compartments in HD tissue[36].

### 4.4. Glutamate Uptake

In addition to disruption of vesicular release of NT, synaptic transmission may also be perturbed at the HD synapse by abnormal clearance of glutamate from the synaptic cleft. *In situ* hybridization studies of astrocytic glutamate transporter (EAAT1/GLT1) mRNA in the neostriatum of post mortem HD brains revealed decreased GLT1 message that was related to reduced expression within astrocytes[37]. GLT1 mRNA and glutamate uptake are also reduced in R6/1 and R6/2 mice after phenotypic onset[20,38]. Such dysfunction may underlie the increases in basal hippocampal transmission that are observed in R6/1 mice long after the onset of phenotype (A. Milnerwood, in press).

### 4.5. Presynaptic Dysfunction and HD Symptom Progression

Several lines of evidence detailed above provide mechanisms by which glutamate release might be perturbed by mhtt. Consistent with these findings, a variety of studies have suggested that progressively reduced presynaptic function occurs at or after the onset of overt symptoms in high polyQ, N-terminal fragment models[16,18,39], that is not observed at early stages in full-length construct YAC72 mice[40]. In light of these changes, it is interesting that a combination drug therapy aimed at increasing NT levels in R6/2 mice delayed cognitive impairment and some alterations in gene expression, in addition to producing a 7% increase in survival; however, motor symptoms continue unabated[41].

## 5. POSTSYNAPTIC DYSFUNCTION

### 5.1. Morphological and Gross Membrane Alterations

Altered cellular morphology is well characterized in human HD tissue, consisting of: increased dendritic complexity, as well as spine number and size in moderate HD grades, and truncated, swollen dendritic processes with marked spine loss in advanced cases[42]. Across many of the HD mouse models, reductions in brain volume and weight are observed when little or no cell death is seen, a finding possibly accounted for by altered cellular morphology, including reduced membrane area and spine numbers, (reviewed in ref. 3). This could be the result of an active process in response to altered afferentation and/or synaptic activity, or a breakdown of the cytoskeletal structure, as has been reported in human HD tissue[24].

The basic integrity of the cell membrane and the rapid passage of ions across it are fundamental to the unique properties of excitable cells. Changes to membrane function have direct consequences for neuronal activity: the ability to receive, process, store, and transmit information. Hippocampal pyramidal cell intrinsic membrane properties and basal transmission are near normal in R6/2 mice, even at the terminal phase[30]. In contrast, the membrane physiology of striatal MSNs is markedly altered in adult R6/2 mice and a subset of cells in CAG knock-in mice[43–46]. Such changes predict that striatal MSNs are hyperexcitable after the onset of symptoms. Similar alterations in membrane properties are observed across several neuronal types at later, postsymptomatic ages in R6/1 mice (K. Murphy, unpublished observations). In MSNs, potassium ($K^+$) channel conductances maintain marked hyperpolarization, regulate AP firing, and confer unique physiological characteristics; importantly, MSNs from symptomatic R6/2 and HD100 mice have reduced $K^+$ currents, and concomitant decreases in $K^+$ channel protein expression[47], findings that may account for altered membrane properties and contribute to latent neuronal vulnerability after the onset of dysfunction.

### 5.2. Excitotoxicity and Glutamate Receptor Function

Intrastriatal injections of excitotoxins produce lesions that resemble the pattern of neurodegeneration seen in human HD. Highly selective lesions that result in the depletion of MSNs have been produced by injections of compounds that act specifically on NMDA receptors; furthermore, such lesions are blocked by NMDAR antagonists (reviewed in ref. 48). NMDAR signaling is critically involved in a variety of fundamental activity-dependent brain processes; notably gating alterations in synaptic weights, gene transcription, and even cell survival. Many seminal reports provide strong evidence that NMDAR receptor dysfunction or aberrant signaling may underlie neuropathology in HD.

Surprisingly, transgenic mice expressing only exon 1 of the HD gene exhibit reduced sensitivity to excitotoxic lesions as compensatory neuroprotective mechanisms develop, correlating with overt symptoms of disease[49,50]. For example, basal $Ca^{2+}$ concentration is higher in R6/2 MSNs, and while quinolinic acid induces larger current changes ($Ca^{2+}$ influx through NMDARs) in R6/2 than in controls, R6/2 MSNs recover to their basal $Ca^{2+}$ concentration whereas WT neurons do not[50]. Such changes would place R6/2 MSNs in a protracted sublethal grade of excitotoxicity that may render them susceptible to further excitotoxic insult over time; in the short term, however, increased $Ca^{2+}$ influx may be responsible for inducing neuroprotective strategies, such as changes in $Ca^{2+}$

buffering[50] and gene transcription that combat NMDAR overactivation[49]. An additional explanation for resistance to excitotoxicity may stem from reduced cortical activity, since intact glutamatergic afferentation is required for excitotoxic lesions to be produced in the striatum[18]. Importantly, intrastriatal injection of quinolinate produces equivalent lesions in WT and larger construct HD100 mice[51], while full-length htt YAC72 mice exhibit increased sensitivity[52]. Thus, the latter models may be representative of earlier disease stages: at the time of, and prior to, induction of neuronal compensatory mechanisms, respectively.

Increased glutamate excitotoxicity could result from increased release, decreased uptake, increased receptor responsiveness, altered downstream receptor signaling, or impaired $Ca^{2+}$ homeostasis following repeated receptor activation. Evidence for a contribution by most of these mechanisms has been demonstrated. Altered release and uptake are implicated at the HD synapse (see Section 4). As well, mitochondrial function (which provides the major $Ca^{2+}$ buffering reservoir following NMDAR stimulation) also appears to be directly impaired by mhtt, while intracellular $Ca^{2+}$ release, in response to mGluR activation, is augmented by altered interaction between the IP3 receptor and expanded htt (reviewed in ref. 53). On the other hand, mGluR desensitization is facilitated by mhtt expression via an interaction between mhtt and the mGluR inhibitory protein optineurin, which could dampen the effects of enhanced IP3 signaling[54]. In addition to these mechanisms that may increase excitotoxic vulnerability, it is increasingly clear that mhtt augments NMDAR activity in striatal MSNs.

Striatal MSNs have high levels of NR2B subunit-containing NMDARs[55], and NMDAR-mediated whole cell currents in cell lines transfected with NR1/NR2B are specifically increased by the presence of mhtt[56]. As one might predict from the study of Chen and colleagues (1999), MSNs in acute slices prepared from R6/2 and CAG knock-in mice exhibit enhanced responsiveness to NMDA application prior to the onset of overt symptoms[43,46], as do a population of MSNs in slices from adult YAC72 and HD100 transgenic mice[43,57]. In cultured striatal MSNs, prepared from YAC mice soon after birth, NMDA-induced whole cell current density and apoptotic cell death are enhanced in a polyQ length-dependent manner[52,58,59]. Also, NMDAR EPSCs are specifically enhanced, whereas AMPA receptor currents are not, at cortico-striatal synapses in slices prepared from 1-month-old YAC72 mice, long before onset of phenotype[40]. The NMDAR activity underlying increased current density, excitotoxicity, and EPSCs is largely mediated by NR2B-containing NMDARs[40,52,58].

Increased NMDAR current may result from increased channel open probability or conductance; however, these channel properties are similar for NR1/NR2B expressed in HEK293 cells with htt containing expanded (138) and normal (15) polyQ[56]. It is possible such findings may not directly correlate with neuronal systems, where the presence of mhtt might alter the channel properties of NR2B-containing heteromeric (NR1/NR2A/NR2B) receptors. Alternatively, increased current could result from increased receptor expression, and increased NR1/NR2B mRNA levels have been observed (prior to large decreases; associated with cell death) in HD post mortem tissue[37]. Striatal immunohistochemistry in symptomatic R6/2 mice has demonstrated increased NR1 NMDAR subunit numbers, but this was accompanied by a decrease in NR2A/B subunits[43]. Other studies have demonstrated no change in striatal NMDAR mRNA and a trend toward decreased membrane bound NR2 in R6/2 mice (see ref. 60). Moreover, expression levels of NR1 and NR2B in the striatum of 2-month-old YAC46 and YAC72 mice (when increased currents are apparent but symptoms are not) are similar to WT controls[61]. Therefore it would appear that mhtt-induced alterations other than transcription or translation of NR1 or NR2B underlie augmented NMDAR current, leaving the possibility that subcellular distribution and/or post-translational modification of NMDARs is altered by mhtt.

### 5.3. NMDA Receptor Trafficking and Surface Expression

A variety of post-translational modifications and protein interactions govern NMDAR function and localization. Recent studies indicate that NMDAR delivery to, and endocytosis from, the surface membrane is regulated by NR2 subunit composition, NR1 splice variants, phosphorylation of specific sites within the C-terminus of NR2B, and NMDAR interactions with SAP102, and postsynaptic density protein-95 (PSD-95), both of which are members of the membrane-associated guanylate kinase (MAGUK) family of scaffolding proteins (reviewed in ref. 62).

PSD-95 binds NR2 subunits, forming clusters in the postsynaptic membrane, in addition to linking NMDARs with cytoskeletal and signaling proteins, including protein kinases and phosphatases. The interaction of PSD-95 with receptors is, unsurprisingly, heavily implicated in the regulation of synaptic plasticity[63,64], and the suppression of PSD-95 attenuates NMDAR-mediated excitotoxicity, despite having no effect on NMDAR current or $Ca^{2+}$ influx[65]. Together these studies demonstrate the importance of physical interactions of NMDARs with other proteins for appropriate downstream responses to their activation. Htt has been shown to associate with both PSD-95 and NMDARs in human cortical tissue, suggesting that htt/PSD-95/NMDA complexes occur[66]. Furthermore, htt/PSD-95 binding is reduced by polyQ expansion and is also reduced in HD patient cortex[66]. The same study found that expression of mhtt in cell lines increased NMDA-mediated excitotoxicity and that the increase was reduced by addition of WT htt. Sun and co-workers (2001) propose that htt/PSD-95 binding negatively regulates NMDAR signaling and that mhtt disrupts this negative regulation. However, it is also possible that mhtt interferes with PSD-95 transcription and that reduced htt/PSD-95 binding is attributable to reduced PSD-95, as has been observed in HD transgenic mice[49,60].

Tyrosine phosphorylation of NMDARs (by nonreceptor Src-family protein tyrosine kinases) increases NMDAR currents, shifts NMDARs from intracellular pools to the surface membrane, and increases NMDAR localization to the postsynaptic membrane (reviewed in ref. 62). Co-expression of mhtt and PSD-95 in cell lines results in increased PSD-95 surface expression and levels of activated Src. Moreover, co-expression of mhtt, PSD-95, NR1, and NR2B results in increased NMDAR-mediated excitotoxicity as well as tyrosine phosphorylation of NR2B; Src inhibitors decreased this phosphorylation and excitotoxicity[67]. As a note of caution, these data may not directly correlate with neurons *in situ*, as PSD-95 is normally targeted not just to the membrane surface, but specifically to synapses by palmitoylation (reviewed in ref. 68), and it has been shown that nonsynaptic NMDARs activate an apoptotic pathway whereas synaptic NMDA receptors preferentially activate a prosurvival pathway[69]. Also, recent studies have demonstrated no increase (but a trend toward decrease) of phosphorylated NR2 in symptomatic HD mouse striatum[49,60].

Serine phosphorylation of NMDARs is also altered in MSNs from HD transgenic mice. Progressive reductions in NR1 phosphorylation at serine 897 (a candidate site for cAMP-induced increases in NMDA activity) were observed in N171 mice[49], which may be related to latent neuroprotective strategies. On the other hand, levels of phosphorylated serine 897 are increased in R6/2 mice at the age at which they exhibit reduced excitotoxicity[70]. Thus, there is no clear correlation between NR1 Ser897 phosphorylation and NMDAR current and excitotoxicity.

NMDARs are internalized by clathrin-dependent mechanisms and HIP1 provides a link between htt and clathrin endocytosis[71]. Moreover, htt/HIP-1 binding is disrupted by polyQ expansion[33,34], and a breakdown in clathrin interactions may contribute to altered NMDAR surface expression or synaptic localization. The binding between another huntingtin-interacting protein (HIP14) and htt is also disrupted by polyglutamine expansion. HIP14 shares sequence homology with proteins involved in endocytosis and is also a palmitoyl acyl-transferase that modulates the trafficking and membrane association of a variety of proteins, including postsynaptic density protein-95 (PSD-95), synaptosomal-associated protein-25 (SNAP-25), and htt itself (reviewed in ref. 68). Altered mhtt/HIP14 function may affect not only endocytosis of NT receptors, but also the trafficking of htt and other synaptic proteins.

### 5.4. Dopamine Receptor Signaling

Dopamine controls the efficacy of fast synaptic transmission by regulating NT release and postsynaptic responsiveness via multiple receptors, second messengers, kinases and phosphatases (reviewed in ref. 72). The striatum receives dopaminergic input from the substantia nigra, and MSN projection pathways are distinguished by heterogeneous dopamine receptor (DAR) expression and projection targets. MSNs of the direct striatonigral pathway express substance P and D1-type DARs, and decreased activity of this pathway (that normally increases thalamocortical excitation) would lead to rigidity and bradykinesia. MSNs of the indirect striatopallidal pathway express met-enkephalin and D2-type DARs, and diminished function of this pathway (that normally reduces thalamocortical excitation) would produce random involuntary movements, including chorea (reviewed in ref. 73). Many post mortem studies suggest that it is the indirect pathway that first shows evidence of depletion in HD (reviewed in ref. 74), and noninvasive brain imaging of living humans has demonstrated reduced DAR labeling correlated with disease duration, cell loss, and cognitive impairment in symptomatic HD patients (reviewed in ref. 75). Moreover, reduced striatal (and cortical) DAR receptor binding has also been reported in asymptomatic HD gene carriers, demonstrating that dopaminergic dysfunction can be detected before clinical diagnosis and cell death[75]. Investigations across many mouse models of HD, after phenotypic onset, demonstrate reductions in striatal DAR protein and mRNA expression as well as diminished met-enkephalin staining with unaltered substance P staining[21,49,60,76,77]. Moreover, specific reductions in striatal D2 binding are seen in knock-in mice that do not exhibit an overt phenotype[78], findings reminiscent of that observed in asymptomatic human carriers and possibly related to dysfunction of the indirect pathway (reviewed in ref. 75).

DAR activity modulates multiple synaptic and plasticity-related proteins (including NMDARs and several molecules that regulate NMDAR localization) via varied cascades, e.g., dopamine and cAMP-regulated phosphoprotein 32kDa (DARPP-32) gating of PKA, PKC, and MAP kinase pathways. Interestingly, D1 receptor agonism and forskolin treatment have been shown to induce the redistribution of htt, HIP1 and clathrin to the membrane[71], which may play a role in regulating surface expression of NMDARs. Notably, DARPP-32 mRNA labeling is decreased in R6/1 and R6/2 mice at about the time of phenotypic onset, and at the same time as D1/DARPP-32 signaling and dopamine regulation of $Ca^{2+}$ channel, GABA receptor, and AMPA receptor function is lost in R6/2 mice[21,79]. Moreover, perturbation of any one of the steps in the dopamine pathway that regulate the activity or subcellular localization of NMDARs may contribute to early enhancement of NMDARs in HD mice.

### 5.5. GABA Receptors and Signaling

Striatal MSNs exert their influence upon target structures through the release of GABA, the major inhibitory NT in the brain. MSNs also receive GABAergic inhibition, primarily from interneurons, but also from other MSNs (reviewed in ref. 80). Stimulated (NMDA-induced) GABA release is increased in R6/1 mice, at about the time of phenotype onset[39]. At the time of symptom onset, striatal MSNs in R6/2 and R6/1 mice exhibit an increased frequency of spontaneous $GABA_A$ receptor ($GABA_AR$)-mediated current; this effect is likely due to elevated presynaptic firing, since miniature IPSC frequencies recorded in tetrodotoxin were not significantly different[44]. Additionally, decay kinetics of $GABA_AR$ currents in R6/2 striatum are also affected, suggesting altered receptor composition. In support of these findings, whole cell $GABA_AR$-mediated current densities and $GABA_A$ $\alpha1$ subunit labeling are increased[44]. Thus, pre- and postsynaptic GABAergic function appears altered in symptomatic R6 mice. On the postsynaptic side, $GABA_AR$ undergo rapid constitutive endocytosis, and internalized receptors are either recycled back to the cell surface or targeted for lysosomal degradation. Notably, the fate of internalized receptors is regulated by a direct interaction between $GABA_ARs$ and HAP1[81]. Overexpression of HAP1 decreases receptor degradation and increases recycling rates, in addition to increasing $GABA_AR$ current amplitude and surface numbers. Thus, HAP1 may play a critical role in the control of fast synaptic inhibition. Direct investigation of the effects of mhtt on $GABA_AR$ cycling may clarify whether increased MSN $GABA_AR$ current in R6/2 mice[44] is attributable to altered htt/HAP1 binding, or other (possibly compensatory) phenomena.

## 6. SYNAPTIC PLASTICITY AND COGNITIVE FUNCTION

Evidence to date suggests that the earliest alteration of neuronal function in mouse models of HD is augmentation of striatal NMDAR function or signaling, to which changes in dopamine signaling may contribute. However, altered NMDAR signaling is not restricted to the striatum. The hippocampus is known to be critical to the formation of episodic memory and, as part of the hippocampo–fronto–striatal pathway, is involved in higher cognitive tasks such as goal-directed behavior and executive function (reviewed in ref. 82). Activity-dependent alterations in synaptic efficacy, such as LTP and long-term depression (LTD), are widely believed to underlie information storage and processing in the brain and may be the best models for cellular substrates of learning and memory (reviewed in ref. 83). Moreover, NMDARs are critical to the induction of many of these modifications. Hippocampal function has been assessed in transgenic[30,84] and knock-in[17] models of HD, and in each study NMDAR-dependent LTP was altered. In addition, the most conspicuous alteration in R6/2 plasticity is the expression of NMDAR-dependent LTD in adult mice, which is entirely absent in controls[30]. In all three models, altered plasticity occurs prior to onset of an overt phenotype, and in R6/2 at a time at which hippocampal-dependent learning is impaired[27,30]. Recent investigations have characterized hippocampal plasticity in R6/1 mice, and following a period of near-normal development, aberrant LTD (mediated by NR2B-containing NMDARs) becomes apparent, long before the onset of an overt phenotype (A. Milnerwood, in press) and prior to reported motor and cognitive and cognitive deficits[85].

Cognitive deficits occur early in both HD patients and mouse models, and it seems likely that the mechanisms that underlie aberrant synaptic plasticity and cellular dysfunction in mouse models may contribute to early cognitive deficits in the human condition. Such dysfunction may, therefore, be the basis of many cognitive and behavioral manifestations of HD. Furthermore the mechanisms responsible for such early changes may directly link to later severe dysfunction that ultimately leads to gross cell death.

## 7. CONCLUSIONS

In human HD, as well as symptomatic stages in HD animal models, there are a multitude of disruptions to processes that are intrinsic to neuronal function and survival. A more complete understanding of the mechanisms for neuronal and synaptic dysfunction at this stage will guide symptomatic therapy in HD and help to develop interventions to slow the course of disease. In order to delay onset of HD, it will be necessary to further investigate the earliest effects of the HD mutation on neuronal function. In this regard, increased NMDA receptor function and toxic signaling, disrupted $Ca^{2+}$ homeostasis, altered dopamine modulation of neurotransmission, and perturbed axonal transport (Figure 31.1; Colorplate 13) have been identified as very early changes that precede onset of the movement disorder. These disturbances may underlie synaptic dysfunction responsible for early cognitive and personality changes, and may also provide triggers for aberrant proteolysis, proteasome dysfunction, and altered gene transcription leading to more severe neuronal dysfunction and death. Genetic testing allows those at risk for HD to be identified early, and biomarkers that will facilitate prediction of an individual's age of disease onset are being developed. With these advances, the possibilities for altering the course of this devastating neurodegenerative disease appear promising.

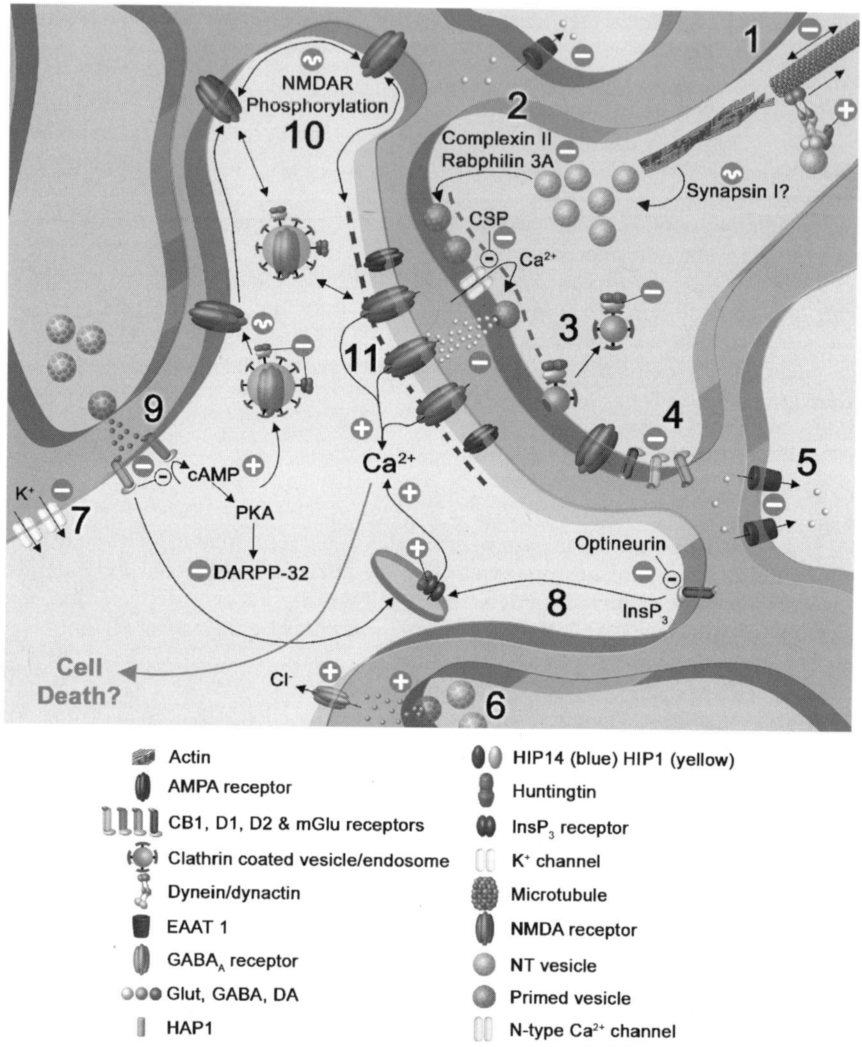


**Figure 31.1.** Summary of Alterations in Cortico-Striatal Synaptic Function Implicated in HD. Simplified figure of a cortico-striatal synapse; a cortical afferent (green) synapses with a spine of a striatal MSN (blue), astrocytes (brown) surround the synaptic cleft, and GABAergic (red) and dopaminergic (purple) terminals are also shown. Red circles indicate where positive, negative, and nonlinear alterations have been reported in HD and HD models. (1) Axonal transport along microtubules is reduced by polyQ expansion[9–12], e.g., disrupted BDNF transport is associated with the detachment of a dynactin p150/HAP1/mhtt complex from microtubules and may be due to increased HAP1/huntingtin binding produced by polyQ expansion[13–15]. (2) Synapsin 1 hyperphosphorylation[20] may alter synaptic vesicle reserve pool regulation; this and decreased levels of the vesicular proteins complexin II[23,25,26,31] and rabphilin 3A[22] may perturb docking and priming of NT vesicles within the active zone (broken green line). $Ca^{2+}$ influx, through N-type $Ca^{2+}$ channels, triggers vesicle fusion and NT release; this influx may be increased in HD by the loss of cysteine string protein (CSP)-mediated inhibition of N-type channels[32]. Earlier increases and later decreases in glutamate release are observed

in some HD mice[18]. (3) HIP1 is involved in clathrin-mediated endocytosis[35], a process which may be disrupted by reduced HIP1/huntingtin binding by polyQ expansion[34]. In addition, htt binding of the PACSIN1 regulator of vesicle recovery increases proportionally to polyQ, and PACSIN1 is abnormally located away from the synapse in HD[36]. (4) Receptors, known to modulate release presynaptically, are reduced in HD[48]. (5) Astrocytic excitatory amino acid transporter 1 (EAAT1/GLT1) mRNA levels are reduced[37] and glutamate uptake is altered in HD[20,38]. (6) Inhibitory GABA release is increased[39,44], as is GABA receptor labeling[44], possibly due to perturbations in HAP1-regulated GABA receptor recycling[81]. (7) $K^+$ channel currents and subunits are decreased in HD models[47]. (8) Activation of group I mGluRs results in the generation of $InsP_3$ and mhtt facilitates the action of the mGluR inhibitory protein optineurin[54] leading to reduced $InsP_3$ signaling; however polyQ-enhanced binding between $InsP_3$ receptor and htt results in increased $Ca^{2+}$ release from intracellular stores (reviewed in ref. 53). (9) Reductions in D1 and D2 dopamine receptors are observed in HD [21,49,60,75,76–78] (reviewed in ref. 74) and increased cAMP labeling[70], decreased DARPP-32 labeling, and a loss of dopaminergic modulation of other NT receptor systems is also observed[21,79]. D1 receptor agonism induces the redistribution of htt, HIP1 and clathrin to the membrane[71] and may play a role in regulating surface expression of NMDARs. (10) Increases in NR1[43] and decreases in NR2A/B protein (43, but see ref. 60), in addition to altered PSD-95 interactions[66,67] and differences in reported NMDAR subunit phosphorylation state[49,60,67,70] together suggest that trafficking of NMDARs is altered in HD. In addition, NMDARs are internalized by clathrin-dependent endocytosis; reduced mhtt/HIP-1 binding[33,34] and a breakdown in clathrin interaction may contribute to altered NMDAR surface expression. Additionally, HIP14/htt binding, which may also be important for endocytosis and protein trafficking, is also reduced by expanded polyQ (reviewed in ref. 68). (11) Altered NMDAR surface expression likely accounts for early increases in NMDAR function seen across HD models[40,43,46,52,56–59]. The resultant increase in $Ca^{2+}$ influx, in combination with increased release from intracellular stores[53], has severe implications for cortico-striatal $Ca^{2+}$ signaling, phosphatase/kinase activity, and consequently synaptic plasticity and the regulation of gene transcription. Moreover, such disruptions in $Ca^{2+}$ signaling and homeostasis may be the triggers for more severe neuronal dysfunction and ultimately cell death. See Colorplate 13.

## 8. REFERENCES


1. Huntington's Disease Collaborative Research Group. (1993) *Cell* **72**, 971–983.
2. Vonsattel, J.P., Myers, R.H., Stevens, T.J., Ferrante, R.J., Bird, E.D., and Richardson, E.P., Jr. (1985) *J Neuropathol Exp Neurol* **44**, 559–577.
3. Levine, M.S., Cepeda, C., Hickey, M.A., Fleming, S.M., and Chesselet, M.F. (2004) *Trends Neurosci* **27**, 691–697.
4. DiFiglia, M., Sapp, E., Chase, K., Schwarz, C., Meloni, A., Young, C., Martin, E., Vonsattel, J.P., Carraway, R., Reeves, S.A., et al. (1995) *Neuron* **14**, 1075–1081.
5. Aronin, N., Chase, K., Young, C., Sapp, E., Schwarz, C., Matta, N., Kornreich, R., Landwehrmeyer, B., Bird, E., Beal, M.F., et al. (1995) *Neuron* **15**, 1193–1201.
6. Landwehrmeyer, G.B., McNeil, S.M., Dure, L.S.T., Ge, P., Aizawa, H., Huang, Q., Ambrose, C.M., Duyao, M.P., Bird, E.D., Bonilla, E., de Young, M., Avila-Gonzales, A.J., Wexler, N.S., DiFiglia, M., Gusella, J.F., MacDonald, M.E., Penney, J.B., Young, A.B., and Vonsattel, J.-P. (1995) *Ann Neurol* **37**, 218–230.
7. Reiner, A., Dragatsis, I., Zeitlin, S., and Goldowitz, D. (2003) *Mol Neurobiol* **28**, 259–276.
8. Li, H., Li, S.H., Yu, Z.X., Shelbourne, P., and Li, X.J. (2001) *J Neurosci* **21**, 8473–8481.
9. Gunawardena, S., Her, L.S., Brusch, R.G., Laymon, R.A., Niesman, I.R., Gordesky-Gold, B., Sintasath, L., Bonini, N.M., and Goldstein, L.S. (2003) *Neuron* **40**, 25–40.
10. Szebenyi, G., Morfini, G.A., Babcock, A., Gould, M., Selkoe, K., Stenoien, D.L., Young, M., Faber, P.W., MacDonald, M.E., McPhaul, M.J., and Brady, S.T. (2003) *Neuron* **40**, 41–52.
11. Trushina, E. et al. (2004) *Mol Cell Biol* **24**, 8195–8209.
12. Li, X.J., Li, S.H., Sharp, A.H., Nucifora, F.C., Jr., Schilling, G., Lanahan, A., Worley, P., Snyder, S.H., and Ross, C.A. (1995) *Nature* **378**, 398–402.
13. Martin, E.J., Kim, M., Velier, J., Sapp, E., Lee, H.S., Laforet, G., Won, L., Chase, K., Bhide, P.G., Heller, A., Aronin, N., and Difiglia, M. (1999) *J Comp Neurol* **403**, 421–430.
14. Engelender, S., Sharp, A.H., Colomer, V., Tokito, M.K., Lanahan, A., Worley, P., Holzbaur, E.L., and Ross, C.A. (1997) *Hum Mol Genet* **6**, 2205–2212.
15. Gauthier, L.R., Charrin, B.C., Borrell-Pages, M., Dompierre, J.P., Rangone, H., Cordelieres, F.P., De Mey, J., MacDonald, M.E., Lessmann, V., Humbert, S., and Saudou, F. (2004) *Cell* **118**, 127–138.
16. Li, H., Wyman, T., Yu, Z.X., Li, S.H., and Li, X.J. (2003) *Hum Mol Genet* **12**, 2021–2030.

17. Usdin, M.T., Shelbourne, P.F., Myers, R.M., and Madison, D.V. (1999) *Hum Mol Genet* **8**, 839–846.
18. Cepeda, C., Hurst, R.S., Calvert, C.R., Hernandez-Echeagaray, E., Nguyen, O.K., Jocoy, E., Christian, L.J., Ariano, M.A., and Levine, M.S. (2003) *J Neurosci* **23**, 961–969.
19. Sudhof, T.C. (2004) *Annu Rev Neurosci* **27**, 509–547.
20. Liévens, J.C., Woodman, B., Mahal, A., and Bates, G.P. (2002) *Mol Cell Neurosci* **20**, 638–648.
21. Bibb, J.A., Yan, Z., Svenningsson, P., Snyder, G.L., Pieribone, V.A., Horiuchi, A., Nairn, A.C., Messer, A., and Greengard, P. (2000) *Proc Natl Acad Sci U S A* **97**, 6809–6814.
22. Smith, R., Petersen, A., Bates, G.P., Brundin, P., and Li, J.Y. (2005) *Neurobiol Dis* **20**, 673–684.
23. Morton, A.J., and Edwardson, J.M. (2001) *J Neurochem* **76**, 166–172.
24. DiProspero, N.A., Chen, E.Y., Charles, V., Plomann, M., Kordower, J.H., and Tagle, D.A. (2004) *J Neurocytol* **33**, 517–533.
25. Edwardson, J.M., Wang, C.T., Gong, B., Wyttenbach, A., Bai, J., Jackson, M.B., Chapman, E.R., and Morton, A.J. (2003) *J Biol Chem* **278**, 30849–30853.
26. Freeman, W., and Morton, A.J. (2004) *Brain Res Bull* **63**, 45–55.
27. Lione, L.A., Carter, R.J., Hunt, M.J., Bates, G.P., Morton, A.J., and Dunnett, S.B. (1999) *J Neurosci* **19**, 10428–10437.
28. Reim, K., Mansour, M., Varoqueaux, F., McMahon, H.T., Sudhof, T.C., Brose, N., and Rosenmund, C. (2001) *Cell* **104**, 71–81.
29. Takahashi, S., Ujihara, H., Huang, G.Z., Yagyu, K.I., Sanbo, M., Kaba, H., and Yagi, T. (1999) *Eur J Neurosci* **11**, 2359–2366.
30. Murphy, K.P.S.J., Carter, R.J., Lione, L.A., Mangiarini, L., Mahal, A., Bates, G.P., Dunnett, S.B., and Morton, A.J. (2000) *J Neurosci* **20**, 5115–5123.
31. Glynn, D., Bortnick, R.A., and Morton, A.J. (2003) *Hum Mol Genet* **12**, 2431–2448.
32. Miller, L.C., Swayne, L.A., Chen, L., Feng, Z.P., Wacker, J.L., Muchowski, P.J., Zamponi, G.W., and Braun, J.E. (2003) *J Biol Chem* **278**, 53072–53081.
33. Wanker, E.E., Rovira, C., Scherzinger, E., Hasenbank, R., Walter, S., Tait, D., Colicelli, J., and Lehrach, H. (1997) *Hum Mol Genet* **6**, 487–495.
34. Kalchman, M.A., Koide, H.B., McCutcheon, K., Graham, R.K., Nichol, K., Nishiyama, K., Kazemi-Esfarjani, P., Lynn, F.C., Wellington, C., Metzler, M., Goldberg, Y.P., Kanazawa, I., Gietz, R.D., and Hayden, M.R. (1997) *Nat Genet* **16**, 44–53.
35. Metzler, M., Legendre-Guillemin, V., Gan, L., Chopra, V., Kwok, A., McPherson, P.S., and Hayden, M.R. (2001) *J Biol Chem* **276**, 39271–39276.
36. Modregger, J., DiProspero, N.A., Charles, V., Tagle, D.A., and Plomann, M. (2002) *Hum Mol Genet* **11**, 2547–2558.
37. Arzberger, T., Krampfl, K., Leimgruber, S., and Weindl, A. (1997) *J Neuropathol Exp Neurol* **56**, 440–454.
38. Behrens, P.F., Franz, P., Woodman, B., Lindenberg, K.S., and Landwehrmeyer, G.B. (2002) *Brain* **125**, 1908–1922.
39. Nicniocaill, B., Haraldsson, B., Hansson, O., O'Connor, W.T., and Brundin, P. (2001) *Eur J Neurosci* **13**, 206–210.
40. Li, L., Murphy, T.H., Hayden, M.R., and Raymond, L.A. (2004) *J Neurophysiol* **92**, 2738–2746.
41. Morton, A.J., Hunt, M.J., Hodges, A.K., Lewis, P.D., Redfern, A.J., Dunnett, S.B., and Jones, L. (2005) *Eur J Neurosci* **21**, 855–870.
42. Ferrante, R.J., Kowall, N.W., and Richardson, E.P., Jr. (1991) *J Neurosci* **11**, 3877–3887.
43. Cepeda, C., Ariano, M.A., Calvert, C.R., Flores-Hernandez, J., Chandler, S.H., Leavitt, B.R., Hayden, M.R., and Levine, M.S. (2001) *J Neurosci Res* **66**, 525–539.
44. Cepeda, C., Starling, A.J., Wu, N., Nguyen, O.K., Uzgil, B., Soda, T., Andre, V.M., Ariano, M.A., and Levine, M.S. (2004) *J Neurosci Res* **78**, 855–867.
45. Klapstein, G.J., Fisher, R.S., Zanjani, H., Cepeda, C., Jokel, E.S., Chesselet, M.F., and Levine, M.S. (2001) *J Neurophysiol* **86**, 2667–2677.
46. Levine, M.S., Klapstein, G.J., Koppel, A., Gruen, E., Cepeda, C., Vargas, M.E., Jokel, E.S., Carpenter, E.M., Zanjani, H., Hurst, R.S., Efstratiadis, A., Zeitlin, S., and Chesselet, M.F. (1999) *J Neurosci Res* **58**, 515–532.
47. Ariano, M.A., Cepeda, C., Calvert, C.R., Flores-Hernandez, J., Hernandez-Echeagaray, E., Klapstein, G.J., Chandler, S.H., Aronin, N., DiFiglia, M., and Levine, M.S. (2005) *J Neurophysiol* **93**, 2565–2574.
48. Yohrling IV, G.J., and Cha, J.H. (2002) in *Huntington's Disease* (Bates, G.P., Harper, P.S., and Jones, L., eds), Third Ed., pp. 276–308, Oxford University Press, Oxford.
49. Jarabek, B.R., Yasuda, R.P., and Wolfe, B.B. (2004) *Brain* **127**, 505–516.

50. Hansson, O., Guatteo, E., Mercuri, N.B., Bernardi, G., Li, X.J., Castilho, R.F., and Brundin, P. (2001) *Eur J Neurosci* **14**, 1492–1504.
51. Petersén, Å., Chase, K., Puschban, Z., DiFiglia, M., Brundin, P., and Aronin, N. (2002) *Exp Neurol* **175**, 297–300.
52. Zeron, M.M., Hansson, O., Chen, N., Wellington, C.L., Leavitt, B.R., Brundin, P., Hayden, M.R., and Raymond, L.A. (2002) *Neuron* **33**, 849–860.
53. Bezprozvanny, I., and Hayden, M.R. (2004) *Biochem Biophys Res Commun* **322**, 1310–1317.
54. Anborgh, P.H., Godin, C., Pampillo, M., Dhami, G.K., Dale, L.B., Cregan, S.P., Truant, R., and Ferguson, S.S. (2005) *J Biol Chem* **280**, 34840–34848.
55. Kuppenbender, K.D., Standaert, D.G., Feuerstein, T.J., Penney, J.B., Jr., Young, A.B., and Landwehrmeyer, G.B. (2000) *J Comp Neurol* **419**, 407–421.
56. Chen, N., Luo, T., Wellington, C., Metzler, M., McCutcheon, K., Hayden, M.R., and Raymond, L.A. (1999) *J Neurochem* **72**, 1890–1898.
57. Laforet, G.A., Sapp, E., Chase, K., McIntyre, C., Boyce, F.M., Campbell, M., Cadigan, B.A., Warzecki, L., Tagle, D.A., Reddy, P.H., Cepeda, C., Calvert, C.R., Jokel, E.S., Klapstein, G.J., Ariano, M.A., Levine, M.S., DiFiglia, M., and Aronin, N. (2001) *J Neurosci* **21**, 9112–9123.
58. Zeron, M.M., Fernandes, H.B., Krebs, C., Shehadeh, J., Wellington, C.L., Leavitt, B.R., Baimbridge, K.G., Hayden, M.R., and Raymond, L.A. (2004) *Mol Cell Neurosci* **25**, 469–479.
59. Shehadeh, J., Fernandes, H.B., Zeron Mullins, M.M., Graham, R.K., Leavitt, B.R., Hayden, M.R., and Raymond, L.A. (2006) *Neurobiol Dis* **21**, 392–403.
60. Luthi-Carter, R., Apostol, B.L., Dunah, A.W., DeJohn, M.M., Farrell, L.A., Bates, G.P., Young, A.B., Standaert, D.G., Thompson, L.M., and Cha, J.H. (2003) *Neurobiol Dis* **14**, 624–636.
61. Li, L., Fan, M., Icton, C.D., Chen, N., Leavitt, B.R., Hayden, M.R., Murphy, T.H., and Raymond, L.A. (2003) *Neurobiol Aging* **24**, 1113–1121.
62. Prybylowski, K., and Wenthold, R.J. (2004) *J Biol Chem* **279**, 9673–9676.
63. Migaud, M., Charlesworth, P., Dempster, M., Webster, L.C., Watabe, A.M., Makhinson, M., He, Y., Ramsay, M.F., Morris, R.G., Morrison, J.H., O'Dell, T.J., and Grant, S.G. (1998) *Nature* **396**, 433–439.
64. Ehrlich, I., and Malinow, R. (2004) *J Neurosci* **24**, 916–927.
65. Sattler, R., Xiong, Z., Lu, W.Y., Hafner, M., MacDonald, J.F., and Tymianski, M. (1999) *Science* **284**, 1845–1848.
66. Sun, Y., Savanenin, A., Reddy, P.H., and Liu, Y.F. (2001) *J Biol Chem* **276**, 24713–24718.
67. Song, C., Zhang, Y., Parsons, C.G., and Liu, Y.F. (2003) *J Biol Chem* **278**, 33364–33369.
68. Huang, K., and El-Husseini, A. (2005) *Curr Opin Neurobiol* **15**, 527–535.
69. Hardingham, G.E., and Bading, H. (2002) *Biochim Biophys Acta* **1600**, 148–153.
70. Ariano, M.A., Wagle, N., and Grissell, A.E. (2005) *J Neurosci Res* **80**, 634–645.
71. Kim, M., Velier, J., Chase, K., Laforet, G., Kalchman, M.A., Hayden, M.R., Won, L., Heller, A., Aronin, N., and Difiglia, M. (1999) *Neuroscience* **89**, 1159–1167.
72. Greengard, P. (2001) *Science* **294**, 1024–1030.
73. Crossman, A.R. (2000) *J Anat* **196**, 519–525.
74. Gutekunst, C., Norflus, F., and Hersch, S.M. (2002) in *Huntington's Disease* (Bates, G.P., Harper, P.S., and Jones, L., eds), Third Ed., pp. 251–275, Oxford University Press, Oxford.
75. Piccini, P. (2004) *Curr Opin Neurol* **17**, 459–466.
76. Ariano, M.A., Aronin, N., Difiglia, M., Tagle, D.A., Sibley, D.R., Leavitt, B.R., Hayden, M.R., and Levine, M.S. (2002) *J Neurosci Res* **68**, 716–729.
77. Zucker, B., Luthi-Carter, R., Kama, J.A., Dunah, A.W., Stern, E.A., Fox, J.H., Standaert, D.G., Young, A.B., and Augood, S.J. (2005) *Hum Mol Genet* **14**, 179–189.
78. Kennedy, L., Shelbourne, P.F., and Dewar, D. (2005) *Brain Res* **1039**, 14–21.
79. Naver, B., Stub, C., Møller, M., Fenger, K., Hansen, A.K., Hasholt, L., and Sørensen, S.A. (2003) *Neuroscience* **122**, 1049–1057.
80. Tepper, J.M., and Bolam, J.P. (2004) *Curr Opin Neurobiol* **14**, 685–692.
81. Kittler, J.T., Thomas, P., Tretter, V., Bogdanov, Y.D., Haucke, V., Smart, T.G., and Moss, S.J. (2004) *Proc Natl Acad Sci U S A* **101**, 12736–12741.
82. Milner, B., Squire, L.R., and Kandel, E.R. (1998) *Neuron* **20**, 445–468.
83. Morris, R. G.M., Moser, E.I., Riedel, G., Martin, S.J., Sandin, J., Day, M., and O'Carroll, C. (2003) *Philos Trans R Soc Lond B Biol Sci* **358**, 773–786.
84. Hodgson, J.G., Agopyan, N., Gutekunst, C.A., Leavitt, B.R., LePiane, F., Singaraja, R., Smith, D.J., Bissada, N., McCutcheon, K., Nasir, J., Jamot, L., Li, X.J., Stevens, M.E., Rosemond, E., Roder, J.C., Phillips, A.G., Rubin, E.M., Hersch, S.M., and Hayden, M.R. (1999) *Neuron* **23**, 181–192.
85. Cybulska-Klosowicz, A., Mazarakis, N.K., Van Dellen, A., Blakemore, C., Hannan, A.J., and Kossut, M. (2004) *Neurobiol Dis* **17**, 427–434.

# 32

# INTERFERENCE PEPTIDES: A NOVEL THERAPEUTIC APPROACH TARGETING SYNAPTIC PLASTICITY IN DRUG ADDICTION

Karen Brebner, Anthony G. Phillips, Yu Tian Wang, and Tak Pan Wong*

## 1. SUMMARY

Synaptic plasticity at excitatory synapses has been proposed as the cellular substrate of information processing and memory formation in the brain under both physiological and pathological conditions, including addiction. There is a growing body of evidence that implicates long-term depression (LTD), particularly in the nucleus accumbens (NAc), as a potential mediator of drug-induced neural plasticity. In animals, behavioral sensitization is used to model enduring changes in neuronal activity and behavior that result from repeated exposure to drugs of abuse. Given the links between behavioral sensitization and enduring drug-induced neuroplasticity, it is possible that compounds that can disrupt LTD may form the basis for a rational drug development strategy for repairing abnormal synaptic functions that are related to exposure to addictive substances. The data reviewed here provide evidence that facilitated AMPAR endocytosis and LTD in the NAc is critically involved in behavioral sensitization associated with drug addiction, indicating that both the expression and possibly the induction phases of LTD may represent promising targets for developing novel therapeutics for the clinical management of drug addicts.

## 2. INTRODUCTION

Signaling and communication between cells is a pervasive and core issue in the biological sciences. Neurons in the mammalian brain accomplish these functions through a process known as synaptic transmission. The strength of

*Department of Psychiatry, University of British Columbia, 2255 Wesbrook Mall, Vancouver, BC, Canada V6T 2A1; brebner@interchange.ubc.ca

synaptic transmission at individual synapses and the number of synapses are not static, but rather are subject to dynamic regulation, and these changes in turn are believed to be intimately associated with development of neuronal circuitry, and behavioral adaptation manifested as learning and memory[1–3]. Many of these adaptive "plastic" changes in synaptic function occur in the region of the postsynaptic density[4]. This is a complex region involving specific neurotransmitter-gated ionic and G-protein-coupled receptors and second messenger cascades involving specific kinase and phosphatase activity, all of which are potential targets for the development of new drugs. Despite this complexity, remarkable progress has been achieved in recent years in identifying molecules essential for prolonged changes in the efficacy of synaptic transmissions (functional alterations), and consequent rewiring of synaptic connections (structural alterations).

A growing body of evidence implicates abnormal synaptic function in a broad spectrum of neurological and neuropsychiatric disorders, including Alzheimer's disease, autism, mental retardation, schizophrenia, and addiction.[5–8] Traditional pharmacotherapies for these disorders have been based primarily on enhancing or reducing extracellular levels of specific neurotransmitters or blocking specific receptors with small molecules. While this approach has alleviated some symptoms of various forms of mental illness, the effects are often nonspecific, and limited by side effects due to interference with the normal function in unaffected regions of the brain and body[9]. Thus, novel approaches with greater specificity in targeting only affected synaptic processes are urgently needed.

Alterations in synaptic connections in certain brain regions may result, at least in part, from prolonged functional and structural alterations arising from abnormal interplay in certain protein–protein interactions within the synapses. This raises the possibility that the therapeutic use of "interference peptides" that can be delivered systemically to block specific protein–protein interactions involved in synaptic dysfunctions may represent a novel pharmacotherapeutic strategy in the clinical management of psychiatric illnesses where synaptic alterations are at the root of the pathogenesis. The successful development of effective interference peptides with therapeutic efficacy could spark the development of similar treatments for many neurological and neuropsychiatric illnesses.

Drug addiction is a complex neuropsychiatric disorder in which repeated self-administration of specific drugs, including amphetamines, cocaine, nicotine, heroin, and alcohol, induces long-lasting changes in neural function and behavior[10]. Addictive behavior is characterized by compulsion centered on the procurement and use of a drug of choice (i.e., *craving*). Drug addiction is also defined as a *chronic relapsing disorder*, as individuals who have successfully abstained from drug use for extended periods are still susceptible to renewed episodes of drug seeking and abuse, following a single exposure to environmental stimuli associated with prior drug use, or a small quantity of the drug itself[11,12]. Studies that have been conducted to investigate the neural adaptation underlying the development and recurrence of addictive behaviors suggest that addiction may involve mechanisms of neural plasticity implicated in learning and memory, and in particular with associative mechanisms which serve to link environmental stimuli with drug reward[12–14].

## 3. LTD AND BEHAVIORAL SENSITIZATION: A MODEL FOR THE ROLE OF SYNAPTIC PLASTICITY IN ADDICTION

Craving is a key feature of relapse to drug-seeking behavior and is modeled in preclinical studies as increased motor activity (i.e., behavioral sensitization) induced by repeated intermittent administration of many drugs of abuse, including amphetamine, cocaine, heroin, and nicotine. Behavioral sensitization in turn is linked to neural adaptations in the mesocorticolimbic regions of the brain, including a terminal region in the NAc that receives dopaminergic projections from the ventral tegmental area (VTA) and excitatory glutamatergic inputs from the prefrontal cortex. Brain-imaging studies with human drug addicts report that craving induced by drug-associated stimuli is accompanied by activation of the prefrontal cortex, anterior cingulate, amygdala, and ventral striatum[15,16]. These brain regions are innervated by the mesocorticolimbic dopamine (DA) system, which subserves the primary reinforcing effects of many drugs of abuse, including psychostimulants and opiates[15–17]. Recent findings also implicate DA in synaptic plasticity and memory[18–20]. Given the importance of DA systems in drug addiction, there have been intensive efforts seeking DA-based therapeutics for the clinical management of patients suffering from drug addictions. However, this approach has met with only limited success, characterized by short periods of efficacy with unacceptable side effects due to interference with the normal function of DA and its receptor systems in unaffected brain regions[9].

Neural adaptations in the VTA play an essential role in the induction of behavioral sensitization, whereas synaptic modifications in the NAc are involved in its long-term maintenance (expression). Initial work on behavioral sensitization focused on pre- and postsynaptic changes in DA systems. However, evidence accumulated recently supports a critical role of synaptic plastic changes in glutamatergic transmission at both the VTA and NAc levels[13]. Consistent with differential roles of the VTA and NAc in mediating sensitization, repeated stimulation of glutamatergic cortical inputs to the VTA triggers sensitization, while the maintenance (expression) of this behavioral adaptation is blocked by inhibition of glutamatergic synapses in the NAc[21–23].

The evidence supporting the importance of the VTA and the NAc in experience-dependent neural plasticity has been accumulating for several years. It was only recently, however, that changes in excitatory synaptic transmission after repeated drug exposure were examined directly. Thomas and colleagues[24] examined excitatory synaptic responses in NAc brain slices from animals that had been sensitized to cocaine. They reported a long-lasting enhancement of LTD of glutamatergic transmission in the shell region of the NAc in sensitized mice. This demonstration of LTD in the NAc following repeated cocaine applications provided the first correlation between acquired LTD and the development of behavioral changes associated with drug addiction, and supports the hypothesis that the neural adaptations that lead to addiction involve the same glutamate-dependent cellular mechanisms that underlie learning and memory.

## 4. GLUTAMATERGIC INVOLVEMENT IN SYNAPTIC PLASTICITY

Glutamate, the principal fast excitatory neurotransmitter in the brain, acts principally via two classes of postsynaptic ionotropic glutamate receptors, NMDARs, and AMPARs. NMDARs make little contribution to basal synaptic transmission due to voltage-dependent $Mg^{2+}$ blockade but are essential for producing different forms of synaptic plasticity in AMPAR-mediated synaptic transmission, including long-term potentiation (LTP) and LTD[1,25]. Both LTP and LTD are believed to be intimately associated with learning and memory and with the refinement of neuronal circuitry during development and under pathological conditions[6,22]. As such, examining the mechanisms mediating the plastic changes in AMPAR-mediated synaptic transmission has been a subject of intensive research in many laboratories around the world.

Although LTP and LTD can be observed in many brain regions, those occurring at hippocampal CA1 synapses are the most well-characterized and best understood examples. It is well accepted that the induction of both LTP and LTD at these synapses requires the activation of NMDARs[1,25]. However there is an emerging consensus that changes in AMPAR function are critically important to the maintenance (expression) of LTP and LTD[2]. Broadly speaking, modification of AMPAR function could be achieved by altering receptor gating properties, such as single channel conductance or mean open time, and/or the number of individual receptors. Indeed, a large body of evidence supports a critical role for rapid changes in the number of postsynaptic AMPARs in the expression of LTP and LTD[2,3] Moreover, these recent studies indicate that, as with most integral plasma membrane proteins, AMPARs undergo constitutive cycling between intracellular compartments and the plasma membrane via vesicle-mediated plasma membrane insertion (*exocytosis*) and internalization (*endocytosis*). Regulation of these processes can lead to rapid increases[2,26,27] or decreases[2,4] in the number of AMPARs expressed in the postsynaptic membrane, thereby providing a powerful means of altering synaptic efficacy. Thus, while the expression of LTP involves a facilitated insertion (*exocytosis*) of AMPARs into the postsynaptic membrane, LTD expression is absolutely dependent on clathrin-mediated AMPAR removal (*endocytosis*) from the synapse[2,3].

## 5. AMPAR ENDOCYTOSIS IN LTD

Structurally, AMPARs are presumed to be heteropentameric structures assembled by combining homologous subunits. The current topology model suggests that each AMPAR subunit (GluR) consists of three transmembrane regions, MD1, MD3, and MD4, (MD, membrane domain), and a re-entrant membrane segment corresponding to MD2. The major intracellular carboxyl-terminal (CT) region contains potential sites for protein phosphorylation and binding motifs for GluR-interacting proteins, including adaptor proteins involved in forming clathrin-coated endocytic vesicles. Based on the mechanisms involved and functional consequences, clathrin-dependent AMPAR endocytosis can be classified into two distinct types: constitutive and regulated (stimulated)[28,29]. *Constitutive endocytosis* occurs across all AMPAR subunits and requires a seven amino acid stretch (EFCYKSR) in the most juxtamembrane region of the CT that is 100% conserved among all AMPAR GluR subunits[30]. Interfering with the constitutive pathway does not lead to detectable alterations in the number of

AMPARs on the cell surface, likely because of the existence of a constitutive AMPAR exocytosis pathway that counteracts the constitutive endocytosis[29].

The *regulated endocytosis*, on the other hand, is specifically associated with AMPARs containing the GluR2 subunit[28]. This GluR2 specificity is mediated by several sequence motifs uniquely present within the CT region of the subunit, including a middle domain that binds to both NSF and AP2 clathrin adaptor[31] and a terminal PDZ-binding domain that interacts with a number of PDZ-containing proteins such as GRIP and PICK1[32]. In addition, systematic mutational analysis has recently identified a novel tyrosine phosphorylation-dependent endocytic motif ($_{869}$YKEGYNVYG$_{877}$) in the GluR2 CT that is absolutely required for insulin-stimulated AMPAR endocytosis (Figure 32.1)[30].

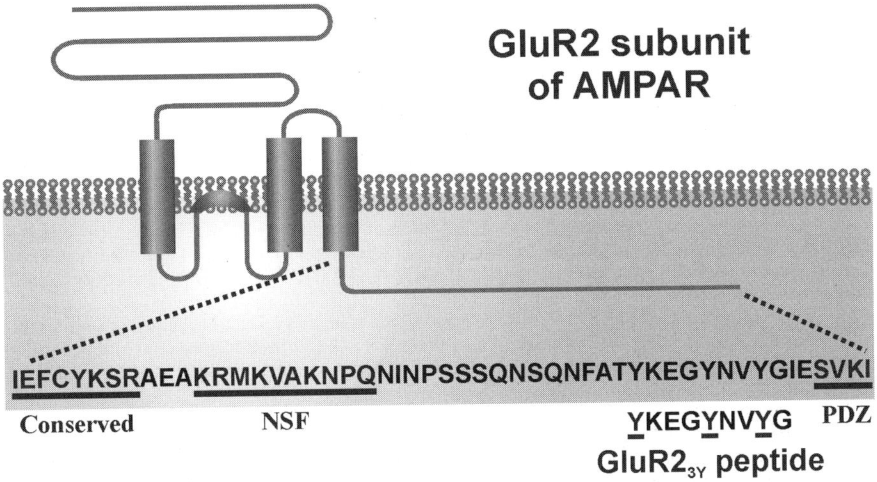


**Figure 32.1.** Schematic Diagram Shows the Amino-Acid Sequence in Carboxyl Terminal (CT) of AMPAR GluR2 Subunit. Sequences conserved among GluR1-4 (conserved) and those involved in its interaction with *N*-ethylmaleimide **S**ensitive **F**actor (NSF) or **PSD-95/D**iscs Large/**Z**ona Occludens-1 domain-containing proteins (PDZ) are highlighted. Note that insulin-induced AMPAR endocytosis can be disrupted only in cells expressing mutants with the three tyrosine residues near the end of the GluR2 CT deleted. The amino acid sequence that covers these three tyrosine residues was used to construct the interference GluR2$_{3Y}$ peptide.

Though the detailed underlying mechanism remains unclear, a recent study has reported that insulin stimulates $\alpha_2$-adrenergic receptor endocytosis by phosphorylating one or more of three tyrosine residues in its CT region, thereby creating a binding site for the Grb2 adaptor protein[33]. By analogy, tyrosine phosphorylation of YKEGYNVYG may create a binding motif for a clathrin-coated adaptor protein such as Grb2. The increased interaction between phosphorylated AMPAR and clathrin-adaptor recruits the receptors to clathrin-coated pits allowing enhanced receptor endocytosis (Figure 32.2).

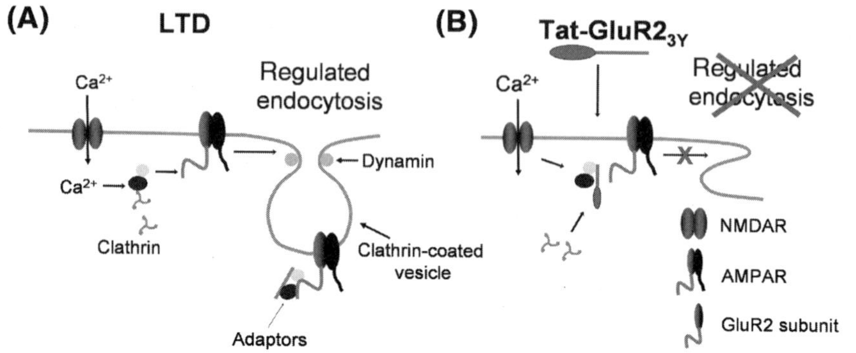


**Figure 32.2.** Proposed Model of Action for Tat-GluR2$_{3Y}$ Interfering Peptide in Blocking AMPAR Endocytosis and LTD. (A) GluR2-dependent endocytosis of AMPARs is a common final step for LTD induction resulting from various stimuli such as low-frequency stimulation, insulin application, and mGluR1/5 activation. These LTD-producing stimuli, via different intracellular cascades, lead to activation of the tyrosine-containing endocytic signal (**Y**KEG**Y**NV**Y**GIE) present in the GluR2 C-terminus (CT), and results in an interaction between the AMPAR and a clathrin-adaptor protein. Since it is the adaptors, and then clathrin, that are recruited to AMPARs, it is the GluR2 CT-adaptor interaction which initiates assembly of AMPAR-containing clathrin-coated vesicles. This leads to regulated endocytosis of the receptor, and hence, LTD. (B) Application of a membrane-permeable peptide containing this GluR2 tyrosine signal (Tat-**Y**KEG**Y**NV**Y**GIE) competitively disrupts the GluR2 CT-adaptor interaction preventing AMPAR endocytosis and LTD. Because the Tat-GluR2$_{3Y}$ peptide disrupts the adaptor interaction in a strict sequence-specific manner without affecting other steps of the signaling cascade, it should be a very specific LTD blocker. This peptide may be used to functionally repair synaptic dysfunctions caused by enhanced LTD such as drug addiction-related behavioral sensitization.

Functionally, unlike *constitutive endocytosis*, stimulating *regulated* AMPAR endocytosis leads to a rapid and persistent reduction in the number of AMPARs expressed on the postsynaptic membrane surface, and hence plays an important role in controlling synaptic strength[28]. The ability to reduce cell-surface AMPARs makes *regulated endocytosis* an ideal candidate mechanism for the expression of certain forms of LTD. Consistent with this conjecture, interfering with either of these GluR2 CT endocytic motifs by postsynaptic application of short "interference peptides" containing these endocytic motif sequences, tyrosine-containing sequences[30], AP2[31], and PICK1[32], can prevent the expression of hippocampal CA1 homosynaptic LTD in brain slices. Moreover, although the facilitated endocytosis of postsynaptic AMPARs is principally observed in conjunction with hippocampal CA1 LTD, it appears to be a common mechanism involved in the expression of certain forms of LTD in many other areas of the brain, including the cerebellum[4] and the NAc[24]. These studies have firmly established a critical role for AMPAR endocytosis in LTD expression. In addition, these results also suggest that these three GluR2 CT endocytic signals are indispensably involved in regulated AMPAR endocytosis and hence LTD expression, and that peptides interfering with the function of these signals may be used as specific inhibitors to probe the functional roles of LTD[30].

## 6. POTENTIAL UTILITY OF INTERFERENCE PEPTIDES IN THE TREATMENT OF DRUG ADDICTION LTD

Given the involvement of the NAc in the maintenance of behavioral sensitization and its association with enhanced LTD, LTD in the NAc may be a critical substrate for the expression of sensitization. LTD observed in the hippocampal CA1 region and LTD in the NAc in animals sensitized to repeated exposure to drugs such as amphetamine or cocaine seem to share a common mechanism, manifested as a consequence of specific reduction of AMPARs, possibly by facilitated endocytosis of AMPARs[24]. Peptides developed to specifically inhibit regulated AMPAR endocytosis, and hence LTD, when delivered into neurons in the NAc, should prevent sensitization-related LTD, thereby blocking the expression of sensitization behavior.

Recently a GluR$2_{3Y}$ peptide that contains the tyrosine-based signal ($_{869}$YKEGYNVYG$_{877}$) required for regulated AMPAR endocytosis has been developed. When delivered into neurons, the GluR$2_{3Y}$ peptide blocks electrically induced LTD in hippocampal slices[30]. Because of its strict sequence requirement, the peptide shows unprecedented specificity, being able to block the expression of LTD without altering AMPAR function and basal synaptic transmission. To facilitate the intracellular delivery of the peptide, both the wild-type GluR$2_{3Y}$ peptide and inactive control GluR$2_{3A}$ peptide were fused to the cell-membrane transduction domain of the human immunodeficiency virus-type 1 Tat protein (YGRKKRRQRRR; ref. 30) to generate Tat-GluR$2_{3Y}$ and Tat-GluR$2_{3A}$ using previously published methods[34]. Tat-GluR$2_{3Y}$ was transduced into neurons and blocked NMDA-induced AMPAR endocytosis in a neuronal cultured model of LTD[35]. Fluorescently tagged Tat-GluR$2_{3Y}$ peptide can cross the blood–brain barrier and reach the cortex of the brain after systemic administration (Figure 32.3).

The potential role of NAc LTD in mediating the expression of behavioral sensitization to psychostimulant drugs was examined using the recently developed Tat-GluR$2_{3Y}$ and Tat-GluR$2_{3A}$ peptides. Persistent depression of cortical glutamatergic inputs to medium spiny neurons can be induced in the NAc[36] (Figure 32.4A) that is similar to a form of hippocampal LTD mediated by facilitated endocytosis of postsynaptic AMPARs[3,37,38]. Several interference peptides for AMPAR endocytosis were delivered to in the postsynaptic neuron in the NAc through pipette solution during whole-cell recording. A dynamin-derived synthetic peptide (Dyn) prevented NAc LTD expression (Figure 32.4B). Dyn inhibits clathrin-mediated endocytosis by blocking the recruitment of dynamin to clathrin-coated pits[39] and prevents expression of certain forms of LTD[4,37]. These findings support an essential role of clathrin-dependent endocytosis of AMPARs for NAc LTD expression.

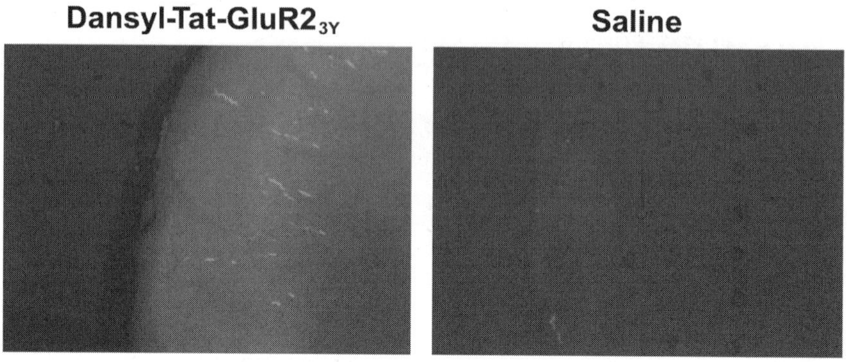


**Figure 32.3.** Dansyl-Labeled Tat-GluR2$_{3Y}$ Crosses the Blood–Brain Barrier and Enters Neural Tissue Following Systemic Application of the Peptide. Two adult male C57-Black/6 mice were given an intraperitoneal injection of either 30 moles/g dansyl-labeled Tat-GluR2$_{3Y}$ or saline. Mice were sacrificed and 40 μ coronal cortical sections were cut with a cryostat. Peptide and saline treated brain sections were visualized with fluorescence microscopy. The images above are of the cortex at –1.0 mm with respect to the bregma.

The required clathrin-dependent endocytosis process in NAc LTD involves GluR2-dependent endocytosis of postsynaptic AMPARs. A synthetic peptide derived from rat GluR2 carboxyl tail (GluR2$_{3Y}$; $_{869}$YKEGYNVYG$_{877}$) that specifically blocks regulated, but not constitutive, AMPAR endocytosis and the expression of CA1 homosynaptic LTD in hippocampal slices[30,40] completely abolished the expression of NAc LTD (Figure 32.4B). These findings support the specific action of GluR2$_{3Y}$ peptide in blocking regulated AMPAR endocytosis, and hence the expression of LTD.

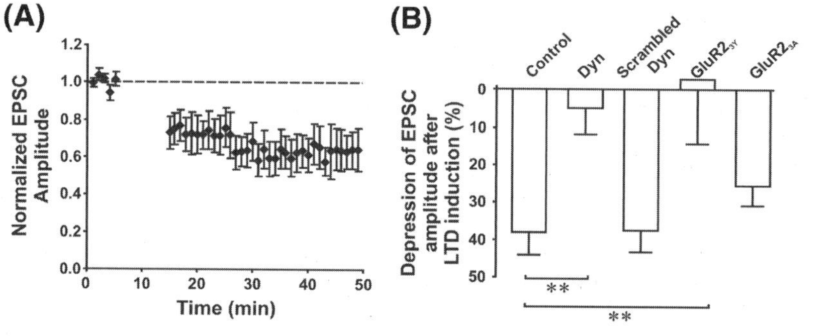


**Figure 32.4.** Clathrin- and GluR2-Dependent AMPAR Endocytosis are Required for the Expression of NAc LTD. (A) LTD of EPSCs evoked by electrical stimulation of glutamatergic cortical inputs projected to medium spiny neurons in the NAc can be induced under whole-cell voltage-clamp mode by pairing presynaptic stimulation (1 Hz, 480 pulses) with postsynaptic depolarization. (B) Histogram shows the depression of EPSC at 30 min after pairing stimulation. Inclusion in the patch pipette of interference peptides for blocking clathrin-dependent AMPAR endocytosis (Dyn), but not the control scrambled Dyn form, abolished the formation of LTD. These findings indicate that NAc LTD is mediated by clathrin-dependent endocytosis of AMPARs. NAc LTD also requires GluR2-dependent endocytosis of postsynaptic AMPARs because the application of GluR2$_{3Y}$ but not the control GluR2$_{3A}$ peptide via recording pipettes inhibited LTD expression. ** $p < 0.01$ versus control (Tukey–Kramer test).

Next, the effects of peptides were tested on behavioral sensitization that was induced in rats with repeated injections of *D*-amphetamine. The expression of behavioral sensitization was evoked by a challenge dose of *D*-amphetamine 21 days later. Pretreatment of sensitized rats with intravenous Tat-GluR2$_{3Y}$ prior to the challenge dose of *D*-amphetamine prevented the expression of behavioral sensitization (Figure 32.5A,B; Colorplate 14).

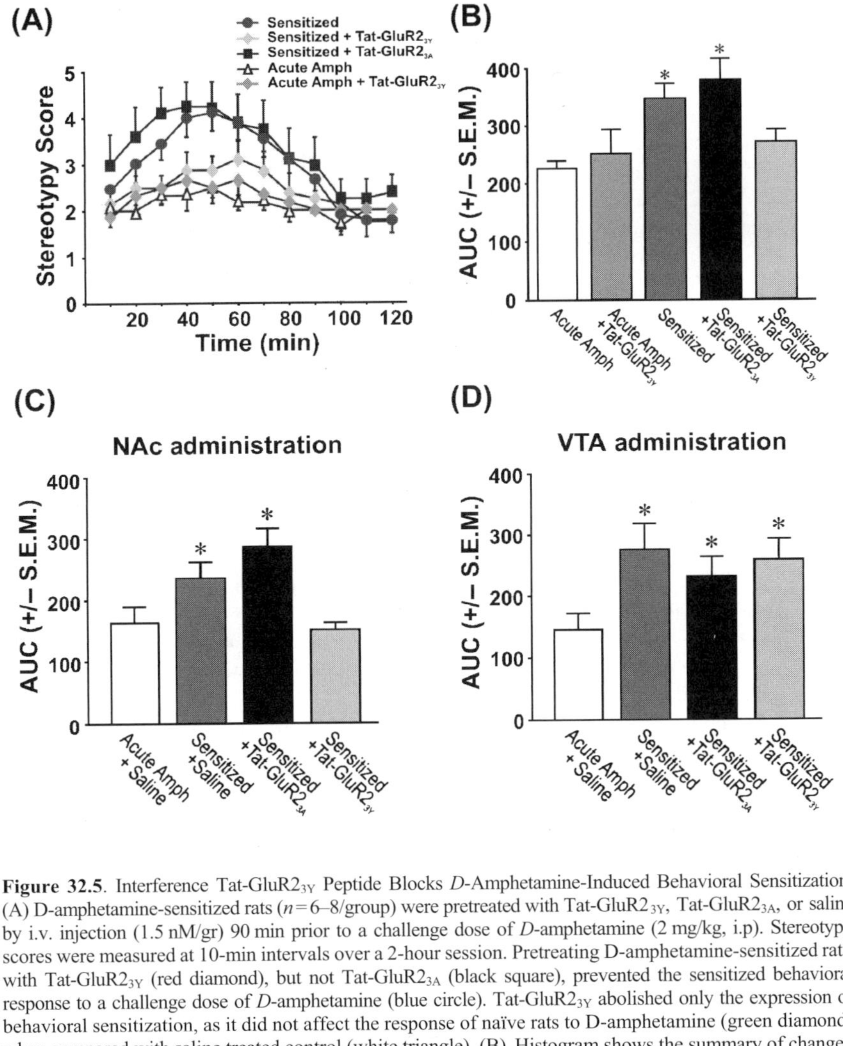


**Figure 32.5**. Interference Tat-GluR2$_{3Y}$ Peptide Blocks *D*-Amphetamine-Induced Behavioral Sensitization. (A) D-amphetamine-sensitized rats ($n = 6$–8/group) were pretreated with Tat-GluR2$_{3Y}$, Tat-GluR2$_{3A}$, or saline by i.v. injection (1.5 nM/gr) 90 min prior to a challenge dose of *D*-amphetamine (2 mg/kg, i.p). Stereotypy scores were measured at 10-min intervals over a 2-hour session. Pretreating D-amphetamine-sensitized rats with Tat-GluR2$_{3Y}$ (red diamond), but not Tat-GluR2$_{3A}$ (black square), prevented the sensitized behavioral response to a challenge dose of *D*-amphetamine (blue circle). Tat-GluR2$_{3Y}$ abolished only the expression of behavioral sensitization, as it did not affect the response of naïve rats to D-amphetamine (green diamond) when compared with saline treated control (white triangle). (B) Histogram shows the summary of changes in stereotypy scores across the 2-hour test session converted to Area Under the Curve (AUC) for individual groups depicted in graph (A). (C) The blockade of behavioral sensitization was reproduced when Tat-GluR2$_{3Y}$ (15 pmol; in 0.3 μl) was microinjected directly into the NAc. Injecting peptides into VTA (D) did not affect behavioral sensitization (* $p < 0.05$ versus acute D-amphetamine group. $n = 6$–7 in each group). Redrawn from ref. 41. See Colorplate 14.

Pretreatment with control Tat-GluR2$_{3A}$ or with saline had no effect. Co-administration of Tat-GluR2$_{3Y}$ with an acute injection of *D*-amphetamine did not significantly change stereotypy responses or locomotor scores relative to acute *D*-amphetamine alone and no noticeable behavioral side effects were associated with peptide pretreatment. To determine where the systemically applied Tat-GluR2$_{3Y}$ peptide was acting to disrupt the expression of behavioral sensitization, the peptide was microinfused directly into the NAc or the VTA. Tat-GluR$2_{3Y}$ prevented the expression of behavioral sensitization when it was delivered into the NAc (Figure 32.5C; Colorplate 14) but not into the VTA (Figure 32.5D; Colorplate 14). The effects of Tat-GluR$2_{3Y}$ did not appear to represent a general disruption of learned motor behavior as systemic treatment with Tat-GluR$2_{3Y}$ failed to disrupt operant responding for food pellets delivered on an FR2 schedule of reinforcement. Responding for intravenous *D*-amphetamine in a more demanding progressive ratio schedule of reinforcement, in which successive deliveries of drug reward require progressively more operant responses[42], was similarly unaffected by treatment with Tat-GluR$2_{3Y}$ or Tat-GluR$2_{3A}$ [41].

The GluR$2_{3Y}$ peptide inhibited the expression of NAc LTD without affecting basal synaptic transmission and other upstream signaling processes involved in LTD. The membrane-permeant form of GluR$2_{3Y}$ peptide made it possible to probe the role of LTD in freely moving rats with unprecedented specificity, providing evidence for the involvement of NAc LTD in the expression of behavioral sensitization. Given the links between behavioral sensitization and enduring drug-induced neuroplasticity, these data raise the possibility that treatment with drugs (such as GluR$2_{3Y}$ peptide) that disrupt LTD may form the basis for a rational drug development strategy for treating maladaptive neuroadaptations related to exposure to addictive substances.

## 7. CONCLUSIONS

Synaptic plasticity at excitatory synapses has been proposed as the cellular substrate of information processing and memory formation in the brain under both physiological[1,43] and pathological conditions, including addiction[22]. Experiments characterizing the nature of synaptic plasticity in sensitized animals have demonstrated both LTP[44,45] and LTD[24] in the NAc. Furthermore, increased excitatory drive in the NAc is observed during drug-seeking behavior[1]. Within a given nucleus such as the NAc, it is suggested that the relative balance between LTP and LTD at different afferent inputs determines the expression of specific patterns of behavior[46]. In sensitized animals the abnormal induction of LTD at PFC inputs to the NAc may disrupt the influence of this pathway on neural activity involved in goal-directed behavior[46].

To date, conclusive evidence for a causal relationship between synaptic plasticity and memory has been lacking due, in part, to the absence of specific drugs that block the expression of LTP or LTD without affecting any upstream signaling. Recent data demonstrate the utility of peptides that disrupt the expression of synaptic plasticity as tools to examine the critical role of LTP and/or LTD in specific aspects of learning and memory in conscious animals. Given the links between behavioral sensitization and enduring drug-induced neuroplasticity, it is possible that drugs that can disrupt LTD may form the basis for a rational drug development strategy for repairing abnormal synaptic functions that are related to exposure to addictive substances.

The evidence that LTD and/or behavioral sensitization may be linked to a number of pathological conditions raises the possibility that interference peptides such as the GluR$2_{3Y}$ peptide may have broad therapeutic potential in the treatment of mental illness. For example, behavioral sensitization has been proposed as a model of psychostimulant-induced psychosis[47] with possible links to schizophrenia. There is also evidence to suggest that LTD triggered by the activation of mGluR1/5 is affected in a number of pathological conditions that involve alterations in the refinement of neuronal circuitry. Evidence for its involvement in drug addiction-related behavioral sensitization is particularly strong[48]. In addition, mGluR1-mediated LTD is enhanced in mice lacking the fragile X mental retardation protein 1 (5; Chapter 30). Taken together, these observations suggest that there are several targets for developing interference peptides aimed at repairing some of the synaptic abnormalities associated not only with drug addiction, but also with other neurological disorders including schizophrenia and certain forms of mental retardation.

## 8. REFERENCES


1. Bliss, T.V.P., and Collingridge, G.L. (1993) *Nature* **361,** 31–39.
2. Malinow, R., and Malenka, R.C. (2002) *Annu Rev Neurosci* **25,** 103–126.
3. Man, H.Y., Ju, W., Ahmadian, G., and Wang, Y.T. (2000) *Cell Mol Life Sci* **57,** 1526–1534.
4. Wang, Y.T., and Linden, D.J. (2000) *Neuron* **25,** 635–647.
5. Huber, K.M., Gallagher, S.M., Warren, S.T., and Bear, M.F. (2002) *Proc Natl Acad Sci U S A* **99,** 7746–7750.
6. Nestler, E.J., Hope, B.T., and Widnell, K.L. (1993) *Neuron* **11,** 995–1006.
7. Rubenstein, J.L., and Merzenich, M.M. (2003) *Genes Brain and Behavior* **2,** 255–267.
8. Honer, W.G. (2002) *Cereb Cortex* **12,** 349–356.
9. Dackis, C.A. (2004) *Curr Psychiatry Rep* 323–331.
10. Leshner, A.I. (1999) *J Am Mental Assoc* **282,** 1314–1316.
11. Connors, G.J., Maisto, S.A., and Donovan, D.M. (1996) *Addiction* **91,** S5–S13.
12. White, F.J. (2002) *J Neurosci* **22,** 3303–3305.
13. Wolf, M.E. (1998) *Prog Neurobiol* **54,** 679–720.
14. Wolf, M.E., Sun, S.X., Mangiavacchi, S., and Chao, S.Z. (2004) *Neuropharmacol* **47,** 61–79.
15. Breiter, H.C., Gollub, R.L., Weisskoff, R.M., Kennedy, D.N., Makris, N., Berke, J.D., Goodman, J.M., Kantor, H.L., Gastfriend, D.R., Riorden, J.P., Mathew, R.T., Rosen B.R, and Hyman, S.E. (1997) *Neuron* **19,** 591–611.
16. Childress, A.R., Mozley, P.D., McElgin, W., Fitzgerald, J., Reivich, M., and O'Brian, C. (1999) *Am J Psychiatry* **156,** 11–18.
17. Wise, R.A., and Bozarth, M.A. (1987) *Psychological Review* **94,** 469–472.
18. Nishiyama, M., Hong, K., Mikoshiba, K., Poo, M.M., and Kato, K. (2000) *Nature* **408,** 584–588.
19. Seamans, J.K., Floresco, S.B., and Phillips, A.G. (1998) *J Neurosci* **18,** 1613–1621.
20. Wang, M., Vijayraghaven, S., and Goldman-Rakic, P.S. (2004) *Science* **303,** 853–856.
21. Cornish, J.L., and Kalivas, P.W. (2001) *J Addict Dis* **20,** 43–54.
22. Hyman, S.E., and Malenka, R.C. (2001) *Nat Rev Neurosci* **2,** 695–703.
23. Kalivas, P.W., Volkow, N., and Seamans, J. (2005) *Neuron* **45,** 647–650.
24. Thomas, M.J., Beurrier, C., Bonci, A., and Malenka, R.C. (2001) *Nat Neurosci* **4,** 1217–1223.
25. Liu, L., Wong, T.P., Pozza, M.F., Lingenhoehl, K., Wang, Y., Sheng, M., Auberson, Y.P., and Wang, Y.T. (2004) *Science* **304,** 1021–1024.
26. Lu, W., Man, H., Ju, W., Trimble, W.S., MacDonald, J.F., and Wang, Y.T. (2001) *Neuron* **29,** 243–254.
27. Man, H.Y., Wang, Q., Lu, W.Y., Ju, W., Ahmadian, G., Liu, L., D'Souza, S., Wong, T.P., Taghibiglou, C., Lu, J., Becker, L.E., Pei, L., Liu, F., Wymann, M.P., MacDonald, J.F., and Wang, Y.T. (2003) *Neuron* **38,** 611–624.
28. Man, H.Y., Lin, J.W., Ju, W.H., Ahmadian, G., Liu, L., Becker, L.E., Sheng, M., and Wang, Y.T. (2000) *Neuron* **25,** 649–662.
29. Liang, F., and Huganir, R.L. (2001) *J Neurochem* **77,** 1626–1631.
30. Ahmadian, G., Ju, W., Liu, L., Wyszynski, M., Lee, S.H., Dunah, A.W., Taghibiglou, C., Wang, Y., Lu, J., Wong, T.P., Sheng, M., and Wang, Y.T. (2004) *EMBO J* **23,** 1040–1050.
31. Lee, S.H., Liu, L., Wang, Y.T., and Sheng, M. (2002) *Neuron* **36,** 661–674.
32. Chung, H.J., Steinberg, J.P., Huganir, R.L., and Linden, D.J. (2003) *Science* **300,** 1751–1755.

33. Karoor, V., Wang, L., Wang, H.Y., and Malbon, C.C. (1998) *J Biol Chem* **273,** 33035–33041.
34. Aarts, M., Liu, Y., Liu, L., Besshoh, S., Arundine, M., Gurd, J.W., Wang, Y.T., Salter, M.W., and Tymianski, M. (2002) *Science* **298,** 846–850.
35. Beattie, E.C., Carroll, R.C., Yu, X., Morishita, W., Yasuda, H., von Zastrow, M., and Malenka, R.C. (2000) *Nat Neurosci* **3,** 1291–1300.
36. Thomas, M.J., Malenka, R.C., and Bonci, A. (2000) *J Neurosci* **20,** 5581–5586.
37. Luscher, C., Xia, H., Beattie, E.C., Carroll, R.C., von Zastrow, M., Malenka, R.C., and Nicoll, R.A. (1999) *Neuron* **24,** 649–658.
38. Collingridge, G.L., Isaac, J.T., and Wang, Y.T. (2004) *Nat Rev Neurosci* **5,** 952–962.
39. Marks, B., and McMahon, H.T. (1998) *Curr Biol* **8,** 740–749.
40. Wang, Y., Ju, W., Liu, L., Fam, S., D'Souza, S., Taghibiglou, C., Salter, M., and Wang, Y.T. (2004) *J Biol Chem* **279,** 41267–41270.
41. Brebner, K., Wong, T.P., Liu, L., Campsall, P., Gray, S., Phelps, L., Phillips, A.G., and Wang, Y.T. (2005) *Science* **310,** 1340–1343.
42. Richardson, N.R., and Roberts, D.C. (1996) *J Neurosci Meth* **66,** 1–11.
43. Bear, M.F., and Abraham, W.C. (1996) *Annu Rev Neurosci* **19,** 437–462.
44. Ungless, M.A., Whistler, J.L., Malenka, R.C., and Bonci, A. (2001) *Nature* **411,** 583–587.
45. Bear, M.F., and Malenka, R.C. (1994) *Curr Opin Neurobiol* **4,** 389–399.
46. Yao, W.D., Gainetdinov, R.R., Arbuckle, M.I., Sotnikova, T.D., Cyr, M., Beaulieu, J.M., Torres, G.E., Grant, S.G., and Caron, M.G. (2004) *Neuron* **41,** 625–638.
47. Tenn, C.C., Fletcher, P.J., and Kapur, S. (2005) *Biol Psychiatry* **57,** 586–593.
48. Kenny, P.J., and Markou, A. (2004) *TIPS* **25,** 265–272.

# Index

**D**